BACTERIOLOGY RESEARCH DEVELOPMENTS SERIES

METHICILLIN-RESISTANT STAPHYLOCOCCUS AUREUS (MRSA): ETIOLOGY, AT-RISK POPULATIONS AND TREATMENT

Bacteriology Research Developments Series

Handbook on Cyanobacteria: Biochemistry, Biotechnology and Applications
Percy M. Gault and Harris J. Marler (Editors)
2009. ISBN: 978-1-60741-092-8

Methicillin-Resistant Staphylococcus Aureus (MRSA): Etiology, At-Risk Populations and Treatment
Charles L. Kolendi (Editos)
2010. ISBN: 978-1-60741-398-1

BACTERIOLOGY RESEARCH DEVELOPMENTS SERIES

METHICILLIN-RESISTANT STAPHYLOCOCCUS AUREUS (MRSA): ETIOLOGY, AT-RISK POPULATIONS AND TREATMENT

CHARLES L. KOLENDI

EDITOR

Nova Science Publishers, Inc.

New York

For permission to use material from this book please contact us:
Telephone 631-231-7269; Fax 631-231-8175
Web Site: http://www.novapublishers.com

LIBRARY OF CONGRESS CATALOGING-IN-PUBLICATION DATA

Methicillin-resistant Staphylococcus aureus (MRSA) : etiology, at-risk populations, and treatment / editor, Charles L. Kolendi.

p. ; cm.

Includes bibliographical references and index.

ISBN 978-1-60741-398-1 (hardcover : alk. paper)

1. Staphylococcus aureus infections. 2. Methicillin resistance. I. Kolendi, Charles L.

[DNLM: 1. Staphylococcal Infections--etiology. 2. Methicillin Resistance--drug effects. 3. Methicillin-Resistant Staphylococcus aureus--pathogenicity. 4. Staphylococcal Infections--drug therapy. WC 250 M9391 2009]

RC116.S8M48 2009

616.9'2071--dc22 2009037440

Published by Nova Science Publishers, Inc. ✢ New York

CONTENTS

PREFACE

Methicillin-Resistant Staphylococcus Aureus (MRSA) is a bacterium responsible for difficult-to-treat infections in humans. It may also be referred to as multidrug-resistant Staphylococcus aureus. MRSA is by definition a strain of Staphylococcus aureus that is resistant to a large group of antibiotics called the beta-lactams, which include the penicillins and the cephalosporins. MRSA was identified as a nosocomial pathogen in the 1960's. MRSA acquisition has traditionally been linked to health care settings and specific patient populations. Established risk factors for MRSA infection include older age, recent hospitalization or surgery, past antimicrobial therapy, residence in a long-term care facility, dialysis, indwelling percutaneous devices and cathethers. This new and important book gathers the latest research from around the globe in this field and focuses on such topics as: risk factors for MRSA in a pediatric surgical ward, MRSA treatments, the anatomy of an outbreak, and others.

Chapter 1 – Chapter 1 - Staphylococci are Gram-positive spherical bacteria that occur in microscopic clusters resembling bunches of grapes. They are aerobic or facultatively anaerobic, nutritionally undemanding and catalase-positive. Today, according to the current List of Bacterial Names with Standing in Nomenclature, the genus Staphylococcus comprises 36 species, eleven of which also contain subdivisions with named subspecies. The classification is still developing.

The growing threat of MRSA is becoming increasingly recognized by the public and health care providers. In 2008, MRSA has evolved to be commonplace in the community and hospital setting. MRSA infections are yet another major factor in the spiraling costs of health care (it is estimated that it costs \$4–7 billion annually in the USA alone to treat resistant infections. It is imperative that clinicians have a high index of suspicion for MRSA so they can promptly initiate the appropriate antibiotic therapy. Failure to act quickly can lead to increased morbidity and mortality for patients. As more virulent strains and increasing resistance occur due to the dynamic nature of this organism, MRSA will remain a growing menace to patients and clinicians. Focused continued education of physicians should improve recognition and treatment, decrease unnecessary antibiotic use, and help reduce the spread of MRSA.

Chapter 2 - *Staphylococcus aureus* causes a wide variety of diseases, ranging from relatively superficial wound infections (e.g. furunculosis) to deep-seated, life-threatening conditions (cardiovascular infections, severe skin infections, osteomyelitis, or pneumonia). Treatment options are, however, becoming limited due to resistance issues. In particular,

Methicillin-Resistant *S. aureus* (MRSA) were first reported in 1961, shortly after introduction of penicillinase-stable β-lactams (such as methicillin). Confined for about 40 years in health care settings, MRSA have also been described in the community (e.g. day care center, schools and sport teams) in various parts of the world. These isolates tend to be highly virulent, causing serious infections and even death in healthy children and adults. Moreover, large-scale CDC surveillance studies have now revealed that hospital acquired (HA-) and community-acquired (CA-) MRSA are no more restricted to specific environments, since strains historically associated with CA-MRSA outbreaks (e.g. mainly US300 isolates) are now recovered in nosocomial infections as well.

Beyond this high capacity to adapt to environment and to develop resistance, the ability of *S. aureus* to survive within host cells can also contribute to the difficulty of eradicating infection with antibiotics. Although primarily known as extracellular, *S. aureus* is now recognized as an opportunistic intracellular pathogen that easily adheres to and invades mammalian cells. This intracellular niche may play an important role in the pathogenesis of several staphylococcal infections, by protecting the bacterium from the lethal action of antibiotics (intracellular forms tend to be poorly susceptible to most available antibiotics) and immune defenses. It may probably explain (i) the recurrent or persistent character of several staphylococcal infections (namely endocarditis, osteomyelitis, complicated skin and soft-tissue infections, or bovine chronic mastitis) even after apparently successful therapy; (ii) the difficulty to eradicate nasal carriage since *S. aureus* can persist in nasal epithelial cells; and (iii) the selection or spreading of resistant mutants due to the poor efficacy of most available antimicrobials. This stresses, therefore, the importance of understanding (i) the intracellular lifestyle and fate of this bacterium, and (ii) whether and to what extent antibiotics may or may not act against intracellular *S. aureus,* with a particular emphasis on multi-drug resistant MRSA. We know, indeed, that cellular accumulation and activity in broth are not, *per se*, predictive factors of intracellular efficacy. This chapter summarizes the authors' current knowledge of the intracellular fate of MRSA and its intracellular susceptibility to antibiotics.

Chapter 3 - For more than a century *Staphylococcus* has been an important human pathogen. "Micrococcus, which, when limited in its extent and activity, causes acute suppurative inflammation (phlegmon), produces, when more extensive and intense in its action on the human system, the most virulent forms of septicaemia and pyaemia…". From these words of Sir Alexander Ogston in 1882, in his "Description of *Staphylococcus*", it can be seen that the bacterium caused serious illness and resulting mortality; this remains true today, more than 100 years later.

S. aureus infections *per se*, and MRSA infections in particular, remain of great concern. For example, CA-MRSA can invade medical institutions, the vancomycin MIC (minimal inhibitory concentration) is increasing, and vancomycin resistance may become more prevalent. Thus, further research on these topics is required.

Chapter 4 - Bone and joint infections (BJIs) are a heterogeneous group of infections in their pathophysiology, clinical presentation, and management. It is felt to be one of the most difficult-to-treat infectious diseases. Although the mortality rate for BJIs may not be so high, morbidity and long-term functional impairment are high (40% to 50% of patients have residual joint dysfunction after septic arthritis). *S. aureus* is the most common virulent pathogen responsible for such infections. Management of staphylococcal BJIs, particularly BJI due to methicillin resistant *S. aureus* (MRSA), is challenging as it is: (i) usually requiring the combination of surgical intervention and long courses of antibiotics; (ii) associated to significant rates of treatment failure and complications.

Today, two categories of MRSA are now recognized: hospital acquired (HA)- and community-acquired (CA)-MRSA. Molecular epidemiological studies show that both HA- and CA-MRSA infections were due to particular pandemic or epidemic MRSA clones. HA-MRSA clones are pandemic and responsible for hospital acquired acute or chronic BJIs, mainly in adult patients with co-morbidities. Conversely, CA-MRSA clones are emergent, epidemic, and responsible for acute and severe BJIs, particularly in the pediatric population, partly due to the production of a particular toxin called Panton-Valentin leukicidin (PVL).

The aim of this chapter was to review the pathogenesis, the clinical presentation, the imaging and the surgical and medical management of HA- and CA-MRSA BJIs.

Chapter 5 - Methicillin resistant *Staphylococcus aureus* is one of the most common pathogens responsible for health-care associated infections as well as community-acquired ones. It is the etiological factor of the wide spectrum of infections. Pathogenesis of those infections is multilevel, thus vaguely described and understood. Virulence factors include components facilitating microbial colonization as well as invasion with severe clinical symptoms, and are often responsible for the chronic state of the disease. Lax hygiene procedures and inappropriate antibiotic policy facilitate the wide spread of this pathogen within a hospital environment. *S. aureus* is also responsible for epidemic infections within human communities. Antibiotic-resistant strains become a serious therapeutic problem, since in such cases a limited spectrum of expensive antibiotics may be used and because of a prolonged hospitalization period resulting in economical losses. Hence, the development of diagnostics enabling faster and proper identification of the etiological factor and its drug-resistance mechanisms, followed by more appropriate clinical treatment, seems to be important. There is also an urgent need for the development of alternative therapeutics, as well as alternative therapeutic methods. The following work is a recapitulative paper on current knowledge concerning staphylococcal diseases, virulence factors, epidemiology, and prospective diagnostics and treatment.

Chapter 6 - Quinolone antibacterial agents are among the most attractive drugs in the field of anti-infective chemotherapy. Quinolones target the essential bacterial type II topoisomerases, which are involved in DNA replication and metabolism. Despite a large number of quinolones approved for the treatment of bacterial infections, there have been progressive efforts for the discovery of new quinolones to overcome the problem of growing bacterial resistance. Over the years, there has been growing concern surrounding the increased prevalence of drug resistance in *Staphylococcus aureus*, especially methicillin-resistant *Staphylococcus aureus* (MRSA). MRSA used to be susceptible to the fluoroquinolones when they were introduced in the early 1980s but a rapid emergence of resistance to these drugs was observed soon after the increasing use of fluoroquinolones from the mid 1980s. Progressive efforts have been made to optimize the biological activity of the quinolones against *S. aureus* and drug-resistant strains including MRSA. Much of the improved potency of modern quinolones against MRSA has been achieved by modification of the N-1, C-7, and C-8 substituents on the quinolone ring system. Several newly developed quinolones and related compounds exhibit enhanced activity against *S. aureus* and have proved useful against MRSA strains and are under rapid preclinical and clinical development.

This chapter will discuss the structural requirements of quinolones for anti-MRSA activity and introduce the newly developed quinolones to manage MRSA infections, their in-vitro effectiveness based on recent studies and structure-activity relationships.

Chapter 7 - After beta-lactamase resistant antibiotics, including methicillin, nafcillin and cephalosporins had been introduced to the clinical use; the first methicillin-resistant *Staphylococcus aureus* (MRSA) isolates were described in 1961. Since then, MRSA clones have become a problem in health-care institutions throughout the world. Community acquired MRSA (CA-MRSA) isolates have also threatened human health for the last two decades. In multicentral surveillance studies, MRSA prevalence was determined as 1-2 % in North European countries but it may reach up to 80 % in some Far East countries.

Many risk factors have been found to be associated with MRSA infections. Recent hospitalization or surgery, residence in long-term care facilities, dialysis, broad-spectrum antibiotic usage and indwelling percutaneous medical devices or catheter are the most common risk factors for nosocomial MRSA infections. The risk factors for CA-MRSA infections have not been well established but several important factors have been described in the literature: HIV infection, recent antibiotic usage, intravenous and intranasal drug usage, living in closed populations, lower socioeconomic status, homelessness, non-hygienic tattoo and health-care workers.

The infection or colonization of health-care workers with MRSA is highly important since it is a cause of transmission and an increase in its incidence. Cellulites, boils and furuncles are among the most common clinical presentations of skin and soft tissue infections (SSTIs) caused by MRSA. These organisms are also responsible for severe infections such as endocarditis, pneumonia and bacteremia. In addition to invasive infections, MRSA produces a variety of toxins that lead to some clinical manifestations such as staphylococcal scalded skin syndrome and toxic shock syndrome. Although CA-MRSA has been traditionally associated with SSTI, there are increasing reports of pneumonia, bacteremia and infective endocarditis caused by CA-MRSA. Hospital acquired MRSA (HA-MRSA) produces more severe clinical conditions associated with increased mortality and morbidity among patients.

Treatment options for both mild SSTI and severe nosocomial infections are limited due to increasing antimicrobial resistance. After MRSA isolates emerged, glycopeptide antibiotics (vancomycin and teicoplanin) have been used more commonly and this resulted in the occurrence of isolates with reduced susceptibility or resistance to vancomycin. Trimetoprim-sulfamethoxazole, rifampin, linezolid, doxycycline, clindamycin and occasionally fluoroquinolones have been used in treating mild SSTIs. In more severe invasive infections, besides vancomycin, other parenteral agents such as linezolid, daptomycin, and tigecycline are commonly used. This review will focus on etiology, at-risk populations and treatment options of MRSA infections.

Chapter 8 - Interest in *Acinetobacter spp.* has been growing for the past 30 years. One of the main reasons for the present increased interest in this genus is the emergence of multiresistant strains—some of which are pan-resistant to antibiotics—that suddenly cause an outbreak of infection.

Multidrug-resistant *Acinetobacter baumannii* is a rapidly emerging pathogen in the health-care setting, where it causes infections that include bacteraemia, pneumonia, meningitis, urinary tract infection, and wound infection. The mortality associated with these infections is high. The crude mortality rate associated with bacteremia is approximately 52% and that associated with pneumonia ranges from 23% to 73%. It is recognized to be among the most difficult antimicrobial-resistant gram negative bacilli to control and treat due to the organism's ability to survive under a wide range of environmental conditions and to persist for extended periods of time on surfaces—which make it a frequent cause of outbreaks of infection and an endemic, health-care-associated pathogen. Recently, an increased rate of infections caused by *A.baumanii* strains resistant to the antibiotics traditionally used in this therapy has been reported. This is an especially severe event in some infections like post-surgical meningitis because the election of the antibiotic depends not only on the sensitivity of *A.baumanii* but also on the penetrability of the chosen antibiotic through the blood-brain barrier. This multidrug-resistance complicates the treatment of these infections and makes imperative the search for new agents and the return of old drugs as we consider the optimal treatment of these multidrug-resistant organisms.

The focus of this chapter is to summarize the current state of knowledge regarding *Acinetobacter baumannii* in relation to this role, including their taxonomy, identification, microbiology, epidemiology, infections, and resistance to antibiotics and the potential therapeutic approaches.

Chapter 9 - As methicillin-resistant *Staphylococcus aureus* (MRSA) has become established in acute care hospitals, high risk or routine newborn care has not been spared. What was once considered community-acquired MRSA has also become established nosocomially and will inevitably affect newborns. An example of a neonatal intensive care unit outbreak with MRSA is highlighted, and the lessons gleaned from the same are detailed. A multi-directional approach to outbreak control is imperative.

The spectrum of MRSA infection in the newborn essentially parallels that due to methicillin-susceptible *Staphylococcus aureus*. Spread to the newborn can occur from maternal, other family, caregiver, and environmental sources. Risk factors for MRSA acquisition include low birth weight, decreased gestational age, multiple gestation, and intensive care. Once established as an outbreak, or for endemic settings otherwise, containment efforts have been quite variable, but some newborn centres have not been successful. MRSA poses several logistical problems for active treatment of infection or for attempted eradication of the carrier state. Prevention strategies for the neonatal intensive care unit are critical in the current context of a global MRSA pandemic in which hypervirulent clones are becoming established in many settings.

With the ability to control various factors given the immobility of the newborn, the scenario of newborn MRSA colonization and infection lends itself to critical related research.

Chapter 10 – Introduction: Methicillin- and also vancomycin (glycopeptide)-resistant Gram-positive organisms have emerged as an increasingly problematic cause of hospital-acquired infections, also spreading into the community. Vancomycin (glycopeptide) resistance has emerged primarily among Enterococci, but the MIC values of vancomycin are also increasing for the entire *Staphylococcus* species, worldwide.

Materials and Methods: Aim of our review is to evaluate the efficacy and tolerability of newer antibiotics with activity against methicillin-resistant and glycopeptide-resistant Gram-positive cocci, on the ground of our experience at a tertiary care metropolitan Hospital, and the most recent literature evidences in this field.

Results: Quinupristin-dalfopristin, linezolid, daptomycin, and tigecycline show an excellent *in vitro* activity, comparable of that of to vancomycin and teicoplanin for methicillin-resistant staphylococci, and superior to that of vancomycin for vancomycin-resistant isolates. Dalbavancin, televancin and oritavancin are new glycopeptide agents with excellent activity against Gram-positive cocci, and have superior pharmacodynamics properties compared to vancomycin. We review the bacterial spectrum, clinical indications and practical use, pharmacologic properties, and expected adverse events and contraindications associated with each of these novel antimicrobial agents, compared with the present standard of care.

Discussion: Quinupristin-dalfopristin is the drug of choice for vancomycin-resistant *Enterococcus faecium* infections, but it does not retain effective activity against *Enterococcus faecalis*. Linezolid acxtivity is substantially comparable to that of vancomycin in patients with methicillin-resistant *Staphylococcus aureus* (MRSA) pneumonia, although its penetration into the respiratory tract is exceptionally elevated. Tigecycline has activity against both *Enterococus* species and MRSA; it is also active against a broad spectrum of Enterobacteriaceae and anaerobes, which allows for use intraabdominal, diabetic foot, and surgical infections. Daptomycin has a rapid bactericidal activity for *Staphylococcus aureus* and it is approved in severe complications, like bacteremia and right-sided endocarditis. It cannot be used to treat pneumonia and respiratory diseases, due to its inactivation in the presence of pulmonary surfactant.

Chapter 11 - Clinically, 23-40% of patients with *S. aureus* infections treated with vancomycin do not respond to it (vancomycin treatment failures). These treatment failures occur in the absence of laboratory-documented vancomycin resistance. We have previously reported that resistance of methicillin-resistant *S. aureus* (MRSA) clinical isolates to vancomycin can be phenotypically increased by exposure to the androgen dehydroepiandosterone (DHEA), a steroid whose chemical structure is similar to that of cholesterol. Herein, we describe an intrinsic phenotypic response to DHEA which results in increased resistance to the host cationic defense peptide, β-1 defensin, as well as vancomycin and other antibiotics that have a positive charge. The increased resistance to vancomycin appears to correlate with a DHEA-mediated alteration in cell surface architecture. DHEA-mediated cell surface changes include alterations in: cell surface charge, surface hydrophobicity, capsule production, and carotenoid production. In addition, exposure to DHEA results in decreased resistance to lysis by Triton X-100 and lysozyme indicating alteration of murien hydrolase activity. We propose that DHEA is an interspecies quorum-like signal that triggers innate phenotypic host survival strategies in *S. aureus* including but not limited to increased carotenoid production. A side effect of this phenotypic change is increased vancomycin resistance. Furthermore this DHEA-mediated survival system may share the cholesterol-squalene pathway shown to be statin sensitive.

Chapter 12 - Nosocomial infection caused by methicillin-resistant *Staphylococcus aureus* (MRSA) is one of the serious worldwide health problems. In order to investigate the risk factors for becoming an MRSA carrier and to prevent MRSA infections, bacteriological surveillance at the time of hospitalization is essential. Bacteriological examination of throat swabs and stool samples in the case of 508 males and 347 females, aged between 0 day and 24 years, who had spent more than 4 days in our pediatric surgical ward were performed within 48 h after hospitalization. In the case of patients with local infections and those with urological diseases, the discharge from the infectious regions and urine, respectively, were also examined. The patients were categorized on the basis of their history of previous hospitalization; neurological disorders, such as cerebral palsy; and age. The carrier rate of MRSA was compared among the groups in each category. We collected 854 throat swabs, 822 stool samples, and 127 other samples. At the time of hospitalization, 8.1% of 855 patients were identified as MRSA carriers. The carrier rate in the case of patients with a history of previous hospitalization within the past year (14.4%: 55/381) was higher than that in the case of patients without any history of previous hospitalization (2.6%: 8/309) and in those who were hospitalized more than a year ago (3.6%: 6/165) ($p < 0.01$). The carrier rate in the case of patients with a neurological disorder (20.9%: 23/110) was higher than that of patients without a neurological disorder (6.2%: 46/745) ($p < 0.01$). The carrier rates differed significantly with age ($p < 0.01$). The carrier rates in the case of patients aged less than 3 years (excluding neonates) (13.4%: 39/291) and patients aged more than 18 years (17.6%: 3/17) were higher than the average carrier rate (8.1%). In the former age group, 66.7% (194/291) had a history of previous hospitalization within the past year. In the latter age group, 29.4% (5/17) had a neurological disorder. Bacteria are known to develop resistance in hospitals and cause serious nosocomial infections. However, MRSA carriers exist not only in hospitals but also in the general community. Therefore, it is very important to identify MRSA carriers at the time of hospitalization. Among our patients, 8.1% were MRSA carriers at the time of hospitalization. A history of previous hospitalization within the past year was identified as a risk factor for being an MRSA carrier. In contrast, the probability of being an MRSA carrier declines after a year of leaving the hospital. MRSA was, however, detected in 2.6% of patients without any history of previous hospitalization and in 3.6% of patients who were hospitalized more than a year ago. Community-acquired MRSA could also cause nosocomial infections. In the case of patients with neurological disorders, 1 out of 5 was an MRSA carrier. With respect to age, MRSA was frequently detected in patients aged less than 3 years (excluding neonates) since many of them had a history of previous hospitalization within the past year. In the case of patients aged more than 18 years, MRSA was frequently detected because many of them had neurological disorders. Bacteriological surveillance at the time of hospitalization is very effective in preventing MRSA infection in a ward. In the case of patients with a history of previous hospitalization within the past year, those with neurological disorders, those aged less than 3 years (excluding neonates), and those more than 18 years, it is particularly important to conduct bacteriological surveillance at the time of hospitalization.

Chapter 13 - Methicillin-resistant *Staphylococcus aureus* (MRSA) was identified as a nosocomial pathogen in the 1960s and later emerged as a cause of community-associated infections in the 1990s. Established risk factors for hospital-associated MRSA infection include older age, recent hospitalization or surgery, past antimicrobial therapy, residence in a long term care facility, dialysis, and indwelling percutaneous devices and catheters. Whereas, risk factors for infection with community-associated MRSA include age less than 2 years and greater than 65 years, participation in contact sports (e.g. rugby, football, and wrestling), injection drug use, residence in a correctional facility, and concurrent influenza-like illness. Identified at-risk populations for community-associated MRSA infection also include racial minorities (i.e., Pacific Islanders, Native Americans/Alaska Natives, and Pacific and Canadian aboriginal), homeless persons, and men who have sex with men. The dichotomy between these two risk factor profiles warrants further study in order to increase understanding of populations at risk for MRSA infection.

Chapter 14 - *Staphylococcus aureus* is a major cause of surgical wounds and infections associated with indwelling medical devices; it is found particularly on the nasal membranes and skin. Naso-oral transmission of *S. aureus* is an important risk factor for development of oral and dental infections. Although prophylactic and postoperative antibiotics are given to patients after oral surgery operations, most strains of *S. aureus* are reported to be resistant to penicillin. After the 1960s, methicillin was replaced by newer penicillin-type antibiotic that were not affected by ß-lactamase, and the resistant *S. aureus* strains became known as *methicillin resistant S. aureus* (MRSA).

There were insufficient data available to determine whether the *S. aureus* isolates were colonizing or infecting the oral cavity and regarding *S. aureus* colonization of the nasal floor in patients with oronasal fistula. Recently our research results showed that bacterial transmission was proven for large oronasal fistulas and a correlation was found with *S. aureus* counts in the children with cleft lip and palate. There were no MRSA strains among our patient population. It appears that as fistula size increases, significantly higher colony numbers of *S. aureus* are found in saliva samples. In addition, *S. aureus* tends to survive in the oral cavity as a result of transmission through the nasal passages as long as an unrepaired cleft exists. Colonization of this site by nasal bacteria seems to occur when the fistulas are sufficiently large to maintain a similar bacterial environment. Therefore, we reported that the high numbers of salivary *S. aureus* count are attributable to poor oral hygiene and dietary factors thereby favoring the outgrowth of *S. aureus,* which can cause wound breakdown. We also recommended that coordination of care and an understanding of special care issues in oral health and maintaining infection control are essential for the health care team for a patient after orodental operations.

Chapter 15 - *Staphylococcus aureus* has developed resistance to many antibiotic agents. Once a source of morbidity and mortality limited to hospitalized patients, drug-resistant *S. aureus* now appears among other populations – often in young people without significant medical conditions. Appropriate control measures, early diagnosis, and proper antibiotic choice can limit its spread and severity.

Chapter 16 – Methicillin-resistant *Staphylococcus aureus* (MRSA) is commonly known for its resistance to methicillin and other antibiotics such as oxacillin, penicillin, and amoxicillin. Hospital-acquired MRSA (HA-MRSA) evolved due to the widespread use of antibiotics in hospitals during the twentieth century. HA-MRSA infections occur most frequently among persons with compromised immune systems in hospitals and other healthcare facilities. Traditionally, most of the attention of the scientific community has been focused on HA-MRSA. The last decade has witnessed the emergence of a genetically different form of MRSA, which has been called Community-acquired MRSA (CA-MRSA) due to the fact that it is acquired from the community and not from hospitals or other healthcare facilities. CA-MRSA infections started to receive attention because they can be highly virulent and can occur in otherwise healthy people, who have no apparent risk factors for MRSA acquisition. The emergence of the first community outbreaks was reported in the 90's. Since then, the prevalence of CA-MRSA infections has been rapidly increasing.

Chapter 17 - There have been few publications on the incidence of infection with methicillin resistant *Staphylococcus aureus (*MRSA) in the head and neck region. In an initial study, over two years, the pattern of infections and the acquisition of MRSA in two groups of patients who had undergone free flap reconstructions of maxillofacial defects were studied. The first group (n=31) were given a 5-day perioperative course of antibiotic prophylaxis (long-term) and the second group (n=33) a 24-hour course (short-term). MRSA was the main infecting organism detected. Seventeen of the 64 patients (27%) developed an infection with MRSA and all infections were acquired after operation. Serious infections delayed healing or discharge in 8% (5) as a result of MRSA and 6% (4) with infections other than MRSA. The median length of hospital admission was significantly greater if MRSA was acquired (p=0.005). There were significantly fewer patients with wounds infected by MRSA in the short-term antibiotic group (4/33 compared with 13/31, p=0.01).

The incidence of acquisition of MRSA and other organisms was then studied over a 6 year period up to 2008. One hundred and fifty-nine patients were reconstructed with a total of 187 free (155) or pedicled (32) flaps. The incidence of MRSA per flap procedure was 19% (36/187). The MRSA acquisition rate decreased from 31% (14/45) with a long-term antibiotic regimen to a rate of 15% (22/142) with the shortened regimen. An infective organism was detected after 85% (159/187) of procedures. The differing groups of organisms were divided in to MRSA alone (2%), "other organisms" (60%), MRSA with "other organisms" (23%), and no growth (15%). The range of organisms increased and these were mainly faecal flora. There was a significant decrease in the incidence of MRSA (p=0.02) and an increase in the detection of Pseudomonas (p=0.02). Age of 70 years or greater (p=0.01) and a pedicled flap (p=0.03) were significant predictors for acquisition of other infectious organisms. The median duration of hospital admission has remained similar throughout the period of the study.

The incidence of MRSA and other infections remains relatively high after major reconstructive maxillofacial surgery with free or pedicled flaps, although the incidence of MRSA is decreasing. A sustained reduction in infections should be achievable through the use of appropriate antibiotic prescribing policies, improved design of specialist wards including the provision of single bedded rooms and isolation facilities together with strict adherence to standard infection control measures. It is not yet known whether the earlier detection of MRSA with rapid screening and preventative isolation will further reduce the incidence of MRSA.

In: Methicillin-Resistant Staphylococcus Aureus... ISBN: 978-1-60741-398-1
Editor: Charles L. Kolendi

Chapter 1

THE CASE FOR MRSA AND TREATMENT OF METHICILLLIN RESISTANT *STAPHYLOCOCCUS AUREUS*

Manal M. Baddour
Microbiology and Immunology Dept., Faculty of Medicine, Alexandria University, Alexandria, Egypt

INTRODUCTION AND TAXONOMY

Staphylococci are Gram-positive spherical bacteria that occur in microscopic clusters resembling bunches of grapes. They are aerobic or facultatively anaerobic, nutritionally undemanding and catalase-positive. Today, according to the current List of Bacterial Names with Standing in Nomenclature (Euzéby, 2004), the genus Staphylococcus comprises 36 species, eleven of which also contain subdivisions with named subspecies. The classification is still developing.

On the basis of comparative 16S rRNA sequence studies, the genus Staphylococcus belongs to the Gram-positive bacteria with a low DNA G+C content. They are closely related to bacilli and other Gram-positive bacteria with low DNA G+C content such as macrococci, enterococci, streptococci, lactobacilli and listeria. Combining Staphylococcus, Gemella, Macrococcus and Salinicoccus within the family Staphylococcaceae has been suggested (Garrity and Holt, 2001).

The Staphylococcus genus is ubiquitously distributed in nature, with some species inhabiting specific ecological niches. Staphylococci are found living naturally on the skin and mucous membranes of warm-blooded animals and humans, but are also isolated from a wide range of foodstuffs such as meat, cheese and milk, and from environmental sources such as soil, sand, air and water (Kloos and Schleifer, 1986).

Most staphylococci species are coagulase-negative. Exceptions are *Staphylococcus aureus, Staphylococcus intermedius, Staphylococcus delphini, Staphylococcus schleiferi* subsp. *coagulans, Staphylococcus lutrae* and some strains of *Staphylococcus hyicus* (Kloos and Bannerman, 1995).

Staphylococcus aureus (*S.aureus*) was discovered in the 1880s. Since then, it has been shown to be a potential pathogenic Gram-positive bacterium, causing such infections as minor skin infections and post-operative wound infections. In the early 1940s, prior to the introduction of penicillin for the treatment of *S.aureus* infections, the mortality rate of individuals with an *S.aureus* infection was about 80% (Skinner and Keefer, 1941). In 1942, only two years after the introduction of penicillin for medical use, the first penicillin resistant *S.aureus* isolate was observed in a hospital. Later on, penicillin-resistant *S.aureus* strains were also observed in the community. Since 1960, around 80% of all *S.aureus* strains are resistant to penicillin. In 1961, 2 years after the introduction of methicillin, a penicillinase-resistant penicillin, *S.aureus* developed methicillin-resistance due to the acquisition of the *mec A* gene (Jevons, 1961).

The prevalence of methicillin resistant *S.aureus* (MRSA) isolates slowly increased over the next three decades, with the majority of isolates being hospital-associated. This started to change in 1982 with an outbreak of MRSA among intravenous drug users (Saravolatx *et al.,* 1982). During the last 45 years, various hospital-associated methicillin resistant *S.aureus* (HA-MRSA) clones disseminated worldwide. The predominant pattern of MRSA in the 1980s and 1990s remained a hospital-acquired disease with traditional risk factors (Brumfit & Hamilton-Miller, 1989; Lowy, 1998). MRSA has emerged epidemiologically in a similar manner to penicillin resistance in the 1950s, occurring first in health care-associated settings then in the community.

Epidemiological studies suggest that hospitals are the primary MRSA reservoir and that hospital contact account for most MRSA infections (Salgado *et al.,* 2003). Three major reservoirs of HA-MRSA within the hospital are patients, health care workers, and inanimate environment. Patients and health care workers may become colonized with MRSA and therefore serve as sources of transmission. The major cause of spread of infection is infection control lapses by health care professionals. Patients with nasal colonization at hospital admission appear to be at increased risk for subsequent infection with the same isolate. Between 10% and 30% of hospitalized patients who acquire MRSA will develop an HA-MRSA infection (Davis *et al.,* 2004; Huang & Platt, 2003).

Not only do MRSA infections lead to higher lengths of stay and more cost then MSSA infections, but more importantly, MRSA infections lead to a higher mortality (Crowcroft & Catchpole, 2002; McHugh & Riley, 2004). A 2003 meta-analysis that included 31 studies showed a significant increase in mortality with MRSA bacteremia compared with MSSA bacteremia (Cosgrove *et al.,* 2003). Mortality among patients with *S.aureus* surgical site infections was increasingly higher when compared with uninfected patients (mortality 2.1%), MSSA-infected patients (mortality 6.7%), and MRSA-infected patients (mortality 20.7%) ($P < 0.001$) (Engeman *et al.,* 2003). A delay in initiation of therapy is associated with a further increase in mortality, making it more important for clinicians to have a heightened awareness for MRSA infection and to be prepared to treat it (Lodise & McKinnon, 2005).

MRSA is largely clonal, with mutation being its main developmental mechanism. Scientists in the field go a long way to de-demonising MRSA as a so-called 'superbug' and reinforce the fact that MRSA are merely pathogenic bacteria which have responded to the intensive antibiotic selective pressure via a classical Darwinian survival process.

PREVALENCE AND EPIDEMIOLOGY

Staphylococcus aureus has three features that make it distinct among most other clinically important bacteria. It can express a variety of virulence factors, it has the ability to develop and expand resistance to a broad spectrum of antimicrobial drug classes, and it is prominent in both hospital and community settings (Flynn & Cohen, 2008; Styers *et al.,* 2006).

The epidemiology of *Staphylococcus aureus* and in particular of MRSA is changing rapidly. Healthcare-associated MRSA is now endemic throughout the world and is a major cause of morbidity and mortality and excess healthcare costs (Gould, 2006; Grundmann *et al.,* 2006). The prevalence of MRSA increased dramatically in many countries during the 1990s. In the UK, MRSA bacteraemia (as a percentage of *S.aureus* bacteraemia) rose from <2% in 1990 to 43% in 2001 (Johnson *et al.,* 2005; Health Protection Agency, 2005) and this trend was mirrored in other countries including the USA and Japan.

Community-acquired MRSA infections have risen dramatically in prevalence in some countries, most notably in the USA (Moran *et al.,* 2006). These infections are caused by MRSA clones that are genetically distinct from those that are endemic in hospitals and they tend to infect younger individuals with few of the risk factors associated with nosocomial MRSA bacteraemia (Vidal *et al.,* 2009). In 2008, MRSA has evolved to be commonplace in the community and hospital setting. Currently, the majority of community-associated strains reported worldwide are methicillin-resistant (Moran *et al.,* 2006). Although the epidemiologic lines are beginning to blur, MRSA infections are still categorized as either health care- or community-associated infections (Byers *et al.,* 2008).

The National Nosocomial Infections Surveillance System reported that the proportion of HA-MRSA infections in U.S. hospital intensive care units (ICUs) increased from 35.9% in 1992 to 64.4% in 2003, an overall increase of 3.1% annually. It is predicted that, if this trend continues, the rate will reach 80% by 2010 (Klevenns *et al.,* 2006).

However, the MRSA prevalence between the countries in Europe is variable. The MRSA prevalence in the northern countries is approximately 1%, whereas the MRSA prevalence in the southern countries is as high as 45% (Tiemersma *et al.,* 2004).

During 2008, some countries have witnessed the first ever significant fall in the rate of healthcare-acquired MRSA bacteraemia, despite concern that this was essentially unachievable (Duckworth, 2005). It causes one-fifth of nosocomial infections in the UK and up to one in six patients in UK intensive care unit (ICU) wards are colonized or infected with MRSA. These strains are usually EMRSA-15 and -16, the most prevalent epidemic MRSA strains in the UK (Hails *et al.,* 2003; Moore & Lindsay, 2002; Johnson *et al.,* 2001; National Audit Office, 2000).

The highest MRSA rates in Europe (ca. 50%) are exceeded in many countries around the world, including the USA, Taiwan, Hong Kong, Singapore and Japan. Much of South America and the Middle East also have a major problem.

The Mediterranean region has been identified as an area of hyper-endemicity for multi-resistant hospital pathogens (Gur & Unal, 2001). This is clearly the case for MRSA within the European continent, where data from the European Antimicrobial Resistance Surveillance System (EARSS) indicate that almost all of the high prevalence countries border the

Mediterranean sea (Tiemersa *et al.,* 2004). Other Mediterranean countries also suffer from this increased prevalence.

A possible measurable factor behind the documented prevalence of MRSA in the southeastern Mediterranean region could be infrastructural deficits related to inadequate bed availability to cope with peak demands, resulting in overcrowding. In addition, inadequate isolation facilities could be a further driver to compound these problems. Furthermore studies show that the mere presence of infection control teams, suitable hand hygiene facilities, as well as antibiotic stewardship initiatives does not necessarily guarantee adequate control of multiresistant infections such as MRSA. This could well be the result of the interplay of other confounding factors as well as possible underlying socio-cultural circumstances (Borg *et al.,* 2008).

HOSPITAL ACQUIRED AND COMMUNITY ACQUIRED

Community acquired MRSA's (CA-MRSA) time in the public eye has been short compared to the 62 years since bacteriologist Mary Barber confirmed the first instances of penicillin-resistant *S.aureus* at London's Hammersmith Hospital (Barber, 1947), or the 48 years since Barber's colleague Patricia Jevons reported the first signs of staphylococcal resistance to the synthetic penicillin replacement methicillin and related beta lactam antibiotics (Jevons, 1961). At the moment, the distinction between CA-MRSA and HA-MRSA is beginning to fade (Lowy, 1998, 2003).

The definitions of HA-MRSA and CA-MRSA are changing, and many different definitions can be found in the literature. In a meta-analysis of 32 studies, at least 8 definitions for CA-MRSA have been reported (Salgado *et al.,* 2003). They are defined by either the epidemiological setting where the infection occurs or by the molecular background of the infecting strains. There appears to be a merging of strains, which makes the definition and distinction even more challenging (Kourbatova *et al.,* 2005).

Two primary factors currently used in the categorization of MRSA infections are *time of infection isolation* and the presence or absence of *MRSA-related risk factors* (Generally, MRSA strains isolated after 48 to 72 hours of admission to a health care facility, or those present at the time of admission in recently discharged patients or residents of long-term care facilities, are interchangeably referred to as *nosocomial*, *hospital acquired*, *hospital-associated*, or *health care–associated* MRSA (HA-MRSA).

Terms used to describe cases of infection not involving a traditional health care setting (CA-MRSA) include *community-acquired*, *community-associated*, and *community-onset.* Of these, *community-onset* is generally used to refer to infections that begin outside of the health care setting (regardless of the presence of risk factors for MRSA), while infections occurring in a community setting in the absence of risk factors for MRSA are considered by some to represent cases of "true" CA-MRSA (Padmanabhan & Fraser, 2005).

Current criteria set forth by the Centers for Disease Control and Prevention (CDC, 2007) for distinguishing CA-MRSA from HA-MRSA state that patients with CA-MRSA infection tend to have all of the following characteristics:

- Diagnosis of MRSA made in the outpatient setting or on the basis of a positive culture for MRSA within 48 hours after hospital admission
- No medical history of MRSA infection or colonization
- No history in the preceding year of hospitalization, dialysis, surgery, or admission to a nursing home, skilled nursing facility, or hospice
- No permanent indwelling catheters or medical devices that pass through the skin into the body.

However, CA-MRSA can also be defined based on genetic markers, such as the presence of *SCCmec* type IV, V or VII and the presence of PVL. In addition, it has been reported that CA-MRSA is associated with several specific *S.aureus* lineages (Tristan *et al.,* 2007). A recent study showed that defining CA-MRSA by the absence of risk factors for healthcare exposure greatly underestimates the burden of epidemic CA-MRSA disease (David *et al.,* 2008).

Community-acquired strains of MRSA are distinct from HA-MRSA strains from genotypic, phenotypic, and epidemiologic perspectives (Hisata *et al.,* 2005; Vandenesch *et al.,* 2003; Daum *et al.,* 2002; Daum, 1998). At a genetic level, CA-MRSA is more similar to methicillin-susceptible *S aureus* (MSSA) than to traditional MRSA, and its emergence appears to be due to the acquisition, by an MSSA strain, of the staphylococcal cassette chromosome (SCC) carrying *mecA*, the gene encoding the methicillin-resistant penicillin binding protein (Robinson *et al.,* 2005). The majority of the studies have described that PVL, together with *SCCmec* type IV or V and a specific genetic background, is a genetic marker for CA-MRSA (Tristan *et al.,* 2007).

In a 2003 meta-analysis of 27 retrospective and 5 prospective studies, CA-MRSA was found to account for 30.2% and 37.3%, respectively, of MRSA isolates from hospitalized patients (Salgado *et al.,* 2003). While a large majority (85%) of these patients had one or more health care-associated risk factors for MRSA (Salgado *et al.,* 2003), the remainder represent cases of "true" CA-MRSA.

In a more recent study that included 319 patients with CA-MRSA infection who presented to one of several rural hospitals in Idaho or Utah, USA, 75% of these patients did not have any identified risk factor for MRSA (Stevenson *et al.,* 2005). Another study from a single medical center in Atlanta evaluated 384 persons with microbiologically confirmed community-onset *S.aureus* skin infections, of which 72% were due to MRSA (King *et al.,* 2006). Among all *S.aureus* isolates, 63% were considered to be community-acquired and 99% were the USA 300 clone. This rate of CA-MRSA represents a much higher percentage than reported in the meta-analysis and suggests that the actual incidence of CA-MRSA is increasing.

Thus the hospital acquired disease pattern associated with MRSA started to change in 1982 with an outbreak of community acquired-MRSA (CA-MRSA) among intravenous drug users. This trend continued with a shift toward more community-acquired infections and was highlighted in the 2006 EMERGEncy ID Net Study, which demonstrated that MRSA was the most common identifiable cause of skin and soft tissue infections among patients who presented to emergency departments in 11 U.S. cities. The overall prevalence of MRSA in this population was 59% (Moran *et al.,* 2006). Community acquired MRSA (CA-MRSA) has rapidly emerged in the community becoming an increasingly common etiologic agent in skin and soft tissue infections and has recently become the most common etiologic agent of these infections in many centers (Moran *et al.,* 2006). This increase in prevalence of CA-MRSA infections is rising both in terms of absolute numbers of cases admitted to hospitals and as a proportion of all cases of MRSA (Fridkin *et al.,* 2005; Eady & Cove, 2003).

Following the four paediatric deaths in Dakota and Minnesota, the emergence of MRSA in the community setting has become a major focal point of epidemiological research (King *et al.,* 2006; Kluytmens *et al.,* 2006; Rossney *et al.,* 2005; Vandenesch *et al.,* 2003; Okuma *et al.,* 2002; Anon, 1999; Lina *et al.,* 1999). MRSA clones of true community origin arise from the horizontal transfer of *mecA* into circulating methicillin-susceptible *S.aureus* strains and are typically characterized by the presence of *SCCmec* IV, susceptibility to non-β-lactams and the gene locus for Panton Valentine leukocidin (PVL) (Tenover *et al.,* 2006; Baba *et al.,* 2002; Lina *et al.,* 1999).

During the last decade, CA-MRSA has emerged worldwide, not only in the community, causing serious infections in people of all ages and backgrounds (Fridkin *et al.,* 2005), but also in healthcare facilities. In general, CA-MRSA is more virulent compared to HA-MRSA due the presence of various virulence factors (Etienne, 2005; Chambers, 2001). The emergence of virulent strains of MRSA in patients in the community who have none of the traditional risk factors for MRSA is extremely worrisome (Fridkin *et al.,* 2005).

The characteristics of HA-MRSA and CA-MRSA differ microbiologically and clinically. These differences are at a molecular level, the presence of specific virulence factors, exotoxin production, and antimicrobial susceptibilities (Diedren & Kluytmans, 2006). Strains of CA-MRSA are more frequently susceptible to a variety of non-beta-lactam antibiotics. CA-MRSA genotypically differs from HA-MRSA in that it carries *SCCmec* types IV or V and is associated with the genes encoding Panton-Valentine leukocidin (PVL) toxin (Deurenberg *et al.,* 2007). Although a small percentage contain *SCCmec* type V, these strains predominantly carry *SCCmec* type IV, which is smaller in size than the gene cassette found in most strains of HA-MRSA (types I, II, and III). This observed difference in SCC size may allow for more efficient transfer of resistance among different bacteria, a factor that may be relevant in the alarmingly rapid emergence of CA-MRSA.

The emergence of CA-MRSA and the growing presence of a community reservoir for methicillin-resistant strains threatens future control of antimicrobial resistance in the health care setting. Since CA-MRSA may now significantly contribute to nosocomial dissemination of MRSA within hospitals, the distinction between CA-MRSA and HA-MRSA within the hospital setting has become blurred. The migration of resistant strains from the community reservoir into hospitals is a potentially troubling development, and gradual increases in this community reservoir can be expected to lead to failure of traditional control measures. Recognition and isolation of symptomatic individuals, along with contact-tracing and quarantining, are two basic measures of control (Fraser *et al.,* 2004) that cannot be used effectively in a community setting. Isolation of infected individuals and carriers is much less manageable in a community setting compared with the relatively closed and controlled environment of the hospital. For this reason, the presence of a community reservoir from which resistant strains can recurrently be transmitted into the health care setting is a significant and growing challenge for the control of MRSA.

Its increase in the community is of additional concern because the CA-MRSA strains appear to be highly virulent, and colonization with CA-MRSA is often undetected in hospitalized patients, which can facilitate spread in the hospital and its potential for becoming resistant to multiple antibiotics.

CA-MRSA could likely have its origins in the health care setting. One source of MRSA in the community is attributed to escaped nosocomial isolates (particularly among people with traditional MRSA risk factors) in addition to the novel community isolates of MRSA. Nosocomial MRSA spreads within the community setting from patients at risk of MRSA carriage to individuals without recent recorded hospital contact. Patients discharged from the hospital often continue to be colonized with MRSA and may become reinfected or may transfer such strains to community contacts. MRSA-colonized patients can persist asymptomatically in 76% of people for up to 12 months. Thus, patients discharged from a hospital can function as reservoirs of MRSA in both the community and the health care-associated setting. Individuals with CA-MRSA then serve as an important reservoir for transmission and dissemination back into the hospital (Eady & Cove, 2003; Seybold *et al.,* 2006). In proof of such a perspective, phenotypical and genotypical analysis demonstrated that MRSA isolates retrieved from the West Midlands community in the UK were not characteristic of CA-MRSA. The majority of the isolates were related to hospital epidemic strains, demonstrating the probable transmission of MRSA from the hospital setting into the surrounding community. The predominance of MRSA isolates harbouring *SCCmec* IV among these demonstrates the transmissible success of *SCCmec* IV outside the hospital setting (Rollason *et al.,* 2008)

One CA-MRSA clone, designated USA300 by the CDC, was first isolated in the San Francisco area in 2000 (McDougal *et al.,* 2003). Despite multiple other strains of CA-MRSA circulating (USA400, USA1000, USA1100), USA300 predominates in outbreaks of skin and soft tissue infection (SSTI) (Kline *et al.,* 2006). Currently, it is unclear why one clone predominates, but most likely USA300 is uniquely adapted for ease of transmission in crowded settings.

It is likely that enzymes such as Toxic Shock Syndrome toxin 1, α-hemolysin, enterotoxins Q and K, staphylokinase, and other proteases are the more significant virulence factors in USA300 rather than PVL historically thought to be responsible for virulence. These toxins are also attributed to the necrotic appearance of CA-MRSA SSTI, which is commonly misdiagnosed as a brown recluse spider bite.

Characteristics of HA-MRSA and CA-MRSA are further differentiated by antimicrobial resistance patterns. Strains of CA-MRSA are more frequently susceptible to a variety of non-beta-lactam antibiotics. Multidrug resistance (MDR) MRSA isolates are a significant concern for HA-MRSA. HA-MRSA is associated with *SCCmec* types II and III and, thus, is significantly more likely to be resistant to >3 antibiotics (Charlebois *et al.,* 2002). Resistance to macrolides, clindamycin, and fluoroquinolones are common along with variable resistance to tetracycline and trimethoprim-sulfamethoxazole. HA-MRSA is usually susceptible to vancomycin and rifampin. Unlike HA-MRSA isolates, many CA-MRSA isolates to date have been resistant only to beta lactams (the antimicrobial class that includes penicillin and cephalosporins) and macrolides/azalides (eg, erythromycin, clarithromycin, azithromycin). However, resistance to other classes of antimicrobial agents, such as fluoroquinolones and tetracycline, occurs and may be increasing in prevalence (Baddour *et al.,* 2006). Most CA-MRSA isolates have been susceptible to trimethorpim-sulfamethoxazole, gentamicin, tetracycline, and clindamycin. Clones resistant to erythromycin can confer inducible resistance to clindamycin and have been reported in USA300 CA-MRSA isolates (Bradley, 2005). The generation of MDR CA-MRSA isolates will make it increasingly difficult for clinicians to choose the appropriate empiric antimicrobial therapy based on epidemiologic information. Although the epidemiologic patterns of HA-MRSA and CA-MRSA may be dissolving, there remain clinical differences that, if unrecognized by the clinician, could cause delay in appropriate treatment and management (Chavez *et al.,* 2008).

Differing Spectrums of Disease

It is important that clinicians be aware of the spectrum of disease caused by CA-MRSA, which differs from that of HA-MRSA in distribution and pattern of infection. Patients infected with CA-MRSA tend to be significantly younger than those infected with traditional strains of MRSA. Unlike traditional MRSA strains, which often are isolated from the bloodstream and the respiratory and urinary tracts, CA-MRSA strains are typically found on skin and in soft tissue and occur in settings that involve crowding, contact, and compromised hygiene (Salgado *et al.,* 2003). Interestingly, because skin infections due to CA-MRSA often have a necrotic center, many have been mistaken for spider bites.

A SSTI develops as a result of a combination of factors: the virulence of the organism, the tropism of the organism to the skin, and the presence of several host risk factors. The Centers for Disease Control and Prevention (CDC) defines CA-MRSA SSTI as a skin or soft tissue lesion occurring in a patient with symptoms such as pain, warmth, or pus, from which MRSA is cultured (CDC, 2001). The most commonly seen CA-MRSA SSTIs are cellulitis, folliculitis, and abscess. However, recent reports have implicated CA-MRSA as causing more invasive SSTI, such as necrotizing fasciitis and pyomyositis (Miller *et al.,* 2007; Fridkin *et al.,* 2005).

Outbreaks have been reported in specific geographic locations (Ma *et al.,* 2005; Peleg *et al.,* 2005; Ribeiro *et al.,* 2005; Schulz *et al.,* 2005; Iyer & Jones, 2004; Salgado *et al.,* 2003) and in several well defined and characteristically "closed" populations, including Alaskan natives, American Indians, children, participants in team sports, military personnel, and correctional facility inmates (Buescher, 2005; Cohen, 2005; Lu & Holton, 2005; Purcell & Fergie, 2005; Rihn *et al.,* 2005; Weber, 2005; Ellis *et al.,* 2004). CA-MRSA is now the predominant cause of community-associated skin infections (Moran *et al.,* 2006).

Although CAMRSA is primarily involved in skin and soft tissue infections, it has also been implicated in sepsis and necrotizing pneumonia (Francis *et al.,* 2005; Mongkolrattabithai *et al.,* 2003). Pneumonia associated with PVL producing MRSA is more frequently associated with sepsis, high fever, leukopenia, hemoptysis, pleural effusions, and death compared with PVL-negative MRSA strains (Boyle-Vavra & Daum, 2007). Recent in vitro work revealed that PVL toxin-producing MRSA binds preferentially to damaged respiratory epithelium, which correlates clinically with data showing that PVL associated pneumonias are associated with prior influenza-like illnesses (Gillet *et al.,* 2002).

Certain clones of MRSA may cause more severe invasive disease. The MRSA clone USA300 is responsible for the dramatic increase in MRSA disease in the United States and is considered the prototypical CA-MRSA strain. Rapid dissemination of this single clone is responsible for this epidemic (Liu *et al.,* 2008; Diep *et al.,* 2006a). Its unique molecular characteristics may also play a key role in pathogenesis. Not only does it possess PVL toxin, but it also has genomic adaptations (*SCCmec* type IV) that can enhance fitness by co-selecting for multidrug resistance and enhance growth and survival on the human skin (Diep *et al.,* 2008, 2006b).

PANTON-VALENTINE LEUKOCIDIN

Since the 1990s, virulent CA-MRSA clones, characterized by the presence of the Panton-Valentine leukocidin (PVL), spread worldwide, first in the community, but later on also in healthcare facilities.

The PVL, having cytolitic activity against human and rabbit monocytes and polymorph nuclear cells, was first reported by Panton and Valentine in 1932 (Panton and Valentine, 1932). PVL is a staphylococcal membrane toxin that targets and kills leukocytes (Kline *et al.,* 2006). Historically, PVL was thought to be the primary virulence factor of CA-MRSA resulting in skin and soft tissue infections (SSTI) (Yamasaki *et al.,* 2005). However, recently it has been proposed that PVL serves as more of a marker of CA-MRSA virulence, and that other toxins played more of a role in the development of SSTI (Voyich *et al.,* 2006).

A total of 40–90% of MRSA strains carrying *SCCmec* type IV express PVL (Kourbatova *et al.,* 2005). This *S.aureus* -specific exotoxin has been associated with skin and soft tissue infections and severe necrotizing pneumonia (Vandenesch *et al.,* 2003). The toxin consists of two polypeptides, S- (slow-eluted) and F- (fast-eluted) components, based on their elution profiles through cation-exchange chromatography (Kaneko and Kamio, 2004; Noda and Kato, 1988; Woodin, 1960). PVL toxin is likely to be involved in severe symptoms such as necrotic pneumonia and furunculosis.

It is postulated that PVL contributes to enhanced community fitness, perhaps from enhanced transmission for draining wounds. It also may enhance the ability to initiate infection in intact skin, directly cause or exacerbate lesions, or facilitate local spread. CA-MRSA strains are also associated with production of other toxins, such as staphylococcal enterotoxin Q and K, α-hemolysin which are capable of causing illness resembling toxic shock syndrome in animal models and may play a role in severe human infections (Labandeira *et al.,* 2007).

Recently, a new class of secreted, short staphylococcal peptides called phenol-soluble modulin peptides, were found to be associated with virulence of CA-MRSA skin infections. These peptides recruit, activate, and lyse neutrophils, the main host defense response against *S.aureus* (Wang *et al.,* 2007).

MECHANISM OF Β-LACTAM RESISTANCE IN MRSA

S.aureus is a dynamic and adaptable bacteria that has a remarkable ability to acquire antibiotic resistance quickly. The cell wall of Gram-positive bacteria is composed of a single, extensively cross-linked peptidoglycan macromolecule, which confers strength and rigidity, maintaining the bacterial shape and protecting against osmotic forces. Structurally, peptidoglycan (murein) is composed of polysaccharide chains of alternating N-acetylglucosamine and N-acetylmuramic acid sugar residues, which are cross-linked by peptide bridges. Its synthesis involves three distinct phases. The first phase, which occurs in the cytoplasm, involves sequentially adding amino acids (L-alanine, D-glutamine, L-lysine, and a dimer of D-alanine) to a UDP-linked N-acetylmuramic acid molecule. In the second phase, this sugar pentapeptide is transferred from the UDP molecule to a lipid carrier (bactoprenol), which transports it across the cytoplasmic membrane, where a further N-acetylglucosamine residue is linked to the N-acetylmuramic acid and, in staphylococci, the ε-amino group of the lysine residue is substituted by pentaglycine.

During the third phase, which occurs at the external surface of the cytoplasmic membrane, the resulting disaccharide-pentapetide is linked onto an existing polysaccharide chain in a reaction termed transglycosylation. Cross-linking of the polysaccharide chains then follows, with attachment via their peptide substituents. This step is catalysed by multiple D-alanyl-D-alanine transpeptidases. They, along with other penicillin susceptible (but non-critical) enzymes, are referred to as penicillin-binding proteins (PBPs) and are the key targets of β-lactam action. β-lactams owe their ability to inhibit these enzymes to a conformational resemblance to D-alanyl-D-alanine; in consequence, they irreversibly inhibit the D-alanyl-D-alanine transpeptidases by covalent acylation of an active site serine. *S.aureus* has three essential PBPs with transpeptidase activity, PBP1, PBP2 and PBP3, and all remain, unaltered, in MRSA (Livermore, 206). Rather, the resistance of MRSA to β-lactams is mediated by a supplementary peptidoglycan transpeptidase PBP, PBP2′ (also known as PBP2a, 78KDa), which continues to function when the normal PBPs have been inactivated by β-lactams (de Lancastre *et al.,* 1994). PBP2′ does bind some available β-lactams, but only with extremely weak affinity. All strains of MRSA produce this unique PBP (PBP2a) that confers resistance ranging from a few cells (heterogeneous resistance) to a majority of cells in a population (homogeneous resistance) (Chambers, 1997; Mulligan *et al.,* 1993).

It is encoded by an acquired gene, *mecA*, 2.1 kb in length ,which is invariably located on a poorly mobile genetic element termed the staphylococcal cassette chromosome mec (*SCCmec*), and which is inserted into the *S.aureus* chromosome at a specific, consistent position (Katayama *et al.,* 2000). The origin of *SCCmec* is not known. Wu et al. suggested that *Staphylococcus sciuri* harbored the ancestor of PBP2a, since it was shown that a PBP was present in *S. sciuri* that had a 87.8% amino acid sequence identity compared to PBP2a. These *S. sciuri* strains were methicillin susceptible, but in the presence of methicillin these strains became resistant to methicillin due to the increased rate of transcription of the *mecA* homologue. It seems that *SCCmec* moved from coagulase-negative staphylococci to *S.aureus* by horizontal transfer (Deurenberg *et al.,* 2007).

The *mecA* gene is regulated by the repressor *MecI* and the transmembrane β-lactam-sensing signal-transducer *MecR1*, which are both divergently transcribed. *MecI* represses both the transcription of *mecA* and *mecR1–mecI* in the absence of a β-lactam antibiotic. However, in the presence of a β-lactam antibiotic, *MecR1* is autocatalytically cleaved, and the metalloprotease domain, which is located in the cytoplasmic part of *MecR1*, becomes active. This metalloprotease cleaves *MecI*, which, in turn, is bound to the *mecA* operator region, allowing the transcription of *mecA*, and the subsequent production of PBP2a to occur (Berger-Bachi and Rohrer, 2002).

There are seven major staphylococcal cassette chromosome variants, differing in size and regulatory genes (Robinson & Enright, 2004). Their restriction to a relatively few clonal complexes of MRSA strains may be because these are the only lineages that are permissive of *mecA* and its product (Ktayama *et al.,* 2000), or may simply reflect the rarity of transfer. Nevertheless, some horizontal transfer does occur, as evidenced by the facts that the same *SCCmec* variants occur in different clonal complexes, and that some MRSA and MSSA strains are clonally identical. Moreover, the same *SCCmec* variants occur in both MRSA and coagulase-negative staphylococci (Enright *et al.,* 2002; Fitzgerald *et al.,* 2001).

At the moment, seven main types of *SCCmec* (type I to VII) are recognized, which vary in size, genetic composition, and antimicrobial resistance patterns. *SCCmec* types range in size from 20.9 to 66.9 kb. *SCCmec* type I (34.3 kb), IV (20.9–24.3 kb), V (28 kb), VI (20.9 Kb), and VII (35.9 kb) cause only β-lactam antibiotic resistance, while *SCCmec* type II (53.0 kb) and III (66.9 kb) cause resistance to multiple classes of antibiotics, due to the additional drug resistance genes integrated into *SCCmec*, i.e. integrated plasmids, e.g. pUB110, pI258 and pT181, and two transposons, e.g. Tn554 and CTn554. Integrated plasmid pUB110 harbors the *ant(4′)* gene, encoding resistance to several aminoglycosides, e.g. kanamycin, tobramycin and bleomycin. Resistance to penicillins and heavy metals, such as mercury, is encoded by pI258, while tetracycline resistance is encoded by pT181. Transposon Tn554 harbors the ermA gene, coding for constitutive and inducible macrolide, lincosamide and streptogramin (MLS) resistance, while CTn554 encodes for resistance to cadmium (Ito *et al.,* 2001, 2003; Leclercq, 2002; Oliveira *et al.,* 2006; Takano *et al.,* 2008). In addition to the resistance genes carried on *SCCmec*, *S.aureus* can also harbor resistance genes on other sites of the genome, such as Tn554, as well as on plasmids (Lindsay & Holden, 2006). It has been shown recently that *SCCmec* type III is a composite element that consists of two SCC elements, i.e. *SCCmec* type III and SCCmercury, harboring ccrC, pI258 and Tn554 (Chongtrakool *et al.,* 2006).

There is a relatively large number of *SCCmec* type IV variants compared to variants of the other *SCCmec* types which could be due to the relative high frequency of horizontal transfer of *SCCmec* type IV, and the genetic plasticity of the MRSA lineages carrying this element (Jansen *et al.,* 2006).

Evidence indicates that *SCCmec* type IV strains may be more fit for survival than other strains (Maree *et al.,* 2007). The smaller size of *SCCmec* type IV types make them more efficient in transfer by phages to MSSA and, along with rapid growth rates, facilitates the spread of MRSA, which may displace other MRSA strains in hospitals (Pan *et al.,* 2005).

Some *S.aureus* strains have a more subtle and less common type of oxacillin resistance that is unrelated to the presence of the *mecA* gene. The resistance mechanism in these isolates is due to either overproduction of β-lactamase or the presence of altered PBP not related to 2a or 2′. These isolates usually have MICs at or just above susceptibility break zones. The isolates are sometimes referred to as borderline-resistant *S.aureus* (BORSA). They are often susceptible to other anti-staphylococcal agents.

MRSA Colonization

The anterior nares are the primary reservoir for *S.aureus* , where replication occurs followed by subsequent dispersal to the skin (Kluytmans *et al.,* 1997). This is supported by studies that show that, if *S.aureus* nasal carriage is eliminated by the use of intranasal mupirocin, colonization often simultaneously resolves from other body sites (Kluytmans *et al.,* 1997; Reagan *et al.,* 1991).

A major principle in the pathogenesis of MRSA infections is that MRSA colonization precedes infection (Mulligan *et al.,* 1993; Miller & Diep, 2008). In proof of this concept of the importance of nasal colonization in the pathogenesis of infection is that, when colonization is eradicated, the risk of clinical infection is lowered (Reagan *et al.,* 1991), which has spurred attempts to eradicate *S.aureus* from the nose to prevent infection among high-risk groups.

However, data from outbreaks of community-acquired MRSA infections suggest that skin-skin and skin-fomite contact represent important and common alternative routes in the acquisition of the infecting MRSA strain and nasal colonization may play a less important role. So the stepwise logical progression of colonization to infection may not occur in community-associated CA-MRSA (Miller & Diep, 2008).

Data that supports non-nasal colonization in the pathogenesis of CA-MRSA infection include outbreak investigations of CA-MRSA infections in football teams, which revealed that the responsible clone of MRSA was not found in the nares of infected players (Kazakova *et al.,* 2005; Begier *et al.,* 2004).

Another investigation found that patients with acute CA-MRSA infection were more likely to have colonization of sites other than nares (axillae, inguinal, and rectum), in contrast to patients with acute CA methicillin-susceptible *S.aureus* (MSSA), HA-MRSA, or HA-MSSA, where non-nasal colonization was rare (Yang *et al.,* 2007). The unique characteristics of CA-MRSA may lead to a different pathogenesis of infection compared with HA-MRSA and MSSA. In outbreaks involving athletic teams, CA-MRSA infections were associated with exposures to various contaminated fomites, such as shared razors, towels, soap bars, and whirlpools. In addition, reported skin-skin contact, especially contact with broken skin, which occurs during games and practices, facilitated the spread of infection (Kazakova *et al.,* 2005; Begier *et al.,* 2004).

Control measures are essential to limit and prevent the spread of MRSA in outbreaks. These measures may include isolation of patients with MRSA, screening of contact patients and staff, and temporary removal of colonized staff from the medical setting (Mulligan *et al.,* 1993). Because the health care staff colonized with MRSA are an important source of MRSA spread, it has been advocated that attempts should be made to decolonize them (Duckworth *et al.,* 1998). Various topical and oral agents have been used to eradicate MRSA colonization, and variable clearance rates have been achieved. In general, the results have been unsatisfactory, and relapse and recolonization have been reported. Because of the consequences of excessive use of vancomycin, it is probably undesirable to use this antibiotic in all colonized patients. One might use oral vancomycin only in patients with digestive tract colonization. However, oral vancomycin is not tolerated by all patients.

It has been emphasized that eradication of MRSA carriage is justified in outbreak situations in health-care settings (Fraise, 1998; Mulligan *et al.,* 1993). Several eradication regimens have been used. Topical agents alone or in combination with oral antibiotics have been used for the elimination of both nasal and extranasal MRSA carriage (Harbarth *et al.,* 1999; Parras *et al.,* 1995; Valls *et al.,* 1994; Walsh *et al.,* 1003; Darouiche *et al.,* 1991; Arathoon *et al.,* 1990; Peterson *et al.,* 1990; Bitar *et al.,* 1987). Intranasal application of mupirocin seems to be the most effective topical agent for the eradication of MRSA nasal carriage (Duckworth *et al.,* 1998; Hill *et al.,* 1998; Kauffman *et al.,* 1993). Relapse and recolonization, however, are still frequently reported, and clearance rates have often been unsatisfactory (Mulligan *et al.,* 1993). It also seems that excessive use of mupirocin has lead to a dramatic drop in susceptibility of MRSA isolates to this antibiotic (Baddour *et al.,* 2006). Several hospitals in the Netherlands use oral vancomycin in combination with topical agents for the eradication of MRSA colonization. One study showed the efficacy of the combination of oral vancomycin, topical intranasal mupirocin, and a bath with povidone-iodine shampoo in eliminating MRSA colonization (Maraha *et al.,* 2002).

In one study, topical application of 5% vancomycin for 2 weeks was successfully used to eradicate MRSA nasal carriage in two nurses (Bitar *et al.,* 1987). In another study, vancomycin ointment failed to eradicate nasal carriage of methicillin sensitive staphylococci (Bryan *et al.,* 1981). In two case reports, decolonization of nasopharyngeal MRSA colonization was achieved using a vancomycin aerosol (Gradon *et al.,* 1992; Weathers *et al.,* 1990).

GISA VISA VRSA

Until now, glycopeptides and especially vancomycin, have been considered as the treatment of choice for MRSA infections. Vancomycin has been used successfully to treat infections with drug-resistant isolates of *S.aureus* for more than three decades. The drug has its limitations. In 1990, Kaatz *et al.,* described the first case of infection with a MSSA isolate with intermediate susceptibility to teicoplanin (TISA). Seven years later, the first infection caused by an MRSA isolate with intermediate susceptibility to vancomycin was reported in Japan (Hiramatsu *et al.,* 1997). Since then, at least 20 cases of infection caused by MRSA with intermediate susceptibility to both vancomycin and teicoplanin (GISA) have been reported worldwide from Japan, the USA, South America and many European countries (Denis *et al.,* 2002; Walsh *et al.,* 2002; Hood *et al.,* 2000; Hiramatsu *et al.,* 1997). Vancomycin-resistant *S.aureus* (VRSA) was first identified in Detroit, Michigan in 2002, mediated by the *vanA* gene complex acquired from vancomycin-resistant enterococci (CDC, 2002). Since then, seven cases of infections caused by VRSA have also been confirmed in the USA (Sievert *et al.,* 2008), and there are other unconfirmed reports from India. Nine VISA and one confirmed VRSA were reported to EARSS in 2006. Although VISA strains have hitherto been thought to be rare, a recent study from Turkey, in which 46 of 256 (18%) MRSA isolates obtained (mainly from blood and pus) between 1998 and 2002 showed the VISA phenotype, suggests that their incidence may be on the rise (Sancak *et al.,* 2005).

In addition to GISA, strains that are borderline-susceptible to glycopeptides, but exhibit low-frequency resistance to glycopeptides $<10^6$ subpopulation; hetero-GISA) have been described more frequently in Europe, Brazil and Asia (Liu & Chambers, 2003; Denis *et al.,* 2002; Walsh *et al.,* 2002; Hiramatsu *et al.,* 1997). Although their clinical relevance is still questioned, such strains appear to be associated with a poor treatment outcome (Howden *et al.,* 2004; Fridkin *et al.,* 2003) and could represent the first step towards the emergence of glycopeptide-resistant mutants following further glycopeptides exposure (Hiramatsu, 2001).

Although the mechanisms underlying vancomycin resistance are not yet fully understood, changes to the bacterial cell wall—the site of action of the glycopeptides—are believed to be key. Vancomycin exerts its antimicrobial effect by inhibiting the cell-wall synthesis of *S aureus*. In contrast to β-lactams, glycopeptides bind to D-alanyl-D-alanine residues of the murein monomer. Binding of glycopeptides to the D-alanyl-D-alanine residues in the completed peptidoglycan layers does not inhibit nascent peptidoglycan synthesis, though it may interfere with crossbridge formation mediated by PBPs. This may be the reason why teicoplanin is synergistic with β-lactam antibiotics. If glycopeptides bind to murein monomers in the cytoplasmic membrane, peptidoglycan synthesis is completely inhibited, and the cells cease to multiply. However, for the glycopeptide molecules to bind to such targets, they have to pass through about 20 peptidoglycan layers without being trapped by the first targets. Since there are many D-alanyl-D-alanine targets in the peptidoglycan layers, many glycopeptide molecules are trapped in the peptidoglycan layers. This compromises the therapeutic effectiveness of glycopeptides. Mechanism of vancomycin resistance is suggested to be production of increased amounts of peptidoglycan. More murein monomers and more layers (probably 30–40 layers) of peptidoglycan are considered to be present in the cell wall. As a result, more vancomycin molecules are trapped in the peptidoglycan layers before reaching the cytoplasmic membrane where peptidoglycan synthesis occurs. This cell wall thickening is a cardinal feature of all vancomycin resistant isolates (Hiramatsu, 2001). This vancomycin trapping is designated "affinity trapping". Moreover, a higher concentration of vancomycin would be required to saturate all the murein monomers. Additionally, the mesh structure of the outer layers of thickened peptidoglycan is destroyed by the trapped vancomycin molecules themselves. This prevents further penetration of vancomycin molecules into the inner part of cell-wall layers (otherwise known as the "clogging" phenomenon) (Cui *et al.,* 2000). It has been suggested that VRSA isolates possess altered peptidoglycan which confers markedly decreased affinity for vancomycin binding (Sakoulas & Moellering, 2008).

A similar resistance mechanism to teicoplanin, the other glycopeptide, is to be expected. However, there appears to be an extra resistance mechanism that might be over-expression of PBP2′ leading to increase in the rate of cross-linking of cell-wall peptidoglycan (Hiramatsu, 2001).

Recent evidence also supports the transfer of genetic material among bacteria as contributing to the development of VRSA. Based on the cases identified to date, risk factors for the development of VRSA may include older age, compromised blood flow to the lower limbs, and the presence of chronic ulcers.

In recent years, the threat of increasing vancomycin resistance in the global *S.aureus* population has led to the development of novel antibiotics for the potential treatment of infections caused by staphylococci.

Clinical Relevance of CA-MRSA

In the mid-1990s, providers in the San Francisco region noted MRSA, an organism historically associated with nosocomial infections, causing SSTI and coined the term community-associated MRSA (CA-MRSA) (Charlebois *et al.,* 2004). Since then, the epidemic of CA-MRSA has spread throughout the United States, currently accounting for more than one-half of all SSTI-related *S.aureus* isolates in the outpatient setting (Moran *et al.,* 2006). A recent longitudinal analysis from 1996 to 2004 of all *S.aureus* infections within a large urban setting showed a dramatic increase in numbers of cases of CA-MRSA from 50 per year to an astonishing 1100 per year (Kline *et al.,* 2006).

Although CA-MRSA's time in the public eye has been short compared to the penicillin-resistant *S.aureus* or to the first report relating to the first signs of staph's resistance to the synthetic penicillin replacement methicillin and related β-lactam antibiotics (Jevons, 1961), it continues to make headlines because of large outbreaks in daycare centers and among members of athletic teams. CA-MRSA infections in children commonly lead to hospitalization. Life-threatening infections, such as necrotizing pneumonitis and brain abscess, can occur. The organism has crossed into hospitals and is now a common cause of hospital-acquired sepsis. Multidrug-resistant strains of MRSA are emerging in Asia, with the resistance based on either a novel gene cassette or a transmissible plasmid. The routine use of antibiotics in livestock seems to be contributing to the emergence of resistant organisms, and some of these have already produced human infection. Fortunately, most cutaneous CA-MRSA infections present as abscesses or furunculosis, and these manifestations generally respond to drainage. The recurrence and attack rates of close contacts are high and relate to persistent colonization. Although CA-MRSA can have dramatic presentations of hemorrhagic necrotizing pneumonia and death, it most commonly occurs as a SSTI in a young, otherwise healthy individual (Garnier *et al.,* 2006).

These data illustrate that the largest burden of illness is identified and treated by primary care and emergency room physicians. It is thus imperative that these clinicians have a high index of suspicion for MRSA so they can promptly initiate the appropriate antibiotic therapy. Failure to act quickly can lead to increased morbidity and mortality for patients. As more virulent strains and increasing resistance occur due to the dynamic nature of this pathogen.

Laboratory Diagnosis and Typing of MRSA

Clinicians are encouraged to collect specimens for culture and antimicrobial susceptibility testing from all patients with suspected MRSA infections, in particular skin abscesses with severe local cellulitis and/or systemic signs of infection or a history suggesting connection to outbreak or cluster of infections. MRSA should be considered in other syndromes compatible with *S.aureus* infections. This is useful in not only the management of the patient but also to determine the local prevalence of MRSA and antimicrobial susceptibility (Gorwitz *et al.,* 2008). At the present time, there is no information to suggest that molecular typing or identification of toxin genes should impact clinical care or management decisions. The Centers for Disease Control states that it is not necessary to routinely collect nasal cultures in all patients presenting with possible MRSA infections.

Conventional methods for the laboratory detection of MRSA include disk diffusion testing, recommended by the National Committee for Clinical Laboratory Standards; broth microdilution minimum inhibitory concentration testing; and oxacillin agar screen test (NCCLS, 1998).

The performance of these tests may be erratic because factors such as inoculum size, incubation temperature, duration of incubation, and variability in culture conditions may affect phenotypic expression of resistance. Most MRSA strains are heterogeneously resistant to methicillin, so only a small proportion of the *S.aureus* colonies that grow from a clinical specimen may show resistance in standard laboratory tests. Although all bacterial cells in a pure culture of MRSA have the basic genetic ability to express their resistance to β-lactam antibiotics, the cells may not behave in a homogenous manner, and frequently only 1 cell in 10^4 to 10^8 display readily detectable resistance in response to oxacillin.

Consequently, if too few cells appear resistant, an oxacillin-resistant strain may go undetected. This property makes detection of MRSA difficult. This phenomenon, called heteroresistance, makes it necessary for labs to use special methods to ensure detection of MRSA. In vitro testing conditions can be modified to enhance the expression of oxacillin resistance, including lower incubation temperatures (30-35°C), final test reading after a full 24 hours of incubation, and higher salt concentrations (supplementation agar with 2-4% sodium chloride).In our hands, we have found that combining 2 of these phenotypic methods greatly improves the test sensitivity and specificity especially in highly heterogeneous populations (Baddour *et al.,* 2007a).

All strains of MRSA have the *mecA* gene, whereas susceptible strains do not. Amplification of genetic determinants by polymerase chain reaction (PCR) assays have been used successfully to amplify and detect *mecA* gene sequences from clinical isolates and to detect MRSA directly from clinical specimens within a few hours (Hueletsky *et al.,* 2004). As they become more widely available, such rapid molecular assays will potentially influence the diagnosis and management of MRSA infections. Isolates identified as borderline-resistant do not grow on oxacillin screen plates.

Recently, a rapid diagnostic test IDI-MRSA (Infectio Diagnostic, Inc., Sainte-Foy, Canada), which uses a real-time polymerase chain reaction (PCR) method to detect MRSA from clinical swabs, has been developed. This method gives a result within 2 hours. The IDI-MRSA_assay utilises real-time PCR for the amplification of an MRSA-specific DNA sequence and uses fluorogenic target-specific hybridization with a molecular probe for the detection of the amplified MRSA DNA. The sequences targeted in this assay are within the *SCCmec* cassette, which carries the *mecA* gene. Target-specific primers within the assay bind to these sequences and generate an MRSA-specific amplicon during the PCR, which can be detected by a molecular probe (Warren *et al.,* 2004). The test has a sensitivity of 92.5% and a specificity of 96.4% (according to the IDI-MRSA manufacturer test assay product insert).

However, whilst technically demanding and more expensive than traditional techniques for patient screening, it did not lead to reduction in transmission of MRSA, possibly due to lapses in infection control practices (Keshtgar *et al.,* 2008).

It must be noted that the results of testing for clindamycin susceptibility in CA-MRSA may be misleading. CA-MRSA strains are often susceptible to clindamycin, but the emergence of resistance during therapy has been reported, especially among erythromycin-resistant strains.Treatment failures have been documented when MRSA isolates were reported as susceptible to clindamycin but resistant to erythromycin (Siberry *et al.,* 2003). A double-disk diffusion test (D zone test) should be performed to identify inducible clindamycin resistance in erythromycin resistant, clindamycin-susceptible *S.aureus* isolates (CLSI, 2007).

The rapid identification of a problem pathogen is imperative to implement appropriate infection control strategies. Controlling the spread of this pathogen by screening patients, personnel and the environment remains a high priority in infection control programs. Tracing the source and transmission routes of MRSA relies on typing methods as tools for the genetic characterization of isolates (Maquelin *et al.,* 2007). Using phenotypic methods, such as antibiotic resistance to type *S.aureus* is disadvantageous because of poor discrimination. The introduction of genotyping techniques has improved the discrimination of *S.aureus*, the most widely used technique being pulsed-field gel electrophoresis (PFGE).

Pulsed-Field Gel Electrophoresis (PFGE)

PFGE is one of the most discriminative typing methods for *S.aureus* and it is therefore considered the golden standard to investigate MRSA outbreaks in hospitals, as well as hospital-to-hospital transmission of MRSA. It is the reference method for molecular strain typing of MRSA. In PFGE for *S.aureus*, the extracted chromosomal DNA is digested with the restriction enzyme *SmaI*, and the resulting DNA fragments (15 to 20 bands) are separated by agarose gel electrophoresis in an electric field with an alternating voltage gradient. The resulting banding patterns are analyzed using a special software package, such as Bionumerics, Quantity one or Gel Compar II, using Dice comparison and unweighted pair group matching analysis (UPGMA) settings according to the criteria of Tenover et al. (1995). Although technically demanding, time consuming and expensive, PFGE is a tool which provides highly reproducible restriction profiles representing the entire bacterial genome and is far more discriminatory than RAPD for staphylococci (Maslow *et al.,* 1993; Raimundo *et al.,* 2002). Several attempts have been made to harmonize PFGE protocols and to establish a common nomenclature, but they were of limited success when assessed by cost of analyses, reproducibility (Baddour *et al.,* 2007b) and speed.

Multilocus Sequence Typing (MLST)

MLST has been proven to be an excellent method to study the molecular evolution of *S.aureus*. This method is based on the sequence analysis of fragments of seven *S.aureus* housekeeping genes, i.e. *arcC, aroE, glpF, gmk, pta, tpi* and *yqiL*, each approximately 500-bp in length. A distinct allele is assigned to each of the different sequences of each housekeeping gene. The alleles of the seven genes define the *S.aureus* lineage, resulting in an allelic profile designated sequence type (ST). The nomenclature of MRSA is currently based on the ST and the type of *SCCmec* element. Disadvantages of MLST are generally that it is expensive, laborious and time consuming. MLST defines unambiguous strain types and results can easily be exchanged between different laboratories (Enright and Spratt, 1999). However, it is a relatively expensive method and therefore not an option for many clinical laboratories (Olive and Bean, 1999).

Typing of the Spa Locus

This single-locus sequence typing technique for *S.aureus* has become increasingly popular during recent years. The method determines the sequence variation of the polymorphic region X of the *S.aureus* protein A (spa) locus (Frenay *et al.,* 1996). The diversity of the spa gene, consisting mainly of a number of repeats of 24 bp in length, is attributed to point mutations, as well as to deletions and duplications of the repeats (Shopsin *et al.,* 1999; Kahl *et al.,* 2005). Spa typing has a discriminative power between that of PFGE and MLST (Malachowa *et al.,* 2005), and, it has been shown that spa typing, in contrast to PFGE and MLST, can be used to study both the molecular evolution as well as hospital outbreaks of MRSA (Koreen *et al.,* 2004). Since spa typing involves sequencing of only a single locus, spa typing, compared to MLST, is less expensive, less laborious and less time consuming.

SCCmec Typing

The structure of *SCCmec* can be determined with a number of PCR-based methods that have been developed during the last few years. However, when investigating the structure of *SCCmec* in the same MRSA isolate, different results were sometimes obtained using these methods (Shore *et al.,* 2005; Kim *et al.,* 2007). A disadvantage of these developed methods is that they determine different structural properties of *SCCmec*. Therefore, the need exists for developing a single, universal assay for the determination of the structure of *SCCmec*.

Repetitive Sequence-Based PCR (rep-PCR)

Repetitive sequence-based PCR (rep-PCR) uses primers that target non-coding repetitive sequences interspersed in bacterial and fungal genomes (Koeuth *et al.,* 1995; Stern *et al.,* 1984; Versalovic *et al.,* 1991). When separated by electrophoresis, the amplified DNA fragments constitute a genomic fingerprint that can be employed for subspecies discrimination and strain delineation of bacteria and fungi (Versalovic and Lupski, 2002). The development of a commercially available, automated rep-PCR assay system, the DiversiLab System (bioMérieux, Boxtel, The Netherlands), offers advances in standardization and reproducibility over manual fingerprint generating systems (Healy *et al.,* 2005). The resulting DNA fingerprint patterns are viewed on a personal DiversiLab website as electropherograms. The software uses the Pearson correlation coefficient to analyze and calculate genetic similarity coefficients among all samples. The unweighted pair-group methods of averages (UPGMA) is employed to automatically compare the rep-PCR profiles and create corresponding dendrograms (Healy *et al.,* 2005). DiversiLab is rapid and non labor-intensive with the advantage of user-friendly web-based software. However, it lacks resolution to differentiate genetically and epidemiologically unique MRSA strains, needed for outbreak analysis.

Randomly Amplified Polymorphic DNA (RAPD)-PCR

Randomly amplified polymorphic DNA (RAPD)-PCR uses short primers with an arbitrary sequence to amplify genomic DNA. These primers randomly hybridize with chromosomal sequences that vary among different strains and that produce different amplification products. These products can be separated by gel electrophoresis to produce fingerprints or patterns characteristic of different epidemiological types (Casey *et al.,* 2007; Tambic *et al.,* 1999; Tang *et al.,* 1997). In comparison to other typing techniques, RAPD is overall a more rapid typing technique, although extracting DNA from staphylococcal isolates for RAPD can be time-consuming.

The method is relatively easy to perform and it is advisable to perform multiple tests with different primers (Cheeseman *et al.,* 2007). It has been claimed to confirm the clonal relationships as determined by PFGE and the Tenover criteria (Casey *et al.,* 2007). The fact that this technique can be used to type many other bacteria and yeasts makes it especially attractive for use in clinical laboratories. Moreover the discriminative power, reproducibility and rapidity of the method and its cost-effectiveness make it a candidate for use as a preventive diagnostic test (Herrera *et al.,* 2004; Hahn *et al.,* 2003; Tang *et al.,* 1997). Nevertheless, the ability of some RAPD methods to discriminate between staphylococcal strains has been debated (Saulinier *et al.,* 1993). Not only that, but acquisition or loss of individual resistance gene clusters or point mutations leading to adaptive resistance will go undetected by RAPD profiling, which only samples a small portion of the genome via PCR amplification to derive a genetic strain type and the reliance on a one-primer approach may not be sufficiently discriminatory.

Recently, an oligonucleotide array called "Staph.Array" has been developed, which targets the manganese-dependant superoxide dismute (*sodA*) gene (Giammarinaro *et al.,* 2005). It is the only tool described to date that distinguishes the 36 validated staphylococcal species in one shot and allows rapid, accurate identification of staphylococcal strains from clinical and food origins.

CA-MRSA has traditionally been distinguished from nosocomial MRSA by distinct epidemiologic and molecular differences noted in the organisms specifically USA300 pulsed-field gel electrophoresis (PFGE) strain type, presence of the PVL gene, and diminished antimicrobial resistance (Baba *et al.,* 2002). Over the last few years, new data on the evolving epidemiology of *S.aureus* and CA-MRSA (Huang *et al.,* 2007, Mwangi *et al.,* 2007), as well as on the appearance and transmission of the organism in the health care setting (Bratu *et al.,* 2005; Saiman *et al.,* 2003), have blurred these distinctions.

Despite some success with MRSA rates in 2008, the original 'superbug' remains one step ahead of us. Its genomic plasticity (Dancer, 2008), ubiquitous existence, longevity and prominent human niche virtually guarantee long-term survival. No biological agent cannot be manipulated by the organism, despite the resilience of certain antibiotics, disinfectants and heavy metal ions (Crawford, 2008). Surveillance remains key for the spur required to search for alternative prevention and control measures.

The Need for MRSA Treatment

The very essence of classifying a *Staphylococcus aureus* isolate as methicillin resistant *Staphylococcus aureus* (MRSA) lies in its resistance to methicillin, a semisynthetic penicillin. However, this is not the only drug to which it is resistant. It actually has the machinery to defy all β-lactam antibiotics, such as penicillins and cephalosporins and carbapenems. MRSA's resistance to all forms of penicillin has earned it the moniker of "superbug." Due to its drug resistance, it is twice as fatal as other staph infections (Cosgrove *et al.,* 2003; Whitby *et al.,* 2001).

It has been contended that the introduction of methicillin put an end to the tremendous amount of research and enthusiasm for the infection control measures which were effective and vigorous at the time when it was first introduced. The medical community experienced a false sense of security against these resistant pathogens. Once you could treat these previously antibiotic-resistant staphylococci, people stopped washing their hands and doing all the other things they now realize they should have been doing all along.

MRSA, like other multidrug resistant organisms (MDR) really seem to be gathering pace, no doubt fuelled by intensive broad-spectrum antibiotic use, poor infection control, highly immunosuppressed patients, and rapid and frequent air travel.

Antibiotic resistance has reached crisis point in many hospitals around the world. Most are swamped with MRSA and many with MDR Gram-negative bacteria. Despite the availability of good treatment alternatives available for serious infections due to MRSA, mortality rates remain high.

Hospital acquired MRSA (HA-MRSA) has traditionally been resistant to other anti-staphylococcal agents, such as clindamycin, erythromycin, and tetracycline, with variable resistance to gentamicin and trimethoprim-sulfamethoxazole (SXT) whereas community-associated MRSA (CA-MRSA) isolates have typically been susceptible to clindamycin, SXT, and gentamicin.

Several studies from the USA have described the need for using empirical MRSA coverage in patients attending emergency departments.

It is imperative that clinicians have a high index of suspicion for MRSA so they can promptly initiate the appropriate antibiotic therapy. Failure to act quickly can lead to increased morbidity and mortality for patients. As more virulent strains and increasing resistance occur due to the dynamic nature of this pathogen, not only is the prevalence of CA-MRSA increasing, but the devastating mortality of these infections continue to be evident since the 4 pediatric deaths from CA-MRSA were reported in 1999 (CDC, 1999).

However, the decision of when to start systemic treatment can be difficult. Features such as fever, underlying illness such as diabetes mellitus, human immunodeficiency virus (HIV) or involvement of deep structures or extensive lesions or if there are infections in other family members support the use of systemic therapy (Enoch *et al.,* 2008).

Antibiotic Drug Development

Despite the continuing emergence of resistant pathogens, development of new classes of antimicrobials has been at a virtual standstill since the late 1970s (Spellberg *et al.,* 2004; IDSA, 2004) probably due to relatively unfavourable returns on investment. Four new classes of antibiotics were introduced in the 1930s and 1940s, including sulphonamides, β-lactams, aminoglycosides, and chloramphenicol. A further six classes were developed and approved in the 1950s and 1960s, including tetracycline, macrolides, glycopeptides, rifamycins, quinolones, and trimethoprim. However, from 1970 to the late 1990s, no new antimicrobial classes were approved and only a few new classes have been approved since 2000 for the treatment of Gram-positive infections; these are the oxazolidinones (linezolid), cyclic lipopeptides (daptomycin), and glycylcyclines (tigecycline). Pharmaceutical companies are discouraged from research and development of new antimicrobials due to high direct costs, risk, and the time associated with animal and in vitro studies (Wenzel, 2004). There are already concerns about resistance to the 2 recently introduced antibiotics in the lipopeptide and oxazolidinone classes. Several reports have described cases of clinically significant resistance to daptomycin (a lipopeptide) for the treatment of MRSA infections (Hayden *et al.,* 2005; Mangili *et al.,* 2005; Vikram *et al.,* 2005), and there are currently 7 case reports of resistance to linezolid (an oxazolidinone) among individuals with *S. aureus* infections (Peters & Sarria, 2005).

Increasing levels of antibiotic resistance, the waning drug development pipeline, and declining numbers of pharmaceutical companies involved with the development of anti-infective agents create a significant public health threat to the effective management of bacterial infections (Wenzel, 2004). Additional factors (Rice, 2003) that establish the imperative for the development of new antibiotics include:

(1) Emergence of bacterial species resistant to new antibiotics.
(2) Intolerance to, or safety concerns about, treatment with currently available antibiotics, such as an increased risk of bleeding associated with linezolid.
(3) Current dosing regimens for most available antimicrobials that are incompatible with outpatient administration, resulting in prolonged hospital stay, increased cost, and increased exposure to nosocomial infections.

TREATMENT OF MRSA INFECTIONS

Vancomycin is currently the first-line agent for treating invasive MRSA infections. Glycopeptide antibiotics, which include vancomycin and teicoplanin, inhibit bacterial cell-wall synthesis. Vancomycin was developed more than 50 years ago and has been the mainstay of treatment for a variety of infections, including endocarditis, pneumonia, and wound infections, with cure rates estimated at 63%, 75%, and 90% of patients, respectively (Esposito & Gleckman, 1977). More recently, however, this role as mainstay of MRSA treatment has been called into question, in part because vancomycin has poorer tissue penetration, especially reported for lung tissue (Rybak, 2006; Cruciani *et al.,* 1996; Lamer *et al.,* 1993) and is less bactericidal with slower bacterial clearance than the anti-staphylococcal penicillins (Gould, 2007a; Chang *et al.,* 2003), it demonstrates poor pharmacokinetic properties, and fluctuating minimum inhibitory concentrations (MICs). Vancomycin monotherapy has demonstrated inferiority to beta-lactam therapy in the treatment of methicillin sensitive *S. aureus* osteomyelitis (Martinez-Aguilar *et al.,* 2003) and bacteraemia (Gould, 2007b; Stryjewski *et al.,* 2007; Siegman-Igra *et al.,* 2005; Chang *et al.,* 2003). Poor pharmacokinetic properties undermining the efficacy of vancomycin include variable tissue penetration and high inoculum sizes (Kollef, 2007; Stevens, 2006).

In a prospective study, 19% (13/70) of MSSA bacteraemia patients treated with vancomycin experienced bacteriological failure, with bacteraemia persisting beyond 7 days or with relapse, compared with none of 18 treated with nafcillin (Chang *et al.,* 2003). Similarly, none of ten patients treated with cloxacillin for MSSA pneumonia died, compared with 47% (8 / 17) of those treated with vancomycin (Gonzalez *et al.,*1999). So generally speaking, there is wide consensus that β-lactams are more efficacious than vancomycin against infections due to susceptible staphylococci.

Vancomycin is also associated with several other limitations, including the development of resistance and associated therapeutic failure, and the potential for serious toxicity. There is growing evidence of vancomycin resistance, including vancomycin-intermediate *S. aureus* (VISA) and heterogeneous VISA (hVISA), in patients with serious infections. hVISA is regarded as a precursor to VISA (Liu & Chambers, 2003): it comprises a heterogeneous population of *S. aureus* cells with an overall MIC in the susceptible range (i.e. < 2 μg/ml), but with non-susceptible subpopulations (i.e. with an overall MIC ≥4 μg/ml) detectable after selection with vancomycin. Heterogeneous VISA infections are associated with high treatment failure rates, prolonged bacteraemia, high bacterial loads, and lower vancomycin trough levels (Kollef, 2007; Stevens, 2006).

An additional concern about vancomycin includes adverse reactions, especially nephrotoxicity, when combined with aminoglycoside agents (Jeffres *et al.,* 2007). Vancomycin has also been shown to provoke hypersensitivity reactions, including anaphylaxis and 'red man syndrome' (an infusion-related reaction involving pruritus and an erythematous rash) (Kollef, 2007).

The standard treatment options for serious MRSA infections are the glycopeptides, vancomycin and teicoplanin, often given together with rifampicin. According to the susceptibility of the particular strain, combinations of two or more of fusidic acid, rifampicin, trimethoprim and minocycline may also be used in less severe infections or as oral follow-on treatment to an intravenous glycopeptide. In addition, there is a growing list of recently developed agents for treatment of severe MRSA infection, with quinupristin–dalfopristin, daptomycin, linezolid and tigecycline already licensed in some parts of the world, and dalbavancin and telavancin anticipated in the next few years (Abbanat *et al*., 2003; Anstead & Owens, 2004). However, all of these agents are significantly more expensive than the older oral agents, and all but linezolid require intravenous administration.

The SENTRY surveillance program (SENTRY is an international antimicrobial surveillance program that documents resistance patterns in bacteria isolated from predominantly hospitalized patients) revealed *S. aureus* to be the leading cause of skin and soft-tissue infections (SSTIs), and the second most common causative organism for bacteraemia in Europe (Moet *et al.,* 2007; Biedenbach *et al.,* 2004). MRSA is of particular concern, as it is associated with greater mortality than MSSA (Cosgrove *et al.,* 2003; Whitby *et al.,* 2001). Although vancomycin is the most widely used agent against Gram positive infections, against *S. aureus*, other agents have demonstrated equivalent and/or superior efficacy to vancomycin in clinical trials of MRSA and MSSA infections (Rehm et al. 2008; Stryjewski *et al.,* 2007; Weigelt *et al.,* 2005; Arbeit *et al.,* 2004; Chang *et al.,* 2003; Stevens *et al.,* 2002; Gentry *et al.,* 1997).

For the treatment of complicated skin and soft tissue infections (cSSTIs), daptomycin showed equivalent efficacy to standard therapy, which comprised vancomycin for suspected MRSA infections and required shorter duration of therapy (Whitby *et al.,* 2001). Daptomycin is a cyclic lipopeptide that depolarizes the bacterial cell membrane. It is a fermentation product of *Streptomyces roseosporus* consisting of a 13-member amino acid cyclic lipopeptide with a decanoyl side chain (Enoch *et al.,* 2008; Micek, 2007). Daptomycin was approved by the United States Food and Drug Administration (USA-FDA) and by the European Medicines Agency (EMEA) for the treatment of cSSSIs caused by methicillin-susceptible and methicillin-resistant *S. aureus*, Daptomycin has shown equivalent efficacy to vancomycin against MRSA bacteraemia/infective endocarditis and might provide an alternative treatment option (Rehm *et al.,* 2008). Although these compounds (daptomycin, linezolid, vancomycin) show similar susceptibility rates, daptomycin was four- to eightfold more active than vancomycin or linezolid (Sader *et al.,* 2009).

Daptomycin is also as effective as standard therapy against MSSA bacteraemia/infective endocarditis (Fowler *et al.,* 2006). In the majority of relevant case reports, daptomycin was administered when treatment of patients with endocarditis or bacteraemia with other potentially effective antibiotics (including glycopeptides and linezolid) had failed (Falagas et al, 2007).

Because of its excellent anti-*S. aureus* spectrum, high potency, and rapid bactericidal activity, daptomycin represents an excellent treatment option for BSI caused by *S. aureus* in at-risk patients such as those on hemodialysis (Sader *et al.,* 2009). In fact, it is the only new antibiotic with a license for use in staphylococcal blood stream infections (Rose *et al.,* 2008). An antibiotic that has high clinical efficacy against both MSSA and MRSA infections is of particular value as an alternative to vancomycin, given that suspected *S. aureus* infections are often treated empirically and the consequences of inappropriate antimicrobial use, such as vancomycin for an infection later confirmed to be MSSA, can be severe (Gould, 2008). Daptomycin does not exhibit cross-resistance with other known classes of antimicrobials and also has a low risk for development of spontaneous mutational resistance (Sader *et al.,* 2009). Resistance to daptomycin remains low although it should always be checked for if there has been prior glycopeptide exposure in a patient (Rose *et al.,* 2008). Indeed, daptomycin has been shown to be active against methicillin resistant *S. aureus* (MRSA) as well as *S. aureus* resistant to linezolid or quinupristin/dalfopristin (Sader *et al.,* 2009).

Linezolid is a synthetic oxazolidinone that inhibits bacterial protein synthesis by binding to the bacterial 23S ribosomal RNA of the 50S subunit. The antibiotic is bacteriostatic against staphylococci. Linezolid received US Food and Drug Administration approval in 2000 and has demonstrated clinical success against indicated Gram-positive organisms causing a variety of infections including serious cutaneous disease and nosocomial pneumonia (Weigelt *et al.*, 2005; Wunderink *et al.*, 2003; Stevens *et al.*, 2002). It is currently approved for the treatment of complicated skin and skin-structure infections (SSSIs) and nosocomial pneumonia caused by MRSA (Vergidis & Falagas, 2008). For the treatment of SSTIs caused by MRSA, linezolid has shown equivalent or greater efficacy to vancomycin in a number of trials (Weigelt *et al.,* 2005; Stevens *et al.,* 2002) and required a shorter duration of hospital stay for complicated SSTIs (Li *et al.,* 2003). There is limited evidence, based on subset analysis, to suggest that linezolid is superior to vancomycin in treating nosocomial pneumonia due to MRSA (Wunderink *et al.,* 2003). Resistance to linezolid remains very uncommon (<0.5%) among surveyed isolates. However, with wider linezolid use, resistance has been recognized (Baddour *et al.*, 2006; Gales *et al.*, 2006; Mutnick *et al.*, 2003; Jones *et al.*, 2002; Marshall *et al.*, 2002; Gonzales *et al.*, 2001; Tsiodras *et al.*, 2001) and characterized as associated with prolonged drug exposure in at-risk patient populations or with breaks in infection control practices (Kainer *et al.*, 2007; Mutnick *et al.*, 2003). Treatment failure typically occurs as a result of deep infection and failure to drain abscesses (Enoch *et al.,* 2008).

Dalfopristin is a derivative of a group A streptogramin, and quinupristin is a group B streptogramin. The streptogramins are a family of compounds isolated from *Streptomyces pristinaespiralis.* These 2 water-soluble streptogramins have been combined in a commercially available injectable form at a 30:70 weight-to-weight ratio (Allington & Rivey, 2001). Individually, quinupristin and dalfopristin demonstrate only modest in vitro bacteriostatic activity. However, the combination of the 2 commonly produces in vitro bactericidal activity that is 8 to 16 times higher than the sum of each component's activity against many gram-positive organisms (Aeschlimann & Rybak, 1998; Rubinstein & Bompart, 1997; Caron et al, 1997). Quinupristin and dalfopristin bind to sequential sites located on the 50s subunit of the bacterial ribosome. Dalfopristin binding causes a conformational change in the ribosome that subsequently increases the binding of quinupristin. The combined actions of the 2 agents create a stable drug-ribosome complex that causes inhibition of protein synthesis by several mechanisms, including prevention of peptide-chain formation, blockade of extrusion of newly formed peptide chains, and, in many instances, bacterial cell death (Thal & Zervos, 1999). Streptogramin resistance is mediated by 3 possible mechanisms: drug inactivation by enzymes, efflux or active transport of the antibiotic out of the cell, and conformational alterations in ribosomal target binding sites. The latter is the most common expression of bacterial resistance to streptogramins. Quinupristin/dalfopristin is active against *S. aureus*, including MRSA, and has been approvedby the FDA for treatment of Gram-positive bacteraemia.

Tigecycline, the first of a new synthetic class of antibiotics known as glycylcyclines, was approved in 2005 (Pankey, 2005). It is a derivative of minocycline and acts by inhibiting the 30S ribosomal subunit and thus protein synthesis. Tigecycline was developed to overcome tetracycline resistance mediated by efflux pumps and ribosomal protection. Tigecycline is a broad-spectrum antibiotic with activity against gram-positive, gram-negative, and atypical anaerobes (Nathwani, 2005). In a multicentre study across Europe testing for comparative antimicrobial susceptibility, tigecycline was the only agent to which all Gram-positive isolates were susceptible (Rodloff *et al.,*2008). Gram-positive bacteria that are susceptible to tigecycline include MRSA, glycopeptide-intermediate *S. aureus* (GISA), vancomycin-resistant enterococci (Cercenado *et al.*, 2003; Petersen *et al.*, 2002) and penicillin-resistant *Streptococcus pneumoniae* (Hoellman *et al.*, 2000). In June 2005, the drug was approved by the FDA for treatment of complicated SSSIs, including those caused by MSSA and MRSA.

TREATMENT OF CA-MRSA

Based on the escalating observance of CA-MRSA in different settings, it has been proposed that CA-MRSA is going to become one of the most challenging infectious disease problems in the 21st century (Deurenberg *et al.,* 2007; Vandenesch *et al.,* 2003).

Physicians evaluating patients with skin and soft tissue infections in the outpatient and emergency departments must be cognizant of the likelihood that CAMRSA is the offending organism. The paradigm that simple abscesses without cellulitis (focal skin/soft tissue infections), require only incision and drainage has not changed (LoVecchio *et al.,* 2009; Ruhe et al., 2007; Martinez-Aguilar et al., 2003).

Clinicians should be aware that the prevalence of CA-MRSA has increased markedly over the past few years. The virulence of these new strains of CA-MRSA is unique not only in the resistance to β-lactam antibiotics but in their capacity to cause tissue necrosis. When approaching a patient with a SSTI, clinicians are required to have a low threshold for sending material for culture and susceptibility testing when an incision and drainage is required. Empiric treatment of a SSTI should take into consideration providing coverage for CA-MRSA, MSSA, and *S. pyogenes* until an organism can be identified through culture. Clinicians should also be aware that, in the empiric selection of antibiotics, trimethoprim-sulfamethoxazole does not adequately cover *S. pyogenes*. Finally, in the setting of more invasive SSTI (pyomyositis or necrotizing fasciitis), empiric intravenous antibiotics directed against CA-MRSA must be included in the initial treatment regimen. There are surprisingly few data regarding oral treatment of the common and serious problem of SSTI caused by MRSA. A case can certainly be made for treating simple and uncomplicated skin infections with surgical drainage alone and reserving the use of systemic antibiotics for patients with fever, large abscesses, complications or underlying risk factors. Drainage of abscesses is required even if antibiotics are used. The latest British Society for Antimicrobial Chemotherapy (BSAC) guideline advises against giving systemic antibiotics to patients with minor SSTIs or small abscesses <5 cm. Thus small abscesses without cellulitis should be incised and drained and antibiotics withheld (Nathwani *et al.*, 2008).

Recent literature has suggested different approaches to the treatment of skin and soft tissue infections. Localized skin infections, including small furuncles and abscesses, often respond to surgical incision and drainage alone, but nonetheless, oral antimicrobials are commonly prescribed. For large soft tissue infection with surrounding cellulitis, lacking a drainable focus, and with systemic symptoms such as fever, most physicians would treat with antimicrobial drugs, although controlled trials are lacking (Moellering RC Jr, 2008; Gorwitz, 2007). Such infections include cellulitis, osteomyelitis, diabetic foot infections, and urinary tract infections [UTIs]). Many such infections receive extended treatment, often in the outpatient setting, which makes intravenous therapy unattractive. Selection of appropriate antibiotics for MRSA infection depends on several factors, including severity of disease (local versus invasive) and strain of MRSA.

Although some reports claim there is no difference between the outcome of septic arthritis caused by MSSA and MRSA (Al-Nammari *et al.*,2007), evidence from other reports show that septic arthritis caused by CA-MRSA could be associated with more severe outcome (Castaldo and Yang, 2007; Arnold *et al.*,2006). In fact, the 1st genome of CA-MRSA sequenced was from a strain that caused fatal septicemia and septic arthritis (Baba *et al.*,2002).

Current guidelines recommend broad-spectrum combination therapy with more than 2 antibiotics as the initial empirical therapy for some types of infections in patients at risk of being infected with resistant organisms (American Thoracic Society, IDSA, 2005). All antibiotics can potentially exert selective pressure and thereby drive resistance Deresinski, 2007b), yet appropriate empiric antimicrobial therapy is important because it can decrease mortality in critically ill patients (Kumar *et al.*, 2006; Ibrahim *et al.*, 2000; Kollef, 2000).

CA-MRSA strains differ in antimicrobial susceptibility compared with hospital-acquired MRSA, because CA-MRSA strains often lack resistance to several various non-ß-lactam antibiotics, a feature most likely attributable to a lower frequency of antibiotic-selective pressures in the community compared with the hospital environment (Chambers, 2001; Suggs *et al.*, 1999). However, with the migration of CA-MRSA strains into hospital settings and movement of these strains back out to the community, an agile CA-MRSA strain with multidrug resistance traits may emerge (Carleton *et al.*, 2004).

Oral agents such as SXT, clindamycin, doxycyline, linezolid, rifampin, and occasionally fluoroquinolones have been used for minor skin infections. In the United States, current susceptibility patterns suggest that all CAMRSA isolates are susceptible to SXT and rifampin, 95% are susceptible to clindamycin, 92% are susceptible to tetracycline, 60% are susceptible to fluoroquinolones, and 6% are susceptible to erythromycin (Moran *et al.*, 2006). But which drug is the first choice, what the regimen should be and whether it should be used in combination are currently more art than science.

In addition to vancomycin, newer parenteral agents for more severe invasive infections include linezolid, daptomycin, and tigecycline; additional potential treatment options are in development.

Treatment of VISA and VRSA

Reduced susceptibility of *S. aureus* to vancomycin and teicoplanin has been definitively observed in recent years with the emergence of glycopeptide-resistant *S. aureus* (GRSA), glycopeptide-intermediate *S. aureus* (GISA) and heterogeneous GISA (hGISA) (Pope & Roecker, 2007; Liu & Chambers, 2003). There is a growing body of evidence that shows the MICs of glycopeptides against *S. aureus* strains to be increasing incrementally, a process referred to as MIC creep.

The prevalence and clinical significance of hVISA cannot be accurately addressed. Furthermore, strains assayed prior to the introduction of reduced CLSI MIC breakpoints in 2006 would be subject to less stringent categorization than they would be today. It is clear from retrospective studies and testing of previously isolated strains that the presence of VISA and hVISA was overlooked prior to 1996. Also, with the variation in effectiveness of susceptibility tests used, and without consensus on how hVISA MICs should be measured, it is difficult to gauge the true extent of vancomycin resistance and reduced susceptibility today.

The *in vitro* activities of a number of antibiotic agents against VISA and hVISA strains have been examined, although the clinical efficacy of such agents has not yet been established in the published literature.

In the SENTRY surveillance program, lower MIC values were measured for daptomycin (0.12–2 μg/ml) compared with vancomycin (1–8 μg/ml) against hVISA and VISA, as well as vancomycin-susceptible MRSA and MSSA (Wootton *et al.,* 2006). In time–kill experiments in two GISA and one hGISA strain, daptomycin at 2 × MIC demonstrated bactericidal activity throughout the 24-hour measurement period and displayed 99.9% kill by 8 hours in all three strains (LaPlante & Rybak, 2004). Arbekacin and tigecycline also showed activity against these strains over 24 hours, whereas in contrast, vancomycin displayed no activity at the 24-hour timepoint.

There has been some concern that *S. aureus* strains with reduced susceptibility to vancomycin may also have reduced susceptibility to daptomycin. In the MIC surveillance study by Steinkraus *et al.* (Steinkraus *et al.,* 2007), daptomycin MICs did not increase over the 5-year period, despite vancomycin MICs incrementally increasing, suggesting that cross-resistance was not occurring. All VRSA strains identified to date have been found susceptible to daptomycin (Patel *et al.,* 2006), but a proportion of VISA isolates have shown reduced daptomycin susceptibility (Moise *et al.,* 2008; Cui *et al.,* 2006; Patel *et al.,* 2006; Sader *et al.,* 2006). Reduced susceptibility in some VISA strains may be a result of adaptive modifications such as thickened cell walls, reducing antibiotic penetration – other resistance mechanisms such as *vanA*-mediated resistance should not influence susceptibility to agents with a different mode of action to vancomycin. In a recent study of 81 patients with MRSA bacteraemia, prior vancomycin exposure within 30 days was associated with significantly higher vancomycin MICs and reduced vancomycin killing activity *in vitro* (Moise *et al.,* 2008). In contrast, there was only a trend towards higher daptomycin MICs with previous vancomycin exposure, and daptomycin retained its full *in vitro* bactericidal activity (Moise *et al.,* 2008). That previous vancomycin exposure does not affect a reduction in the bactericidal activity of daptomycin is supported by a separate study of VISA isolates (Sader *et al.,* 2006).

Other antibiotics have also shown activity in VISA and hVISA; in a comparison of eight licensed antibiotics, linezolid and quinupristin/dalfopristin were the only agents to demonstrate potent activity against VISA and hVISA. The MICs for linezolid were consistently low (0.25–2 μg/ml), whereas the MICs for quinupristin/dalfopristin were more variable (0.25–16 μg/ml) (Howe *et al.,* 2003).

Vancomycin Treatment Failures

MRSA infections can be difficult to treat. Clinical failures, even with appropriate treatment, occur regularly with vancomycin monotherapy, resulting in recommendations that vancomycin be used in combination with other antimicrobial agents to yield the best clinical results.

Appropriate treatment could be defined as intravenous treatment with an antibiotic to which the infecting bacterial strain is susceptible for at least 1 week in pneumonia, 2 weeks in bloodstream infections, 4 weeks in endocarditis, epidural abscess, and joint infection; and 6 weeks in osteomyelitis (Gilbert *et al.,* 2007). In cases of bloodstream infection, removal of intravascular catheters is considered a necessary part of appropriate treatment.

In 2006, increasing reports of vancomycin treatment failure and decreased susceptibility of MRSA to vancomycin prompted the Clinical and Laboratory Standards Institute (CLSI) to decrease the vancomycin breakpoint for *S. aureus* from 4 to 2 mg/L (Clinical and Laboratory Standards Institute, 2006).

Some clinical failures among MRSA infections treated with vancomycin may result from decreased susceptibility to the drug. MRSA strains with intermediate susceptibility to vancomycin (VISA; vancomycin minimum inhibitory concentration [MIC] = 4-8 μg/mL) and strains with frank resistance to vancomycin (VRSA; MIC >16 μ/mL) remain uncommon. However, some portion of vancomycin failures may be due to unrecognized heterogeneously resistant *S. aureus* (hVISA) (Liu & Chambers, 2003). Heteroresistance does not correlate well with MIC, and is thus not readily detected by standard clinical laboratory methods (Tenover & Moellering, 2007; Baddour *et al.,* 2007). Furthermore, infections due to MRSA which are still considered susceptible to vancomycin but have MICs at the higher end of the susceptible range (MIC=2) are at higher risk for clinical failure (Tenover & Moellering, 2007).

Years of glycopeptide under-dosing owing to fears of toxicity coupled with the poor penetration of glycopeptides into tissues (e.g. lung and endocardial tissue), is thought to have played a part in the emergence of reduced susceptibility by exposing *S. aureus* to sub-inhibitory concentrations of vancomycin (Deresinski, 2007a; Gould, 2007a; Sakoulas *et al.,* 2006). Glycopeptides are also only weakly bactericidal, and this might also have contributed to the development of strains with reduced susceptibility.

Interestingly, some studies claim that there was no association between patient outcomes and the susceptibility of the pathogen to the prescribed antimicrobial agents (Enoch *et al.,* 2008; Fridkin *et al.,* 2005; Lee *et al.,* 2004).

Apart from the intrinsic pharmacologic properties of vancomycin and emerging resistance, other factors may affect MRSA treatment failure rates. These include the site of infection, characteristics of the patient, antibiotic dosing and serum drug concentrations, and the presence and quality of concomitant surgical therapy where indicated. Such factors, however, are not well studied.

For MRSA bacteraemia/infective endocarditis, vancomycin is considered the gold-standard treatment. However, there have been recent reports of treatment failures against vancomycin-susceptible strains (Howden *et al.,* 2004; Charles *et al.,* 2004). In one study, glycopeptide therapy failed in 19/25 patients with infections caused by MRSA that conformed to definitions of VISA or hVISA by population analysis profiling (Howden *et al.,* 2004). In another prospective study, 19% (13/70) of MSSA bacteraemia patients treated with vancomycin experienced bacteriological failure, with bacteraemia persisting beyond 7 days or with relapse, compared with none of 18 treated with nafcillin (Chang *et al.,* 2003). Similarly, none of ten patients treated with cloxacillin for MSSA pneumonia died, compared with 47% (8/17) of those treated with vancomycin (Gonzalez *et al.,*1999). A separate study showed vancomycin treatment to fail in 5/5 MRSA patients with VISA, compared with 1/48 patients with vancomycin-susceptible MRSA (Charles *et al.,* 2004). Alternative treatments to vancomycin may be considered for the empiric therapy of suspected MRSA bacteraemia/infective endocarditis before the MIC and the presence of VISA or hVISA have been determined (Gould, 2008).

Another site of infection where this reported clinical failure occurs frequently despite appropriate therapy is the bones. Daver and colleagues recently reported a failure rate of 35% in patients with MRSA osteomyelitis treated with either >4 weeks of parenteral antibiotics or <4 weeks of parenteral therapy followed by oral therapy, with no difference in the relapse rates between the two groups (Daver *et al.,* 2007). Dombrowski and Winston, 2008, also reported a 25% clinical failure rate in an appropriately treated (Gilbert *et al.,* 2007) cohort of patients with diverse clinical conditions. They report that factors associated with an increased risk of failure at a p-value of <0.05 in bivariate analysis included diabetes, admission to the orthopedics service, osteomyelitis, and hospitalization in the intensive care unit at the time of antibiotic initiation. Pneumonia and endocarditis were associated with decreased risk of failure. In multivariate analysis, only a diagnosis of osteomyelitis was independently associated with failure ($p < 0.001$). They conclude justifiably that failure in cases of osteomyelitis was particularly common (46%) (Dombrowski and Winston, 2008).

It seems that regimens other than vancomycin monotherapy and/or longer duration of monotherapy may be more appropriate for treatment of MRSA osteomyelitis. In animal models of *S. aureus* osteomyelitis, vancomycin monotherapy has demonstrated inferiority to moxifloxacin (Kalteis et al, 2006), tigecycline (Yin et al, 2005), and rifampin-vancomycin combination therapy (Norden & Shaffer). Use of surgical debridement followed by local implantation of gatifloxacin implants had a favourable outcome in experimental MRSA osteomyelitis (ElKamel & Baddour, 2007).

The substantial failure rates in cases of MRSA osteomyelitis that are reported may also indicate poor performance of vancomycin in bone. Vancomycin has higher penetration into infected bone than non-infected bone, but it may function more poorly in the more anaerobic environment of osteomyelitic bone (Garazzino *et al.,* 2008).

The empirical use of vancomycin against Gram-positive infections should be carefully considered in the context of local vancomycin MIC data and, where possible, local MIC trends should be taken into account. Where vancomycin MIC is >1 μg/ml, alternative therapies should be considered to avoid the possibility of treatment failure. Precise determination of MIC, as well as regular surveillance are therefore of key importance because even small changes in MIC can have clinical significance (Gould, 2008).

TREATMENT OPTIONS AND NEW DIRECTIONS

MRSA are inherently resistant to all β-lactam antibiotics, but some lineages (clones) have additionally evolved resistance to multiple antibiotic classes. Within the species, resistance to all known antibiotic classes has occurred due to mutation and horizontal gene transfer and this has led to anxiety regarding the future availability of effective chemotherapeutic options.

Fortunately, most MRSA isolates remain susceptible to 1 or more of several older oral antimicrobial agents, including SXT, clindamycin, tetracyclines, and fluoroquinolones (Klevens et al., 2006; Moran et al., 2006).

Use of older oral agents could potentially slow the development of resistance to newer antimicrobials, avoid the morbidity and expense of intravenous access, and reduce drug costs.

This has prompted interest in such compounds for oral therapy for non-life-threatening MRSA infections (Stevens *et al.,* 2005). However, studies examining tetracycline (Ruhe and Menon, 2007; Ruhe *et al.,* 2005; Yuk *et al.,* 1991), SXT (Cenizal *et al.,* 2007; Markowitz *et al.,* 1992), or clindamycin (Martinez-Aguilar *et al.,* 2003) for treating MRSA infections were relatively small, involved only intravenous therapy, and/or were limited to a single type of infection.

In a study assessing the efficacy of alternative older antibiotics against MRSA, treatment was clinically successful in 89% of cases. The most commonly used alternative agent was SXT (49%), followed by long-acting tetracyclines (doxycycline and minocycline, 35%). SXT has received comparatively little attention in this role, and the attention it has received includes cautionary comments regarding the paucity of clinical data. Although Markowitz et al. (1992) showed good efficacy with intravenous SXT for serious MRSA infections (including bacteremia and endocarditis), evidence supporting oral SXT therapy for MRSA infections is quite limited (Stein et al., 1998; Honda et al., 1996). This has been the subject of a recent review by Proctor, who states that the new strains of community-acquired MRSA show decreased resistance to SXT. Clinical and experimental reports show a mixture of successes and failures with SXT. A reason for failure might be the amount of thymidine released from damaged host tissues and bacteria, a concept strengthened by the fact that *S. aureus* thermonuclease releases thymidine from DNA. Thus, success or failure with SXT may well depend on the amount of tissue damage and organism burden, rather than acquisition of a resistance gene. Clinical trials and experimental animal studies show high failure rates, perhaps because of released thymidine. He concludes that the use of SXT for community-acquired MRSA infection should not be endorsed without further research (Proctor, 2008).

A more recent retrospective assessment of alternative (non-vancomycin, non-linezolid) oral antimicrobial therapy for non-life-threatening MRSA infections for a 2-year period at a Veterans Affairs Medical Center demonstrated that this strategy was generally successful and well tolerated. Nearly 90% of subjects had satisfactory outcomes, and nearly half (40%) of the failures resulted from medication intolerance rather than frank clinical failure (Drekonja *et al.*, 2009).

There are several new classes with activity against MRSA which are currently in development. New glycopeptide agents include dalbavancin and telavancin, which are suitable for the treatment of MRSA and VISA strains and exhibit activity against a variety of other organisms (Billeter *et al.*, 2008; Jauregui *et al.*, 2005). Dalbavancin is a semi-synthetic lipoglycopeptide derived from a teicoplanin-like glycopeptide agent (Billeter *et al.*, 2008). It inhibits cell wall synthesis and has in vitro activity against MRSA with a long half-life of 6-10 d that allows once-weekly Dosing (Micek, 2007). Dalbavancin demonstrated similar cure rates when compared with linezolid for treating cSSTIs (Jauregui *et al.*, 2005). Dalbavancin seems to be a safe drug, with the most commonly reported adverse events including nausea, diarrhoea, and constipation with some evidence suggesting an increased risk of hypotension, hypokalemia, and increased levels of alanine aminotransferase and aspartate aminotransferase (Billeter *et al.*, 2008; Micek, 2007).

Telavancin is an investigational lipoglycopeptide antimicrobial agent with a dual mechanism of action that inhibits cell wall synthesis and disrupts membrane barrier function. Its half life ranges from 7 to 9 h, which permits once-daily dosing at 7.5 10 mg/kg/day. Disruption of the membrane barrier function permits rapid bactericidal activity for a variety of pathogens including MRSA and other drug-resistant strains. In vivo studies have demonstrated that telavancin may be effective for the treatment of Gram-positive soft tissue infections, bacteraemia, endocarditis, meningitis, and pneumonia. Furthermore, the MIC of telavancin is generally 2-8 times less than that of vancomycin for many Gram-positive bacteria, including MRSA (Attwood & LaPlante, 2007). Telavancin also appears to be more effective for the treatment of MRSA SSTIs infections than vancomycin (Stryjewski *et al.*, 2006).

A major coming challenge will be to determine to what extent these agents offer absolute and cost-benefit advantages over vancomycin and to define their relative merits. The lack of wider evidence of superiority is perhaps surprising because vancomycin is not considered to be an especially potent antibiotic.

For severe clinical infections caused by MRSA, vancomycin remains the pivotal parenteral agent. However, numerous reports have confirmed the often slow and suboptimal treatment outcome when using vancomycin in single drug or combination therapy (Fowler *et al.*, 1999; Levine *et al.*, 1991). Thus, it seems essential to evaluate new antistaphylococcal drugs and regimens.

In principle, it should be possible to redesign β-lactams so that they bind to PBP2′, thereby gaining activity against MRSA. This task should be facilitated because the crystal structure of PBP2′ is known, permitting molecular modeling (Lim & Strynadka, 2002). Moreover, PBP2′ is exposed on the cell surface, meaning that the antibiotic need not be a good permeant of bacterial membranes, whereas this would be a sine qua non for an anti-Gram-negative agent. The problem lies in the fact that the active site of PBP2′ is located in an extended, narrow groove. The structure of this groove is such that the β-lactam is displaced and misaligned with respect to the active site serine residue, militating against acylation (Chambers, 2003; Lim & Strynadka, 2002). Many attempts have been made to produce β-lactams with modified side group structures to promote association with PBP2′ and to make acylation more energetically favourable. Modifications known to enhance activity involve acyl substituent groups with greater length and hydrophobicity, which result in increased interaction with the active site groove. Such research has yielded several compounds with MIC values of 2-8 mg/L for MRSA (Chambers, 2003), but it has proved difficult to develop these for clinical use. One of the earliest such analogues, an aminothiazolylpenicillin called BRL-44154 (Beecham), was investigated in the 1980s (Mizen *et al.,* 1995) and had MICs of 4–8 mg/L for MRSA, compared with 0.5–2 mg/L for MSSA. It was effective against MRSA infections in mice but was inactivated by rapid metabolism in humans. A later carbapenem, L-695 256 (Merck), was even more active in vitro, with MICs of 2 mg/L for MRSA, and was effective against MRSA endocarditis in rabbits (Chambers, 1995). However, it readily formed insoluble crystals, making it unsuitable for use in humans or for commercial manufacture (Sundelof *et al.,* 1996). Numerous anti-MRSA cephalosporins (e.g., TOC-50) (Nomura *et al.,* 1996) were presented at one or more ICAAC meetings during the 1990s before fading from view, usually without publicly available explanation. More recent anti-MRSA cephalosporins include RWJ-54428, now abandoned (Chamberland *et al.,* 2001), CAB-175 (Iizawa *et al.,* 2004; Koga *et al.,* 2005) and PPI-0903 (CX- 0903, T-91825, TAK-599) (Iizawa *et al.,* 2004), which is in phase I development, as is an anti-MRSA carbapenem, CS-023 (RO4908463) (Koga *et al.,* 2005).

One further promising anti-MRSA broad-spectrum cephalosporin, ceftobiprole (Ro 63-9141, BAL9141), has progressed much further. Ceftobiprole displays a wide range of activity both against Gram-positive and Gram-negative pathogens, including several resistant species such as MRSA, penicillin-resistant *Streptococcus pneumoniae* (Fritsche *et al.,* 2008; Rouse *et al.,* 2007; Jones *et al.,* 2002; Hebeisen *et al.,* 2001) and even heteroresistant vancomycin-intermediate *S. aureus* (hVISA) (MICs$\leq$2mg/L) (Deshpande *et al.,* 2004). It is approved in Canada with the indication of cSSTIs. Ceftobiprole has an oxyimino aminothiazolyl substituent linked to the 7-amino group of the cephalosporin nucleus, conferring stability to many β-lactamases, and a vinylpyrrolidinone moiety at position 3, promoting association with PBP2′ and facilitating the subsequent acylation reaction (Livermore, 2006). PBP assays, which reflect both the initial binding and the acylation, indicate that ceftobiprole produces a 50% saturation of PBP2′ at an extremely low concentration compared with those needed for ceftriaxone, imipenem and methicillin (Hebeisen *et al.,* 2001). For improved solubility, ceftobiprole is administered in vivo as a water-soluble ester prodrug, ceftobiprole medocaril, which is rapidly cleaved in plasma to yield the active compound (Hebeisen *et al.,* 2001). The pharmacokinetics of ceftobiprole is similar to other β-lactams in that it displays a two-compartment model, distributes into extracellular fluid, does not cross membranes well and is primarily cleared renally. Ceftobiprole also displays a safety profile similar to other β-lactams, with the exception of dysgeusia being a common adverse effect. The potential for resistance development is low and therefore ceftobiprole seems to be good first-line therapeutic option against a variety of infections. This compound is not active against all pathogens and is subject to hydrolysis by some ESBLs and carbapenemases (Barbour *et al.,* 2009). It is currently under review by the FDA for approval in the USA.

Iclaprim is a selective dihydrofolate reductase inhibitor from a class of diaminopyrimidines that is currently in phase 3 trials. Iclaprim at concentrations close to the MIC has exhibited antimicrobial efficacy against Gram-positive pathogens including resistant *S. aureus* (MRSA, VISA, GISA) (Schneider *et al.,* 2003). The antibacterial profile of iclaprim compared with linezolid and vancomycin for treatment of SSTIs is promising (Morgan *et al.,* 2009).

The Microbiology Laboratory

In the microbiology laboratories, a few points have to be kept in mind, sometimes MRSA appears susceptible *in vitro* to other β-lactam agents, such as cephalosporins; however, these are not clinically effective. MRSA must be reported by the microbiology laboratory as resistant to all β-lactam agents (including cephalosporins, β-lactam and β-lactamase inhibitor combinations, and carbapenems), regardless of the *in vitro* test results.

Microbiology laboratories should also be aware that the results of testing for clindamycin susceptibility in CA-MRSA may be misleading. Treatment failures have been documented when MRSA isolates were reported as susceptible to clindamycin but resistant to erythromycin. A double-disk diffusion test (D zone test) should be performed to identify inducible clindamycin resistance in erythromycin resistant, clindamycin-susceptible *S. aureus* isolates (Yilmaz *et al.*, 2007).

Clinical microbiologists and infectious disease clinicians should be aware that staphylococci showing reduced susceptibility to vancomycin may also demonstrate reduced susceptibility to daptomycin. Such strains must be tested in the clinical microbiology laboratory for their susceptibility to daptomycin by a reliable and easy to use method. E-test strips are available and have been reported to give results comparable with the CLSI reference method (Fuchs *et al.,* 2001).

In conclusion, this growing threat of MRSA is becoming increasingly recognized by the public and health care providers. In 2008, MRSA has evolved to be commonplace in the community and hospital setting. MRSA infections are yet another major factor in the spiraling costs of health care (it is estimated that it costs $4–7 billion annually in the USA alone to treat resistant infections (US Congress, Office of Technology Assessment, 1995)). It is imperative that clinicians have a high index of suspicion for MRSA so they can promptly initiate the appropriate antibiotic therapy. Failure to act quickly can lead to increased morbidity and mortality for patients. As more virulent strains and increasing resistance occur due to the dynamic nature of this organism, MRSA will remain a growing menace to patients and clinicians. Focused continued education of physicians should improve recognition and treatment, decrease unnecessary antibiotic use, and help reduce the spread of MRSA.

Nevertheless, there is a need for new antimicrobial agents for the treatment of human pathogens that are not susceptible to currently available antibiotics (Wenzel, 2004; Rice, 2003). These development efforts must be coordinated with aggressive infection control efforts and rational use of currently available and emerging antimicrobial agents (Rice, 2003).

References

Abbanat, D; Macielag, M; Bush, K. Novel antibacterial agents for the treatment of serious Gram-positive infections. *Expert Opin Invest Drugs*, 2003, 12, 379–399.

Aeschlimann, JR; Rybak, MJ. Pharmacodynamic analysis of the activity of quinupristin-dalfopristin against vancomycin resistant *Enrerococcus fuecium* with differing MBCs via time-kill-curve and postantibiotic effect methods. *Antimicrob Agents Chemothe*, 1998, 42, 2188-2192.

Allington, DR; Rivey, MR. Quinupristin Dalfopristin: A Therapeutic Review. *Clin Therapeut*, 2001, 23(1), 24-44.

Al-Nammari, SS; Bobak, P; Venkatesh, R. Methicillin resistant *Staphylococcus aureus* versus methicillin sensitive *Staphylococcus aureus* adult haematogenous septic arthritis. *Arch Orthop Trauma Surg*., 2007, 127, 537-542.

American Thoracic Society, Infectious Diseases Society of America. Guidelines for the management of adults with hospital-acquired, ventilator-associated, and healthcare-associated pneumonia. *Am J Respir Crit Care Med., 2005*, 171, 388-416.

Anon. Four pediatric deaths from community-acquired methicillin-resistant *Staphylococcus aureus* Minnesota and North Dakota. *Morb Mortal Wkly Rep 1999;48*:707-710.

Anstead, GM; Owens, AD. Recent advances in the treatment of infections due to resistant *Staphylococcus aureus*. *Curr Opin Infect Dis*., 2004, 17, 549–555.

Arathoon EG, Hamilton JR, Hench CE, Stevens DA. Efficacy of short courses of oral novobiocin-rifampin in eradicating carrier state of methicillin resistant *Staphylococcus aureus* and in vitro killing studies of clinical isolates. *Antimicrob Agents Chemother* 1990; 34: 1655-1659.

Arbeit, RD; Maki, D; Tally, FP; Campanaro, E; Eisenstein, BI. The safety and efficacy of daptomycin for the treatment of complicated skin and skin-structure infections. *Clin Infect Dis*., 2004, 38, 1673-1681.

Arnold, SR; Elias, D; Buckingham, SC; Thomas, ED; Novais, E; Arkader, A; Howard, C. Changing patterns of acute hematogenous osteomyelitis and septic arthritis: emergence of community-associated methicillin resistant *Staphylococcus aureus*. *J. Pediatr Orthop*, 2006, 26, 703-708.

Attwood, RJ; LaPlante, KL. Telavancin: a novel lipoglycopeptide antimicrobial agent. *Am J Health Syst Pharm.*, 2007, 64, 2335-2348.

Baba T, Takeuchi F, Kuroda M, Yuzawa H, Aoki K, Oguchi A, Nagai Y, Iwama N, Asano K, Naimi T, Kuroda H, Cui L, Yamamoto K, Hiramatsu K. Genome and virulence determinants of high-virulence community acquired MRSA. *Lancet* 2002;359:1819-1827.

Baba, T; Takeuchi, F; Kuroda, M; Yuzawa, H; Aoki, K; Oguchi, A; Nagai, Y; Iwama, N; Asano, K; Naimi, T; Kuroda, H; Cui, L; Yamamoto, K; Hiramatsu, K. Genome and virulence determinants of high virulence community-acquired MRSA. *Lancet*, 2002, 359, 1819-1827.

Baddour MM, Abuelkheir MM, Fatani AJ, Bohol M, Alahdal M. Molecular epidemiology of Methicillin Resistant *Staphylococcus aureus* (MRSA) isolates from major hospitals in Riyadh, Saudi Arabia. *Canadian Journal of Microbiology* 2007; 53 (8): 931-936.

Baddour MM, Abuelkheir MM, Fatani AJ. Comparison of *mecA* PCR and phenotypic methods for the detection of MRSA. *Curr Microbiol* 2007a; 55(6): 473-479.

Baddour MM, Abuelkheir MM, Fatani AJ. Trends in antibiotic susceptibility patterns and epidemiology of MRSA isolates from several hospitals in Riyadh, Saudi Arabia. *Annals of Clinical Microbiology and Antimicrobials* 2006; 5: 30.

Barber M. Staphylococcal infection due to penicillin-resistant strains. *Br Med J.* 1947;863-865.

Barbour, A; Schmidt, S; Rand, KH; Derendorf, H. Ceftobiprole: a novel cephalosporin with activity against Gram-positive and Gram-negative pathogens, including methicillin-resistant *Staphylococcus aureus* (MRSA). *International Journal of Antimicrobial Agents* 2009; 34(1):1-7.

Begier EM, Frenette K, Barrett NL, Mshar P, Petit S, Boxrud DJ, Watkins-Colwell K, Wheeler S, Cebelinski EA, Glennen A, Nguyen D, Hadler JL; Connecticut Bioterrorism Field Epidemiology Response Team. A high-morbidity outbreak of methicillin resistant *Staphylococcus aureus* among players on a college football team, facilitated by cosmetic body shaving and turf burns. *Clin Infect Dis* 2004;39: 1446-1453.

Berger-Bachi B, Rohrer S. Factors influencing methicillin resistance in Staphylococci. *Arch Microbiol* 2002; 178:165-171.

Biedenbach, DJ; Moet, GJ; Jones, RN. Occurrence and antimicrobial resistance pattern comparisons among bloodstream infection isolates from the SENTRY Antimicrobial Surveillance Program (1997–2002). *Diagn Microbiol Infect Dis.*, 2004, 50, 59-69.

Billeter, M; Zervos, MJ; Chen, AY; Dalovisio, JR; Kurukularatne, C. Dalbavancin: a novel once-weekly lipoglycopeptide antibiotic. *Clin Infect Dis.*, 2008, 46, 577-583.

Bitar CM, Mayhall CG, Lamb VA, Bradshaw TJ, Spadora AC, Dalton HP. Outbreak due to methicillin-resistant *Staphylococcus aureus*: epidemiology and eradication of the resistant strain from the hospital. *Infect Control* 1987; 8: 15-23.

Borg MA, Cookson BD, Rasslan O, Ben Redjeb S, Benbachir M, Rahal K, Bagatzouni DP, Elnasser Z, Daoud Z, Scicluna EA. Correlation between methicillin-resistant *Staphylococcus aureus* prevalence and infection control initiatives within southern and eastern Mediterranean hospitals. *J Hosp Infect* 2009;71(1):36-42.

Boyle-Vavra S, Daum RS. Community-acquired methicillin-resistant *Staphylococcus aureus*: the role of Panton-Valentine leukocidin. *Lab Invest* 2007;87:3-9.

Bradley SF. *Staphylococcus aureus* pneumonia: emergence of MRSA in the community. *Semin Respir Crit Care Med* 2005;26:643-649.

Bratu S, Eramo A, Kopec R, Coughlin E, Ghitan M, Yost R, Chapnick EK, Landman D, Quale J. Community-associated methicillin-resistant *Staphylococcus aureus* in hospital nursery and maternity units. *Emerg Infec Dis* 2005;11:808-813.

Bryan CS, Wilson RS, Meade P, Sill LG. Topical antibiotic ointments for staphylococcal nasal carriers: survey of current practices and comparison of bacitracin and vancomycin ointments. *Infect Control* 1981; 1: 153-156.

Buescher ES. Community-acquired methicillin-resistant *Staphylococcus aureus* in pediatrics. *Curr Opin Pediatr* 2005; 17:67-70.

Byers DK, Decker CF. The Changing Epidemiology of Methicillin-Resistant *Staphylococcus aureus Dis Mon* 2008;54:756-762

Carleton, HA; Diep, BA; Charlebois, ED; Sensabaugh, GF; Perdreau-Remington, F. Community-adapted methicillin-resistant *Staphylococcus aureus* (MRSA): population dynamics of an expanding community reservoir of MRSA. *J Infect Dis.*, 2004, 190, 1730-1738.

Caron, F; Gold, HS; Wennersten, CB; Farris, MG; Moellering, RC; Jr, Eliopoulos, GM. Influence of erythromycin resistance, in-oculum growth phase, and incubation time on assessment of the bactericidal activity of RP 59500 (quinupristin-dalfopristin) against vancomycin-resistant *Enterococcus faecium.* Antimicrob Agents Chemother, 1997, 41, 2749-2753.

Casey AL, Worthington T, Caddick JM, Hilton AC, Lambert PA, Elliott TSJ. RAPD for the typing of *Staphylococcus aureus* implicated in nosocomial Infection. *J Hosp Infect* 2007;66: 192-193.

Castaldo, ET; Yang, EY. Severe sepsis attributable to community associated methicillin-resistant *Staphylococcus aureus*: an emerging fatal problem. *Am Surg*., 2007, 73, 684-687.

Catherine F. Decker Pathogenesis of MRSA Infections. *Dis Mon* 2008;54:774-779

Cenizal, MJ; Skiest, D; Luber, S; Bedimo, R; Davis, P; Fox, P; Delaney, K; Hardy, RD. Prospective randomized trial of empiric therapy with trimethoprim–sulfamethoxazole or doxycycline for outpatient skin and soft tissue infections in an area of high prevalence of methicillin-resistant *Staphylococcus aureus*. *Antimicrob Agents Chemother*, 2007, 51, 2628-2630.

Centers for Disease Control and Prevention. Community-associated MRSA information for clinicians. Available at: http://www.cdc.gov/ncidod/dhqp/ar_mrsa_ca_clinicians.html. Accessed April 2007.

Centers for Disease Control and Prevention. Four pediatric deaths from community-acquired methicillin-resistant *Staphylococcus aureus* -Minnesota and North Dakota, 1997-1999. JAMA 1999, 282, 1123-5.

Centers for Disease Control and Prevention. *Staphylococcus aureus* resistant to vancomycin: United States, 2002. MMWR *Morb Mortal Wkly Rep* 2002;51: 565-567.

Centers for Disease Control and Prevention: Methicillin-resistant *Staphylococcus aureus* skin or soft tissue infections in a state prison: Mississippi, 2000. MMWR *Morb Mortal Wkly Rep* 2001;50:919-922.

Cercenado, E; Cercenado, S; Gomez, JA; Bouza, E. In vitro activity of tigecycline (GAR-936), a novel glycylcycline, against vancomycin-resistant enterococci and staphylococci with diminished susceptibility to glycopeptides. *J Antimicrob Chemother*, 2003, 52, 138–9.

Chamberland, S; Blais, J; Hoang, M; Park, C; Chan, C; Mathias, K; Hakem, S; Dupree, K; Liu, E; Nguyen, T; Dudley, MN. In vitro activities of RWJ-54428 (MC-02,479) against multiresistant gram-positive bacteria. *Antimicrob Agents Chemother*, 2001, 45, 1422– 1430.

Chambers HF. Methicillin resistance in Staphylococci: molecular and biochemical basis and clinical implications. *Clin Microbiol Rev* 1997;10:781-791.

Chambers, HF. In vitro and in vivo antistaphylococcal activities of L-695,256, a carbapenem with high affinity for the penicillin-binding protein PBP 2a. *Antimicrob Agents Chemother*, 1995, 39, 462–466.

Chambers, HF. Solving staphylococcal resistance to betalactams. *Trends Microbiol*, 2003, 11, 145–148.

Chambers, HF. The changing epidemiology of *Staphylococcus aureus*? *Emerg Infect Dis* 2001; 7: 178-182.

Chang, FY; Peacock, JE; Jr., Musher, DM; Triplett, P; MacDonald, BB; Mylotte, JM; O'Donnell, A; Wagener, MM; Yu, VL. *Staphylococcus aureus* bacteremia: recurrence and the impact of antibiotic treatment in a prospective multicenter study. *Medicine (Baltimore)*, 2003; 82: 333-339.

Charlebois ED, Bangsberg DR, Moss NJ, Moore MR, Moss AR, Chambers HF, Perdreau□Remington F. Population-based community prevalence of methicillin-resistant

Staphylococcus aureus in the urban poor of San Francisco. *Clin Infect Dis* 2002;34:425-433.

Charlebois ED, Perdreau-Reminton F, Kreiswirth B, Bangsberg DR, Ciccarone D, Diep B, Valerie L Ng, Chansky K, Edlin B, Chambers HF. Origins of community strains of methicillin-resistant *Staphyloccus aureus. Clin Infect Dis* 2004;39:47-54.

Charles, PG; Ward, PB; Johnson, PD; Howden, BP; Grayson, ML. Clinical features associated with bacteremia due to heterogeneous vancomycin-intermediate *Staphylococcus aureus. Clin Infect Dis.*, 2004, 38, 448-451.

Chavez TT, Decker CF. Health Care-Associated MRSA Versus Community-Associated MRSA. *Dis Mon* 2008;54:763-768.

Cheeseman KE, Williams GJ, Maillard J-Y, Denyer, SP, Mahenthiralingam E, Typing of *Staphylococcus aureus* clinical isolates using random amplification of polymorphic DNA method and comparison with antibiotic susceptibility typing. *J Hosp Infect* 2007;67(4):388-90.

Chongtrakool P, Ito T, Ma XX, Kondo Y, Trakulsomboon S, Tiensasitorn C, Jamklang M, Chavalit T, Song JH, Hiramatsu K. Staphylococcal cassette chromosome mec (*SCCmec*) typing of methicillin-resistant *Staphylococcus aureus* strains isolated in 11 Asian countries: a proposal for a new nomenclature for *SCCmec* elements. *Antimicrob. Agents Chemother* 2006; 50: 1001-1012.

Clinical and Laboratory Standards Institute. Performance standards for antimicrobial susceptibility testing: 16th informational supplement. Wayne, PA: *Clinical and Laboratory Standards Institute;* 2006, M100-S16

CLSI Performance standards for antimicrobial susceptibility testing: seventeenth informational supplement. M100–S17. Wayne, PA: Clinical and Laboratory Standards Institute, 2007;52.

Cohen PR. Cutaneous community-acquired methicillin-resistant *Staphylococcus aureus* infection in participants of athletic activities. *South Med J* 2005; 98:596-602.

Cosgrove, SE; Sakoulas, G; Perencevich, EN; Schwaber, MJ; Karchmer, AW; Carmeli, Y. Comparison of mortality associated with methicillin resistant and methicillin-susceptible *Staphylococcus aureus* bacteremia: a meta-analysis. *Clin Infect Dis.*, 2003, 36, 53-59.

Crawford DH. *Deadly companions: How microbes shaped our history*. Oxford: Oxford University Press; 2008.

Crowcroft NS, Catchpole M. Mortality from methicillin resistant *Staphylococcus aureus* in England and Wales: analysis of death certificates. *Br Med J* 2002;325:1390-1391.

Cruciani, M; Gatti, G; Lazzarini, L; Furlan, G; Broccali, G; Malena, M; Franchini, C; Concia, E. Penetration of vancomycin into human lung tissue. *J Antimicrob Chemother*, 1996; 38: 865-869.

Cui L, Murakami H, Kuwahara-Arai K, Hanaki H, Hiramatsu K. Contribution of a thickened cell wall and its gultamine nonamidated component to the vancomycin resistance expressed by *Staphylococcus aureus* Mu50. *Antimicrob Agents Chemother* 2000; 44: 2276-2285.

Cui, L; Tominaga, E; Neoh, HM; Hiramatsu, K. Correlation between reduced daptomycin susceptibility and vancomycin resistance in vancomycin-intermediate *Staphylococcus aureus. Antimicrob Agents Chemother*, 2006; 50: 1079-1082.

Dancer SJ. Keeping watch over the Staphylococcus. *J Hosp Infect* 2008; 70: 297.

Darouiche R, Wright C, Hamill R, Koza M, Lewis D, Markowski J. Eradication of colonization by methicillin-resistant *Staphylococcus aureus* by using oral minocycline–rifampin and topical mupirocin. *Antimicrob Agents Chemother* 1991; 35: 1612-1615.

Daum RS, Ito T, Hiramatsu K, Hussain F, Mongkolrattanothai K, Jamklang M, Boyle-Vavra S. A novel methicillin-resistance cassette in community-acquired methicillin-resistant *Staphylococcus aureus* isolates of diverse genetic backgrounds. *J Infect Dis* 2002; 186:1344-1347.

Daum RS. Community-acquired methicillin-resistant *Staphylococcus aureus* infections. *Pediatr Infect Dis J* 1998; 17:745-746.

Daver, NG; Shelburne, SA; Atmar, RL; Giordano, TP; Stager, CE; Reitman, CA, White, AC. Oral step-down therapy is comparable to intravenous therapy for *Staphylococcus aureus* osteomyelitis. *Journal of Infection*, 2007; 54(6): 539-544.

David MZ, Glikman D, Crawford SE, Peng J, King KJ, Hostetler MA, Boyle-Vavra S, Daum RS. What Is community-associated methicillin-resistant *Staphylococcus aureus*? *J Infect Dis* 2008; 197: 1235-1243.

Davis KA, Stewart JJ, Crouch HK, Florez CE, Hospenthal DR. Methicillin-resistant *Staphylococcus aureus* (MRSA) nares colonization at hospital admission and its effect on subsequent MRSA infection. *Clin Infect Dis* 2004;39 (6):776-782.

Decker BS, Babel CF. Microbiology and Laboratory Diagnosis of MRSA. *Dis Mon* 2008;54:769-773

Denis O, Nonhoff C, Byl B, Knoop C, Bobin-Dubreux S, Struelens MJ. Emergence of vancomycin-intermediate *Staphylococcus aureus* in a Belgian hospital: microbiological and clinical features. *J Antimicrob Chemother* 2002; 50: 383-391.

Deresinski, S. Counterpoint: Vancomycin and *Staphylococcus aureus* an antibiotic enters obsolescence. *Clin Infect Dis.*, 2007a; 44: 1543-1548.

Deresinski, S. Principles of antibiotic therapy in severe infections: optimizing the therapeutic approach by use of laboratory and clinical data. *Clin Infect Dis.*, 2007b; 45: 177-183.

Deshpande, L; Rhomberg, PR; Fritsche, TR; Sader, HS; Jones, RN. Bactericidal activity of BAL9141, a novel parenteral cephalosporin against contemporary Gram-positive and Gram-negative isolates. *Diagn Microbiol Infect Dis.*, 2004; 50: 73–5.

Deurenberg RH, Vink C, Kalenic S, Friedrich AW, Bruggeman CA, Stobberingh EE. The molecular evolution of methicillin resistant *Staphylococcus aureus. Clin Microbiol Infect* 2007;13 (3):222-235.

Diederen BM, Kluytmans JA. The emergence of infections with community associated methicillin resistant *Staphylococcus aureus. J Infect* 2006;52:157-168.

Diep BA, Carleton HA, Chang RF, Friedrich AW, Bruggeman CA, Stobberingh EE. Role of 34 virulence genes in the evolution of hospital-and community-associated strains of methicillin-resistant *Staphylococcus aureus. J Infect Dis* 2006a;193:1495-1503.

Diep BA, Gill SR, Chang RF, Phan TH, Chen JH, Davidson MG, Lin F, Lin J, Carleton HA, Mongodin EF, SensabaughGF, Perdreau-Remington F. Complete genome sequence of USA300, an epidemic clone of community-acquired methicillin-resistant *Staphylococcus aureus. Lancet* 2006b;367:731-739.

Diep BA, Stone GG, Basuino L, Graber CJ, Miller A, des Etages SA, Jones A, Palazzolo-Ballance AM, Perdreau-Remington F, Sensabaugh GF, DeLeo FR, Chambers HF. The arginine catabolic mobile element and staphylococcal chromosomal cassette mec linkage:

convergence of virulence and resistance in the USA300clone of methicillin-resistant *Staphylococcus aureus. J Infect Dis* 2008;97(11):1523-1530.

Dombrowski, JC; Winston, LG. Clinical failures of appropriately-treated methicillin-resistant *Staphylococcus aureus* infections. *Journal of Infection*, 2008; 57: 110-115.

Drekonja, DM; Traynor, LM; DeCarolis, DD; Crossley, KB; Johnson, JR. Treatment of non–life-threatening methicillin-resistant *Staphylococcus aureus* infections with alternative antimicrobial agents: a 2-year retrospective review. *Diagnostic Microbiology and Infectious Disease*, 2009; 63(2): 201-207.

Duckworth G, Cookson B, Humphreys H, Heathcock R. Working party report. Revised guidelines for the control of methicillin-resistant *Staphylococcus aureus* infection in hospitals. *J Hosp Infect* 1998; 39: 253-290.

Duckworth G. Improving surveillance of MRSA bacteraemia. Br Med J 2005;331:976-977.

Eady EA, Cove JH. Staphylococcal resistance revisited: community-acquired methicillin resistant *Staphylococcus aureus*-an emerging problem for the management of skin and soft tissue infections. *Curr Opin Infect Dis* 2003;16:103-124.

ElKamel, A; Baddour, MM. Gatifloxacin-biodegradable implant for treatment of experimental osteomyelitis. In vitro and in vivo evaluation. *Drug Delivery*, 2007; 14 (6): 349-356.

Ellis MW, Hospenthal DR, Dooley DP, Gray PJ, Murray CK. Natural history of community-acquired methicillin-resistant *Staphylococcus aureus* colonization and infection in soldiers. *Clin Infect Dis* 2004; 39:971-979.

Enoch, DA; Karas, JA; Aliyu, SH. Oral antimicrobial options for the treatment of skin and soft-tissue infections caused by methicillin-resistant *Staphylococcus aureus* (MRSA) in the UK. *Int. J. Antimicrob. Agents* (2008), doi:10.1016/j.ijantimicag.2008.10.014

Enright MC, Spratt BG. Multilocus sequence typing. *Trends in Microbiology* 1999; 7: 482-487.

Esposito, AL; Gleckman, RA. Vancomycin: a second look. *J Am Med Assoc.*, 1977, 238, 1756-1757.

Etienne J. Panton-Valentine leukocidin: a marker of severity for *Staphylococcus aureus* infection? *Clin Infect Dis* 2005; 41: 591-593.

Euzéby JP. Definitions and abbreviatons. J.P. Euzéby: List of Bacteria Names with Standing in Nomenclature. 2004 http://www.bacterio.cict.fr2004.

Falagas, ME; Giannopoulou, KP; Ntziora, F; Vardakas, KZ. Daptomycin for endocarditis and/or bacteraemia: a systematic review of the experimental and clinical evidence. *J Antimicrob Chemother*, 2007, 60, 7-19.

Flynn N, Cohen SH. The continuing saga of MRSA. *J Infect Dis* 2008;197:1217-1219.

Fowler, Jr VG; Kong, LK; Corey, GR; Gottlieb, GS; McClelland, RS; Sexton, DJ; Gesty-Palmer, D; Harrell, LJ. Recurrent *Staphylococcus aureus* bacteremia: pulsed-field gel electrophoresis findings in 29 patients. *J Infect Dis.*, 1999, 179, 1157-1161.

Fowler, VG; Jr., Boucher, HW; Corey, GR; Abrutyn, E; Karchmer, AW; Rupp, ME; Levine, DP; Chambers, HF; Tally, FP; Vigliani, GA; Cabell, CH; Link, AS; DeMeyer, I; Filler, SG; Zervos, M; Cook, P; Parsonnet, J; Bernstein, JM; Price, CS; Forrest, GN; Fatkenheuer, G; Gareca, M; Rehm, SJ; Brodt, HR; Tice, A; Cosgrove, SE. Daptomycin versus standard therapy for bacteremia and endocarditis caused by *Staphylococcus aureus*. *N Engl J Med.*, 2006, 355, 653-665.

Fraise AP. Guidelines for the control of methicillin resistant *Staphylococcus aureus*. *J Antimicrob Chemother* 1998; 42: 287-289.

Francis JS, Doherty MC, Lopatin U, Johnston CP, Sinha G, Ross T, Cai M, Hansel NN, Perl T, Ticehurst JR, Carroll K, Thomas DL, Nuermberger E, Bartlett JG. Severe community-onset pneumonia in healthy adults caused by methicillin-resistant *Staphylococcus aureus* carrying the Panton-Valentine leukocidin genes. *Clin Infect Dis* 2005;40:100-107.

Fraser C, Riley S, Anderson RM, Ferguson NM. Factors that make an infectious disease outbreak controllable. *Proc Natl Acad Sci U S A* 2004; 101:6146-6151.

Frenay HM, Bunschoten AE, Schouls LM, van Leeuwen WJ, Vandenbroucke-Grauls CM, Verhoef J, Mooi FR. Molecular typing of methicillin resistant *Staphylococcus aureus* on the basis of protein A gene polymorphism. *Eur J Clin Microbiol Infect Dis* 1996; 15: 60-64.

Fridkin SK, Hageman J, McDougal LK , Mohammed J, Jarvis WR, Perl TM, Tenover FC; Vancomycin-Intermediate *Staphylococcus*. Epidemiological and microbiological characterization of infections caused by *Staphylococcus aureus* with reduced susceptibility to vancomycin, United States, 1997–2001. *Clin Infect Dis* 2003; 36: 429-439.

Fridkin SK, Hageman JC, Morrison M, Sanza LT, Como-Sabetti K, Jernigan JA, Harriman K, Harrison LH, Lynfield R, Farley MM; Active Bacterial Core Surveillance Program of the Emerging Infections Program Network. Methicillin-resistant *Staphylococcus aureus* disease in three communities. *N Engl J Med* 2005;352:1436-1444.

Fritsche, TR; Sader, HS; Jones, RN. Antimicrobial activity of ceftobiprole, a novel anti-methicillin-resistant *Staphylococcus aureus* cephalosporin, tested against contemporary pathogens: results from the SENTRY Antimicrobial Surveillance Program (2005–2006). *Diagn Microbiol Infect Dis*., 2008, 61, 86–95.

Fuchs, PC; Barry, AL; Brown, SD. Evaluation of daptomycin susceptibility testing by E-test and the effect of different batches of media. *J Antimicrob Chemother*, 2001, 48, 557-561.

Gales, AC; Sader, HS; Andrade, SS; Lutz, L; Machado, A; Barth, AL. Emergence of linezolid-resistant *Staphylococcus aureus* during treatment of pulmonary infection in a patient with cystic fibrosis. *Int J Antimicrob Agents*, 2006, 27, 300-302.

Garazzino, S; Aprato, A; Baietto, L; D'Avolio, A; Maiello, A; De Rosa, FG; Aloj, D; Siccardi, M; Biasibetti, A; Massè, A; Di Perri, G. Glycopeptide bone penetration in patients with septic pseudoarthrosis of the tibia. *Clin Pharmacokinet*, 2008, 47(12), 793-805.

Garnier F, Tristan A, Francois B, Etienne J, Delage-Corre M, Martin C, Liassine N, Wannet W, Denis F, Ploy MC. Pneumonia and new methicillin-resistant *Staphylococcus aureus* clone. *Emerg Infect Dis* 2006;12:498-500.

Garrity GM, Holt JG. 2001. The road map to the manual. In: Boone, R.D., Castenholz, R.W., Garrity, G.M. (Eds.), *Bergey's Manual of Systematic Bacteriology*, 2nd ed. Springer-Verlag, New York, NY, pp. 119-166.

Gentry, CA; Rodvold, KA; Novak, RM; Hershow, RC; Naderer, OJ. Retrospective evaluation of therapies for *Staphylococcus aureus* endocarditis. *Pharmacotherapy*, 1997; 17, 990-997.

Giammarinaro P, Leroy S, Chacornac JP, Delmas J, Talon R. Development of a new oligonucleotide array to identify staphylococcal strains at species level. *J Clin Microbiol* 2005; 43 (8): 3673-3680.

Gilbert, DN; Mollering, RC; Eliopoulos, GM; Sande, MA; editors. The Sanford guide to antimicrobial therapy. Sperryville, VA: *Antimicrobial Therapy*, Inc.; 2007.

Gillett Y, Issartel B, Vanhems P, Fournet JC, Lina G, Bes M, Vandenesch F, Piémont Y, Brousse N, Floret D, Etienne J. Association between *Staphylococcus aureus* strains carrying gene for Panton-Valentine leukocidin and highly lethal necrotizing pneumonia in young immunocompetent patients. *Lancet* 2002;359:753-759.

Gonzales, RD; Schreckenberger, PC; Graham, MB; Kelkar, S; DenBesten, K; Quinn, JP. Infections due to vancomycin-resistant *Enterococcus faecium* resistant to linezolid. *Lancet*, 2001; 357, 1179.

Gonzalez, C; Rubio, M; Romero-Vivas, J; Gonzalez, M; Picazo, JJ. Bacteremic pneumonia due to *Staphylococcus aureus*: a comparison of disease caused by methicillin-resistant and methicillin-susceptible organisms. *Clin Infect Dis*., 1999, 29, 1171–1177.

Gorwitz RJ, Jernigan DB, Powers JH. Strategies for clinical management of MRSA in the community: summary of an experts' meeting convened by the Centers for Disease Control and Prevention, 2006. Available at: http://www.cdc.gov/ncidod/dhqp/ar_mrsa_ca.html. Accessed May 15, 2008.

Gorwitz, RJ. The role of ancillary antimicrobial therapy for treatment of uncomplicated skin infections in the era of community-associated methicillin-resistant *Staphylococcus aureus*. *Clin Infect Dis*., 2007, 44, 785-787.

Gould IM. Costs of hospital-acquired methicillin-resistant *Staphylococcus aureus* (MRSA) and its control. *Int J Antimicrob Agents* 2006;28:379-384.

Gould, I. Clinical relevance of increasing glycopeptide MICs against *Staphylococcus aureus*. *Int J Antimicrob Agents*, 2008, 31, 1-9.

Gould, IM. MRSA bacteraemia. *Int J Antimicrob Agents*, 2007b, 30 Suppl 1, S66-S70.

Gould, IM. The problem with glycopeptides. *Int J Antimicrob Agents*, 2007a, 30, 1-3.

Gradon JD, Wu EH, Lutwick LI. Aerosolized vancomycin therapy facilitating nursing home placement. *Ann Pharmacother* 1992; 26: 209-210.

Grundmann H, Aires-de-Sousa M, Boyce J, Tiemersma E. Emergence and resurgence of methicillin-resistant *Staphylococcus aureus* as a public-health threat. *Lancet* 2006; 368(9538):874-885.

Gur D, Unal S. Resistance to antimicrobial agents in Mediterranean countries. *Int J Antimicrob Agents* 2001; 17:21-26.

Hahn RC, Macedo AM, Fontes CJF, Batista RD, Santos NL, Hamdan S. Randomly amplified polymorphic DNA as a valuable tool for epidemiological studies of *Paracoccidioides brasiliensis*. *J Clin Microbiol* 2003;41:2849-2854.

Hails J, Kwaku F, Wilson AP, Bellingan G, Singer M. Large variation in MRSA policies, procedures and prevalence in English intensive care units: a questionnaire analysis. *Intensive Care Med* 2003;29:481-483.

Harbarth S, Dharan S, Liassine N, Herrault P, Auckenthaler R, Pittet D. Randomized, placebo controlled, double-blind trial to evaluate the efficacy of mupirocin for eradicating carriage of methicillin-resistant *Staphylococcus aureus*. *Antimicrob Agents Chemother* 1999; 43: 1412-1416.

Hayden, MK; Rezai, K; Hayes, RA; Lolans, K; Quinn, JP; Weinstein, RA. Development of Daptomycin resistance in vivo in methicillin-resistant *Staphylococcus aureus*. *J Clin Microbiol*, 2005, 43, 5285-5287.

Health Protection Agency, 2005. *Staphylococcus aureus* bacteraemia laboratory reports and methicillin susceptibility (voluntary reporting scheme): England and Wales, 1990–2003. http://www.hpa.org.uk/infections topics az/staphylo/lab data staphyl.htm. [Accessed May 2007].

Healy M, Huong J, Bittner T, Lising M, Frye S, Raza S, Schrock R, Manry J, Renwick A, Nieto R, Woods C, Versalovic J, Lupski JR. Microbial DNA typing by automated repetitive-sequence-based PCR. *J Clin Microbiol* 2005; 43:199-207.

Hebeisen, P; Heinze-Krauss, I; Angehrn, P; Hohl, P; Page, MG; Then, RL. In vitro and in vivo properties of Ro 63- 9141, a novel broad-spectrum cephalosporin with activity against methicillin-resistant staphylococci. *Antimicrob Agents Chemother*, 2001, 45, 825–836.

Herrera MO, Gallegos AG, Schacht FC, Fernandez LP, Jimenez RC. RAPD-PCR characterization of *Pseudomonas aeruginosa* strains obtained from cystic fibrosis patients. *Salud Publica Mex* 2004;46:149-157.

Hill RLR, Duckworth GJ, Casewell NW. Elimination of nasal carriage of methicillin resistant *Staphylococcus aureus* with mupirocin during a hospital outbreak. *J Antimicrob Chemother* 1998; 22: 377-384.

Hiramatsu K, Aritaka N, Hanaki H, Kawasaki S, Hosoda Y, Hori S, Fukuchi Y, Kobayashi I. Dissemination in Japanese hospitals of strains of *Staphylococcus aureus* heterogeneously resistant to vancomycin. *Lancet* 1997;350:1670-1673.

Hiramatsu K, Hanaki H, Ino T, Yabuta K, Oguri T, Tenover FC. Methicillin-resistant *Staphylococcus aureus* clinical strain with reduced vancomycin susceptibility. *J Antimicrob Chemother* 1997; 40: 135-136.

Hiramatsu K. Vancomycin-resistant *Staphylococcus aureus*: a new model of antibiotic resistance. *Lancet Infect Dis* 2001; 1: 147-155

Hisata K, Kuwahara-Arai K, Yamanoto M, Ito T, Nakatomi Y, Cui L, Baba T, Terasawa M, Sotozono C, Kinoshita S, Yamashiro Y, Hiramatsu K. Dissemination of methicillin-resistant staphylococci among healthy Japanese children. *J Clin Microbiol* 2005; 43:3364-3372.

Hoellman, DB; Pankuch, GA; Jacobs, MR; Appelbaum, PC. Antipneumococcal activities of GAR-936 (a new glycylcycline) compared to those of nine other agents against penicillin-susceptible and -resistant pneumococci. *Antimicrob Agents Chemother*, 2000, 44, 1085–8.

Hood J, Edwards GFS, Cosgrove B, Curran E, Morrison D, Gemmell CG. Vancomycin-intermediate *Staphylococcus aureus* at a Scottish hospital. *J Infect* 2000;40:A11.

Howden BP, Ward PB, Charles PG Korman TM, Fuller A, du Cros P, Grabsch EA, Roberts SA, Robson J, Read K, Bak N, Hurley J, Johnson PD, Morris AJ, Mayall BC, Grayson ML. Treatment outcomes for serious infections caused by methicillin-resistant *Staphylococcus aureus* with reduced vancomycin susceptibility. *Clin Infect Dis* 2004; 38: 521-528.

Howe, RA; Wootton, M; Noel, AR; Bowker, KE; Walsh, TR; MacGowan, AP. Activity of AZD2563, a novel oxazolidinone, against *Staphylococcus aureus* strains with reduced susceptibility to vancomycin or linezolid. *Antimicrob Agents Chemother*, 2003, 47, 3651-3652.

Huang SS, Platt R. Risk of methicillin-resistant *Staphylococcus aureus* infection after previous infection or colonization. *Clin Infect Dis* 2003;36:281-285.

Huang YH, Tseng SP, Hu JM, Tsai JC, Hsueh PR, Teng LJ. Clonal spread of *SCCmec* type IV methicillin-resistant *Staphylococcus aureus* between community and hospital. *Clin Microbiol Infect* 2007;13:717-724.

Hueletsky A, Giroux R, Rossbach V, et al. New real-time PCR assay for rapid detection of methicillin resistant *Staphylococcus aureus* directly from specimens containing a mixture of staphylococci. *J Clin Microbiol* 2004;42:1835-1841.

Ibrahim, EH; Sherman, G; Ward, S; Fraser, VJ; Kollef, MH. The influence of inadequate antimicrobial treatment of bloodstream infections on patient outcomes in the ICU setting. *Chest*, 2000, 118, 9-11.

Iizawa, Y; Nagai, J; Ishikawa, T; Hashiguchi, S; Nakao, M; Miyake, A; Okonogi, K. In vitro antimicrobial activity of T-91825, a novel anti-MRSA cephalosporin, and in vivo anti-MRSA activity of its prodrug, TAK-599. *J Infect Chemother*, 2004, 10, 146–156.

Infectious Disease Society of America (IDSA). Bad Bugs, No Drugs. As Antibiotic Discovery Stagnates. A Public Health Crisis Brews. Alexandria, Va: IDSA, 2004.

Ito T, Katayama Y, Asada K, Mori N, Tsutsumimoto K, Tiensasitorn C, Hiramatsu K. Structural comparison of three types of staphylococcal cassette chromosome mec integrated in the chromosome in methicillin-resistant *Staphylococcus aureus*. *Antimicrob Agents Chemother* 2001;45: 1323-1336.

Ito T, Ma XX, Takeuchi F, Okuma K, Yuzawa H, Hiramatsu K. Novel type V staphylococcal cassette chromosome mec driven by a novel cassette chromosome recombinase, ccrC. *Antimicrob Agents Chemother* 2004; 48 :2637-2651.

Ito T, Okuma K, Ma XX, Yuzawa H, Hiramatsu K. Insights on antibiotic resistance of *Staphylococcus aureus* from its whole genome: genomic island SCC. *Drug Resist Updat* 2003; 6:41-52.

Iyer S, Jones DH. Community-acquired methicillin-resistant *Staphylococcus aureus* skin infection: a retrospective analysis of clinical presentation and treatment of a local outbreak. *J Am Acad Dermatol* 2004; 50:854-858.

Jansen WT, Beitsma MM, Koeman CJ, van Wamel WJ, Verhoef J, Fluit AC. Novel mobile variants of staphylococcal cassette chromosome mec in *Staphylococcus aureus*. Antimicrob. *Agents Chemother* 2006: 50: 2072-2078.

Jauregui, LE; Babazadeh, S; Seltzer, E; Goldberg, L; Krievins, D; Frederick, M; Krause, D; Satilovs, I; Endzinas, Z; Breaux, J; O'Riordan, W. Randomized, double-blind comparison of once-weekly dalbavancin versus twice-daily linezolid therapy for the treatment of complicated skin and skin structure infections. *Clin Infect Dis*., 2005, 41, 1407 15.

Jeffres, MN; Isakow, W; Doherty, JA; Micek, ST; Kollef, MH. A retrospective study of possible renal toxicity in patients with healthcare-associated meticillin resistant *Staphylococcus aureus* pneumonia treated with vancomycin. *Clin Ther*., 2007, 29, 1107-1115.

Jevons M. "Celbenin"-resistant staphylococci. *Br Med J*. 1961:124-125.

Johnson AP, Aucken HM, Cavendish S, et al. Dominance of EMRSA-15 and -16 among MRSA causing nosocomial bacteraemia in the UK: analysis of isolates from the European Antimicrobial Resistance Surveillance System (EARSS). *J Antimicrob Chemother* 2001;48:143-144.

Johnson AP, Pearson A, Duckworth G. Surveillance and epidemiology of MRSA bacteraemia in the UK. *J Antimicrob Chemother* 2005;56:455–62.

Jones, RN; Della-Latta, P; Lee, LV; Biedenbach, DJ. Linezolid resistant *Enterococcus faecium* isolated from a patient without prior exposure to an oxazolidinone: report from the SENTRY Antimicrobial Surveillance Program. *Diagn Microbiol Infect Dis.*, 2002, 42.

Jones, RN; Deshpande, LM; Mutnick, AH; Biedenbach, DJ. In vitro evaluation of BAL9141, a novel parenteral cephalosporin active against oxacillin-resistant staphylococci. *J Antimicrob Chemother*, 2002, 50, 915–32.

Kaatz GW, Seo SM, Dorman NJ, Lerner SA. Emergence of teicoplanin resistance during therapy of *Staphylococcus aureus* endocarditis. *J Infect Dis* 1990; 162: 103-108.

Kahl BC, Mellmann A, Deiwick S, Peters G, Harmsen D. Variation of the polymorphic region X of the protein A gene during persistent airway infection of cystic fibrosis patients reflects two independent mechanisms of genetic change in *Staphylococcus aureus*. *J Clin Microbiol* 2005; 43: 502-505.

Kainer, MA; Devasia, RA; Jones, TF; Simmons, BP; Melton, K; Chow, S; Broyles, J; Moore, KL; Craig, AS; Schaffner, W. Response to emerging infection leading to outbreak of linezolid-resistant enterococci. *Emerg Infect Dis.*, 2007, 13, 1024-103.

Kalteis, T; Beckmann, J; Schröder, HJ; Handel, M; Grifka, J; Lehn, N; Lerch, K. Acta Orthop. Moxifloxacin superior to vancomycin for treatment of bone infections--a study in rats, 2006, 77(2), 315-319.

Kaneko J, Kamio Y. Bacterial two-component and hetero-heptameric poreforming cytolytic toxins: structures, pore-forming mechanism, and organization of the genes. *Biosci Biotechnol Biochem* 2004; 68: 981-1003.

Katayama Y, Ito T, Hiramatsu K. Genetic organization of the chromosome region surrounding *mecA* in clinical staphylococcal strains: role of IS431-mediated mecI deletion in expression of resistance in *mecA*-carrying, low-level methicillin-resistant Staphylococcus haemolyticus. *Antimicrob Agents Chemother* 2001;45: 1955-1963.

Kauffman CA, Terpenning MS, He X, Zarins LT, Ramsey MA, Jorgensen KA, Sottile WS, Bradley SF. Attempts to eradicate methicillin-resistant *Staphylococcus aureus* from a long-term-care facility with the use of mupirocin ointment. *Am J Med* 1993; 94: 371-378.

Kazakova SV, Hageman JC, Matava ME, Srinivasan A, Phelan L, Garfinkel B, Boo T, McAllister S, Anderson J, Jensen B, Dodson D, Lonsway D, McDougal LK, Arduino M, Fraser VJ, Killgore G, Tenover FC, Cody S, Jernigan DB. A clone of methicillin-resistant *Staphylococcus aureus* among professional football players. *N Engl J Med* 2005;352:468-475.

Keshtgar MR, Khalili A, Coen PG, *et al.*,.. Impact of rapid molecular screening for methicillin-resistant *Staphylococcus aureus* in surgical wards. *Br J Surg* 2008;95:381-386.

Kim J, Jeong JH, Cha HY, Jin JS, Lee JC, Lee YC, Seol SY, Cho DT. Detection of diverse *SCCmec* variants in methicillin-resistant *Staphylococcus aureus* and comparison of *SCCmec* typing methods. *Clin Microbiol Infect* 2007; 13: 1128-1130.

King MD, Humphrey BJ, Wang YF, Kourbatova EV, Ray SM, Blumberg HM. Emergence of community-acquired methicillin-resistant *Staphylococcus aureus* USA 300 clone as the predominant cause of skin and soft-tissue infections. *Ann Intern Med* 2006; 144:309-317.

Klevens, RM; Edwards, JR; Tenover, FC; McDonald, LC; Horan, T; Gaynes, R. Changes in the epidemiology of methicillin-resistant *Staphylococcus aureus* in intensive care units in US hospitals, 1992–2003. *Clin Infect Dis.*, 2006, 42, 389-391.

Kline MD, Humphrey BS, Wang YF, Kourbatova EV, Ray SM, Blumberg HM. Emergence of community-acquired methicillin-resistant *Staphylococcus aureus* USA 300 clone as the predominant cause of skin and soft tissue infections. *Ann Intern Med* 2006;144:309-317.

Kloos WE, Bannerman TL. 1995. Staphylococcus and Micrococcus. In: Murray, P.R., Baron, E.J., Pfaller, M.A., Tenover, F.C., Yolken, R.H. (Eds.), Manual of Clinical Microbiology, 6th ed. ASM Press, Washington, D.C., USA, pp. 282-298.

Kluytmans J, Van Belkum A, Verbrugh H. Nasal carriage of *Staphylococcus aureus*: epidemiology, underlying mechanism, and associated risks. *Clin Microbiol Rev* 1997;10:505-20.

Kluytmans-Vandenbergh MF, Kluytmans JA. Community-acquired methicillin-resistant *Staphylococcus aureus*: current perspectives. *Clin Microbiol Infect* 2006;12:697-698.

Koga, T; Abe, T; Inoue, H; Takenouchi, T; Kitayama, A; Yoshida, T; Masuda, N; Sugihara, C; Kakuta, M; Nakagawa, M; Shibayama, T; Matsushita, Y; Hirota, T; Ohya, S; Utsui, Y; Fukuoka, T; Kuwahara, S. In vitro and in vivo antibacterial activities of CS-023 (RO4908463), a novel parenteral carbapenem. *Antimicrob Agents Chemother*, 2005, 49, 3239–3250.

Kollef, MH. Limitations of vancomycin in the management of resistant staphylococcal infections. *Clin Infect Dis*., 2007, 45(Suppl 3):S191-195.

Kollef, MH; Inadequate antimicrobial treatment: an important determinant of outcome for hospitalized patients. *Clin Infect Dis*., 2000, 31, 131-138.

Koreen L, Ramaswamy SV, Graviss EA, Naidich S, Musser JM, Kreiswirth BN. spa typing method for discriminating among *Staphylococcus aureus* isolates: implications for use of a single marker to detect genetic micro- and macrovariation. *J Clin Microbiol* 2004;42: 792-799.

Kourbatova EV, Halvosa JS, King MD, Ray SM, White N, Blumberg HM. Emergence of community-associated methicillin-resistant *Staphylococcus aureus* USA300 clone as a cause of healthcare-associated infections among patients with prosthetic joint infections. *Am J Infect Control* 2005;33:385-391.

Kumar, A; Roberts, D; Wood, KE; Light, B; Parrillo, JE; Sharma, S; Suppes, R; Feinstein, D; Zanotti, S; Taiberg, L; Gurka, D; Kumar, A; Cheang, M. Duration of hypotension before initiation of effective antimicrobial therapy is the critical determinant of survival in human septic shock. *Crit Care Med*., 2006, 34, 1589-1596.

Labandeira-Rey M, Couzon F, Boisset S, Brown EL, Bes M, Benito Y, Barbu EM, Vazquez V, Höök M, Etienne J, Vandenesch F, Bowden MG. *Staphylococcus aureus* Panton-Valentine leukocidin causes necrotizing pneumonia. *Science* 2007;315:1130-1133.

Lamer, C; De Beco, V; Soler, P; Calvat, S; Fagon, JY; Dombret, MC; Farinotti, R; Chastre, J; Gibert, C. Analysis of vancomycin entry into pulmonary lining fluid by bronchoalveolar lavage in critically ill patients. *Antimicrob Agents Chemother*, 1993, 37, 281-286.

LaPlante, KL; Rybak, MJ. Clinical glycopeptide-intermediate staphylococci tested against arbekacin, daptomycin, and tigecycline. *Diagn Microbiol Infect Dis*., 2004, 50, 125-130.

Leclercq R. Mechanisms of resistance to macrolides and lincosamides: nature of the resistance elements and their clinical implications. *Clin Infect Dis* 2002;34: 482-492.

Lee, MC; Rios, AM; Aten, MF; Mejias, A; Cavuoti, D; McCracken, Jr GH; Hardy, RD. Management and outcome of children with skin and soft tissue abscesses caused by community-acquired methicillin-resistant *Staphylococcus aureus*. *Pediatr Infect Dis J.*, 2004, 23, 123-127.

Levine, DP; Fromm, BS; Reddy, BR. Slow response to vancomycin or vancomycin plus rifampin in methicillin-resistant *Staphylococcus aureus* endocarditis. *Ann Intern Med.*, 1991, 115, 674-680.

Li, JZ; Willke, RJ; Rittenhouse, BE; Rybak, MJ. Effect of linezolid versus vancomycin on length of hospital stay in patients with complicated skin and soft tissue infections caused by known or suspected methicillin-resistant staphylococci: results from a randomized clinical trial. *Surg Infect (Larchmt)*, 2003, 4, 57-70.

Lim, D; Strynadka, NC. Structural basis for the beta lactam resistance of PBP2a from methicillin-resistant *Staphylococcus aureus*. *Nat Struct Biol.*, 2002, 9, 870–876.

Lina G, Piemont Y, Godail-Gamot F, Bes M, Peter MO, Gauduchon V, Vandenesch F, Etienne J. Involvement of PantoneValentine leukocidin-producing *Staphylococcus aureus* in primary skin infections and pneumonia. *Clin Infect Dis* 1999;29:1128-1132.

Lindsay JA, Holden MT. Understanding the rise of the superbug: investigation of the evolution and genomic variation of *Staphylococcus aureus*. *Funct Integr Genomics* 2006;6: 186-201.

Liu C, Chambers HF. *Staphylococcus aureus* with heterogeneous resistance to vancomycin: epidemiology, clinical significance, and critical assessment of diagnostic methods. *Antimicrob Agents Chemother* 2003; 47: 3040-3045.

Liu C, Graber CJ, Karr M, Diep BA, Basuino L, Schwartz BS, Enright MC, O'Hanlon SJ, Thomas JC, Perdreau-Remington F, Gordon S, Gunthorpe H, Jacobs R, Jensen P, Leoung G, Rumack JS, Chambers HF. A population-based study of the incidence and molecular epidemiology of methicillin-resistant *Staphylococcus aureus* disease in San Francisco, 2004-2005. *Clin Infect Dis* 2008;46(11):1637-1646.

Livermore DM. Can β-lactams be re-engineered to beat MRSA? *Clin Microbiol Infect* 2006;12(Suppl.2):11-16.

LoVecchio, F; Perera, N; Casanova, L; Mulrow, M; Pohl, A. Board-certified emergency physicians' treatment of skin and soft tissue infections in the community-acquired methicillin-resistant Staphylococcus aureus era. *Am J Emerg Med.*, 2009, 27(1), 68-70.

Lowy FD. *Staphylococcus aureus* infections. *N Engl J Med* 1998;339:520-532.

Lu D, Holtom P. Community-acquired methicillin-resistant *Staphylococcus aureus*, a new player in sports medicine. *Curr Sports Med Rep* 2005; 4:265-270.

Ma XX, Galiana A, Pedreira W, Mowszowicz M, Christophersen I, Machiavello S, Lope L, Benaderet S, Buela F, Vincentino W, Albini M, Bertaux O, Constenla I, Bagnulo H, Llosa L, Ito T, Hiramatsu K. Community-acquired methicillin-resistant *Staphylococcus aureus*, Uruguay. *Emerg Infect Dis* 2005; 11:973-976.

Malachowa N, Sabat A, Gniadkowski M, Krzyszton-Russjan J, Empel J, Miedzobrodzki J, Kosowska-Shick K, Appelbaum PC, Hryniewicz W, 2005. Comparison of multiple-locus variable-number tandem-repeat analysis with pulsed-field gel electrophoresis, spa typing, and multilocus sequence typing for clonal characterization of *Staphylococcus aureus* isolates. *J Clin Microbiol* 2005; 43: 3095-3100.

Mangili, A; Bica, I; Snydman, DR; Hamer, DH. Daptomycin-resistant, methicillin-resistant *Staphylococcus aureus* bacteremia. *Clin Infect Dis.*, 2005, 40, 1058-1060.

Maquelin K, Cookson B, Tassios P, van Belkum A. Current trends in the epidemiological typing of clinically relevant microbes in Europe. *J Microbiol Methods* 2007; 69: 222-226.

Maraha B, van Halteren J, Verzij JM, Wintermans RGF, Buiting AGM. Decolonization of methicillin-resistant *Staphylococcus aureus* using oral vancomycin and topical mupirocin. *Clin Microbiol Infect* 2002; 8: 671- 675.

Maree CL, Daum RS, Boyle-Vavra S, Matayoshi K, Miller LG. Community-associated methicillin resistant *Staphylococcus aureus* isolates causing health care-associated infections. *Emerg Infect Dis* 2007;13:236-242.

Markowitz, N; Quinn, EL; Saravolatz, LD Trimethoprim–sulfamethoxazole compared with vancomycin for the treatment of *Staphylococcus aureus* infection. *Ann Intern Med*., 2002, 117, 390-398.

Marshall, SH; Donskey, CJ; Hutton-Thomas, R; Salata, RA; Rice, LB. Gene dosage and linezolid resistance in *Enterococcus faecium* and *Enterococcus faecalis*. *Antimicrob Agents Chemother*, 2002, 46, 3334-3336.

Martinez-Aguilar, G; Hammerman, WA; Mason, Jr EO; Kaplan, SL. Clindamycin treatment of invasive infections caused by community acquired, methicillin-resistant and methicillin-susceptible *Staphylococcus aureus* in children. *Pediatr Infect Dis J*., 2003, 22, 593-598.

Maslow JN, Mulligan ME, Arbeit RD. Molecular epidemiology: application of contemporary techniques to the typing of microorganisms. *Clin Infect Dis* 1993; 17(2): 153-164.

McDougal LK, Steward CD, Killgore GE, Chaitram JM, McAllister SK, Tenover FC. Pulsed-field gel electrophoresis typing of oxacillin-resistant *Staphylococcus aureus* isolates from the United States; establishing a national database. *J Clin Microbiol* 2003;41:5113-5120.

McHugh CG, Riley LW. Risk factors and costs associated with methicillin-resistant *Staphylococcus aureus* bloodstream infections. *Infect Control Hosp Epidemiol* 2004; 25: 425-430.

Micek, ST. Alternatives to vancomycin for the treatment of meticillin resistant *Staphylococcus aureus* infections. *Clin Infect Dis*., 2007, 45(Suppl 3), S184-190.

Miller LG, Diep BA. Colonization, fomites and virulence: rethinking the pathogenesis of community-associated methicillin-resistant *Staphylococcus aureus* infections. *Clin Infect Dis* 2008;46:752-60.

Miller LG, Perdreau-Remington F, Bayer AS, Diep B, Tan N, Bharadwa K, Tsui J, Perlroth J, Shay A, Tagudar G, Ibebuogu U, Spellberg B. Clinical and epidemiologic characteristics cannot distinguish community-associated methicillin-resistant *Staphylococcus aureus* infection from methicillin-susceptible *S.aureus* infection: a prospective investigation. *Clin Infect Dis* 2007;44:471-482.

Mizen, L; Berry, V; Woodnutt, G. The influence of uptake from the gastrointestinal tract and first-pass effect on oral bioavailability of (Z)-alkyloxyimino penicillins. *J Pharm Pharmacol*, 1995, 47, 725–730.

Moellering, RC; Jr. Current treatment options for community-acquired methicillin resistant *Staphylococcus aureus* infection. *Clin Infect Dis*., 2008, 46, 1032-1037.

Moet, GJ; Jones, RN; Biedenbach, DJ; Stilwell, MG; Fritsche, TR. Contemporary causes of skin and soft tissue infections in North America, Latin America, and Europe: Report from the SENTRY Antimicrobial Surveillance Program (1998–2004). *Diagn Microbiol Infect Dis*., 2007, 57, 7-13.

Moise, PA; Smyth, DS; El, Fawal, N; Robinson, DA; Holden, PN; Forrest, A; Sakoulas, G. Microbiological effects of prior vancomycin use in patients with methicillin-resistant *Staphylococcus aureus* bacteraemia. *J Antimicrob Chemother*, 2008, 61, 85-90.

Mongkolrattabithai K, Boyle S, Kahana MD, Daum RS. Severe *Staphylococcus aureus* infections caused by clonally related communityacquired methicillin-susceptible and methicillin-resistant isolates. *Clin Infect Dis* 2003;37:1050-1058.

Moore PC, Lindsay JA. Molecular characterisation of the dominant UK methicillin-resistant *Staphylococcus aureus* strains, EMRSA-15 and EMRSA-16. *J Med Microbiol* 2002;51:516-521.

Moran GJ, Krishnadasan A, Gorwitz RJ, Fosheim GE, McDougal LK, Carey RB, Talan DA; EMERGEncy ID Net Study Group. Methicillin-resistant *S.aureus* infections among patients in the emergency department. *N Engl J Med* 2006; 355:666-674.

Moran, GJ; Krishnadasan, A; Gorwitz, RJ; Fosheim, GE; McDougal, LK; Carey, RB, Talan DA. Methicillin-resistant *S. aureus* infections among patients in the emergency department. *N Engl J Med.*, 2006, 355, 666-674.

Morgan, A; Cofer, C; Stevens, DL. Iclaprim: a novel dihydrofolate reductase inhibitor for skin and soft tissue infections. *Future Microbiol*, 2009, 4, 131-144.

Mulligan ME, Murray-Leisure KA, Ribner BS, Standiford HC, John JF, Korvick JA, Kauffman CA, Yu VL. Methicillin-resistant *Staphylococcus aureus*: a consensus review of microbiology, pathogenesis, and epidemiology with implications for prevention and management. *Am J Med* 1993; 94: 313-328.

Mutnick, AH; Biedenbach, DJ; Tunbridge, JD; Jones, RN. Spectrum and potency evaluation of a new oxazolidinone, linezolid: report from the SENTRY Antimicrobial Surveillance Program, 1998-2000. *Diagn Microbiol Infect*, 2002, 43, 65-73.

Mwangi MM, Wu SW, Zhou Y, Sieradzki K, de Lencastre H, Richardson P, Bruce D, Rubin E, Myers E, Siggia ED, Tomasz A. Tracking the in vivo evolution of multidrug resistance in *Staphylococcus aureus* by whole-genome sequencing. *Proc Natl Acad Sci U S A* 2007;104:9451-9456.

Nathwani, D. Tygecycline: clinical evidence and formulary positioning. Int *J Antimicrob Agents*, 2005, 25, 185-192.

Nathwani, D; Morgan, M; Masterton, RG; Dryden, M; Cookson, BD; French, G; Lewis, D. British Society for Antimicrobial Chemotherapy Working Party on Community-onset MRSA Infections. Guidelines for UK practice for the diagnosis and management of methicillin resistant *Staphylococcus aureus* (MRSA) infections presenting in the community. *J Antimicrob Chemother*, 2008, 61, 976-994.

National Audit Office. The management and control of hospital acquired infection in acute NHS trusts in England: report by Controller and Auditor General. London: Stationery Office; 2000. Paragraphs 15-16.

National Committee for Clinical Laboratory Standards. Performance standards for antimicrobial susceptibility testing; eighth informational supplement. NCCLS document MS100-SB. Wayne, PA: National Committee for Clinical Laboratory Standards, 1998.

Noda M, Kato I. Purification and crystallization of staphylococcal leukocidin. *Methods Enzymol* 1988; 165: 22-32.

Nomura, S; Hanaki, H; Unemi, N. In vitro antibacterial activity of TOC-50, a new parenteral cephalosporin against methicillin-resistant *Staphylococcus aureus* and Staphylococcus epidermidis. *Chemotherapy*, 1996, 42, 253–258.

Norden, CW; Shaffer, M. Treatment of experimental chronic osteomyelitis due to *Staphylococcus aureus* with vancomycin and rifampin. *J Infect Dis.*, 1983, 147(2), 352-357.

Okuma K, Iwakawa K, Turnidge JD, Grubb WB, Bell JM, O'Brien FG, Coombs GW, Pearman JW, Tenover FC, Kapi M, Tiensasitorn C, Ito T, Hiramatsu K. Dissemination of new methicillin-resistant *Staphylococcus aureus* clones in the community. *J Clin Microbiol* 2002;40:4289-4294.

Olive DM, Bean P. Principles and applications of methods for DNA-based typing of microbial organisms. *J Clin Microbiol* 1999; 37: 1661-1669.

Oliveira DC, Milheirico C, de Lencastre H. Redefining a structural variant of staphylococcal cassette chromosome mec, *SCCmec* type VI. *Antimicrob Agents Chemother* 2006; 50: 3457-3459.

Padmanabhan RA, Fraser TG. The emergence of methicillin-resistant *Staphylococcus aureus* in the community. *Cleve Clin J Med* 2005; 72:235–241.

Pan ES, Diep BA, Charlebois ED, Auerswald C, Carleton HA, Sensabaugh GF, Perdreau-Remington F. Population dynamics of nasal strains of methicillin-resistant *Staphylococcus aureus*-and their relation to community-associated disease activity. *J Infect Dis* 2005;192:811-818.

Pankey, GA. Tigecycline. *J Antimicrob Chemother*, 2005, 56, 470-480.

Panton PN, Valentine FCO. Staphylococcal toxin. *Lancet* 1932; 222: 506-508.

Parras F, Del Carmen Guerrero M, Bouza E, Blazquez MJ, Moreno S, Menarguez MC, Cercenado E. Comparative study of mupirocin and oral cotrimoxazole plus topical fusidic acid in eradication of nasal carriage of methicillin-resistant *Staphylococcus aureus*. *Antimicrob Agents Chemother* 1995; 39: 175-179.

Patel, JB; Jevitt, LA; Hageman, J; McDonald, LC; Tenover, FC. An association between reduced susceptibility to daptomycin and reduced susceptibility to vancomycin in *Staphylococcus aureus*. *Clin Infect Dis*., 2006, 42, 1652-1653.

Paterson, DL. Clinical experience with recently approved antibiotics. *Cur Op Pharmacol*, 2006, 6, 486-490.

Peleg AY, Munckhof WJ, Kleinschmidt SL, Stephens AJ, Huygens F. Life-threatening community-acquired methicillin-resistant *Staphylococcus aureus* infection in Australia. *Eur J Clin Microbiol Infect Dis* 2005; 24:384-387.

Peters, MJ; Sarria, JC. Clinical characteristics of linezolid-resistant *Staphylococcus aureus* infections. *Am J Med Sci*., 2005, 330, 102-104.

Petersen, PJ; Bradford, PA; Weiss, WJ; Murphy, TM; Sum, PE; Projan, SJ. In vitro and in vivo activities of tigecycline (GAR-936), daptomycin, and comparative antimicrobial agents against glycopeptide-intermediate *Staphylococcus aureus* and other resistant gram-positive pathogens. *Antimicrob Agents Chemother*, 2002, 46, 2595–601.

Peterson LR, Quick JN, Jensen B, Homann S, Johnson S, Tenquist J, Shanholtzer C, Petzel RA, Sinn L, Gerding DN. Emergence of ciprofloxacin resistance in nosocomial methicillin resistant *Staphylococcus aureus* isolates. *Arch Intern Med* 1990; 150: 2151-2155.

Pope, SD; Roecker, AM. Vancomycin for treatment of invasive, multi-drug resistant *Staphylococcus aureus* infections. *Expert Opin Pharmacother*, 2007, 8, 1245-1261.

Proctor RA. Role of folate antagonists in the treatment of methicillin-resistant *Staphylococcus aureus* infection. *Clin Infect Dis*., 2008, 46(4), 584-593.

Purcell K, Fergie J. Epidemic of community-acquired methicillin resistant *Staphylococcus aureus* infections: a 14-year study at Driscoll Children's Hospital. *Arch Pediatr Adolesc Med* 2005; 159:980-985.

Raimundo O, Heussler H, Bruhn JB, Suntrarachun S, Kelly Deighton MA, Garland SM. Molecular epidemiology of coagulase negative staphylococcal bacteramia in a newborn intensive care unit. *J Hosp Infect* 2002; 51: 33-42.

Reagan DR, Doebbeling BN, Pfaller MA, Sheetz CT, Houston AK, Hollis RJ, Wenzel RP. Elimination of coincident *Staphylococcus aureus* nasal and hand carriage with intranasal application of mupirocin calcium ointment. *Ann Intern Med* 1991;114:101-6.

Rehm, SJ; Boucher, H; Levine, D; Campion, M; Eisenstein, BI; Vigliani, GA; Corey, GR; Abrutyn, E. Daptomycin versus vancomycin plus gentamicin for treatment of bacteraemia and endocarditis due to Staphylococcus aureus: subset analysis of patients infected with methicillin-resistant isolates. *J Antimicrob Chemother*, 2008, Dec;62(6), 1413-1421.

Ribeiro A, Dias C, Silva-Carvalho MC, Berquó L, Ferreira FA, Santos RN, Ferreira-Carvalho BT, Figueiredo AM. First report of infection with community-acquired methicillin-resistant *Staphylococcus aureus* in South America. *J Clin Microbiol* 2005; 43:1985-1988.

Rice, LB. Do we really need new anti-infective drugs? *Curr Opin Pharmacol*, 2003, 3, 459-463.

Rihn JA, Posfay-Barbe K, Harner CD, Macurak A, Farley A, Greenawalt K, Michaels MG. Community-acquired methicillin-resistant *Staphylococcus aureus* outbreak in a local high school football team: unsuccessful interventions. *Pediatr Infect Dis J* 2005; 24:841-843.

Robinson DA, Enright MC. Multilocus sequence typing and the evolution of methicillin-resistant *Staphylococcus aureus*. *Clin Microbiol Infect* 2004; 10: 92-97.

Robinson DA, Kearns AM, Holmes A, Morrison D, Grundmann H, Edwards G, O'Brien FG, Tenover FC, McDougal LK, Monk AB, Enright MC. Re-emergence of early pandemic *Staphylococcus aureus* as a community-acquired methicillin-resistant clone. *Lancet* 2005; 365:1256-1258.

Rodloff, AC; Leclercq, R; Debbia, EA; Cantón, R; Oppenheim, BA; Dowzicky, MJ. Comparative analysis of antimicrobial susceptibility among organisms from France, Germany, Italy, Spain and the UK as part of the tigecycline evaluation and surveillance trial. 1: *Clin Microbiol Infect.*, 2008, 14(4), 307-314.

Rollason J, Bastin L, Hilton AC, Pillay DG, Worthington T, McKeon C, Dec P, Burrows K, Lambert PA. Epidemiology of community-acquired methicillin resistant *Staphylococcus aureus* obtained from the UK West Midlands region. *J Hosp Infect* 2008; 70(4):314-320.

Rose, WE; Leonard, SN; Sakoulas, G; Kaatz, GW; Zervos, MJ; Sheth, A; Carpenter, CF; Rybak, MJ. Daptomycin activity against *Staphylococcus aureus* following vancomycin exposure in an in vitro pharmacodynamic model with simulated endocardial vegetations. *Antimicrob Agents Chemother*, 2008, 52, 831-836.

Rossney A, Morgan P, O'Connell B. Community-acquired PVL+MRSA in Ireland: a preliminary report. *Euro Surveill* 2005; 10(4):E050421.1.

Rouse MS, Steckelberg JM, Patel R. In vitro activity of ceftobiprole, daptomycin, linezolid, and vancomycin against methicillin-resistant staphylococci associated with endocarditis and bone and joint infection. *Diagn Microbiol Infect Dis.*, 2007, 58, 363–5.

Rubinstein, E; Bompart, F. Activity of quinupristin/dalfopristin against gram-positive bacteria: clinical applications and therapeutic potential. *J Antimicrob Chemother*, 1997, 39(Suppl A), 139-143.

Ruhe, JJ; Menon, A. Tetracyclines as an oral treatment option for patients with community onset skin and soft tissue infections caused by methicillin-resistant *Staphylococcus aureus*. *Antimicrob Agents Chemother*, 2007, 51, 3298-3303.

Ruhe, JJ; Monson, T; Bradsher, RW; Menon, A. Use of long-acting tetracyclines for methicillin-resistant *Staphylococcus aureus* infections: case series and review of the literature. *Clin Infect Dis*., 2005, 40, 1429-1434.

Ruhe, JJ; Smith, N; Bradsher, RW; Menon, A. Community-onset methicillin-resistant *Staphylococcus aureus* skin and soft-tissue infections: impact of antimicrobial therapy on outcome. *Clin Infect Dis*., 2007, 44, 777-784.

Rybak, MJ. The pharmacokinetic and pharmacodynamic properties of vancomycin. *Clin Infect Dis*., 2006, 42(Suppl 1), S35-39.

Sader, HS; Fritsche, TR; Jones, RN. Antimicrobial activity of daptomycin and selected comparators tested against bloodstream *Staphylococcus aureus* isolates from hemodialysis patients. *Int J Infect Dis*., 2009, 13, 291-295

Sader, HS; Fritsche, TR; Jones, RN. Daptomycin bactericidal activity and correlation between disk and broth microdilution method results in testing of *Staphylococcus aureus* strains with decreased susceptibility to vancomycin. *Antimicrob Agents Chemother*, 2006, 50(7), 2330-2336.

Saiman L, O'Keefe M, Graham PL. Hospital transmission of community acquired methicillin-resistant *Staphylococcus aureus* among postpartum women. *Clin Infect Dis* 2003;37:1313-1319.

Sakoulas G, Moellering RC Jr. Increasing antibiotics resistance among methicillin resistant *Staphylococcus aureus* strains. *Clin Infect Dis* 2008;46:S360-S367.

Sakoulas, G; Moellering, RC; Jr., Eliopoulos, GM. Adaptation of methicillin-resistant *Staphylococcus aureus*in the face of vancomycin therapy. *Clin Infect Dis*., 2006, 42 Suppl 1, S40-S50.

Salgado CD, Farr BM, Calfee DP. Community-acquired methicillin-resistant *Staphylococcus aureus*: a meta-analysis of prevalence and risk factors. *Clin Infect Dis* 2003; 36:131-139.

Sancak B, Ercis S, Menemenlioglu D, Colakoglu S, Hascelik G. Methicillin-resistant *Staphylococcus aureus* heterogeneously resistant to vancomycin in a Turkish university hospital. *J Antimicrob Chemother* 2005; 56: 519-523.

Saravolatx LD, Markowitz N, Arking L, Pohlod D, Fisher E. Methicillin-resistant *Staphylococcus aureus*: epidemiologic observations during a community acquired outbreak. *Ann Intern Med* 1982;96(1):11-6.

Saulnier P, Bourneix C, Prevost G, Andremont A. Random amplified polymorphic DNA assay is less discriminant than pulsed-field gel electrophoresis for typing strains of methicillin-resistant *Staphylococcus aureus*, *J Clin Microbiol* 1993; 31: 982-985.

Schneider, P; Hawser, S; Islam, K. Iclaprim, a novel diaminopyrimidine with potent activity on trimethoprim sensitive and resistant bacteria. *Bioorg Med Chem Lett*., 2003, 13, 4217 21.

Schulz P, Allen M, Murray Q, Smith SA, Goss L, Carrico R, Ramirez J. Infections due to community acquired methicillin-resistant *Staphylococcus aureus*: an emergent epidemic in Kentucky. *J Ky Med Assoc* 2005; 103:194-203.

Seybold U, Kourbatova EV, Johnson JG, Halvosa SJ, Wang YF, King MD, Ray SM, Blumberg HM. Emergence of community associated methicillin-resistant

Staphylococcus aureus USA300 genotype as a major cause of health-care associated blood stream infections. *Clin Infect Dis* 2006;42:647-656.

Shopsin B, Gomez M, Montgomery SO, Smith DH, Waddington M, Dodge DE, Bost DA, Riehman M, Naidich S, Kreiswirth BN. Evaluation of protein A gene polymorphic region DNA sequencing for typing of *Staphylococcus aureus* strains. *J Clin Microbiol* 1999; 37: 3556-3563.

Shore A, Rossney AS, Keane CT, Enright MC, Coleman DC. Seven novel variants of the staphylococcal chromosomal cassette *mec* in methicillin-resistant *Staphylococcus aureus* isolates from Ireland. *Antimicrob Agents Chemother* 2005; 49: 2070-2083.

Siberry GK, Tekle T, Carroll K, Dick J. Failure of clindamycin treatment of methicillin-resistant *Staphylococcus aureus* expressing inducible clindamycin resistance in vitro. *Clin Infect Dis* 2003;37:1257-1260.

Siegman-Igra, Y; Reich, P; Orni-Wasserlauf, R; Schwartz, D; Giladi, M. The role of vancomycin in the persistence or recurrence of *Staphylococcus aureus* bacteraemia. *Scandinavian Journal of Infectious Diseases*, 2005, 37(8), 572-578.

Sievert DM, Rudrik JT, Patel JP, McDonald LC, Wilkins MJ, Hageman JC. Vancomycin-resistant *Staphylococcus aureus* in the United States, 2002-2006. *Clin Infect Dis* 2008;46:668-674.

Spellberg, B; Powers, JH; Brass, EP; Miller, LG; Edwards, JE; Jr. Trends in antimicrobial drug development: implications for the future. *Clin Infect Dis*., 2004, 38, 1279-1286.

Steinkraus, G; White, R; Friedrich, L. Vancomycin MIC creep in nonvancomycin-intermediate *Staphylococcus aureus* (VISA), vancomycin-susceptible clinical methicillin-resistant *S. aureus* (MRSA) blood isolates from 2001–05. *J Antimicrob Chemother*, 2007, 60, 788-794.

Stevens, DL. The role of vancomycin in the treatment paradigm. *Clin Infect Dis*., 2006, 42(Suppl 1), S51-57.

Stevens, DL; Bisno, AL; Chambers, HF; Everett, ED; Dellinger, P; Goldstein, EJ; Gorbach, SL; Hirschmann, JV; Kaplan, EL; Montoya, JG; Wade, JC. Practice guidelines for the diagnosis and management of skin and soft-tissue infections. *Clin Infect Dis*., 2005, 41, 1373-1406.

Stevens, DL; Herr, D; Lampiris, H; Hunt, JL; Batts, DH; Hafkin, B. Linezolid versus vancomycin for the treatment of methicillin-resistant *Staphylococcus aureus* infections. *Clin Infect Dis*., 2002, 34, 1481-1490.

Stevenson KB, Searle K, Stoddard GJ, Samore M. Methicillin-resistant *Staphylococcus aureus* and vancomycin-resistant enterococci in rural communities, western United States. *Emerg Infect Dis* 2005; 11:895–903.

Stryjewski, ME; Chu, VH; O'Riordan, WD; Warren, BL; Dunbar, LM; Young, DM; Vallée, M; Fowler, VG; Jr, Morganroth, J; Barriere, SL; Kitt, MM; Corey, GR. FAST 2 Investigator Group. Telavancin versus standard therapy for treatment of complicated skin and skin structure infections caused by Gram-positive bacteria: FAST 2 study. *Antimicrob Agents Chemother*, 2006, 50, 862 7.

Stryjewski, ME; Szczech, LA; Benjamin, DK; Jr., Inrig, JK; Kanafani, ZA; Engemann, JJ; Chu, VH; Joyce, MJ; Reller, LB; Corey, GR; Fowler, VG; Jr. Use of vancomycin or first-generation cephalosporins for the treatment of hemodialysis-dependent patients with methicillin susceptible *Staphylococcus aureus* bacteremia. *Clin Infect Dis*., 2007, 44 (2), 190-196.

Styers D, Sheehan DJ, Hogan P, Sahm DF. Laboratory-based surveillance of current antimicrobial resistance patterns and trends among *Staphylococcus aureus*: 2005 status in the United States. *Ann Clin Microbiol Antimicrob* 2006;5:2.

Suggs, AH; Maranan, MC; Boyle-Vavra, S; Daum, RS. Methicillin-resistant and borderline methicillin-resistant asymptomatic *Staphylococcus aureus* colonization in children without identifiable risk factors. *Pediatr Infect Dis J.*, 1999, 18, 410-414.

Sundelof, JG; Thompson, R; White, KM; Sasor, MW; Cama, L; Kropp, H. Pharmacokinetics in nonhuman primates of a prototype carbapenem active against methicillin-resistant *Staphylococcus aureus*. *Antimicrob Agents Chemother*, 1996, 40, 795–798.

Takano T, Higuchi W, Otsuka T, Baranovich T, Enany S, Saito K, Isobe H, Dohmae S, Ozaki K, Takano M, Iwao Y, Shibuya M, Okubo T, Yabe S, Shi D, Reva I, Teng LJ, Yamamoto T. Novel characteristics of community-acquired methicillin-resistant *Staphylococcus aureus* belonging to multilocus sequence type 59 in Taiwan. *Antimicrob Agents Chemother* 2008; 52: 837-845.

Tambic A, Power EGM, Tambic T, Snur I, French GL. Epidemiological analysis of meticillin-resistant *Staphylococcus aureus* in a Zagreb trauma hospital using a randomly amplified polymorphic DNA-typing method. *Eur J Microbiol Infect Dis* 1999;18:335-340.

Tang YW, Procop GW, Persing DH. Molecular diagnostics of infectious diseases. *Clin Chem* 1997;43:2021-2038.

Tenover FC, Arbeit RD, Goering RV, Mickelsen PA, Murray BE, Persing DH, Swaminathan B. Interpreting chromosomal DNA restriction patterns produced by pulsed-field gel electrophoresis: criteria for bacterial strain typing. *J Clin Microbiol* 1995; 33: 2233-2239.

Tenover FC, McDougal LK, Goering RV, Killgore G, Projan SJ, Patel JB, Dunman PM. Characterization of a strain of community-associated methicillin-resistant *Staphylococcus aureus* widely disseminated in the United States. *J Clin Microbiol* 2006;44:108-118.

Tenover FC, Moellering Jr RC. The rationale for revising the Clinical and Laboratory Standards Institute vancomycin minimal inhibitory concentration interpretive criteria for *Staphylococcus aureus*. *Clinical Infectious Diseases*, 2007 May 1, 44(9), 1208-1215.

Thal, LA; Zervos, MJ. Occurrence and epidemiology of resistance to virginiamycin and streptogramins. *J Antimicrob Chemother*, 1999, 43, 171-176.

Tiemersma EW, Bronzwaer SL, Lyytikainen O, Degener JE, Schrijnemakers P, Bruinsma N, Monen J, Witte W, Grundman H. Methicillin resistant *Staphylococcus aureus* in Europe, 1999–2002. *Emerg Infect Dis* 2004; 10: 1627-1634.

Torell E, Molin D, Tano E, Ehrenborg C, Ryden C. Community-acquired pneumonia and bacteraemia in a healthy young woman caused by methicillin resistant *Staphylococcus aureus* (MRSA) carrying the genes encoding Panton–Valentine leukocidin (PVL). *Scand J Infect Dis* 2005; 37: 902-904.

Tristan A, Bes M, Meugnier H, Lina G, Bozdogan B, Courvalin P, Reverdy ME, Enright MC, Vandenesch F, Etienne J. Global distribution of Panton-Valentine leukocidin-positive methicillin-resistant *Staphylococcus aureus*, 2006. *Emerg Infect Dis* 2007; 13: 594-600.

Tsiodras, S; Gold, HS; Sakoulas, G; Eliopoulos, GM; Wennersten, C; Venkataraman, L; Moellering, RC; Ferraro, MJ. Linezolid resistance in a clinical isolate of *Staphylococcus aureus*. *Lancet*, 2001, 358, 207-208.

US. Congress, Office of Technology Assessment. Impacts of antibiotic-resistant bacteria. Washington, DC: US Government Printing Office; September, 1995.

Valls V, Gomez-Herruz P, Gonzalez-Palacios R, Cuadros JA, Romanyk JP, Ena J. Long-term efficacy of a program to control methicillin-resistant *Staphylococcus aureus*. *Eur J Clin Microbiol Infect Dis* 1994; 13: 90-95.

Vandenesch F, Naimi T, Enright MC, Lina G, Nimmo GR, Heffernan H, Liassine N, Bes M, Greenland, Reverdy ME, Etienne J. Community-acquired methicillin-resistant *Staphylococcus aureus* carrying Panton-Valentine leukocidin genes: worldwide emergence. *Emerg Infect Dis* 2003; 9:978-984.

Vergidis, PI; Falagas, ME. New antibiotic agents for bloodstream infections. International *Journal of Antimicrobial Agents*, 32S (2008) S60–S65.

Vidal PM, Trindade PA, Garcia TO, Pacheco RL, Costa SF, Reinert C, Hiramatsu K, Mamizuka EM, Garcia CP, Levin AS. Differences between "classical" risk factors for infections caused by methicillin-resistant *Staphylococcus aureus* (MRSA) and risk factors for nosocomial bloodstream infections caused by multiple clones of the staphylococcal cassette chromosome mec type IV MRSA strain. *Infect Control Hosp Epidemiol* 2009;30(2):139-45.

Vikram, HR; Havill, NL; Koeth, LM; Boyce, JM. Clinical progression of methicillin-resistant *Staphylococcus aureus* vertebral osteomyelitis associated with reduced susceptibility to daptomycin. *J Clin Microbiol,* 2005, 43, 5384-5387.

Voyich JM, Otto M, Mathema B, Braughton KR, Whitney AR, Welty D, Long RD, Dorward DW, Gardner DJ, Lina G, Kreiswirth BN, DeLeo FR. Is Panton-Valentine Leukocidin the major virulence determinant in community-associated methicillin resistant *Staphylococcus aureus* disease? *J Infect Dis* 2006;194:1761-1770.

Walsh TJ, Standiford HC, Reboli AC, John JF, Mulligan ME, Ribner BS, Montgomerie JZ, Goetz MB, Mayhall CG, Rimland D. Randomized double-blinded trial of rifampin with either novobiocin or trimethoprim–sulfamethoxazole against methicillin-resistant *Staphylococcus aureus* colonization: prevention of antimicrobial resistance and effect of host factors on outcome. *Antimicrob Agents Chemother* 1993; 37: 1334-1342.

Walsh TR, Howe RA. The prevalence and mechanisms of vancomycin resistance in *Staphylococcus aureus*. *Ann Rev Microbiol* 2002; 56: 657-675.

Wang R, Braughton KR, Kretschmer D, Bach TH, Queck SY, Li M, Kennedy AD, Dorward DW, Klebanoff SJ, Peschel A, DeLeo FR, Otto M. Identificaiton of novel cytolytic peptides as key virulence determinants for community-associated MRSA. *Nat Med* 2007;13:1510-1514.

Warren DK, Liao RS, Merz LR, Eveland M, Dunne Jr WM. Detection of methicillin-resistant *Staphylococcus aureus* directly from nasal swab specimens by a real-time PCR assay. *J Clin Microbiol* 2004;42:5578-5581.

Weathers L, Riggs D, Santeiro M, Weibley RE. Aerosolized vancomycin for treatment of airway colonization by methicillin-resistant *Staphylococcus aureus*. *Pediatr Infect Dis J* 1990; 9: 220-221.

Weber JT. Community-associated methicillin-resistant *Staphylococcus aureus*. *Clin Infect Dis* 2005; 41(Suppl 4):S269-S272.

Weigelt, J; Itani, K; Stevens, D; Lau, W; Dryden, M; Knirsch, C. Linezolid versus vancomycin in treatment of complicated skin and soft tissue infections. *Antimicrob Agents Chemother*, 2005, 49, 2260-2266.

Wenzel, RP. The antibiotic pipeline challenges, costs, and values. *N Engl J Med.*, 2004, 351, 523-526.

Whitby, M; McLaws, ML; Berry, G. Risk of death from methicillin resistant *Staphylococcus aureus* bacteraemia: a meta-analysis. *Med J Aust.*, 2001, 175, 264-267.

Witt Rt, Kanhai V, vanLeeuwen WB. Comparison of theDiversiLab™system, Pulsed-Field Gel Electrophoresis and Multi-Locus Sequence Typing for the characterization of epidemic reference MRSA strains. *J Microbiol Methods* 2009; 77: 130-133.

Woodin AM. Purification of the two components of leucocidin fron *Staphylococcus aureus. Biochem J* 1960; 75:158-165.

Wootton, M; MacGowan, AP; Walsh, TR. Comparative bactericidal activities of daptomycin and vancomycin against glycopeptides intermediate *Staphylococcus aureus* (GISA) and heterogeneous GISA isolates. *Antimicrob Agents Chemother*, 2006, 50, 4195-4197.

Wu SW, de Lencastre H, Tomasz A. Recruitment of the *mecA* gene homologue of Staphylococcus sciuri into a resistance determinant and expression of the resistant phenotype in *Staphylococcus aureus. J Bacteriol* 2001; 183: 2417-2424.

Wunderink, RG; Rello, J; Cammarata, SK; Croos-Dabrera, RV; Kollef, MH. Linezolid vs vancomycin: analysis of two double-blind studies of patients with methicillin-resistant *Staphylococcus aureus* nosocomial pneumonia. *Chest*, 2003, 124, 1789–1797.

Yamasaki O, Kaneko J, Morizane S, Akiyama H, Arata J, Narita S, Chiba J, Kamio Y, Iwatsuki K. The association between *Staphylococcus aureus* strains carrying Panton-Valentine leukocidin genes and the development of deep-seated follicular infection. *Clin Infect Dis* 2005;40:381-385.

Yang ES, Tan J, Rieg G. Body site colonization prevalence in patients with community-associated methicillin resistant *Staphylococcus aureus* infections [abstract 285]. In: Program and Abstracts of the 45th Annual Meeting of the Infectious Diseases Society of America.

Yilmaz, G; Aydin, K; Iskender, S; Caylan, R; Koksal, I. Detection and prevalence of inducible clindamycin resistance in staphylococci. *J Med Microbiol*, 2007, 56, 342-345.

Yin, LY; Lazzarini, L; Li, F; Stevens, CM; Calhoun, JH. Comparative evaluation of tigecycline and vancomycin, with and without rifampicin, in the treatment of methicillin-resistant *Staphylococcus aureus* experimental osteomyelitis in a rabbit model. *J Antimicrob Chemother*, 2005, 55(6), 995-1002.

Yuk, JH; Dignani, MC; Harris, RL; Bradshaw, MW; Williams, Jr TW. Minocycline as an alternative antistaphylococcal agent. *Rev Infect Dis.*, 1991, 13, 1023-1024.

In: Methicillin-Resistant Staphylococcus Aureus...
Editor: Charles L. Kolendi
ISBN: 978-1-60741-398-1

Chapter 2

THERAPEUTIC OPTIONS FOR INTRACELLULAR MRSA: CURRENT VIEWS AND PERSPECTIVES

Sandrine Lemaire, Paul M. Tulkens and Françoise Van Bambeke
Louvain Drug Research Institute
Université Catholique de Louvain, Brussels, Belgium.

INTRODUCTION

Staphylococcus aureus causes a wide variety of diseases, ranging from relatively superficial wound infections (e.g. furunculosis) to deep-seated, life-threatening conditions (cardiovascular infections, severe skin infections, osteomyelitis, or pneumonia). Treatment options are, however, becoming limited due to resistance issues. In particular, Methicillin-Resistant *S. aureus* (MRSA) were first reported in 1961, shortly after introduction of penicillinase-stable β-lactams (such as methicillin). Confined for about 40 years in health care settings, MRSA have also been described in the community (e.g. day care center, schools and sport teams) in various parts of the world. These isolates tend to be highly virulent [1-3], causing serious infections and even death in healthy children and adults [4;5]. Moreover, large-scale CDC surveillance studies have now revealed that hospital acquired (HA-) and community-acquired (CA-) MRSA are no more restricted to specific environments, since strains historically associated with CA-MRSA outbreaks (e.g. mainly US300 isolates) are now recovered in nosocomial infections as well [6].

Beyond this high capacity to adapt to environment and to develop resistance, the ability of *S. aureus* to survive within host cells can also contribute to the difficulty of eradicating infection with antibiotics. Although primarily known as extracellular, *S. aureus* is now recognized as an opportunistic intracellular pathogen that easily adheres to and invades mammalian cells [7-10]. This intracellular niche may play an important role in the pathogenesis of several staphylococcal infections, by protecting the bacterium from the lethal action of antibiotics (intracellular forms tend to be poorly susceptible to most available antibiotics) and immune defenses. It may probably explain (i) the recurrent or persistent character of several staphylococcal infections (namely endocarditis, osteomyelitis, complicated skin and soft-tissue infections, or bovine chronic mastitis) even after apparently successful therapy; (ii) the difficulty to eradicate nasal carriage since *S. aureus* can persist in nasal epithelial cells [11]; and (iii) the selection or spreading of resistant mutants due to the poor efficacy of most available antimicrobials. This stresses, therefore, the importance of understanding (i) the intracellular lifestyle and fate of this bacterium, and (ii) whether and to what extent antibiotics may or may not act against intracellular *S. aureus,* with a particular emphasis on multi-drug resistant MRSA. We know, indeed, that cellular accumulation and activity in broth are not, *per se*, predictive factors of intracellular efficacy [12;13]. This chapter summarized our current knowledge of the intracellular fate of MRSA and its intracellular susceptibility to antibiotics.

1. Intracellular Survival of *S. Aureus*

1.1. Intracellular Life-Style

Over the past decades, it has been shown that S. aureus invades both phagocytic (macrophages [14-16]), polymorphonuclear neutrophils [PMN, [7;16]]) and non phagocytic cells (endothelial cells [17], osteoblasts [18;19], skin keratinocytes [20], fibroblasts [21], and bovine mammary epithelial cells [22]) (the intracellular fate of S. aureus is illustrated in figure 1). In non phagocytic cells, host cell invasion depends on the fibronectin-mediated bridging between staphylococcal fibronectin-binding proteins (FnBPs, a family of adhesins expressed at the cell surface of the bacteria) and host cell integrins [23], which induces actin cytoskeleton rearrangements though zipper-type mechanism [24]. In PMN and macrophages, which constitute a first line of defense in the clearance of microorganisms, internalization process seems to involve both phagocytosis and FnBPs-mediated internalization since fibronectin-bound S. aureus are more efficiently ingested than non opsonized S. aureus [25;26].

After internalization, the fate of the bacterium varies depending on cell type or on the strain. In macrophages and PMN [7;27;28], electron microscopy studies have shown that S. aureus remains confined in membrane-limited vacuoles, suggesting that this organism can resist to bactericidal mechanisms within phagolysosomes. In contrast, a few organisms are found entirely free in the cytosol of infected endothelial cells [29] or epithelial cells [30].

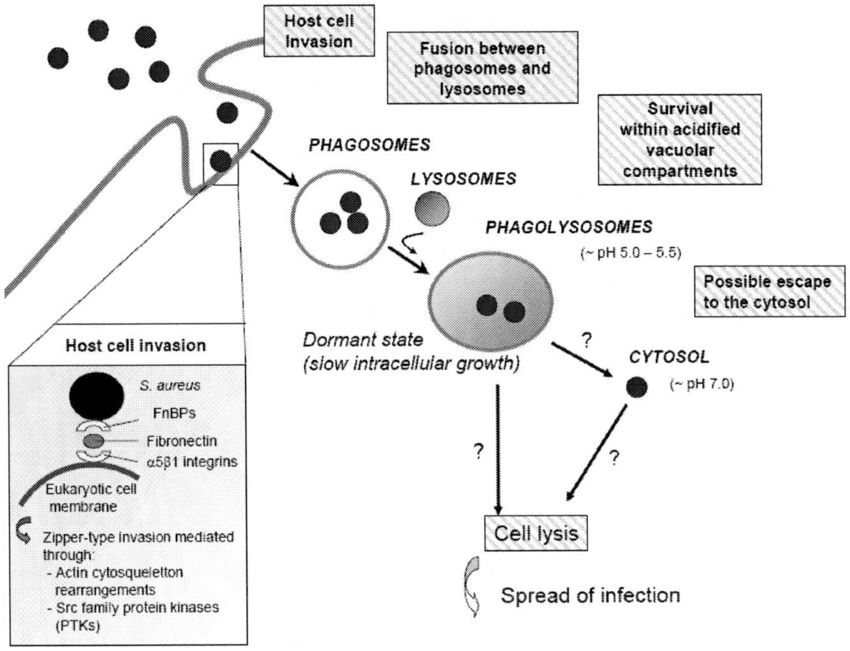

Figure 1. Pictorial description of the infectious cycle of *S. aureus* within eukaryotic cells, illustrating adhesion to host cells, residence and multiplication within phagolysosomal apparatus, and the possible escape into the cytosic compartment in endothelial and epithelial cells.

1.2. Post-Phagocytosis Changes in Response to the Intracellular Environment

In order to survive intracellularly, bacteria have to cope with the eukaryotic environment. Using a microarray procedure, Garzoni and colleagues have demonstrated that, following internalization by human lung epithelial cells [31], *S. aureus* extensively reprograms its transcriptome to maintain a metabolically dormant state (instead of the active bacterial replication observed in broth [see figure 2]) until the intracellular environment becomes suitable for escape. Hence, the expression of genes involved in bacterial metabolism, nutrient transport and cell-wall synthesis is markedly down-regulated. This is noticeably different from what is observed with intracellular *Salmonella* [32] or *Shigella* [33], which show up-regulation of genes involved in division processes. In contrast, there is an increased expression of genes involved in host cell invasion (e.g. gene coding for FnBPs proteins), neutralization of free radicals and iron scavenging. Of particular importance also, the expression of virulence factors that damage the host (such as *hla* [haemolysin alpha] [31]) is reduced in intracellular *S. aureus* (to preserve host cell integrity [27]) while expression of genes involved in resistance to host defense appears critical. In particular, it has been shown recently that bacteria expressing high levels of staphyloxanthin (a carotenoid pigment blocking attack by host reactive oxygen species [34;35]) shows increased tolerance to H_2O_2, which is reflected intracellularly by an increased bacterial growth [36].

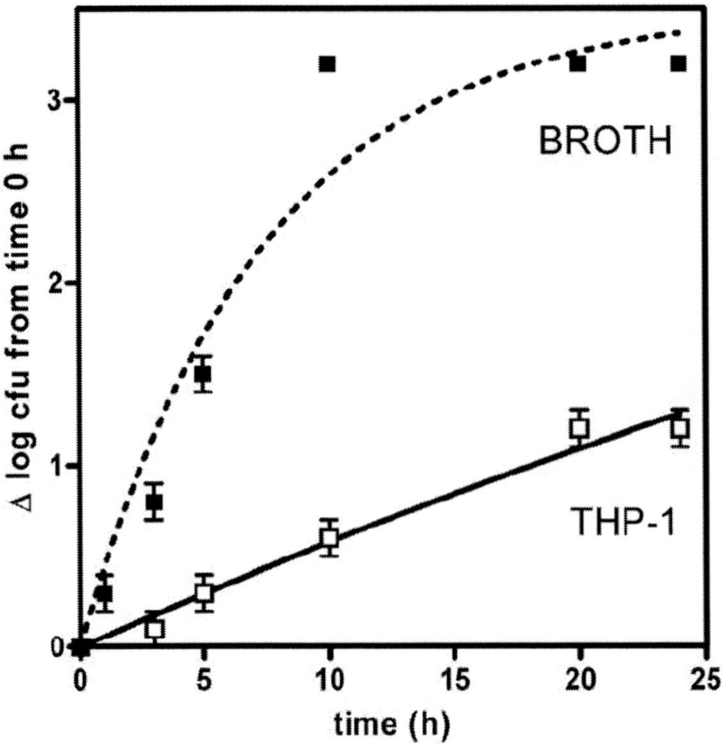

Figure 2. Kinetics of MRSA (strain ATCC 33591) bacterial growth in Mueller Hinton broth (extracellular bacteria) or THP-1 cells (intracellular bacteria).

1.3. Intracellular Persistence of Small Colony Variants of *S. Aureus*

First described in the earlier 1900's, Small Colony Variants (SCVs) represent naturally occurring subpopulations of *S. aureus* that are often associated with persistent and recurrent illnesses, namely pulmonary infections associated with cystic fibrosis, chronic osteomyelitis or keratosis follicularis [Darier-White disease])[37;38]. These variants are characterized by a slow growth due to auxotrophism by defects in hemin/menadione [39] or thymidine biosynthesis [40]. These metabolic alterations make also them respectively more resistant to aminoglycosides (decreased antibiotic uptake) or sulfonamides (bypass of the pathway inhibited by trimethoprim-sulfamethoxazole due to the thymidine dependence of this process). Being relatively unstable, SCVs can also revert to a highly virulent phenotype. Combined with their increased ability to invade eukaryotic cells (due to enhanced expression of FnBPs [41]), all these characteristics make SCV-related infections difficult to diagnose and to treat.

1.4. Clinical relevance of *S. Aureus* Intracellular Persistence

Despite clear evidences of *in vitro* persistence, the ability of *S. aureus* to reside and multiply intracellularly *in vivo* is still an ongoing debate. Yet, intracellular foci were documented in patients suffering from recurrent rhinosinusitis [11;42], and seem to constitute a significant risk factor for antibiotic failure. Likewise, intracellular bacteria were evidenced in fibroblasts of biopsy specimens from periprosthetic tissues [43], or surrounded by glycocalyx in osteoblasts from surgical bone specimens [44]. The ability of S. aureus to secrete glycocalyx further contributes to difficulty in eradication, since this extracellular polymeric material offers significant advantage for the bacteria (increased bacterial adherence to bone matrix, tissues and biomaterials [45;46]; protection from the lethal action of antibiotics [47]). Similar evidences of in vivo intracellular persistence are, however, still lacking for life-threatening cardiovascular infections, severe skin and soft-tissues diseases and pulmonary infections associated with cystic fibrosis, probably due to technical difficulties in localizing such reservoirs.

2. Intracellular Activity of Antibiotics

Over the past decades, cellular accumulation of antibiotics has been the subject of a large number of studies (see [12;13;48] for detailed description of mechanisms of antibiotics uptake, distribution and efflux in eukaryotic cells), but its importance for activity against intracellular bacteria remains largely controversial. Studies examining a series of antibiotics belonging to the main pharmacological classes have convincingly shown that accumulation *per se* is not a predictive factor of antibiotic efficacy. Interestingly, antibiotics belonging to pharmacological classes known for their low cellular accumulation (such as β-lactams) are not necessarily inactive against intracellular forms of susceptible bacteria (as demonstrated for *S. aureus* or *L. monocytogenes)*. Conversely, and contrary to most original assumptions, macrolides failed to show activity towards intraphagocytic *S. aureus* despite their massive cellular accumulation. In order to be effective intracellularly, it is therefore critical to select antibiotics reaching the infected subcellular compartment (phagolysosomes and cytosol) in some cell types at concentration levels far above the minimal inhibitory concentration (MIC) for a reasonable period of time, but also being able to express activity therein.

Routine evaluation of antibiotic activity is performed on extracellular bacteria, ignoring cellular pharmacokinetics (PK) and the modulation of pharmacodynamics (PD) by host cell environment. In this context, a complete evaluation of antistaphylococcal antibiotics should include an assessment of their ability to control intracellular infections. While animal models are currently being developed [49], models of cultured cells remain valuable tools because they offer the possibility to explore in details the pharmacological parameters governing the response of intracellular bacteria to the drug in the absence of host factors. Figure 3 shows a summarized view of the intracellular activity of antibiotics representative of the main pharmacological classes compared at their respective human *Cmax* (see table 1, with a full description of their intrinsic activity at both neutral and acidic pH) against a MSSA and a MRSA (MLS_B). As a whole, activity is grossly similar against both strains for all antibiotics tested except for azithromycin to which the MRSA is resistant. Interestingly, activity of cloxacillin (as a representative of β-lactams) is not affected, suggesting that the MRSA phenotype does not express it-self intracellularly (see 2.5 for more details). Figure 4 shows the effect reached in the same conditions against a series of MRSA with variable MICs to these agents. With most antibiotics, activity decreases gradually when MICs increases, corroborating a concentration-dependent profile of activity. The β-lactam cloxacillin makes exception, with a similar effect obtained whatever the MIC of the strain. This is in accordance with the non-concentration pharmacodynamic profile of this class of drug. Linezolid, also known as a poorly concentration-dependent antibiotic, becomes inactive as soon as the bacteria show elevated MICs.

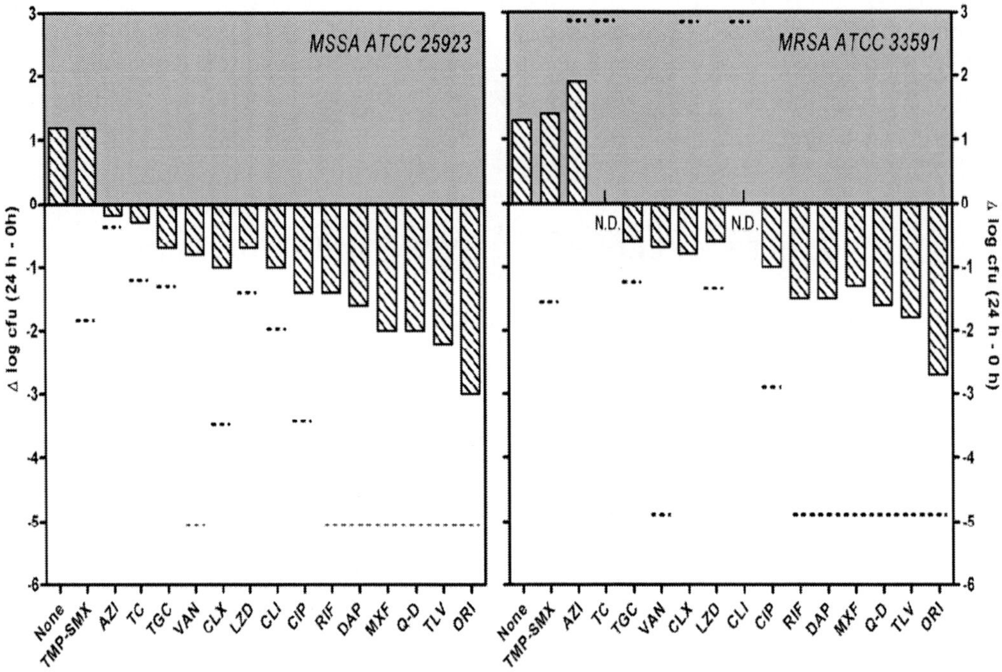

Figure 3. Comparison of the intracellular activities of antibiotics against MSSA and MRSA. The graphs show the activity of a series of antibiotics against MSSA strain ATCC 25923 (fully susceptible) and MRSA strain ATCC 33591 (resistance in broth to β-lactams, azithromycin and clindamycin) phagocytized by human THP-1 macrophages. Activity is expressed as the change in bacterial counts following 24 h of exposure to each of the selected antibiotics using an extracellular concentration corresponding to their respective human C*max*. The limit of detection is set at – 5 $\log_{10}$ cfu. The dotted lines represent the extracellular activities of each antibiotic (broth), while the gray zone represents the bacterial growth. AZI, azithromycin; CIP, ciprofloxacin; CLI, clindamycin; CLX, cloxacillin; DAP, daptomycin; LZD, linezolid; MXF, moxifloxacin; N.D., Not determined; ORI, oritavancin; Q-D, quinupristin-dalfopristin (vol/vol, 30/70); RIF, rifampicin; TET, tetracycline; TGC, tigecycline; TLV, telavancin; STX, trimethoprim-sulfamethoxazole; VAN, vancomycin.

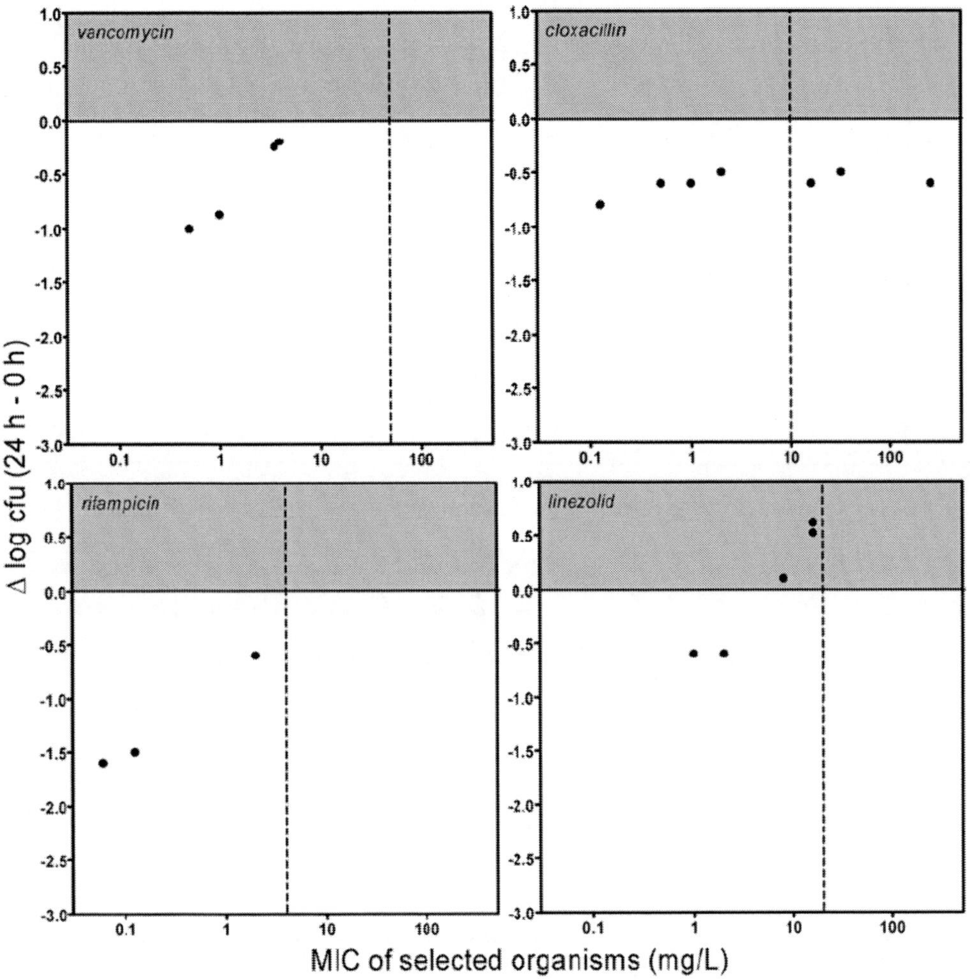

Figure 4A. Correlation between intrinsic and intracellular activities of a series of antibiotics (vancomycin, cloxacillin, rifampicin and linezolid) against *S. aureus* isolates (including MRSA strains of clinical epidemiological interest). The abscissas show antibacterial intrinsic activities, as determined by standard MIC assays. The ordinates show intraphagocytic activities (human THP-1 macrophages) expressed as the change in bacterial counts following 24 h of exposure to each of the selected antibiotics (using an extracellular concentration corresponding to their respective human C*max [as represented by the vertical dotted line])*. The gray zone corresponds to the bacterial growth.

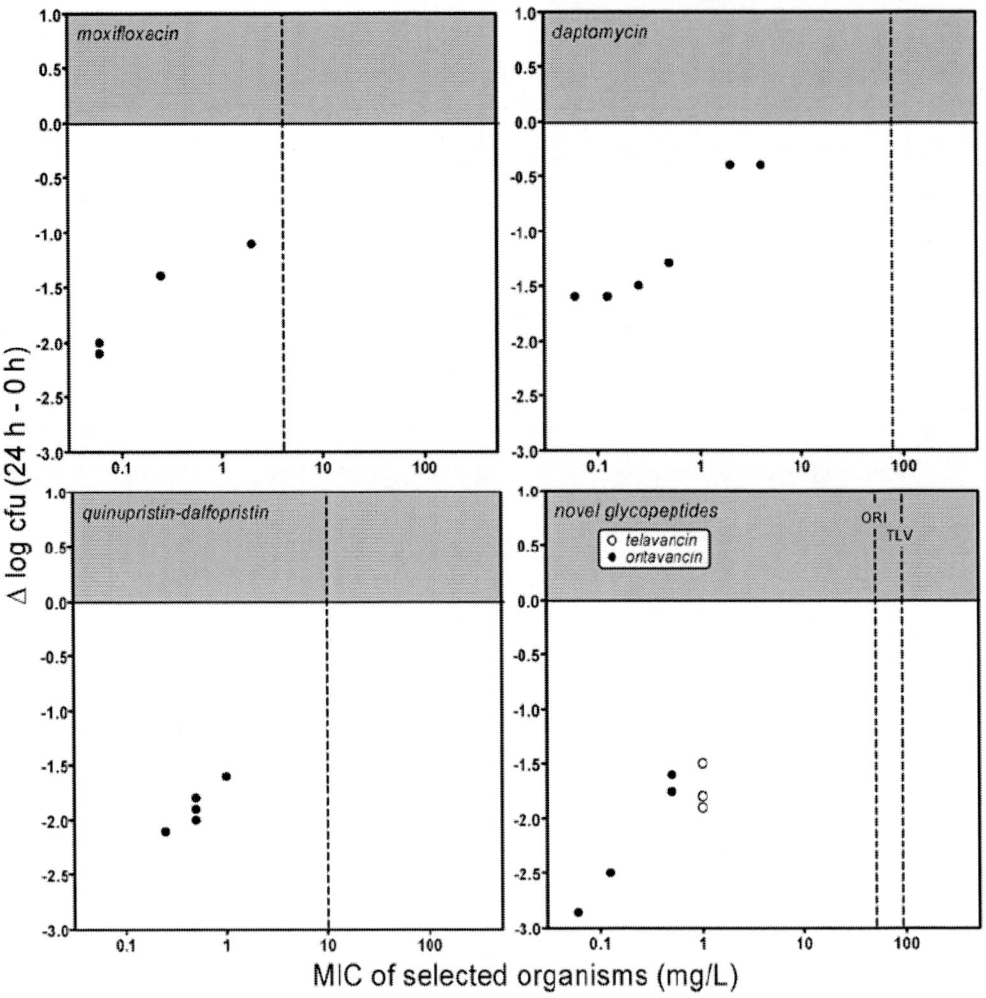

Figure 4B. Correlation between intrinsic and intracellular activities of a series of antibiotics (moxifloxacin, daptomycin, quinupristin-dalfopristin, and novel glycopeptides antibiotics [namely oritavancin and telavancin]) against *S. aureus* isolates (including MRSA strains of clinical epidemiological interest). The abscissas show antibacterial intrinsic activities, as determined by standard MIC assays. The ordinates show intraphagocytic activities (human THP-1 macrophages) expressed as the change in bacterial counts following 24 h of exposure to each of the selected antibiotics (using an extracellular concentration corresponding to their respective human C*max [as represented by the vertical dotted line])*. The gray zone corresponds to the bacterial growth.

Table 1. Comparative susceptibilities of antibiotics at neutral and acidic pH

Antibiotics	Human Cmax (mg/L)	MICs (mg/L)			
		MSSA (ATCC 25923)		MRSA (ATCC 33591)	
		pH 7.4	pH 5.5	pH 7.4	pH 5.5
Cell-wall/membrane					
Oxacillin (OXA)	8	0.25	0.03	256	0.25
Cloxacillin (CLX)	10	0.125	0.06	16	0.125
Meropenem (MEM)	50	0.125	0.06	16-32	0.125
Daptomycin (DAP)	77	0.125	0.25	0.125-0.25	0.5
Vancomycin (VAN)	50	1	1	2	2
Telavancin (TLV)	90	1	1	1	2
Oritavancin (ORI)	50	0.06	0.125	0.25	0.5
Nucleic acid synthesis					
Ciprofloxacin (CIP)	4	0.125	0.5	0.25	0.5-1
Moxifloxacin (MXF)	4	0.06	0.125-0.25	0.125	0.25
Rifampicin (RIF)	4	0.03	0.01	0.06	0.03
Metabolism					
Trimethoprim-sulfamethoxazole (STX)	25	1	4	1	2
Protein synthesis					
Gentamicin (GEN)	18	0.25	16	2	16
Tetracycline (TET)	5	0.5	0.25	256	64
Tigecycline (TGC)	1	0.125-0.25	1-2	0.5	2
Azithromycin (AZI)	0.5	0.5	256	> 256	> 256
Clindamycin (CLI)	20	0.125	8-16	> 256	16
Quinupristin-dalfopristin (Q-D)	10	0.25	0.5	0.5	0.5
Linezolid (LZD)	20	1-2	2	1-2	2

2.1. Trimethoprim-Sulfamethoxazole

Along with clindamycin, trimethoprim-sulfamethoxazole is the most commonly used antimicrobial agents for outpatient treatment of CA-MRSA infections. Despite its marked activity towards both extracellular HA-MRSA and CA-MRSA, this combination of antibiotics is not effective for the treatment of experimental intracellular *S. aureus*, even when using high extracellular concentrations (up to 100 mg/L).

2.2. Azithromycin

A large activity of azitromycin towards intracellular *S. aureus* is expected based on its massive cellular accumulation [50] and its co-localization with the bacteria in acidic vacuoles [51]. However, azithromycin never yields a truly intracellular effect towards intraphagocytic *S. aureus* [28;52], probably in relation with the deleterious effect of acidic pH on its antibacterial efficacy, and its bacteriostatic character. Improved potency (lower static concentrations) can, however, be obtained in the presence of verapamil [53], which blocks azithromycin efflux by P-glycoprotein [54;55].

2.3. Tetracyclines

For a long time, tetracyclines were considered as the drugs of choice for the treatment of several intracellular infections. Despite their apparent cellular accumulation (10- to 60-fold [56;57]), these antibiotics reduces only modestly the intracellular inoculum of *S. aureus* [28], again probably in relation with their bacteriostatic character.

2.4. Vancomycin

Vancomycin often remains the treatment of choice against MRSA infections. Unfortunately, it only shows a modest effect on intracellular *S. aureus* [28;58]. Moreover, a further loss of intracellular efficacy (*Emax*) is obtained when cells are infected by less susceptible strains, such as heteroresistant subpopulations of vancomycin-intermediate *S. aureus* [hVISA] or VISA isolates [59]. Poorer outcome in patients treated with vancomycin for staphylococcal infections implicating less susceptible isolates may thus result from insufficient eradication of both extracellular and intracellular bacteria. An illustrative case was reported by Julian and colleagues [60], who described a chronic infective endocarditis that did not respond to vancomycin. Bacteriemia was documented over a 33-day period with isolates showing progressive elevated MIC values to vancomycin. Resolution was finally obtained by surgical procedures, indicating that persistent foci were the cause of the therapeutic failure.

2.5. Glycylcyclines

Tigecycline achieves quickly high intracellular accumulation within PMN but shows only a modest effect towards intracellular *S. aureus*, probably again in relation with its bacteriostatic character [61;62].

2.6. β–Lactam Antibiotics

For a long time, it has been stated that β-lactams would not be active against intracellular organisms because they do not accumulate within eukaryotic cells. Clinical data tend to prove the contrary, since β-lactams are highly effective in the treatment of listeriosis [63], an infection caused by the intracellular organism *Listeria monocytogenes*. Cellular models have now rationalized this observation by showing that appropriate doses (i.e. high concentrations such as serum levels) and prolonged time of exposure (imposed by the time-dependent activity of these agents in extracellular models of infection) allow to obtain sustained cellular concentrations of the same order of magnitude as those reached in the serum. This explains a certain degree of intracellular activity, including against *S. aureus* [28;64].

A still more unanticipated observation is that MRSA regain almost full susceptibility to cloxacillin or meropenem after phagocytosis by human THP-1 macrophages (figure 5) or upon internalization by keratinocytes [65;66]. This is clearly due to the acidic pH [67] prevailing in the phagolysosomes, since neutralization of these organelles (using lysosomotropic agent such as ammonium chloride) renders intraphagocytic MRSA again insensitive to the action of β-lactams. These findings have been rationalized by the observation that acidic pH caused a conformational change of the Penicillin-Binding Protein 2a (PBP2a, a unique transpeptidase poorly inhibited by β-lactams and responsible for the MRSA phenotype [68-70]), improving the access of penicillins to the active site and the subsequent acylation of the protein required for antibiotic activity [71]. Lack of effective inhibition of PBP2a at neutral pH has indeed been ascribed to the closed conformation of its active site [69;72].

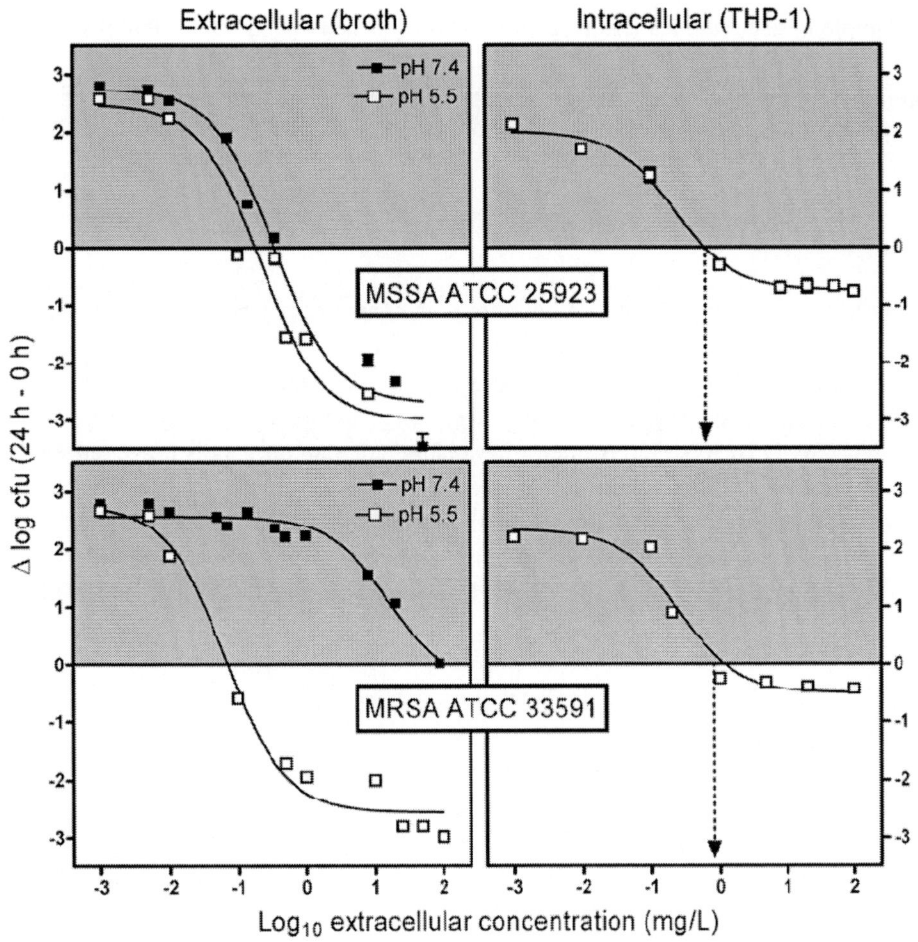

Figure 5. Influence of concentration on the extracellular (Mueller Hinton Broth adjusted to pH 7.4 and pH 5.5) and intracellular activities of cloxacillin against MSSA strain ATCC 25923 and MRSA strain ATCC 33591. The graph shows antibiotic activity over a 24 h period against bacteria in broth (pH 7.4 *vs.* pH 5.5) or post-phagocytized by human THP-1 macrophages. Activity is expressed as the change in bacterial counts from time 0 h. The gray zone corresponds to the bacterial growth. Adapted from data presented in reference [65].

2.7. Oxazolidinones

It has been shown previously that linezolid kills more efficiently MRSA compared to vancomycin [73]. In the intracellular environment, however, this bacteriostatic agent [28;59] kills only modestly intracellular *S. aureus* with a maximal effect similar to that of vancomycin.

2.8. Ansamycins

Rifampicin accumulates from 2- to 5-fold in cells [74], and demonstrates potent intracellular killing of rifampicin-susceptible *S. aureus* [28;59], yielding even truly intracellular bactericidal effect towards SCVs (< 3 log CFU decrease, as described by the CLSI criteria) for prolonged time exposure (72 h) [62].

2.9. Daptomycin

This last resort drug for MRSA infections is characterized by (i) extremely fast killing effects towards MRSA at all growth phases (which may be particularly useful for the treatment of indolent, deep-seated infections such as endocarditis or osteomyelitis in which bacteria reach stationary phase [75]); and (ii) marked intraphagocytic killing of both MSSA [76] and MRSA [77]. As demonstrated for vancomycin, daptomycin is however less efficient against intracellular VISA isolates due to reduced intrinsic activity [59;77]. This intracellular effect is obtained in spite of a low cellular accumulation (Cc/Ce < 1 [76]) and is further increased in the presence of P-glycoprotein inhibitors (verapamil or elacridar) since this antibiotic is substrate of these transporters, as demonstrated in THP-1 macrophages and MDCK (Madin-Darby Canine Kidney) cells [76].

2.10. Fluoroquinolones

These antibiotics accumulate to variable levels in eukaryotic cells (5- to 20-fold (*78*)) and are suspected to easily diffuse in all subcellular compartments, as they show activity against cytosolic (*L. monocytogenes* [79-81]), phagosomal (*L. pneumophila* [82;83*]*) or phagolysosomal (*S. aureus [(28;52;84)*) bacteria. Within this class, molecules with higher level of accumulation but also higher intrinsic activity like moxifloxacin are more active intracellularly than derivatives with lower cellular accumulation and intrinsic activities (such as ciprofloxacin)[28]. These properties, combined with a bactericidal character, make of moxifloxacin one of the most effective antibiotics against intracellular *S. aureus,* yielding potent intracellular killing in a concentration- and time-dependent fashion (-2 and -3 log cfu after 24 h and 48 h, respectively) against both MSSA [28;52;85] and CA-MRSA isolates [86;87].

Cellular accumulation of fluoroquinolones is highly modulable but surprinsingly, changes in accumulation do not necessarily translate in improved efficacy. In J774 macrophages, ciprofloxacin, but not moxifloxacin, is substrate of the multidrug transporter MRP4 [88]. Inhibition of this transporter increases ciprofloxacin cellular content to values close to those reached by moxifloxacin [53;89;90]. However, this is not accompagnied by an improved activity against intracellular *S. aureus* but well on *L. monocytogenes* (*53;89*), because the additional amount of drug accumulates in the cytosol. In human THP-1 macrophages, differentiation by PMA specifically increases the accumulation of moxifloxacin but not of levofloxacin by a still unknown mechanism [84]. Again, this specifically increases the intracellular activity against *L. monocytogenes* but not on *S. aureus*, presumably for the same reason.

2.11. Quinupristin-Dalfopristin

This combination of streptogramins accumulates within cells [91], and demonstrate potent intraphagocytic killing of *S. aureus* causing typically a 2 – 2.5 log decrease in intracellular inoculum [91]. This effect is also observed for MRSA isolates of current clinical and epidemiological interest, including VISA and Vancomycin Resistant *S. aureus* (VRSA) strains [59;92].

2.12. Novel Investigational Antibiotics

2.12.1. Novel Glycopeptides

Loss of susceptibility to vancomycin in MRSA stimulated the search for alternative therapy. In this context, novel glycopeptides have been successfully designed to keep activity against Vancomycin-resistant *S. aureus* (VRSA) isolates and Vancomycin-intermediate *S. aureus* (VISA) to a lower extent (see for review: [93-95], because of a dual mode of action implying both inhibition of cell-wall synthesis and membrane destabilization [96;97]. Telavancin [98;99] and oritavancin [100;101] accumulate to high levels (~ 50- and 200-fold after 24 hours, respectively) in the lysosomes of murine macrophages [102;103], show superior *in vitro* activity in broth, and are highly bactericidal towards both extracellular and intracellular forms of vancomycin-susceptible and vancomycin-resistant *S. aureus* (figure 2, figure 4) [58;59;104] compared to vancomycin. Oritavancin is also highly efficient against intracellular SCV [62], and this activity is still reinforced when combined with other bactericidal agents such as rifampicin or moxifloxacin [105].

2.12.2. Anti-MRSA Cephalosporins

Ceftobiprole and ceftaroline are two cephalosporins under development that show similar activity against both MSSA and MRSA [106-108]. This is due to the fact that they strongly bind to PBP2a, being able to induce the conformational change of the protein required for antibiotic activity [109], as did acidic pH for conventional β-lactams. Ceftobiprole shows similar intrinsic activities against MSSA and MRSA (HA- or CA-MRSA) at neutral and acidic pH, as well as a similar potency against intracellular MSSA or MRSA in models of human macrophages or keratinocytes [110].

2.12.3. Macrolide / Ketolide

CEM-101 is a novel ketolide that, in contrast to azithromycin, keeps activity under acidic conditions and is not substrate for P-glycoprotein. Accordingly, it is considerably more active against intracellular forms of *S. aureus* [111], as well against intraphagocytic *Listeria monocytogenes* and *Legionella pneumophila.*

2.12.4. Oxazolidinones

Novel investigational oxazolidinones (such as radezolid [RX-1741] [112;113] or torezolid [TR-700] [114;115]) have been designed to keep activity against linezolid-resistant *S. aureus* strains. As compared to linezolid, they show increased intrinsic activities against linezolid-resistant *S. aureus* (lower MIC values), and higher intracellular accumulation accompanied by an increased potency towards both linezolid-susceptible and linezolid-resistant *S. aureus* [116-118].

2.12.5. Iclaprim

Iclaprim is a diaminopyrimidine dihydrofolate reductase inhibitor that has completed clinical development for the treatment of skin and skin structure infections (cSSSI). This compound demonstrates potent intraphagocytic activity against different intracellular organisms, including *L. monocytogenes* [119] *and L. pneumophila* [120], compared to trimethoprim-sulfamethoxazole. Data are however still lacking for intracellular *S. aureus*.

2.12.6. Linopristin-Flopristin

Linopristin-flopristin (NXL103, formerly XRP2868) is a novel semi-synthetic combination of streptogramins (vol/vol, 30/70). Compared to quinupristin-dalfopristin, it demonstrates slightly lower MICs values against MRSA [121], but similar activity against intracellular *S. aureus* (approx. -2 log CFU [122]).

CONCLUSIONS

Gaining an intracellular niche, even briefly, may afford several opportunities to *S. aureus* to survive and promote disease. Because antibiotics are considerably less effective intracellularly than extracellularly, treatment of relapsing and persistent *S. aureus* infections is particularly challenging. In this context, the possibility that conventional anti-MRSA antibiotics (vancomycin, tetracyclines, linezolid, azithromycin, trimethoprim-sulfamethoxazole) may have limited intracellular activity should be borne in mind when selecting appropriate treatment for infections in which intracellular foci are suspected. In this respect, cellular models may help in selecting for further development molecules with low MICs values and high efficacy (*Emax*) against both extracellular and intracellular MRSA.

BOX 1. Intraphagocytic model of infection

Gentamicin protection assays are widely used for the determination of antibiotic intracellular activities. Briefly, phagocytes are incubated in the presence of pre-opsonized *S. aureus* (e.g. multiplicity of infection [MOI] equivalent to 4 bacteria per cells) during 1-2 h [28;52] before extensive washing in pre-warmed phosphate buffered saline (PBS) and incubation in the presence of gentamicin during 45 min-1 h. Cells are then incubated in the presence of increasing concentrations of antibiotics [28;64;123] and, after different time intervals, cells are washed free from antibiotics, lysed by resuspension in water, and the corresponding samples then processed for CFU counting and protein content determination. In order to prevent the development of extracellular bacteria, control cells systematically include exposition to gentamicin (1/2- to 1-fold the MIC, avoiding marked acidification and subsequent cell death)[28;52]. Complete sterilization of the extracellular medium has been also reported in lysostaphin protection assay, but a significant amount of lysostaphin is able to enter the cells, which may interfere with the expression of antibiotic activity or CFU counting [52].

REFERENCES

[1] Amador-Miranda, R., Bertran-Pasarell, J., Gonzalez, M., and Conde, A. (2008) *Bol. Asoc. Med. P. R. 100*, 21-23.

[2] Boyle-Vavra, S. and Daum, R. S. (2007) *Lab Invest 87*, 3-9.

[3] Diep, B. A., Palazzolo-Ballance, A. M., Tattevin, P., Basuino, L., Braughton, K. R., Whitney, A. R., Chen, L., Kreiswirth, B. N., Otto, M., DeLeo, F. R., and Chambers, H. F. (2008) *PLoS. ONE. 3*, e3198.

[4] Corriere, M. D. and Decker, C. F. (2008) *Dis. Mon. 54*, 751-755.

[5] Nourse, C., Starr, M., and Munckhof, W. (2007) *J. Paediatr. Child Health 43*, 656-661.

[6] Klevens, R. M., Morrison, M. A., Nadle, J., Petit, S., Gershman, K., Ray, S., Harrison, L. H., Lynfield, R., Dumyati, G., Townes, J. M., Craig, A. S., Zell, E. R., Fosheim, G. E., McDougal, L. K., Carey, R. B., and Fridkin, S. K. (2007) *JAMA 298*, 1763-1771.

[7] Gresham, H. D., Lowrance, J. H., Caver, T. E., Wilson, B. S., Cheung, A. L., and Lindberg, F. P. (2000) *J. Immunol. 164*, 3713-3722.

[8] Lowy, F. D. (2000) *Trends Microbiol. 8*, 341-343.

[9] Ellington, J. K., Harris, M., Webb, L., Smith, B., Smith, T., Tan, K., and Hudson, M. (2003) *J. Bone Joint Surg. Br. 85*, 918-921.

[10] Garzoni, C. and Kelley, W. L. (2009) *Trends Microbiol. 17*, 59-65.

[11] Clement, S., Vaudaux, P., Francois, P., Schrenzel, J., Huggler, E., Kampf, S., Chaponnier, C., Lew, D., and Lacroix, J. S. (2005) *J. Infect. Dis. 192*, 1023-1028.

[12] Carryn, S., Chanteux, H., Seral, C., Mingeot-Leclercq, M. P., Van Bambeke, F., and Tulkens, P. M. (2003) *Infect. Dis. Clin. North Am. 17*, 615-634.

[13] Van Bambeke, F., Barcia-Macay, M., Lemaire, S., and Tulkens, P. M. (2006) *Curr. Opin. Drug Discov. Devel. 9*, 218-230.

[14] Whaley, K. and Singh, H. (1973) *Immunology 24*, 25-35.
[15] Baughn, R. and Bonventre, P. F. (1975) *Infect. Immun. 12*, 346-352.
[16] Hof, D. G., Repine, J. E., Peterson, P. K., and Hoidal, J. R. (1980) *Am. Rev. Respir. Dis. 121*, 65-71.
[17] Hamill, R. J., Vann, J. M., and Proctor, R. A. (1986) *Infect. Immun. 54*, 833-836.
[18] Hudson, M. C., Ramp, W. K., Nicholson, N. C., Williams, A. S., and Nousiainen, M. T. (1995) *Microb. Pathog. 19*, 409-419.
[19] Jevon, M., Guo, C., Ma, B., Mordan, N., Nair, S. P., Harris, M., Henderson, B., Bentley, G., and Meghji, S. (1999) *Infect. Immun. 67*, 2677-2681.
[20] Mempel, M., Schnopp, C., Hojka, M., Fesq, H., Weidinger, S., Schaller, M., Korting, H. C., Ring, J., and Abeck, D. (2002) *Br. J. Dermatol. 146*, 943-951.
[21] Usui, A., Murai, M., Seki, K., Sakurada, J., and Masuda, S. (1992) *Microbiol. Immunol. 36*, 545-550.
[22] Hebert, A., Sayasith, K., Senechal, S., Dubreuil, P., and Lagace, J. (2000) *FEMS Microbiol. Lett. 193*, 57-62.
[23] Sinha, B., Francois, P. P., Nusse, O., Foti, M., Hartford, O. M., Vaudaux, P., Foster, T. J., Lew, D. P., Herrmann, M., and Krause, K. H. (1999) *Cell Microbiol. 1*, 101-117.
[24] Schwarz-Linek, U., Werner, J. M., Pickford, A. R., Gurusiddappa, S., Kim, J. H., Pilka, E. S., Briggs, J. A., Gough, T. S., Hook, M., Campbell, I. D., and Potts, J. R. (2003) *Nature 423*, 177-181.
[25] Shinji, H., Kamada, M., Seki, K., Tajima, A., Iwase, T., and Masuda, S. (2007) *Microbiol. Immunol. 51*, 63-71.
[26] Yonemasu, K., Sasaki, T., Hashimoto, H., and Kashiba, S. (1988) *Microbiol. Immunol. 32*, 795-805.
[27] Kubica, M., Guzik, K., Koziel, J., Zarebski, M., Richter, W., Gajkowska, B., Golda, A., Maciag-Gudowska, A., Brix, K., Shaw, L., Foster, T., and Potempa, J. (2008) *PLoS. ONE. 3*, e1409.
[28] Barcia-Macay, M., Seral, C., Mingeot-Leclercq, M. P., Tulkens, P. M., and Van Bambeke, F. (2006) *Antimicrob. Agents Chemother. 50*, 841-851.
[29] Menzies, B. E. and Kourteva, I. (2000) *FEMS Immunol. Med. Microbiol. 29*, 39-45.
[30] Qazi, S. N., Counil, E., Morrissey, J., Rees, C. E., Cockayne, A., Winzer, K., Chan, W. C., Williams, P., and Hill, P. J. (2001) *Infect. Immun. 69*, 7074-7082.
[31] Garzoni, C., Francois, P., Huyghe, A., Couzinet, S., Tapparel, C., Charbonnier, Y., Renzoni, A., Lucchini, S., Lew, D. P., Vaudaux, P., Kelley, W. L., and Schrenzel, J. (2007) *BMC. Genomics 8*, 171.
[32] Eriksson, S., Lucchini, S., Thompson, A., Rhen, M., and Hinton, J. C. (2003) *Mol. Microbiol. 47*, 103-118.
[33] Lucchini, S., Liu, H., Jin, Q., Hinton, J. C., and Yu, J. (2005) *Infect. Immun. 73*, 88-102.
[34] Pelz, A., Wieland, K. P., Putzbach, K., Hentschel, P., Albert, K., and Gotz, F. (2005) *J. Biol. Chem. 280*, 32493-32498.
[35] Liu, C. I., Liu, G. Y., Song, Y., Yin, F., Hensler, M. E., Jeng, W. Y., Nizet, V., Wang, A. H., and Oldfield, E. (2008) *Science 319*, 1391-1394.
[36] Olivier, A., Lemaire, S., Van Bambeke, F., Tulkens, P. M., and Oldfield, E. Role of *rsbU* and staphyloxanthin in phagocytosis and intracellular growth of *Staphylococcus aureus* in human macrophages and endothelial cells. J.Infect.Dis. In press. 2009.
[37] Von Eiff, C., Peters, G., and Becker, K. (2006) *Injury 37 Suppl 2*, S26-S33.

[38] Von Eiff, C., Proctor, R. A., and Peters, G. (2000) *Berl Munch. Tierarztl. Wochenschr. 113*, 321-325.

[39] Von Eiff, C., McNamara, P., Becker, K., Bates, D., Lei, X. H., Ziman, M., Bochner, B. R., Peters, G., and Proctor, R. A. (2006) *J. Bacteriol. 188*, 687-693.

[40] Chatterjee, I., Kriegeskorte, A., Fischer, A., Deiwick, S., Theimann, N., Proctor, R. A., Peters, G., Herrmann, M., and Kahl, B. C. (2008) *J. Bacteriol. 190*, 834-842.

[41] Vaudaux, P., Francois, P., Bisognano, C., Kelley, W. L., Lew, D. P., Schrenzel, J., Proctor, R. A., McNamara, P. J., Peters, G., and Von Eiff, C. (2002) *Infect. Immun. 70*, 5428-5437.

[42] Plouin-Gaudon, I., Clement, S., Huggler, E., Chaponnier, C., Francois, P., Lew, D., Schrenzel, J., Vaudaux, P., and Lacroix, J. S. (2006) *Rhinology 44*, 249-254.

[43] Sendi, P., Rohrbach, M., Graber, P., Frei, R., Ochsner, P. E., and Zimmerli, W. (2006) *Clin. Infect. Dis. 43*, 961-967.

[44] Bosse, M. J., Gruber, H. E., and Ramp, W. K. (2005) *J. Bone Joint Surg. Am. 87*, 1343-1347.

[45] Gristina, A. G. and Costerton, J. W. (1985) *J. Bone Joint Surg. Am. 67*, 264-273.

[46] Gristina, A. G., Shibata, Y., Giridhar, G., Kreger, A., and Myrvik, Q. N. (1994) *Semin. Arthroplasty 5*, 160-170.

[47] Naylor, P. T., Myrvik, Q. N., and Gristina, A. (1990) *Clin. Orthop. Relat Res.* 126-133.

[48] Tulkens, P. M. (1991) *Eur. J. Clin. Microbiol. Infect. Dis. 10*, 100-106.

[49] Sandberg, A., Hessler, J. H., Skov, R. L., Blom, J., and Frimodt-Moller, N. (2009) *Antimicrob. Agents Chemother. 53*, 1874-1883.

[50] McDonald, P. J. and Pruul, H. (1991) *Eur. J. Clin. Microbiol. Infect. Dis. 10*, 828-833.

[51] Carlier, M. B., Garcia-Luque, I., Montenez, J. P., Tulkens, P. M., and Piret, J. (1994) *Int. J. Tissue React. 16*, 211-220.

[52] Seral, C., Van Bambeke, F., and Tulkens, P. M. (2003) *Antimicrob. Agents Chemother. 47*, 2283-2292.

[53] Seral, C., Carryn, S., Tulkens, P. M., and Van Bambeke, F. (2003) *J. Antimicrob. Chemother. 51*, 1167-1173.

[54] Seral, C., Michot, J. M., Chanteux, H., Mingeot-Leclercq, M. P., Tulkens, P. M., and Van Bambeke, F. (2003) *Antimicrob. Agents Chemother. 47*, 1047-1051.

[55] Vazifeh, D., Abdelghaffar, H., and Labro, M. T. (1997) *Antimicrob. Agents Chemother. 41*, 2099-2107.

[56] Forsgren, A. and Bellahsene, A. (1985) *Scand. J. Infect. Dis. Suppl 44*, 16-23.

[57] Yang, Q., Nakkula, R. J., and Walters, J. D. (2002) *J. Dent. Res. 81*, 836-840.

[58] Barcia-Macay, M., Lemaire, S., Mingeot-Leclercq, M. P., Tulkens, P. M., and Van Bambeke, F. (2006) *J. Antimicrob. Chemother. 58*, 1177-1184.

[59] Lemaire, S., Kosowska-Shick, K., Julian, K., Tulkens, P. M., Van Bambeke, F., and Appelbaum, P. C. (2008) *Clin. Microbiol. Infect. 14*, 766-777.

[60] Julian, K., Kosowska-Shick, K., Whitener, C., Roos, M., Labischinski, H., Rubio, A., Parent, L., Ednie, L., Koeth, L., Bogdanovich, T., and Appelbaum, P. C. (2007) *Antimicrob. Agents Chemother. 51*, 3445-3448.

[61] Ong, C. T., Babalola, C. P., Nightingale, C. H., and Nicolau, D. P. (2005) *J. Antimicrob. Chemother. 56*, 498-501.

[62] Nguyen, H. A., Denis, O., Vergison, A., Theunis, A., Tulkens, P. M., Struelens, M. J., and Van Bambeke, F. (2009) *Antimicrob. Agents Chemother.*

[63] Crum, N. F. (2002) *Curr. Gastroenterol. Rep. 4*, 287-296.
[64] Lemaire, S., Van Bambeke, F., Mingeot-Leclercq, M. P., and Tulkens, P. M. (2005) *J. Antimicrob. Chemother. 55*, 897-904.
[65] Lemaire, S., Van Bambeke, F., Mingeot-Leclercq, M. P., Glupczynski, Y., and Tulkens, P. M. (2007) *Antimicrob. Agents Chemother. 51*, 1627-1632.
[66] Lemaire, S., Olivier, A., Van Bambeke, F., Tulkens, P. M., Appelbaum, P. C., and Glupczynski, Y. (2008) *Antimicrob. Agents Chemother. 52*, 2797-2805.
[67] Sabath, L. D., Wallace, S. J., and Gerstein, D. A. (1972) *Antimicrob. Agents Chemother. 2*, 350-355.
[68] Hartman, B. J. and Tomasz, A. (1984) *J. Bacteriol. 158*, 513-516.
[69] Fuda, C., Suvorov, M., Vakulenko, S. B., and Mobashery, S. (2004) *J. Biol. Chem. 279*, 40802-40806.
[70] Roychoudhury, S., Dotzlaf, J. E., Ghag, S., and Yeh, W. K. (1994) *J. Biol. Chem. 269*, 12067-12073.
[71] Lemaire, S., Fuda, C., Van Bambeke, F., Tulkens, P. M., and Mobashery, S. (2008) *J. Biol. Chem. 283*, 12769-12776.
[72] Lim, D. and Strynadka, N. C. (2002) *Nat. Struct. Biol. 9*, 870-876.
[73] Weigelt, J., Kaafarani, H. M., Itani, K. M., and Swanson, R. N. (2004) *Am. J. Surg. 188*, 760-766.
[74] Mor, N., Simon, B., Mezo, N., and Heifets, L. (1995) *Antimicrob. Agents Chemother. 39*, 2073-2077.
[75] Mascio, C. T., Alder, J. D., and Silverman, J. A. (2007) *Antimicrob. Agents Chemother. 51*, 4255-4260.
[76] Lemaire, S., Van Bambeke, F., Mingeot-Leclercq, M. P., and Tulkens, P. M. (2007) *Antimicrob. Agents Chemother. 51*, 2748-2757.
[77] Lemaire, S., Van Bambeke, F., and Tulkens, P. M. Intracellular activity of daptomycin against methicillin-sensitive, methicillin-resistant, and vancomycin-intermediate S. aureus. 17th European Congress of Clinical Microbiology and Infectious Diseases Poster 829. 2007.
[78] Ballesta, S., Pascual, A., Garcia, I., and Perea, E. J. (1996) *Enferm. Infecc. Microbiol. Clin. 14*, 470-473.
[79] Carryn, S., Van Bambeke, F., Mingeot-Leclercq, M. P., and Tulkens, P. M. (2003) *J. Antimicrob. Chemother. 51*, 1051-1052.
[80] Carryn, S., Van Bambeke, F., Mingeot-Leclercq, M. P., and Tulkens, P. M. (2002) *Antimicrob. Agents Chemother. 46*, 2095-2103.
[81] Grayo, S., Join-Lambert, O., Desroches, M. C., and Le Monnier, A. (2008) *Antimicrob. Agents Chemother. 52*, 1697-1702.
[82] Carlier, M. B., Scorneaux, B., Zenebergh, A., Desnottes, J. F., and Tulkens, P. M. (1990) *J. Antimicrob. Chemother. 26 Suppl B*, 27-39.
[83] Baltch, A. L., Bopp, L. H., Smith, R. P., Michelsen, P. B., and Ritz, W. J. (2005) *J. Antimicrob. Chemother. 56*, 104-109.
[84] Van, d., V, Nguyen, H. A., Van Bambeke, F., Tulkens, P. M., Grellet, J., Dubois, V., Quentin, C., and Saux, M. C. (2008) *J. Antimicrob. Chemother. 62*, 518-521.
[85] Ulrich, M., Berger, J., Moller, J. G., and Doring, G. (2005) *Infection 33 Suppl 2*, 50-54.
[86] Lemaire, S., Van Bambeke, F., Glupczynski, Y., and Tulkens, P. M. Intracellular activity of moxifloxacin against European, American and Asian clinical isolates of

community-acquired methicillin-resistant S. aureus (CA-MRSA). 27ème Réunion Interdisciplinaire de Chimiothérapie Anti-Infectieuse Poster 487. 2009.
[87] Lemaire, S., Van Bambeke, F., Tulkens, P. M., and Glupczynski, Y. Comparative activity of moxifloxacin, linezolid, cloxacillin, clindamycin, trimethoprim-sulfamethoxazole, and ciprofloxacin against intracellular Methicillin-sensitive and Community-Acquired Methicillin-Resistant *S. aureus*. 17th European Congress of Clinical Microbiology and Infectious Diseases Poster 703. 2007.
[88] Marquez, B., Caceres, N. E., Mingeot-Leclercq, M. P., Tulkens, P. M., and Van Bambeke, F. (2009) *Antimicrob. Agents Chemother. 53*, 2410-2416.
[89] Lismond, A., Tulkens, P. M., Mingeot-Leclercq, M. P., Courvalin, P., and Van Bambeke, F. (2008) *Antimicrob. Agents Chemother. 52*, 3040-3046.
[90] Michot, J. M., Van Bambeke, F., Mingeot-Leclercq, M. P., and Tulkens, P. M. (2004) *Antimicrob. Agents Chemother. 48*, 2673-2682.
[91] Bebear, C. and Bouanchaud, D. H. (1997) *J. Antimicrob. Chemother. 39 Suppl A*, 59-62.
[92] Baudoux, P., Glupczynski, Y., Tulkens, P. M., and Van Bambeke, F. Extracellular and intracellular activities of Quinupristin-Dalfopristin (Synercid) against S. aureus, with different resistant phenotypes (MSSA, MRSA, VISA). 17th European Congress of Clinical Microbiology and Infectious Diseases Poster 2058. 2007.
[93] Van Bambeke, F., Mingeot-Leclercq, M. P., Struelens, M. J., and Tulkens, P. M. (2008) *Trends Pharmacol. Sci. 29*, 124-134.
[94] Van Bambeke, F., Van Laethem, Y., Courvalin, P., and Tulkens, P. M. (2004) *Drugs 64*, 913-936.
[95] Linden, P. K. (2008) *Expert. Rev. Anti. Infect. Ther. 6*, 917-928.
[96] Leonard, S. N. and Rybak, M. J. (2008) *Pharmacotherapy 28*, 458-468.
[97] Belley, A., Neesham-Grenon, E., McKay, G., Arhin, F. F., Harris, R., Beveridge, T., Parr, T. R., Jr., and Moeck, G. (2009) *Antimicrob. Agents Chemother. 53*, 918-925.
[98] Laohavaleeson, S., Kuti, J. L., and Nicolau, D. P. (2007) *Expert. Opin. Investig. Drugs 16*, 347-357.
[99] Nannini, E. C. and Stryjewski, M. E. (2008) *Expert. Opin. Pharmacother. 9*, 2197-2207.
[100] Mercier, R. C. and Hrebickova, L. (2005) *Expert. Rev. Anti. Infect. Ther. 3*, 325-332.
[101] Crandon, J. and Nicolau, D. P. (2008) *Future. Microbiol. 3*, 251-263.
[102] Barcia-Macay, M., Mouaden, F., Mingeot-Leclercq, M. P., Tulkens, P. M., and Van Bambeke, F. (2008) *J. Antimicrob. Chemother. 61*, 1288-1294.
[103] Van Bambeke, F., Carryn, S., Seral, C., Chanteux, H., Tyteca, D., Mingeot-Leclercq, M. P., and Tulkens, P. M. (2004) *Antimicrob. Agents Chemother. 48*, 2853-2860.
[104] Baudoux, P., Tulkens, P. M., and Van Bambeke, F. Intracellular activity of oritavancin against MSSA, MRSA and VISA strains in a model of human macrophages. 48th Interscience Conference on Antimicrobial Agents and Chemotherapy Poster no. A-971. 2008.
[105] Nguyen, H. A., Denis, O., Vergison, A., Tulkens, P. M., Struelens, M. J., and Van Bambeke, F. (2009) *Antimicrob. Agents Chemother. 53*, 1443-1449.
[106] Vidaillac, C. and Rybak, M. J. (2009) *Pharmacotherapy 29*, 511-525.
[107] Jacqueline, C., Caillon, J., Le, M., V, Miegeville, A. F., Hamel, A., Bugnon, D., Ge, J. Y., and Potel, G. (2007) *Antimicrob. Agents Chemother. 51*, 3397-3400.

[108] Mushtaq, S., Warner, M., Ge, Y., Kaniga, K., and Livermore, D. M. (2007) *J. Antimicrob. Chemother. 60*, 300-311.

[109] Lovering, A., Danel, F., Page, M. G. P., and Strynadka, N. J. Mechanism of action of ceftobiprole: structural basis for the anti-MRSA activity. 17th European Congress of Clinical Microbiology and Infectious Diseases Program Abstr. 2006.

[110] Lemaire, S., Glupczynski, Y., Duval, V., Joris, B., Tulkens, P. M., and Van Bambeke, F. (2009) *Antimicrob. Agents Chemother. 53*, 2289-2297.

[111] Lemaire, S., Van Bambeke, F., and Tulkens, P. M. (2009) *Antimicrob. Agents Chemother.*

[112] Skripkin, E., McConnell, T. S., DeVito, J., Lawrence, L., Ippolito, J. A., Duffy, E. M., Sutcliffe, J., and Franceschi, F. (2008) *Antimicrob. Agents Chemother. 52*, 3550-3557.

[113] Lawrence, L., Danese, P., DeVito, J., Franceschi, F., and Sutcliffe, J. (2008) *Antimicrob. Agents Chemother. 52*, 1653-1662.

[114] Shaw, K. J., Poppe, S., Schaadt, R., Brown-Driver, V., Finn, J., Pillar, C. M., Shinabarger, D., and Zurenko, G. (2008) *Antimicrob. Agents Chemother. 52*, 4442-4447.

[115] Jones, R. N., Moet, G. J., Sader, H. S., Mendes, R. E., and Castanheira, M. (2009) *J. Antimicrob. Chemother.*

[116] Lemaire, S., Van Bambeke, F., Appelbaum, P. C., and Tulkens, P. M. Cellular pharmacokinetics and intracellular activity of Torezolid (TR-700) in phagocytic and non-phagocytic cells. J.Antimicrob.Chemother. 2009.

[117] Lemaire, S., Tulkens, P. M., and Van Bambeke, F. Radezolid (RX-1741), a novel oxazolidinone, is active against intracellular *S. aureus*, *L. monocytogenes* and *L. pneumophila* phagocytosed by human THP-1 macrophages. 19th European Congress of Clinical Microbiology and Infectious Diseases O-30. 2009.

[118] Lemaire, S., Tulkens, P. M., and Van Bambeke, F. Radezolid (RX-1741), a novel oxazolidinone, accumulates extensively within human macrophages and PMNs and shows activity towards intracellular linezolid-sensitive and linezolid-resistant Staphylococcus aureus. 26th International Conference on Chemotherapy . 2009.

[119] Fischer, M., Weiss, L., Scheidegger, D., Bernardini, F., and Islam, K. Iclaprim, a Novel Diaminopyrimidine Antibiotic: In Vitro and Intracellular Activity against Listeria monocytogenes. 43rd Interscience Conference on Antimicrobial Agents and Chemotherapy E-2004. 2003.

[120] Morrissey, I. and Hawser, S. (2007) *J. Antimicrob. Chemother. 60*, 905-906.

[121] McGhee, P., Lin, G., Pankuch, G., Bryskier, A., and Appelbaum, P. Activity of linopristin-flopristin (NXL103) Against *Streptococcus pneumoniae, Haemophilus influenzae,* and *Staphylococcus aureus* . 19th European Congress of Clinical Microbiology and Infectious Diseases . 2009.

[122] Pace, J., Levasseur, P., Girard, A., and Lavallade, L. Antibacterial Activity of NXL103 (linopristin/flopristin), against Intracellular *Staphylococcus aureus* and Efficacy Following *per oral* Administration in Murine Models of Systemic Infection. 19th European Congress of Clinical Microbiology and Infectious Diseases. 2009.

[123] Baudoux, P., Bles, N., Lemaire, S., Mingeot-Leclercq, M. P., Tulkens, P. M., and Van Bambeke, F. (2007) *J. Antimicrob. Chemother. 59*, 246-253.

In: Methicillin-Resistant Staphylococcus Aureus...
Editor: Charles L. Kolendi

ISBN: 978-1-60741-398-1

Chapter 3

METHICILLIN-RESISTANT STAPHYLOCOCCUS AUREUS (MRSA)

Dong Choon Park[1], Sun Kyu Lee[2] and Seung Geun Yeo[2*]

[1]Department of Obstetrics and Gynecology, The Catholic University of Korea
[2]Department of Otorhinolaryngology and Head and Neck Surgery, School of Medicine, KyungHee University, Seoul, Korea

1. INTRODUCTION

For more than a century *Staphylococcus* has been an important human pathogen. "Micrococcus, which, when limited in its extent and activity, causes acute suppurative inflammation (phlegmon), produces, when more extensive and intense in its action on the human system, the most virulent forms of septicaemia and pyaemia… [1]". From these words of Sir Alexander Ogston in 1882, in his "Description of *Staphylococcus*", it can be seen that the bacterium caused serious illness and resulting mortality; this remains true today, more than 100 years later.

Prior to antibiotic development, *Staphylococcus aureus* (*S. aureus*) infection was associated with very high mortality rates. After penicillin was developed in 1942, the number of deaths from *S. aureus* infections started to decrease rapidly. However, only 2 years later, in 1944, a strain showing resistance to penicillin was discovered [2]. Until the 1960s, penicillin-resistant *S. aureus* were widespread in the community and healthcare centers. The development of methicillin, a penicillinase-resistant semisynthetic penicillin analog, in 1960, quickly resolved the problem. However less than 1 year later, in 1961, methicillin-resistant *S. aureus* (MRSA) was reported for the first time [3].

MRSA strains show resistance to the semisynthetic penicillins methicillin and oxacillin, and to cephalosporins and carbapenems such as β-lactam antibiotics. MRSAs are also resistant to macrolides, clindamycin, aminoglycosides, and other common antibiotics.

* Corresponding authors: #1 Hoegi-dong, dongdaemun-gu, Seoul 130-702, Korea, TEL; 82-2-958-8980, FAX; 82-2-958-8470, E-mail; yeo2park@yahoo.co.kr

In 1956, the glycopeptide antibiotic vancomycin was introduced and was an important component of MRSA treatment for more than 30 years. In 1997, vancomycin-intermediate-susceptible *S. aureus* (VISA) was reported for the first time; resistance to vancomycin began to increase [4, 5]; and, ultimately, fully vancomycin-resistant *S. aureus* (VRSA) was identified in 2002 [6, 7].

Despite the emergence of VRSA, most staphylococcal infections are still caused by MRSA. After the first such infections were identified in 1961, the proportion of staphylococcal infections caused by MRSA has gradually increased. In 1974, MRSA strains accounted for 2% of all hospital staphylococcal infections but in 1997 this level had increased to 50% [8]. According to National Nosocomial Infection Surveillance (NNIS) data, nosocomial infections caused by MRSA in intensive care units in the United States comprise at least 60% of all *S. aureus* infections [9]. Compared with treatment of MSSA (methicillin-sensitive *S. aureus*) infections, resolution of MRSA infections is costlier, more prolonged, and associated with higher mortality rates [10]. Surgical site MRSA infection is associated with a 20.7% mortality rate, which is significantly higher than that of MSSA infection (6.7%) or non-infection (2.1%) [11]. MRSA infections kill 19,000 hospitalized patients each year in the United States, similar to the number of deaths from AIDS, tuberculosis, or viral hepatitis [12].

MRSA usually evolve in hospitals where antibiotics are heavily used, and the MRSA infection rate is low among healthy community subjects; infection is rarely reported even among drug addicts [13]. However, in Minnesota and North Dakota in the United States, between 1997 and 1999, four healthy children died of severe septicemia caused by MRSA [14]. Thereafter, interest in community-associated MRSA (CA-MRSA) began to increase. The most important feature of CA-MRSA infections is that they are not associated with underlying disease. The incidence of CA-MRSA infections is increasing; there were 110, 259, and 995 cases per 100,000 pediatric patients in 1988, 1990, and 1993, respectively [15]. CA-MRSA strains show features distinct from nosocomial isolates and are a serious public health problem.

S. aureus infections *per se*, and MRSA infections in particular, remain of great concern. For example, CA-MRSA can invade medical institutions, the vancomycin MIC (minimal inhibitory concentration) is increasing, and vancomycin resistance may become more prevalent. Thus, further research on these topics is required.

REFERENCES

Ogston, A. Micrococcus poisoning. *J Anl Physiol*, 1883,17, 24-58.

Barber, M; Rozwadowska-Dowzenko, M. Infection by penicillin-resistant Staphylococci. *Lancet*, 1948, ii, 641-4.

Jevons, MP. Celbenin-resistant Staphylococci. *BMJ*, 1961, 1, 124-5.

Hiramatsu, K; Hanaki, H; Ino, T; Yabuta, K; Oguri, T; Teover, FC. Methicillin-resistanct Staphylococcus aureus clinical strain with reduced vancomycin susceptibility. *J Antimicrob Chemother*, 1997, 40, 135-6.

From Centers for Disease Control and Prevention. Update: Staphylocccus aureus with reduced susceptibility to vancomycin-United States, 1997. *MMWR Morb Mortal Wkly Rep.*, 1997, 46, 813-15.

From Centers for Disease Control and Prevention. Vancomycin-resistant: Staphylococcus aureus-Pennsylvania, 2002. *MMWR Morb Mortal Wkly Rep.*, 2002, 51, 902.

From Centers for Disease Control and Prevention. Staphylococcus aureus resistant to vancomycin-United States, 2002. *MMWR Morb Mortal Wkly Rep.*, 2002, 51, 565-7.

Lowy, F. Staphylococcus aureus infections. *N Engl J Med.*, 1998, 339, 520-32.

National Nosocomial Infections Surveillance System. National Nosocomial Infections Surveillance (NNIS) system report, data summary from January 1992 through June 2004, issued October 2004. *Am J Infect Control*, 2004, 32, 470-85.

Cosgrove, SE; Sakoulus, G; Parencevich, EN; Schwaber, MJ; Karchmer, AW; Carmeli, Y. Comparison of mortality associated with methicillin-resistant and methicillin-susceptible Staphylococcus aureus bacteremia: a meta-analysis. *Clin Infect Dis.*, 2003, 36, 53-9.

Engeman, JJ; Carmilli, Y; Cosgrove, SE; Fowler, VG; Bronstein, MZ; Trivette, SL; et al. Adverse clinical and economic outcomes attributable to methicillin resistance among patients with Staphylococcus aureus surgical site infection. *Clin Infect Dis.*, 2003, 36, 592-9.

Bouche, HW; Corey, GR. Epidemiology of methicillin-resistant Staphylococcus aureus. *Clin Infect Dis.*, 2008, 46, S344-9.

From the Centers for Disease Control and Prevention. Community-acquired methicillin-resistant Staphylococcus aureus infections-Michigan. *MMWR Morb Mortal Wkly Rep,.* 1981, 30,185-7.

From the Centers for Disease Control and Prevention. Four pediatric deaths from community-acquired methicillin-resistant Staphylococcus aureus-Minnesota and North Dakota. *JAMA*, 1999, 282, 1123-5.

Herold, BC; Immergluck, LC; Maranan, MC; Lauderdale, DS; Gaskin, RE; Boyle-Vavra, S; et al. Community-acquired methicillin-resistant Staphylococcus aureus in children with no identified predisposing risk. *JAMA*, 1998, 279, 593-8.

2. Microbiological and Laboratory Diagnosis of *Staphylococcus* and Methicillin-Resistant *S. Aureus*. Microbiological Characteristics of *S. Aureus*.

Staphylococcus is a Gram-positive bacterium that is globular in shape, and the colonies resemble bunches of grapes (Figure 1).

Staphylococcus grows well in aerobic or anaerobic peptone-containing nutrient medium. Sometimes mannitol and 7.5–10% (w/v) NaCl, are used as selective components for *S. aureus* growth. In general, *S. aureus* colonies grow to 1–3 mm in diameter after 24–72 hours on blood agar, nutrient agar, tryptic soy agar, or brain heart infusion agar at 34–37°C, and can reach a diameter of 8 mm after 72 hours growth [1].

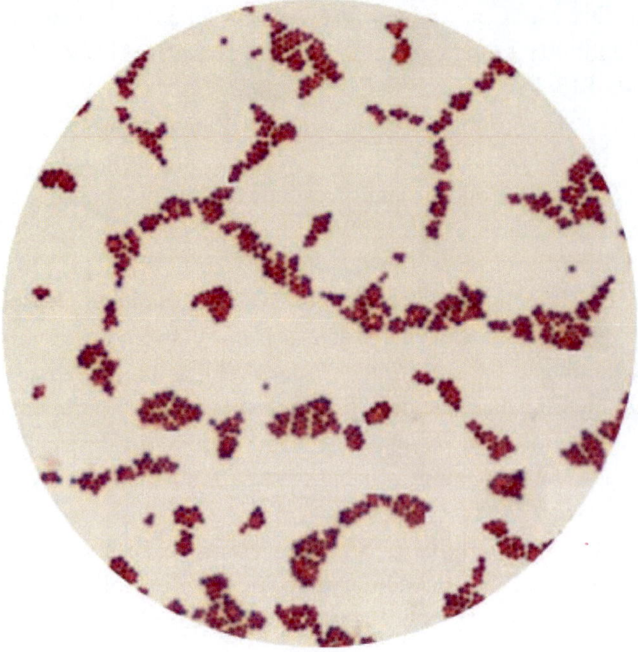

Figure 1. Micrograph of *S. aureus* after Gram staining.

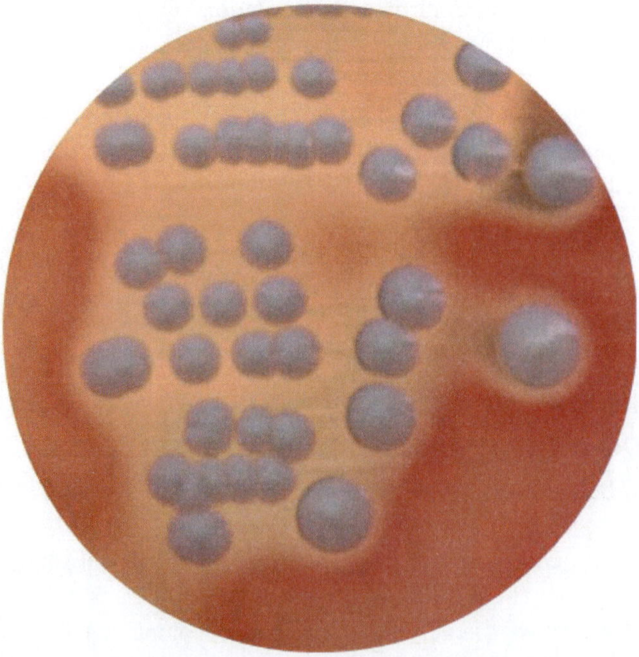

Figure 2. Hemolytic growth of *S. aureus* on blood agar.

The typical *S. aureus* colony is creamy-yellow to orange in color, smooth, thus not fissured, round, and slightly elevated. When growing on blood agar, hemolysis can be observed (Figure 2).

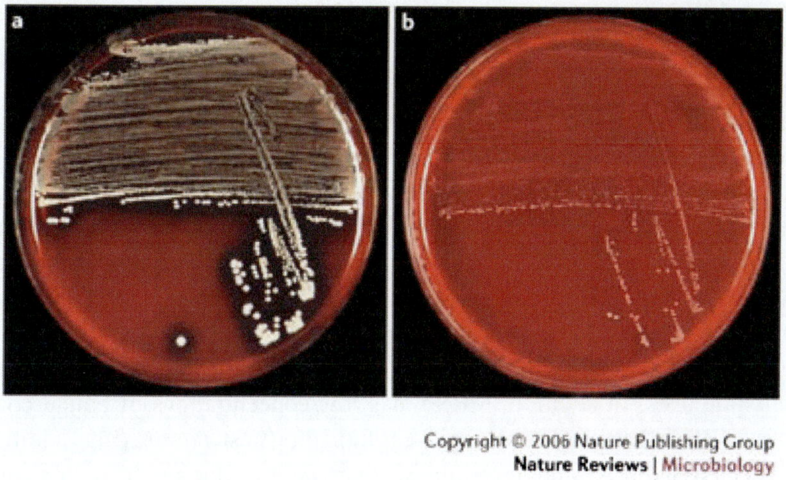

Figure 3. Phenotypic differences between wild-type and small colony variants (SCVs) of *S. aureus*. The blood agar plate shows a normal strain (a) and a small colony variant (b).

With small colony variants (SCVs) of *S. aureus*, colonies on agar can be seen no earlier than 48 hours after culture commencement. The colony size is 10-fold smaller than that of the wild type and pigmentation is lacking (Figure 3). Usually, SCVs show auxotrophism for heme and menadione, reduced carbohydrate utilization, and gentamicin resistance [2]. SCVs can be isolated alone or as mixtures with wild-type strains. When SCVs are subcultured, the SCV morphology can be maintained or SCVs may change rapidly to a normal (wild-type) growth form. SCVs are often found in cases of persistent infection, such as in cystic fibrosis patients, those with chronic osteomyelitis, and patients who have had long-term exposure to aminoglycoside or trimethoprim–sulfamethoxazole antibiotics [2, 3]. If isolated SCV colonies are supplied with menadione, hemin, and CO_2, normal growth is restored [4].

Laboratory Diagnosis of *S. Aureus*

Staphylococcus may be distinguished form other micrococci by the lysostaphin susceptibility test, aerobic acid production from glycerol, anaerobic acid production from glucose, the lysozyme susceptibility test, the bacitracin susceptibility test, and the presence of a modified oxidase [5]. The presence of catalase is diagnostic and easily distinguishes *Streptococcus or Enterococcus*.

After cultivation, *Staphylococcus* can be distinguished from other bacteria by coagulase production, a slide agglutination test, heat-soluble nuclease production, phosphatase activity, pyrrolidonyl arylamidase production, ornithine decarboxylase activity, urease synthesis, β-galactosidase activity, acetoin production, novobiocin resistance, resistance to polymyxin B, and acid production from carbohydrates [1]. The simplest method to distinguish different *Staphylococcus* types is by coagulase detection. Coagulase is antigenically classified into two distinct forms. The first is a secreted enzyme, thus not attached to the cell wall, and is easily detected using the tube coagulase test (or a coagulase production test). This test is used to detect acute infection with *S. aureus* and is also used to identify *Staphylococcus intermedius* and all most kinds of *Staphylococcus hyicus*. Other strains secrete coagulase but the enzyme remains attached to the cell wall to form the so-called clumping factor. A slide coagulase test detects this material. A positive slide test is given by *Staphylococcus lugdunensis* or *Staphylococcus schleiferi*. A rare form of *S. intermedius* also reacts in this test. The biochemical discrimination of *S. aureus* from other *Staphylococcus* strains is summarized below (Figure 4).

A commercial kit has been developed that detects the clumping factor, protein A, protein S, and other *Staphylococcus* surface antigens. It has been reported that the latex agglutination test shows higher sensitivity and specificity compared with other conventional slide tests [6]. However, the test is unsuitable for identification of *S. lugdunensis*, and some other species of *Staphylococcus*, such as *Staphylococcus saprophyticus* and *Staphylococcus sciuri*; some *Macrococcus* species score positively in the latex agglutination test but are negative in conventional slide tests. In addition, MRSA has low concentrations of bound coagulase and protein A, so MRSA detection by a rapid agglutination test requires the additional use of antibodies directed against staphylococcal capsular polysaccharide. Thus, recently, a study using MSSA and MRSA of serotypes 5 and 8 found that the latex agglutination test showed positive results [7] but false-positives can also be found. If *S. aureus* is strongly suspected but the slide test is negative, a tube coagulase test should confirm diagnosis. Currently, many commercial latex agglutination MRSA test kits are under development and are predicted to be both faster and more accurate than traditional procedures. These will be discussed in the Chapter entitled "Early diagnosis of MRSA infection".

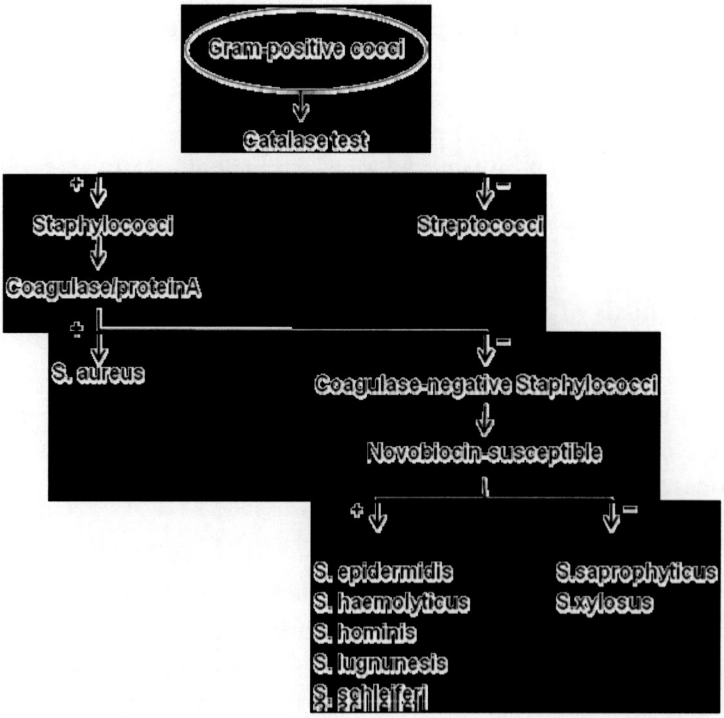

Figure 4. Biochemical characterization of *Staphylococci*; an algorithm of biochemical tests is used to discriminate among clinically important strains. Additional tests are necessary to identify all particular species (adapted from Harrison's Principles of Internal Medicine, 17th Edition; http://www.accessmedicine.com).

Antibiotic Sensitivity Tests

If a pathogen is detected on culture, antibiotic sensitivity tests must be conducted. Owing to the great variety of available antibiotics the NCCLS (National Committee for Clinical Laboratory Standards) disc diffusion method is used. In this method, antibiotics with similar antibacterial activities are grouped and a representative drug is tested. In addition, we test related drugs and combine the results. In this manner, unnecessary duplication is prevented, costs are reduced, and confusion minimized.

A *Staphylococcus* strain showing penicillin sensitivity is generally sensitive to a series of penicillin, cephem, and carbapenem. If the strain is resistant to penicillin and sensitive to oxacillin the bacterium will be resistant to the types of penicillins destroyed by β-lactamase but sensitive to the drugs cephem and carbapenem that are resistant to β-lactamase. Oxacillin (methicillin)-resistant *Staphylococci* are resistant to all drugs that can be inactivated by β-lactamase. If a strain is resistant to penicillin but sensitive to oxacillin, this may readily be determined using a 10-unit penicillin disk. Penicillin is used for sensitivity testing as a surrogate for ampicillin, amoxicillin, azlocillin, carbenicillin, mezlocillin, piperacillin, and ticarcillin, all of which are unstable. Thus, if a strain expresses a β-lactamase, resistance to the drugs listed can be predicted. Oxacillin is used as a representative drug not inactivated by β-lactamase. Oxacillin is stable so is easy to store, and is thus useful to identify heteroresistant *S. aureus* strains. Penicillin resistance involves an inducible plasmid-encoded β-lactamase, so penicillin sensitivity is assessed after induction of the enzyme [8].

The incidence of resistance to a penicillinase-resistant penicillin (methicillin, oxacillin, nafcillin) is 80% in coagulase-negative *Staphylococci* (CoNS) and 50% in *S. aureus*. Penicillin resistance is mediated by the *mecA* gene, which encodes an altered penicillin-binding protein, PBP2a (or PBP2'). In general, resistance is heterogeneous. In other words, only a small number of cells (1 in $10^{4\text{-}8}$) is resistant. For this reason, special guidelines for resistance detection are needed. Traditionally, a 1-μg oxacillin disk is used in the diffusion test (Figure 5). Suppression of growth to less than 10 mm in diameter indicates resistance, 11–12 mm indicates an intermediate state, and 13 mm or more indicates susceptibility. In the MIC method, inhibition by ≥4 μg/mL indicates resistance, whereas inhibition by ≤2 μg/mL is considered to indicate susceptibility [9].

Figure 5. Methicillin-resistant *S. aureus* in a disk diffusion test.

Some *S. aureus* strains show oxacillin resistance due to synthesis of a modified PBP (not PBP2a) not correlated with *mecA* gene presence or overproduction of β-lactamase. Most such strains are mildly susceptible in MIC tests. Such strains are sometimes termed borderline-resistant *S. aureus* (BORSA), or are alternatively considered to be truly resistant *S. aureus* isolates. These strains do not grow on oxacillin screening plates and, in an oxacillin MIC test for a β-lactamase inhibitor (clavulinic acid), MecA-mediated resistance and β-lactamase overproduction can be distinguished. In addition, these strains are all sensitive to other anti-staphylococcal agents, unlike other stains resistant due to β-lactamase production or showing traditional resistance caused by expression of modified PBPs encoded by *mecA*.

In recent years, *mecA*-mediated oxacillin resistance has been investigated in several studies that used cefoxitin as an oxacillin surrogate. Cefoxitin is a stronger inducer of the *mecA* regulatory system than are the other penicillins [9]. The CLSI (Clinical and Laboratory Standards Institute) considered that the sensitivity and specificity of the cefoxitin disk diffusion method was the same as that of the oxacillin disk diffusion test [10]. The specificity of the cefoxitin disk diffusion method was, however, greater than that of the oxacillin method [11]. The biggest disadvantage of the cefoxitin disk diffusion test used to assess antibiotic resistance in *mecA*-positive *S. aureus* is that the gene is not induced [10].

Except for modified interpretive criteria, the cefoxitin disk diffusion test does not differ from the routine test. *S. aureus* strains with inhibition zones less than 21 mm in diameter are considered oxacillin-resistant and those with zones of at least 22-mm diameter are considered susceptible. However, CoNS strains yielding inhibition zones of no more than 24-mm diameter are considered resistant, and those with zones of at least 25-mm are considered susceptible [9]. The cefoxitin disk diffusion test is easier to interpret because reflected light is used for visualization, and it is not necessary to check the disk diffusion zone for surrounding light growth or small colonies. In addition, the cefoxitin disk diffusion test is useful to detect *S. lugdunensis*, which is negative for *mecA* but is reportedly oxacillin-resistant [12]. The oxacillin interpretive criteria for *S. lugdunensis* were thus excluded from the CLSI M100 guidelines.

D-Test

If a *S. aureus* strain shows erythromycin resistance and sensitivity to clindamycin in the MIC test, the "D test" should be performed to check for inducible resistance to clindamycin. At a spacing of 15 mm, a disc with 15 μg erythromycin and another with 2 μg clindamycin are placed on an inoculated blood agar plate After overnight culture, if the test strain can express inducible clindamycin resistance, a curved D-shape can be seen in the inhibition zone around the clindamycin disk adjacent to the erythromycin disk [13] (Figure 6). The inducible resistance results from expression of the MRSA *erm* (erythromycin ribosome methylase) gene. The enzyme alters the ribosomal features, causing macrolide sensitivity and resulting in cross-resistance to all macrolide-lincosamide-streptogramin B (MLSB) antibiotics [14].

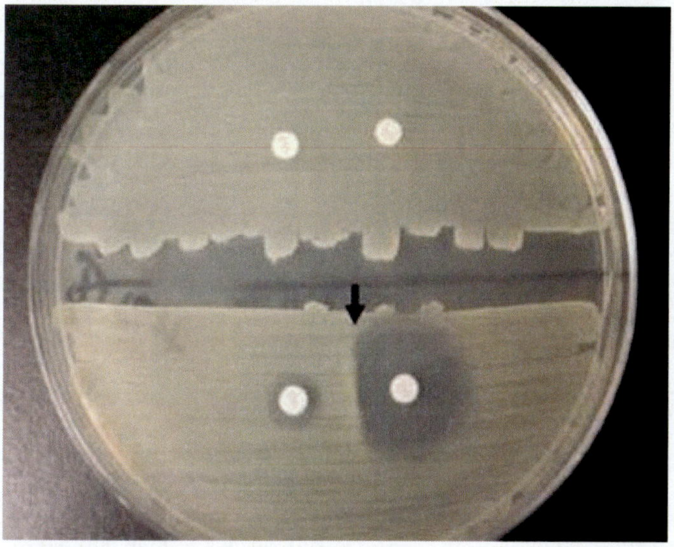

Figure 6. D-test for erythromycin-resistant, clindamycin-susceptible *S. aureus*.

Oxacillin-Salt Agar Screening Test

The CLSI recommends that 4% (w/v) sodium chloride and 6 μg oxacillin/mL should be added to MHA (Mueller-Hinton agar). [11, 15]. The plate does not contain methicillin. After one night of culture on nonselective agar a sample colony is transferred to broth (for example, tryptic soy broth) or saline prior to adjustment with the "0.5 McFarland standard" for inoculum preparation. The suspension is transferred to an oxacillin agar screening plate as follows. A cotton swab is dipped in the test suspension and excess liquid is removed. Next, the swab is used to inoculate an agar plate with a circle 10–15 mm in diameter or by streaking on a quarter segment of the plate. A 1-μl loop may also be used [16]. The test plate is incubated at 35°C under circulating air for at least 24 hours. If even one colony is seen, antibiotic resistance is confirmed.

The oxacillin-salt agar screen test has been widely used but sensitivity, especially for heterogeneously resistant strains, has been of concern [17]. The test affords lower sensitivity than the cefoxitin disk diffusion assay [18-20] and, because the oxacillin level is 6 μg/mL, the test cannot be used to investigate oxacillin resistance of CoNS [21]. Therefore, CLSI considers the agar screening test inappropriate for CoNS [9].

DETECTION OF *S. AUREUS* AND IDENTIFICATION OF METHICILLIN-RESISTANT *S. AUREUS*

Early Diagnosis of MRSA Infection

Many tests are available to diagnose MRSA infection, based on antibody–antigen interaction, microbial residue detection, and gene detection and microbial cultivation. For rapid diagnosis, methods other than traditional culture approaches (which take 2–3 days) are used but culture is required for diagnosis confirmation and antibiotic sensitivity tests. *S. aureus* in serum or urine is difficult to detect by antibody–antigen interaction or analysis of bacterial residues. Therefore, traditional microbiological culture and sensitivity tests are used to detect early infections with MRSA, and are followed by additional tests. Speed, accuracy, and cost-effectiveness are always considerations.

Table 1. Detection and identification methods for MRSA.

Microscopic examination	Gram stain; acid-fast bacillus test; FISH
Culture methods	Blood agar plate followed by morphological examination Chromogenic agar Selective agar; mannitol-salt agar
Antigen-antibody interactions	Anti-surface protein antibody Anti-secretory protein antibody
Molecular methods (*mecA* gene detection or *mecA* gene + *S. aureus*-specific sequence detection)	PCR RT-PCR Real-time PCR Multiplex PCR

Early MRSA diagnosis is particularly important, and several tests have been developed to this end. The slide coagulase test was traditionally used to view agglutination with serum; the test is rapid but has a false-negative rate of 10–15%. The tube coagulase test takes at least 4 hours and sometimes up to 24 hours [5]. The slide coagulase latex agglutination test, which was designed to detect clumping factor, surface protein A and other surface antigens, gives immediate results [6]. Such kits include the Pastorex Staph-Plus (Sanofi Diagnostic Pasteur), Staphaurex Plus (Murex Diagnostic), and Staphylo-LA (Denka-Seiken) [7]. However, *S. lugdunensis* can yield false-positive results, and some *S. aureus* strains are protein A- and clumping factor-negative. Thus, the tube coagulase test should be conducted to confirm diagnosis [5]. Many rapid identification kits are used in hospitals. Although results are available within 2–4 hours, kit accuracy is poor and the kits are expensive.

For early MRSA diagnosis, tests detecting the low-affinity PBP2a or the *mecA* gene are used. The MRSA screen (Denka-Seiken) and the Oxoid PBP latex agglutination test (Oxoid) detect PBP2a [22]. These methods take approximately 10–15 minutes. Around 3% of MRSA strains may show false-negative results and *mecA*-containing CoNS may score as false-positives. Therefore, kit diagnoses must be accompanied by bacterial isolation.

MRSA is easily confirmed by the presence of the *mecA* gene. An optimal test would achieve this cheaply and easily. However, if existing kits are used on a regular basis, the cost is relatively high and the tests are tedious. Multiplex PCR can be used to diagnose MRSA by simultaneously detecting genes unique to *S. aureus* and the *mecA* gene [23]. The unique genes of *S. aureus* include those encoding a nuclease (*nuc*), a coagulase (*coa*), protein A (*spa*), and a factor essential for methicillin resistance (*fem*) [24-27]. Many efforts have been made to detect these genes and *mecA* simultaneously. However, if methicillin-resistant coagulase-negative *Staphylococcus* (MRCoNS) occurs together with *S. aureus,* a false-positive MRSA diagnosis can result, which is highly problematic.

Kits available for MRSA screening tests involve multiplex PCR technology or are certified products such as the Becton Dickinson GeneOhm MRSA (BD), BD GeneOhm StaphSR (BD), Xpert MRSA (Cepheid), Xpert MRSA/SA SSTI (Cepheid), MRSA Velogene (ID Biomedical), and EASY Q MRSA (Biomerieux).

MRSA is detected by PCR amplification using the BD GeneOhm MRSA kit, which detects the *mecA* gene in the staphylococcal chromosomal cassette containing *mec* (SCC*mec*), and the unique genes of *S. aureus*, using a smart cycler and molecular labeling. In a test that does not detect the *mecA* gene but instead uses multiple primers and molecular labeling, there is no longer a risk that MRCoNS shows a false-positive result [28]. In theory, the test is well designed but in practice yields a false-negative result in strains with SCC*mec* mutations and a false-positive signal from MRSA with a novel type of SCC*mec*. In screening of nasal swabs the sensitivity and specificity have been reported to be 100% and 98.6%, respectively, and the positive predictive value and negative predictive value have been reported as 95.8% and 100%, respectively [29]. The Xpert MRSA test (Cepheid) takes 1 hour to obtain results from nasal swabs and is fast and easy to use. The sensitivity and specificity of this test are 94.3% and 93.2%, respectively. The positive predictive value is 73.0% and the negative predictive value about 98.8% [30]. NASBA NucliSENS EASY Q MRSA (Biomerieux) can simultaneously detect the *mecA* gene and SCC*mec*. This reduces the false-positive rate, but the test takes about 3 hours and more work is needed to optimize the procedure.

After the pathogen is cultured from blood it takes 2 days of further culture to confirm MRSA, but MRSA tests using molecular genetic methods are also available. To examine blood cultures the BD GeneOhm StaphSR test takes 2 hours and the Xpert MRSA/SA BC test (Cepheid) takes 1 hour to determine whether MRSA or MSSA has been grown [28, 31]. The *mecA* EVIGENE (AdvanDx) test, which uses a capture probe for *mecA* and detects gene presence with a color change, takes 4 hours to detect MRSA. However, the *mecA* EVIGENE test (AdvanDx) can be used after defection of the *nuc* gene using the *S. aureus* EVIGENE or *S. aureus* PA FISH kits (AdvanDx). The EVIGENE kit can evaluate resistance to mupirocin, vancomycin, Panton-Valentine leukocidin (PVL), and toxic shock syndrome toxin 1 (TSST-1) [32].

In soft tissue-infected patients, direct analysis of lesion swabs can be achieved by the Xpert MRSA/SA SSTI kit (Cepheid). With this kit, MRSA detection takes about 1 hour, [33], at an accuracy of up to 95%, but the kit is relatively expensive.

Efforts have been made to directly detect the pathogen and to evaluate drug resistance in patients with septicemia. The LightCycler SeptiFast kit (Roche) can report within 6 hours on six species of Gram-positive and eight species of Gram-negative bacteria, and six fungal species ; a 3-mL blood sample is used. The test involves real-time PCR assay that detects three types of internally transcribed spacers (ITS) using several probes. The *mecA* gene can be detected within 6 hours, which is in contrast to other kits that identify *mecA* in 2 hours 30 minutes. The Roche test offers many advantages in the detection of septicemia etiologies, but the success rate is only 80% and the kit is expensive [34].

Many techniques for early MRSA diagnosis have been developed and are used in parallel with traditional bacterial culture tests. Selective markers are useful for management of MRSA infection and MRSA carrier detection. Laboratories can report positivity within 1 hour, so such tests are used in many institutions. In addition, MRSA detection by molecular genetic testing can be achieved within 1 day in the United States. Direct MRSA detection from a clinical sample is usually performed after a positive result has been reported for a blood culture. The cost of molecular genetic tests and the labor involved suggest that the clinical applicability of such tests requires additional research.

Table 2. Traditional and recent methods of MRSA identification.

Traditional identification step	Newer early identification tools
Gram stain Catalase test Coagulase test (slide coagulase test, tube coagulase test) DNase test Mannitol fermentation	Latex agglutination for protein A, clumping factor, and other surface antigens Rapid ID kits (biochemical methods) PCR, RT-PCR, multiplex PCR for *mecA*, genes encoding toxins and surface antigens, and the SCC*mec-orfX* junction

REFERENCES

[1] Bannerman, T. Staphylococcus, Micrococcus, and other catalse positive cocci that grow aerobically. In: Murray, PR; Baron, EJ; Jorgensen, JH; Landry, ML; Pfaller, MA; editors. Manual of Clinical Microbiology. 8th ed. Washington, DC: *American Society for Microbiology*, 2003, p.384-404.

[2] Proctor, RA; Peters, G. Small colony variants in Staphylococcal infection: diagnostic and therapeutic implications. *Clin Infect Dis*., 1998, 27, 419-22.

[3] Looney, WJ. Small-colony variants of Staphylococcus aureus. *Br J Biomed*, 2000, 57, 317-22.

[4] Proctor, RA; Van Langevelde, P; Kristijansson, M; Maslow, JM. Persistent and relapsing infections associated with with small-colony variants of Staphylococcus aureus. *Clin Infect Dis*., 1995, 20, 95-102.

[5] Catalase positive cocci. In: Murray, PR; Baron, EJ; Jorgensen, JH; Landry, ML; Pfaller, MA; editors. Manual of clinical microbiology. 9th ed. Washington, DC: *American Society of Microbiology*, 2007, p. -411.

[6] Chediac-Tannoury, R; Araj, GF. Rapid MRSA detection by Latex kit. *Clin Lab Sci*., 2003, 16, 198-202.

[7] Fournier, JM; Bouvet, A; Mathieu, D; Nato, F; Boutonnier, A; Gerbal, R; et al. New latex reagent using monoclonal antibodies to capsular polysaccharide for reliable identification of both oxacillin-susceptible and oxacillin-resistant Staphylococcus aureus. *J Clin Microbiol*, 1993, 31, 1342-4.

[8] Clinical and Laboratory Standards Institute/NCCLS, 2003

[9] Clinical and Laboratory Standards Institute. Performance standards for antimicrobial susceptibility testing; 18th informational supplement. Document M100-S18. Wayne, PA: National Committee for Clinical Laboratory Standards, 2008.

[10] Swenson, JM; Tenover, FC; and the Cefoxitin Disk Study Group. Results of disk diffusion testing with cefoxitin correlate with presence of mecA in Staphylococcus spp. *J Clin Microbiol*, 2005, 43, 3818-23.

[11] Clinical and Laboratory Standards Institute Performance Standards for Antimicrobial Disk Susceptibility Test; Approved Strandard-8th edition. Document M2-A9. Wayne, PA: *National Committee for Clinical Laboratory Standards*, 2006.

[12] Clinical and Laboratory Standards Institute. Performance Standards for Antimicrobial Susceptibility Testing; 15th Information supplement. Document M100-S15. Wayne, PA: *National Committee for Clinical Laboratory Standards*, 2006.

[13] Clinical and Laboratory Standards Institute. Performance standards for antimicrobial susceptibility testing; 17th informational supplement. Document M100–S17. Wayne, PA: *National Committee for Clinical Laboratory Standards*, 2007.

[14] Leclercq R. Mechanisms of resistance to macrolides and lincosamides: nature of the resistance elements and their clinical implications. *Clin Infect Dis*., 2002, 34, 482-9.

[15] Clinical and Laboratory Standards Institute. Methods for Dilution Antimicrobial Susceptibility Tests for Bacteria That Grow Aerobically; Approved Standard-6th edition. Document M7-A7. Wayne, PA: *National Committee for Clinical Laboratory Standards*, 2006.

[16] Swenson, JM; Spargo, J; Tenover, FC; Ferraro, MJ. Optimal inoculation methods and quality control for the NCCLS oxacillin agar screen test for detection of oxacillin resistance in Staphylococcus aureus. *J CLin Microbiol*, 2001, 39, 3781-4.

[17] Cavassini, MA; Wenger, A; Jaton, K; Blanc, DS; Bille, J. Evaluation of MRSA-screen, a simple anti-PBP 2a slide latex agglutination kit, for rapid detection of methicillin resistance in Staphylococcus aureus. *J Clin Microbiol*, 1999, 37, 1591-4.

[18] Caywekuer, B; Gordts, B; Cescheemaecker, P. Van Landuyt H. Evaluation of a disk diffusion method with cefoxitin (30 ug) for detection of methicillin-resistant Staphylococcus aureus. *Eur J Clin Microbiol Infec Dis*., 2004, 23, 389-92.

[19] Felten, A; Grandry, B; Lagrange, PH; Casin, I. Evaluation of three technique for detection of low level methicillin-resistant Staphylococcus aureus (MRSA): a disk diffusion method with cefoxitin and moxalctam, the Vitek 2 system, and the MRSA-Screen latex agglutination test. *J Clin Microbiol*, 2002, 40, 2766-71.

[20] Swenson, JM; Williams, PP; Killgore, G; O'Hara, CM; Tenover, FC. Performance of eight methods, including two rapid methods, for detection of oxacillin resistance in a challenge set of Staphylococcus aureus organisms. *J Clin Microbiol*, 2001, 39, 3785-8.

[21] Tenover, FC; Jones, RN; Swenson, JM; Zimmer, B; McAllister, S; Jorgensen, JH. Methods for improved detection of oxacillin resistance in coagulase-negative staphylococci: results of a multicenter study. *J Clin Microbiol*, 1999, 37, 4051-8.

[22] Brown, DF; Walpole, E. Evaluation of the Mastalex latex agglutination test for methicillin resistance in Staphylococcus aureus grown on different screening media. *J Antimicrob Chemother*, 2001, 47, 187-9.

[23] Perez-Roth, E; Claverie-Martin, F; Villar, J; Mendez-Alvarez, S. Multiplex PCR for simultaneous identification of Staphylococcus aureus and detection of methicillin and mupirocin resistance. *J Clin Microbiol*, 2001, 39, 4037-41.

[24] Huluetsky, A; Giroux, R; Rossbach, V; Gagnon, M; Vaillancourt, M; Bernier, M; et al. New real-time PCR assay for rapid detection of methicillin-resistant Staphylococcus aureus directly from specimens containing a mixture of staphylococci. *J Clin Microbiol*, 2004, 42, 1875-84.

[25] Rohrer, M; Tschierske, M; Zbinden, R; Berger Bachi, B. Improved methods for detection of methicillin-resistant Staphylococcus aureus. *Eur J Clin Microbiol Infect Dis.*, 2001, 20, 267-70.

[26] Shrestha, NK; Tuohy, MJ; Hall, GS; Isada, CM; Procop, GW. Rapid identification of Staphylococcus aureus and the mecA gene from BacT/ALERT blood culture bottles by using the Lightcycler systems. *J Clin Microbiol*, 2002, 40, 2659-61.

[27] Costa, AM; Kay, I; Palladino, S. Rapid detection of mecA and nuc genes in staphylococci by real-time multiplex polymerase chain reaction. *Diagn Microbiol Infect*, 2005, 51, 13-7.

[28] Stamper, PD; Cai, M; Howard, T; Speser, S; Carroll, KC. Clinical validation of the molecular BD GeneOhm StaphSR assay for direct detection of Staphylococcus aureus and methicillin-resistant Staphylococcus aureus in positive blood cultures. *J Clin Microbiol*, 2007, 45, 2191-6.

[29] Boyce, JM; Havill, NL. Comparison of BD GeneOhm Methicillin-Resistant Staphylococcus aureus (MRSA) PCR versus the CHROMagar MRSA Assay for Screening Patients for the Presence of MRSA Strains. *J Clin Microbiol*, 2008, 46, 350–1.

[30] Wolk, DM; Picton, E; Johnson, D; Davis, T; Pancholi, P; Ginocchio, CC; et al. Multicenter evaluation of the Cepheid Xpert methicillin-resistant Staphylococcus aureus (MRSA) test as a rapid screening method for detection of MRSA in nares. *J Clin Microbiol*, 2009, 47, 758-64.

[31] Rossney, AS; Herra, CM; Brennan, GI; Morgan, PM; O'Connell, B. Evaluation of the Xpert methicillin-resistant Staphylococcus aureus (MRSA) assay using the GeneXpert real-time PCR platform for rapid detection of MRSA from screening specimens. *J Clin Microbiol*, 2008, 46, 3285-90.

[32] Levi, K; Towner, KJ. Detection methicillin-resistant Staphylococcus aureus (MRSA) in blood with the EVIGENE MRSA detection kit. *J Clin Microbiol*, 2003, 41, 3890-2.

[33] Wolk, DM; Struelens, MJ; Pancholi, P, Davis, T; Della-Latta, P; Fuller, D; et al. Rapid detection of Staphylococcus aureus and methicillin-resistant S. aureus (MRSA) in wound specimens and blood cultures: multicenter preclinical evaluation of the Cepheid Xpert MRSA/SA skin and soft tissue and blood culture assays. *J Clin Microbiol*, 2009, 47, 823-6.

[34] Casalta, JP; Gouriet, F; Roux, V; Thuny, F; Habib, G; Raoult, D. Evaluation of the LightCycler(R) SeptiFast test in the rapid etiologic diagnostic of infectious endocarditis. Eur J Clin Microbiol Infect Dis. 2008 Dec 2. [Epub ahead of print]

3. MRSA Pathogenesis

To invade and infect, *S. aureus* has to colonize and penetrate a body surface, avoid host immune reactions, and establish a metastatic infection. To be susceptible to *S. aureus* infection, a patient must have a break in the skin or a mucosal barrier. A colonizing strain, perhaps from another infected patient, can pass through damaged skin into the bloodstream. Unless there is a skin/mucosal break, *S. aureus* is a harmless, common, surface-resident bacterium.

The anterior nares is the main site of *S. aureus* colonization. *S. aureus* continuously colonizes the anterior nares in about 20% of the general population, and temporarily colonizes it in a further 30% of subjects [1-4]. The biological mechanism of colonization is poorly understood, although *S. aureus* is known to attach to the skin or mucosa. The type and density of resident bacteria, possible nasal cavity oromucosal damage, the antimicrobial properties of nasal secretions, and host genetic factors (particularly human leukocyte antigen [HLA] type) affect colonization [5].

S. aureus invasion through abrasions, injections, or insertion of intravenous catheters is followed by internal dissemination and colonization. The various *S. aureus* surface proteins are termed matrix molecules (MSCRAMMs), and have important roles in facilitating *S. aureus* attachment to various tissues [6]. MSCRAMMs such as fibronectin-binding protein, clumping factor, and collagen-binding protein allow *S. aureus* to grow on the surfaces of fibronectin, fibrinogen, and collagen, thereby facilitating endovascular infection, residence on bones and in joints, and prosthetic-device colonization [6-9].

Artificial biofilm formation by CoNS and a high frequency of colony formation in artificial environments are well known. However, *S. aureus* also has a biofilm formation gene, the intercellular adhesion (*ica*) locus, and can form biofilms on the surfaces of artificial inserts; these biofilms allow the bacteria to escape both the host immune system and antibiotic action [10].

After colonization, *S. aureus* produces many enzymes such as serine protease, hyaluronidase, thermonuclease, and lipase. These enzymes destroy host tissues and allow bacterial spread. In fat-rich areas, such as hair follicles, lipase facilitates *S. aureus* survival and initiates infection spread. The PVL toxin, rarely found in hospital-associated methicillin-resistant S. aureus (HA-MRSA) and MSSA, is involved in CA-MRSA infection. This toxin affects polymorphonuclear leukocytes, macrophages, and monocytes [11, 12]. Strains expressing this toxin are epidemiologically related to those causing faruncles, carbuncles, and other soft-tissue infections, and to strains responsible for post-influenza necrotizing pneumonia [13, 14]. A detailed description of the PVL toxin can be found in "Microbiological characteristics of CA-MRSA" (Chapter 6-2).

Other than peptidoglycan, both lipoteichoic acid and alpha-toxin interact with the activated host immune system and initiate inflammatory reactions including septic shock syndrome [15-17].

Escape from the host defense mechanism is critical for invasion. *S. aureus* has a polysaccharide microcapsule with anti-phagocytosis activity. Most clinical *S. aureus* strains have capsules of types 5 or 8. The capsule of *S. aureus* has an important role in induction of abscess formation. The polysaccharide has a zwitterionic charge (positive and negative charges together), which is critical in abscess formation [18, 19]. Of the MSCRAMMs, protein A is unique to *S. aureus*. The protein attaches to the Fc portion of IgG1, IgG2, and IgG4 antibodies, in the manner of an Fc receptor, and inhibits both opsonization and phagocytosis by polymorphic nuclear leukocytes [20].

The ability of *S. aureus* to avoid the host defense mechanism involves cell internalization. Both professional and nonprofessional phagocytes can internalize the bacterium. Internalization by endothelial cells, a form of nonprofessional phagocyte, can provide sanctuary from the host defense mechanism. Such internalization results in expression of the Fc receptor and integrin, and secretion of cytokines. These cellular changes contribute to systemic disease manifestations including septicemia or vasculitis [21-24].

S. aureus SCV, strains auxotrophic for menadione and hemin, and strains lacking alpha-toxin typically survive in endothelial cells. SCVs often appear after the use of aminoglycoside antibiotics and are found more frequently in chronic infection sites, such as those of chronic osteomyelitis or in the laryngeal secretions of cystic fibrosis patients. Phase shifting to SCV allows *S. aureus* to survive, and later recover to wild-type, which is another mechanism of re-infection. Also, SCVs residing in polymorphic leukocytes use the cells to transport *S. aureus* to other bodily sites [25-27].

S. aureus produces three kinds of toxin, a cytotoxin, a pyrogenic toxin superantigen, and an exfoliative toxin. Epidemiologic reports and animal experiments have shown that if a host has antibodies to the toxins, these may function as defenses to toxic shock syndrome (TSS), food poisoning, staphylococcal scalded skin syndrome (SSSS), and other diseases [28]. Pyrogenic toxin superantigen, of small molecular weight, is a member of a family of structurally similar proteins and causes TSS and food poisoning. TSS results when enterotoxin and TSS toxin promote the division of T cells. In normal antigen presentation, an antigen is processed within a phagocytic cell. The processed peptides of the antigen are presented to the cleft of major histocompatibility complex class 2 proteins, thereby initiating T-cell reactions. However, enterotoxin binds directly to the constant region of major histocompatibility complex class 2 proteins and contacts the T-cell receptor through the V-chain, inducing T-cell overexpression. Next, inflammatory mediators produced by T cells, such as interferon (IFN)-r, interleukin (IL)-1, IL-6, and tumor necrosis factor (TNF)-β, are released from T cells to create the "cytokine storm" [28, 29]. Multi-organ disorders, including muscle pain, fever, rash, and low blood pressure, follow. The symptoms are similar to those of septicemia caused by endotoxins, but the mechanisms differ.

Exfoliative toxin is the cause of SSSS. The exfoliative toxins can be divided into two types, ETA and ETB. These toxins destroy desmosomes connecting adjacent cells. The exact mechanism remains unclear but some studies have suggested that the toxin has a serine protease activity that triggers exfoliation. Thus, a granular sheet is split from the desmosome layer, which causes elimination of the epidermis, a characteristic of the disease [29, 30].

REFERENCES

[1] Klutymans, J; Van Belkum, A; Verbrugh, H. Nasal carriage of Staphylococcusaureus: epidemiology, underlying mechanisms, and associated risks. *Clin Microbiol Rev.*, 1997, 10, 505-20.

[2] Eriken, NH; Espersen, F; Rosdahl, VT; Jensen, K. Carriage of Staphylococcus aureus among 104 healthy persons during a 19-month period. *Epidemiol Infect.*, 1995, 115, 51-60.

[3] Nouwen, JL; Ott, A; Klutymans-Vandenbergh, MF; Boelens, HA; Hofman, A, Van Belkum, A; et al. Predicting the Staphylococcus aureus nasal carrier state: derivation and validation of a "culture rule". *Clin Infect Dis.*, 2004, 39, 806-11.

[4] Hu, L; Umeda, A; Kondo, S; Amako K. Typing of Staphylococcus aureus colonizing human nasal carriers by pulsed-field gel electrophoresis. *J Med Microbiol*, 1995, 42, 127-32.

[5] Wertheim, HF; Melles, DC; Vos, MC; Van Leeuwen, W; Van Belkum, A; Verbrugh, HA; et al. The role of nasal carriage in Staphylococcus aureus infections. *Lancet infect Dis.*, 2005, 5, 751-62.
[6] Menzies, BE. The role of fibronectin binding proteins in the pathogenesis of Staphylococcus aureus infections. *Curr Opin Infect Dis.*, 2003, 16, 225–9.
[7] Patti, JM; Allen, BL; McGavin, MJ; Hook, M. MSCRAMM-mediated adherence of microorganisms to host tissues. *Annu Rev Microbiol*, 1994, 48, 585–617.
[8] Foster, TJ; Hook, M. Surface protein adhesins of Staphylococcus aureus. *Trends Microbiol*, 1998, 6, 484–8.
[9] Tung, H; Guss, B; Hellman, U; Persson, L; Rubin, K; Ryden, C. A bone sialoprotein-binding protein from Staphylococcus aureus: a member of the staphylococcal Sdr family. *Biochem J.*, 2000, 345, 611–9.
[10] Ogawa, SK; Yurberg, ER; Hatcher, VB; Levitt, MA; Lowy, FD. Bacterial adherence to human endothelial cells in vitro. *Infect Immun*, 1985, 50, 218–24.
[11] Menestrina, G; Dalla, SM; Comai, M; Coraiola, M; Viero, G; Werner, S; et al. Ion channels and bacterial infection: the case of b-barrel pore-forming protein toxins of Staphylococcus aureus. *FEBS Lett.*, 2003, 552, 54–60.
[12] Prevost, G; Cribier, B; Couppie, P; et al. Panton-Valentine leucocidin and gamma-hemolysin from Staphylococcus aureus ATCC 49775 are encoded by distinct genetic loci and have different biological activities. *Infect Immun*, 1995, 63, 4121–9.
[13] Seybold, U; Kourbatova, EV; Johnson, JG; Halvosa, SJ; Wang, YF; King, MD; et al. Emergence of community associated methicillin-resistant Staphylococcus aureus USA300 genotype as a major cause of health care-associated blood stream infections. *Clin Infect Dis.*, 2006, 42, 647-56.
[14] Huang, H; Flynn, NM; King, JH; et al. Comparisons of community-associated methicillin-resistant Staphylococcus aureus (MRSA) and hospital-associated MSRA infections in Sacramento, California. *J Clin Microbiol*, 2006, 44, 2423-7.
[15] Timmerman, CP; Mattsson, E; Martinez-Martinez, L; De Graaf, L; Van Striip, JA; Verbrugh, HA; et al. Induction of release of tumor necrosis factor from human monocytes by staphylococci and staphylococcal peptidoglycans. *Infect Immun*, 1993, 61, 4167–72.
[16] Heumann, D; Barras, C; Severin, A; Glauser, MP; Tomasz, A. Grampositive cell walls stimulate synthesis of tumor necrosis factor alpha and interleukin-6 by human monocytes. *Infect Immun*, 1994, 62, 2715–21.
[17] Bhakdi, S; Tranum-Jensen, J. Alpha-toxin of Staphylococcus aureus. *Microbiol Rev.*, 1991, 55, 733–51.
[18] O'Riordan, K; Lee, JC. Staphylococcus aureus capsular polysaccharides. *Clin Microbiol Rev.*, 2004, 17, 218–34.
[19] Tzianabos, AO; Wang, JY; Lee, JC. Structural rationale for the modulation of abscess formation by Staphylococcus aureus capsular polysaccharides. *Proc Natl Acad Sci USA*, 2001, 98, 9365–70.
[20] Deisenhofer, J. Crystallographic refinement and atomic models of a human Fc fragment and its complex with fragment B of protein A from Staphylococcus aureus at 2.9- and 2.8-A resolution. *Biochemistry*, 1981, 20, 2361–70.
[21] Ogawa, SK; Yurberg, ER; Hatcher, VB; Levitt, MA; Lowy, FD. Bacterial adherence to human endothelial cells in vitro. *Infect Immun*, 1985, 50, 218–24.

[22] Hamill, RJ; Vann, JM; Proctor, RA. Phagocytosis of Staphylococcus aureus by cultured bovine aortic endothelial cells: model for postadherence events in endovascular infections. *Infect Immun*, 1986, 54, 833–6.

[23] Arrecubieta, C; Lowy, FD. Staphylococcus aureus–eukaryotic cell interactions. In: Fischetti, VA; Novick, RP; Ferretti, JJ; Portnoy, DA; Rood, JI; editors. Gram-positive pathogens. 2nd ed. Washington, DC: ASM Press; 2006, p.517–25.3

]24] Moreillon, P; Que, YA; Bayer, AS. Pathogenesis of streptococcal and staphylococcal endocarditis. *Infect Dis Clin North Am.*, 2002, 16, 297–318.

[25] Kahl, B; Herrmann, M; Everding, AS; et al. Persistent infection with small colony variant strains of Staphylococcus aureus in patients with cystic fibrosis. *J Infect Dis.*, 1998, 177, 1023–9.

[26] Proctor, RA; Peters, G. Small colony variants in staphylococcal infections: diagnostic and therapeutic implications. *Clin Infect Dis.*, 1998, 27, 419–22.

[27] Proctor, RA; Van Langevelde, P; Kristjansson, M; Maslow, JN; Arbeit, RD. Persistent and relapsing infections associated with small-colony variants of Staphylococcus aureus. *Clin Infect Dis.*, 1995, 20, 95–102.

[28] McCormick, JK; Yarwood, JM; Schlievert, PM. Toxic shock syndrome and bacterial superantigens: an update. *Annu Rev Microbiol*, 2001, 55, 77–104.

[29] Dinges, MM; Orwin, PM; Schlievert, PM. Exotoxins of Staphylococcus aureus. *Clin Microbiol Rev.*, 2000, 13, 16–34.

[30] Prevost, G; Couppie, P; Monteil, H. Staphylococcal epidermolysins. *Curr Opin Infect Dis.*, 2003, 16, 71–6.

4. Antimicrobial Resistance of MRSA

Resistance to Penicillin

Mechanisms of penicillin resistance include [1]:

(1) Inactivation of antibiotics by β-lactamase
(2) Alteration of target PBPs
(3) Formation of permeability barriers that prevent antibiotics from invading bacterial targets.

Of these, β-lactamase production is the most frequently encountered resistance mechanism. To date, more than 100 types of β-lactamase have been reported. Several β-lactamases synthesized by *S. aureus, Haemophilus* spp., and *Escherichia coli* have narrow active substrate spectra, degrading only penicillin and not cephalosporin. However, other enzymes produced by *Pseudomonas aeruginosa* and *Enterobacter* spp. have broad active substrate spectra and can degrade both cephalosporin and penicillin.

Carbapenems show high resistance to penicillinase and cephalosporinase; these drugs are degraded by metallo-β-lactamase.

Resistance due to inhibition of drug penetration through cytoplasmic membranes is based on outer-membrane barriers, and is most common in Gram-negative bacteria. Although drug penetration can be attenuated, the mechanism cannot provide complete resistance because the antibiotics finally enter the cells and suppress growth. However, if a β-lactamase is concurrently present, antibiotics may be effectively degraded during gradual penetration and the permeability barrier can have a key resistance role. A recent report found that morphological cell-wall thickening due to decreased peptidoglycan cross-linking, lower cell-wall turnover, and decreased autolysis might contribute to vancomycin resistance in *S. aureus*, which has recently posed a problem [2, 3].

Alterations in PBPs cause antibiotic resistance in MRSA and penicillin-resistant *Pneumococci*. Such strains synthesize PBPs with lower affinities for β-lactam antibiotics, thus inducing drug resistance. At higher drug doses, however, the resistance cannot be maintained.

mecA – The Methicillin-Resistance Gene

The resistance of *S. aureus* to methicillin is due to the presence of the *mecA* gene that encodes the 78-kDa PBP2a (or PBP2'). β-lactam antibiotics normally bind to PBPs in the cell wall and inhibit formation of the peptidoglycan layer, thus inducing bacterial death. PBP2a does not bind to β-lactam antibiotics, thus making bacteria resistant to methicillin. Expression of the *mecA* gene is controlled by the repressor MecI and a trans-membrane β-lactam-sensing signal transducer MecRI. If β-lactam antibiotics are absent, however, MecI suppresses the transcription of both *mecA* and *mecRI-mecI.* When β-lactam antibiotics are present, MecI is autocatalytically cleaved, which activates a metalloprotease domain present in the cytoplasmic MecRI region. The metalloprotease cleaves MecI attached to the *mecA* operon, thus initiating *mecA* transcription and the subsequent synthesis of PBP2a [4]. The termini of *mecI* and *mecRI* are truncated by the insertion sequences IS431 or IS1272, which can suppress *mecA* gene expression [5].

The 2.1-kb *mecA* gene is located in a mobile genetic element termed the staphylococcal chromosomal cassette *mec* (SCC*mec*) [6]. Until recently, five main types of SCCmec (types Ⅰ–Ⅴ) have been identified. SCCmec types Ⅰ (34.3 kb), Ⅳ (20.9–24.3 kb), and Ⅴ (28 kb) all provide resistance to β-lactam antibiotics. Conversely, SCCmec types Ⅱ (53.0 kb) and Ⅲ (66.9 kb) contain integrated plasmids (pUB110, pI258, or pT181) and a transposon (Tn554) that include genes mediating resistance to other drugs. Each plasmid and transposon contains genes that provide resistance to antibiotics other than methicillin or to heavy metals. In addition, SCC*mec* contains *mecRI* (SCC*mec* types Ⅰ, Ⅳ, and Ⅴ), which regulates *mecA* transcription, or *mecRI* and *mecI* (SCC*mec* types Ⅱ and Ⅲ), as well as insertion sequences such as IS431 [6-9]. These genes are located in *mec* complexes, the five major classes (AE) of which have recently been described and used in diagnosis [5-7].

For integration and excision of SCC*mec* into/from specific genetic loci, genes encoding cassette chromosome recombinases (*ccr* genes) are present in SCC*mec* elements. These genes include *ccrA1, ccrB1* (in SCC*mec* types Ⅱ and Ⅳ, respectively), *ccrA3* and *ccrB3* (in SCC*mec* type Ⅲ), *ccrA4* and *ccrB4* (in MRSA strain HDE288 of SCC*mec* type Ⅳ) and *cork* (in SCC*mec* type Ⅴ). The area between *mec* and *ccr* is termed the J (junkyard) region [5, 8-11], and any SCC*mec* element J region can be divided into three subregions: J1, J2, and J3

[12, 13]. J1 runs from the chromosomal right junction to *ccr*, J2 lies between the *ccr* gene and the *mec* complex, and J3 runs between the *mec* complex and the left extremity of SCC*mec*. In addition to these five major SCC*mec* types, several variant SCC*mec* constructs, with or without plasmids and transposons, and/or with transformed J regions, are known [6, 11, 12, 14-18].

SCC*mec* is also present in *Staphylococcus* species (methicillin-resistant CoNS; MRCoNS) other than MRSA. SCC*mec* types I–IV have been identified from MRCoNS strains isolated in the 1970s [19]. However, these genes have not been found in bacteria related to *Staphylococcus*; the origin the staphylococcal *mecA* therefore remains unknown.

CoNS with greater levels of SCC*mec* elements have been proposed to be reservoirs of resistance genes in *Staphylococcus.*

S. sciuri expresses a novel PBP (which may be the intrinsic PBP) that has 87.7% amino acid homology to PBP2a. Thus, the PBP of *S. sciuri* may be the precursor of *S. aureus* PBP2a [20]. In addition, *S. hemolyticus* is known to be multidrug resistant and has also been assumed to be a resistant precursor of the homologous *S. aureus* [1].

Resistance to Other Antimicrobial Agents

Resistance to antibiotics other than methicillin can be provided by plasmids or transposons integrated into SCC*mec*. Plasmid pUB110 contains the *ant* (4') gene, which mediates resistance to kanamycin, tobramycin, and bleomycin. In addition, pI258 encodes resistance to penicillin and heavy metals, plasmid pT181 carries resistance genes for tetracycline, and transposon Tn554 carries the *ermA* (erythromycin ribosome methylase) gene responsible for inducible resistance to the macrolides lincosamide and streptogramin [6, 22]. The *erm* gene triggers activation site modification of the 30S ribosome, which is the target site of the macrolides. This leads to resistance to all types of MLSB antibiotics. When MLSB antibiotics are present, the *msrA* (macrolide-streptogramin resistance) gene product provides resistance by forming an efflux mechanism [22]. The resistance mechanisms of erythromycin-resistant and clindamycin-sensitive bacterial strains can involve an inducible MLSB system or macrolide efflux. A constitutive MLSB-resistance system is assumed to be present in strains showing resistance to both antibiotics [23]. Furthermore, point mutations in *gyrA, grlA*, the DNA gyrase gene, or the topoisomerase IV gene, the products of which are affected by quinolone antibiotics, modify antibiotic interactions and contribute to quinolone resistance [24-27]. It is known that an aminoglycoside-inactivating enzyme encoded by plasmid R mediates aminoglycoside resistance [28].

References

[1] Chmbers, HF; Hadley, WK; Jawetz, E. Beta-lactam antibiotics and other inhibitors of cell wall synthesis. In: Katzung, BG; editors. Basic and clinical pharmacology. 7th ed. Stanford, Connecticut: *Appleton and Lange*, 1998, p.724-42.

[2] Sieradzki, K; Tomasz, A. Alterations of cell wall structure and metabolism accompany reduced susceptibility to vancomycin in an isogenic weries of clinical isolates of Staphylococcus aureus. *J Bacteriol*, 2003, 185, 7103-10.

[3] Reipert, A; Ehlert, K; Kast, T; Bierbaum, G. Morphological and genetic differences in two isogenic Staphylococcus aureus strains with decreased susceptibilities to vancomycin. *Antimicrob Agents Chemother*, 2003, 47, 568-76.
[4] Berger-Bachi, B; Rohrer, S. Factors influencing methicillin resistance in staphylococci. *Arch Microbiol*, 2002, 178, 165–71.
[5] Katayama, Y; Ito, T; Hiramatsu, K. Genetic organization of the chromosome region surrounding mecA in clinical staphylococcal strains: role of IS431-mediated mecI deletion in expression of resistance in mecA-carrying, low level methicillin-resistant Staphylococcus haemolyticus. *Antimicrob Agents Chemother*, 2001, 45, 1955–63.
[6] Ito, T; Okuma, K; Ma, XX; Yuzawa, H; Hiramatsu, K. Insights on antibiotic resistance of Staphylococcus aureus from its whole genome: genomic island SCC. *Drug Resist Update*, 2003, 6, 41–52.
[7] Ito, T; Katayama, Y; Asada, K; et al. Structural comparison of three types of staphylococcal cassette chromosome mec integrated in the chromosome in methicillin-resistant Staphylococcus aureus. *Antimicrob Agents Chemother*, 2001, 45, 1323–36.
[8] Ito, T; Ma, XX; Takeuchi, F; et al. Novel type V staphylococcal cassette chromosome mec driven by a novel cassette chromosome recombinase ccrC. *Antimicrob Agents Chemother*, 2004, 48, 2637–51.
[9] Daum, RS; Ito, T; Hiramatsu, K; et al. A novel methicillin-resistance cassette in community-acquired methicillin-resistant Staphylococcus aureus isolates of diverse genetic backgrounds. *J Infect Dis*., 2002, 186, 1344–47.
[10] Hiramatsu, K; Cui, L; Kuroda, M; Ito, T. The emergence and evolution of methicillin-resistant Staphylococcus aureus. *Trends Microbiol*, 2001, 9, 486–93.
[11] Oliveira, DC; Tomasz, A; De Lencastre, H. The evolution of pandemic clones of methicillin-resistant Staphylococcus aureus: identification of two ancestral genetic backgrounds and the associated mec elements. *Microb Drug Resist*, 2001, 7, 349–61.
[12] Shore, A; Rossney, AS; Keane, CT; Enright, MC; Coleman, DC. Seven novel variants of the staphylococcal chromosomal cassette mec in methicillin-resistant Staphylococcus aureus isolates from Ireland. *Antimicrob Agents Chemother*, 2005, 49, 2070–2083.
[13] Chongtrakool, P; Ito, T; Ma, XX; et al. Staphylococcal cassette chromosome mec (SCCmec) typing of methicillin-resistant Staphylococcus aureus strains isolated in 11 Asian countries: a proposal for a new nomenclature for SCCmec elements. *Antimicrob Agents Chemother*, 2006, 50, 1001–12.
[14] Hisata, K; Kuwahara-Arai, K; Yamanoto, M; et al. Dissemination of methicillin-resistant staphylococci among healthy Japanese children. *J Clin Microbiol*, 2005, 43, 3364–72.
[15] Ma, XX; Ito, T; Tiensasitorn, C; et al. Novel type of staphylococcal cassette chromosome mec identified in community-acquired methicillin-resistant Staphylococcus aureus strains. *Antimicrob Agents Chemother*, 2002, 46, 1147–52.
[16] Oliveira, DC; De Lencastre, H. Multiplex PCR strategy for rapid identification of structural types and variants of the mec element in methicillin-resistant Staphylococcus aureus. *Antimicrob Agents Chemother*, 2002, 46, 2155–61.
[17] Boyle-Vavra, S; Ereshefsky, B; Wang, CC; Daum, RS. Successful multiresistant community-associated methicillin-resistant Staphylococcus aureus lineage from Taipei,

Taiwan, that carries either the novel staphylococcal chromosome cassette mec (SCCmec) type V_T or SCCmec type IV. *J Clin Microbiol*, 2005, 43, 4719–30.
[18] Kwon, NH; Park, KT; Moon, JS; et al. Staphylococcal cassette chromosome mec (SCCmec) characterization and molecular analysis for methicillin-resistant Staphylococcus aureus and novel SCCmec subtype IVg isolated from bovine bovine milk in Korea. *J Antimicrob Chemother*, 2005, 56, 624–32.
[19] Wisplinghoff, H; Rosato, AE; Enright, MC; et al. Related clones containing SCCmec type IV predominate among clinically significant Staphylococcus epidermidis isolates. *Antimicrob Agents Chemother*, 2003, 47, 3574–9.
[20] Wu, S; Piscitelli, C; De Lencastre, H; Tomasz, A. Tracking the evolutionary origin of the methicillin resistance gene: cloning and sequencing of a homologue of mecA from a methicillin-susceptible strain of Staphylococcus sciuri. *Microb Drug Resist*, 1996, 2, 435-41.
[21] Takeuchi, F; Watanabe, S; BaBa, T; Yuzawa, H; Ito, T; Morimoto, Y; Kuroda, M; et al. Whole-genome sequencing of staphylococcus haemolyticus uncovers the extreme plasticity of its genome and the evolution of human-colonizing staphylococcal species. *J Bacgteriol*, 2005, 187, 7292-308.
[22] Leclercq, R. Mechanisms of resistance to macrolides and lincosamides: nature of the resistance elements and their clinical implications. *Clin Infect Dis*., 2002, 34, 482-9.
[23] Livermore, DM; Winstanley, TG; Shannon, KP. Interpretative reading: recognizing the unusual and inferring resistance mechanisms from resistance phenotypes. *J Antimicrob Chemother*, 2001, 48, S87-102.
[24] Dubin, DT; Fitzgibbon, JE; Nahvi, MD; John, JF. Topoisomerase sequence of coagulase-negative Staphylococcus isolates resistant to ciprofloxacin or trovafloxacin. *Antimicrob Agents Chemother*, 1999, 43, 1631-7.
[25] Fey, PD; Climo, MW; Archer, GL. Determination of the chromosomal relationship between *mecA* and *gyrA* in methicillin-resistant coagulase-negative Staphylococci. *Antimicrob Agents Chemother*, 1998, 42, 306-12.
[26] Peterson, LR; Willard, KE; Sinn, LM; Fasching, CE; Gerding, DN. gyrA sequence analysis of Staphylococcus aureus and methicillin-resistant S. aureus strains selected, in vitro, for high-level ciprofloxacin resistance. *Diagn Microbiol Infect Dis*., 1993, 17, 97-101.
[27] Takahata, M; Yonezawa, M; Kurose, S; Futakuchi, N; Matsubara, N; Watanabe, Y; Narita, H. Mutations in the *gyrA* and *grlA* genes of quinolone-resistant clinical isolates of methicillin-resistant Staphylococcus aureus. *J Antimicrob Chemother*, 1996, 38, 543-6.
[28] Culebras, E; Martine, JL. Amonoglycoside resistance mediated by the bifunctional enzyme 6'-N-aminoglycoside acetyltransferase-2"-O-aminoglycoside phosphotransferase. *Front Biosci*., 1999, 4, D1-8.

5. GENETIC EVOLUTION OF MRSA

Prior to the discovery of penicillin, the mortality rate from *S. aureus* infection was 80%. After penicillin was developed for clinical use in the early 1940s, treatment of staphylococcal infections was assumed to be solved. With the emergence of penicillin-resistant *S. aureus* in 1942, however, hope faded. Moreover, bacterial strains resistant to semisynthetic penicillins (methicillin, oxacillin, and nafcillin) and resistant to glycopeptides, the only treatment agents effective against MRSA, subsequently emerged.

In MRSA, antibiotic resistance is based on *mecA* gene expression; the gene encodes a modified PBP (PBP2a or PBP2') of 78-kDa molecular weight. Modified PBP does not bind to β-lactam antibiotics and peptidoglycan synthesis is thus not impaired. Therefore, a normal cell wall is formed and survival is possible in the presence of β-lactam antibiotics. To date, five major types of SCC*mec* (types I – V) have been identified. The origin of SCC*mec* is unknown. Based on the finding that 87.8% of PBP amino acids in *S. sciuri* are identical to those of PBP2a, it has been suggested that the *S. sciuri* gene is the ancestor of the gene encoding PBP2a [1]. *S. sciuri* is normally methicillin-sensitive, but, if grown with methicillin, the strain shows methicillin resistance by increasing the transcription rate of the *mecA* homolog. In addition, if this homolog is introduced into MSSA, such strains become methicillin-resistant, as is seen in MRSA [1]. In light of these findings, MRSA strains may have been generated by horizontal transfer of *mecA* to generate two types of *Staphylococcus* [2].

After the emergence of the first MRSA by horizontal transfer of *mecA*, and the subsequent development of epidemic MRSA (EMRSA), two hypotheses have been proposed to explain MRSA spread and evolution. First, it has been suggested that all MRSA clones originated from one common ancestor through the insertion of SCC*mec* into a single, early *S. aureus* strain; this is the single-clone theory [3]. Second, MRSA clones may have been derived from various genetic lineages of ancestral *S. aureus,* into each of which a SCC*mec* cassette was inserted; this is the multi-clone theory [4-7]. The latter view is gaining credence [4-7].

Many molecular-typing methods have been used in efforts to understand MRSA origin, propagation, and evolution to yield EMRSA. Among these, the most common methods include PFGE (pulsed-field gel electrophoresis), MLST (multilocus sequence typing), and SCC*mec* typing.

PFGE is performed as follows. After purified chromosomal DNA is digested with *Sma*I restriction enzyme, agarose gel electrophoresis is performed. The DNA fragment pattern is analyzed using the Dice coefficient and unweighted pair-group matching analysis (UPGMA) according to defined criteria [8]. PFGE is one of the most commonly used research methods in studies of MRSA outbreaks and hospital-to-hospital transmission. However, the disadvantages include poor between-laboratory reproducibility, long experimental time, and expense [9-11]. No common PFGE terminology has yet been internationally established.

The MLST approach has been recognized as the best method for study of MRSA clonal evolution. MLST analyzes seven housekeeping genes (*arcC, aroE, glpF, gmk, pta, tpi,* and *yqiL*) of *S. aureus* and compares housekeeping gene sequences to those of *S. aureus* alleles stored in the MLST database; various alleles are thus defined. Strains of different allelic profiles are defined as sequence types (STs) and each *S. aureus* genotype is accurately and easily categorized [4, 12]. If SCC*mec* type data are included, each strain can be individually named. For example, a MRSA strain termed the "Iberian clone" had an MLST profile of 3-3-1-12-4-4-16 and was identified as ST247 (http://www.mlst.net), and, because the strain had SCC*mec* type I, it was finally given the descriptor ST247-MRSA- I . Each MLST profile or sequence type is subdivided into clonal complexes by BURST analysis, with reference to the seven loci (the housekeeping genes). If five or more of the seven loci are identical in two strains, the bacteria are classified as belonging to the same clonal complex (CC). Both initial grouping and CC classification predict ancestral allelic profiles and this is the simplest method to predict the evolutionary descent pattern from a predicted ancestral allelic profile [13]. The disadvantages of MLST are that the approach is time-consuming and requires good laboratory facilities.

SCC*mec* typing detects the *mec* complex and the *ccr* gene using PCR and classifies strains into SCC*mec* types Ⅰ–Ⅴ by difference analysis [13]. In addition, *spa* typing, a single-locus sequence typing method, is based on the polymorphic region X sequence of the *S. aureus* protein A-encoding gene (*spa*).

Enright and colleagues studied 553 MSSA isolates from 20 countries purified from 1961 to 1999, and examined 359 using the MLST and SCC*mec* typing methods [4]. First, five clonal complexes (CC 5, 8, 22, 30, 45) were adequate to include all studied strains and, even in strains with the same ST, different SCC*mec* types were apparent. The various SCC*mec* types are thought to originate from acquisition by *S. aureus* of DNA from different genetic backgrounds; thus, each SCC*mec* was inserted into a different *S. aureus* genetic lineage. In the case of CC8, the ST8-MSSA strain is probably the earliest ancestor of MRSA. ST8-MSSA is well known as a common cause of epidemic MSSA infections. This strain may have evolved to yield ST8-MRSA-Ⅰ, -Ⅱ, and -Ⅳ by acquisition of SCC*mec* types Ⅰ, Ⅱ, and Ⅳ. ST250-MSSA may have arisen from a point mutation of the *yqiL* gene in ST8-MSSA and, after acquisition of SCCmec type Ⅰ, eventually became ST250-MRSA-Ⅰ. ST250-MRSA-Ⅰ is the first known strain of MRSA discovered in the 1960s, also termed the “Archaic” clone [15]. ST247-MRSA-Ⅰ evolved from a point mutation in the *gmk* locus of ST250-MRSA-Ⅰ and is a major cause of MRSA infection in European hospitals. One of the major CC8 strains, ST239-MRSA-Ⅲ originated from ST8-MRSA-Ⅲ by transfer of part of the ST30 chromosome by homologous recombination, and is identical to a Brazilian clone (Figure 7). It can be inferred that CC5, CC22, CC30, and CC45, in addition to CC8, evolved by insertion of SCC*mec* DNA into different epidemic MSSA lineages.

Decreased vancomycin susceptibility of MRSA strains occurs through: increased synthesis of D-Ala-D-Ala dipeptide residues in the cell wall, thereby reducing vancomycin binding; by vancomycin excretion; and by increasing cell-wall thickness, thus preventing vancomycin access to the cell-wall formation area [16]. Although the exact mechanism of vancomycin resistance is not known, it can be inferred that resistance occurs by spontaneous mutation of genes associated with cell-wall synthesis because resistance develops gradually during treatment with vancomycin. MLST and SCC*mec* typing of GISA strains showed that such strains originated from ST5-MRSA-Ⅱ, a well-known New York/Japanese clone [4] (Figure 8).

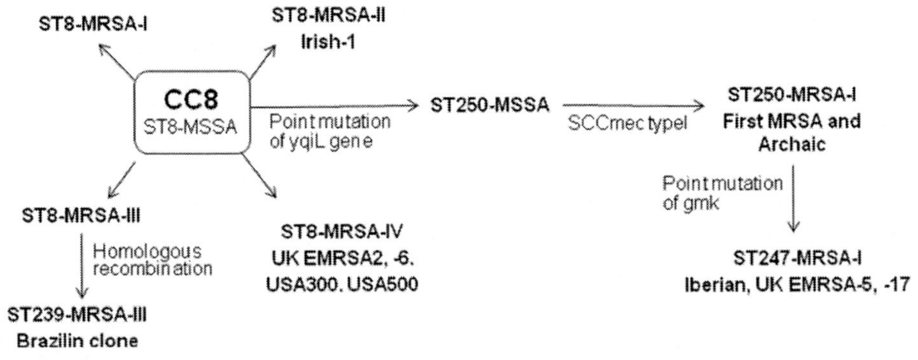

Figure 7. Evolutionary origins and patterns of development in CC8. Alterations in the allelic profile and acquisition of the SCC*mec* types are shown. The earlier names of the five major EMRSA clones are indicated under the current MLST-SCC*mec* nomenclature. The previous names are adapted from Enright et al. [4], McDougal et al. [26], Tenover et al. [27], and Melles et al. [28].
CC: clonal complex, EMRSA: epidemic MRSA, MRSA: methicillin-resistant *S. aureus*, SCC*mec*: staphylococcal chromosomal *mec* cassette, ST: sequence type.

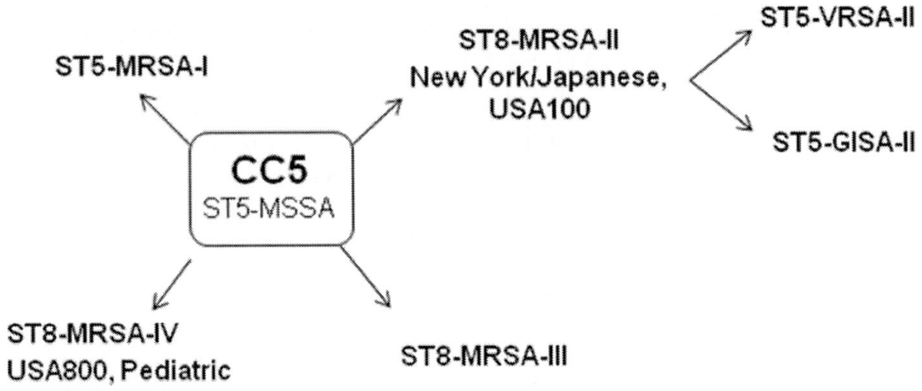

Figure 8. Evolutionary origins and patterns of development in CC5. Alterations in the allelic profile and acquisition of the SCC*mec* types are shown. Earlier names of major EMRSA clones are indicated under the current MLST-SCC*mec* nomenclature. Previous names are adapted from Enright et al. [4], McDougal et al. [26], Tenover et al. [27], and Melles et al. [28]. ST5-MRSA- II has evolved both intermediate and full resistance to vancomycin.
CC: clonal complex, EMRSA: epidemic MRSA, GISA: glycopeptide-intermediate-resistant-*S. aureus*; MRSA: methicillin-resistant *S. aureus*, SCC*mec*: staphylococcal chromosomal *mec* cassette, ST: sequence type, VRSA: vancomycin-resistant *S. aureus*.

CA-MRSA is distinct from HA-MRSA in that CA-MRSA strains are typically only resistant to β-lactam antibiotics and have one of the simplest, smallest, SCC*mec* forms, type IV [17]. CA-MRSA is also genetically different from HA-MRSA. First, CA-MRSA strains are of different clonal types, as detected in hospitals [18, 19], and demonstrate more MLST clonal diversity than do HA-MRSA strains [4, 20]. Most CA-MRSA strains have SCC*mec* types IV and V, but SCCmec types I, II, and III are also found [21, 22]. SCC*mec* type IV is also seen in HA-MRSA [23], so any distinction between CA-MRSA and HA-MRSA made by SCC*mec* typing is unreliable. It is not known whether SCC*mec*

DNA of CA-MRSA arose by novel acquisition of SCC*mec* IV from community MSSA strains [20] or was derived from HA-MRSA [24]. In a recent study, it was hypothesized that one penicillin-resistant *S. aureus* clone (ST30-MSSA) became dormant after the invention of methicillin, and emerged again only after acquiring the PVL gene and SCC*mec* type IV [25]. In addition, this strain (ST30-MSSA) may have evolved to become ST36-MRSA-II by obtaining SCC*mec* type II, and thereafter caused a pandemic outbreak [25] (Figure 9).

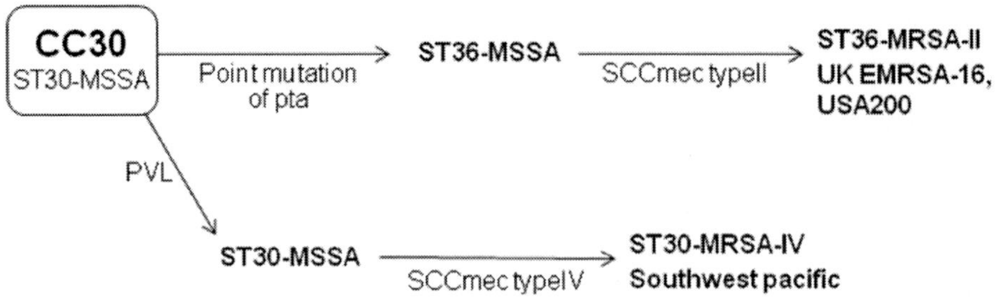

Figure 9. Evolutionary origins and patterns of development in CC30 and a possible relationship with community-acquired MRSA. Alterations in the allelic profile and acquisition of the SCC*mec* types are shown. Earlier names of major EMRSA clones are indicated under the current MLST-SCC*mec* nomenclature. Previous names are adapted from Enright et al. [4], McDougal et al. [26], Tenover et al. [27], and Melles et al. [28].
CC: clonal complex, EMRSA: epidemic MRSA, MRSA: methicillin-resistant *S. aureus*, SCC*mec*: staphylococcal chromosomal *mec* cassette, ST: sequence type, PVL: Panton-Valentine leukocidin.

References

[1] Wu, SW; De Lencastre, H; Tomasz, A. Recruitment of the mecA gene homologue of Staphylococcus sciuri into a resistance determinant and expression of the resistant phenotype in Staphylococcus aureus. *J Bacteriol*, 2001,183:2417–24.

[2] Wielders, CL; Vriens, MR; Brisse, S; De Graaf-Miltenburg, LA; Troelstra, A; Fleer, A; et al. In-vivo transfer of mecA DNA to Staphylococcus aureus. *Lancet*, 2001, 357, 1674–5.

[3] Kreiswirth, B; Kornblum, J; Arbeit, RD; Eisner, W; Maslow, JN; McGeer, A; et al. Evidence for a clonal origin of methicillin resistance in Staphylococcus aureus. *Science*, 1993, 259, 227–30.

[4] Enright, MC; Robinson, DA; Randle, G; Feil, EJ; Grundmann, H; Spratt, BG. The evolutionary history of methicillin-resistant Staphylococcus aureus (MRSA). *Proc Natl Acad Sci USA*, 2002, 99, 7687–92.

[5] Qi, W; Ender, M; O'Brien, F; Imhof, A; Ruef, C; McCallum, N; et al. Molecular epidemiology of methicillin-resistant Staphylococcus aureus in Zurich, Switzerland (2003): prevalence of type IV SCCmec and a new SCCmec element associated with isolates from intravenous drug users. *J Clin Microbiol*, 2005, 43, 5164–70.

[6] Musser, JM; Kapur, V. Clonal analysis of methicillin-resistant Staphylococcus aureus strains from intercontinental sources: association of the mec gene with divergent

phylogenetic lineages implies dissemination by horizontal transfer and recombination. *J Clin Microbiol*, 1992, 30, 2058–63.

[7] Fitzgerald, JR; Sturdevant, DE; Mackie, SM; Gill, SR; Musser, JM. Evolutionary genomics of Staphylococcus aureus: insights into the origin of methicillin-resistant strains and the toxic shock syndrome epidemic. *Proc Natl Acad Sci USA.*, 2001, 98, 8821–6.

[8] Tenover, FC; Arbeit, RD; Goering, RV; Mickelsen, PA; Murray, BE; Persing, DH; et al. Interpreting chromosomal DNA restriction patterns produced by pulsed-field gel electrophoresis: criteria for bacterial strain typing. *J Clin Microbiol*, 1995, 33, 2233–9.

[9] Murchan, S; Kaufmann, ME; Deplano, A; De Ryck, R; Struelens, M; Zinn, CE; et al. Harmonization of pulsed-field gel electrophoresis protocols for epidemiological typing of strains of methicillin-resistant Staphylococcus aureus: a single approach developed by consensus in 10 European laboratories and its application for tracing the spread of related strains. *J Clin Microbiol*, 2003, 41, 1574–85.

[10] Tenover, FC; Arbeit, R; Archer, G; Biddle, J; Byrne, S; Goering, R; et al. Comparison of traditional and molecular methods of typing isolates of Staphylococcus aureus. *J Clin Microbiol*, 1994, 32, 407–15.

[11] Van Belkum, A; Van Leeuwen, W; Kaufmann, ME; Cookson, B; Forey, F; Etienne, J; et al. Assessment of resolution and intercenter reproducibility of results of genotyping Staphylococcus aureus by pulsedfield gel electrophoresis of SmaI macrorestriction fragments: a multicenter study. *J Clin Microbiol*, 1998, 36, 1653–9.

[12] Enright, MC, Day, NPJ; Davies, CE; Peacock, SJ; Spratt, BG. Multilocus sequence typing for characterization of methicillin-resistant and methicillin-susceptible clones of staphylococcus aureus. *J Clin Microbiol*, 2000, 38, 1008-15,

[13] Hiramatsu, K; Kuroda, M; Baba, T; Okuma, K. In: Persing D, editor. Diagnostic molecular microbiology. Washington, DC: *American Society for Microbiology*, 2002. p.

[14] Frenay, HM; Bunschoten, AE; Schouls, LM; Van Leeuwen, WJ; Vandenbroucke-Grauls, CM; Verhoef, J; et al. Molecular typing of methicillin-resistant Staphylococcus aureus on the basis of protein A gene polymorphism. *Eur J Clin Microbiol Infect Dis.*, 1996, 15, 60–4.

[15] De Lencastre, H; De Lencastre, A; Tomasz, A. Methicillin-resistant Staphylococcus aureus isolates recovered from a New York City hospital: analysis by molecular fingerprinting techniques. *J Clin Microbiol*, 1996, 34, 2121-4.

[16] Cui, L; Murakami, H; Kuwahara-Arai, K; Hanaki, H; Hiramatsu, K. Contribution of a thickened cell wall and its glutamine nonamidated component to the vancomycin resistance expressed by Staphylococcus aureus Mu50. *Antimicrob Agents Chemother*, 2000, 44, 2276-85.

[17] Ma, XX; Ito, T; Tiensasitorn, C; Jamklang, M; Chongtrakool, P; Boyle-Vavra, S; et al. Novel type of staphylococcal cassette chromosome mec identified in community-acquired methicillin-resistant Staphylococcus aureus strains. *Antimicrob Agents Chemother*, 2002, 46, 1147-52.

[18] Groom, AV; Wolsey, DH; Naimi, TS; Smith, K; Johson, S; Boxrud, D; et al. Community-acquired methicillin-resistant Staphylococcus aureus in a rural American Indian community. *JAMA*, 2001, 286, 1201–5.

[19] Naimi, TS; LeDell, KH; Boxrud, DJ; Groom, AV; Steward, CD; Johnson, SK; et al. Epidemiology and clonality of community-acquired methicillin-resistant Staphylococcus aureus in Minnesota, 1996–1998. *Clin Infect Dis*., 2001, 33, 990–6.
[20] Okuma, K; Iwakawa, K; Turnidge, JD; Grubb, WB; Bell, JM; O'Brien, FG; et al. Dissemination of new methicillin-resistant Staphylococcus aureus clones in the community. *J Clin Microbiol*, 2002, 40, 4289–94.
[21] Chung, M; Dickinson, G; De Lencastre, H; Tomasz, A. International clones of methicillin-resistant Staphylococcus aureus in two hospitals in Miami, Florida. *J Clin Microbiol*, 2004, 42, 542–7.
[22] Wannet, WJ; Spalburg, E; Heck, ME; Pluister, GN; Tiemersma, E; Willems, RJ; et al. Emergence of virulent methicillin-resistant Staphylococcus aureus strains carrying Panton–Valentine leucocidin genes in The Netherlands. *J Clin Microbiol*, 2005, 43, 3341–5.
[23] De A Trindade, P; Pancheco, RL; Costa, SF; Rossi, F; Barone, AA; Mamizuka, EM; et al. Prevalence of SCCmec type IV in nosocomial bloodstream isolates of methicillin-resistant Staphylococcus aureus. *J Clin Microbiol*, 2005, 43, 3435–7.
[24] Aires De Sousa, M; De Lencastre, H. Evolution of sporadic isolates of methicillin-resistant Staphylococcus aureus (MRSA) in hospitals and their similarities to isolates of community-acquired MRSA. *J Clin Microbiol*, 2003, 41, 3806–15.
[25] Robinson, DA; Kearns, AM; Holmes, A; Morrison, D; Grundmann, H; Edwards, G; et al. Re-emergence of early pandemic Staphylococcus aureus as a community-acquired methicillin-resistant clone. *Lancet*, 2005, 365, 1256–8.
[26] McDougal, LK; Steward, CD; Killgore, GE; Chaitram, JM; McAllister, SK; Tenover, FC. Pulsed-field gel electrophoresis typing of oxacillin-resistant Staphylococcus aureus isolates from the United States: establishing a national database. *J Clin Microbiol*, 2003, 41, 5113–20.
[27] Tenover, FC; McDougal, LK; Goering, RV; et al. Characterization of a strain of community-associated methicillin-resistant Staphylococcus aureus widely disseminated in the United States. *J Clin Microbiol*., 2006, 44, 108–18.
[28] Melles, DC; Gorkink, RF; Boelens, HA; et al. Natural population dynamics and expansion of pathogenic clones of Staphylococcus aureus. *J Clin Invest*, 2004, 114, 1732–40.

6-1. Healthcare-Associated (HA)-MRSA

Traditionally, MRSA has been thought to have a nosocomial origin. Methicillin was introduced during 1959–1960. One year later, a methicillin-resistant strain was reported [1]. MRSA was first found in medical institutions rather than in local communities. In the mid-1980s, MRSA was most commonly detected in large-volume centers and medical institutions where patients received long-term hospitalization. From the 1990s, the MRSA prevalence rate exceeded 20% even in medium- and small-sized hospitals with fewer than 200 beds. In urban hospitals, the prevalence rate has been two-fold higher [2]. In 2003, 60% of *S. aureus* isolated from intensive care units (ICUs) showed methicillin resistance [3].

According to the National Nosocomial Infections Surveillance System, ICUs in the United States reported an HA-MRSA infection rate of 35.9% in 1992 and 64.4% in 2003. These figures show that infection increased by 3.1% on a yearly basis and, if these trends continue, would reach 80% by 2010 [4]. MRSA infection rates increase with patient age and are highest in patients aged 65 years or older [5].

With respect to HA-MRSA infection, risk factors include long-term hospitalization, residency in a nursing home, dialysis, indwelling catheter use, ICU exposure, prior antimicrobial therapy (in particular, with cephalosporin or fluoroquinolones), surgical procedures, a decubitus ulcer, close proximity to a hospital patient colonized with MRSA, intravenous drug use, and a history of enteral feeding [6-8].

Major mechanisms of β-lactam resistance in HA-MRSA strains are based on PBP2a expression; the protein has a low-binding affinity for β-lactam antibiotics (9). PBP2a is encoded by the *mecA* gene; the mobile *mecA* gene complex is composed of a genomic island termed SCC*mec* together with regulator genes such as *mecI* and *mecR* [10-12]. Transmission of resistance mediated by *mecA* occurs through exchange of plasmid and chromosomal DNA between pathogens, and these resistant pathogens infect other patients. It is well known that the most important mechanism of bacterial transmission involves transmission on the hands of hospital staff [13].

Many studies have investigated the origin of HA-MRSA. MLST showed that MRSA originated from five ancestral genotypes of different clonal complexes [14]. ST 8 MSSA bacterial strains evolved to ST 250 MSSA by point mutation, and it is assumed that this strain was the ancestor of the first clinically discovered MRSA strain (termed ST 250-MRSA- I) after acquiring SCC*mec* type I [14].

HA-MRSA became internationally disseminated through clonal transmission. Five principal clones (the Iberian, Brazilian, Hungarian, New York/Japanese, and pediatric clones), identified by sequencing, have been assigned to various clonal groups. These multidrug resistant clones were transmitted throughout the world and became important causes of HA-MRSA infections in several regions. It remains unclear why the transmission of some specific clones was facile and why these clones caused HA-MRSA infection only in particular countries. It is at least certain that resistance to multiple antibacterial agents has a key role in niche establishment in medical institutions where antibiotic use is prevalent. In addition, the transmitted clones may be of high virulence.

The virulence of HA-MRSA strains is not fully understood. A Brazilian clone (Brazilian epidemic clonal complex [BECC], also known as PFGE type A1) was the main cause of a serious staphylococcal infection in Brazil's Joao Barro Barreto University Hospital in the 1990s. When this clone was compared with MSSA and sporadic MRSA strains, it was found that the Brazilian strain more readily formed biofilms, showed high affinity for polystyrene and bronchial epithelial cells, and was capable of efficient tissue infiltration. An accessible fibronectin-binding domain probably contributed to the high level of infiltration. These properties allowed the bacterium to readily attach to host cells, to remain host-associated, and to penetrate and become a major infection problem [15]. However, it is not known whether such changes also occurred in other HA-MRSA strains.

Both direct bacterial contact and airborne bacteria may be routes of infection for MRSA. According to a report on the rapidly increasing MRSA infection rate in hospitals from 1976 to 1980 [13], 74 of 140 (53%) patients from whom MRSA was isolated showed a true infection, whereas 66 (47%) were merely colonized. Therefore, patients and healthcare professionals who have bacteria colonizing the anterior nasal cavity or hands, without true infection, could be an important source of nosocomial infection. The most important cause of infection spread is disinfection failure by healthcare professionals. Patients with bacteria colonizing the anterior nasal cavity are known to be at increased infection risk. In hospitals where MRSA is present, 10–30% of inpatients develop MRSA infection [16, 17].

Unlike CA-MRSA, HA-MRSA usually causes only pneumonia, urinary tract infection, sepsis, and operative wound infection.

REFERENCES

[1] Jevons, MP. "Celbenin"-resistant staphylococci. *Br Med J.*, 1961, 1, 124–5.

[2] Chambers, HF. The changing epidemiology of Staphylococcus aureus? *Emerg Infect Dis.*, 2001, 7, 178-82.

[3] National Nosocomial Infections Surveillance System. National Nosocomial Infections Surveillance (NNIS) System Report, data summary from January 1992 to June 2004, issued October 2004. *Am J Infect Control*, 2004, 32, 470-85.

[4] Klevens, RM; Edwards, JR; Tenover, FC; McDonald, LC; Horan, T; Gaynes, R. National Nosocomial Infections Surveillance System. Changes in epidemiology of methicillin-resistant Staphylococcus aureus in intensive care units in US hospitals, 1992-2003. *Clin Infect Dis.*, 2006, 42, 389-91.

[5] Kuehnert, MJ; Hill, HA; Kupronis, BA; Tokars, JI; Solomon, SL; Jernigan, DB. Methicillin-resistant Staphylococcus aureus hospitalizations, United States. *Emerg Infect Dis.*, 2005, 11, 868-72.

[6] Lodise, TP; Jr, McKinnon, PS; Rybak, M. Prediction model to identify patients with Staphylococcus aureus bacteremia at risk for methicillin-resistance. *Infect Control Hosp Epidemiol*, 2003, 24, 655-71.

[7] Karchmer, AW. Nosocomial bloodstream infections: organisms, risk factors and implications. *Clin Infect Dis.*, 2000, 31, S13-43.

[8] Graffunder, EM; Venezia, RA. Risk factors associated with noscomial methicillin-resistant Staphylococcus aureus (MRSA) infection including previous use of antimicrobials. *J Antimicrob Chemother*, 2002, 49, 999-1005.

[9] Hartman, A; Tomasz, B. Altered penicillin-binding proteins in methicillin-resistant strains of Staphylococcus aureus. *Antimicrob Agents Chemother*, 1981, 19, 726-35.

[10] Katayama, Y; Ito, T; Hiramatsu, K. A new class of genetic element, Staphylococcus cassette chromosome mec, encodes methicillin resistance in Staphylococcus aureus. *Antimicrob Agents Chemother*, 2000, 44, 1549-55.

[11] Hiramatsu, K. Elucidation of the mechanism of antibiotics resistance acquisition of methicillin-resistant Staphylococcus aureus (MRSA) and determination of its whole genome nucleotide sequence. *JMAJ*, 2004, 47, 153-9.

[12] Robinson, DA; Enright, MC. Multilocus sequence typing and the evolution of methicillin-resistant Staphylococcus aureus. *Clin Microbiol Infect*, 2004, 10, 92-7.

[13] Thompson, RL; Cabezudo, I; Wenzel, RP. Epidemiology of nosocomial infections caused by methicillin-resistant Staphylococcus aureus. *Ann Intern Med.*, 1982, 97, 309-17.
[14] Enright, MC; Robinson, DA; Randle, G; Feil, EJ; Grundmann, H; Spratt, BG. The evolutionary history of methicillin-resistant Staphylococcus aureus (MRSA). *Proc Natl Acad Sci USA.*, 2002, 99, 7687-92.
[15] Amaral, MM; Coelho, LR; Flores, RP; Souza, RR; Silva-Carvalho, MC; Teixeira, LA; et al. The predominant variant of the Brazilian epidemic clonal complex of methicillin-resistant Staphylococcus aureus has an enhanced ability to produce biofilm and to adhere to and invade airway epithelial cells. *J infect Dis.*, 2005, 192, 801-10.
[16] Davis, KA; Stewart, JJ; Crouch, HK; Florez, CE; Hospenthal, DR. Methicillin-resistant Staphylococcus aureus (MRSA) nasal colonization at hospital admission and its effect on subsequent MRSA infection. *Clin Infect Dis.*, 2004, 39, 776-82.
[17] Huang, SS; Platt, R. Risk of methicillin-resistant Staphylococcus aureus infection after previous infection or colonization. *Clin Infect Dis.*, 2003, 36, 281-5.

6-2. Community Associated Methicillin-Resistant *S. Aureus* (CA-MRSA)

A definition of CA-MRSA may be helpful. As outlined above, *Staphylococcus* strains have acquired resistance to antibiotics ever since penicillin was developed in the 1940s; MRSA appeared in 1961 and, more recently, VRSA strains have been isolated. These findings can be regarded as a chronology of HA-MRSA development. The first discovery of MRSA outside a hospital was in intravenous drug users in 1982 [1]. There was no reported case of MRSA infection in community members not exposed to medical institutions until the 1990s. However, in 1989 and 1991, MRSA infections occurred in the indigenous population of Western Australia with no history of medical institution contact [2], and in the 1990s, severe MRSA infections in patients not recognizably at risk were reported in the United States [3-5]. In Minnesota and North Dakota between 1997 and 1999, severe MRSA infections associated with necrotizing pneumonia, pulmonary abscess, and sepsis, were reported, and four pediatric patients died. Thereafter, interest in CA-MRSA has increased rapidly [6]. More recently, clonal outbreaks of skin/soft tissue infection caused by CA-MRSA among convicts, homosexual men, soldiers, and athletes (particularly soccer players) have been reported [7-10]. CA-MRSA infections are now increasing in number and frequency, and such strains also occur in medical institutions, thus obscuring the distinction between HA and CA bacteria [11, 12].

From the perspective of epidemiology, "community acquired" means that the bacterium is isolated within 48–72 hours of admission or from patients in the outpatient clinic or emergency room. However, CA infection sometimes occurs together with HA infection and classification by isolation time may artificially increase the prevalence of CA-MRSA infection. Thus, when MRSA infection occurs in the community, if the infection is seen in a community associated with a hospital, the term “healthcare-associated infection” is used, and any other type of community infection is considered to be a “community-associated infection” [13]. However, confusion in terminology remains [14]. The Active Bacterial Core (ABC) Surveillance Program of the United States Centers for Disease Control and Prevention

(CDC) defined CA-MRSA as a strain isolated in the outpatient clinic, the emergency room, or within 48 hours of admission if there was no risk factor for MRSA infection. Established MRSA infection risk factors are: 1) recent hospitalization, surgery, or residence in a nursing home for less than 1 year; 2) dialysis; 3) a history of prior MRSA isolation; and, 4) use of a central line, a Foley catheter, a tracheotomy tube, or percutaneous gastrostomy at the time of culture [15-18]. All other infections are classified as HA-MRSA [19]. However, this definition is also imprecise, and the detailed location of infection cannot be easily defined. Thus, some researchers classify MRSA according to isolation location, thus retaining the community-onset MRSA and hospital-onset MRSA distinctions.

CA-MRSA is narrowly defined as follows [20]: 1) a multidrug-resistant phenotype is absent; 2) a specific exotoxin virulence factor (such as PVL toxin) is present; 3) a type □ SCC is present; and, 4) there are molecular distinctions from nosocomial strains. Among these criteria, only confirmation of absence of a multidrug-resistant phenotype can be clinically applied. However, the pattern of resistance to antibiotics is dynamic and can vary by region. In addition, and unlike previous reports, CA-MRSA can show resistance to antibacterial agents other than □-lactams [19], and a nosocomial outbreak caused by CA-MRSA in a medical institution has been reported [21]. It may thus be unreasonable to distinguish CA-MRSA only by a pattern of antibiotic resistance.

Epidemiologic Features of CA-MRSA

Many CA-MRSA patients do not have specific risk factors [22]. Outbreaks have been reported in certain groups such as a wrestling team, football players, rugby players, fencing athletes, homosexual men, prisoners, children in a daycare center, and homeless individuals [23-29]. Bodily contact, crowding, contaminated items and environmental surfaces, compromised skin integrity, and cleanliness could have important roles in the spread of the pathogen [20].

Populations at Risk

Young children [17]
Daycare attendees [28]
Prison inmates [29]
Indigenous populations (Pacific islanders, native Americans/Alaskan natives, Pacific and Canadian aboriginals) [30]
Homosexual men [26]
Those with HIV infection [17]
Members of football teams [23-25]
Wrestlers [25]
Fencing teams [25]
Intravenous drug users [29]
Homeless [29]
Tattoo recipients [31]
Military recruits [27]

According to a large-scale study in the United States, 131 (12%) of 1,100 cases of MRSA infection were CA-MRSA and patients showed no risk factors [19]. Most of these CA-MRSA strains (85%) were of SCC*mec* type IV, and the CA-MRSA strains were sensitive to several drugs. Moreover, such strains of MRSA accounted for 10% of hospital infections, and the strains were believed to spread within the hospital. Recent data from community residents and 12 hospital laboratories showed that 17% of all isolated strains were CA-MRSA [19]. In other studies, however, patients with CA-MRSA infection had risk factors for MRSA infection and the bacterial isolates were of various PFGE types. The possibility that HA-MRSA may have propagated to the local community has also been proposed.

A recently published meta-analysis on CA-MRSA frequency [14] reviewed 27 retrospective studies, and found that, although CA-MRSA comprised 30% of all *Staphylococcus* infections, 86% of patients had risk factors for MRSA infection. Also, a review of five prospective studies on CA-MRSA infection revealed that CA-MRSA comprised 37% of all cases and 87% of patients had risk factors for MRSA infection. These results indicate that most CA-MRSA infections originate in hospitals.

Various frequencies of MRSA infection are regionally reported, but definitions and research methods differ. Nevertheless, increasing numbers of MRSA infections have been recently reported both in the community and in hospitals not only in the United States [32], but also in Japan, Taiwan, Finland, Uruguay, Canada, and Australia [33-37].

Origin of CA-MRSA

There are two theories on the origin of CA-MRSA. One possibility is that HA-MRSA spread among patients after infected subjects entered a community, and the other explanation is that an MSSA strain naturally gained a methicillin-resistance gene (*mecA*) due to increased use of antibiotics in the community [38, 39]. If no outbreak of MRSA is seen, antibiotic use in medical institutions may be an important cause of emerging CA-MRSA. However, if it is considered that CA-MRSA may differ in genotype from HA-MRSA, CA-MRSA may originate in the community. Recently, as outbreaks of CA-MRSA in hospitals have been reported, the distinction between the community and the hospital environment from an epidemiological viewpoint becomes blurred.

Microbiologic Caracteristics of CA-MRSA

CA-MRSA strains have many features different from those of HA-MRSA strains [43-45]. First, CA-MRSA strains have SCC*mec* of type IV, which does not include genes mediating resistance to antibiotics other than β-lactams. The SCC*mec* type IV is smaller than other forms of SCC*mec,* and can thus be easily transferred. In addition, CA-MRSA strains have greater mitosis rates than do HA-MRSA strains [40]. This biological fitness of CA-MRSA makes it possible for CA-MRSA to substitute for other MRSA strains under conditions of antimicrobial pressure and to successfully propagate in hospitals. One report claimed that one-third of MRSA-induced sepsis was caused by CA-MRSA clones in a hospital in the United States [46]. Several lines of evidence from recent molecular genetic studies using MLST have indicated that CA-MRSA strains evolve from insertion of the mobile SCC*mec* type IV into MSSA strains [47, 48]. It has been reported that several such clones have caused community outbreaks in the United States.

Recent genetic analysis of CA-MRSA found 19 virulence genes absent from HA-MRSA [49]. These included superantigens such as enterotoxins B/C (SEA/SEB) and PVL genes encoding leukotoxins. In particular, PVL genes were found in most CA-MRSA strains but in fewer than 5% of HA-MRSA isolates [50]. Skin and soft tissue infections and necrotizing pneumonia are known to be caused by PVL-positive strains [51, 52]. In 172 *S. aureus* strains [52], PVL was associated with community-acquired pneumonia (85%) rather than with hospital-acquired pneumonia (0%), and with invasive skin infections such as furunculosis (93%) and cutaneous abscess (50%) rather than with superficial abscesses (0%). PVL was not detected in strains associated with infective endocarditis, urinary tract infection, TSS, or mediastinitis.

PVL contains an S protein and an F protein, and has leukocytolytic, erythrocytolytic, and dermonecrotic properties depending on the combination of these two proteins [53, 54]. PVL is composed of Luk-PV and LukF-PV; the four units form an octameric □-barrel, thus creating a pore in the leukocyte membrane and eventually causing cell lysis [55-57]. In addition, PVL has been shown to cause neutrophil apoptosis through the mitochondrial pathway and induce necrosis when present at high concentration [58]. PVL may contribute to the increased virulence of CA-MRSA. However, there was no difference in virulence between a PVL-positive strain and a PVL-negative strain in a recent study [59, 60]. These results indicate that the effect of PVL under experimental conditions may differ from the *in vivo* PVL effect; PVL may only be a marker for another virulence factor or may be one of many important virulence factors.

In recent years, increased upregulation of phenol-soluble modulin in CA-MRSA, compared with HA-MRSA, has been suggested to increase virulence [61]. Moreover, in the USA400 strain, staphylococcal enterotoxin H (She), Bsa (a potential bacteriocin) [62], and, in the USA300 strain, ACME (arginine catabolic mobile element) have been identified as new virulence factors [63]. It is thought that ACME increases virulence by increasing the cellular infiltration ability under conditions of low oxygen tension, thus improving survival at low pH levels, suppressing the growth of peripheral monocytes, and inhibiting formation of nitric oxide.

Clinical Features of CA-MRSA Infection

The epidemiology of CA-MRSA infection is not completely understood. CA-MRSA infections reported in North America differ in many clinical features from HA-MRSA infections. Whereas HA-MRSA frequently infect the bloodstream, urinary tract, and respiratory tract, CA-MRSA usually infect skin and soft tissue [64, 65] and were associated with younger individuals; infection characteristics also differ with race and socioeconomic status [19]. Most CA-MRSA infections provoke furunculosis, cutaneous skin abscesses, or simple cellulitis and repetitive furunculosis, although transmission through a skin lesion after direct contact is common [66]. MRSA infections must be considered in patients at risk for such infections and who show slow responses to β-lactam treatment. In addition, MRSA infection should be discriminated from a dermonecrotic lesion caused by a spider bite [7].

CA-MRSA also causes lethal community-acquired pneumonias (CAPs). Pneumonia caused by *S. aureus* is a rare disease except in the case of postinfluenza infection. There were two reports of CA-MRSA CAP in United Statesduring the influenza seasons of 2003 and 2004 [67], and 19 cases were reported in the United States. In all MRSA infection patients, previous influenza viral infection was identified or suspected. Although most patients were young and healthy, the mortality rate reached 25% and survivors remained in the ICU for a considerable period. PVL toxin was detected in 14 of 15 patients. Leukopenia, hemoptysis, influenza prodrome, and concomitant furunculosis were clinical features of PVL-positive necrotizing pneumonia, with final rapid progression to fatality [68].

REFERENCES

[1] Saravolatz, LD; Markowitz, N; Arking, L; Pohlod, D; Fisher, E. Methicillin-resistant Staphylcoccus aureus: epidemiologic observation during a community-acquired outbreakt. *Ann Intern Med.*, 1982, 96, 11-6.

[2] Udo, EE; Pearman, JW; Grubb, WB. Genetic analysis of community isolates of methicillin-resistant Staphylococcus aureus in western Austrailia. *J Hosp Infect.*, 1993, 25, 97-108.

[3] Riley, TV; Rouse, IL. Methicillin-resistant Staphylcoccus aureus in Western Austrailia, 1983-1992. *J Hosp Infect*, 1995, 29, 177-88.

[4] Adcock, PM; Pastor, P; Medley, F; Patterson, JE; Murphy, TV. Methicillin-resistant Staphylococcus aureus in two child care centers. *J Infect Dis.*, 1998, 178, 577-80.

[5] Herold, BC; Immergluck, LC; Maranan, MC; Lauderdale, DS; Gaskin, RE; Boyle-Vavra, S; et al. Community-acquired methicillin-resistant Staphylococcus aureus in children with no identified predisposing risk. *JAMA*, 1998, 279, 593-8.

[6] From the Centers for Disease Control and Prevention. Four pediatric deaths from community-acquired methicillin-resistant Staphylococcus aures-Minnesota and North Dakota, 1997-1999. MMWR Morb Mortal Wkly Rep., 1999, 48, 707-10.

[7] From the Centers for Disease Control and Prevention. Outbreaks of community-associated methicillin-resistant Staphylococcus aureus skin infections—Los Angeles County, California, 2002–2003. MMWR *Morb Mortal Wkly Rep.*, 2003, 52, 88.

[8] From the Centers for Disease Control and Prevention. Methicillin-resistant Staphylococcus aureus infections among competitive sports participants—Colorado, Indiana, Pennsylvania, and Los Angeles County, 2000–2003. *MMWR Morb Mortal Wkly Rep.*, 2003, 52, 793–5.

[9] From the Centers for Disease Control and Prevention. Methicillin-resistant Staphylococcus aureus skin or soft tissue infections in a state prison—Mississippi, 2000. *MMWR Morb Mortal Wkly Rep.*, 2001, 50, 919–22.

[10] Kazakova, SV; Hageman, JC; Matava, M; Srinivasan, A; Phelan, L; Garfinkel, B; et al. A clone of methicillin-resistant Staphylococcus aureus among professional football players. *N Engl J Med.*, 2005, 352, 468–75.

[11] Seybold, U; Kourbatova, EV; Johnson, JG; Halvosa, SJ; Wang, YF; King, MD; et al. Emergence of community-associated methicillin-resistant Staphylococcus aureus

USA300 genotype as a major cause of health care–associated blood stream infections. *Clin Infect Dis*., 2006, 42, 647–56.

[12] Gonzalez, BE; Rueda, AM; Shelburne, SA; III, Musher, DM; Hamill, RJ; Hulten, KG. Community-associated strains of methicillin-resistant Staphylococccus aureus as the cause of healthcare-associated infection. *Infect Control Hosp Epidemiol*, 2006, 27, 1051–6.

[13] Maltezou, HC; Giamarellou, H. Community-acquired methicillin-resistant Staphylococcus aureus infections. *Int J Antimicrob Agents*., 2006, 27, 87-96.

[14] Salgado, CD; Farr, BM; Calfee, DP. Community-acquired methicillin-resistant Staphylococcus aureus: a meta-analysis of prevalence and risk factors. *Clin Infect Dis*., 2003, 36, 131-139.

[15] Lowy, FD. Staphylococcus aureus infections. *N Engl J Med*., 1998, 339, 520-32.

[16] Brumfitt, W; Hamilton-Miller, J. Methicillin-resistant Staphylococcus aureus. *N Engl J Med*., 1989, 320, 1188-96.

[17] Fridkin, SK; Hageman, JC; Morrison, M; Sanza, LT; Como-Sabetti, K; Jernigan, JA; ea al. Methicillin-resistant Staphylococcus aureus disease in three communities. *N Engl J Med*., 2005, 352, 1436-44.

[18] Naimi, TS; LeDell, KH; Boxrud, DJ; Groom, AV; Steward, CD; Johnson, SK; et al. Epidemiology and clonality of community-acquired methicillin-resistant Staphylococcus aureus in Minnesota, 1996-1998. *Clin Infect Dis*., 2001, 33, 990-6.

[19] Naimi, TS; LeDell, KH; Como-Sabetti, K; Borchardt, SM; Boxrud, DJ; Etienne, J; et al. Comparison of community- and health care-associated methicillin-resistant Staphylococcus aureus infection. *JAMA* 2003, 290, 2976-84.

[20] Kowalski, TJ; Berbari, EF; Osmon, DR. Epidemiology, treatment, and prevention of community-acquired methicillin-resistant Staphylococcus aureus infections. *Mayo Clin Proc*., 2005, 80, 1201-8.

[21] Saiman, L; O'Keefe, M; Graham, PL; 3rd, Wu, F; Said-Salim, B; Kreiswirth, B; et al. Hospital transmission of community-acquired methicillin-resistant Staphylococcus aureus among postpartum women. *Clin Infect Dis*., 2003, 37, 1313-9.

[22] Moran, GJ; Krishnadasan, A; Gorwitz, RJ; Fosheim, GE; McDougal, LK; Carey, RB; et al. Methicillin-resistant S. aureus infections among patients in the emergency department. *N Engl J Med*., 2006, 355, 666-74.

[23] Begier, EM; Frenette, K; Barrett, NL; Mshar, P; Petit, S; Boxrud, DJ; et al. A high-morbidity outbreak of methicillin-resistant Staphylococcus aureus among players on a college football team, facilitated by cosmetic body shaving and turf burns. *Clin Infect Dis*., 2004, 39, 1446-53.

[24] Kazakova, SV; Hageman, JC; Matava, M; Srinivasan, A; Phelan, L; Garfinkel, B; et al. A clone of methicillin-resistant Staphylococcus aureus among professional football players. *N Engl J Med*., 2005, 352, 468-75.

[25] From the Centers for Disease Control and Prevention. Methicillin-resistant Staphylococcus aureus infections among competitive sports participants—Colorado, Indiana, Pennsylvania, and Los Angeles county, 200-2003. *MMWR Morb Mortal Wkly Rep*., 2003, 52, 793-5.

[26] Diep, BA; Chambers, HF; Graber, CJ; Szumowski, JD; Miller, LG; Han, LL; et al. Emergence of multidrug-resistant, community-associated, methicillin-resistant

Staphylococcus aureus clone USA300 in men who have sex with men. *Ann Intern Med.*, 2008, 148, 249-57.

[27] Aiello, AE; Lowy, FD; Wright, LN; Larson, EL. Methicillin-resistant Staphylococcus aureus among US prisoners and military personnel: review and recommendations for future studies. *Lancet Infect Dis.*, 2006, 6, 335-41.

[28] Adcock, PM; Pastor, P; Medley, F; Patterson, JE; Murphy, TV. Methicillin-resistant Staphylococcus aureus in two child care centers. *J Infect Dis.*, 1998, 178, 577-80.

[29] Gilbert, M; MacDonald, J; Gregson, D; Siushansian, J; Zhang, K; Elsayed, S; et al. Outbreak in Alberta of community-acquired methicllin-resistant Staphylococcus aureus in people with a history of drug abuse, homelessness or incarceration. *CMAJ*, 2006, 175, 149-54.

[30] Center for Disease Control and Prevention. Community-associated methicillin-resistant Staphylococcus aureus infections in Pacific Islanders-Hawaii, 2001-2003. *MMWR Morb Mortal Wkly Rep.*, 2004, 53, 767-70.

[31] Centers for Disease Control and Prevention. Methicillin resistant Staphylococcus aureus skin infections among tattoo recipients—Ohio, Kentucky and Vermont, 2004-2005. *MMWR Morb Mortal Wkly Rep.*, 2006, 55, 677-9.

[32] Charlebois, ED; Perdreau-Remington, F; Kreiswirth, B; Bangsberg, DR; Ciccarone, D; Diep, BA; et al. Origins of community strains of methicillin-resistant Staphylococcus aureus. *Clin Infect Dis.*, 2004, 39, 47-54.

[33] Hsu, LY; Koh, YL; Chlebicka, NL; Tan, TY; Krishnan, P; Lin, RT; et al. Establishment of ST30 as the predominant clonal type among community-associated methicillin-resistant Staphylococcus aureus isolates in Singapore. *J Clin Microbiol*, 2006, 44, 1090-3.

[34] Ma, XX; Galiana, A; Pedreira, W; Mowszowicz, M; Christophersen, I; Machiavello, S; et al. Community-acquired methicillin-resistant Staphylococcus aureus, Uruguay. *Emerg Infect Dis.*, 2005, 11, 973- 6.

[35] Mulvey, MR; MacDougall, L; Cholin, B; Horsman, G; Fidyk, M; Woods, S; Community-associated methicillin-resistant Staphylococcus aureus, *Canada. Emerg Infect Dis,*. 2005, 11, 844-50.

[36] Chen, FJ; Lauderdale, TL; Huang, IW; Lo, HJ; Lai, JF; Wang, HY; et al. Methicillin-resistant Staphylococcus aureus in Taiwan. *Emerg Infect Dis.*, 2005, 11, 1760-3.

[37] Coombs, GW; Pearson, JC; O'Brien, FG; Murray, RJ; Grubb, WB; Christiansen, KJ. Methicillin-resistant Staphylococcus aureus clones, Western Australia. *Emerg Infect Dis.*, 2006, 12, 241-7.

[38] Fey, PD; Said-Salim, B; Rupp, ME; Hinrichs, SH; Boxrud, DJ; Davis, CC; et al. Comparative molecular analysis of community- or hospital-acquired methicillin-resistant Staphylococcus aureus. *Antimicrob Agents Chemother*, 2003, 47, 196-203.

[39] Daum, RS; Ito, T; Hiramatsu, K; Hussain, F; Mongkolrattanothai, K; Jamklang, M; et al. A novel methicillin-resistance cassette in community-acquired methicillin-resistant Staphylococcus aureus isolates of diverse genetic backgrounds. *J Infect Dis.*, 2002, 186, 1344-7.

[40] Okuma, K; Iwakawa, K; Turnidge, JD; Grubb, WB; Bell, JM; O'Brien, FG; et al. Dissemination of new methicillin-resistant Staphylococcus aureus clones in the community. *J Clin Microbiol.*, 2002, 40, 4289-94.

[41] Bratu, S; Eramo, A; Kopec, R; Coughlin, E; Ghitan, M; Yost, R; et al. Community-associated methicillin-resistant Staphylococcus aureus in hospital nursery and maternity units. *Emerg Infect Dis*., 2005, 11, 808-13.

[42] Regev-Yochay, G; Rubinstein, E; Barzilai, A; Carmeli, Y; Kuint, J; Etienne, J; et al. Methicillin-resistant Staphylococcus aureus in neonatal intensive care unit. *Emerg Infect Dis*., 2005, 11, 453-6.

[43] Diederen, BM; Kluytmans, JA. The emergence of infections with community-associated methicillin resistant Staphylococcus aureus. *J Infect*., 2006, 52, 157-68.

[44] Kluytmans-Vandenbergh, MF; Kluytmans, JA. Community-acquired methicillin-resistant Staphylococcus aureus: current perspectives. *Clin Microbiol Infect*., 2006, 12, S9-S15.

[45] Deresinski, S Methicillin-resistant Staphylococcus aureus: an evolutionary, epidemiologic, and therapeutic odyssey. *Clin Infect Dis*., 2005, 40, 562-73.

[46] Seybold, U; Kourbatova, EV; Johnson, JG; Halvosa, SJ; Wang, YF; King, MD; et al. Emergence of community-associated methicillinresistant Staphylococcus aureus USA300 genotype as a major cause of health care-associated blood stream infections. *Clin Infect Dis*., 2006, 42, 647-56.

[47] Enright, MC; Day, NP; Davies, CE; Peacock, SJ; Spratt, BG. Multilocus sequence typing for characterization of methicillin-resistant and methicillin-susceptible clones of Staphylococcus aureus. *J Clin Microbiol*, 2000, 38, 1008-15.

[48] Enright, MC; Robinson, DA; Randle, G; Feil, EJ; Grundmann, H; Spratt, BG. The evolutionary history of methicillin-resistant Staphylococcus aureus (MRSA). *Proc Natl Acad Sci USA*., 2002, 99, 7687-92.

[49] Baba, T; Takeuchi, F; Kuroda, M; Yuzawa, H; Aoki, K; Oguchi, A; et al. Genome and virulence determinants of high virulence community-acquired MRSA. *Lancet*, 2002, 359, 1819-27.

[50] Vandenesch, F; Naimi, T; Enright, MC; Lina, G; Nimmo, GR; Heffernan, H; et al. Community-acquired methicillin-resistant Staphylococcus aureus carrying Panton-Valentine leukocidin genes: worldwide emergence. *Emerg Infect Dis*., 2003, 9, 978-84.

[51] Gillet, Y; Issartel, B; Vanhems, P; Fournet, JC; Lina, G; Bes, M; et al. Association between Staphylococcus aureus strains carrying gene for Panton-Valentine leukocidin and highly lethal necrotising pneumonia in young immunocompetent patients. *Lancet*, 2002, 359, 753-759.

[52] Lina, G; Piemont, Y; Godail-Gamot, F; Bes, M; Peter, MO; Gauduchon, V; et al. Involvement of Panton-Valentine leukocidin-producing Staphylococcus aureus in primary skin infections and pneumonia. *Clin Infect Dis*., 1999, 29, 1128-32.

[53] Menestrina, G; Dalla Serra, M; Comai, M; Coraiola, M; Viero, G; Werner, S; et al. Ion channels and bacterial infection: the case of b-barrel pore-forming protein toxins of Staphylococcus aureus. *FEBS Lett*., 2003, 552, 54–60.

[54] Prevost, G; Cribier, B; Couppie, P; Petiau, P; Supersac, G; Finck-Barbacon, V; et al. Panton-Valentine leucocidin and gamma-hemolysin from Staphylococcus aureus ATCC 49775 are encoded by distinct genetic loci and have different biological activities. *Infect Immun*, 1995, 63, 4121–9.

[55] Szmigielski, S; Prevost, G; Monteil, H; Colin, DA; Jeljaszewicz, J. Leukocidal toxins of staphylococci. *Zentralbl Bakteriol*, 1999, 289, 185–201.

[56] Miles, G; Movileanu, L; Bayley, H. Subunit composition of a bicomponent toxin: staphylococcal leukocidin forms an octameric transmembrane pore. *Protein Sci.*, 2002, 11, 894–902.

[57] Colin, DA; Mazurier, I; Sire, S; Finck-Barbancon, V. Interaction of the two components of leukocidin from Staphylococcus aureus with human polymorphonuclear leukocyte membranes: sequential binding and subsequent activation. *Infect Immun*, 1994, 62, 3184–8.

[58] Genestier, AL; Michallet, MC; Prevost, G; Bellot, G; Chalabreysse, L; Peyrol, S; et al. Staphylococcus aureus Panton-Valentine leukocidin directly targets mitochondria and induces Bax-independent apoptosis of human neutrophils. *J Clin Invest.*, 2005, 115, 3117–27.

[59] Said-Salim, B; Mathema, B; Braughton, K; Davis, S; Sinsimer, D; Eisner, W; et al. Differential distribution and expression of Panton-Valentine leucocidin among community-acquired methicillin-resistant Staphylococcus aureus strains. *J Clin Microbiol*, 2005, 43, 3373–9.

[60] Voyich, JM; Otto, M; Mathema, B; Braughton, KR; Whitney, AR; Welty, D; et al. Is Panton-Valentine leukocidin the major virulence determinant in community-associated methicillin-resistant Staphylococcus aureus disease? *J Infect Dis.*, 2006, 194, 1761–70.

[61] Wang, R; Braughton, KR; Kretschmer, D; Bach, TH; Queck, SY; Li, M; et al. Identification of novel cytolytic peptides as key virulence determinants for community-associated MRSA. *Nat Med.*, 2007, 13, 1510–4.

[62] Baba, T; Takeuchi, F; Kuroda, M; Yuzawa, H; Aoki, K; Oguchi, A; et al. Genome and virulence determinants of high virulence community-acquired MRSA. *Lancet*, 2002, 359, 1819–27.

[63] Diep, BA; Gill, SR; Chang, RF; Phan, TH; Chen, JH; Davidson, MG; et al. Complete genome sequence of USA300, an epidemic clone of community-acquired meticillin-resistant Staphylococcus aureus. *Lancet*, 2006, 367, 731-739.

[64] Eady, EA; Cove, JH. Staphylococcal resistance revisited: community-acquired methicillin resistant Staphylococcus aureus--an emerging problem for the management of skin and soft tissue infections. *Curr Opin Infect Dis.*, 2003, 16, 103-124.

[65] King, MD; Humphrey, BJ; Wang, YF; Kourbatova, EV; Ray, SM; Blumberg, HM. Emergence of community-acquired methicillin-resistant Staphylococcus aureus USA 300 clone as the predominant cause of skin and soft-tissue infections. *Ann Intern Med.*, 2006, 144, 309-317.

[66] Faden, H; Ferguson, S. Community-acquired methicillin-resistant Staphylococcus aureus and intrafamily spread of pustular disease[letter]. *Pediatr Infect Dis J.*, 2001, 20, 554-5.

[67] Hageman, JC; Uyeki, TM; Francis, JS; Jernigan, DB; Wheeler, JG; Bridge, CB; et al. Severe community-acquired pneumonia due to Staphylococcus aureus, 2003-04 influenza season. *Emerg Infect Dis.*, 2006, 12, 894-9.

[68] Gillet, Y; Issartel, B; Vanhems, P; Fournet, JC; Lina, G; Bes, M; et al. Association between Staphylococcus aureus strains carrying gene for Panton-Valentine leukocidin and highly lethal necrotising pneumonia in young immunocompetent patients. *Lancet*, 2002, 359, 753-9.

6-3. Comparison of CA-MRSA and HA-MRSA

Differences between CA-MRSA and HA-MRSA (adapted from references 1–11).

Characteristic	CA-MRSA	HA-MRSA
At-risk populations	Children, athletes, prisoners, soldiers, selected ethnic populations (native American, Alaskan natives, Pacific islanders), intravenous drug users, men who have sex with men	Outbreaks are healthcare associated
Clinical infections	Skin and soft tissue infections (furuncles, skin abscesses); postinfluenza necrotizing pneumonia	Pneumonia, urinary tract infections, catheter-related or bloodstream infections, surgical site infections
Underlying conditions	None	Healthcare-associated risk factors*
Age group	Younger	Older
Race/ethnicity	Non-white	White
Antimicrobial resistance		
Chloramphenicol	Usually susceptible	Frequently resistant
Clindamycin	Usually susceptible	Frequently resistant
Erythromycin	Usually resistant	Usually resistant
Fluoroquinolone	Variable	Usually resistant
TMP-SMZ	Usually susceptible	Usually susceptible
SCC*mec* type	SCCmec type IV, V	SCCmec type I, II, III
Toxin production	More	Less
PVL gene	Frequent	Rare

SCC; staphylococcal chromosome cassette, PVL; Panton-Valentine leukocidin, TMP-SMZ; trimethoprim-sulfamethoxazole. * Isolation of MRSA ≥2 days post-hospitalization, history of hospitalization, surgery, dialysis, residence in long-term care facility, permanent indwelling catheter or percutaneous device, previous MRSA isolation.

References

[1] Diederen, BM; Kluytmans, JA. The emergence of infections with community-associated methicillin resistant Staphylococcus aureus. *J Infect*, 2006, 52, 157-68.

[2] Kowalski, TJ; Berbari, EF; Osmon, DR. Epidemiology, treatment, and prevention of community-acquired methicillin-resistant Staphylococcus aureus infections. *Mayo Clin Proc.*, 2005, 80, 1201-1207.

[3] Chavez, TT; Decker, DF. Health care-associated MRSA versus community-associated MRSA. *Dis Mon.*, 2008, 54, 763-8.

[4] Riley, TV; Rouse, IL. Methicillin-resistant Staphylococcus aureus in Western Australia, 1983-1992. *J Hosp Infect.*, 1995, 29, 177-88.
[5] Adcock, PM; Pastor, P; Medley, F; Patterson, JE; Murphy, TV. Methicillin-resistant Staphylococcus in two childcare centers. *J Infect Dis.*, 1998, 178, 577-80.
[6] Herold, BC; Immergluck, LC; Maranan, MC; Lauderdale, DS; Gaskin, RE; Boyle-Vavra S, et al. Community-acquired methicillin-resistant Staphylococcus aureus in children with no identified predisposing risk. *JAMA*, 1998, 279, 593-8.
[7] Frank, AL; Marcinak, JF; Mangat, PD; Schreckenberger, PC. Community-acquired and clindamycin-susceptible methicillin-resistant Staphylococcus aureus in children. *Pediatr Infect Dis J.*, 1999, 18, 993-1000.
[8] Naimi, TS; LeDell, KH; Como-Sabetti, K; Borchardt, SM; Boxrud, DJ; Etienne, J; et al. Comparison of community- and health care-associated methicillin-resistant Staphylococcus aureus infection. *JAMA*, 2003, 290, 2976-84.
[9] Vandenesch, F; Naimi, T; Enright, MC; Lina, G; Nimmo, GR; Heffernan, H; et al. Community-acquired methicillin-resistant Staphylococcus aureus carrying Panton-Valentine leukocidin genes: worldwide emergence. *Emerg Infect Dis.*, 2003, 9, 978-84.
[10]Moreillon, P; Que, YA; Glauser, MP. Staphylococcus aureus (including staphylococcal toxic shock). In: Mandell, GL; Bennett, JE; Dolin, R; editors. *Mandell, Couglas and Bennett's Principles and Practice of Infectious Disease*. Vol 2. 6th ed. Philadelphia, Pa: Elsevier; 2005. p.2321-51.
[11]Center for Disease Control and Prevention. Community-associated methicillin-resistant Staphylococcus aureus infections in Pacific Islanders-Hawaii, 2001-2003. *MMWR Morb Mortal Wkly Rep.*, 2004, 53, 767-70.

7. Antimicrobial Agents in Treatment of MRSA Infection

In the past 50 years, vancomycin has been preferred for the treatment of MRSA infection because the material is more stable than other antibiotics, resistance acquisition is low, and no other economical substitute has shown an equivalent therapeutic effect. However, vancomycin also has several disadvantages, including poor tissue penetration and the fact that it is difficult to attain therapeutic drug levels in the lung or cerebrospinal fluid (CSF) [1-3]. Moreover, because the drug is of high molecular weight and has low bioavailability, vancomycin cannot be administered orally and intravenous administration is required. Therefore, economic and time burdens have been associated with vancomycin use because each patient prescribed the drug has to be admitted to hospital [4]. Vancomycin also causes anaphylactic reactions such as hypotension, dyspnea, and pruritus. Serum drug levels must be monitored in some patients to prevent complications such as pseudomembranous colitis, deafness, vestibular disorder, neutropenia, and vasculitis.

Recently, clinical failure of vancomycin treatment has been frequently reported and vancomycin resistance has developed in MRSA; the phenomenon is termed "vancomycin MIC creep", which means that vancomycin MIC levels increase in antibiotic sensitivity tests, and is thought to explain treatment failure [5]. With MIC creep, even if a strain is susceptible to a high drug level clinical failure will eventually result [6]. Strains such as vancomycin-resistant *S. aureus* (VRSA), glycopeptide-intermediate-resistant *S. aureus* (GISA), and heterogeneous glycopeptide-intermediately-resistant *S. aureus* (hGISA) are known to be insensitive to vancomycin. Such strains are of little clinical significance to date, but they are increasing in number annually according to recent reports [7]. It is assumed that VRSA arose by acquisition of the *vanA* resistance operon from vancomycin-resistant *Enterococci* (VRE) when *S. aureus* combined with VRE [8]. It is known that the incidence of GISA infections increase after long-term vancomycin treatment. The thickened cell-wall characteristics of such strains seems to explain the resistance [9]. If hGISA infection is treated with glycopeptide, the glycopeptide-susceptible population that was dominant is killed by the antibiotic, and the glycopeptide-resistant subpopulation becomes the major source of infection. Thus, patients with hGISA infection have poorer prognosis compared with those with MRSA [10]. When the problems caused by such resistant strains are considered, glycopeptide may no longer be the treatment of choice for resistant strains. It is critical to develop more effective and versatile antibiotics to treat VRSA, GISA, and hGISA. However, the speed of resistance acquisition is much greater than the speed at which new antibiotic agents can be developed. The newly developed antibiotics have mechanisms of action distinct from those of current antibiotics, such as inhibition of protein synthesis, formation of DNA conjugates, inhibition of membrane synthesis, formation of protein antibodies, or reconstruction of existing antibiotic actions. Thus, present antibiotic-resistant strains are susceptible to the newer compounds.

In this chapter, we will review linezolid, daptomycin, and tigecycline; these are drugs approved by the FDA in 2001–2003 for the treatment of MRSA infection. We will also examine recently approved drugs or that are currently undergoing clinical trials to combat MRSA infection.

Linezolid

- Origin: Oxazolidinone.
- mechanism: Bacteriostatic; by conjugation with the 23S subunit of the 50S ribosome, which eventually blocks 70S ribosome complex formation and inhibits protein synthesis initiation.
- Coverage: Bacteria.

  G(+): Bacteriostatic action on MSSA, MRSA, CoNS, *Enterococci* (also VRE strains), *Streptococci* (also penicillin-resistant and multidrug-resistant *Pneumococci*).

  G(-): Affected some Gram-negative bacteria in an *in vitro* test, but is thought to be generally ineffective due to the efflux pump system common to such bacteria [11].

  anaerobes: Effective against *Clostridium difficile, Clostridium perfringens, Bacteroides,* and *Fuseobacterium.*
- Clinical use and characteristics: Used to treat pneumonia caused by Gram-positive bacteria and complicated and uncomplicated skin and soft-tissue infections (SSTI) . Although the antibiotic is not approved for treatment of MRSA bacteremia, some reports have described successful results when the drug was used alone or in combination with rifampicin or fusidic acid [12, 13]. The antibiotic is used to treat central nervous system (CNS) infections caused by CoNS or other Gram-positive organisms as the drug has excellent tissue and Cerebrospinal fluid (CSF) infiltration ability. Linezolid is indicated for the treatment of MRSA endocarditis, but both

failure and success have been reported [14-17]. Linezolid is also useful in combination with clindamycin for the treatment of toxic shock, necrotizing fascitis, and necrotizing pneumonia caused by Gram-positive strains due to reduced synthesis of PVL, α-hemolysin, and TSS toxin [18].

- Pharmacodynamics and pharmacokinetics: 100% bioavailability when administered at the recommended dose of 600 mg twice daily. The serum peak level is attained 1–2 hours after oral administration and the half-life is 4.26–5.4 hours. The drug is readily distributed in tissues, and only 31% binds to protein [20, 21]. The antibiotic is usually excreted by the kidney, but can be eliminated by a nonrenal route. Thus, dose adjustment is not needed even in patients with renal or hepatic impairment.
- Adverse effects: When the drug is used chronically (for more than 14 days), reversible bone marrow (BM) suppression can develop [20, 23]. Thus, regular complete blood cell count (CBC) monitoring is necessary. If BM suppression develops, the drug must be discontinued. If the antibiotic is used for more than 28 days, peripheral/optic neuritis may occur. Lactic acidosis is seen at very low incidences. Linezolid has a reversible and non-selective monoamine oxidase inhibitor action, so serotonin syndrome can develop [20, 23-25]. According to reports by Pfizer and the United States FDA in 2007, linezolid treatment of critically ill patients resulted in higher mortality rates compared with treatment with vancomycin, oxacillin, or dicloxacillin in patients with intravascular catheter-related bloodstream infections, catheter site infections caused by Gram-negative bacteria alone or from a combination of Gram-negative/Gram-positive strains. Therefore, linezolid is not recommended for the treatment of catheter-related bloodstream infections, catheter site infections, or Gram-negative bacterial infections [26]. Linezolid alone is also not recommended for treatment of mixed infections with MRSA and Gram-negative bacteria.

Daptomycin

- Origin: A semisynthetic cyclic lipopeptide originating from a fermentation product of *Streptomyces roseoporus*. Composed of a 13-amino-acid cyclic lipopeptide with a decanoyl side chain (a lipophilic tail).
- Mechanism: The lipophilic decanoyl side chain irreversibly binds to the bacterial cell wall and is inserted in a calcium-dependent manner, but not as far as the cytoplasm. Thus, rapid potassium efflux results from destruction of the cell membrane; depolarization eventually occurs and synthesis of macromolecules such as RNA, DNA, and proteins is inhibited. The drug shows rapid bactericidal activity [29].
- Coverage: Bacteria.
  G(+): Effective only on Gram-positive bacteria such as MRSA, MSSA, and VRE, but also kills linezolid- and quinupristin/dalfopristin-resistant VRE [29-32].
- Clinical use and characteristics: Used mainly used to treat SSTI complications [33-35]. Use in patients with bacteremia and endocarditis is also approved [36]. Effectiveness in pneumonia patients is limited because the drug is inhibited by pulmonary surfactant [37].

- Pharmacokinetics and pharmacodynamics: The half-life is 7–9 hours [38]. The updated recommended dosage is 6 mg/kg once-daily in patients with bacteremia or endovascular infection, and 4 mg/kg once-daily in those with SSTI complications.
- Adverse effects: Myopathy occurs in a dose-dependent manner and increases creatinine kinase levels or abnormalities in hepatic transaminase function [34, 39]. Dose adjustment is needed in patients with renal failure.
- Resistance: The precise resistance mechanism is unknown. However, it is assumed that a thickened cell wall inhibits drug entry, as indicated by the increase in daptomycin MIC seen in patients previously prescribed vancomycin [40-42].

Tigecycline

- Origin: A new antibiotic derived from minocycline 9-*t* butylglycylamido.
- Mechanism: Inhibition of protein synthesis because the minocycline side chain combines effectively with the 30S ribosomal unit, resulting in bacteriostasis.
- Coverage: A very broad-spectrum antibiotic.

  G(+): Combats MRSA, penicillin-resistant *Pneumococci*, vancomycin-susceptible/resistant *Enterococci*, □-hemolytic *Streptococci*, *Listeria monocytogenes*, and tetracycline-resistant MRSA.

  G(-): Kills *Hemophilus influenzae, Neisseria* spp., *Acinetobacter* spp., and *Enterobacteriacae* spp. [43-45].

  Effective *in vitro* on the following bacteria: atypical pathogens, *Mycobacterium chelonae, Mycobacterium abscessus, Mycobacterium fortuitum* group, *Mycobacterium marinum, Mycoplasma* spp., and *Ureaplasma urealyticum* [46].
- Clinical use and characteristics: Indicated for the treatment of complicated SSRI, multidrug- resistant abdominal infections, and is currently being considered for the treatment of hospital-acquired/ventilator-associated pneumonia [47].
- Pharmacokinetics and pharmacodynamics: More than 68% of the drug binds to proteins and the distribution volume is large (10 L/kg). The half-life is 36 hours. Less than 15% of unmetabolized drug is excreted in the urine. The recommended dose is 100 mg and the maintenance dosage is 50 mg, both twice-daily by intravenous administration.
- Adverse effects: No major complication has been reported, although dose-dependent gastrointestinal effects such as nausea and vomiting have been documented [44].
- Resistance: No MRSA resistance has been reported. Resistance has been seen in *Pseudomonas aeruginosa* and *Proteus* spp., caused by expression of efflux pumps encoded by the *mexAB-oprM* and *acrAB* genes, respectively [48-50].

Quinupristin–Dalfopristin

- Origin: A streptogramin combination of quinupristin and dalfopristin (both are derivatives of semisynthetic pristinamycin) at a weight ratio of 30:70.
- Mechanism: Inhibition of protein synthesis by the 50S ribosomal subunit.
- Coverage: Bacteria.

G(+): Bactericidal effect on MRSA, MSSA, and *Streptococcus pyogenes,* and a bacteriostatic effect on vancomycin-resistant *Enterococcus faecium* (VREF) [24,25].

- Clinical use and characteristics: Used to treat severe Gram-positive infections, hospital-acquired pneumonia, SSTI, and VREF. Sometimes used to treat VREF bacteremia, but must be used with care in endocarditis patients.
- Pharmacokinetics and pharmacodynamics: The half-life is 0.7–0.85 hours [51]. Owing to low bioavailability, the drug must be administered into a central vein [21, 25]. The antibiotic is excreted mainly into the bile, but clearance may decrease in patients with renal insufficiency. Interacts with other drugs by inhibiting cytochrome P450 3A4.
- Adverse effects: Hepatotoxicity, hyperbilirubinemia, myalgia, and arthralgia have been reported, and thrombophlebitis has been reported in 75% of treated patients [21, 23, 25]. When combined with other drugs, quinupristin–dalfopristin may induce QTc prolongation; electrocardiogram monitoring is thus needed.
- Resistance: The SNTRY antimicrobial surveillance program showed 10% resistance of *E. faecium* in Europe and 0.6% in North America.

Ceftobiprole

- Origin: Pyrrolidinone cephalosporin.
- Mechanism: Conventional β-lactam antibiotics are irreversibly conjugated with PBPs by covalent acylation, but do not bind to PBP2a. However, the new cephalosporin (ceftobiprole) has high affinity for PBP2a through the vinylpyrrolidone moiety at position 3 [53-38]. The drug also has a high affinity for PBP2x, which is associated with β-lactam-resistance of *Pneumococci* [58,59]. The antibiotic is bactericidal in character.
- Coverage: Bacteria. The drug has a very broad-spectrum activity [56, 60-65].

  G(+): Excellent effect on MRSA, VISA, penicillin-resistant *Pneumococci, Enterococcus faecalis* (vancomycin-susceptible/resistant). However, shows little effect on *E. faecium.*

  G(-): Effective against *H. influenzae, Neisseria gonorrhoeae, Moraxella catarrhalis,* extended-spectrum β-lactamase-negative *Enterobacteriaceae, P. aeruginosa,* and *Acinetobacter baumanii.*

  Anaerobes: Has some effect on *Actinomyces, Clostridium, Fusibacterium,* and *Veillonella* spp., but little effect on Gram-negative anaerobes.
- Clinical use and characteristics: Effective in the treatment of healthcare-associated pneumonia and complicated SSTI [66-68].
- Pharmacokinetics and pharmacodynamics: Administered intravenously. The half-life is about 3–4 hours. The recommended dose in patients with Gram-positive infections is a twice-daily 500 mg infusion over 1 hour. In cases of polymicrobial infection, 500-mg infusion over 2 hours, three times a day is suggested.
- Adverse effects: Nausea and taste disturbances are the most common side effects reported to date.

- Resistance: No resistant strain has been isolated to date. However, ceftobiprole is not stable in the presence of metallo-β-lactamase and is degraded by expanded spectrum β-lactamase (ESBL)-positive organisms.

Ceftaroline

- Origin: Novel cephalosporin.
- Mechanism: High affinity to and selectivity for PBP2s.
- Coverage: Bacteria.
  G(+): Combats MRSA, VRS, *Streptococci*, *Enterobacteriaceae*, and *E. fecalis*.
- Clinical use and characteristics: Studies on treatment of complicated superficial surgical wound infection (SSSI) [69,70] and community-acquired pneumonia [71] are ongoing. An animal endocarditis study has been successfully concluded [72]. The drug must be administered intravenously due to low bioavailability [21].
- Pharmacokinetics and pharmacodynamics: As the drug has a short half-life of 0.6–2.4 hours, the antibiotic must be frequently infused [21, 72]. There is as yet no confirmed recommended dose guideline.
- Resistance: It has been reported that ceftaroline is degraded by ESBLs, class C β-lactamases, and class B metallo-β-lactamase. However, a combination of the drug with clavulanate can inhibit such processes [73].

Dalbavancin

- Origin: Second-generation glycopeptide made by modifying the functional groups of a teicoplanin-like glycopeptide and the sugar moieties.
- Mechanism: Inhibition of bacterial cell-wall formation.
- Coverage: Bacteria. The drug has a similar activity spectrum to that of other glycopeptides, including teicoplanin.
  G(+): Combats MSSA, MRSA, CoNS, *Enterococci* (excluding *vanA*-positive strains), GISA, linezolid-resistant *S. aureus*, penicillin-resistant/susceptible *Pneumococci*, and *Streptococcus viridans* [74,75].
  G(-): Almost completely ineffective [75].
  Anaerobes: Although the drug is known to affect Gram-positive anaerobes such as *Clostridium* spp., the antibiotic has no effect on *Clostridium clostridioforme*.
- Clinical use and characteristics: Ongoing studies are investigating effectiveness in catheter-related bloodstream infections [77], in complicated SSSI cases [76], and in infections caused by Gram-positive organisms.
- Pharmacokinetics and pharmacodynamics: Drug half-life is 123–210 hours. Due to this very long half-life, the drug can administered weekly. The antibiotic must be administered intravenously due to low bioavailability. The drug has a broad distribution volume (10 L/kg), shows a high protein-binding level (>95%), and slow systemic clearance (0.05 L/hour). About 33% of the drug is excreted unchanged in the urine [74, 78].

- Adverse effects: The only reports are of slight headache and gastrointestinal-tract symptoms [70]. Hypokalemia and hypotension, although rare, have also been reported [77].

Oritavancin

- Origin: Made by replacing the vancosamine of the vancomycin disaccharide moiety with epivancosamine [25, 79].
- Mechanism: Has a rapid bactericidal effect that involves destruction of the transmembrane potential [24].
- Coverage: Bacteria [79-84].
  G(+): Combats MRSA, glycopeptide-resistant *Enterococci* (including *vanA*-positive strains), *Streptococci*, *Peptostreptococcus* spp., *Propionibacterium* spp., *Clostridium perfringens*, and *Corynebacterium jeikeium*. Also effective against penicillin-intermediate-resistant/resistant *Streptococcus pneumoniae*, methicillin-resistant coagulase-positive *Staphylococci*, and methicillin-resistant CoNS.
  G(-) and anaerobes: No effect.
- Clinical use and characteristics: A study in treatment of complicated superficial surgical wound infection (cSSSI) [84,85] is in progress and a single report of efficacy in *S. aureus* bacteremia treatment has been published [86].
- Pharmacokinetics and pharmacodynamics: The drug has a long half-life of nearly 100 hours [82]. The antibiotic can be easily prescribed in the outpatient clinic because the drug is administered once daily or on alternate days.
- Adverse effects: Only rash and phlebitis have been reported to date.
- Resistance: None reported.

Telavancin

- Origin: A semisynthetic analog of vancomycin with a hydrophobic decylaminoethyl side- chain on vancosamine sugar and a hydrophilic phosphonomethylaminomethyl group at the 4' position of amino acid 7.
- Mechanism: The drug blocks the transglycosylation and transpeptidation steps in peptidoglycan-chain formation, in a manner similar to that of vancomycin. Also, the antibiotic acts directly to depolarize the cell membrane, thus affecting permeability [21, 24, 79, 87].
- Coverage: Bacteria [24, 79, 87-91].
  G(+): The drug combats *Streptococci* (including penicillin-resistant/sensitive *Streptococcus pneumoniae*) and is also more effective against MRSA, MSSA, and GISA compared with vancomycin. Moreover, the antibiotic is effective against *Enterococci*, including *vanA*-positive organisms.
  G(-): No known effect.
- Clinical use and characteristics: A study on drug use in cases of cSSSI caused by Gram-positive organisms is in progress.

- Pharmacokinetics and pharmacodynamics: The known half-life is 7–11 hours [21, 91].
  Due to a low bioavailability when administered orally, the drug is administered intravenously. The recommended dosage is 10 mg/kg once-daily.
- Adverse effects: The reported side effects are nausea, taste disturbance, insomnia, headache, and a QTc prolongation time of 12.5 ms; the clinical importance of these features is under study [88].

Iclaprim

- Origin: This drug is a diaminopyrimidine dihydrofolate reductase inhibitor, similar to trimethoprim.
- Mechanism: The antibiotic selectively inhibits bacterial dihydrofolate reductase [92-94].
- Coverage: Bacteria. The antibiotic has broad-spectrum activity.
  G(+): The drug combats *S. aureus*; not only MRSA, MSSA, and VRSA, but also strains resistant to macrolides, quinolones, trimethoprim, and other drugs. The antibiotic has varying effects on *Streptococci*.
  G(-): The antibiotic combats *Enterobacter* spp., *Neisseria* spp., *H. influenzae, M. catarrhalis,* and *Chlamydia pneumoniae* [92-95]. The drug is not effective against *Pseudomonas*.
- Clinical use and characteristics: Use in cSSSI and HAP (hospital-acquired pneumonia) patients is being studied.
- Pharmacokinetics and pharmacodynamics: The drug can be administered either orally or intravenously because the bioavailability is 40%. The half-life is about 4 hours and 70% of the unmetabolized drug is excreted by the kidney [95].
- Adverse effects: Nausea, headache, constipation, and diarrhea have been reported.

Ranbezolid

This drug is a new oxazolidinone and may be useful for the treatment of MRSA and penicillin-resistant pneumococcal infections. The antibiotic is bacteriostatic against MRSA, MSSA, *Enterococci, M. catarrhalis, Pneumococci*, and Gram-negative and Gram-positive anaerobes [79, 96-98]. The drug may inhibit biofilm production by MRSA, MSSA, and CoNS [99].

Tomopenem

This new 1-β-methylcarbapenem is under development. Addition of a guanidine-pyrrolidine side chain affords high affinity for PBP1, PBP2, and PBP4 of *S. aureus* [100]. Unlike existing carbapenems, the drug is active against MRSA, but also kills MSSA, ESBL-producing *E. coli, Klebsiella* spp., and *P. aeruginosa* [100, 101].

Lysostaphin

This drug is active against zinc metalloenzymes, and may therefore be selective for *S. aureus*. The bacterium has three zinc metalloenzymes, an endopeptidase, endo-ß-N-acetyl glucosamidase, and N-acetyl-muramyl-L-alanine amidase. Among these, the glycylglycine endopeptidase acts on the glycine–glycine bonds specific to the *S. aureus* cell membrane. It is thus expected that the drug will be valuable in the treatment of multidrug-resistant *S. aureus* because all strains have glycine–glycine bonds [102].

In addition, many other drugs are now in clinical trials or under development. Research on new medicines is important but treatment with currently available medications and methods remains critical. It is essential to seek to inhibit resistant strains with the right combination of existing medicines and infection control using mixtures of currently available medications, and appropriate surgical intervention (to reduce bacterial load by incision and drainage). These interventions remain important and appropriate to prevent development of strains resistant to new drugs.

References

[1] Deresinski, S. Counterpoint: vancomycin and Staphylococcus aureus-an antibiotic enters obsolescence. *Clin Infect Dis*., 2007, 44, 1543-8.

[2] Scheetz, MH; Wunderink, RG; Postelnick, MJ; Noskin, GA. Potential impact of vancomycin pulmonary distribution on treatment outcomes in patients with methicillin-resistant Staphylococcus aureus pneumonia. *Pharmacotherapy*, 2006, 26, 539-50.

[3] Ricard, JD; Wolff, M; Lacherade, JC; Mourvillier, B; Hidri, N; Barnaud, G; et al. Levels of vancomycin in cerebrospinal fluid of adult patients receiving adjunctive corticosteroids to treat pneumococcal observational study. *Clin Infect Dis*., 2007, 44, 250-255.

[4] Santos, F; Mankarious, LA; Eavey, RD. Methicillin-resistant Staphylococcus aureus: pediatric otitis. *Arch Otolaryngol Head Neck Surg*., 2000, 126, 1383-5.

[5] Steinkraus, G; White, R; Friedrich, L. Vancomycin MIC creep in non-vancomycin-intermediated Staphylococcus aureus (VISA), vancomycin-susceptibile clinical methicillin-resistant S. aureus (MRSA) blood isolates from 2001-05. *J Antimicrob Chemother*, 2007, 60, 788-94.

[6] Hidayat, LK; Hsu, DI; Quist, R; Chriner, KA; Wong-Beringer, A. High-dose vancomycin therapy for methicillin-resistant Staphylococcus aureus infections: efficacy and toxicity. *Arch Intern Med*., 2006, 166, 2138-44.

[7] Appelbaum, PC. Reduced glycopeptide susceptibility in methicillin-resistant Staphylococcus aureus (MRSA). *Int J Antimicrob Agents*, 2007, 30, 398-408.

[8] Weigel, LM; Clewell, DB; Gill, SR; Clark, NC; McDougal, LK; Flannagan, SE; et al. Genetic analysis of a high-level vancomycin-resistant isolates of Staphylococcus aureus. *Science*, 2003, 302, 1569-71.

[9] Naimi, TS; Anderson, D; O'Boyle, C; Boxrud, DJ; Johnson, SK; Tenover, FC; et al. Vancomycin-intermediate Staphylococcus aureus with phenotypic susceptibility to methicillin in a patient with recurrent bacteremia. *Clin Infect Dis*., 2003, 36, 1609-12.

[10] Neoh, HM; Hori, S; Komatsu, M; Oguri, T; Takeuchi, F; Cui, L; et al. Impact of reduced vancomycin susceptibility on the therapeutic outcome of MRSA bloodstream infections. *Ann Clin Microbiol Antimicrob*, 2007, 6, 13.

[11] Livermore, DM. Linezolid in vitro: mechanism and antibacterial spectrum. *J Antimicrob Chemother*, 2003, 51,S9-16.

[12] Huang, YT; Hsiao, CH; Liao, CH; LeeCW, Hsueh, PR. Bacteremia and infective endocarditis caused by a non-daptomycin-susceptible, vancomycin-intermediate, and methicillin-resistant Staphylococcus aureus strain in Taiwan. *J Clin Microbiol*, 2008, 46, 1132-6.

[13] Schwalm, JD; El-Helou, P; Lee, CH. Clinical outcome with oral linezolid and rifampin following recurrent methicillin-resistant Staphylococcus aureus bacteremia despite prolonged vancomycin treatment. *Can J Infect Dis*., 2004, 15, 97-100.

[14] Chesi, G; Colli, A; Mestres, CA; Gambarati, G; Boni, F; Gherli, T. Multiresistant-MRSA tricuspid valve infective endocarditis with ancient osteomyelitis locus. *BMC Infect Dis*., 2006, 6, 124.

[15] Nathani, N; Iles, P; Elliott, TS. Successful treatment of MRSA native valve endocarditis with oral linezolid therapy: a case report. *J Infect*, 2005, 51, e213-5.

[16] Falagas, ME; Manta, KG; Ntziora, F; Vardakas, KZ. Linezolid for the treatment of patients with endocarditis: a systematic review of the published evidence. *J Antimicrob Chemother*, 2006, 58, 273–80.

[17] Munoz, P; Rodriguez-Creixems, M; Moreno, M; Marin, M; Ramallo, V; Bouza, E. Linezolid therapy for infective endocarditis. *Clin Microbiol Infect*., 2007, 13, 211–5.

[18] Micek, ST. Alternatives to vancomycin for the treatment of methicillin-resistant Staphylococcus aureus infections. *Clin Infect Dis*., 2007, 45, S184–90.

[19] Pfizer. Zyvox (Linezolid) Datasheet (2007).

[20] French, G. Safety and tolerability of linezolid. *J Antimicrob Chemother*, 2003, 51, S45–53.

[21] Pan, A; Lorenzotti, S; Zoncada, A. Registered and investigational drugs for the treatment of methicillin-resistant Staphylococcus aureus infection. *Recent Patents Anti-Infect Drug Disc*., 2008, 3; 10–33.

[22] Macgowan, AP. Pharmacokinetic and pharmacodynamic profile of linezolid in healthy volunteers and patients with Gram-positive infections. *J Antimicrob Chemother*, 2003, 51, S17–25.

[23] Anstead, GM; Quinones-Nazario, G; Lewis, JS. Treatment of infections caused by resistant Staphylococcus aureus. *Methods Mol Biol*., 2007, 391, 227–58.

[24] Lentino, JR; Narita, M; Yu, VL. New antimicrobial agents as therapy for resistant Gram-positive cocci. *Eur J Clin Microbiol Infect Dis*., 2008, 27, 3–15.

[25] Aksoy, DY; Unal, S. New antimicrobial agents for the treatment of Gram-positive bacterial infections. *Clin Microbiol Infect*, 2008, 14, 411–20.

[26] U.S. Food and Drug Administration [homepage on the internet]. Available from: http://www.fda.gov/medwatch/

[27] Woodford, N. Biological counterstrike: antibiotic resistance mechanisms of Gram-positive cocci. *Clin Microbiol Infect*., 2005, 11, S2–21.

[28] 28=88

[29] Tedesco, KL; Rybak, MJ. Daptomycin. *Pharmacotherapy*, 2004, 24, 41–57.

[30] Shah, PM. The need for new therapeutic agents: what is the pipeline? *Clin Microbiol Infect*, 2005, 11, S36–42.

[31] Denis, O; Deplano, A; Nonhoff, C; Hallin, M; De Ryck, R; Vanhoof, R; et al. In vitro activities of ceftobiprole, tigecycline, daptomycin, and 19 other antimicrobials against methicillin-resistant Staphylococcus aureus strains from a national survey of Belgian hospitals. *Antimicrob Agents Chemother*, 2006, 50, 2680–5.

[32] Falagas, ME; Giannopoulou, KP; Ntziora, F; Papagelopoulos, PJ. Daptomycin for treatment of patients with bone and joint infections: a systematic review of the clinical evidence. *Int J Antimicrob Agents*, 2007, 30, 202–9.

[33] Eisenstein, BI. Treatment challenges in the management of complicated skin and soft-tissue infections. *Clin Microbiol Infect*, 2008, 14, S17–25.

[34] Arbeit, RD; Maki, D; Tally, FP; Campanaro, E; Eisenstein, BI. The safety and efficacy of daptomycin for the treatment of complicated skin and skin-structure infections. *Clin Infect Dis*., 2004, 38, 1673–81.

[35] Lipsky, BA; Stoutenburgh, U. Daptomycin for treating infected diabetic foot ulcers: evidence from a randomized, controlled trial comparing daptomycin with vancomycin or semi-synthetic penicillins for complicated skin and skin-structure infections. *J Antimicrob Chemother*, 2005, 55, 240–5.

[36] Falagas, ME; Giannopoulou, KP; Ntziora, F; Vardakas, KZ. Daptomycin for endocarditis and/or bacteraemia: a systematic review of the experimental and clinical evidence. *J Antimicrob Chemother*, 2007, 60, 7–19 .

[37] Silverman, JA; Mortin, LI; Vanpraagh, AD; Li, T; Alder, J. Inhibition of daptomycin by pulmonary surfactant: in vitro modeling and clinical impact. *J Infect Dis*., 2005, 191, 2149–52.

[38] Cubist Pharmaceuticals. Daptomycin (Cubicin) Datasheet (2007).

[39] Fowler, VG; Jr, Boucher, HW; Corey, GR; Abrutyn, E; Karchmer, AW; Rupp, ME; et al. Daptomycin versus standard therapy for bacteremia and endocarditis caused by Staphylococcus aureus. *N Engl J Med*., 2006, 355, 653–65.

[40] Sakoulas, G; Alder, J; Thauvin-Eliopoulos, C; Moellering, RC; Jr, Eliopoulos, GM. Induction of daptomycin heterogeneous susceptibility in Staphylococcus aureus by exposure to vancomycin. Antimicrob. *Agents Chemother*, 2006, 50, 1581–5.

[41] Wootton, M; Macgowan, AP; Walsh, TR. Comparative bactericidal activities of daptomycin and vancomycin against glycopeptide-intermediate Staphylococcus aureus (GISA) and heterogeneous GISA isolates. *Antimicrob Agents Chemother*, 2006, 50, 4195–7.

[42] Cui, L; Tominaga, E; Neoh, HM; Hiramatsu, K. Correlation between reduced daptomycin susceptibility and vancomycin resistance in vancomycin-intermediate Staphylococcus aureus. *Antimicrob Agents Chemother*, 2006, 50, 1079–82.

[43] Livermore, DM. Tigecycline: what is it, and where should it be used? *J Antimicrob Chemother*, 2005, 56, 611–4.

[44] Pankey, GA. Tigecycline. *J Antimicrob Chemother*, 2005, 56, 470–80.

[45] Pankey, GA; Steele, RW. Tigecycline: a single antibiotic for polymicrobial infections. *Pediatr Infect Dis J*., 2007, 26, 77–8.

[46] Al-Tatari, H; Abdel-Haq, N; Chearskul, P; Asmar, B. Antibiotics for treatment of resistant Gram-positive coccal infections. *Indian J Pediatr*, 2006, 73, 323–4.

[47] Maroko, R. Results of Phase 3 study comparing a tigecycline (TGC) regimen with an imipenem/cilastatin (IMI) regimen in treatment of patients (pts) with hospital-acquired pneumonia (HAP). Presented at: 47th *Interscience Conference on Antimicrobial Agents and Chemotherapy,* Chicago, IL, USA, 17–20 September 2007.
[48] Ruzin, A; Keeney, D; Bradford, PA. AcrAB efflux pump plays a role in decreased susceptibility to tigecycline in Morganella morganii. *Antimicrob Agents Chemother*, 2005, 49, 791–3.
[49] Dean, CR; Visalli, MA; Projan, SJ; Sum, PE; Bradford, PA. Efflux-mediated resistance to tigecycline (GAR-936) in Pseudomonas aeruginosa PAO1. *Antimicrob Agents Chemother*, 2003, 47, 972–8.
[50] Hawkey, P; Finch, R. Tigecycline: in-vitro performance as a predictor of clinical efficacy. *Clin Microbiol Infect*, 2007, 13, 354–62.
[51] King Pharmaceuticals. Syndercid Datasheet (2003).
[52] Woodford, N. Biological counterstrike: antibiotic resistance mechanisms of Gram-positive cocci. *Clin Microbiol Infect*, 2005, 11, S2-21.
[53] Noel, GJ. Clinical profile of ceftobiprole, a novel β-lactam antibiotic. *Clin Microbiol Infect*, 2007, 13, S25–9.
[54] Chambers, HF. Ceftobiprole: in-vivo profile of a bactericidal cephalosporin. *Clin Microbiol Infect*, 2006, 12, S17–22.
[55] Bogdanovich, T; Ednie, LM; Shapiro, S; Appelbaum, PC. Antistaphylococcal activity of ceftobiprole, a new broad-spectrum cephalosporin. *Antimicrob Agents Chemother*, 2005, 49, 4210–9.
[56] Lodise, TP; Jr, Pypstra, R; Kahn, JB; Murthy, BP; Kimko, HC; Bush, K; et al. Probability of target attainment for ceftobiprole as derived from a population pharmacokinetic analysis of 150 subjects. *Antimicrob Agents Chemother*, 2007, 51, 2378–87.
[57] Lodise, TP; Patel, N; Renaud-Mutart, A; Gorodecky, E; Fritsche, TR; Jones, RN. Pharmacokinetic and pharmacodynamic profile of ceftobiprole.Diagn. *Microbiol Infect Dis*., 2008, 61, 96–102.
[58] Murthy, B; Schmitt-Hoffmann, A. Pharmacokinetics and pharmacodynamics of ceftobiprole, an anti-MRSA cephalosporin with broad-spectrum activity. *Clin Pharmacokinet*, 2008, 47, 21–33.
[59] Appelbaum, PC. MRSA–the tip of the iceberg. *Clin Microbiol Infect*, 2006, 12, S3–10.
[60] Noel, GJ. Clinical profile of ceftobiprole, a novel β-lactam antibiotic. *Clin Microbiol Infect*, 2007, 13, S25–9.
[61] Chambers, HF. Ceftobiprole: in-vivo profile of a bactericidal cephalosporin. *Clin Microbiol Infect*, 2006, 12, S17–22.
[62] Bogdanovich, T; Ednie, LM; Shapiro, S; Appelbaum, PC. Antistaphylococcal activity of ceftobiprole, a new broad-spectrum cephalosporin. *Antimicrob Agents Chemother*, 2005, 49, 4210–9.
[63] Lodise, TP; Patel, N; Renaud-Mutart, A; Gorodecky, E; Fritsche, TR; Jones, RN. Pharmacokinetic and pharmacodynamic profile of ceftobiprole. *Diagn Microbiol Infect Dis*., 2008, 61, 96–102.
[64] Jones, ME. In-vitro profile of a new β-lactam, ceftobiprole, with activity against methicillin-resistant Staphylococcus aureus. *Clin Microbiol Infect*, 2007, 13, S17–24.

[65] Livermore, DM. Can β-lactams be re-engineered to beat MRSA? *Clin Microbiol Infect*, 2006, 12, S11–6.

[66] Noel, GJ; Bush, K; Bagchi, P; Ianus, J; Strauss, RS. A randomized, double-blind trial comparing ceftobiprole medocaril with vancomycin plus ceftazidime for the treatment of patients with complicated skin and skin-structure infections. *Clin Infect Dis*., 2008, 46, 647-55.

[67] Noel, GJ; Strauss, RS; Amsler, K; Heep, M; Pypstra, R; Solomkin, JS. Results of a double-blind, randomized trial of ceftobiprole treatment of complicated skin and skin-structure infection caused by Gram-positive bacteria. *Antimicrob Agents Chemother*, 2008, 52, 37-44.

[68] Basilea Pharmaceutica Ltd. Basilea announces positive top-line data from phase III study of ceftobiprole in hospital-acquired pneumonia. (2008) www.basilea.com//template_loader.php?tplpage_id=34&mode=details&id=144.

[69] Parish, D; Scheinfeld, N. Ceftaroline fosamil, a cephalosporin derivative for the potential treatment of MRSA infection. *Curr Opin Investig Drugs*, 2008, 9, 201–9.

[70] Talbot, GH; Thye, D; Das, A; Ge, Y. Phase 2 study of ceftaroline versus standard therapy in treatment of complicated skin and skin structure infections. *Antimicrob Agents Chemother*, 2007, 51, 3612–16.

[71] Morrissey, I; Ge, Y; Janes, R. Activity of the new cephalosporin ceftaroline against bacteremia isolates from patents with community-acquired pneumonia. *Int J Antimicrob Agents*, 2009, 7 [Epub ahead of print].

[72] Jacqueline, C; Caillon, J; Le Mabecque, V; Miegeville, AF; Hamel, A; Bugnon, D; et al. In vivo efficacy of ceftaroline (PPI-0903), a new broad-spectrum cephalosporin, compared with linezolid and vancomycin against methicillin-resistant and vancomycin-intermediate Staphylococcus aureus in a rabbit endocarditis model. *Antimicrob Agents Chemother*, 2007, 51, 3397–400.

[73] Mushtaq, S; Warner, M; Ge, Y; Kaniga, K; Livermore, DM. In vitro activity of ceftaroline (PPI-0903M, T-91825) against bacteria with defined resistance mechanisms and phenotypes. *J Antimicrob Chemother*, 2007, 60, 300–11.

[74] Leighton, A; Gottlieb, AB; Dorr, MB; Jabes, D; Mosconi, G; VanSaders, C; et al. Tolerability, pharmacokinetics, and serum bactericidal activity of intravenous dalbavancin in healthy volunteers. *Antimicrob Agents Chemother*, 2004, 48, 940–5.

[75] Billeter, M; Zervos, MJ; Chen, AY; Dalovisio, JR; Kurukularatne, C. Dalbavancin: a novel once-weekly lipoglycopeptide antibiotic. *Clin Infect Dis*., 2008, 46, 577–83.

[76] Jauregui, LE; Babazadeh, S; Seltzer, E; Goldberg, L; Krievins, D; Frederick, M; et al. Randomized, double-blind comparison of once-weekly dalbavancin versus twice-daily linezolid therapy for the treatment of complicated skin and skin structure infections. *Clin Infect Dis*., 2005, 41, 1407–15.

[77] Raad, I; Darouiche, R; Vazquez, J; Lentnek, A; Hachem, R; Hanna, H; et al. Efficacy and safety of weekly dalbavancin therapy for catheter-related bloodstream infection caused by Gram-positive pathogens. *Clin Infect Dis*., 2005, 40, 374–80.

[78] Andes, D; Craig, WA. In vivo pharmacodynamic activity of the glycopeptide dalbavancin. *Antimicrob Agents Chemother*, 2007, 51, 1633–42.

[79] Bush, K; Macielag, M; Weidner-Wells, M. Taking inventory: antibacterial agents currently at or beyond Phase 1. *Curr Opin Microbiol*, 2004, 7, 466–76.

[80] Poulakou, G; Giamarellou, H. Oritavancin: a new promising agent in the treatment of infections due to Gram-positive pathogens. *Expert Opin Investig Drugs*, 2008, 17, 225–43.

[81] Bhavnani, SM; Passarell, JA; Owen, JS; Loutit, JS; Porter, SB; Ambrose, PG. Pharmacokinetic–pharmacodynamic relationships describing the efficacy of oritavancin in patients with Staphylococcus aureus bacteremia. *Antimicrob Agents Chemother*, 2006, 50, 994–1000.

[82] Bhavnani, SM; Owen, JS; Loutit, JS; Porter, SB; Ambrose, PG. Pharmacokinetics, safety, and tolerability of ascending single intravenous doses of oritavancin administered to healthy human subjects. *Diagn Microbiol Infect Dis.*, 2004, 50, 95–102.

[83] Fetterly, GJ; Ong, CM; Bhavnani, SM; Loutit, JS; Porter, SB; Morello, LG; et al. Pharmacokinetics of oritavancin in plasma and skin blister fluid following administration of a 200-milligram dose for 3 days or a single 800-milligram dose. *Antimicrob Agents Chemother*, 2005, 49, 148–52.

[84] Wasilewski, MM; Disch, D; McGill, JM; Harris, HW; O'Riordan, W; Zeckel, ML. Equivalency of shorter course therapy with oritavancin compared with vancomycin/cephalexin in complicated skin/skin structure infections. Presented at: 41st Interscience Conference on *Antimicrobial Agents and Chemotherapy*, Washington, DC, USA, 16–19 December 2001.

[85] Giamarellou, H; O'Riordan, W; Harris, H; Porter, S; Loutit, J. Phase III trial comparing 3–7 days of oritavancin versus 10–14 days of vancomycin/cephalexin in the treatment of patients with complicated skin/skin structure infections (cSSSI). Presented at: 43rd Interscience Conference on *Antimicrobial Agents and Chemotherapy*, Chicago, IL, USA, 14–17 September 2003.

[86] Loutit, JS; O'Riordan, W; San Juan, J; et al. Phase 2 trial comparing four regimens of oritavancin vs comparator in the treatment of patients with S. aureus bacteraemia. *Clin Microbiol Infect*, 2008, 10, S122.

[87] Kanafani, ZA. Telavancin: a new lipoglycopeptide with multiple mechanisms of action. *Expert Rev Anti Infect Ther.*, 2006, 4, 743–9.

[88] Stryjewski, ME; Graham, DR; Wilson, SE; O'Riordan, W; Young, D; Lentnek, A; et al. Telavancin versus vancomycin for the treatment of complicated skin and skin-structure infections caused by Gram-positive organisms. *Clin Infect Dis.*, 2008, 46, 1683–93.

[89] Stryjewski, ME; Chu, VH; O'Riordan, WD; Warren, BL; Dunbar, LM; Young, DM; et al. Telavancin versus standard therapy for treatment of complicated skin and skin structure infections caused by Gram-positive bacteria: FAST 2 study. *Antimicrob Agents Chemother*, 2006, 50, 862–7.

[90] Stryjewski, ME; O'Riordan, WD; Lau, WK; Pien, FD; Dunbar, LM; Vallee, M; et al. Telavancin versus standard therapy for treatment of complicated skin and soft-tissue infections due to Gram-positive bacteria. Clin. *Infect Dis.*, 2005, 40, 1601–7.

[91] Hegde, SS; Reyes, N; Wiens, T; Vanasse, N; Skinner, R; McCullough, J; et al. Pharmacodynamics of telavancin (TD-6424), a novel bactericidal agent, against Gram-positive bacteria. *Antimicrob Agents Chemother*, 2004, 48, 3043–50.

[92] Schneider, P; Hawser, S; Islam, K. Iclaprim, a novel diaminopyrimidine with potent activity on trimethoprim sensitive and resistant bacteria. *Bioorg Med Chem Lett.*, 2003, 13, 4217–21.
[93] Laue, H; Weiss, L; Bernardi, A; Hawser, S; Lociuro, S; Islam, K. In vitro activity of the novel diaminopyrimidine, iclaprim, in combination with folate inhibitors and other antimicrobials with different mechanisms of action. *J Antimicrob Chemother*, 2007, 60, 1391–4.
[94] Kohlhoff, SA; Sharma, R. Iclaprim. *Expert Opin Investig Drugs*, 2007, 16, 1441–8.
[95] Peppard, WJ; Schuenke, CD. Iclaprim, a diaminopyrimidine dihydrofolate reductase inhibitor for the potential treatment of antibiotic-resistant staphylococcal infections. *Curr Opin Investig Drugs*, 2008, 9, 210–25.
[96] Hoellman, DB; Lin, G; Ednie, LM; Rattan, A; Jacobs, MR; Appelbaum, PC. Antipneumococcal and antistaphylococcal activities of ranbezolid (RBX 7644), a new oxazolidinone, compared with those of other agents. *Antimicrob Agents Chemother*, 2003, 47, 1148–50.
[97] Ednie, LM; Rattan, A; Jacobs, MR; Appelbaum, PC. Antianaerobe activity of RBX 7644 (ranbezolid), a new oxazolidinone, compared with those of eight other agents. *Antimicrob Agents Chemother*, 2003, 47, 1143–7.
[98] Bozdogan, B; Appelbaum, PC. Oxazolidinones: activity, mode of action, and mechanism of resistance. *Int J Antimicrob Agents*, 2004, 23, 113–9.
[99] Mathur, T; Bhateja, P; Pandya, M; Fatma, T; Rattan, A. In vitro activity of RBx 7644 (ranbezolid) on biofilm producing bacteria. *Int J Antimicrob Agents*, 2004, 24, 369–73.
[100] Macgowan, AP; Bowker, KE; Noel, AR. Pharmacodynamics of the antibacterial effect and emergence of resistance to tomopenem, formerly RO4908463/CS-023, in an in vitro pharmacokinetic model of Staphylococcus aureus infection. *Antimicrob Agents Chemother*, 2008, 52, 1401–6.
[101] Thomson, KS; Moland, ES. CS-023 (R-115685), a novel carbapenem with enhanced in vitro activity against oxacillin-resistant staphylococci and Pseudomonas aeruginosa. *J Antimicrob Chemother*, 2004, 54, 557–62.
[102] Kumar, JK. Lysostaphin: an antistaphylococcal agent. *Appl Microbiol Biotechnol*, 2008, 80, 555-61.

8. Prevention and Control of MRSA Infection

Control and Prevention of HA-MRSA Infection

In the United Kingdom MRSA constituted only 5% of *S. aureus* samples isolated from blood in the early 1990s [1]. However, in the next 10 years, the picture changed markedly. From the 1990s to the early 2000s, MRSA increased from 2% of bloodborne infections to more than 40% and the mean incidence of MRSA bacteremia increased from 0.10/1,000 beds to 0.19/1,000 beds [2]. However, MRSA incidence did not increase in all countries. In Europe, marked differences in MRSA prevalence were observed. Scandinavian countries reported low infection incidences, whereas England, Ireland, Spain, and Italy, among other countries, reported high MRSA infection rates [3]. The control and prevention of MRSA infection aims to not only decrease the primary incidence of MRSA, but also to prevent MRSA from spreading to lower-prevalence areas. Several researchers have expressed skepticism about the control and prevention of MRSA because MRSA does not spread easily, is not of high virulence, and preventative measures are not cost-effective [4-7]. In Finland, however, although two MRSA outbreaks occurred in the early 1990s, successful control and infection management made it possible to inhibit MRSA transfer from long-term healthcare institutions to other environments and to control MRSA infection. This is a good example of control of MRSA transfer and eradication of MRSA in a non-endemic situation.

Control of MRSA is also helpful for the overall management of multidrug-resistant bacterial levels [10]. One reason for MRSA control failure is a lack of compliance with practice guidelines among healthcare workers (HCWs); for example, not following bacterial isolation precautions and avoiding handwashing [11]. Transfers from patients to the hands of HCW hands occurred during 17% of interactions between MRSA-colonized patients and HCWs. But only 75% of all HCWs and only 27% of doctors routinely wore gloves [12]. In a study on risk factors for MRSA spread in ICUs, the only significant factor was carelessness of the medical team; only 59% of personnel followed hand-disinfection recommendations when in contact with patients [13]. According to Society for Healthcare Epidemiology of America (SHEA) guidelines issued in 2003, it is critical to obey hand hygiene protocols to control and prevent the spread of MRSA and multidrug resistant *Enterococci*, although disinfection, active surveillance culture (ASC), barrier precautions, and other transfer-inhibiting measures were also shown to be important [14]. One report claimed that the nosocomial infection rate decreased from 16.9% to 9.9%, and MRSA transmission also decreased, by simple implementation of a hospital-wide hand-hygiene program [15].

Active screening or ASC of MRSA carriage to inpatients also has an important role in the control of MRSA. It remains controversial to screen all inpatients, but regular screening of high-risk patients is essential. High-risk patients may be identified as follows [15]: 1) history of infection with and colonization by MRSA; 2) frequent re-admission to hospital; 3) direct transfer between hospitals; 4) history of admittance to a hospital in a high-MRSA-prevalence area; and 5) residence in a medical institution with a high level of MRSA carriage. In addition to screening of such high-risk patients, active screening is necessary in units where patients are particularly susceptible to MRSA infection including, for example, ICUs, neonatal ICUs (NICUs), burns units, transplantation units, cardiothoracic units, orthopedic units, trauma units, vascular surgery units, renal units, and referral centers. In one study, universal screening of all patients was performed when patients were admitted and weekly thereafter; the program decreased the incidence of bacteremia from 3.6 cases/month to 1.8 cases/month [17]. Samples should be obtained from the anterior nares, any skin lesion, any wound, catheter insertion sites, urine catheters, the groin/perineum area, tracheostomy orifices, any other skin breaks, sputum from patients with productive cough, and umbilical blood of neonates.

Decolonization is thought to decrease the spread and incidence of MRSA. First, epidemiological control was achieved by treating colonized/infected HCWs associated with MRSA outbreaks [18], and by administration of an intranasal mupirocin ointment to healthy HCWs twice daily for 5 days. Under such a regimen, the prevalence of *S. aureus* carriage decreased by 91% [19]. However, after 4 weeks, recolonization was noted in 26% of decolonized HCWs, suggesting relapse or reinfection [19].

Decolonization successfully decreased the risk of *S. aureus* infection in several studies [20-22]. However, in other work, decolonization did not have a clear effect [20, 23, 24]. According to a Cochrane review of randomized, controlled trials, four of six such studies showed no MRSA eradication efficacy, and 25% of patients were reported to become free of MRSA by taking a placebo. Nevertheless, selective and short-term use of intranasal mupirocin in specific groups (patients scheduled for major surgery [25], or to prevent relapsing MRSA infections in patients in whom standard prevention methods failed) were effective to some degree [26].

There are various methods of decolonization. In general, topical mupirocin can be spread into the nasal cavity twice daily for 5 days (mupirocin should be applied until a patient feels the taste on the tongue); a chlorohexidine bath can also be used. One method involves dilution of a bleaching agent (3.4 g) in water (3.8 L) (1 United States gallon) [27]. In some cases of recurrent infection, TMP-SMX or tetracycline (with or without rifampin) is used over an extended period. When long-term intranasal mupirocin is used (more than twice daily over 5 days), mupirocin resistance can occur and this possibility should be monitored [28].

Not only hand hygiene, but also identification and contact isolation of MRSA carriers, patient decolonization, and correct and appropriate use of antibiotics have major roles in MRSA infection prevention. Overdosing of antibiotics can cause selective increases in levels of resistant bacteria, and further propagation of preexisting MRSA could thus occur [29]. It is known that use of various antibiotics is associated with MRSA infection and colonization; the drugs include broad-spectrum cephalosporins and fluoroquinolones [30-34]. Also, it was found that colonization and infection with drug-resistant/intermediate *S. aureus* was strongly associated with long-term exposure of preexisting MRSA to glycopeptide antibiotics [35-37]. Appropriate guidelines for antibiotic use are: 1) prohibition of misuse and overuse of antibiotics for the purposes of treatment and/or prevention; 2) carefully selected antibiotics should be prescribed in accurate doses and for the proper period; 3) glycopeptide use should be restricted to appropriate patients, and long-term glycopeptide use should be limited; 4) if clinically not indicated, the use of broad-spectrum cephalosporins (especially third-generation cephalosporins) and fluoroquinolones should be restricted.

In recent years, the appearance of heteroresistant vancomycin-intermediate-resistant *S. aureus* (hVISA) and vancomycin-resistant *S. aureus* (VRSA) have indicated that few treatment options for *S. aureus* infection remain. Thus, control and prevention of MRSA infection are important for the treatment of infection and prevention of VRSA outbreaks.

Prevention and Control of CA-MRSA

More than 10% of CA-MRSA infections are skin infection relapses [27]. In most relapsed patients, the infectious species is usually identical to that of the prior infection [38]. One study reported that more than 60% of family members had MRSA infection histories in the 6 months before an infant acquired a CA-MRSA infection [27]. The known main transmission routes are skin-to-skin and skin-to-fomite [39]; thus good wound care, hand washing, good personal hygiene, and frequent laundering of clothes, blankets, and towels are recommended by the CDC [26].

Anti-Staphylococcal Vaccination

Although infectious disease control has developed, mortality rates have not changed markedly. In addition, MRSA constituted 2% of nosocomial infections in 1974 and 50% in 1997 [40]. It is now understood that an effective vaccine is more important than development of a new antibiotic. However, the effectiveness and safety of existing vaccines remain controversial.

S. aureus virulence is determined by the degree of adherence, the extent of host-cell penetration, and the extent to which the immune mechanism of the host is subverted. These processes are affected by variation in MSCRAMMs such as surface proteins and secreted products, which makes vaccine development difficult. For example, the microcapsule is important for invasive infection of *S. aureus* and provides the bacterium with the ability to resist opsonization by polymorphonuclear cells and phagocytosis [41]. However, the microcapsule is a poor vaccine candidate because the microcapsule is only weakly immunogenic [42]. Proteins that are significantly immunogenic include α-toxoid, β-toxoid, and protein A. Notably, "Staphycan", made from alpha-toxin and killed whole bacteria, has been developed. This vaccine has been used in peritoneal dialysis patients, who developed positive antibody reactions, but the infection rate was not significantly lower than in patients given placebo [43]. Later, a vaccine composed of capsular polysaccharides 5 and 8, and recombinant exotoxin A (StaphVAX ™) was reported to show reasonable safety and effectiveness in a clinical trial [44]. Although the vaccine did not decrease the short-term infection rate of hemodialysis patients, a decrease in bacteremia was seen in a 40-week follow-up study [45]. However, after 54 weeks there was no significant difference in infection rates between the two groups. Eventually, the phase III clinical trial was terminated on January 11, 2005, because test patient data showed no significant difference compared with placebo control individuals.

A human immunoglobulin G preparation termed INH-A21 (Veronate) increased antibody levels to ClfA *S. aureus* clumping factor A (a staphylococcal surface adhesion protein), and SdrG (*S. epidermidis* fibrin-binding protein), but no clinical benefit was seen in neonates and vaccine development has ceased.

Although *S. aureus* vaccination has been studied for more than 10 years, unsolved questions remain [47]. Is it possible to prevent *S. aureus* infection by vaccination? What are the useful vaccine candidates? Which part of the host immune system must be stimulated by a successful *S. aureus* vaccine? Which staphylococcal antigens are suitable for inclusion in a multicomponent staphylococcal vaccine? Will such a vaccine work in the environment of ever-changing *S. aureus* infection profiles? Will the vaccine prevent prosthesis-related infections? Will pre-clinical animal studies really be predictive of human responses? Will a vaccine that reduces *S. aureus* colonization effectively limit *Staphylococcus* infections?

Answers to these questions are being vigorously sought. First, candidates for a multicomponent vaccine include CP5, CO8, PNAG, cell-wall-anchored proteins (ClfA, IsdB), and the alpha-hemolysis H135A protein. Vaccines based not on a single protein but rather on multiple proteins are being studied. Vaccine candidates will be evaluated in immunocompromised patients.

Vaccine work should be accompanied by treatment of *S. aureus* infection. Passive antistaphylococcal immunization is thought to have an accessory role in combating severe infections such as endocarditis. Vaccination will be used to block re-acquisition after removal of nasally carried bacteria. Vaccination *per se* is important but must be used in association with conventional treatment.

REFERENCES

[1] Communicable Disease Surveillance Centre, Public Health Laboratory Service. Antimicrobial resistance in 2000. *England and Wales.* London: Public Health Laboratory Service, 2002.

[2] Johnson, AP; Pearson, A; Duckworth, G. Surveillance and epidemiology of MRSA bacteraemia in the UK. *J Antimicrob Chemother*, 2005, 56, 455-62.

[3] Tiemersma, EW; Bronzwaer, SL; Lyytikainen, O; Degener, JE; Schrijnemakers, P; Bruinsma, N; et al. Methicillin-resistant Staphylococcus aureus in Europe, 1999-2002. *Emerg Infect Dis.*, 2004, 10, 1627-34.

[4] Lacey, RW. Multi-resistant Staphylococcus aureus--a suitable case for inactivity? *J Hosp Infect*, 1987, 9, 103-5.

[5] Barrett, SP; Mummery, RV; Chattopadhyay, B. Trying to control MRSA causes more problems than it solves. *J Hosp Infect*, 1998, 39, 85-93.

[6] Teare, EL; Barret, SP. Stop the ritual of tracing colonized people. *BMJ*, 1997, 314, 665-6.

[7] Farrington, M; Redpath, C; Trundle, C; Coomber, S; Brown, NM. Winning the battle but losing the war: methicillin-resistant Staphylococcus aureus (MRSA) infection at a teaching hospital. *Q J Med.*, 1998, 91, 539-48.

[8] Kotilainen, P; Routamaa, M; Peltonen, R; et al. Elimination of epidemic methicillin-resistant Staphylococcus aureus from a university hospital and district institutions, *Finland. Emerg Infect Dis.*, 2003, 9, 169-75.

[9] Kerttula, AM; Lyytika¨inen, O; Salmenlinna, S; Vuopio-Varkila, J. Changing epidemiology of methicillin-resistant Staphylococcus aureus in Finland. *J Hosp Infect*, 2004, 58, 109-14.

[10] Rubinovitch, B; Pittet, D. Screening for methicillin-resistant Staphylococcus aureus in the endemic hospital: what have we learned? *J Hosp Infect.*, 2001, 47, 9-18.

[11] Boyce, JM; Havill, NL; Kohan, C; Dumigan, DG; Ligi, CE. Do infection control measures work for methicillin-resistant Staphylococcus aureus? *Infect Control Hosp Epidemiol*, 2004, 25, 395-401.

[12] McBryde, ES; Bradley, LC; Whitby, M; McElwain, DLS. An investigation of contact transmission of methicillin-resistant Staphylococcus aureus. *J Hosp Infect*, 2004, 58, 104-8.

[13] Grundmann, H; Hori, S; Winter, B; Tami, A; Austin, DJ. Risk factors for the transmission of methicillin-resistant Staphylococcus aureus in an adult intensive care unit: fitting a model to the data. *J Infect Dis.*, 2002, 185, 481-88.

[14] Muto, CA; Jernigan, JA; Ostrowsky, BE; Richet, HM; Jarvis, WR; Boyce, JM; et al. SHEA guidelines for preventing nosocomial transmission of multi-drug-resistant strains of Staphylococcus aureus and Enterococcus. *Infect Control Hosp Epidemiol*, 2003, 24, 362-86.

[15] Pittet, D; Hugonnet, S; Harbarth, S; Mourouga, P; Sauvan, V; Touveneau, S; et al. Effectiveness of a hospital-wide program to improve compliance with hand hygiene. *Lancet*, 2000, 356, 1307-12.
[16] Coia, JE; Duckworth, GJ; Edwards, DI; Farrington, M; Fry, C; Humphreys, H; et al. Guidelines for the control and prevention of meticillin-resistant Staphylococcus aureus (MRSA) in healthcare facilities. *J Hosp Infect*, 2006, 63, S1-44.
[17] Shitrit, P; Gottesman, BS; Katzir, M; Kilman, A; Ben-Nissan, Y; Chowers, M. Active surveillance for methicillin-resistant Staphylococcus aureus (MRSA) decreases the incidence of MRSA bacteremia. *Infect Control Hosp Epidemiol*, 2006, 27, 1004-8.
[18] The AGREE Collaboration. Appraisal of guidelines for research and evaluation (AGREE) instrument. London: St. Georges Hospital Medical School; 2001. Available from: http://www.agreecollaboration.org
[19] Doebbeling, BN; Breneman, DL; Neu, HC; Aly, R; Yangco, BG; Holley, HP; et al. Elimination of Staphylococcus aureus nasal carriage in health care workers: analysis of six clinical trials with mupirocin calcium ointment. The Mupirocin Collaborative Study Group. *Clin Infect Dis.*, 1993, 17, 466-74.
[20] Perl, TM; Cullen, JJ; Wenzel, RP; Zimmerman, MB; Pfaller, MA; Sheppard, D; et al. Intranasal mupirocin to prevent postoperative Staphylococcus aureus infections. *N Engl J Med.*, 2002, 346, 1871-7.
[21] Mody, L; Kauffman, CA; McNeil, SA; Galecki, AT; Bradley, SF. Mupirocin-based decolonization of Staphylococcus aureus carriers in residents of 2 long-term care facilities: a randomized, double-blind, placebo-controlled trial. *Clin Infect Dis.*, 2003, 37, 1467-74.
[22] Tacconelli, E; Carmeli, Y; Aizer, A; Ferreira, G; Foreman, MG; D'Agata, EM. Mupirocin prophylaxis to prevent Staphylococcus aureus infection in patients undergoing dialysis: a meta-analysis. *Clin Infect Dis.*, 2003, 37, 1629-38.
[23] Kalmeijer, MD; Coertjens, H; Van Nieuwland-Bollen, PM; Bogaers-Hofman, D; De Baere, GA; Stuurman, A; et al. Surgical site infections in orthopedic surgery: the effect of mupirocin nasal ointment in a double-blind randomized, placebo-controlled study. *Clin Infect Dis.*, 2002, 35, 353-8.
[24] Wertheim, H; Vos, MC; Ott, A. Mupirocin prophylaxis against nosocomial Staphylococcus aureus infections in nonsurgical patients: a randomized study. *Ann Intern Med.*, 2004, 140, 419-25.
[25] Loeb, M; Main, C; Walker-Dilks, C; Eady, A. Antimicrobial drugs for treating meticillin resistant Staphylococcus aureus colonization. *Cochrane Database Syst Rev.*, 2003, (4), CD003340.
[26] Gorwitz, RJ; Jemigan, DB; Powers, JH. Strategies for clinical management of MRSA in the community: summary of an experts' meeting convened by the Centers for Disease Control and Prevention. March 2006. Available from: http://www.cdc.gov/ncidod/dhqp/pdf/ar/CAMRSA_ExpMtgStrategies.pdf. Accessed May 20, 2008.
[27] Daum, RS. Skin and soft-tissue infections caused by methicillin-resistant Staphylococcus aureus. *N Engl J Med.*, 2007, 357, 280-90.
[28] Jones, JC; Rogers, TJ; Brookmeyer, P; Dunne, WM; Storch, GA; Coopersmith, CM; et al. Mupirocin resistance in patients colonized with methicillin-resistant Staphylococcus aureus in a surgical intensive care unit. *Clin Infect Dis.*, 007, 45, 541-7.

[29] Enright, MC; Robinson, DA; Randle, G; Feil, EJ; Grundmann, H; Spratt, BG. The evolutionary history of methicillin-resistant Staphylococcus aureus (MRSA). *Proc Natl Acad Sci.*, USA. 2002, 99, 7687-92.
[30] Washio, M; Mizoue, T; Kajioka, T; Yoshimitsu, T; Okayama, M; Hamada, T; et al. Risk factors for methicillin-resistant Staphylococcus aureus (MRSA) infection in a Japanese geriatric hospital. *Public Health*, 1997, 111, 187-90.
[31] Dziekan, G; Hahn, A; Thune, K; Schwarzer, G; Schafer, K; Daschner, FD; et al. Methicillin-resistant Staphylococcus aureus in a teaching hospital: investigation of nosocomial transmission using a matched case-control study. *J Hosp Infect.*, 2000, 46, 263-70.
[32] Hori, S; Sunley, R; Tami, A; Grundmann, H. The Nottingham Staphylococcus aureus population study: prevalence of MRSA among the elderly in a university hospital. *J Hosp Infect*, 2002, 50, 25-9.
[33] Crowcroft, NS; Ronveaux, O; Monnet, DL; Mertens, R. Methicillin-resistant Staphylococcus aureus and antimicrobial use in Belgian hospitals. *Infect Control Hosp Epidemiol*, 1999, 20, 31-6.
[34] Campillo, B; Dupeyron, C; Richardet, JP. Epidemiology of hospital-acquired infections in cirrhotic patients: effect of carriage of methicillin-resistant Staphylococcus aureus and influence of previous antibiotic therapy and norfloxacin prophylaxis. *Epidemiol Infect*, 2001, 127, 443-50.
[35] Srinivasan, A; Dick, JD; Perl, TM. Vancomycin resistance in staphylococci. *Clin Microbiol Rev.*, 2002, 15, 430-8.
[36] Stelfox, HT; Bates, DW; Redelmeier, DA. Safety of patients isolated for infection control. *JAMA*, 2003, 290, 1899-905.
[37] Paterson, DL. Reduced susceptibility of Staphylococcus aureus to vancomycin e a review of current knowledge. *Commun Dis Intell*, 1999, 23, 69-73.
[38] Huang, SS, Diekema, DJ; Warren, DK; Zuccotti, G; Winokur, PL; Tendolkar, S; et al. Strain-relatedness of methicillin-resistant Staphylococcus aureus isolates recovered from patients with repeated infection. *Clin Infect Dis.*, 2008, 46, 1241-7.
[39] Miller, LG; Diep, BA. Colonization, fomites, and virulence: rethinking the pathogenesis of community-acquired methicillin-resistant Staphylococcus aureus infection. *Clin Infect Dis.*, 2008, 46, 752-60.
[40] Lowy, F. Staphylococcus aureus infections. *N Engl J Med.*, 1998, 339, 520-32.
[41] Deisenhofer, J. Crystallographic refinement and atomic models of a human Fc fragment and its complex with fragment B of pretein A from Staphylococcus aureus at 2.9- and 2.8A resolution. *Biochemistry*, 1981, 20, 2361-70.
[42] Projan, SJ; Nesin, M; Dunman, PM. Staphylococcal vaccines and immunotherapy: to dream the impossible dream? *Curr Opin Pharmacol*, 2006, 6, 473-9.
[43] Poole-Warren, LA; Hallett, MD; Hone, PW; Burden, SH; Farrell, PC. Vaccination for prevention of CAPD associated staphylococcal infection: results of a prospective multicentre clinical trial. *Clin Nephrol*, 1991, 35, 198-206.
[44] Shinefield, H; Black, S; Fattom, A; Horwith, G; Rasgon, S; Ordonez, J; et al. Use of a Staphylococcus aureus conjugate vaccine in patients receiving hemodialysis. *N Engl J Med.*, 2002, 346, 491-6.
[45] Fattom, A; Fuller, S; Propst, M; Winston, S; Muenz, L; He, D; et al. Safety and immunogenicity of a booster dose of Staphylococcus aureus types 5 and 8 capsular

polysaccharide conjugate vaccine (StaphVAX) in hemodialysis patients. *Vaccine*. 2004, 23, 656-63.
[46] Schaffer, AC; Lee, JC. Staphylococcal vaccines and immunotherapies. *Infect Dis Clin North Am*., 2009, 23, 153-71.
[47] Schaffer, AC; Lee, JC. Vaccination and passive immunization against Staphylococcus aureus. *Int J Antimicrob Agent*, 2008, 32, S71-8.

9. Overview of Glycopeptide-Intermediately-Resistant/Resistant *S. Aureus*

The past century can be described as a war between *Staphylococci* and humans. Penicillin was developed in 1942 and just 1 year later penicillin-resistant *Staphylococci* were found. In 1960, semisynthetic penicillinase-resistant penicillins (methicillin, oxacillin, and nafcillin) were discovered, and 1 year later, in 1961, MRSA appeared. MRSA was not only resistant to methicillin, oxacillin, and nafcillin but also to cephalosporin, carbapenem, macrolide, clindamycin, and aminoglycosides. However, the strains remained highly sensitive to trimethoprim–sulfamethoxazole, tetracycline, rifampin and the newly developed antibiotics linezolid, quinupristin–dalfopristin, and daptomycin.

Vancomycin was developed in 1956 and has been used for the treatment of MRSA infections for more than 50 years. However, GISA (MIC for two glycopeptides 4–16 μg/mL) was first reported in Japan in 1996 [1]. In 2002, the *vanA* gene complex, derived from vancomycin-resistant *Enterococci*, spread to *S. aureus* and conferred a high level of vancomycin resistance (MIC >16–32 μg/mL) in an outbreak in Detroit, United States [2]. The appearance of GISA, followed by VRSA, posed a worldwide threat to human health.

Unlike β-lactam antibiotics, which bind to PBPs, vancomycin binds to a peptidoglycan pentapeptide that is an early intermediate in cell-wall synthesis. Vancomycin binding blocks the actions of transglycosylase, transpeptidase, and D,D-carboxypeptidase to inhibit peptidoglycan elongation and crosslinking [3]. As a result, the peptidoglycan becomes weaker, and the cell becomes more susceptible to lysis. Eventually, the cell membrane collapses; this is how vancomycin exerts a bactericidal effect.

Resistance to vancomycin in VRE and VRSA strains occurs due to such enzymes, which produces the low-affinity vancomycin-binding peptidoglycan precursor or degrades high-affinity vancomycin-binding proteoglycan. However, this enzyme is absent from GISA strains. Instead, accumulation of mutations and modulation of regulatory systems alter cellular physiology. The cell-wall composition is changed, and high levels of residual D-Ala-D-Ala components are synthesized, binding to vancomycin and reducing the diffusion coefficient of the antibiotic. This false target sequestrates vancomycin and blocks penetration to the target [3]. When *S. aureus* has reduced vancomycin sensitivity, the degree of peptidoglycan crosslinking is reduced, cell-wall turnover is lower, and autolysis decreases owing to structural and metabolic changes in teichoic acid [4]. Such alterations cause significant changes in cell-wall morphology [5] (Figure 10).

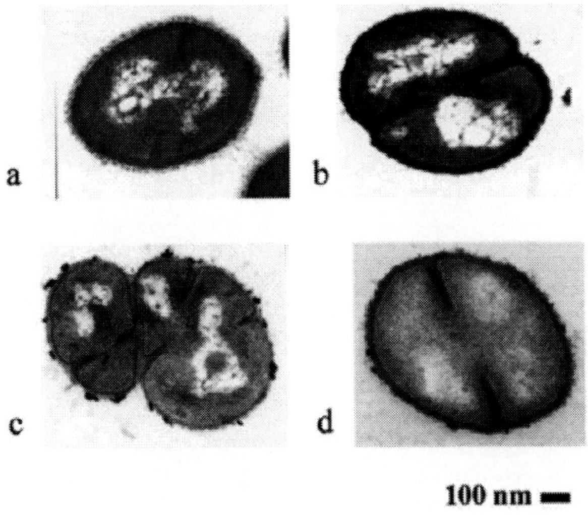

Figure 10. Morphological cell-wall thickening in vancomycin-intermediate-resistant *S. aureus* strains (a and b), a vancomycin-susceptible revertant strain (c), and control NCTC 8324 (d) (magnification x63,000) [5]. Preprint from [5].

The number of strains with vancomycin MIC >2 μg/mL is increasing annually by 0–0.1% [6]. One study showed a small but significant vancomycin MIC increase among MRSA and MSSA [7]. A study that examined more than 6,000 *S. aureus* strains found no VRSA strainand only one VRSA strain [8]. However, strains with vancomycin MICs more than 1.0 μg/mL have also increased in number, constituting 19.9% of all strains in 2000 and 70.4% in 2004 [8]. CA-MRSA strains have lower vancomycin MICs because such strains are less exposed to glycopeptides than are HA-MRSA strains. However, considering the current invasion of healthcare facilities by CA-MRSA, the CA-MRSA vancomycin MIC might be higher than thought.

A recent study showed an increased risk of treatment failure when the vancomycin MIC of a vancomycin-susceptible MRSA increased [10-12]. If the vancomycin MIC was <0.5 μg/mL during MRSA treatment, the success rate was 56%, but when the MIC was 1–2 μg/mL, the success rate was less than 10% [10]. Other MRSA-related infections that developed when the vancomycin MIC was <2 μg/mL showed treatment response and mortality rates of 85% and 10%, respectively, but when the MIC was >2 μg/mL the figures were 62% and 24%, respectively; the differences were statistically significant [13].

S. aureus can be classified into four groups by vancomycin MIC: susceptible, heteroresistant, intermediate, and resistant (Table 1):

hVISA strains include a large number of vancomycin-susceptible isolates and an extremely small number of vancomycin-resistant strains. Vancomycin-resistant subpopulations grow slowly. Clinically, using a common culture medium, it is almost impossible to identify hVISA. It is nonetheless important to identify such strains because they are associated with poorer clinical outcomes [15] and increase the vancomycin MIC [16]. One study found that when the vancomycin MIC was 1, 2, or 4 μg/mL the hVISA levels were 57%, 81%, and 100%, respectively [16]. Until clinical microbiology laboratories can easily identify hVISA, vancomycin should be used with caution when the MIC is greater than 1 μg/mL. The CLSI recommends careful reduction of the vancomycin MIC breakpoint to 2 μg/mL. An algorithm for vancomycin use to combat *S. aureus* infections is shown below (Figure 11).

Table 1. Vancomycin susceptibility among *S. aureus* strains (data adapted from the Clinical Laboratory Standards Institute) [14].

Vancomycin susceptibility	Name	MIC, μg/mL
Susceptible	VSSA	≤2*
Heteroresistant	hVISA/hGISA	1–2
Intermediate	VISA or GISA	4–8
Resistant	VRSA	≥16**

GISA: intermediate-glycopeptide-resistant S. aureus, hGISA, heteroresistant intermediate-glycopeptide-resistant S. aureus, hVISA: heteroresistant vancomycin-intermediate-resistant S. aureus, MIC: minimal inhibitory concentration, VISA: vancomycin intermediate-resistant S. aureus, VRSA: vancomycin-resistant S. aureus, VSSA: vancomycin-susceptible S. aureus.

*consisting of subpopulations ($<10^{-6}$) that may grow in media containing >2 μg/mL vancomycin.

**requires confirmatory testing by evaluation of overnight growth on a plate with 6 μg/mL vancomycin.

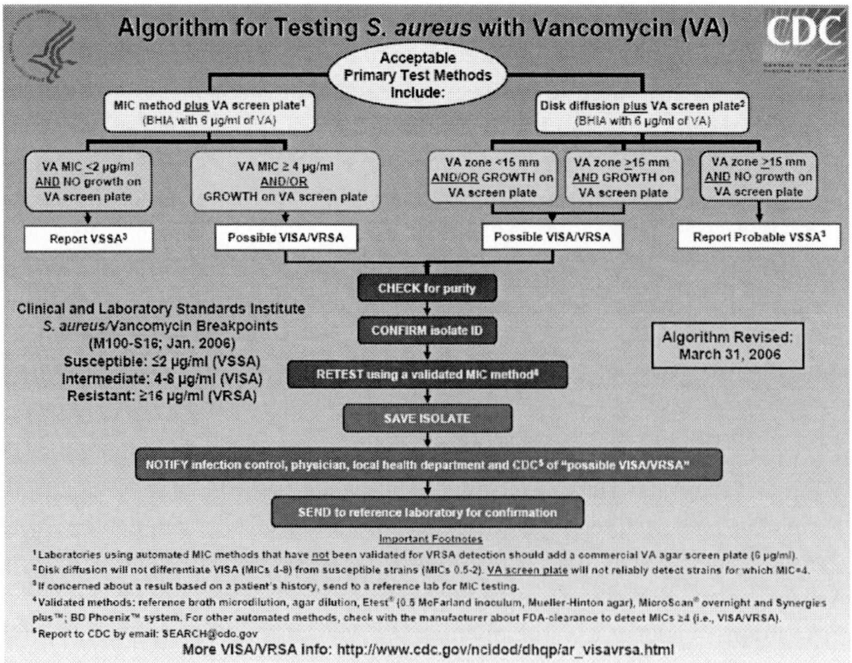

Figure 11. Algorithm for testing *S. aureus* susceptibility to vancomycin.

It is known that if vancomycin dosage is increased, drug effectiveness also increases [17, 18]. Thus, in patients with severe HAP caused by MRSA, the bioconcentration of vancomycin should be at least 15–20 μg/mL [19]. However, in a retrospective study of 102 patients with HAP, vancomycin trough concentration and area-under-the-concentration curve were not associated with survival [20]. A recent case-control study showed an increase in drug effectiveness until the vancomycin serum trough level was four-fold greater than the MIC, but low effectiveness when the MIC was more than 2 μg/mL [13] . This means that in cases of hVISA or VISA infection vancomycin will not be useful and may be associated with a high risk of complications owing to the increased dosage. Although use of high-dose vancomycin may not have a role in treatment, an increased dose may contribute to lowering GISA or hGISA levels by increasing the vancomycin MIC [21,22].

Drugs other than vancomycin, including linezolid, daptomycin, and tigecycline, are approved by the FDA for the treatment of MRSA [23, 24]. There is no official CLSI microbiological definition of daptomycin resistance; however, a daptomycin MIC of 2 μg/mL in *S. aureus* was thought to result in clinical treatment failure [25, 26]. When the daptomycin MIC was greater than the susceptible range shown by MRSA, *mprF* (encoding lysophosphatidylglycerol synthetase), *yycG* (encoding a sensor histidine kinase), and *rpoB* and *rpoC* (encoding the B and B' subunits of RNA polymerase, respectively) mutations were discovered. When vancomycin MIC was increased, daptomycin MIC also increased. There is no genetic relationship between the two resistance mechanisms. It was thought that the thickened cell wall prevented daptomycin from penetrating the target, thus increasing the MIC [27, 28].

In clinically isolated MRSA, linezolid resistance arises due to mutations in the central loop of 23S ribosomal RNA (rRNA) domain V [29, 30]. Linezolid resistance in MRSA is extremely rare, but in MRSA bloodstream isolates linezolid MICs have been noted to increase [31]. The linezolid-resistant CoNS isolated in some hospitals are a clinical problem requiring future research.

REFERENCES

[1] Hiramatsu, K; Hanaki, H; Ino, T; Yabuta, K; Oguri, T; Tenover, FC. Methicillin-resistant Staphylococcus aureus clinical strain with reduced vancomycin susceptibility. *J Antimicrob Chemother*, 1997, 40, 135–6.

[2] From the Centers for Disease Control and Prevention. Staphylococcus aureus resistant to vancomycin—United States, 2002. *MMWR Morb Mortal Wkly Rep.*, 2002, 51, 565–7.

[3] Courvalin, P. Vancomycin resistance in gram-positive cocci. *Clin Infect Dis.*, 2006, 42, S25–34.

[4] Sieradzki, K; Tomasz, A. Alterations of cell wall structure and metabolism accompany reduced susceptibility to vancomycin in an isogenic series of clinical isolates of Staphylococcus aureus. *J Bacteriol*, 2003, 185, 7103–10.

[5] Reipert, A; Ehlert, K; Kast, T; Bierbaum, G. Morphological and genetic differences in two isogenic Staphylococcus aureus strains with decreased susceptibilities to vancomycin. *Antimicrob Agents Chemother*, 2003, 47, 568–76.

[6] Jones, RN. Microbiological features of vancomycin in the 21st century: minimum inhibitory concentration creep, bactericidal/static activity, and applied breakpoints to predict clinical outcomes or detect resistant strains. *Clin Infect Dis.*, 2006, 42, S13–24.

[7] Rhee, KY; Gardiner, DF; Charles, M. Decreasing in vitro susceptibility of clinical Staphylococcus aureus isolates to vancomycin at the New York Hospital: quantitative testing redux. *Clin Infect Dis.*, 2005, 40, 1705–6.

[8] Wang, G; Hindler, JF; Ward, KW; Bruckner, DA. Increased vancomycin MICs for Staphylococcus aureus clinical isolates from a university hospital during a 5-year period. *J Clin Microbiol*, 2006, 44, 3883–6.

[9] Tsuji, BT; Rybak, MJ; Cheung, CM; Amjad, M; Kaatz, GW. Community and health care-associated methicillin-resistant Staphylococcus aureus: a comparison of molecular epidemiology and antimicrobial activities of various agents. *Diagn Microbiol Infect Dis.*, 2007, 58, 41–7.

[10] Sakoulas, G; Moise-Broder, PA; Schentag, J; Forrest, A; Moellering, RC; Jr, Eliopoulos GM. Relationship of MIC and bactericidal activity to efficacy of vancomycin for treatment of methicillin-resistant Staphylococcus aureus bacteremia. *J Clin Microbiol*, 2004, 42, 2398–402.

[11] Moise, PA; Sakoulas, G; Forrest, A; Schenag, JJ. Vancomycin in vitro bactericidal activity and its relationship to efficacy in clearance of methicillin-resistant Staphylococcus aureus bacteremia. *Antimicrob Agents Chemother*, 2007, 57, 2582–6.

[12] Moise, PA; Smyth, DS; El-Fawal, N; Robinson, DA; Holden, PN; Forrest A; et al. Microbiological effects of prior vancomycin use in patients with methicillin-resistant Staphylococcus aureus bacteraemia. *J Antimicrob Chemother*, 2008, 61, 85–90.

[13] Hidayat, LK; Hsu, DI; Quist, R; Shriner, KA; Wong-Beringer, A. High dose vancomycin therapy for methicillin-resistant Staphylococcusaureus infections: efficacy and toxicity. *Arch Intern Med.*, 2006, 166, 2138–44.

[14] Clinical Laboratory Standards Institute. Performance standards for antimicrobial disc susceptibility tests: 16th informational supplement. Document M100-S16. Wayne, PA: *National Committee for Clinical Laboratory Standards*, 2006.

[15] Liu, C; Chambers, HF. Staphylococcus aureus with heterogeneous resistance to vancomycin: epidemiology, clinical significance, and critical assessment of diagnostic methods. *Antimicrob Agents Chemother*, 2003, 47, 3040–5.

[16] Bell, JM; Chitsaz, M; Turnidge, JD; Barton, M; Walters, LJ; Jones, RN. Prevalence and significance of a negative extended-spectrum beta-lactamase (ESBL) confirmation test result after a positive ESBL screening test result for isolates of Escherichia coli and Klebsiella pneumoniae: results from the SENTRY Asia-Pacific Surveillance Program. *J Clin Microbiol* 2007, 45, 1478-82.

[17] Rybak, MJ. The pharmacokinetic and pharmacodynamic properties of vancomycin. *Clin Infect Dis.*, 2006, 42, S35–9.

[18] Moise-Broder, PA; Forrest, A; Birmingham, MC; Schentag, JJ; Pharmacodynamics of vancomycin and other antimicrobials in patients with Staphylococcus aureus lower respiratory tract infections. *Clin Pharmacokinet*, 2004, 43, 925–42.

[19] American Thoracic Society, Infectious Diseases Society of America. Guidelines for the management of adults with hospital-acquired, ventilator-associated, and healthcare-associated pneumonia. *Am J Respir Crit Care Med.*, 2005, 171, 388–416.

[20] Jeffres, MN; Isakow, W; Doherty, JA; McKinnon, PS; Ritchie, DJ; Micek, ST; et al. Predictors of mortality for methicillin-resistant Staphylococcus aureus health-care–associated pneumonia: specific evaluation of vancomycin pharmacokinetic indices. *Chest*, 2006, 130, 947–55.

[21] Tsuji, BT; Rybak, MJ; Lau, KL; Sakoulas, G. Evaluation of accessory gene regulator (agr) group and function in the proclivity towards vancomycin intermediate resistance in Staphylococcus aureus. *Antimicrob Agents Chemother*, 2007, 51, 1089–91.

[22] Sakoulas, G; Gold, HS; Cohen, RA; Venkataraman, L; Moellering, RC; Eliopoulos, GM. Effects of prolonged vancomycin administration on methicillin-resistant Staphylococcus aureus (MRSA) in a patient with recurrent bacteraemia. *J Antimicrob Chemother*, 2006, 57, 699–704.

[23] ZYVOX prescribing information. In: Physicians' desk reference. 59th ed. Montvale, NJ: Thomson PDR; 2005. p. 2773–81.

[24] Cubicin (daptomycin for injection) [package insert]. Lexington, MA: *Cubist Pharmaceuticals*, 2006.

[25] Hayden, MK; Rezai, K; Hayes, RA; Lolans, K; Quinn, JP; Weinstein, RA. Development of daptomycin resistance in vivo in methicillin-resistant Staphylococcus aureus. *J Clin Microbiol*, 2005, 43, 5285–7.

[26] Hirschwerk, D; Ginocchio, CC; Bythrow, M; Condon, S. Diminished susceptibility to daptomycin accompanied by clinical failure in a patient with methicillin-resistant Staphylococcus aureus bacteremia. *Infect Control Hosp Epidemiol*, 2006, 27, 315–7.

[27] Cui, L; Tominaga, E; Neoh, HM; Hiramatsu, K. Correlation between reduced daptomycin susceptibility and vancomycin resistance in vancomycin-intermediate Staphylococcus aureus. *Antimicrob Agents Chemother*, 2006, 50, 1079–82.

[28] Pillai, SK; Gold, HS; Sakoulas, G; Wennersten, C; Moellering, RC; Jr, Eliopoulos, GM. Daptomycin nonsusceptibility in Staphylococcus aureus with reduced vancomycin susceptibility is independent of alterations in MprF. *Antimicrob Agents Chemother*, 2007, 51, 2223–5.

[29] Tsiodras, S; Gold, HS; Sakoulas, G; Eliopoulos, GM; Wennersten, C; Venkataraman, L; et al. Linezolid resistance in a clinical isolate of Staphylococcus aureus. *Lancet*, 2001, 358, 207–8.

[30] Meka, VG; Pillai, SK; Sakoulas, G; Wennersten, C; Venkataraman, L; DeGirolami, PC; et al. Linezolid resistance in sequential Staphylococcus aureus isolates associated with a T2500A mutation in the 23S rRNA gene and loss of a single copy of rRNA. *J Infect Dis.,* 2004, 190, 311–7.

[31] Steinkraus, G; White, R; Friedrich, L. Vancomycin MIC creep in nonvancomycin-intermediate Staphylococcus aureus (VISA), vancomycin-susceptible clinical methicillin-resistant S. aureus (MRSA) blood isolates from 2001–05. *J Antimicrob Chemother*, 2007, 60, 788–94.

[32] Potoski, BA; Adams, J; Clarke, L; et al. Epidemiological profile of linezolid-resistant coagulase-negative staphylococci. *Clin Infect Dis.*, 2006, 43, 165–71.

In: Methicillin-Resistant Staphylococcus Aureus... ISBN: 978-1-60741-398-1
Editor: Charles L. Kolendi

Chapter 4

HOSPITAL AND COMMUNITY-ACQUIRED METHICILLIN-RESISTANT *STAPHYLOCOCCUS AUREUS* BONE AND JOINT INFECTIONS

Ilker Uçkay[1,2], Frederic Laurent[3,4], Sébastien Lustig[4,5], Jean-Philippe Rasigade[3,4], Vivien Thomson[4,6], Vincent Paoli[4,7], Thomas Perpoint[4,8], Richard Stern[9], André Boibieux[8], Florence Ader[3,4,8], Gerard Lina[3,4], Daniel Lew[1,2], François Vandenesch[3,4], Dominique Peyramond[4,8], Christian Chidiac[3,4,,8], Jerôme Etienne[3,4] and Tristan Ferry[3,4,8*]

[1]Service de Maladies Infectieuses, Hôpitaux Universitaires de Genève, Suisse
[2]Université de Genève, Suisse
[3]INSERM U851, Equipe pathogénie des staphylocoques, Centre National de référence des staphylocoques, Lyon, France
[4]Université Claude Bernard Lyon 1, France
[5]Service de Chirurgie Orthopédique, Centre Livet, Hospices Civils de Lyon, France
[6]Département de Radiologie, Hôpital de la Croix-Rousse, Hospices Civils de Lyon, France
[7]Département de Radiologie Pédiatrique et Gynécologique, Hôpital Femme-Mère-Enfant, Hospices Civils de Lyon, France
[8]Service de Maladies Infectieuses et Tropicales, Hôpital de la Croix-Rousse, Hospices Civils de Lyon, France
[9]Service de l'Orthopédie et de Traumatologie de l'appareil moteur, Hôpitaux Universitaires de Genève.

* Corresponding author: Service de Maladies Infectieuses et Tropicales, Hôpital de la Croix-Rousse, Hospices Civils de Lyon, 93 Grande rue de la Croix-Rousse, 69004 Lyon, France, e-mail: tristan.ferry@univ-lyon1.fr

INTRODUCTION

Bone and joint infections (BJIs) are a heterogeneous group of infections in their pathophysiology, clinical presentation, and management. It is felt to be one of the most difficult-to-treat infectious diseases. Although the mortality rate for BJIs may not be so high, morbidity and long-term functional impairment are high (40% to 50% of patients have residual joint dysfunction after septic arthritis). *S. aureus* is the most common virulent pathogen responsible for such infections. Management of staphylococcal BJIs, particularly BJI due to methicillin resistant *S. aureus* (MRSA), is challenging as it is: (i) usually requiring the combination of surgical intervention and long courses of antibiotics; (ii) associated to significant rates of treatment failure and complications.

Today, two categories of MRSA are now recognized: hospital acquired (HA)- and community-acquired (CA)-MRSA. Molecular epidemiological studies show that both HA- and CA-MRSA infections were due to particular pandemic or epidemic MRSA clones. HA-MRSA clones are pandemic and responsible for hospital acquired acute or chronic BJIs, mainly in adult patients with co-morbidities. Conversely, CA-MRSA clones are emergent, epidemic, and responsible for acute and severe BJIs, particularly in the pediatric population, partly due to the production of a particular toxin called Panton-Valentin leukicidin (PVL).

The aim of this chapter was to review the pathogenesis, the clinical presentation, the imaging and the surgical and medical management of HA- and CA-MRSA BJIs.

FIRST PART: HA-MRSA BJI

During the 1960s, methicillin-resistant *S. aureus* (MRSA) strains emerged rapidly in hospitals after the introduction of methicillin. MRSA strains have now spread among hospitals worldwide and are the leading cause of hospital-acquired infections, which are particularly frequent in intensive care units (1). Molecular epidemiological studies show that a few pandemic clones are responsible for most of HA-MRSA infections worldwide (2, 3). These pandemic HA-MRSA clones are mainly defined on the basis of the multilocus sequence type (MLST) (4-6), the *spa* type (after sequencing of the polymorphic domain x of the protein A gene) (7, 8) and the type of genetic cassette encoding resistance to methicillin (SCC*mec*) (9-11). The pandemic Iberian clone, first described in Spain in 1989, was the predominant clone in Southern Europe. The pediatric clone, first reported in pediatric hospitals in Portugal in 1992, has also been found in Poland, the USA, Argentina and Colombia. The Brazilian clone has mainly been described in South America, and the New York/Japan clone (NY/J clone) predominates in hospitals in metropolitan New York, New Jersey, Pennsylvania, and Connecticut. The New York clone V (NY clone V), a single-locus variant of the Iberian clone, was initially identified in Europe (Czech Republic), before being shown to be an epidemic MRSA clone in the USA. Recent data revealed that the two HA-MRSA clones predominate in France, the "Lyon clone", and the "Géraldine clone". Most of these clones belong to only five lineages determined by MLST, corresponding to five clonal complexes (CC) designated CC1, CC5, CC8, CC30 and CC45 (2, 12, 13).

Patients at Risk for HA-MRSA BJI

Particular populations of patients, especially patients with hemodialysis, cancer, illicit intravenous drug abuse, diabetes mellitus and rheumatoid arthritis, have a significantly greater risk of invasive staphylococcal infection than the normal population (14). Focusing on HA-MRSA BJIs, the most important factors that independently adds to these predisposing conditions is chronic HA-MRSA carriage and orthopaedic surgical procedure (15).

Risk factors for HA-MRSA carriage include advanced age, prior antibiotic exposure, history of hospitalization (particularly in an intensive care unit), living in a nursing home, and close proximity to a patient colonized or infected with HA-MRSA (16, 17). HA-MRSA carriage then facilitates invasive MRSA infections such as pneumonia, urinary tract infection and bacteremia with or without endocarditis. These invasive HA-MRSA infections may facilitate, through the bloodstream, the deposition of bacteria in bone and joints that may lead to acute or latent BJIs (18, 19).

After orthopaedic surgical procedure, risk factors of HA-MRSA surgical site infection include the MRSA nasal carriage and various comorbidities such as vascular diseases, chronic obstructive pulmonary disease, poor functional status and dementia (20, 21). Then, the surgical site infection is acquired in the operation theatre from the patient's own skin. However, a significant proportion of patients who developed HA-MRSA surgical site infection has no evidence of MRSA prior to surgery, and may acquire the MRSA during the surgery, from the environment, particularly in institution with a high HA-MRSA endemicity (15).

Different Forms of HA-MRSA BJI

HA-MRSA BJIs include a heterogeneous group of MRSA infections. Multiple different forms of HA-MRSA BJI have been described and all bones (long, short and flat bones) and/or joints could be affected. Currently, native joint arthritis is distinguished to osteomyelitis (18). Native joint arthritis, also called septic arthritis, is defined by an acute infection of the synovial membrane and the synovial fluid of a joint (22). Native joint arthritis could be subclassified on the basis of the route of infection (mainly hematogeneous) and the anatomic location of the infection. Osteomyelitis is defined by the infection of the bone and/or the bone marrow (18). Osteomyelitis could be subclassified on the basis of the route of infection (hematogeneous or direct inoculation), the duration of illness (acute versus chronic), the presence of orthopaedic implant (prosthetic joints or internal fixation device), the anatomic location of the infection and the type of infected bone (vertebra, long bone) (18). Most frequent HA-MRSA osteomyelitis include spondylodiscitis, chronic osteomyelitis and implant-associated osteomyelitis. Spondylodiscitis is defined by the infection of the intervertebral disk and the adjacent vertebrae (vertebral osteomyelitis is an infection of the vertebrae only) that could be associated with epidural or paravertebral abscesses (23). Chronic osteomyelitis includes long bone (typically post-traumatic chonic osteomyelitis) and diabetic foot infections (not described in this chapter, as diabetic foot infections are often plurimicrobial) (18). Multiple forms of long bone osteomyelitis have been described and two staging systems, the

Waldvogel and the Cierny-Mader, have been proposed to classified these infections (24, 25).

Pathophysiology of HA-MRSA BJI

Routes of infection and pathological examination

Native joint infections and spondylodiscitis are usually hematogenously acquired (18, 22). Indeed, the synovial membrane is highly vascularized and is particularly susceptible to the deposition of bacteria during occult or clinically apparent bacteremia (22). In the same way, during spondylodiscitis and vertebral osteomyelitis, it is hypothesized that the bacteria is deposited in the segmental arteries supplying the intervertebral disc and the two adjacent vertebrae. However, direct inoculation of bacteria into a joint, into an intervertebral disk or into a vertebra is possible through percutaneous puncture or during a surgical procedure (23, 26). Native joint infections are associated with bacterial proliferation in the synovial fluid with large recruitment of polymorphonuclear neutrophils (PMNs) (18, 22). Conversely, chronic osteomyelitis usually occurred after direct inoculation of the bacteria (during trauma or surgery) or during adjacent soft tissues infections as seen in diabetic foot infection. During chronic osteomyelitis, after an early proinflammatory phase characterized by edema, vascular congestion and leukocytes recruitment, large areas of dead bone, also called sequestrum, may be formed due to intramedullary ischemia (18). Sequestrum is a common pathologic feature of chronic osteomyelitis and may facilitate the bacterial persistence. Sequestrum is also frequently associated with a surrounded production of new bone, called involucrum, which may be associated with pus forming a chronic sinus tract that drained the infected bone to the skin surface. Implant-associated osteomyelitis is frequently chronic and due to a direct inoculation of the bacteria during a surgery, but hematogeneous seeding, frequently associated with an acute presentation, may also occur. Orthopaedic implants remain devoid of a microcirculation and their surface is rapidly covered after implantation by extracellular matrix proteins such as fibrinogen and fibronectin. Direct inoculation of HA-MRSA during the surgical procedure can lead to acute or latent implant-associated osteomyelitis (18).

Involvement of HA-MRSA virulence factors

S. aureus owns a variety of receptors, named microbial surface components recognizing adhesive matrix molecules (MSCRAMMs), for host proteins such as fibronectin, fibronectin, laminin, elastin and collagen (27). In a substantial number of animal studies, mutations in these receptors strongly reduced the ability of *S. aureus* to produce infection (28-31). MSCRAMMs are considered to be determinant by enabling colonisation, which is the first step of joint infection. Hurt bone and post-operative hematoma facilitate the adhesion of the bacterium and in this context MSCRAMMs are also considered to be crucial (18, 19, 32). In acute form of the disease, microscopy reveals an acute suppurative inflammation in which bacteria are embedded. In this situation, *S. aureus* is able to invade host tissue by producing multiple toxins and degrading components of extracellular matrix (18, 19, 32). After the acute phase of infection, persistent HA-MRSA infections may occur and is facilitated by the formation of sequestrum, which is a segment of bone resulting from localized ischaemia and bone necrosis. Sequestrum is often associated with chronic osteomyelitis, which is characterized by a long-standing bone infection (18). *S. aureus* has two strategies to facilitate its in situ persistence. The production of biofilm is the first way for the bacteria to persist (33). In biofilm, microorganism lives clustered together in a highly hydrated extracellular matrix, which is often attached to some surface such as orthopaedic devices. Within biofilm, *S. aureus* develops into organized and complex communities with structural and functional heterogeneity resembling multicellular organism. In this environment, the bacterium is protected from antimicrobial agents and host immune responses. Moreover, depletion of metabolic substances causes microbes to enter a slow or nongrowing phase state that reduces also the susceptibility to antibiotics. The second strategy for *S. aureus* to persist is to promote endocytic uptake by epithelial cells or osteoblast. Intracellular survival may be a key factor for *S. aureus* persistence and probably occur early in the process of the disease. Inside cells, particularly into non-professional phagocytes, a subpopulation of *S. aureus*, namely small colony variants (SCVs), could emerge (34). SCVs reveal many changed physiological characteristics comparing with the normal *S. aureus* phenotype, such as reduced pigmentation and hemolysis. This altered phenotype has been clearly associated with persistent infections such as chronic staphylococcal osteomyelitis, and may be selectively enriched during antibiotic therapy (35).

To date, few studies have compared the capacity to adhere or to persist of different HA-MRSA, of HA-MRSA with sporadic MRSA, and of HA-MRSA with methicillin-susceptible *S. aureus* (MSSA). Recently, it has been suggested that each MRSA pandemic clones might have specific virulence properties. In a recent study comparing the adhesion in vitro onto epithelial cells, a clear difference appeared between the two predominant HA-MRSA clones in France, the "Lyon clone" and the "Géraldine clone" (36). Moreover, in another study comparing in vitro characteristics of the predominant HA-MRSA clone in Brazil, also called Brazilian clone, with sporadic MRSA isolates and MSSA isolates, authors conclude to a higher ability of the Brazilian clone to adhere and to produce biofilm in vitro (37). These data, collectively, seem to indicate that particular HA-MRSA clones might be particularly adapted to produce chronic osteomyelitis, especially when orthopaedic materials are associated.

Clinical Presentation of HA-MRSA BJI

Native joint arthritis

The classical clinical presentation of native joint arthritis due to HA-MRSA is common with that observed with other organisms, with the exceptions of gonococcal arthritis (which is acute with erythematous papules or pustules on the affected joint), and mycobacterial or fungal arthritis (which are usually subacute or chronic). The patient frequently experiences acute onset of fever with local findings of pain, warmth, swelling and decreased range of motion in the involved joint. An underlying HA-MRSA bacteremia may be associated and blood cultures have to be performed. In this latter case, the patient has to be screened for infective endocarditis, especially if multiple joints are injured (22).

Spondylodiscitis and vertebral osteomyelitis

The classical clinical presentation of vertebral HA-MRSA spondylodiscitis and vertebral osteomyelitis is subacute. The onset of symptoms is often insidious with a delay in diagnosis of weeks to months after the initial symptoms. Patient usually complains of progressive back pain (mainly located in the lumbar or thoracic spine), which increases on straining, with or without fever. Patients may have history of HA-MRSA bloodstream infection or concomitant HA-MRSA blood cultures may be associated (23, 26, 38). During postoperative spondylodiscitis and vertebral osteomyelitis, external signs of incisional infection could be detected (23). Patients with HA-MRSA spondylodiscitis have to be screened for motor or sensory deficits as spinal cord or nerve root compression may occur. Compression of the spinal cord is usually associated with large epidural involvement (23, 39).

Chronic osteomyelitis

The classical presentation of long-bone HA-MRSA chronic osteomyelitis includes chronic mild pain and drainage trough a sinus tract (Figure 1). If fever is present, it is low grade (18, 40). Sometimes, an obstruction of the sinus tract may occur and may lead to the formation of localized abscess and/or acute soft-tissue infection.

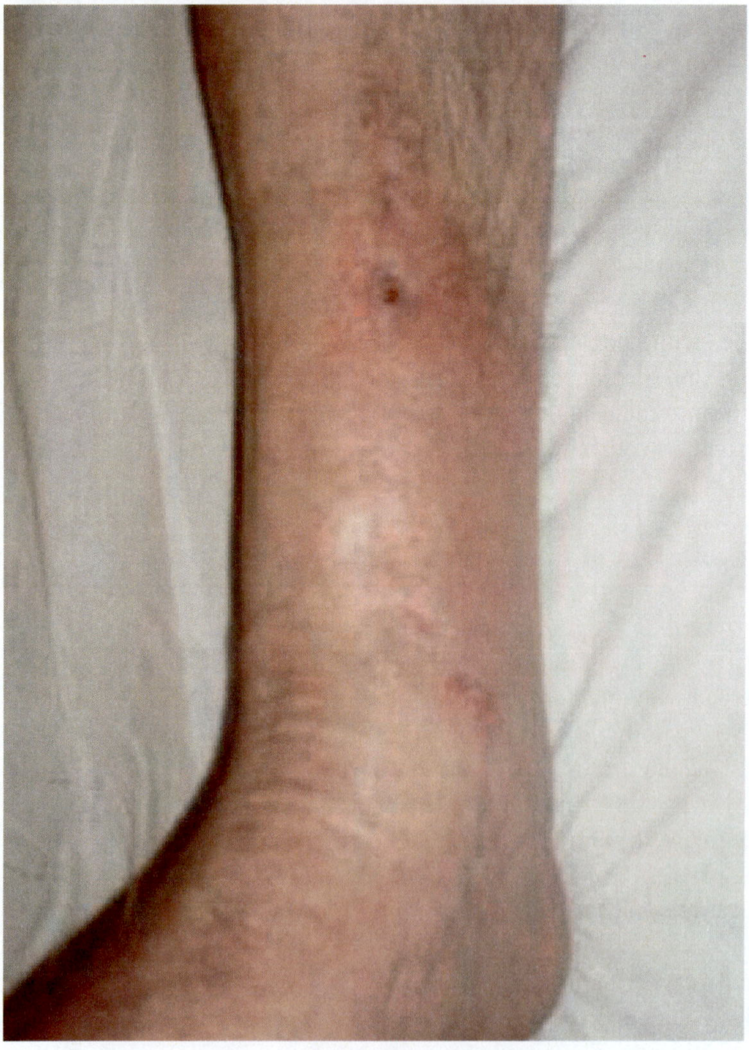

Figure 1. Sinus tract in a 49 year-old patient with HA-MRSA chronic osteomyelitis of the tibia.

Implant-associated osteomyelitis

Patients with implant-associated osteomyelitis can present with clinical signs of acute or chronic infection, depending on many factors (18, 40-42). In cases of early infection (< 2 weeks after surgery in patients with osteosynthesis and <3 months in patients with joint prosthesis), the patient typically presents with systemic acute signs, such as fever and tachycardia, associated with local signs of postoperative wound infection, such as persisting local pain, wound healing disturbance, persistent hematoma, and/or erythema. In cases of delayed infection (2-10 weeks after surgery in patients with osteosynthesis and 3-24 months in patients with prosthesis joint), clinical signs and symptoms of infection are different and could be entirely lacking. They are mainly subacute, as during long bone chronic osteomyelitis (see above), and one of the subtle sign of delayed infection is a persistent or increasing post-surgical pain (with mild signs of arthritis in patients with joint prosthetis) associated with a sinus tract. Some patients developed late infection (>10 weeks after surgery in patients with osteosynthesis and >2 years in patients with prosthesis joint). In this situation, focusing only on *S. aureus* infections, the onset of clinical signs is frequently acute, as during early infection. Indeed, late infections, particularly in patients with joint prosthesis, are frequently due to *S. aureus* hematogenous spreading from another primary (detectable or not) focus of infection (18, 40-42).

Imaging of HA-MRSA BJI

Different radiologic modalities could be used for the imaging of MRSA BJIs, as during BJIs due to other pathogen. If an active BJI is suspected, it is essential to delineate exactly its anatomical extent and possible involvement of other surrounding tissues to decide on the best therapeutic modality.

Plain radiographs are always the basic first line examination in the assessment of any clinical suspicion of bone infection. Whatever the type of bone infection, plain-film radiographies are insensitive for early BJIs, and remain normal during at least the first 10 days of infection. Plain radiographs can also help in detecting and characterizing the possible presence of metallic surgical material and are also important for follow-up. Newly discovered abnormalities are highly suspicious for active infection. However, even when modifications of plain radiographs are considered, reported sensitivity and specificity were only of 14 and 70% in Tumeh et al. study (43).

Ultrasound (US), which can provide rapid and useful information for the diagnosis of musculoskeletal infections, can facilitate the microbiological diagnosis of infection, as guided puncture could be performed with US. However, negative US examination never excludes the diagnosis of BJI and further examinations should be performed when an osteomyelitis is suspected. As a matter of fact, US accuracy is particularly effective when performed on children in whom subperiosteal abscesses are frequent, due to a weak attachment to the subjacent bone, and for whom computed tomography and magnetic resonance imaging usually requires sedation or anesthesia (44).

Computed tomography (CT) scan has proved to be a useful examination in BJIs. It allows the delineation of subtile osseous changes, the detection of cortical sequestra, cortical defects, and intraosseous fistula. The use of intravenous iodinated contrast also allows the delineation of abscess. CT scan offers a good evaluation of the intra and extraosseous extent of the lesion, as well as guidance to biopsy.

Magnetic resonance imaging (MRI), as CT scan, is a cross-sectional imaging. MRI has an important role in the diagnosis of musculoskeletal infections, due to its capability to demonstrate anatomical details as well as pathological changes of bone marrow and surrounding soft tissues with high sensitivity and excellent spatial resolution. MRI provides a superior analysis of compartmental anatomy and a superior sensitivity to detect acute BJIs and soft-tissue abnormalities, in comparison with CT. Nevertheless it is not always the first cross-sectional test available and is more time consuming than CT. Marrow abnormality on MRI appears much earlier in the course of the disease than lytic changes seen radiographically. MRI screening usually includes T1-weighted, T2-weighted (with or without fat suppression) sequences as well as T1-weighted sequences with adjunction of gadolinium intravenous contrast. Contrast media enhancement, due to hyperemia and increased endothelial permeability, may help to define area of bone marrow inflammation and to abscess formation.

Native joint arthritis

During the course of native joint arthritis, the earliest change is accumulation of intra-articular fluid. US is very sensitive in detecting joint effusions during septic arthritis. Moreover, thickened synovial walls could be detected (45, 46).

Spondyliscitis and vertebral osteomyelitis

Lateral vertebral radiographs can show disc space narrowing, as well as erosion and destruction of vertebral body. Even if these findings are consistent with vertebral osteomyelitis, they usually are detected many weeks after beginning of clinical symptoms. MRI is the imaging modality of choice for detecting spondylodiscitis and vertebral osteomyelitis, as their complications, such as epidural abscess. MRI can show intervertebral disk and/or epidural enhancement, with vertebral marrow edema. End-plates erosion or destruction, with or without bony sequestra could be detected, especially with CT scan (Figure 2).

Distortion of local spine architecture consecutive to surgery is source of confusion with infection-related lesions. A particular type of infection, potentially due to HA-MRSA, is represented by post-surgical spondylodiscitis (47-49). Many changes in lumbar MRI, that appear to reflect accelerated disk degeneration, can be founded, mimicking infectious complication. Up to 6 months after discectomy, at an early post-operative stage, a high signal intensity band is seen to extend from the nucleus pulposus to the site of annular disruption on T2-weighted images. Ross et al. described in their study 20% of postoperative intervertebral disk annular enhancement that was not present on the preoperative MRI exam [47]. The intervertebral disc space enhances in approximately 67% of cases for up to 6 weeks after surgery. The marrow may develop an appearance similar to Type 1 (inflammatory) end-plate changes, resulting in low SI on T1-weighted images and high signal intensity on T2-weighted images (50). Nerve root enhancement in the immediate post-operative period may be a normal finding and can be attributed to breakdown of the blood-nerve barrier consecutive to surgery. In opposition, enhancement after 6 months is always considered pathological. Facet joint enhancement is also reported in more than 65% of patients at 6 weeks. This may represent a local response to surgical dissection (51). After 6 months, maturation of scar tissue in the laminectomy space results in a well-defined area of intermediate signal intensity on T1-weighted and hypointensity on T2-weighted sequences. The incidence of para-spinal soft tissue enhancement decreases from 100% in immediate post-operative stage to 18% of cases 6 months after surgery (51). Enhanced MRI is the technique of choice in detecting pathological post-operative changes, like discitis and its associated complications such as epidural and paraspinal abscess. Changes in MRI signal include loss of disc height, intradiscal high signal on T2-weighted images and loss of the intra-nuclear cleft. The end-plates usually show loss of definition, erosion as far as the vertebral marrow which is edematous. The major difference between a normal post-operative degenerative disk and an infected disk consist in the persistent low T2-weighted intensity of the degenerate disk. The most reliable pattern for infection remains the combination of diffuse enhancement of the intervertebral disc, annulus and vertebral body marrow.

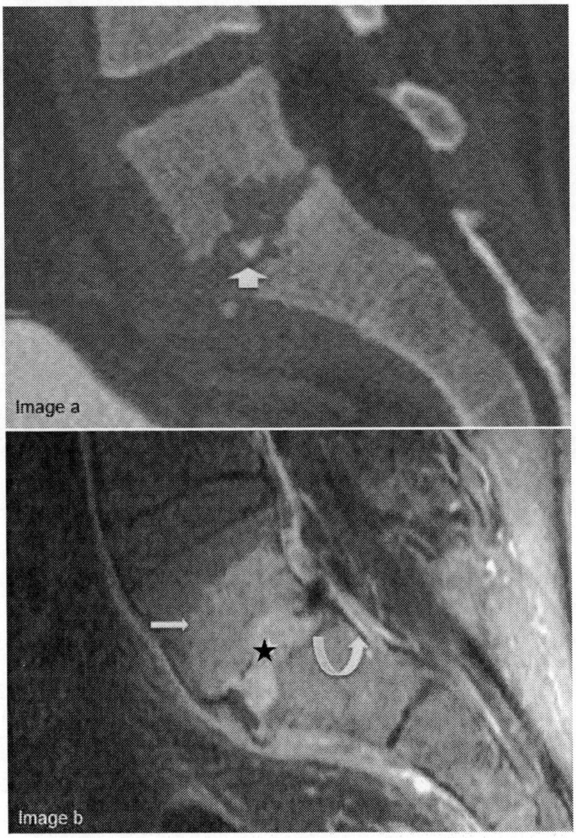

Figure 2. 25 year-old man with acute spondylodiscitis after HA-MRSA bacteremia. CT scan (image a) can show cortical sequestra (large arrow) as well as end-plate destruction T1 post gadolinium weighted MRI (image b) demonstrate inter-vertebral disk enhancement (black star) as well as epidural (curved arrow) and paraspinal soft tissue enhancement. It is difficult to detect bony sequestra with MRI, but vertebral marrow edema especially for L5 (thin arrow) is evident.

Long-bone chronic osteomyelitis

In their study, Tumeh et al. described that the only reliable conventional radiographs finding of bone infection was presence of a sequestrum. The sequestered bone fragment is visualized as a piece of osseous segment with increased density, demarcated from the adjacent osteopenic bone (43). Those signs are followed by periosteal thickening and osteopenia that can be sometimes detected early in the evolution of the infection. Cortical destruction appears later as small holes, which become larger coalescent lesions. US also contribute for the diagnosis of osteomyelitis, as it is able to detect early changes like oedema or fluid effusion in soft tissues surrounding suspected infected areas. During acute osteomyelitis, a subtle swelling close to the bone surface could be detected (45, 46). In case of subperiosteal abscess, an hyper or hypo echoic fluid collection could be seen (Figure 3).

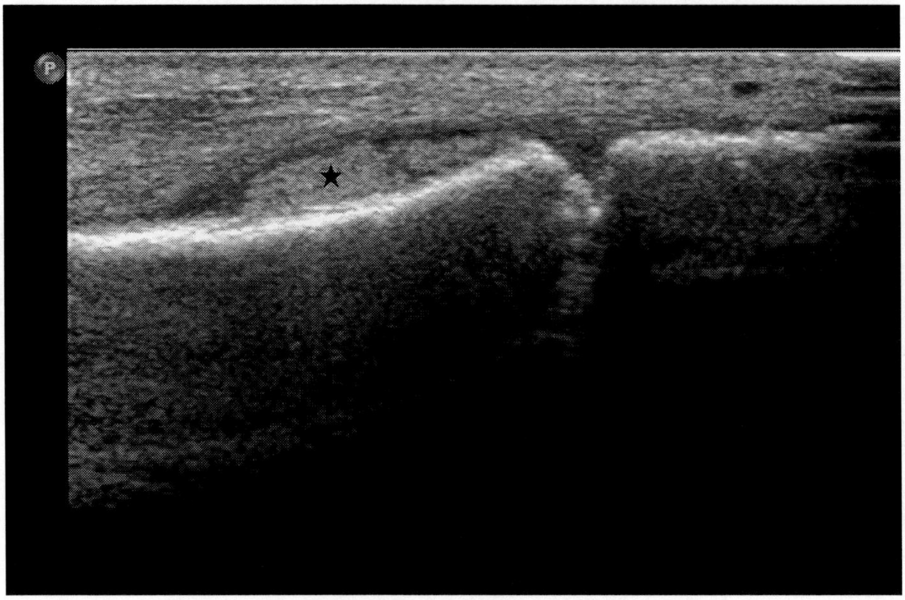

Figure 3. US of the lower extremity of the tibia of a child demonstrating a hyperechoic superiosteal abscess (black star) with mild inflammation of the surrounding soft tissues.

During chronic osteomyelitis, cortical erosions may be found by well-trained radiologists using a high frequency probe (52). With CT scan, as with MRI, contrast enhancement is strongly recommended during chronic osteomyelitis. A strong peripheral enhancement within medullar canal or soft tissue characterizes abscess. Sequestra or involucra appear hypo intense on T1- or T2-weighted sequences, without enhancement after adjunction of gadolinium. In chronic osteomyelitis, low signal intensity areas may be seen traducing devascularized fibrotic tissue and a relative well-defined interface between the normal and abnormal marrow is commonly observed. Nevertheless, detection of active infection may be very difficult in bone, particularly in bone that has been altered by prior traumatism or surgical procedure. In postoperative conditions, inflammatory changes such as edema pattern on T2-weighted sequences and contrast enhancement are constant, and may lead to an inaccuracy diagnosis of infection. This is due to similar signal intensities and contrast media enhancement during reparative fibrovascular scar tissue as during infection. Those signs may persist for up to 12 months after surgical intervention. One criterion that may help to discriminate is the pattern of contrast enhancement. Ledermann et al. have shown that granulation tissue in surgical bone defects is characterized by a homogeneous enhancement that change over time, up to leave a less vascularized scar in the center of the bone cavity. This process progresses faster in small bony defects, but equivocal findings, particularly in large defects, may persist up to 1 year (53). That is the reason why only a specificity of 60% of enhanced MRI, was documented in the study of Kaim et al. in a highly selective patient group with longstanding recurrent osteomyelitis (54). In this situation, bony condensation is frequently observed (figure 4). Finally, secondary signs of osteomyelitis, such as cutaneous ulcera, sinus tracts, abscess formation, and cortical interruption, are very useful in suspected chronic osteomyelitis.

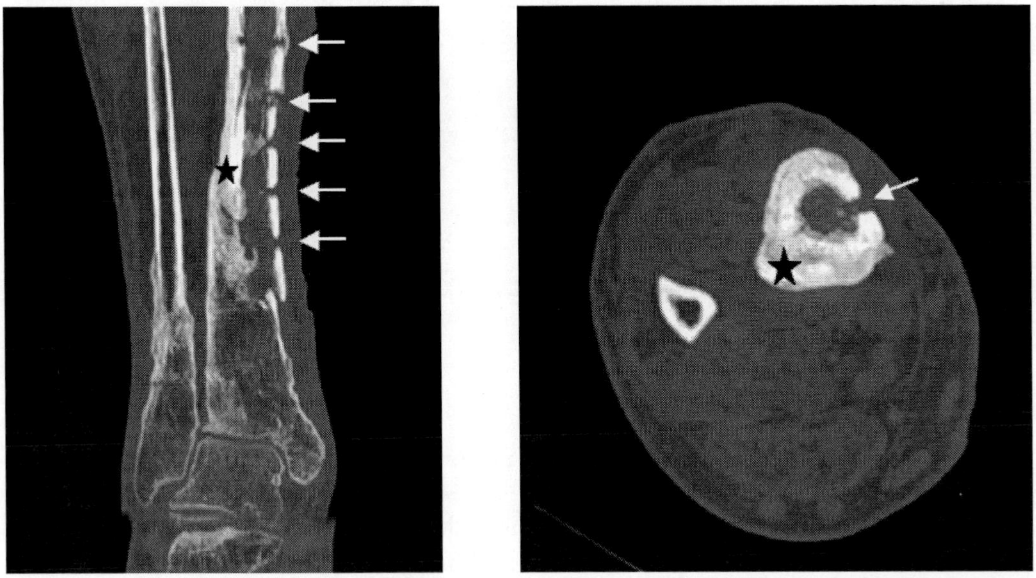

Figure 4. 59 year-old man, with chronic ostemyelitis of the tibia 3 years after osteosynthesis for leg fracture and 6 months after explantation of the material devices. Multiplanar reconstructions of CT scan show post-surgical sequella (white arrows) and very important bony condensation (black stars) due to chronic ostemyelitis. No abnormal soft tissue enhancement was observed.

Implant-associated osteomyelitis

Severe changes in bony architecture with localized osseous defects after fractures or surgery may be confused with infection-related osteolysis. US is also helpful to evaluate musculoskeletal soft tissues next to orthopedic hardware. It can depict a suggestive fluid collection and sinus tracts in the soft tissues close to the surgical hardware, indicating an active or a reactivated infectious process (55, 56). An important technical limitation of CT scan and MRI is occurrence of field inhomogeneities around a metallic foreign body. Such signal alterations are found in a high percentage, since patients frequently have undergone multiple surgical procedures with implantation and removal of surgical implants. Both over- and underestimation of lesion size may occur and can lead to equivocal findings on MRI images. In their series of 18 patients, Kaim et al. founded susceptibility artefacts in 15 of 18 patents that severely limited analysis in 3 patients only (54). Indeed, the area immediately surrounding metallic fragments, even if they are small, remain obscure. This is always a problem because those foreign bodies are potential source of infection. However, CT scan and MRI can show, despite very significant artefacts, significant soft-tissue collection adjacent to bone (Figure 5).

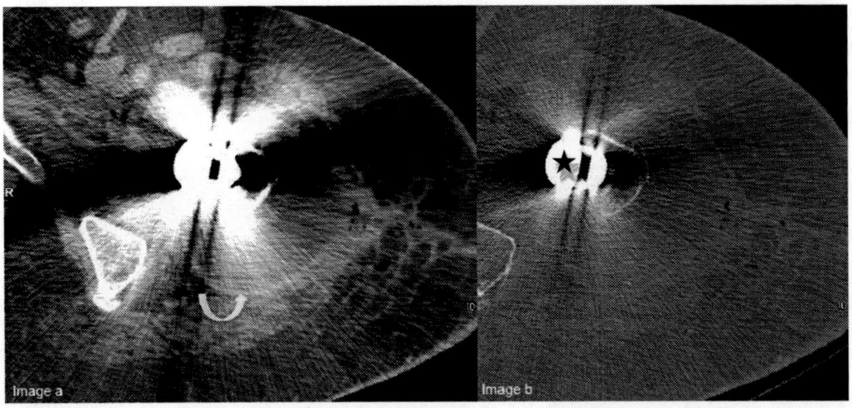

Figure 5. 75 year-old woman, 40 days after hip prosthesis. CT scan with iodinated enhancement shows, despite very significant artefacts due to prostheses (black star), an important soft tissue collection due to HA-MRSA (curved arrow), adjacent to bone.

LABORATORY DIAGNOSIS OF HA-MRSA BJI

The accurate diagnosis of BJIs including arthritis, osteomyelitis, and infections on orthopaedic implants due to HA-MRSA is usually made on the combined basis of clinical, imaging and laboratory techniques.

The Gold Standard remains histological and microbiological examination of the bone specimens and/or synovial fluid. Just a few data, specifically focused on BJIs due to MRSA, are available in the literature. But a large part of the general recommendations commonly reported for the diagnosis of BJIs can be applied to HA-MRSA. We chose to describe the various stages of the diagnosis and to stress the specific features of them for MRSA BJIs.

Analytical data

Some analytical data have been described as associated with the presence of BJIs but in fact they are not essential and must be just considered as one item in the evidence pool for diagnosis. The erythrocyte sedimentation rate, the level of C-reactive protein or leukocytosis lack sensitivity and specificity and are not reliable enough to be of routine value in the diagnosis of BJIs (57).

Gram staining

After collecting samples, Gram staining (usually best seen on a concentrated sediment after centrifugation in the case of synovial fuid) has a high specificity (more than 97 percent) but generally has a low sensitivity (less than 26 percent in some series) (58-60). Detection of the presence of Gram positive cocci is highly suggestive of staphylococcal infection but the identification of MRSA will require culture, identification tests and susceptibility tests.

Examination of synovial liquid

Data from synovial liquid have been proposed to evaluate joint disorders even if synovial aspiration is not always possible. Chemistries (lactate, glucose, protein, etc.) have been shown to lack both sensitivity and specificity for the diagnosis of BJI (61). The American College of Rheumatology has defined interpretative criteria of cell count and differential for synovial fluid in patients without orthopedic implants (Table 1). These criteria cannot be applied to patients with orthopaedic implants, and specific criteria for this kind of patients remind to be definitively established (62).

The cut-off values for diagnosing prosthetic joint infection are considerably lower than those for septic arthritis in native joints. In prosthetic knee infections, synovial fluid leukocyte count of more than 1.7×10^9/l and differential of more than 65% neutrophils had a sensitivity for infection of 94% and 97%, and specificity of 88% and 98%, respectively (63).

Table 1. Synovial fluid cell count criteria in patients without orthopaedic implants.

	Normal	Non inflammatory	inflammatory	purulent
Leukocytes (x10^3 cells/l)	<150	< 3 000	3 000 - 50 000	≥50 000
Neutrophils (%)	< 25	< 25	≥70	> 90

Culture

Microbiological cultures are currently regarded as the reference for identification of pathogenic bacteria. However, the sensitivity of these cultures is highly variable and depends both on the context in which clinical samples are taken, and on the pathogen involved.

As recently demonstrated for periprosthetic joint infections, prolonged microbiological culture at least for 2 weeks is necessary because it yields signs of periprosthetic infection in a significant proportion of patients that would otherwise remain unidentified (64). Even if *S. aureus* emerged predominantly during the first week, this delay appears reasonable to be sure to detect all *S. aureus* strain including SCVs (see below). Rich media such as blood agar, polyvitex agar in CO_2 atmosphere have to be preferentially used as well as an anaerobic agar plate combined with liquid medium such as Schaedler liquid medium or blood culture bottles (preferentially filled directly when sampled). This latter have the advantage to allow the direct inoculation of bacteria in liquid medium, dilution of potential remaining antibiotics from samples, continued detection of bacterial growth.

The rate of global bacterial detection varies according to infection type (50 to 70% for infectious spondylodiscitis, 65 to 95% for prosthetic-joint infections, 50% for gonococcal arthritis and 90% for non-gonococcal arthritis) and bacteria involved. It must be noticed that patients undergoing antibiotic treatment during sampling, and bacteria enclosed within biofilms on an orthopaedic implant are two situations that frequently explain false-negative cultures. It is the reason why, if possible, antimicrobial therapy should be discontinued at least 2 weeks prior to tissue sampling and perioperative prophylaxis at revision surgery should not be started until after tissue specimens have been collected for culture (65, 66).

Different kinds of sample (superficial wound, sinus tract, synovial fluid aspiration, bone, periprosthetic tissue) can be performed for the BJI diagnosis depending on the infection site and nature of BJIs. Cultures of a superficial wound or sinus tract are often positive because of microbial colonization from the surrounding skin. Only the isolation of *S. aureus* (including MRSA) from sinus tracts has been described as predictive of the causative pathogen isolated from deep intra-operative tissue specimens; other organisms had a low predictive value (67). Because of the difficulty of interpretation what can be misleading results, such samples should be avoided. In aspiratedsynovial fluid, the pathogen can be detected in 45 to 100 percent of cases (60). Cultures of periprosthetic tissue provide the most reliable means of detecting the pathogen in PJIs and are frequently used as the reference standard for diagnosing infections associated with prosthetic joints. The sensitivity of these cultures ranges from 65 to 94 percent, depending on the definition of infection (60).

In a study evaluating criteria for microbiological diagnosis of prosthetic-joint infection at revision arthroplasty, the isolation of an indistinguishable microorganism from three or more independent specimens is highly predictive of infection (sensitivity, 65%; specificity, 99.6%), while Gram staining was less useful (sensitivity, 12%; specificity, 98%) (58). Based on a simple mathematical model developed to predict the performance of the diagnostic test, the authors recommended five or six specimens be sent, that the cutoff for a definite diagnosis of infection be three or more operative specimens that yield an indistinguishable organism, and that because of its low level of sensitivity, Gram staining should be abandoned as a diagnostic tool at elective revision arthroplasty.

Recently, some studies showed that the culture of samples obtained by sonication of prostheses (allowing the dislodgement of microorganisms from the surface) was more sensitive than conventional periprosthetic-tissue culture for the microbiologic diagnosis of prosthetic hip and knee infection, especially in patients who had received antimicrobial therapy within 14 days before surgery (68, 69). The culture sensitivity appeared superior to that of standard periprosthetic tissue (75% versus 54%) whereas the specificity was 87% and 98% (70). Sonication in bags lacked specificity due to bag leakage. Given the significant risk of contamination during processing, the time-consuming and difficulty of the process more data are necessary before a generalization of this approach.

Cultures may be negative because of prior antimicrobial exposure, a low number of organisms, an inappropriate culture medium, fastidious organisms, or prolonged transport time to the microbiology laboratory. Some microorganisms, such as *Abiotrophia defectiva* or *Granulicatella adiacens* (formerly classified as nutritionally variant streptococci) or small-colony variants of staphylococci (see below), are difficult to detect. To detect cases of low-grade infection, antimicrobial therapy should be discontinued at least two weeks before tissue specimens are obtained. If revision surgery is planned, per-operative prophylaxis should not be administered until after tissue specimens have been collected for culture. If the prosthesis is removed, the implant or its components can be cultured in enrichment broth. However, the risk of contamination during specimen processing is high.

Identification

Staphylococci are catalase-positive, Gram-positive and cluster forming cocci. The genus contains more than 40 species. Speciation of isolates is essential to distinguish *S. aureus* from coagulase-negative staphylococci (CoNS), all the more for BJIs because CoNS that are often normal skin flora and classically considered as non-significant human pathogens, are conversely frequently associated with BJIs, especially from foreign bodies like prosthetic devices.

A variety of tests are used to differentiate *S. aureus* from the other staphylococcal species. *S. aureus* has traditionally been identified by the coagulase assay which detects free coagulase. But the detection of surface proteins such as clumping factor (CF) (also named bound coagulase) and/or protein A also enables rapid identification with commercialized kits. In these test systems (e.g, Dry Spot Staphytect Plus® (Oxoid), Pastorex Staph Plus® (Bio-Rad), Slidex Staph-Kit® and Slidex Staph Plus® (bioMérieux), Staphaurex Plus (Remel) and Staphylase Test® (Oxoid)), particles coated with specific antibodies precipitate with one or multiple *S. aureus* surface antigens, and allow *S. aureus* and CNS isolates to be distinguished within a few seconds. The majority of these tests are based on slide agglutination. Unfortunately, the accuracy of these test systems is not perfect. In particular, MRSA may be misclassified in agglutination tests because of false-negative results (71), which might be due to changes in various surface components, such as capsular polysaccharides, the clumping factor or protein A (72), and certain strains are therefore misidentified as coagulase negative staphylococci (CNS) (73). In an attempt to improve the accuracy (high sensitivity and high specificity) of the first commercialized *S. aureus* identification kits, the last third generation tests used latex particles coated with fibrinogen and antibodies to detect CF and protein A as well as with polyclonal and monoclonal antibodies specific for capsular antigens, especially type 5 elaborate by some MRSAs and not detected by previous latex (74). The presence of a DNase activity could also be used as a surrogate marker for the identification of coagulase-positive staphylococci. However, the specificity of this criterion or of the more laborious test for thermonucleases is not entirely satisfactory. A combination of coagulase test or DNase test with one of the latex agglutination is classically performed and allows the accurate and rapid presumptive identification of *S. aureus* from colonies obtained on culture plates. Then, commercial kit identification systems and automated instruments, based on biochemical features of isolates, can be used to confirm the identification. The most common reagents are: API STAPH gallery (bioMérieux), VITEK card (bioMérieux), MicroScan panel (Dade MicroScan), Phoenix Panel (Becton Dickinson). In case of doubt, molecular identification can be performed using homebrew PCR or commercial kits. A range of primers designed to amplify species-specific targets have now been developed. Such targets include the nuclease (*nuc*), coagulase (*coa*), protein A (*spa*), *femA* and *femB*, *Sa442*, 16S rRNA and surface-associated fibrinogen-binding protein genes.

The identification of *S. aureus* must be completed by the characterization of the methicillin resistance. Staphylococcal resistance to methicillin occurs when an isolate carries an altered penicillin-binding protein, PBP2a, which is encoded by the *mec*A gene. The new penicillin-binding protein (PBP) binds beta-lactams with lower avidity, which results in resistance to all molecules belonging to this class of antimicrobial agents. Standardized disc diffusion methods have been recommended by almost all national society for antimicrobial Chemotherapy (CLSI, BSAC, CA-SFM, etc.) and represent the most common way to detect the production of PBP2a. When resistance was first described in 1961, methicillin was used to test and treat infections caused by *S. aureus*. However, oxacillin, which is in the same class of drugs as methicillin, was chosen as the agent of choice for testing staphylococci in the early 1990s. Finally more recently, cefoxitin was adopted because it is an even better inducer of the mecA gene and disk diffusion tests using cefoxitin give clearer endpoints and are easier to read than tests with oxacillin. Nevertheless, the acronym MRSA for "methicillin resistant *Staphylococcus aureus*" is still used to describe these isolates because of its historic role.

A 10-min rapid slide latex agglutination test based on detection of PBP2a (75). The method involves extraction of PBP2a from suspensions of colonies and detection by agglutination with latex particles coated with monoclonal antibodies to PBP2a. The test is sensitive and specific with *S. aureus* (76). The method requires no special equipment and is suitable for confirmation of resistance or equivocal tests in routine clinical laboratories, but incubation times and experiment instructions must be followed accurately. Isolates producing small amounts of PBP2a may give weak agglutination reactions or agglutinate slowly. New immunochromatographic dipsticks using the same monoclonal antibodies should be available before the end of 2009. Finally, automated systems including Vitek2 (bioMérieux), Phoenix (Becton Dickinson) and Microscan (Dade Behring) include tests for oxacillin and cefoxitin susceptibility and are generally reported to be reliable for *S. aureus* although a small number of incorrect results, have been reported. In case of doubts, disc diffusion methods or molecular methods must be applied.

Nucleic acid amplification tests have been also developped to detect the *mec*A gene coding for PBP2a. In addition to classic homebrew PCR, numerous variations in combination with a variety of detection chemistries and automated systems are being developed. Currently available PCR-based molecular tests for MRSA include the LightCycler *Staphylococcus* and MRSA detection kit (LC assay; Roche Diagnostics, Mannheim, Germany), the Hyplex StaphyloResist PCR (BAG, Lich, Germany), the GenoType MRSA direct assay (Hain Lifescience, Nehren, Germany), the IDI-MRSA assay (GenOhm, San Diego, CA; BD Diagnostics), and the recently introduced GeneXpert MRSA assay (Cepheid, Sunnyvale, CA).

Molecular methods for BJI diagnosis

To overcome the drawbacks of microbiological culture, PCR based methods to detect and discriminate bacteria been used. These methods are of interest because of their high sensitivities and the potential for rapid detection of bacterial genome in a sample without prior knowledge of wether or not the bacteria can be cultured in vitro. Different For osteoarticular infections, few reliable studies have been reported (77-79). Different approachs have been proposed. Broad-range PCRs, using primers targeting 16S rRNA gene, the 23S rRNA and the *rpoB* gene are particularly suitable as they are ubiquitous to all bacteria (80). After DNA extraction, that is a clue step based on automated mechanical and/or chemical disruption, targeted genes is amplified in a first step. In a second step, the amplified product is sequenced. In a third step, the obtained bacterial sequence is analyzed and compared with that given in a database, mainly Genbank or Bibi, to find the closest similar sequences, allowing an accurate identification. Most studies have shown that the use of broad-range PCR and sequencing in BJIs increased mainly the identification of specific bacteria such as Kingalla, anaerobic bacteria, Streptococcus. For MRSA infections, the interest of such approach is limited to cases i) in which SCVs are involved and exhibited unusual growth requirements or failed to grow because of unfavorable environmental conditions ii) after antimicrobial exposure. The limit of conventional broad-range PCR is that it may lack both specificity and sensitivity. Moreover, these PCR techniques are cost intensive, methodically cumbersome, and always debated. In some studies, no noteworthy advantages over culture have been demonstrated (81-83). A significant advance is quantitative real-time PCR, in which amplification and detection of amplified products are coupled in a single vessel allowing a quicker results and a lower risk of contamination. Primers are designed to detect unique specific DNA sequence of a single species such as *S. aureus*. Pathogen-specific PCR has been shown to be more sensitive than 16S rDNA PCR in BJIs due to *S. aureus* (77, 84, 85). The data using such a specific PCR are seldom but the results appeared consistent and should be more extensively explored. The use of molecular biology based on DNA amplification provides a complementary assay to culture to identify microorganisms from bone or joint samples. This tool should be used for culture-negative cases when infection is suspected on the basis of clinical signs and symptoms or when inflammatory syndrome is present as highlighted by blood test results and purulent samples. Nevertheless, the significance of genotypic characterization is severely restricted with regard to the evaluation of antibiotic susceptibility.

S. aureus small colony variants (SCVs)

A variant subpopulation of *S. aureus* was first described about 100 years ago as a slow growing subpopulation by Kolle and Hetsch H. Since then, the significance of *S. aureus* SCVs has been demonstrated and explored (34). In contrast to the normal staphylococcal phenotype, SCVs are characterized by enhanced ability to persist intracellularly, fastidious growth characteristics, nonpigmented and nonhemolytic pinpoint colonies after 24–72 hours of incubation on blood agar, and absence or reduction of biochemical reactions (eg, mannitol salt agar negative). Most *S. aureus* SCVs are coagulase-positive on the tube test only after incubation for > 18 hours. Thus, these uncommon morphological and physiological features of SCVs present a challenge to clinical microbiologists in terms of recovery and identification (34).

To avoid misidentification and non-diagnosis, a prerequisite is the application of extended conventional culture and identification techniques (86, 87). If possible, the most accurate and rapid method to detect both the *S. aureus* species and the SCV phenotype is to inoculate specimens on both Columbia blood agar and the new chromogenic agar, *S. aureus* ID agar (bioMérieux, La Balme Les Grottes, France) (87). SCVs are rapidly overgrown and are easily missed when the normal *S. aureus* is present because they divide about nine times slower than normal-phenotype *S. aureus*. Isolates suspicious for *S. aureus* SCVs, which may give a false negative coagulase test, should be confirmed as *S. aureus* by testing the species-specific *nuc* and *coa* genes or by diagnostic sequencing of 16S rDNA sequences (86).

The intracellular location of *S. aureus* SCVs may provide a survival niche for these bacteria because the microorganisms are protected against antibiotic therapy and host defenses. This phenotype, inducing persistent and recurrent infections has been identified especially in BJIs (34). Recently, according to a model based on the study of clinical isolates associated with clinical patterns of persistence and relapse, microbiological and clinical features of *S. aureus* SCVs were related to alterations in electron transport inducing menadione or/and hemin auxotrophism. The *S. aureus* SCVs reverts to normal colony forms in the presence of these substances and/or CO2 supplement (34).

For several reasons, SCVs also present a challenge with regard to susceptibility testing: i) SCVs are often in mixed populations with normal *S. aureus*, thus, even a small percentage of normally growing organisms will rapidly replace the SCVs in liquid medium in an overnight culture, thereby making susceptibility testing (as well as recovery) of the SCVs difficult (88); ii) the slow growth makes standardization of testing difficult because slow growth alters diffusion tests and times for measuring susceptibility; iii) irrespective of their auxotrophism, errors may occur when these variants are resistant to oxacillin when tested by disc-diffusion test, Etest, microdilution test, and automated susceptibility-testing systems as well as anti-PBP2a slide-latex agglutination tests (87, 89). In consequence, *mec*A gene detection by molecular methods or the use of an anti-PBP2a slide-latex agglutination test using a drastically increased inoculum (approximately a loopful with 100–200 SCV colonies) should be applied for reliable diagnosis or validation of MRSA-SCVs (89).

In any case, it has been demonstrated that interruption of electron transport (see below) reduces the electrochemical gradient across the bacterial membrane, resulting in a decreased uptake of antimicrobial agents that require a charge differential to be active, such as aminoglycosides. Therefore, these substances should not be used, although single strains with an SCV phenotype might be susceptible to gentamicin or other aminoglycosides. Finally, *S. aureus* SCVs are often also more resistant to cell wall active antibiotics (34). Slow growth, and hence reduced cell-wall division, reduces the efficacy of beta-lactam antibiotics (90). Optimal therapy for infections due to *S. aureus* SCVs has not yet been defined. In the past, it was found that trimethoprim-sulfamethoxazole combined with rifampicin was the most active therapeutic regimen in a tissue-culture system with SCVs being inside endothelial cells; but more studies are necessary to define the optimal therapy, especially to include the therapeutic place of new antibiotics such as tigecycline, linezolid or daptomycin (34).

TREATMENT OF HA-MRSA BJI

Surgical Treatment of HA-MRSA BJI

Surgery may go hand in hand with antibiotic treatment, but surgery can accomplish in one hour what the antibiotic treatment may require days or weeks to do. The purpose of surgery is to perform debridement that reduces the inoculum and removes necrotic and avascular bone, bacteria, and harmful bacterial products.

Native joint arthritis

There are three essential principles in the management of acute septic arthritis: (i) the joint must be adequately drained; (ii) antibiotics must be given to diminish the systemic effects of sepsis; and (iii) the joint must be rested in a stable position (91). Prompt drainage and evaluation of purulent joint fluid seems to be crucial for preservation of articular cartilage and for resolution of the infection (92). Few questions remain concerning the beneficial effects of early drainage of an infected joint, but controversy continues about the advantages of open surgical drainage, arthroscopic drainage and multiple aspirations. Excellent results have been reported with all three methods and clinical comparisons of aspiration with arthroscopy or arthrotomy are inconclusive (93). Drainage should be performed for all infectious arthritis. Adjuvant therapies, such as synovectomy, salicylate administration, and constant passive motion, although theoretically helpful, have not been shown to influence the results.

If the diagnosis is made early and the involved joint is superficial, such as the elbow or ankle; aspiration should be performed and repeated if necessary. Appropriate antibiotics should be administrated, and the joint should be splinted in a position of function. The patient should be observed for a decrease in pain, swelling, and temperature and for improved joint mobility. If the response is not favorable and repeat aspiration does not show a decrease in the synovial leukocyte count within 24 to 48 hours, open surgical drainage is necessary (94). If purulent material is deeply situated in a joint, such as the shoulder or hip, open surgical drainage should be done. Arthroscopic drainage is a good alternative to open drainage in many instances, especially for infections involving the knee, elbow, shoulder or ankle.

The ankle may be drained through any approaches: anterolateral, anteromedial, posterolateral or posteromedial (95). After the capsule has been incised, the wound is closed loosely over drains. A posterior splint is applied with the foot in neutral position and the ankle at 90 degrees. The splint is worn until the wound has healed; the graduated weight bearing and active range-of motion exercises are begun.

In acute knee septic arthritis, usually anteromedial arthrotomy or arthroscopic drainage and antibiotic treatment are adequate. In more difficult cases, if the posterior compartment of the knee is distended (with sometimes a popliteal abcess) parallel anterior incisions combined with posterolateral and posteromedial incisions usually are best. A posterior midline approach should not be used to drain an infected knee because it exposes the popliteal vessels to pus and to pressure from the drain and creates a potentially contracting scar across the joint. Several studies have reported good results with arthroscopic drainage for acute arthritis of the knee in adults (96). It combines the advantages and avoids the disadvantages of needle aspiration and arthrotomy. With arthroscopy, purulent material can be removed, and the joint can be irrigated. The joint cartilage can be inspected, and loculations or adhesions can be removed with the arthoscope. A partial synovectomy can be performed if necessary. Drains can be placed into the joint through the portal sites for drainage or for a continuous suction drainage system. Arthroscopy also has the advantage of allowing much earlier range of motion and rehabilitation of the knee joint compared with arthrotomy (97).

During hip septic arthritis, drainage may be accomplished through a posterior, medial, lateral or anterior approach. In an adult, the posterior approach allows dependent drainage and is a more familiar approach for most orthopaedic surgeons. Some authors reported good results with arthroscopic treatment too (98). After treatment patients are confined to bed rest in Buck traction until the wound has healed and the patient can control the leg. Protective weight bearing using crutches is permitted and active range-of-motion exercises are started.

In adults, septic arthritis of the shoulder usually is associated with a debilitating disease and rarely responds well to treatment. The joint should be aspired whenever an infection is suspected, and early surgical drainage is indicated if franc pus is obtained. The shoulder may be drained through an anterior incision. In a review of adult patients treated for septic arthritis of the shouder, Leslie et al. found that arthrotomy yielded better results than repeated aspirations (99). Arthrsoscopy drainage has been reported and is seen in increasing frequency for the treatment of acute septic arthritis of the shoulder. With arthroscopy, washout is done under direct vision ensuring complete removal of purulent material. The advantages are the same than for the knee. After treatment, the shoulder is supported on a splint at 45 degrees of abduction until the wound has healed. Then active and active-assisted range-of-motion exercises are started.

Vertebral osteomyelitis

The primary method of treatment of vertebral osteomyelitis is antibiotic therapy. However, a surgical intervention is required in case of large abscess formation, neurological deficit due to epidural abscess with stenosis, instability or severe deformity.

Autologous bone grafting and instrumentation if instability occurs after complete infected tissue removal. Many authors reported good results with early and "aggressive" surgical treatment (100).

The method of surgical treatment requires an accurate assessment of the location of the abscess and the presence of an associated osteomyelitis (101).

Acute or chronic isolated dorsal (posterior), lateral, and some ventral (anterior) infections are best treated with total laminectomy for drainage and closure over drains. Epidural infections associated with osteomyelitis are best exposed by anterior or postero-lateral exposures that allow treatment of the osteomyelitis and the epidural infection.

If the cervical spine is involved, the abscess may be present retropharyngeally in the posterior triangle of the neck or supraclavicular area. Drainage can be performed in an emergency through an incision in the posterior wall of the pharynx (anterior approach), but most of the time drainage should be through an extraoral antero-lateral approach. Most abscesses caused by disease of the dorsal spine can be evacuated by costotransversectomy (dorsal approach). The drainage of lumber spine has no particularity and can be performed with a longitudinal incision lateral to the midline parallel to the spinous processes (dorsal approach).

Recently some authors propose endoscopic surgery with promising results, but this is only indicated in case of very serious comorbidity (102).

Long-bone chronic osteomyelitis

Long-bone chronic osteomyelitis generally cannot be eradicated without surgical treatment. Antibiotics alone rarely can eradicate the infection for numerous reasons: Bacteria are able to adhere to bone matrix through various receptors and can form a slimy coat that protects them from phagocytic cells and antibiotics. Surgery for chronic osteomyelitis consists of sequestrectomy and resection of scarred and infected bone and soft tissue (18, 24). The goal of surgery is eradication of the infection by achieving a viable and vascular environment. Radical debridement may be required to achieve this goal. Inadequate debridement may be one reason for recurrence in chronic osteomyelitis. Simpson et al. prospectively evaluated the extent of surgical resection margins for the treatment of chronic osteomyelitis in 50 patients (103). There was significant difference between wide resection (>5 mm) and marginal resection (<5 mm). Adequate debridement often leaves a large dead space that must be managed to prevent recurrence but bone loss may result in bony instability. Appropriate reconstruction of the bone and soft-tissue defects may be needed after proper identification of the infecting organism and appropriate antibiotic therapy. Reconstruction should be undertaken only after careful planning and identification of sequestra and intraosseous abcesses by plain radiographs, CT scan and MRI (104).

Sequestrectomy and curettage require time to perform and result in considerable bone loss. Consequently, appropriate preparation should be made before surgery. Sinus tracks can be injected with methylene blue 24 hours before surgery to make them easier to locate and excise. The infected area of bone should be exposed with complete excision of all sinus tracks. A drill can be use to outline a cortical window at the appropriate site, that can be then removed. All sequestra, purulent material and necrotic tissue have to be removed.

If a cavity cannot be filled by the surrounding soft tissue, a local muscle flap or a free tissue transfer can be used to obliterate the dead space (105). If there is a nonunion present with any bony instability, the bone must be stabilized. If possible, the skin should be closed lossely over drains, and ensure that no excessive skin tension is present. If closure is impossible, a delayed closure or skin grafting is planed.

Management of bony and soft-tissue defects

Bony and soft-tissue defects must be filled to reduce the chance of continued infection and loss of function. Several techniques have been described for the management of such defects and have proved successful when properly performed, but they require meticulous surgical technique.

The methods described to eliminate this dead space are: (i) bone grafting with primary or secondary closure; (ii) use of antibiotic polymethyl methacrylate (PMMA) beads as a temporary filler of the dead space before reconstruction (106); (iii) local muscle flaps and skin grafting with or without bone grafting (107); (iv) microvascular transfer of muscle, myocutaneous, osseous, and osteocutaneous flaps (108); and (v) use of bone transport (Ilizarov technique) (109).

Amputation for osteomyelitis

Amputation is performed infrequently for osteomyelitis. In certain patients, this form of treatment may be preferable, however, to multiple operations that are unlikely to eradicate the infection.

Implant-associated osteomyelitis

When the diagnosis of infection is established, treatment options are debridement with prosthesis retention, resection arthroplasty, one-stage or two-stage reimplantation or arthrodesis.

Joint debridement with prosthesis retention

It is limited to a small subset of patients: patients with an early (< 4 weeks postoperatively, acte onset of symptoms) with a well-fixed prosthesis without signs of loosening. For some authors, infection with *S. aureus* is a relative contraindication to debridement and component retention (110).

Some authors emphasized several points that could lead to higher success rates for this approach: (i) infectious disease consultation and antibiotic monitoring; (ii) diagnosis and treatment of hematogenous sources of infection; (iii) newer antibiotics; (iv) six-week duration of postoperative intravenous antibiotics; (v) repeated cultures within 2 weeks of the initial debridement and re-debridement if these cultures were positive; and (vi) polyethylene exchange at the time of debridement; exchange of grown, gloves, and instruments; and redraping at the time of wound closure (111).

According to their experience, all of the early postoperative infections and 70% of the acute hematogenous infections could be treated successfully.

Resesection arthroplasty

It was for example described by Kaufer and Matthews for the knee (112), consisting of removal of the infected prosthesis and ciment and debridement of the synovium. Resection arthroplasty is ideal for a patient with an infected TKA and severe polyarticular rheumatoid arthritis with limited ambulation. Such treatment is preferable to fusion in these patients because it allows some mobility for sitting.

Many authors have confirmed that infection can be controlled in most patients, but function is poor because of instability and mild to moderate residual pain with standing (113).

Arthrodesis

Arthrodesis as treatment of an infected arthroplasty can provide a stable, generally painless limb with some expected shortening (114). The indications are high functional demands, disease involving a single joint, young age, deficient extensor mechanism (for the knee), poor soft tissue coverage, immunocompromised patient, and infection with a highly virulent microorganism. Various techniques of arthrodesis have been used, including external fixation, plating, and intramedullary nailing (for the knee).

Exchange arthroplasty

Exchange arthroplasty in one or two-stages offers the greatest chance of functional recovery after an infected arthroplasty.

One-stage exchange arthroplasty is performed essentially as a revision arthroplasty with an emphasis on debridement and surgically rendering the wound a close to sterile as possible. Some authors reported excellent results with such techniques (89% success rate for Silva et l. and 88% success for Göksan et al.) (115, 116). The series are characterized by susceptible organisms, the use of antibiotics-impregnated cement and prolonged postoperative antibiotic therapy.

More commonly, exchange arthroplasty is performed in two stages: initial prosthesis removal and debridement followed by a period of intravenous antibiotics and later reimplantation. The most commonly accepted protocol calls for 6 weeks of intravenous antibiotics, followed by reimplantation of an new prosthesis (117).

Antibiotic-impregnated PMMA spacers are used by many surgeons to maintain soft tissue tension of the articulation during the interval between debridement and reimplantation in two-stage procedures. Some authors suggested benefits of this technique include high levels of local antibiotic delivery, improved exposure at the time of reimplantation, and the ability to maintain weight bearing during the interval period (118). Of course some concern can be expressed about leaving a foreign body within the wound, along with potential bone loss from weight-bearing on the spacer block. Nevertheless, success rates of 88% to 96% have been reported (119).

Amputation

The last option for the treatment of the infected arthroplasty is amputation. It essentially concerns infected knee arthroplasty (according to Rand, after reviewing multiple series of infected TKAs, the frequency of amputation was 5.5% (120)). Amputation is indicated only for life-threatening infection or persistent local infection with massive bone loss, not amenable to arthrodesis.

Medical Treatment of HA-MRSA BJIs

Generalities

As a general principle, nosocomial soft tissue and osteo-articular infections due to healthcare-associated MRSA (HA-MRSA) are preventable to a high proportion. They represent a failure in the chain of quality of care, are expensive and ultimately compromise patients' safety (121). Hence, every attempt should be done to reduce further transmission and infection of HA-MRSA. A good surveillance of MRSA cross-transmission / infection, or of surgical site infections in general belong to quality requisites of an Orthopaedic Service.

The relatively high failure rate following antibiotic treatment of bone infection is well documented. Chronic osteomyelitis and prosthetic joint infections are particularly difficult to cure and it has been suggested that 'arrest' is a more appropriate term than 'cure' for defining outcome in chronic osteomyelitis as relapse may occur many months or years later (122).

The antibiotic resistance of Staphylococcus sp or of any other pathogen does not alter the general management of osteo-articular infections; there is a common need for obtainment of good quality culture specimens, adequate surgical treatment, sufficient duration of antibiotic treatment, assessment of the patients' adherence to therapy, and an adequate general management. Differences among various pathogens lie in different pathogenic microorganisms, the choice of antibiotic agents, and regarding infection control policies.

Obtainment of culture specimens is important not only for documentation of the presence of MRSA but also for susceptibility testing to guide therapy. The resistance pattern of MRSA isolates may vary from one region to another.

Antibiotic agents per se are not always necessary. Their indication is clinical and depends of the severity of disease and the co-morbidities of the host. This attitude does not rely on the antibiotic resistance of the isolate. In healthy patients with small purulent lesions with localised abscesses of soft tissues and no systemic signs of infection, drainage alone without concomitant antibiotic treatment may be sufficient (123).

Table 2. Antibiotic treatment of bone and joint infections due to healthcare-associated MRSA (concomitant to surgery)

	Parenteral treatment			Oral treatment			
	Antibiotic	Alternatives	Duration	Antibiotic	Alternatives	Duration	Remarks
Arthroplasty infection with implant retention	Vancomycin[a] + rifampin[b]	Teicoplanin[c]	~2 weeks	Fusidic acid[g] + rifampin[b]	Ciprofloxacin[h] + rifampin[b]	2½ months	Teicoplanin and fusidic acid not available in the U.S.
		Daptomycin[n]	~2 weeks		Levofloxacin[i] + rifampin[b]	2½ months	Total duration: 3 months. For knee joint prostheses, up to six months can be considered
		Tigecycline[d]	~2 weeks		Doxycyclin[k] + rifampin[b]	2½ months	
		Linezolid[e]	~2 weeks		Minocyclin[l] + rifampin[b]	2½ months	
		Ceftobripole[f]	~2 weeks		Cotrimoxazole[m] + rifampin[b]	2½ months	
Arthroscopy infection with implant removal	Vancomycin[a] + rifampin[b]	Teicoplanin[c]	~2 weeks	Fusidic acid[g] + rifampin[b]	Ciprofloxacin[h] + rifampin[b]	~4 weeks	Teicoplanin and fusidic acid not available in the U.S.
Other implant-related with implant retention		Daptomycin[n]	~2 weeks		Levofloxacin[i] + rifampin[b]	~4 weeks	Total duration: 6-12 weeks
		Tigecycline[d]	~2 weeks		Doxycyclin[k] + rifampin[b]	~4 weeks	
		Linezolid[e]	~2 weeks		Minocyclin[l] + rifampin[b]	~4 weeks	
		Ceftobripole[f]	~2 weeks		Cotrimoxazole[m] + rifampin[b]	~4 weeks	
Acute osteomyelitis without implant	Vancomycin[a]	Teicoplanin[c]	~1-2 weeks	Fusidic acid[g] + rifampin[b]	Ciprofloxacin[h] + rifampin[b]	~4-6 weeks	Teicoplanin and fusidic acid not available in the U.S.
		Daptomycin[n]	~1-2 weeks		Levofloxacin[i] + rifampin[b]	~4-6 weeks	Total duration: 4-6 weeks For chronic osteomyelitis without surgical removal: 3 months
		Tigecycline[d]	~1-2 weeks		Doxycyclin[k]	~4-6 weeks	
		Linezolid[e]	~1-2 weeks		Minocyclin[l]	~4-6 weeks	
		Ceftobripole[f]	~1-2 weeks		Cotrimoxazole[m]	~4-6 weeks	
Native joint arthritis	Vancomycin[a]	Teicoplanin[c]	~2-3 weeks	Fusidic acid[g] + rifampin[b]	Ciprofloxacin[h] + rifampin[b]	~0-1 weeks	Teicoplanin and fusidic acid not available in the U.S.
		Daptomycin[n]	~2-3 weeks		Levofloxacin[i] + rifampin[b]	~0-1 weeks	Total duration: probably 2-4 weeks. In short durations, only iv

Table 2. (Continued)

Parenteral treatment				Oral treatment			
		Tigecycline[d]	~2-3 weeks		Doxycyclin[k]	~0-1 weeks	
		Linezolid[e]	~2-3 weeks		Minocyclin[l]	~0-1 weeks	
		Ceftobripole[f]	~2-3 weeks		Cotrimoxazole[m]	~0-1 weeks	
Spondylodiscitis	Vancomycin[a]	Teicoplanin[c]	~2-4 weeks	Fusidic acid[g] + rifampin[b]	Ciprofloxacin[h] + rifampin[b]	~2-4 weeks	Teicoplanin and fusidic acid not available in the U.S.
		Daptomycin[n]	~2-4 weeks		Levofloxacin[i] + rifampin[b]	~2-4 weeks	Total duration: 6 weeks in absence of epidural abscess
		Tigecycline[d]	~2-4 weeks		Doxycyclin[k]	~2-4 weeks	
		Linezolid[e]	~2-4 weeks		Minocyclin[l]	~2-4 weeks	
		Ceftobripole[f]	~2-4 weeks		Cotrimoxazole[m]	~2-4 weeks	

[a]Vancomycin 2 x 15 mg/kg iv or 30 mg/kg/d in continuous infusion. Targeted serum vancomycinemia in steady state: ~25 mg/L

[b]Rifampin 600-1200 mg/d. Parenteral medication not necessary. Always in combination, never alone (development of resistance). In absence of implants, rifampin is not indicated, but may be used in combination therapy because of good bone penetration.

[c]Teicoplanin 1st day 2 x 400 mg intravenously. From 2nd day 1 x 400 mg iv. It can also be given by intramuscular route.

[d]Tigecycline 100 mg iv once, thereafter 2 x 50 mg/d iv. Mostly experimental so far.

[e]Linezolid 2 x 600 mg/d. In non-bacteremic cases, linezolid orally. Interaction with MAO-inhibitors, myelosuppression and polyneuropathy.

[f]Ceftobripol 2-3 x 500 mg/d. No studies for osteomyelitis so far.

[g]Fusidic acid 3 x 500 mg/d. Always in combination (possible development of resistance during monotherapy)

[h]Ciprofloxacin 2 x 500 mg/d. Only if the MRSA is susceptible. This is rarely the case.

[i]Levofloxacin 2 x 500 mg/d. Only if the MRSA is susceptible. This is rarely the case.

[k]Doxycyclin 2 x 100 mg/d.

[l]Minocyclin 2 x 100 mg/d.

[m]Cotrimoxazole 2 double-strenght tablets (800mg trimethroprim, 160mg sulfadiaxozide) per day. May have failure when high inoculum of MRSA.

[n]Daptomycine 6-10 mg/kg/d once daily intravenously. Few data on human osteo-articular infections available so far.

The impact of methicillin-resistance in the outcome of localized staphylococcal osteo-articular infections

Contrary to staphylococcal bloodstream infections, it is an unresolved topic if staphylococcal methicillin-resistance increases the likelihood of treatment failure in localized soft tissue or osteo-articular infections. Al-Nammari et al. reported the same duration of antimicrobial therapy and the same number of surgical interventions required for the treatment of septic arthritis due to MRSA or susceptible *S. aureus* (MSSA) (124). Volin et al proved that methicillin-resistance did not influence cure in patients with two-staged re-implantation after periprosthetic total joint infection (125). In contrast, Kilgus et al showed that hip replacements infected with MRSA were treated successfully only in 48% of cases compared to 81% with MSSA (126). In a more recent study, HA-MRSA in peri-prosthetic tissue culture resulted in a higher failure rate than MSSA when controlling for joint hip arthroplasty and removal of hardware (127).

Choice of antibiotic agents

Single-agent antibiotic therapy is usually adequate for intravenous treatment in absence of implant material. Antibiotics should be chosen based on the susceptibility of the isolated MRSA, bone penetration and, in later stage, on the best oral bioavailability (Table 2). In patients with complicated infections requiring hospitalization or intravenous therapy, vancomycin is the drug of choice because of the low cost and long-term experience. Teicoplanin, linezolid, daptomycin, tigecycline, and ceftobiprole are alternative agents. The latter three are new agents and yet in experimental status for osteo-articular infections due to MRSA.

New agents versus older glycopeptides for parenteral treatment

In general, cure rates for skin, soft-tissue and osteo-articular infections due to HA-MRSA vary between 60-90% with any antibiotic. The newer anti-MRSA agents are not adequately studied yet for bone infections and do not necessarily yield better outcomes than the older glycopeptide antibiotics vancomycin and teicoplanin in terms of microbiological eradication and clinical cure. Better pharmacokinetic and -dynamic properties do not need to show a direct correlation with clinical outcome. Clinical trials with the newer agents have been designed as non-inferiority studies and only in fact in skin and soft tissue infections. Moreover, the majority of trials did not require a vancomycin dose adjustment or report the vancomycin serum levels. The newer molecules have nevertheless other undeniable advantages over glycopeptide antibiotics. In particular, tigecycline and ceftobripole also a wide variety of Gram-negative pathogens, which is beyond the topic of this chapter?

Parenteral versus peroral treatment, duration of treatment

As a general principle for acute osteomyelitis, arthritis and prosthetic joint infections, antibiotics are recommended parenterally for 2 weeks, followed by several weeks of oral therapy (128). This initial two weeks of parenteral medication is based on experts' opinion rather than on clinical trials. Due to lack of prospective randomized trials, it remains difficult to change this established opinion. Without doubt, the bone penetration of antibiotic agents in parenteral administration is good and the bioavailability per definition 100% (129). However, recent retrospective data suggest that regimens with an early switch to oral antibiotics are as effective as prolonged parenteral regimens at least for osteomyelitis due to methicillin-susceptible *S. aureus* (MSSA) (130). Whether this finding can be applied for HA-MRSA infection, remains to be elucidated.

In case of concomitant endocarditis, spondylodiscitis or undrained abscesses, this intravenous application may be prolonged to four or six weeks.

Glycopeptides

Vancomycin

Vancomycin is a glycopeptide that inhibits cell-wall synthesis and has a serum half-life of 6h. Oral bioavailability is 2% at best. It is not dialysed. In dialysed patients, it can be administered during dialysis three times a week. Since the bone penetration of vancomycin is only about 15-30% of the serum concentration, minimal serum through levels of 20 mg/ml are believed to treat bone infections best (131). Fortunately, penetration in infected bone is higher than in uninfected one (132). Often, nephrotoxicity concerns argue against a high dose. This issue has been exaggerated in the past, partly also due to incapacities concerning vancomycin fabrication and the use of concomitant nephrotoxic medication. As an example of recent literature, the nephrotoxicity rate in patients receiving vancomycin dropped from 12 to 2% if patients with concomitant nephrotoxic medication were analyzed separately (133). Vancomycin should be administered over 1 hour, to prevent a histamine-mediated "red-man" syndrome, which should not be confounded with a true, but rare, allergy or pseudoallergy.

In continuous perfusion of vancomycin, the changes in serum concentrations are much lower than in intermittent application. The target concentrations are achieved quicker with less adverse drug effects (131). In continuous perfusion, vancomycin is practically always in steady state and above the required minimal inhibitory concentration. However, continuous perfusion does not guarantee a better outcome in term of cure of infection.

Teicoplanin

Teicoplanin is available in Europe and elsewhere, but not in the United States. It is a glycopeptide with a long serum half-life of 72 hours (134). It can be administered within half an hour; generally at a dose of 400 mg once a day parentally (after a charging dose of 2 x 400 mg the first day). Alternatively, teicoplanin can be given intramuscularly or be administered three times a week, even for the treatment of bone and joint infections (135, 136). For arthritis and staphylococcal bone infections, a high dose of serum concentrations appear necessary to treat bone infection, but the ideal serum through level or daily dose remains to be established. Le Frock et al. used doses of 6-12 mg/kg to treat bone and joint infections. After an average of 6 weeks therapy for osteomyelitis and 3 weeks for septic arthritis, cure rates were 90%, 88% and 100% for acute and chronic osteomyelitis and septic arthritis, respectively (135).

Novel antibiotic agents

Daptomycin

Daptomycin depolarizes membranes and yields a rapid, dose-dependant bactericidal effect. The serum half-life is 9h. It is only available in parenteral form and is administered once a day at a dose of 6 mg/kg in absence of renal dysfunction. This makes it suitable for an outpatient treatment. Daptomycin has shown its efficacy in clinical trials in the field of osteo-articular, skin and soft tissue infection, although so far it remains FDA-approved only for the treatment of complicated skin and skin-structure infections (137). Few adverse events are known, among them muscular toxicity with an elevation of creatinin-phosphokinases. Clinicians should keep in mind that emergence of a daptomycin-resistant *S. aureus* isolate during treatment of initially daptomycin-susceptible MRSA bacteraemia and osteomyelitis has been decribed (138). Trials with higher doses up to 10 mg/kg are being tried to overcome this problem.

Tigecycline

Tigecycline belongs to the glycylcyclines, a further development of tetracycline antibiotics. Today, it has to be considered as an experimental drug for osteo-articular infections (139). It inhibits ribosomal protein synthesis, and is only available in parenteral form: charging dose of 100mg, followed by 50mg bid i.v. As for daptomycin, FDA approval for management of MRSA infections is limited to skin and soft tissue infections. A trial in diabetic foot infections with osteomyelitis is ongoing, but studies in other forms of osteomyelitis due to MRSA are clearly required. Major adverse effects include nausea and vomiting.

Ceftrobripole

Ceftobiprole is the first cephalosporin that is bactericidal against MRSA and most Gram-negative organisms. It is adapted to renal function and inhibits cell-wall synthesis. With a serum half-life of 4 hours, it is administered at a dose of 500 mg bid i.v. for MRSA infections (tid i.v. for severe Gram-negative infections) (140, 141). Ceftrobiprole has been tested in clinical trials of skin and soft tissue infections as well as healthcare-acquired infections (142). As a beta-lactam antibiotic, it is a very attractive compound that requires further testing in osteo-articular infections.

Second line intravenous drugs

Aminogylcosides

Aminoglycosides might be indicated in combination therapy for sustained bacteraemia, but are not indicated for osteo-articular infections: They are less active in synovial fluid or in bone (143). Furthermore, staphylococcal small-colony variants, a hallmark of chronic pre-treated osteo-articular *S. aureus* infections, are generally resistant to aminoglycosides (144). However, in desperate situations and in low-income countries, aminoglycosides might be an option (i.v. or intramuscularly) if the HA-MRSA is susceptible.

Outpatient parenteral antibiotic therapy (OPAT)

OPAT was widespread during the last two decades to treat patient with continuous intravenous antibiotics outside of the hospital setting (145). However, for most situations, OPAT may be obsolete, because we dispose of a good per oral armentarium against HA-MRSA. In special circumstances such as impossibility of oral medication and early hospital discharge, teicoplanin is an ideal candidate for OPAT (134). OPAT with vancomycin is also feasible, although vancomycin was associated with a greater number of side effects than other molecules (131, 146). Vancomycin can be stored during 7 days at 25°C (147).

Oral antibiotic agents for MRSA

For reasons of cost-savings, prevention of catheter-related problems, patient and nurses' comfort, parenteral medication should be kept for a minimum. According to todays' expert opinion, this minimal duration is 2 weeks for acute osteo-articular infection. When it comes to oral medication, certain rules have to be imperatively followed: Not every antibiotic agent has a good oral bioavailability and good penetration in tissues like bone and joints. Moreover, though some antibiotics are monitored by serum drug concentrations, bone concentrations would be more important to monitor, but here are no standardized methods for determination of bone concentrations of antibiotics. As for every medicament, the pharmacokinetic bioavailabilty is important. However, patients' co-morbidities, e.g. malabsorption, bad compliance, etc. must equally be taken into account when prescribing oral medication. Fortunately, most oral agents for HA-MRSA infections have over 90% bioavailability and have proved their clinical efficacy (132).

Linezolid

Linezolid belongs to a new class of antibiotics, the oxazolinidones. It inhibits ribosomal protein synthesis and can be administered parenterally or orally at a dose of 600 mg bid, without adjustment for renal insufficiency. It is bacteriostatic with no cross-resistance to other antibiotics. Due to its excellent bioavailability of 100% it is a good choice for outpatient treatments of HA-MRSA (148). Nevertheless, it also features some inconveniences: Besides an expensive price, it is associated with reversible with bone marrow suppression, and in particular thrombopenia during administration of more than two weeks. Regular control of the hematogram is mandatory. Optic neuropathy and non-reversible peripheral neuropathy have been reported in 2% to 4% of patients with prolonged administration (149, 150). A severe serotonin syndrome in co-medication with certain antidepressive drugs, such as monoamine oxidase inhibitors, has been described.

Co-trimoxazole

With the epidemic of HA-MRSA infection, there is an increasing off-label use of antibiotics, such as co-trimoxazole, an inexpensive bactericidal folate antagonist. Clinical experience shows that this molecule can heal small MRSA soft-tissue infections and was reported efficacious in infections of orthopaedic implants (151, 152). However, it has not been FDA-approved for severe *S. aureus* infections (153). One reason for failure in severe infections might be the amount of thymidine released from damaged host tissues and bacteria, a concept strengthened by the fact that *S. aureus* thermonuclease releases thymidine from DNA. Thymidine antagonizes the anti-staphylococcal effects of both trimethoprim and sulfamethoxazole, the two compounds of co-trimoxazole. Thus, failure with co-trimoxazole may well depend on the amount of tissue damage and organism burden (154). We recommend its use as monotherapy only in localised small skin and soft tissue infections due to MRSA (155). Alternatively, co-trimoxazole can be combined with rifampin. The dose of co-trimoxazole is not standardized for deep MRSA infections in osteo-articular tissues. It ranges from 1-2 double-strength tablets every 8 to 12 hours to 20 mg/kg of trimethoprim per day (152). Main adverse events during prolonged medication are nausea, rash, myelosuppression, allergy, and hepatitis.

Tetracyclines

Tetracyclines (doxycycline and minocycline; both 100mg bid) are lipophilic, facilitating the passage into tissues. Evidence with tissue infections due to MRSA is quite limited, and exist essentially for skin and soft tissue infections (156). As for co-trimoxazole, we recommend its use only in localised small skin and soft tissue infections due to MRSA (157). Tetracylines are often combined with rifampin, although firm data are lacking (155). Whether the combination is also superior to minocycline or doxycycline alone, has been questioned (158). Main adverse events of tetracyclines for adults are nausea and the danger of photosensitivity during summer.

Fusidic acid

Oral fusidic acid (regular dose 3 x 500mg) has demonstrated efficacy in acute osteomyelitis, chronic osteomyelitis, vertebral infection, septic arthritis, prosthetic and other device-related infections. It inhibits protein synthesis. Although a small number of studies has also examined the use of fusidic acid alone for the treatment of bone infections most experts do not recommend monotherapy because of development of (potentially reversible) resistance (159-161). The time delay under current therapy until the appearance of resistance is unknown and might be variable. The antibiotic can be combined with rifampin (122). Hepatic failure has been reported with use of fusidic acid and rifampin combinations for MRSA, so monitoring of liver function is advisable. Fusidic acid is available in some (not all) European countries, Australia, but not in the United States.

Streptogramins

Streptogramins such as quinupristin-dalfopristin (intravenous) or pristinamycin (peroral) inhibit protein synthesis by binding to bacterial ribosomes. Quinupristin-dalfopristin administration requires central venous access and dextrose infusion. Streptogramins might be useful for treatment of MRSA infections of bone, joint and soft tissue infections in patients intolerant of or failing to vancomycin, but are not considered as first line drugs. Adverse effects limit the use of quinupristin-dalfopristin such as hyperbilirubinemia, myalgias, arthralgias, and nausea.

Clindamycin

For community-acquired MRSA and some few HA-MRSA isolates bacterial protein synthesis inhibition by clindamycin 600-900mg tid may be an excellent option, if the HA-MRSA reveals to be susceptible to clindamycin and to macrolide antibiotics. The clinical efficacy of clindamycin in bone infection can be explicated by its excellent penetration despite its classification as a bacteriostatic agent (132). However, there is a potential pitfall: For isolates routinely tested as susceptible for clindamycin but resistant to erythromycin, clindamycin resistance may be inducible during ongoing treatment. To detect this relatively frequent inducible clindamycin resistance, a D-zone test should be performed by the laboratory (162). However, the relationship between inducible resistance to clindamycin and treatment failure is poorly defined (163).

Fosfomycine and chloramphenicol

Although HA-MRSA may be susceptible to fosfomycin and chloramphenicol, these antibiotics have not been much used or approved for the treatment of osteo-articular and should be avoided (164). Physicians fear agranulocytosis under chloramphenicol medication.

Combination therapy with rifampin

Rifampin exerts a bactericidal effect by inhibiting the RNA polymerase. Rifampin has excellent oral bioavailability, can penetrate white blood cells to kill phagocytosed bacteria and can eradicate adherent organisms in the stationary phase making it the ideal antibiotic for bone infection (122). However, in staphylococcal infections, it must always be combined with another drug to prevent emergence of resistance, which can emerge after only two days of monotherapy (165). Quinolones are the traditional combination drugs. Unfortunately, many HA-MRSA are resistant to quinolones. Therefore, other anti-MRSA drugs have to be combined with rifampin, such as glycopeptides, linezolid, co-trimoxazole, doxycycline, minocycline, or fusidic acid (166). In summary, almost every agent described before could be feasible in combination with rifampin with the exception of clindamycin (167). The dose of rifampin in combination therapy is a matter of debate. In contrast to rodents in which higher doses yielded some better cure rates, no prospective randomized trials exist in humans. Doses of 1 x 600 mg, 2 x 450 mg or 2 x 600 mg are used in routine practice in several parts of the world. When administered with rifampin, the standard dose of the combination antibiotic is not reduced, an attitude which does not base on evidence. Theoretically, the standard dose of the combined drug could be reduced, but this has to be confirmed in prospective trials. As a counter-argument, reduction of any dose should not be attempted for bone tissue where antibiotic penetration is already reduced.

The classical indication for combined rifampin use is staphylococcal implant-infections. This concept does again not base on strong evidence. A recent meta-analysis found few human studies addressing the role of adjunctive rifampin therapy. Adjunctive therapy seems most promising for the treatment of osteomyelitis and prosthetic device-related infections, although studies were underpowered and benefits were not always seen in vivo or in vitro (168). The debate continues. Meanwhile, for most patients with MRSA infections in absence of implant material, adjunctive therapy with rifampin cannot be recommended except for fusidic acid that in turn needs a combination in order to prevent the development of resistance during monotherapy. When prescribing rifampin, physicians have to remember potential interactions with other medications such as warfarin, antiepileptic drugs, anti-HIV treatment, contraceptiva, or steroids. Rifampin can color body liquids in orange or red and can provoke nausea and hepatitis.

Empirical treatment

A true empirical treatment for HA-MRSA is seldom necessary, since the identification of a clinical isolate of MRSA normally leads to an antibiogram or the standard resistance pattern of the hospitals' predominant strain is already known. Awaiting the first results of the antibiogram, vancomycin is the drug of choice for empirical treatment if an infection due to

HA-MRSA is suspected. Waiting for an antibiogram before the start of antibiotics is also an option and depends on the clinical emergency of (empirical) antibiotic treatment. Intermediately-resistant or fully vancomycin-resistant MRSA isolates are rarities.

The presence of cutaneous MRSA carriage does not oblige to cover it immediately for the treatment of skin and soft tissue infections. In mild to moderate infections, the physicians may cover other susceptible pathogens such as Streptococcus sp and only switch to anti-MRSA therapy in case of adverse evolution under initial empiric antibiotic treatment.

Antimicrobial therapy during distinct clinical presentation of HA-MRSA BJIs

Native joint arthritis

Native joint arthritis due to HA-MRSA is uncommon in adults, but continues to present a challenge to medical and surgical treatment. It is both a surgical and medical emergency to prevent irreversible cartilage damage (169). No randomized controlled studies have evaluated antibiotic regimens for bacterial arthritis. Par default, the choice of antibiotics is the same as for all osteo-articular infections due to MRSA. The recommended duration of antibiotic medication is typically 2-3 weeks for uncomplicated infection though this parenteral throughout (143, 170). Some experts advocate 2 weeks i.v. for streptococci, and 3 weeks for staphylococci (171). Others favor an antibiotic treatment of two weeks parenterally and of another two weeks orally. Provided that surgical lavage is adequately performed, theoretically even septic arthritis due to MRSA could be treated by peroral antibiotics, but so far there is no data supporting this approach. Intraarticular antibiotics are not recommended, as effective parenteral or oral therapy produce adequate levels of antimicrobial agents in joint fluid (172). Furthermore, direct instillation of antibiotics into a joint may provoke an inflammatory response.

Vertebral osteomyelitis

Unless it is a nosocomial infection after spine surgery, haematogenous spread is generally the most common origin of vertebral osteomyelitis and/or spondylodiscitis. In bacteraemic disease, endocarditis must be excluded. The infection is acute and there are exceptional reasons to prolong a 6-week course of antibiotic treatment in uncomplicated infection, even if recurrence risk of spodylodiscitis due to HA-MRSA may be 30%: extensive bone destruction and/or paravertebral infection such as non-drained abscesses in the epidural space or psoas muscles (173). In these cases, the decision for the duration of antibiotic treatment is individual.

No randomized controlled studies have evaluated antibiotic regimens for vertebral osteomyelitis. Practically, the choice of antibiotic agents is not different from any other osteo-articular infections, but a parenteral treatment of at least 3–4 weeks is suggested (173). This may not be true if antibiotics with excellent oral bioavailability can be used. However, the role of these regimens has not been evaluated so far in patients with spondylodiscitis and epidural abscesses (174). Basing on today's expert's opinion, a switch to a peroral treatment should only be made if a suitable drug with proven susceptibility to the MRSA is available, and if compliance with oral therapy can be carefully monitored.

Osteomyelitis without implants

The choice of antibiotic agents is not different from the rational mentioned above. The optimal duration of antibiotic treatment is unknown. The acuity of infection probably has the biggest impact on the decision of the duration. As a general principle for acute osteomyelitis, antibiotics should be given for 4-6 weeks, of which the initial 2 weeks by the parenteral route (18, 171). Chronic osteomyelitis, characterized by sequesters, poor local vascularisation and the potential presence of small-colony variants, is almost impossible to cure without major surgery. Antibiotics are secondary in the curative approach, and maybe the duration is not so important. Practically, patients receive three months of concomitant antibiotic treatment, if they support it. Otherwise the treatment is interrupted earlier.

When surgery cannot be performed (e.g. large infected area of the pelvis), anti-MRSA antibiotics are suppressive. In this case, they are only occasionally prescribed during one to several weeks upon clinical indication; mostly to calm down ongoing severe flares of inflammation due to chronic infection. These flares might be the presence of fever or increased purulent discharge out of the sinus tract. Often the patients decide themselves for how long they need antibiotics to calm down the purulent discharge. Importantly, the sinus tract should not be attempted to close, because it represents the spontaneous drainage of a chronic infection. In case of retention the situation would get worse. Chronic osteomyelitis without possibility of surgery is life-long burden.

Arthroplasty infections with removal of the implant

Treatment of arthroplasty infections is not easy to standardize due to the lack of randomized trials. According to the traditional rules for MRSA infections and other pathogens, standard treatment includes two-stage exchange with meticulous removal of all foreign material combined with a 6-week course of parenteral antibiotics (42). Afterwards, reimplantation should be delayed by 2 more weeks in order to obtain reliable tissue specimens for culture and to document treatment success during reimplantation. The optimal duration of this "antibiotic window" is unknown. Two weeks are considered as minimal for most experts. After reimplantation, antimicrobial therapy with vancomycin is reinstated. If cultures of intraoperative specimens remain negative, treatment is stopped. Otherwise it is continued for 6 to 12 weeks (175).

Arthroplasty infections with implant kept in place

In MRSA arthroplasty infections without removal of the implant, a rifampin combination should be used (175). In theses cases, the ideal duration of antibiotic treatment has never been tested in comparative trials. Per precaution, experts suggest duration of 3 months for hip prostheses and 6 months for knee prostheses, although the excessive duration of six months for knee prostheses have been questioned (42, 175-177).

Arthroplasty infections with removal of the implant and use of antibiotic-releasing cement or spacers

Antibiotic-containing cement is used for treatment of bone and joint infections, but still remains controversial (66). Many antibiotics are released from the cement in potentially efficacious amounts, but the duration of time over which these antibiotics continue to be released is less certain. Moreover, the advantage appears minimal in two-stage procedures

(178). Thus, a lowering of the failure rate is often difficult to attribute to the use of antibiotic-impregnated cement alone. It is also worth noting that HA-MRSA is often resistant to gentamicin, the most widely used compound. Even if not, gentamicin may lead to development of small colony variants (144).

Suppressive antibiotic treatment

Exchange surgery is fastidious, time-consuming and can be depletory for patients with multiple co-morbidities. As an alternative, lifelong suppressive oral antimicrobial treatment without surgical intervention is an option. For MSSA, the overall success rate in prosthetic joint infection was 69% with suppressive therapy after a mean follow-up of 5 years, but is unknown if such a period would be achieved also for HA-MRSA (179). Even if suppressive treatment is a palliative treatment, most experts recommend a lavage of the infected joint in order to reduce the inoculum of MRSA. The use of rifampin for suppressive purposes in discouraged due to adverse events and the need of combination therapy in an already multi-morbid patient population.

Non-arthroplasty implant infections

The goals of treating infection associated with internal fixation devices are consolidation of the fracture and prevention of chronic osteomyelitis. Thus, in contrast to prosthetic joint associated infection, complete eradication of infection is not the primary goal, since the device can theoretically be removed after consolidation or temporary stabilization can be achieved in worse case scenarios with external fixation and removal of internal fixation (41).

The choice of antibiotic agents is the same as for all osteo-articular infections and the ideal duration of antibiotic treatment is unknown due to lack of randomized prospective trials. Some experts favor continuing antibiotics until implant removal and some weeks (usually 4-6 weeks) beyond (41). Others prescribe the same antibiotics only for six weeks despite the presence of implant material. In chronic infections, defined by onset of infectious symptoms after ten weeks post-implant, the duration may be prolonged to three months (41). Other experts treat for 3 months when the device is retained and 6 weeks when there is complete removal of all hardware (41). In any case the duration has to be decided individually. It is also unclear if rifampin should be added to the antibiotic regimen from the start, even if the implant will be removed later on. The authors of this chapter personally prefer to treat acute infections with combined rifampin during 6 weeks post-surgery, independently if the device is removed or planned to be removed. In case of chronic infection, device retention, and good tolerability of the antibiotics, the regimen can be prolonged to 3 months. Provided that there is no documented bacteraemia, the antibiotic regimen can be peroral from the start.

Future aspects

The future will probably provide a firm place in the antibiotic armentarium for some of today's investigational agents: Dalbavancin is not used yet in clinical routine. It inhibits bacterial cell-wall synthesis and has the longest serum half-life of all antibiotics known: 50 hours. Thus would enable a once-weekly parenteral administration with a charging dose of 1000 mg on day 1, followed by 500 mg on day 8 (180). However, the molecule has to show its clinical efficacy and suitability in human trials (181). Many other drugs are in the pipeline and may be in clinical use in immediate future: Orbitracin, telbavancin, iclaprim, oritavancin, lysostaphin, RWJ-416457. It should not be forgotten that most, if not all, newer molecules are and will be equivalent to glycopeptide antibiotics in terms of cure rates in vivo. As for the old glycopeptide vancomycin, the increase of the minimum inhibitory concentration (MIC) may become a concern with a theoretical rise in treatment failures despite correct administration, which is already discussed for bacteraemic disease (182). The potential benefice of a continuous perfusion of vancomycin against the strains with higher MICs remains to be investigated (131).

Despite significant progress in the understanding of foreign body infections due to HA-MRSA, improved knowledge of metabolic properties of biofilm-grown bacteria is still needed. The discovery of specific genomic elements promoting its invasiveness may offer the possibility to develop new drug targets. For example, by using a quorum-sensing inhibitor and other innovative approaches such as urokinase, bacteriophages, ultrasonically enhanced antibiotic activity or electric current treatment of biofilm may prove to be useful in future. Anti-staphylococcal vaccines could become clinical reality (166). Epidemiologically speaking, randomized prospective trials are needed to determine the ideal duration and the optimal form of administration of antibiotic agents in the field of osteo-articular and soft tissue infections due to HA-MRSA. Finally, we will surely not lack of MRSA-infected patients. Hence, prevention of HA-MRSA transmission will always remain a cornerstone (167).

Second Part: CA-MRSA BJI

Since the mid-1990s, MRSA infections began to be detected in the community in persons who did not have contact with the health care system. Molecular typing such as pulsed-field gel electrophoresis (PFGE) and MLST revealed distinct genetic backgrounds from HA-MRSA suggesting that they arose de novo (183). These new MRSA strains, called CA-MRSA, frequently carry the genes encoding PVL (which is a pore-forming toxin involved in the process of abscess formation) and staphylococcal chromosomal cassette (SCC*mec* element types IV or V (183, 184). These latter elements code for methicillin resistance, but lack other antibiotics resistance genes. CA-MRSA isolates have been detected worldwide with continent-specific clones (183). One of the most prevalent PVL-positive CA-MRSA clone in the USA, called USA300, belong to ST8; the main European clone belongs to ST80; the major clone in Asia and Oceania belongs to ST30; and in international travel locations such as Singapore or Geneva, several clones have been detected (ST80, ST30, ST59 and ST80, ST88, ST5, respectively) (183, 185, 186).

Patients at Risk for CA-MRSA BJI

Numerous initial reports have suggested the easy transmission of CA-MRSA isolates in settings where people are in close contact, including soldiers, jail detainees, athletes, men who have sex with men, native Americans, Pacific Islanders and intravenous drug users (187). During the last 10 years, a dramatic increase of the CA-MRSA infections occurrence has been observed. Whereas CA-MRSA infections was uncommon before 2000, the study of

Moran et al. in 11 US cities in August 2004 revealed that CA-MRSA became the most common identifiable cause of skin and soft-tissue infections among adults patients presenting to emergency department (188). In the multivariate analysis, history of CA-MRSA infection, close contact with person with similar infection and previous use of any antibiotic in the past month were the three main risk factors for CA-MRSA infection. The pediatric population was not a population at risk for CA-MRSA infection at the emergence of CA-MRSA strains. Nowadays, children, particularly in the USA, became at risk for developing such infections. In this population, the incidence of CA-MRSA infection gradually rises. For instance, in the Driscoll Children's Hospital located in Texas, the number of CA-MRSA infections was 36 in 2000 and reached 589 in 2004 (189). In the same way, Mishaan et al. recorded a 62% increase from 2001-2002 to 2002-2003 of the incidence of CA-MRSA infection in the Texas Children's Hospital of Houston (190). In Johns Hopkins Hospital, 64% of all *S. aureus* skin and soft-tissue infection were due to CA-MRSA in the 2003-2005 period (191). Risk factors for developing CA-MRSA infection in children are not currently known. Although skin and soft tissue infections predominated in the initial descriptions of the clinical spectrum of infections caused by CA-MRSA, increasing numbers of children with invasive infection such as CA-MRSA BJIs have been reported.

Different Forms of CA-MRSA BJI

We previously described different forms of HA-MRSA BJIs which include various different and heterogeneous forms of infections. CA-MRSA BJIs are considered to be more homogeneous. Contrary to HA-MRSA BJIs, CA-MRSA BJIs mainly occurred in the pediatric population and the initial presentation is always acute (192, 193). Children with CA-MRSA BJIs frequently present with hematogeneous femur or tibia osteomyelitis and/or with arthritis of knee, hip or shoulder. In comparison with MSSA BJIs, CA-MRSA BJIs are particularly associated with musculoskeletal complications such as subperiosteal or muscle abscesses, necrotizing fasciitis or septic shock (192, 193). In Europe, it is notheworthy that most of PVL-associated *S. aureus* BJIs are due to MSSA and not CA-MRSA (192).

Pathophysiology of CA-MRSA BJI

Routes of infection and pathological examination

As during MSSA BJIs in children, CA-MRSA infections are acquired hematogenously. The origin of the bacteremia is evident when CA-MRSA skin and soft-tissue is associated, but could be absent.

As during osteomyelitis due to MSSA, the metaphyse of long bone is the most frequent site for CA-MRSA osteomyelitis in children. For adolescents, ilio-pubial osteomyelitis has also been reported. The predilection of infection to the metaphyse of long bone is explained by the particular anatomy of bone in contact with the growth plate. Indeed, under the growth plate, capillary ends of the nutrient artery make sharp loops, leading to blood flow slowing down and turbulence. Any minor trauma can lead to the development of a small hematoma, vascular obstruction, and subsequent bone necrosis which could be seeded from a transient bacteremia. The acute infection initially produces an inflammatory infiltration that compromises the medullary circulation and enhances the spread of infection.

Native joint hematogenous arthritis is the second clinical entity of CA-MRSA BJIs. Hip, knee and shoulder are the most frequent infected joint. As in adults, the joint synovial membrane is highly vascularized and is particularly susceptible to the deposition during bacteremia. In children with long bone osteomyelitis, direct extension of the infection to the joint is possible (192, 193).

CA-MRSA virulence factors

As during HA-MRSA BJI, MSCRAMMs and production of classical toxins probably favour the adhesion and the invasion of CA-MRSA. In situ production of PVL might participate to the local inflammatory response and abscess formation.

MSCRAMMs and classical toxins

Adhesion and invasion determinants involved in HA-MRSA BJIs, i.e. MSCRAMMs and the ability to produce biofilm, are also found in CA-MRSA (194-197), but it is not known whether CA-MRSA can undergo endocytic uptake like HA-MRSA to gain prolonged intracellular survival. However, besides these classical adhesion and invasion mechanisms, CA-MRSA carry several additional virulence or fitness determinants whose contribution to the pathogenesis will be discussed below: (i) PVL; (ii) staphylococcal chromosomal cassette mec (SCC*mec*) types IV-VII (195); (iii) arginine catabolic mobile element (ACME) (198); and (iv) α-type phenol-soluble modulins (PSMs) (199).

Panton-Valentine leukocidin

PVL is a pore-forming protein exotoxin initially described by Van de Velde in 1894, who demonstrated its ability to lyse leukocytes and designated it "substance leukocidine" (200). Panton and Valentine first associated the leukocidin with skin and soft tissue infection in 1932. Subsequent identification of the genes encoding this toxin strengthened its association with severe skin and soft tissue infection and necrotizing pneumonia in MSSA isolates (184, 201), and later in CA-MRSA isolates (183). Numerous epidemiological studies have enlightened the strong association between PVL and CA-MRSA, and PVL is now postulated widely as a major virulence determinant of CA-MRSA, and not solely a surrogate marker of these strains (183, 202, 203). As a matter of fact, whereas *S. aureus* strains harbour an extensive repertoire of transferable toxin genes, those encoding PVL are the most consistently present in CA-MRSA (183, 204-206).

PVL belongs to a family of bi-component leukocidal toxins (synergohymenotropic toxins) produced by staphylococci (206). These toxins consist of two polypeptides, S-(slow eluted) and F-(fast eluted) components, based on their elution profiles through cation-exchange chromatography. Both PVL components LukF-PV and LukS-PV are encoded by the *lukPV* operon. This operon is carried by several transferable phages designated ΦSLT, ΦPVL, ΦSA2MW or ΦSA2usa (198, 207, 208). After LukF-PV and LukS-PV are secreted, the latter binds to a yet unidentified receptor on leukocytes, and dimerizes with LukF-PV. Subsequently, LukF-PV and LukS-PV components bind successively until they assemble into a heptameric β-barrel, forming pores in the membranes of polymorphonuclear cells (PMNs) and monocytes. This pore-forming mechanism is similar to that used by α-hemolysin, but in contrast to PVL, α-hemolysin does not lyse neutrophils (206, 209). The molecular determinants of the target-cell specificity of these structurally similar toxins are not completely elucidated (210).

Depending on its concentration, PVL can cause either leukocyte lysis or apoptosis. Leukocyte apoptosis is presumably associated with the formation of pores in the mitochondrial membrane (211). Once bound to PMN membrane, LukS-PV is phosphorylated by a host protein kinase A or C (206). The following activation of Ca++ ion channels points toward the induction of transduction signaling events able to trigger production of inflammatory mediators. Indeed, PVL present at sublytic concentrations induces secretion by PMNs of chemotactic components and reactive oxygen species, contributing to an enhanced local inflammatory response and leukocyte recruitment (212-214). Although PMN lysis and apoptosis are clearly involved in PVL-mediated pathogenesis, the mechanisms setting off tissue necrosis and severe sepsis are unclarified. Indeed, purified PVL exerts no effect on cultured epithelial cells, which makes it unlikely to be directly responsible for tissue necrosis (215). It is thus more probable that PVL-associated tissue necrosis and sepsis are mediated indirectly by the release of cytotoxic components from lysed PMNs, and by the inflammatory cascade initiated by such a release (200).

PVL as a purified toxin causes tissue necrosis

The role of PVL in pathogenesis has been studied in a number of in vitro experiments and animal models. In vitro, PVL lyses human and rabbit PMNs, and to a lesser extent, mouse PMNs (207, 216). Purified PVL injected intravenously into rabbits causes a transient depletion of PMNs followed by significant granulocytosis, without being lethal (217). Purified PVL also causes dose-dependant dermonecrotic lesions in rabbits (207, 218), and lethal necrotic lesions in mice lungs through nasal instillation (219). Along with the fact that PVL is expressed at toxic levels in human cutaneous abscesses, these data bring compelling arguments for the role of PVL in local tissue necrosis (220). Moreover, in a rabbit osteomyelitis model, bone deformation and muscle or joint involvement are significantly more frequent in rabbits infected with PVL positive USA300 CA-MRSA than with its PVL negative isogenic strain (Saleh-Mghir A. et al., unpublished).

Regulation of virulence factors

LukS-PV signal peptide enhances adhesion to extracellular matrix components. Besides their ability to cause local tissue necrosis, PVL positive strains display an increased adhesion to matrix molecules, notably type I collagen and laminin (215). Type I collagen is the main organic component of bone extracellular matrix (221), while laminin is part of the bone marrow stroma (222). In order to understand the mechanisms involved in this enhanced adhesion of PVL positive strains, Tristan et al. compared the ability of PVL positive strains and their isogenic PVL negative strains to bind various immobilized extracellular matrix components (ECMC). Through studying the effects of recombinant fragments of the PVL subunits and synthetic PVL peptides, they showed that the processed LukS-PV signal peptide was sufficient to significantly enhance the adhesion of *S. aureus* to ECMC. This adhesion was inhibited by heparin and heparan sulfates, suggesting that in vivo, heparan sulfates could function as a molecular bridge between the matrix and *S. aureus* expressing LukS-PV signal peptide (223).

Controversy about the gene regulatory effect of PVL

Wether PVL has a role in regulating other virulence factors remains controversial. A global effect on gene regulation has been initially attributed to PVL, whose expression was associated, among others, with a strong up-regulation of protein A in a laboratory strain lysogenized with ΦSLT (219). Because protein A plays a major role in both immune evasion and the triggering of inflammation (224), it has been hypothesized that PVL, besides its own pro-inflammatory properties, indirectly enhances inflammation and necrosis through the up-regulation of protein A (219). However, in a later study, Diep et al. aimed at addressing this question and found that PVL had no effect on global gene and protein expression in prevalent USA300 and USA400 CA-MRSA strains, when comparing transcriptional profiling of wild-type parental and isogenic Δpvl strains (225). The inconsistency between these studies may be ascribed to methodological differences, and the experiments being conducted on different *S. aureus* genetic backgrounds. Moreover, the question of whether this potential gene regulatory effect of PVL occurs in vivo is still to be addressed.

PVL contribution to BJIs pathogenesis

Despite experimental data are lacking about the role of PVL per se in CA-MRSA BJIs, a local production of PVL is most likely responsible for severe cell damage and acute complications. Clinical data bring compelling arguments for this hypothesis, with CA-MRSA BJIs being frequently associated with severe sepsis and extensive soft-tissue involvement (193, 226). In the same way, PVL production in joint fluid may enhance the pro-inflammatory response, PMN recruitment, and clinical signs of arthritis.

SCCmec IV

Beta-lactam resistance gene mecA is carried by distinct allotypes of the SCCmec element, whose presence in HA-MRSA is generally thought to reduce their biological fitness in the absence of antibiotic selection pressure. However, CA-MRSA mostly carry type IV SCC*mec* (SCC*mec* IV) elements that differ in size and gene content from the classical type I-III elements found in most HA-MRSA. SCC*mec* IV, when compared to SCC*mec* I-III, is up to 3-fold smaller in size, do not encode resistance determinants other than mecA and thus might impose a reduced fitness burden (227). Through comparing isogenic strains complemented with different allotypes of the SCC*mec* element, Lee et al. demonstrated that SCC*mec* I, but not SCC*mec* IV, inflicts a significant fitness cost to their host, causing decreased in vitro growth rate and cell yield per mole of ATP consumed (228). Moreover, USA300 CA-MRSA and its isogenic strain with deleted SCC*mec* IV showed no difference in infectivity in a rabbit co-infection model (229). Collectively, these observations indicate that SCC*mec* IV imposes minimal, if any, biological fitness cost to CA-MRSA, whereas HA-MRSA probably depend on high antibiotic selection pressure to balance the fitness cost inflicted by their carrying multiple resistance determinants (230).

Virulence attenuator gene fudoh is present in HA-MRSA-associated SCCmec elements but not in CA-MRSA

Type II and III SCC*mec* elements found in most HA-MRSA have been shown recently to carry a newly identified gene, *fudoh*, which has been associated with decreased in vitro exotoxin production and in vivo virulence in a mouse bacteremia model (231). Noteworthy, fudoh was not found in CA-MRSA-associated SCC*mec* IV. These observations suggest that SCC*mec* elements found in HA-MRSA are involved in decreasing the virulence of their host, whereas the absence of *fudoh* in CA-MRSA is a plausible cause for their conserved exotoxin production and virulence (231).

Arginine catabolic mobile element

USA300 is a geographically widespread CA-MRSA lineage, accounting for the majority (67%) of invasive CA-MRSA infections in the United States (232). In this prevalent lineage, SCC*mec* IV is physically linked to the arginine catabolic mobile element (ACME), a pathogenicity island probably acquired from *Staphylococcus epidermidis* (198). In a rabbit co-infection model, deletion of ACME in USA300 significantly decreased its virulence and fitness, whereas deletion of SCC*mec* IV in the same strain had no effect on competitive fitness (229). ACME carries the *arc* and *opp3* gene clusters that encode for an arginine deiminase pathway and an oligopeptide permease system, respectively (198). The arginine deiminase system *arc* could play a role in virulence through depleting L-arginine, which is essential for the production of nitric oxide, a free radical produced by macrophages in response to bacterial infections (230, 233). Although the precise function of *opp3* is unknown, this operon belongs to the ATP-binding cassette (ABC) transporter family. ABC transporters exhibit various functions such as peptide nutrient uptake, quorum sensing, pheromone transport, chemotaxis, eukaryotic cell adhesion, binding of serum components and resistance to antimicrobial peptides. Taken together, these data indicate that ACME plays a role in the pathogenicity of USA300 (230). Some other CA-MRSA lineages such as ST5 and ST59 carry ACME-like elements, which contain the arc gene cluster but lack the *opp3* operon (229). Whether these ACME-like elements lacking *opp3* also confer pathogenicity or fitness to their hosts is still to be addressed.

Phenol-soluble modulins

Among the numerous cytolytic peptides harbored by *S. aureus*, it is unclear which molecules play the most significant role in lysing leukocytes in vivo. Indeed, the major lytic activity exerted by CA-MRSA on human leukocytes is unlikely to be due to PVL alone (230). A novel class of cytolytic peptides designated α-type phenol-soluble modulins (PSMs), has been recently described in *S. aureus* (199). PSMs are endowed with considerable leukocidal, proinflammatory and chemotactic activities, which have been demonstrated both in vitro and in vivo. Although the operon encoding PSMs is present in all sequenced genomes of *S. aureus*, HA-MRSA produce PSMs only at reduced levels, if at all, whereas CA-MRSA express them at high levels. Noteworthy, over-expression of PSMs in a HA-MRSA strain increased its leukocidal activity to a level comparable with that observed in CA-MRSA (199). The difference in expression levels of PSMs between HA- and CA-MRSA is, along with PVL, likely to contribute for the pronounced difference in leukocidal activity between these pathogens. As a matter of fact, bacteremia and abscess mouse infection models comparing CA-MRSA MW2 and its isogenic ΔPSM strain, confirmed that α-type PSMs exert a crucial influence on the infectivity of their host (199).

Collectively, these findings indicate that differences in virulence between HA- and CA-MRSA are at least partially related to differential gene expression. For example, expression levels of α-hemolysin, a strong cytolytic and proinflammatory toxin, differ considerably among different strains of *S. aureus*. Although α-hemolysin does not target leukocytes, its expression levels impact the outcome of CA-MRSA necrotizing pneumonia (234). This toxin and the PSMs are under control of the global regulator *agr*. Notably, the *agr* regulatory molecule RNAIII is strongly expressed in CA-MRSA, whereas HA-MRSA are often *agr*-null (i.e. they carry non functional *agr*) (36). Hence, this virulence regulator might play an important, although largely unexplored, role in the pathogenicity of CA-MRSA (199).

Concluding remarks

The epidemic spread of CA-MRSA strains in recent years is one of the most challenging infectious disease issues of the last decade. Among CA-MRSA strains, the pandemic clone USA300 is probably the major contributor to this epidemic, due to its unusual virulence and transmission capacities in the community. Recent observations have revealed that besides the horizontal acquisition of novel pathogenicity determinants, CA-MRSA pathogenicity also resides in the up-regulation of virulence factors present in all *S. aureus* genomes (199). Despite we lack experimental data about the specific role of CA-MRSA virulence determinants in BJIs, their relative contributions to the disease process have to be clarified in order to point out potential drug development targets.

Clinical Signs of CA-MRSA BJI

CA-MRSA infections are usually acute with a short interval between symptoms and hospital admission. Most of patients have primary or secondary skin and soft-tissue infection during the previous days. Patients usually presented with high fever, local pain and severe disability. Severe sepsis and septic shock are frequently associated. At the site of infection, the bone or the joint are usually widely warmth. A rapid spread of swelling with pain out of proportion to the physical findings, a necrotic aspect (even small), a local anesthesia, and an occasional crepitus should raise the clinical suspicion of necrotizing fasciitis. Occurrence of a rash should raise the diagnosis of toxic shock syndrome (192, 193).

Imaging of CA-MRSA BJI

Usual imaging presentation and MRI findings

Musculoskeletal involvement may be the most common site of CA-MRSA invasive infection providing osteomyelitis, osteoarthritis, and bacterial myositis. High clinical virulence is usually associated with this bacterium, indeed a greater frequency of complications will have to be detected by imaging (235).

Acute osteoarticular infection normally requires a rapid radiologic confirmation in order to manage the best therapeutical modality. Therefore, role of imaging is not only to provide an accurate assessment of the infectious process but also to display its complications. Thus, this helps to determine whether surgery is necessary and where it should be performed. In case of CA-MRSA osteomyelitis, it has been clearly established that MRI is the best imaging to focus on the development of infections (236). Indeed the high incidence of complications associated with osseous and soft tissues which are statistically linked with CA-MRSA BJI mean that MRI is becoming essential in cases where such a micro organism is implicated (32, 237).

In the peculiar case of CA-MRSA, subjacent affected bone has normal characteristics, as far as affected people are primary healthy and no promoting factors may be found in their medical history (238). On the MRI, osteomyelitis is manifested as bone marrow signal intensity alterations consistent with inflammation and hyperemia: decreased T1 signal, increased T2 signal and highly increased T1 after gadolinium enhancement. Some authors have assessed sensitivity for CA-MRSA osteomyelitis as being up to 99%, which is quite superior to other radiologic devices (236). Bone marrow oedema remains the main radiologic sign for osteomyelitis but is often associated, in the case of CA-MRSA involvement, with osseous complications including marrow suppuration, periosteal reaction and cortical destruction. Marrow oedema in CA-MRSA is widely extended and pandiaphysitis is frequent. Occurrence of an epiphyseal attempt and its association with osteoarthritis may be common; indeed MRI displays fluid effusion and strong synovial enhancement in the adjacent joint. Such a complication should be carefully attended in children because of a higher risk of later developing epiphysiodesis and bone growth abnormalities (239). Bone abscesses are demonstrated by T1 gadolinium imaging with a localized intramedullary fluid area of low signal, surrounded by strong peripheral enhancement. Subperiosteal abscess occurrence is higher with CA-MRSA than with other Staphylococci and are of a great size and can involve multiple sites, especially in juvenile tubular bone because of the relative weak attachment of the periosteum to the subjacent bone (240). Some authors have evaluated the occurrence of this complications up to 71 %, with a greater rate of surgery requirement making MRI essential (241). Cortical bone can be disrupted but the periosteum usually remains thin because of the severity of the infectious process ant its acute evolution (Figure 6, 7).

Surrounding soft tissues development is extended and bacterial myositis and pyomyositis are very frequent complications of CA-MRSA osteoarticular infections. Intramuscular abscess is a large fluid-full intramuscular suppurative area, mean diameter of which has been evaluated as up to 3.5 cm for CA-MRSA, sometimes requiring drainage procedure (242) (Figure 6, 7). High prevalence of venous thrombosis is associated with CA-MRSA osteomyelitis and complications may be detected on MRI to prevent septic emboly (243).

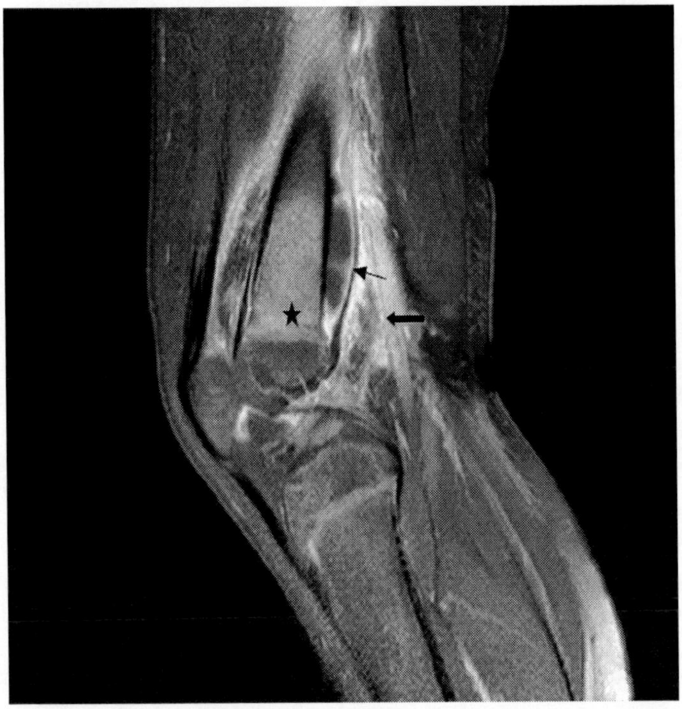

Figure 6. 15 year-old child with CA-MRSA osteomyelitis of the lower femoral extremity. Sagittal MR T1 FS gadolinium enhanced imaging demonstrates bone marrow oedema (black star), extended subperiosteal absccess (thin arrow) and adjacent myositis (large arrow) without soft tissues abscess.

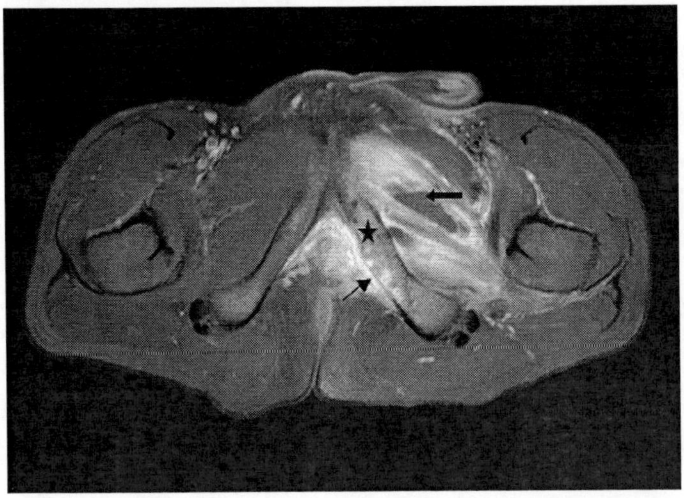

Figure 7. 9 year-old boy with CA-MRSA osteomyelitis of the left ischiopubic branch. Axial MR T1 FS gadolinium enhanced imaging demonstrates strong enhancement of the bone marrow oedema (black star) in association with subperiostel abscess (thin arrow), myositis and an intramuscular abscess within (large arrow).

All these radiologic features are quite unspecific but acute onset of the disease, in association with extended initial bone marrow oedema and severe osseous and soft tissues complications, may be indicative of a CA-MRSA origin. Repeat imaging should be performed in order to detect the occurrence of inflammation extension and abscess formation to guide any surgical drainage or to adapt antibiotherapy (193).

Others Radiologic Imagings

Due to its high virulence and its sudden evolution, suspected CA-MRSA BJI usually requires MRI to be performed, therefore other radiologic devices have a minor role in that case even they can provide some useful informations.

X-Rays

Plain radiographs are always performed for differential diagnosis exclusion but signs are delayed and unspecific. Also sensitivity is low, insofar as CA-MRSA BJI onset is acute and evolution includes early complications (244).

Ultra-sound

US is a good way to detect soft tissue damages cellulitis, myositis, pyomyostis and abscesses which are frequently associated with CA-MRSA. This assessment is likely to be more effective in children who have suspected osteomyelitis as it can help detect subperiosteal abscess and differentiate isolated cellulitis from bone infection. In the incidence of isolated arthritis, an ultra-sound may be sufficient but occurrence of such situations is not usual with CA-MRSA involvement and MRI is a necessary requirement in the diagnosis and treatment of severe cases. Ultra-sound is also helpful as it can be used for needle guided targeting of detected abscesses.

CT- Scan

Lower efficiency in diagnosis of BJI has been longer established in comparison with high MRI contrast resolution. The CT scan has few indications: (i) it can be performed when the CA-MRSA BJI became chronic (requiring looking for Brodie abscess or sequestrum); and (ii) it can guided puncture procedure.

Bone scintigraphy

CA-MRSA BJIs are characterized by virulence and a potential involvement a of multiples sites, therefore bone scans can be performed to provide a whole-body examination. This might be specially indicated in newborns and young children in which clinical examination is sometimes uneasy (245).

Microbial Diagnosis of CA-MRSA BJI

As during HA-MRSA BJIs, the microbioal diagnosis of CA-MRSA BJIs is firstly based on the identification of MRSA on culture. Direct detection of the PVL on clinical sample is possible but nowadays, molecular tools allowing the characterization of the strain are essential.

Genotypic detection of the pvl locus

Molecular testing for the presence of PVL genes is seldom routinely performed in hospital laboratories, and clinical suspicion has to be aroused to trigger sending the isolate to a reference laboratory for toxin testing. The presence of PVL genes is associated with the specific clinical features and severity of CA-MRSA infections. These genes are always expressed when present, although the amount of PVL produced varies from one strain to another.

Conventional PCR

Conventional PCR testing for the PVL genes lukS-PV and lukF-PV (GenBank accession n° AB006796) is the gold standard method for PVL positive *S. aureus* diagnosis. Primers sequences are as follows: PVL-1 (forward), ATCATTAGGTAAAATGTCTGGACAT GATCCA; NPVL-2 (reverse), GCATCAASTGTATTGGATAGCAAAAGC. The yielded PCR product is 433 bp long. Reference strain ATCC49775 is used as positive control strain. This method is performed on pure *S. aureus* culture, which involves a minimum delay of 24-48 h between sample collection and result availability

Multiplex PCR

In order to shorten diagnosis delay through bypassing the step of culture. microbiological diagnosis of CA-MRSA infection from wound or pus swab can be carried out using multiplex real-time PCR. This PCR uses: (i) five forward primers to amplify the most common SCC*mec* elements (types II-V and VII) and a reverse primer in the *orfX* flanking region of the *S. aureus* chromosome, and (ii) primers targeting a conserved area across the *lukS-PV* and *lukF-PV* genes. Sensitivity and specificity for SCC*mec* detection are 95% and 99%, respectively. PVL detection displays 100% sensitivity and specificity, but only a limited number of positive samples were tested. The high throughput of real-time PCR, with automated extraction and no requirement for post-amplification analysis, makes it well-suited for routine and emergency diagnosis, at least in areas of high CA-MRSA prevalence.

Phenotypic detection of PVL

Phenotypic toxin detection assays, compared to genotypic assays, determine actual toxin production rather than inferring it from the presence of toxin genes. However, with respect to PVL this advantage is weakened because: (i) as mentioned above, PVL genes are always expressed when present; and (ii) although the amount of PVL produced in vitro varies from strain to strain, this amount does not correlate with the severity of infection.

To date (early 2009), there is no commercially available phenotypic assay to detect PVL. In-house quantitative assays have been reported, including an ELISA using LukF-PV polyclonal rabbit antiserum, and a latex agglutination assay using antibodies to both LukF-PV and LukS-PV components. Sensitivity and specificity were not evaluated for the ELISA. The agglutination assay displayed 100% sensitivity and specificity, with 1 ng/ml detection limit.

SCCmec typing

Methicillin resistance gene mecA is located on a transferable genetic element designated SCC*mec*. Several variants or allotypes of this element exist, with substantial differences in the size and nucleotide sequence. To date, SCC*mec* types I to VII are recognized. The most common method for SCC*mec* typing uses a multiplex PCR strategy with primers targeting multiple loci in SCC*mec*, the combination of which is specific for a given SCC*mec* type.

Most CA-MRSA strains harbour SCC*mec* element type IV, with sporadic reports of SCC*mec* V to VII. Conversely, SCC*mec* types I to III have been historically associated with HA-MRSA and are not encountered in CA-MRSA. Therefore, SCC*mec* IV has been initially considered a candidate marker for CA-MRSA. However, the epidemiology of HA-MRSA is rapidly changing: (i) in high CA-MRSA prevalence areas such as the United States or North Africa, PVL positive CA-MRSA have spread to the hospital setting and have become a major cause of inpatient and surgical site infection; and (ii) in low CA-MRSA prevalence areas such as Switzerland, Denmark or France, SCC*mec* type IV is gradually replacing historical SCC*mec* I to III elements among PVL negative HA-MRSA. Because SCC*mec* IV is present in both CA-MRSA and HA-MRSA, it should not be used as a marker for CA-MRSA.

Molecular typing methods

Genotyping techniques such as PFGE, MLST or *spa* typing have been initially developped for epidemiological and outbreak surveillance purposes. These expensive and time-consuming methods, although unvaluable for population structure investigations, are not suited for routine detection of CA-MRSA.

PVL production is the main identified microbiological marker linked to the specific clinical features and severity of CA-MRSA BJIs. PVL status of all *S. aureus* isolates from severe, acute-onset BJIs should be requested, as this will alert clinicians and orthopaedic surgeons to the need for early surgical intervention and prolonged intravenous therapy.

Surgical Treatment of CA-MRSA BJI

Surgical treatment of CA-MRSA bone and joint infections do not differ of the basic principles for treatment of BJIs. Nevertheless, CA-MRSA infections have some particularities influencing indications and modalities of the surgical treatment. Actually, difficulties encountered in the medical treatment of such infections and high virulence of some strains (i.e. secretion of PVL) can lead to very severe deep seated necrotizing process. Increasing prevalence of MRSA in North America and Europe justify multidisciplinary approach of the treatment facing the severity of these infections (246).

Basic principles

Some authors suggested that acute hematogenous osteomyelitis could be cure with a single course of antibiotics if the treatment was initiated within 1 or 2 days after the early symptoms (247). Conversely others suggested an early and aggressive procedure (248). Today, it is wildly accepted that surgical treatment of hematogenous bone and joint infections required a measured approach based on the failure of antibiotic treatment or when the diagnosis of infection is done at advanced stage. In these situations the presence of deep seated abscesses (soft tissues, subperiosteal and intra osseous) requires surgical drainage and excision of infected tissues.

No study compared the results of the various procedures suggested neither in terms of healing nor in terms of late after-effects. Nevertheless, whatever the bone and joint infections presented with deep seated abscesses, basic principles of the surgical treatment include first: puncture of abscesses and biopsy of infected bone for culture and organism identification. At the same time swabs must be taken for pathology. In a second time, debridement of abscesses, excision of infected tissues and drainages with irrigation of infected joints must complete the treatment (249, 250).

The initial treatment of septic arthritis is aspiration and lavage of the joint, followed, if necessary by repeated lavages using puncture, arthroscopy or arthrotomy. Serial joint aspiration or arthroscopic drainage and lavages are now well documented with good results (251, 252). I has to be underlined that initial samples as blood, bone, or pus must set up before antibiotic treatment when joint effusion, subperiosteal abscess or intra osseous abscess are present at initial examination. After initial antibiotic treatment, early organism identification leads to adapt medical treatment to an appropriate antibiotic. After antibiotic treatment has been started, surgery must set up for drainage and excision if joint effusion, subperiosteal abscess or intra osseous abscess are present. These circumstances can result of pre-existing abscesses when diagnosis is late, when antibiotic treatment fails or in case of high virulent organisms.

Septic arthritis and osteomyelitis in children

Septic arthritis and osteomyelitis are usually described separately, especially for surgical procedures. In most cases, during severe bone and joint infection, the two localizations coexist. Actually, in children most of the main joints are susceptible to inoculation from the intracapsular portion of the infected bone metaphysis. Concomitant soft tissues infections are common in severe cases in children (253). Spreading infections in bone and joint increase the risks of after-effects, especially in children, leading to limb discrepancy or limb deviation as well as joint destruction.

Surgery in CA-MRSA bone and joint infections

Increasing incidence of CA-MRSA BJI leads to recent reports underlying the severity of these infections (254). CA-MRSA are described as requiring long courses antibiotics, combination of medical and surgical treatments and are associated with a high risk of treatment failure, complications and after-effects (241, 254). Dombrowski et al. had reported a failure rate of 46% in cases of MRSA osteomyelitis with completed recommended therapy (255). The diagnosis of osteomyelitis was significantly associated with relapse. On the other hand, the risk of failure didn't increase even if surgical debridement has been required.

Some strains of CA-MRSA produce virulent toxins, especially the PVL. This necrotizing toxin increase inflammatory response and spreading of the bone and joint infection. These severe infections can lead to septic shocks, multi visceral failures, and very severe deep seated necrotizing processes (192, 193, 256). Martinez-Aguilar et al. noted the high percentage of patients who required surgical procedure in such invasive infections however they didn't noted significant difference in the percentage of surgical procedures performed in MRSA and invasive MSSA infections in children (257). Conversely, some authors noted the efficacy of surgical drainage in case of CA-MRSA infections in children (192, 258-260). Lamp KC et al. reported a study about treatment of ostetomyelitis with 70% of positives cultures for *S aureus* (261). MRSA were 43% and MSSA 18% of the positives cultures. Studying the results of daptomycin in this condition, he noted that failures were more likely to occur if surgical debridement was not performed (24% vs. 5%, $p=0.045$).

Deep seated musculoskeletal infections are frequent in MRSA infections and especially in case of production of virulence factors (i.e. PVL). Pyomyositis is an abscess of muscle, primary or secondary to local spread of a progressive BJI. These lesions impose surgical drainage. In some cases CT or ultrasonography guidance could be assistance. Necrotizing fasciitis is an infection of subcutaneous and fascia tissues which require surgical excision and drainage. Soft tissues abscesses could be primary or secondary to the spread of intra-osseous abscess or subperiosteal abscess. Debridement and excision of necrotizing tissues must be preformed.

Hematogenous MRSA spondylitis has been reported in few cases until now. Al-Nammari reported 13 cases. In 54% of these spondylitis, surgical procedure would have been suitable. Thirty eight per cent had surgery, others have been considered with high risk of death if surgery. Surgical treatment was decided in case of bone destruction with progressive spinal deformity or abscesses. These results confirm the high rate of surgical procedures required in such CA-MRSA BJIs (173).

Indications

We reported in a study 14 cases of BJIs related to *S. aureus* producing PVL (192). We can consider these BJIs as similarly severe as CA-MRSA BJIs, actually these organisms are frequently PVL producing (193, 254, 256). In our study we focused discussion on surgical treatment among others topics. We noted a significant difference in term of surgical drainage in case of PVL positive infection versus PVL negative BJI (a mean of 2.14 vs 0.17, $p<0.001$). We noted the efficiency of MRI to investigate the spreading of the necrotizing process in bone and soft tissues, Hawkshead et al. confirmed this observation (254).

Few studies focused on the improvement of results in the treatment of BJI by using surgical drainage when abscesses or necrotizing tissues were associated. A comparative study can't be performed in such pathology and most of the time; surgery is performed in complicated cases only. There is a lack of literature about this situation; nevertheless, in case of CA-MRSA BJIs we recommend to look for deep seated abscesses (in bone as in soft tissues) when the antibiotic treatment fails with relapsing clinical and biological inflammatory syndrome or when initial features include severe clinical and biological inflammatory signs (which could be related to septic gathering). The best way to indentify these deep abscesses appears to be MRI.

Extensive surgical drainage, debridement and excision of the necrotized tissues are the recommended procedures to treat these BJI in association with an appropriate antibiotic treatment.

Medical Treatment of CA-MRSA BJI

Generalities

CA-MRSA BJIs are frequently acute with bacteremia, systemic clinical signs of infection, and septic shock (that occurs approximately in half of the cases) (192, 193, 254). As a result, in association with aggressive surgical treatment, the medical treatment of CA-MRSA BJIs have to include a combination of effective intravenous antibiotics. At least one antibiotic has to have an excellent bone and joint diffusion. In the absence of clinical trials in the field of CA-MRSA BJIs, data are lacking about the optimal treatment of CA-MRSA infections. In skin and soft-tissue infections, some recommendations have been proposed, depending on the clinical severity and on the drug susceptibility of CA-MRSA isolates and the in vitro antitoxinic activity of particular antibiotics (262).

Drug susceptibility of CA-MRSA isolates

In comparison with multidrug resistant HA-MRSA isolates, initial PVL-positive CA-MRSA isolates were susceptible to most antimicrobial agents (most of initial isolates were only resistant to penicillin and oxacillin alone). Since 2003, new antimicrobial resistance determinants have been acquired. Consequently, recognition of the PVL status of a strain by its antibiotics resistance profile has become generally unreliable (183, 186). However, there are certain recognized susceptibility patterns associated with the main circulating CA-MRSA lineages, or with different geographical locations (263). Although a definitive diagnosis of CA-MRSA infection cannot be inferred from such susceptibility patterns, they should raise suspicion and lead to further examinations when associated with consistent clinical features. USA300 strains are the most prevalent CA-MRSA strains in the United States (232). Most of these strains are resistant only to β-lactams, or to β-lactams and macrolides without inducible resistance to clindamycin (226). However, resistance to kanamycin, tetracyclin, clindamycin, mupirocin, daptomycin and ofloxacin has been reported, as well as intermediate susceptibility to vancomycin (186, 264). Strains of the worldwide pandemic ST30 lineage (initially described as the Southwest Pacific Clone) are mostly resistant only to β-lactams, with sporadic reports of resistance to kanamycin, tobramycin, gentamicin, macrolides and clindamycin (183, 186). The majority of CA-MRSA from Europe and North Africa belong to the ST80 lineage, which displays the most specific susceptibility pattern, i.e. resistance to β-lactams, kanamycin, tetracyclin and fusidic acid. However, multidrug resistant isolates have been increasingly reported, especially from North Africa (186).

Antibiotic agents with antitoxinic activity

PVL is associated with intense necrosis in vivo, possibly leading to poor antibiotic diffusion and suboptimal concentration at the site of infection. Subinhibitory antibiotic concentrations modulate the production of PVL, depending on the drug and its site of action. Dumitrescu et al. examined the influence of subinhibitory concentration of various antibiotics (alone and in combination) on the level PVL production. It has been demonstrated that oxacillin enhanced the PVL production of CA-MRSA isolates. This could be related to transcriptional activation in the presence of beta-lactmams, as it has been also observed with other staphylococcal toxins, such as alpha-toxin (265). In contrast, clindamycin, linezolid, and fusidic acid, reduced significantly the PVL release (266, 267). These antibiotics have previously been shown to reduce the production of several staphylococcal toxins, possibly through their impact on bacterial protein synthesis and transcription (268). The vancomycin had roughly no effect (266).

Choice of antibiotic agent for the initial intravenous treatment

As a part of severe *S. aureus* BJIs may also be due to MSSA, initial antimicrobial therapy may include an antistaphylococcal beta-lactam agent (100 mg/kg/day) and vancomycin (30 mg/kg/day). In addition, clindamycin (30 mg/kg/day) appeared to be the drug of choice as: (i) it is effective on MSSA and on most of CA-MRSA, even if some of CA-MRSA exhibit erythromycin resistance, which is a marker of inductible resistance to clindamycin; (ii) it has an excellent diffusion in bone and joint tissue; and (iii) it blocks the production of PVL in vitro. Finally, gentamicin could be added if a bacteremia and/or a septic shock occurred.

Choice of antibiotic agent for the secondary oral treatment

Intravenous antimicrobial therapy (i.e. vancomycin) has to be continued longer as necessary (usually 2 to 6 weeks) to obtain significant clinical and local improvement. As the bioavailability of clindamycin is excellent, peroral administration could be kept as soon as possible. Trimethoprim-sulfamethoxazole (10 and 50 mg/kg/day) or fusidic acid (60 mg/kg/day, always in combination) could be used. As most of CA-MRSA isolates are susceptible to fluoroquinolones, this therapeutic class has to be considered as a treatment option, as the biodisponibility and the bone and joint diffusion of these drugs are excellent (186). However, data from animal studies demonstrate that fluoroquinolones can cause cartilage damage and the limited in vivo data in children have made pediatricians reluctant to use these agents. Ciprofloxacin is the most frequently used fluoroquinolone in children, most often in the treatment of pulmonary infection in cystic fibrosis. Safety data of fluoroquinolones in children appear to be similar to those in adults (269). The total duration of antimicrobial therapy depends on the clinical form of infection and the occurrence of long-term complications. A treatment of six weeks to three months seems to be acceptable.

Conclusions

HA and CA-MRSA BJIs are two heterogeneous groups of infections with distinct epidemiology, pathophysiology, clinical presentation, and management. Improving the surveillance, explaining the virulence and evaluating different treatment options in both groups are required to limit the morbidity and the mortality of these severe infections.

References

[1] Brun-Buisson C, Meshaka P, Pinton P, Vallet B. EPISEPSIS: a reappraisal of the epidemiology and outcome of severe sepsis in French intensive care units. Intensive Care Med. 2004 Apr;30(4):580-8.

[2] Oliveira DC, Tomasz A, de Lencastre H. The evolution of pandemic clones of methicillin-resistant Staphylococcus aureus: identification of two ancestral genetic backgrounds and the associated mec elements. Microb Drug Resist. 2001 Winter;7(4):349-61.

[3] Robinson DA, Enright MC. Multilocus sequence typing and the evolution of methicillin-resistant Staphylococcus aureus. Clin Microbiol Infect. 2004 Feb;10(2):92-7.

[4] Enright MC, Day NP, Davies CE, Peacock SJ, Spratt BG. Multilocus sequence typing for characterization of methicillin-resistant and methicillin-susceptible clones of Staphylococcus aureus. J Clin Microbiol. 2000 Mar;38(3):1008-15.

[5] Enright MC, Spratt BG. Multilocus sequence typing. Trends Microbiol. 1999 Dec;7(12):482-7.

[6] Maiden MC, Bygraves JA, Feil E, Morelli G, Russell JE, Urwin R, et al. Multilocus sequence typing: a portable approach to the identification of clones within populations of pathogenic microorganisms. Proc Natl Acad Sci U S A. 1998 Mar 17;95(6):3140-5.

[7] Frenay HM, Bunschoten AE, Schouls LM, van Leeuwen WJ, Vandenbroucke-Grauls CM, Verhoef J, et al. Molecular typing of methicillin-resistant Staphylococcus aureus on the basis of protein A gene polymorphism. Eur J Clin Microbiol Infect Dis. 1996 Jan;15(1):60-4.

[8] Shopsin B, Gomez M, Montgomery SO, Smith DH, Waddington M, Dodge DE, et al. Evaluation of protein A gene polymorphic region DNA sequencing for typing of Staphylococcus aureus strains. J Clin Microbiol. 1999 Nov;37(11):3556-63.

[9] Milheirico C, Oliveira DC, de Lencastre H. Multiplex PCR strategy for subtyping the staphylococcal cassette chromosome mec type IV in methicillin-resistant Staphylococcus aureus: 'SCCmec IV multiplex'. J Antimicrob Chemother. 2007 Jul;60(1):42-8.

[10] Milheirico C, Oliveira DC, de Lencastre H. Update to the multiplex PCR strategy for assignment of mec element types in Staphylococcus aureus. Antimicrob Agents Chemother. 2007 Sep;51(9):3374-7.

[11] Oliveira DC, de Lencastre H. Multiplex PCR strategy for rapid identification of structural types and variants of the mec element in methicillin-resistant Staphylococcus aureus. Antimicrob Agents Chemother. 2002 Jul;46(7):2155-61.

[12] Ferry T, Bes M, Dauwalder O, Meugnier H, Lina G, Forey F, et al. Toxin gene content of the Lyon methicillin-resistant Staphylococcus aureus clone compared with that of other pandemic clones. J Clin Microbiol. 2006 Jul;44(7):2642-4.

[13] Melter O, Santos Sanches I, Schindler J, Aires de Sousa M, Mato R, Kovarova V, et al. Methicillin-resistant Staphylococcus aureus clonal types in the Czech Republic. J Clin Microbiol. 1999 Sep;37(9):2798-803.

[14] Laupland KB, Church DL, Mucenski M, Sutherland LR, Davies HD. Population-based study of the epidemiology of and the risk factors for invasive Staphylococcus aureus infections. J Infect Dis. 2003 May 1;187(9):1452-9.

[15] Harbarth S, Fankhauser C, Schrenzel J, Christenson J, Gervaz P, Bandiera-Clerc C, et al. Universal screening for methicillin-resistant Staphylococcus aureus at hospital admission and nosocomial infection in surgical patients. Jama. 2008 Mar 12;299(10):1149-57.

[16] Boyce JM. Methicillin-resistant Staphylococcus aureus. Detection, epidemiology, and control measures. Infect Dis Clin North Am. 1989 Dec;3(4):901-13.

[17] Harbarth S, Sax H, Uckay I, Fankhauser C, Agostinho A, Christenson JT, et al. A predictive model for identifying surgical patients at risk of methicillin-resistant Staphylococcus aureus carriage on admission. J Am Coll Surg. 2008 Nov;207(5):683-9.

[18] Lew DP, Waldvogel FA. Osteomyelitis. Lancet. 2004 Jul 24-30;364(9431):369-79.

[19] Lowy FD. Staphylococcus aureus infections. N Engl J Med. 1998 Aug 20;339(8):520-32.

[20] Anderson DJ, Chen LF, Schmader KE, Sexton DJ, Choi Y, Link K, et al. Poor functional status as a risk factor for surgical site infection due to methicillin-resistant Staphylococcus aureus. Infect Control Hosp Epidemiol. 2008 Sep;29(9):832-9.

[21] Shukla S, Nixon M, Acharya M, Korim MT, Pandey R. Incidence of MRSA surgical-site infection in MRSA carriers in an orthopaedic trauma unit. J Bone Joint Surg Br. 2009 Feb;91(2):225-8.
[22] Shirtliff ME, Mader JT. Acute septic arthritis. Clin Microbiol Rev. 2002 Oct;15(4):527-44.
[23] Ozuna RM, Delamarter RB. Pyogenic vertebral osteomyelitis and postsurgical disc space infections. Orthop Clin North Am. 1996 Jan;27(1):87-94.
[24] Cierny G, 3rd, Mader JT, Penninck JJ. A clinical staging system for adult osteomyelitis. Clin Orthop Relat Res. 2003 Sep(414):7-24.
[25] Waldvogel FA, Medoff G, Swartz MN. Osteomyelitis: a review of clinical features, therapeutic considerations and unusual aspects. 3. Osteomyelitis associated with vascular insufficiency. N Engl J Med. 1970 Feb 5;282(6):316-22.
[26] Cottle L, Riordan T. Infectious spondylodiscitis. J Infect. 2008 Jun;56(6):401-12.
[27] Foster TJ, Hook M. Surface protein adhesins of Staphylococcus aureus. Trends Microbiol. 1998 Dec;6(12):484-8.
[28] Elasri MO, Thomas JR, Skinner RA, Blevins JS, Beenken KE, Nelson CL, et al. Staphylococcus aureus collagen adhesin contributes to the pathogenesis of osteomyelitis. Bone. 2002 Jan;30(1):275-80.
[29] Gemmell CG, Goutcher SC, Reid R, Sturrock RD. Role of certain virulence factors in a murine model of Staphylococcus aureus arthritis. J Med Microbiol. 1997 Mar;46(3):208-13.
[30] Palmqvist N, Foster T, Tarkowski A, Josefsson E. Protein A is a virulence factor in Staphylococcus aureus arthritis and septic death. Microb Pathog. 2002 Nov;33(5):239-49.
[31] Palmqvist N, Josefsson E, Tarkowski A. Clumping factor A-mediated virulence during Staphylococcus aureus infection is retained despite fibrinogen depletion. Microbes Infect. 2004 Feb;6(2):196-201.
[32] Ferry T, Perpoint T, Vandenesch F, Etienne J. Virulence determinants in Staphylococcus aureus and their involvement in clinical syndromes. Curr Infect Dis Rep. 2005 Nov;7(6):420-8.
[33] Brady RA, Leid JG, Calhoun JH, Costerton JW, Shirtliff ME. Osteomyelitis and the role of biofilms in chronic infection. FEMS Immunol Med Microbiol. 2008 Jan;52(1):13-22.
[34] Proctor RA, von Eiff C, Kahl BC, Becker K, McNamara P, Herrmann M, et al. Small colony variants: a pathogenic form of bacteria that facilitates persistent and recurrent infections. Nat Rev Microbiol. 2006 Apr;4(4):295-305.
[35] Sendi P, Proctor RA. Staphylococcus aureus as an intracellular pathogen: the role of small colony variants. Trends Microbiol. 2009 Feb;17(2):54-8.
[36] Karauzum H, Ferry T, de Bentzmann S, Lina G, Bes M, Vandenesch F, et al. Comparison of adhesion and virulence of two predominant hospital-acquired methicillin-resistant Staphylococcus aureus clones and clonal methicillin-susceptible S. aureus isolates. Infect Immun. 2008 Nov;76(11):5133-8.
[37] Amaral MM, Coelho LR, Flores RP, Souza RR, Silva-Carvalho MC, Teixeira LA, et al. The predominant variant of the Brazilian epidemic clonal complex of methicillin-resistant Staphylococcus aureus has an enhanced ability to produce biofilm and to adhere to and invade airway epithelial cells. J Infect Dis. 2005 Sep 1;192(5):801-10.

[38] Livorsi DJ, Daver NG, Atmar RL, Shelburne SA, White AC, Jr., Musher DM. Outcomes of treatment for hematogenous Staphylococcus aureus vertebral osteomyelitis in the MRSA ERA. J Infect. 2008 Aug;57(2):128-31.

[39] Rigamonti D, Metellus P. Spinal epidural abscess. N Engl J Med. 2007 Feb 8;356(6):638; author reply -9.

[40] Lazzarini L, Mader JT, Calhoun JH. Osteomyelitis in long bones. J Bone Joint Surg Am. 2004 Oct;86-A(10):2305-18.

[41] Trampuz A, Zimmerli W. Diagnosis and treatment of infections associated with fracture-fixation devices. Injury. 2006 May;37 Suppl 2:S59-66.

[42] Zimmerli W, Trampuz A, Ochsner PE. Prosthetic-joint infections. N Engl J Med. 2004 Oct 14;351(16):1645-54.

[43] Tumeh SS, Aliabadi P, Weissman BN, McNeil BJ. Disease activity in osteomyelitis: role of radiography. Radiology. 1987 Dec;165(3):781-4.

[44] Cardinal E, Bureau NJ, Aubin B, Chhem RK. Role of ultrasound in musculoskeletal infections. Radiol Clin North Am. 2001 Mar;39(2):191-201.

[45] Harcke HT, Grissom LE, Finkelstein MS. Evaluation of the musculoskeletal system with sonography. AJR Am J Roentgenol. 1988 Jun;150(6):1253-61.

[46] Mah ET, LeQuesne GW, Gent RJ, Paterson DC. Ultrasonic features of acute osteomyelitis in children. J Bone Joint Surg Br. 1994 Nov;76(6):969-74.

[47] Babar S, Saifuddin A. MRI of the post-discectomy lumbar spine. Clin Radiol. 2002 Nov;57(11):969-81.

[48] Longo M, Granata F, Ricciardi K, Gaeta M, Blandino A. Contrast-enhanced MR imaging with fat suppression in adult-onset septic spondylodiscitis. Eur Radiol. 2003 Mar;13(3):626-37.

[49] Ross JS, Ruggieri P, Tkach J, Obuchowski N, Dillinger J, Masaryk TJ, et al. Lumbar degenerative disk disease: prospective comparison of conventional T2-weighted spin-echo imaging and T2-weighted rapid acquisition relaxation-enhanced imaging. AJNR Am J Neuroradiol. 1993 Sep-Oct;14(5):1215-23.

[50] Grand CM, Bank WO, Baleriaux D, Matos C, Levivier M, Brotchi J. Gadolinium enhancement of vertebral endplates following lumbar disc surgery. Neuroradiology. 1993;35(7):503-5.

[51] Boden SD, Davis DO, Dina TS, Sunner JL, Wiesel SW. Postoperative diskitis: distinguishing early MR imaging findings from normal postoperative disk space changes. Radiology. 1992 Sep;184(3):765-71.

[52] Pineda C, Vargas A, Rodriguez AV. Imaging of osteomyelitis: current concepts. Infect Dis Clin North Am. 2006 Dec;20(4):789-825.

[53] Ledermann HP, Kaim A, Bongartz G, Steinbrich W. Pitfalls and limitations of magnetic resonance imaging in chronic posttraumatic osteomyelitis. Eur Radiol. 2000;10(11):1815-23.

[54] Kaim AH, Gross T, von Schulthess GK. Imaging of chronic posttraumatic osteomyelitis. Eur Radiol. 2002 May;12(5):1193-202.

[55] Bureau NJ, Chhem RK, Cardinal E. Musculoskeletal infections: US manifestations. Radiographics. 1999 Nov-Dec;19(6):1585-92.

[56] Venkatesh SK, Riederer B, Chhem RK, Cardinal E, Wang SC. Reactivation in post-traumatic chronic osteomyelitis: ultrasonographic findings. Can Assoc Radiol J. 2003 Jun;54(3):163-8.

[57] Morrey BF, Westholm F, Schoifet S, Rand JA, Bryan RS. Long-term results of various treatment options for infected total knee arthroplasty. Clin Orthop Relat Res. 1989 Nov(248):120-8.

[58] Atkins BL, Athanasou N, Deeks JJ, Crook DW, Simpson H, Peto TE, et al. Prospective evaluation of criteria for microbiological diagnosis of prosthetic-joint infection at revision arthroplasty. The OSIRIS Collaborative Study Group. J Clin Microbiol. 1998 Oct;36(10):2932-9.

[59] Ghanem E, Ketonis C, Restrepo C, Joshi A, Barrack R, Parvizi J. Periprosthetic infection: where do we stand with regard to Gram stain? Acta Orthop. 2009 Feb;80(1):37-40.

[60] Trampuz A, Osmon DR, Hanssen AD, Steckelberg JM, Patel R. Molecular and antibiofilm approaches to prosthetic joint infection. Clin Orthop Relat Res. 2003 Sep(414):69-88.

[61] Shmerling RH. Synovial fluid analysis. A critical reappraisal. Rheum Dis Clin North Am. 1994 May;20(2):503-12.

[62] Kersey R, Benjamin J, Marson B. White blood cell counts and differential in synovial fluid of aseptically failed total knee arthroplasty. J Arthroplasty. 2000 Apr;15(3):301-4.

[63] Trampuz A, Hanssen AD, Osmon DR, Mandrekar J, Steckelberg JM, Patel R. Synovial fluid leukocyte count and differential for the diagnosis of prosthetic knee infection. Am J Med. 2004 Oct 15;117(8):556-62.

[64] Schafer P, Fink B, Sandow D, Margull A, Berger I, Frommelt L. Prolonged bacterial culture to identify late periprosthetic joint infection: a promising strategy. Clin Infect Dis. 2008 Dec 1;47(11):1403-9.

[65] Spangehl MJ, Masri BA, O'Connell JX, Duncan CP. Prospective analysis of preoperative and intraoperative investigations for the diagnosis of infection at the sites of two hundred and two revision total hip arthroplasties. J Bone Joint Surg Am. 1999 May;81(5):672-83.

[66] Widmer AF. New developments in diagnosis and treatment of infection in orthopedic implants. Clin Infect Dis. 2001 Sep 1;33 Suppl 2:S94-106.

[67] Mackowiak PA, Jones SR, Smith JW. Diagnostic value of sinus-tract cultures in chronic osteomyelitis. Jama. 1978 Jun 30;239(26):2772-5.

[68] Esteban J, Gomez-Barrena E, Cordero J, Martin-de-Hijas NZ, Kinnari TJ, Fernandez-Roblas R. Evaluation of quantitative analysis of cultures from sonicated retrieved orthopedic implants in diagnosis of orthopedic infection. J Clin Microbiol. 2008 Feb;46(2):488-92.

[69] Trampuz A, Piper KE, Jacobson MJ, Hanssen AD, Unni KK, Osmon DR, et al. Sonication of removed hip and knee prostheses for diagnosis of infection. N Engl J Med. 2007 Aug 16;357(7):654-63.

[70] Trampuz A, Piper KE, Hanssen AD, Osmon DR, Cockerill FR, Steckelberg JM, et al. Sonication of explanted prosthetic components in bags for diagnosis of prosthetic joint infection is associated with risk of contamination. J Clin Microbiol. 2006 Feb;44(2):628-31.

[71] Croize J, Gialanella P, Monnet D, Okada J, Orsi A, Voss A, et al. Improved identification of Staphylococcus aureus using a new agglutination test. Results of an international study. Apmis. 1993 Jun;101(6):487-91.

[72] Kuusela P, Hilden P, Savolainen K, Vuento M, Lyytikainen O, Vuopio-Varkila J. Rapid detection of methicillin-resistant Staphylococcus aureus strains not identified by slide agglutination tests. J Clin Microbiol. 1994 Jan;32(1):143-7.
[73] Neville LO, Billington OJ, Kibbler CC, Gillespie SH. Methicillin resistant Staphylococcus aureus without clumping factor, protein A, and DNAse. Lancet. 1991 Aug 24;338(8765):518.
[74] Fournier JM, Boutonnier A, Bouvet A. Staphylococcus aureus strains which are not identified by rapid agglutination methods are of capsular serotype 5. J Clin Microbiol. 1989 Jun;27(6):1372-4.
[75] Nakatomi Y, Sugiyama J. A rapid latex agglutination assay for the detection of penicillin-binding protein 2'. Microbiol Immunol. 1998;42(11):739-43.
[76] van Griethuysen A, Pouw M, van Leeuwen N, Heck M, Willemse P, Buiting A, et al. Rapid slide latex agglutination test for detection of methicillin resistance in Staphylococcus aureus. J Clin Microbiol. 1999 Sep;37(9):2789-92.
[77] Fenollar F, Roux V, Stein A, Drancourt M, Raoult D. Analysis of 525 samples to determine the usefulness of PCR amplification and sequencing of the 16S rRNA gene for diagnosis of bone and joint infections. J Clin Microbiol. 2006 Mar;44(3):1018-28.
[78] Fihman V, Hannouche D, Bousson V, Bardin T, Liote F, Raskine L, et al. Improved diagnosis specificity in bone and joint infections using molecular techniques. J Infect. 2007 Oct 26.
[79] Verdier I, Gayet-Ageron A, Ploton C, Taylor P, Benito Y, Freydiere AM, et al. Contribution of a broad range polymerase chain reaction to the diagnosis of osteoarticular infections caused by Kingella kingae: description of twenty-four recent pediatric diagnoses. Pediatr Infect Dis J. 2005 Aug;24(8):692-6.
[80] Fenollar F, Fournier PE, Robert C, Raoult D. Use of genome selected repeated sequences increases the sensitivity of PCR detection of Tropheryma whipplei. J Clin Microbiol. 2004 Jan;42(1):401-3.
[81] Dempsey KE, Riggio MP, Lennon A, Hannah VE, Ramage G, Allan D, et al. Identification of bacteria on the surface of clinically infected and non-infected prosthetic hip joints removed during revision arthroplasties by 16S rRNA gene sequencing and by microbiological culture. Arthritis Res Ther. 2007;9(3):R46.
[82] Ince A, Rupp J, Frommelt L, Katzer A, Gille J, Lohr JF. Is "aseptic" loosening of the prosthetic cup after total hip replacement due to nonculturable bacterial pathogens in patients with low-grade infection? Clin Infect Dis. 2004 Dec 1;39(11):1599-603.
[83] Tunney MM, Patrick S, Curran MD, Ramage G, Hanna D, Nixon JR, et al. Detection of prosthetic hip infection at revision arthroplasty by immunofluorescence microscopy and PCR amplification of the bacterial 16S rRNA gene. J Clin Microbiol. 1999 Oct;37(10):3281-90.
[84] Chometon S, Benito Y, Chaker M, Boisset S, Ploton C, Berard J, et al. Specific real-time polymerase chain reaction places Kingella kingae as the most common cause of osteoarticular infections in young children. Pediatr Infect Dis J. 2007 May;26(5):377-81.
[85] Lecouvet F, Irenge L, Vandercam B, Nzeusseu A, Hamels S, Gala JL. The etiologic diagnosis of infectious discitis is improved by amplification-based DNA analysis. Arthritis Rheum. 2004 Sep;50(9):2985-94.

[86] Becker K, Harmsen D, Mellmann A, Meier C, Schumann P, Peters G, et al. Development and evaluation of a quality-controlled ribosomal sequence database for 16S ribosomal DNA-based identification of Staphylococcus species. J Clin Microbiol. 2004 Nov;42(11):4988-95.
[87] Kipp F, Kahl BC, Becker K, Baron EJ, Proctor RA, Peters G, et al. Evaluation of two chromogenic agar media for recovery and identification of Staphylococcus aureus small-colony variants. J Clin Microbiol. 2005 Apr;43(4):1956-9.
[88] Chuard C, Vaudaux PE, Proctor RA, Lew DP. Decreased susceptibility to antibiotic killing of a stable small colony variant of Staphylococcus aureus in fluid phase and on fibronectin-coated surfaces. J Antimicrob Chemother. 1997 May;39(5):603-8.
[89] Kipp F, Becker K, Peters G, von Eiff C. Evaluation of different methods to detect methicillin resistance in small-colony variants of Staphylococcus aureus. J Clin Microbiol. 2004 Mar;42(3):1277-9.
[90] Proctor RA, Kahl B, von Eiff C, Vaudaux PE, Lew DP, Peters G. Staphylococcal small colony variants have novel mechanisms for antibiotic resistance. Clin Infect Dis. 1998 Aug;27 Suppl 1:S68-74.
[91] Nade S. Acute haematogenous osteomyelitis in infancy and childhood. J Bone Joint Surg Br. 1983 Mar;65(2):109-19.
[92] Esterhai JL, Jr., Gelb I. Adult septic arthritis. Orthop Clin North Am. 1991 Jul;22(3):503-14.
[93] Ivey M, Clark R. Arthroscopic debridement of the knee for septic arthritis. Clin Orthop Relat Res. 1985 Oct(199):201-6.
[94] Perry CR. Septic arthritis. Am J Orthop. 1999 Mar;28(3):168-78.
[95] Holtom PD, Borges L, Zalavras CG. Hematogenous septic ankle arthritis. Clin Orthop Relat Res. 2008 Jun;466(6):1388-91.
[96] Parisien JS. Arthroscopic management of pyarthrosis. An overview. Bull Hosp Jt Dis Orthop Inst. 1987 Spring;47(1):52-7.
[97] Thiery JA. Arthroscopic drainage in septic arthritides of the knee: a multicenter study. Arthroscopy. 1989;5(1):65-9.
[98] Kim SJ, Choi NH, Ko SH, Linton JA, Park HW. Arthroscopic treatment of septic arthritis of the hip. Clin Orthop Relat Res. 2003 Feb(407):211-4.
[99] Leslie BM, Harris JM, 3rd, Driscoll D. Septic arthritis of the shoulder in adults. J Bone Joint Surg Am. 1989 Dec;71(10):1516-22.
[100] Heyde CE, Boehm H, El Saghir H, Tschoke SK, Kayser R. Surgical treatment of spondylodiscitis in the cervical spine: a minimum 2-year follow-up. Eur Spine J. 2006 Sep;15(9):1380-7.
[101] Hadjipavlou AG, Mader JT, Necessary JT, Muffoletto AJ. Hematogenous pyogenic spinal infections and their surgical management. Spine. 2000 Jul 1;25(13):1668-79.
[102] Ito M, Abumi K, Kotani Y, Kadoya K, Minami A. Clinical outcome of posterolateral endoscopic surgery for pyogenic spondylodiscitis: results of 15 patients with serious comorbid conditions. Spine. 2007 Jan 15;32(2):200-6.
[103] Simpson AH, Deakin M, Latham JM. Chronic osteomyelitis. The effect of the extent of surgical resection on infection-free survival. J Bone Joint Surg Br. 2001 Apr;83(3):403-7.
[104] Seltzer SE. Value of computed tomography in planning medical and surgical treatment of chronic osteomyelitis. J Comput Assist Tomogr. 1984 Jun;8(3):482-7.

[105] Fitzgerald RH, Jr., Ruttle PE, Arnold PG, Kelly PJ, Irons GB. Local muscle flaps in the treatment of chronic osteomyelitis. J Bone Joint Surg Am. 1985 Feb;67(2):175-85.
[106] Levy M, Goldberg I. Treatment of chronic osteomyelitis by insertion of gentamicin-containing methyl methacrylate. Contemp Orthop. 1981;3:1114.
[107] Patzakis MJ, Abdollahi K, Sherman R, Holtom PD, Wilkins J. Treatment of chronic osteomyelitis with muscle flaps. Orthop Clin North Am. 1993 Jul;24(3):505-9.
[108] Mattar R, Azze R, Ferreira M. Vascularized fibular graft for management of severe osteomyelitis of the upper extremity. Microsurgery. 1994;15:22.
[109] Catagni MA, Ottaviani G, Camagni M. Treatment of massive tibial bone loss due to chronic draining osteomyelitis: fibula transport using the Ilizarov frame. Orthopedics. 2007 Aug;30(8):608-11.
[110] Deirmengian C, Greenbaum J, Stern J, Braffman M, Lotke PA, Booth RE, Jr., et al. Open debridement of acute gram-positive infections after total knee arthroplasty. Clin Orthop Relat Res. 2003 Nov(416):129-34.
[111] Mont MA, Waldman B, Banerjee C, Pacheco IH, Hungerford DS. Multiple irrigation, debridement, and retention of components in infected total knee arthroplasty. J Arthroplasty. 1997 Jun;12(4):426-33.
[112] Kaufer H, Matthews LS. Resection arthroplasty: an alternative to arthrodesis for salvage of the infected total knee arthroplasty. Instr Course Lect. 1986;35:283-9.
[113] Falahee MH, Matthews LS, Kaufer H. Resection arthroplasty as a salvage procedure for a knee with infection after a total arthroplasty. J Bone Joint Surg Am. 1987 Sep;69(7):1013-21.
[114] Damron TA, McBeath AA. Arthrodesis following failed total knee arthroplasty: comprehensive review and meta-analysis of recent literature. Orthopedics. 1995 Apr;18(4):361-8.
[115] Goksan SB, Freeman MA. One-stage reimplantation for infected total knee arthroplasty. J Bone Joint Surg Br. 1992 Jan;74(1):78-82.
[116] Silva M, Tharani R, Schmalzried TP. Results of direct exchange or debridement of the infected total knee arthroplasty. Clin Orthop Relat Res. 2002 Nov(404):125-31.
[117] Insall JN, Thompson FM, Brause BD. Two-stage reimplantation for the salvage of infected total knee arthroplasty. J Bone Joint Surg Am. 1983 Oct;65(8):1087-98.
[118] Durbhakula SM, Czajka J, Fuchs MD, Uhl RL. Antibiotic-loaded articulating cement spacer in the 2-stage exchange of infected total knee arthroplasty. J Arthroplasty. 2004 Sep;19(6):768-74.
[119] McPherson EJ, Lewonowski K, Dorr LD. Techniques in arthroplasty. Use of an articulated PMMA spacer in the infected total knee arthroplasty. J Arthroplasty. 1995 Feb;10(1):87-9.
[120] Hanssen AD, Rand JA. Evaluation and treatment of infection at the site of a total hip or knee arthroplasty. Instr Course Lect. 1999;48:111-22.
[121] Pittet D, Donaldson L. Clean Care is Safer Care: a worldwide priority. Lancet. 2005 Oct 8;366(9493):1246-7.
[122] Darley ES, MacGowan AP. Antibiotic treatment of gram-positive bone and joint infections. J Antimicrob Chemother. 2004 Jun;53(6):928-35.
[123] Stryjewski ME, Chambers HF. Skin and soft-tissue infections caused by community-acquired methicillin-resistant Staphylococcus aureus. Clin Infect Dis. 2008 Jun 1;46 Suppl 5:S368-77.

[124] Al-Nammari SS, Bobak P, Venkatesh R. Methicillin resistant Staphylococcus aureus versus methicillin sensitive Staphylococcus aureus adult haematogenous septic arthritis. Arch Orthop Trauma Surg. 2007 Sep;127(7):537-42.

[125] Volin SJ, Hinrichs SH, Garvin KL. Two-stage reimplantation of total joint infections: a comparison of resistant and non-resistant organisms. Clin Orthop Relat Res. 2004 Oct(427):94-100.

[126] Kilgus DJ, Howe DJ, Strang A. Results of periprosthetic hip and knee infections caused by resistant bacteria. Clin Orthop Relat Res. 2002 Nov(404):116-24.

[127] Salgado CD, Dash S, Cantey JR, Marculescu CE. Higher risk of failure of methicillin-resistant Staphylococcus aureus prosthetic joint infections. Clin Orthop Relat Res. 2007 Aug;461:48-53.

[128] Uçkay I, Lew D. Infections in Skeletal Prostheses. Bennet & Brachman's Hospital Infections 2007;5th ed:665.

[129] Smilack JD, Flittie WH, Williams TW, Jr. Bone concentrations of antimicrobial agents after parenteral administration. Antimicrob Agents Chemother. 1976 Jan;9(1):169-71.

[130] Daver NG, Shelburne SA, Atmar RL, Giordano TP, Stager CE, Reitman CA, et al. Oral step-down therapy is comparable to intravenous therapy for Staphylococcus aureus osteomyelitis. J Infect. 2007 Jun;54(6):539-44.

[131] Vuagnat A, Stern R, Lotthe A, Schuhmacher H, Duong M, Hoffmeyer P, et al. High dose vancomycin for osteomyelitis: continuous vs. intermittent infusion. J Clin Pharm Ther. 2004 Aug;29(4):351-7.

[132] Toma MB, Smith KM, Martin CA, Rapp RP. Pharmacokinetic considerations in the treatment of methicillin-resistant Staphylococcus aureus osteomyelitis. Orthopedics. 2006 Jun;29(6):497-501.

[133] Hidayat LK, Hsu DI, Quist R, Shriner KA, Wong-Beringer A. High-dose vancomycin therapy for methicillin-resistant Staphylococcus aureus infections: efficacy and toxicity. Arch Intern Med. 2006 Oct 23;166(19):2138-44.

[134] Lazzarini L, Tramarin A, Bragagnolo L, Tositti G, Manfrin V, de LF. Three-times weekly teicoplanin in the outpatient treatment of acute methicillin-resistant staphylococcal osteomyelitis: a pilot study. J Chemother. 2002 Feb;14(1):71-5.

[135] LeFrock JL, Ristuccia AM, Ristuccia PA, Quenzer RW, Haggerty PG, Allen JE, et al. Teicoplanin in the treatment of bone and joint infections. Teicoplanin Bone and Joint Cooperative Study Group, USA. Eur J Surg Suppl. 1992(567):9-13.

[136] Pensotti C, Nacinovich F, Vidiella G, Carbone E, Marin M, Di Stefano C, et al. [Teicoplanin in the treatment of bone and joint infections due to methicillin resistant staphylococci. Experience in adult patients]. Medicina (B Aires). 2002;62 Suppl 2:40-7.

[137] Arbeit RD, Maki D, Tally FP, Campanaro E, Eisenstein BI. The safety and efficacy of daptomycin for the treatment of complicated skin and skin-structure infections. Clin Infect Dis. 2004 Jun 15;38(12):1673-81.

[138] Marty FM, Yeh WW, Wennersten CB, Venkataraman L, Albano E, Alyea EP, et al. Emergence of a clinical daptomycin-resistant Staphylococcus aureus isolate during treatment of methicillin-resistant Staphylococcus aureus bacteremia and osteomyelitis. J Clin Microbiol. 2006 Feb;44(2):595-7.

[139] Ellis-Grosse EJ, Babinchak T, Dartois N, Rose G, Loh E. The efficacy and safety of tigecycline in the treatment of skin and skin-structure infections: results of 2 double-

blind phase 3 comparison studies with vancomycin-aztreonam. Clin Infect Dis. 2005 Sep 1;41 Suppl 5:S341-53.

[140] Noel GJ, Strauss RS, Amsler K, Heep M, Pypstra R, Solomkin JS. Results of a double-blind, randomized trial of ceftobiprole treatment of complicated skin and skin structure infections caused by gram-positive bacteria. Antimicrob Agents Chemother. 2008 Jan;52(1):37-44.

[141] Rouse MS, Steckelberg JM, Patel R. In vitro activity of ceftobiprole, daptomycin, linezolid, and vancomycin against methicillin-resistant staphylococci associated with endocarditis and bone and joint infection. Diagn Microbiol Infect Dis. 2007 Jul;58(3):363-5.

[142] Deresinski SC. The efficacy and safety of ceftobiprole in the treatment of complicated skin and skin structure infections: evidence from 2 clinical trials. Diagn Microbiol Infect Dis. 2008 May;61(1):103-9.

[143] Cunha B. Antibiotics in orthopaedic infections. In Orthopaedic Infection Schlossberg, D, Ed. 1998:156–74.

[144] Vaudaux P, Kelley WL, Lew DP. Staphylococcus aureus small colony variants: difficult to diagnose and difficult to treat. Clin Infect Dis. 2006 Oct 15;43(8):968-70.

[145] Matthews PC, Conlon CP, Berendt AR, Kayley J, Jefferies L, Atkins BL, et al. Outpatient parenteral antimicrobial therapy (OPAT): is it safe for selected patients to self-administer at home? A retrospective analysis of a large cohort over 13 years. J Antimicrob Chemother. 2007 Aug;60(2):356-62.

[146] Wynn M, Dalovisio JR, Tice AD, Jiang X. Evaluation of the efficacy and safety of outpatient parenteral antimicrobial therapy for infections with methicillin-sensitive Staphylococcus aureus. South Med J. 2005 Jun;98(6):590-5.

[147] Galperine T, Ader F, Piriou P, Judet T, Perronne C, Bernard L. [Outpatient parenteral antimicrobial therapy (OPAT) in bone and joint infections]. Med Mal Infect. 2006 Mar;36(3):132-7.

[148] Weigelt J, Itani K, Stevens D, Lau W, Dryden M, Knirsch C. Linezolid versus vancomycin in treatment of complicated skin and soft tissue infections. Antimicrob Agents Chemother. 2005 Jun;49(6):2260-6.

[149] Bishop E, Melvani S, Howden BP, Charles PG, Grayson ML. Good clinical outcomes but high rates of adverse reactions during linezolid therapy for serious infections: a proposed protocol for monitoring therapy in complex patients. Antimicrob Agents Chemother. 2006 Apr;50(4):1599-602.

[150] Ferry T, Ponceau B, Simon M, Issartel B, Petiot P, Boibieux A, et al. Possibly linezolid-induced peripheral and central neurotoxicity: report of four cases. Infection. 2005 Jun;33(3):151-4.

[151] Markowitz N, Quinn EL, Saravolatz LD. Trimethoprim-sulfamethoxazole compared with vancomycin for the treatment of Staphylococcus aureus infection. Ann Intern Med. 1992 Sep 1;117(5):390-8.

[152] Stein A, Bataille JF, Drancourt M, Curvale G, Argenson JN, Groulier P, et al. Ambulatory treatment of multidrug-resistant Staphylococcus-infected orthopedic implants with high-dose oral co-trimoxazole (trimethoprim-sulfamethoxazole). Antimicrob Agents Chemother. 1998 Dec;42(12):3086-91.

[153] Gorwitz RJ, Jernigan DB, Powers JH, Jernigan JA, Community PitCfDCaPCEMoMoMit. Strategies for clinical management of MRSA in the

community: summary of an experts' meeting convened by the Centers for Disease Control and Prevention. . Department of Health and Human Services, Centers for Disease Control and Prevention. 2006.
[154] Proctor RA. Role of folate antagonists in the treatment of methicillin-resistant Staphylococcus aureus infection. Clin Infect Dis. 2008 Feb 15;46(4):584-93.
[155] Johnson MD, Decker CF. Antimicrobial agents in treatment of MRSA infections. Dis Mon. 2008 Dec;54(12):793-800.
[156] Ruhe JJ, Monson T, Bradsher RW, Menon A. Use of long-acting tetracyclines for methicillin-resistant Staphylococcus aureus infections: case series and review of the literature. Clin Infect Dis. 2005 May 15;40(10):1429-34.
[157] Cenizal MJ, Skiest D, Luber S, Bedimo R, Davis P, Fox P, et al. Prospective randomized trial of empiric therapy with trimethoprim-sulfamethoxazole or doxycycline for outpatient skin and soft tissue infections in an area of high prevalence of methicillin-resistant Staphylococcus aureus. Antimicrob Agents Chemother. 2007 Jul;51(7):2628-30.
[158] Yourassowsky E, van der Linden MP, Lismont MJ, Crokaert F. Combination of minocycline and rifampicin against methicillin- and gentamicin-resistant Staphylococcus aureus. J Clin Pathol. 1981 May;34(5):559-63.
[159] Atkins B, Gottlieb T. Fusidic acid in bone and joint infections. Int J Antimicrob Agents. 1999 Aug;12 Suppl 2:S79-93.
[160] Drancourt M, Stein A, Argenson JN, Roiron R, Groulier P, Raoult D. Oral treatment of Staphylococcus spp. infected orthopaedic implants with fusidic acid or ofloxacin in combination with rifampicin. J Antimicrob Chemother. 1997 Feb;39(2):235-40.
[161] Turnidge J, Collignon P. Resistance to fusidic acid. Int J Antimicrob Agents. 1999 Aug;12 Suppl 2:S35-44.
[162] Yilmaz G, Aydin K, Iskender S, Caylan R, Koksal I. Detection and prevalence of inducible clindamycin resistance in staphylococci. J Med Microbiol. 2007 Mar;56(Pt 3):342-5.
[163] Panagea S, Perry JD, Gould FK. Should clindamycin be used as treatment of patients with infections caused by erythromycin-resistant staphylococci? J Antimicrob Chemother. 1999 Oct;44(4):581-2.
[164] Falagas ME, Giannopoulou KP, Kokolakis GN, Rafailidis PI. Fosfomycin: use beyond urinary tract and gastrointestinal infections. Clin Infect Dis. 2008 Apr 1;46(7):1069-77.
[165] Tuazon CU, Lin MY, Sheagren JN. In vitro activity of rifampin alone and in combination with nafcillin and Vancomycin against pathogenic strains of Staphylococcus aureus. Antimicrob Agents Chemother. 1978 May;13(5):759-61.
[166] Uckay I, Pittet D, Vaudaux P, Sax H, Lew D, Waldvogel F. Foreign body infections due to Staphylococcus epidermidis. Ann Med. 2009;41(2):109-19.
[167] Trampuz A, Zimmerli W. Antimicrobial agents in orthopaedic surgery: Prophylaxis and treatment. Drugs. 2006;66(8):1089-105.
[168] Perlroth J, Kuo M, Tan J, Bayer AS, Miller LG. Adjunctive use of rifampin for the treatment of Staphylococcus aureus infections: a systematic review of the literature. Arch Intern Med. 2008 Apr 28;168(8):805-19.
[169] Matthews PC, Dean BJ, Medagoda K, Gundle R, Atkins BL, Berendt AR, et al. Native hip joint septic arthritis in 20 adults: delayed presentation beyond three weeks predicts need for excision arthroplasty. J Infect. 2008 Sep;57(3):185-90.

[170] Smith JW, Hasan MS. Bone and joint infections. In Mandell, Douglas and Bennett's Principles and Practice of Infectious Diseases 2000;5th edn (Mandell, G. L., Bennett, J. E. & Dolin, R., Eds).1175–82.

[171] Berendt AR, Lipsky B. Is this bone infected or not? Differentiating neuro-osteoarthropathy from osteomyelitis in the diabetic foot. Curr Diab Rep. 2004 Dec;4(6):424-9.

[172] Mader JT, Mohan D, Calhoun J. A practical guide to the diagnosis and management of bone and joint infections. Drugs. 1997 Aug;54(2):253-64.

[173] Al-Nammari SS, Lucas JD, Lam KS. Hematogenous methicillin-resistant Staphylococcus aureus spondylodiscitis. Spine. 2007 Oct 15;32(22):2480-6.

[174] Sendi P, Bregenzer T, Zimmerli W. Spinal epidural abscess in clinical practice. Qjm. 2008 Jan;101(1):1-12.

[175] Zimmerli W. Infection and musculoskeletal conditions: Prosthetic-joint-associated infections. Best Pract Res Clin Rheumatol. 2006 Dec;20(6):1045-63.

[176] Laffer RR, Graber P, Ochsner PE, Zimmerli W. Outcome of prosthetic knee-associated infection: evaluation of 40 consecutive episodes at a single centre. Clin Microbiol Infect. 2006 May;12(5):433-9.

[177] Zimmerli W, Widmer AF, Blatter M, Frei R, Ochsner PE. Role of rifampin for treatment of orthopedic implant-related staphylococcal infections: a randomized controlled trial. Foreign-Body Infection (FBI) Study Group. Jama. 1998 May 20;279(19):1537-41.

[178] Duncan CP, Masri BA. The role of antibiotic-loaded cement in the treatment of an infection after a hip replacement. Instr Course Lect. 1995;44:305-13.

[179] Rao N, Crossett LS, Sinha RK, Le Frock JL. Long-term suppression of infection in total joint arthroplasty. Clin Orthop Relat Res. 2003 Sep(414):55-60.

[180] Billeter M, Zervos MJ, Chen AY, Dalovisio JR, Kurukularatne C. Dalbavancin: a novel once-weekly lipoglycopeptide antibiotic. Clin Infect Dis. 2008 Feb 15;46(4):577-83.

[181] Jauregui LE, Babazadeh S, Seltzer E, Goldberg L, Krievins D, Frederick M, et al. Randomized, double-blind comparison of once-weekly dalbavancin versus twice-daily linezolid therapy for the treatment of complicated skin and skin structure infections. Clin Infect Dis. 2005 Nov 15;41(10):1407-15.

[182] Sakoulas G, Moise-Broder PA, Schentag J, Forrest A, Moellering RC, Jr., Eliopoulos GM. Relationship of MIC and bactericidal activity to efficacy of vancomycin for treatment of methicillin-resistant Staphylococcus aureus bacteremia. J Clin Microbiol. 2004 Jun;42(6):2398-402.

[183] Vandenesch F, Naimi T, Enright MC, Lina G, Nimmo GR, Heffernan H, et al. Community-acquired methicillin-resistant Staphylococcus aureus carrying Panton-Valentine leukocidin genes: worldwide emergence. Emerg Infect Dis. 2003 Aug;9(8):978-84.

[184] Lina G, Piemont Y, Godail-Gamot F, Bes M, Peter MO, Gauduchon V, et al. Involvement of Panton-Valentine leukocidin-producing Staphylococcus aureus in primary skin infections and pneumonia. Clin Infect Dis. 1999 Nov;29(5):1128-32.

[185] Francois P, Harbarth S, Huyghe A, Renzi G, Bento M, Gervaix A, et al. Methicillin-resistant Staphylococcus aureus, Geneva, Switzerland, 1993-2005. Emerg Infect Dis. 2008 Feb;14(2):304-7.

[186] Tristan A, Bes M, Meugnier H, Lina G, Bozdogan B, Courvalin P, et al. Global distribution of Panton-Valentine leukocidin--positive methicillin-resistant Staphylococcus aureus, 2006. Emerg Infect Dis. 2007 Apr;13(4):594-600.
[187] Chambers HF. The changing epidemiology of Staphylococcus aureus? Emerg Infect Dis. 2001 Mar-Apr;7(2):178-82.
[188] Moran GJ, Krishnadasan A, Gorwitz RJ, Fosheim GE, McDougal LK, Carey RB, et al. Methicillin-resistant S. aureus infections among patients in the emergency department. N Engl J Med. 2006 Aug 17;355(7):666-74.
[189] Purcell K, Fergie J. Epidemic of community-acquired methicillin-resistant Staphylococcus aureus infections: a 14-year study at Driscoll Children's Hospital. Arch Pediatr Adolesc Med. 2005 Oct;159(10):980-5.
[190] Mishaan AM, Mason EO, Jr., Martinez-Aguilar G, Hammerman W, Propst JJ, Lupski JR, et al. Emergence of a predominant clone of community-acquired Staphylococcus aureus among children in Houston, Texas. Pediatr Infect Dis J. 2005 Mar;24(3):201-6.
[191] Szczesiul JM, Shermock KM, Murtaza UI, Siberry GK. No decrease in clindamycin susceptibility despite increased use of clindamycin for pediatric community-associated methicillin-resistant Staphylococcus aureus skin infections. Pediatr Infect Dis J. 2007 Sep;26(9):852-4.
[192] Dohin B, Gillet Y, Kohler R, Lina G, Vandenesch F, Vanhems P, et al. Pediatric bone and joint infections caused by Panton-Valentine leukocidin-positive Staphylococcus aureus. Pediatr Infect Dis J. 2007 Nov;26(11):1042-8.
[193] Martinez-Aguilar G, Avalos-Mishaan A, Hulten K, Hammerman W, Mason EO, Jr., Kaplan SL. Community-acquired, methicillin-resistant and methicillin-susceptible Staphylococcus aureus musculoskeletal infections in children. Pediatr Infect Dis J. 2004 Aug;23(8):701-6.
[194] Stefani S, Bongiorno D, Cafiso V, Campanile F, Crapis M, Cristini F, et al. Pathotype and susceptibility profile of a community-acquired methicillin-resistant Staphylococcus aureus strain responsible for a case of severe pneumonia. Diagn Microbiol Infect Dis. 2009 Jan;63(1):100-4.
[195] Baba T, Takeuchi F, Kuroda M, Yuzawa H, Aoki K, Oguchi A, et al. Genome and virulence determinants of high virulence community-acquired MRSA. Lancet. 2002 May 25;359(9320):1819-27.
[196] Souza RR, Coelho LR, Botelho AM, Ribeiro A, Rito PN, Vieira VV, et al. Biofilm formation and prevalence of lukF-pv, seb, sec and tst genes among hospital- and community-acquired isolates of some international methicillin-resistant Staphylococcus aureus lineages. Clin Microbiol Infect. 2009 Jan 31.
[197] Zhu Y, Weiss EC, Otto M, Fey PD, Smeltzer MS, Somerville GA. Staphylococcus aureus biofilm metabolism and the influence of arginine on polysaccharide intercellular adhesin synthesis, biofilm formation, and pathogenesis. Infect Immun. 2007 Sep;75(9):4219-26.
[198] Diep BA, Gill SR, Chang RF, Phan TH, Chen JH, Davidson MG, et al. Complete genome sequence of USA300, an epidemic clone of community-acquired meticillin-resistant Staphylococcus aureus. Lancet. 2006 Mar 4;367(9512):731-9.
[199] Wang R, Braughton KR, Kretschmer D, Bach TH, Queck SY, Li M, et al. Identification of novel cytolytic peptides as key virulence determinants for community-associated MRSA. Nat Med. 2007 Dec;13(12):1510-4.

[200] Boyle-Vavra S, Daum RS. Community-acquired methicillin-resistant Staphylococcus aureus: the role of Panton-Valentine leukocidin. Lab Invest. 2007 Jan;87(1):3-9.

[201] Gillet Y, Issartel B, Vanhems P, Fournet JC, Lina G, Bes M, et al. Association between Staphylococcus aureus strains carrying gene for Panton-Valentine leukocidin and highly lethal necrotising pneumonia in young immunocompetent patients. Lancet. 2002 Mar 2;359(9308):753-9.

[202] Tristan A, Ferry T, Durand G, Dauwalder O, Bes M, Lina G, et al. Virulence determinants in community and hospital meticillin-resistant Staphylococcus aureus. J Hosp Infect. 2007 Jun;65 Suppl 2:105-9.

[203] Maree CL, Daum RS, Boyle-Vavra S, Matayoshi K, Miller LG. Community-associated methicillin-resistant Staphylococcus aureus isolates causing healthcare-associated infections. Emerg Infect Dis. 2007 Feb;13(2):236-42.

[204] Miller LG, Perdreau-Remington F, Rieg G, Mehdi S, Perlroth J, Bayer AS, et al. Necrotizing fasciitis caused by community-associated methicillin-resistant Staphylococcus aureus in Los Angeles. N Engl J Med. 2005 Apr 7;352(14):1445-53.

[205] Diep BA, Carleton HA, Chang RF, Sensabaugh GF, Perdreau-Remington F. Roles of 34 virulence genes in the evolution of hospital- and community-associated strains of methicillin-resistant Staphylococcus aureus. J Infect Dis. 2006 Jun 1;193(11):1495-503.

[206] Kaneko J, Kamio Y. Bacterial two-component and hetero-heptameric pore-forming cytolytic toxins: structures, pore-forming mechanism, and organization of the genes. Biosci Biotechnol Biochem. 2004 May;68(5):981-1003.

[207] Szmigielski S, Prevost G, Monteil H, Colin DA, Jeljaszewicz J. Leukocidal toxins of staphylococci. Zentralbl Bakteriol. 1999 Apr;289(2):185-201.

[208] Ma XX, Ito T, Chongtrakool P, Hiramatsu K. Predominance of clones carrying Panton-Valentine leukocidin genes among methicillin-resistant Staphylococcus aureus strains isolated in Japanese hospitals from 1979 to 1985. J Clin Microbiol. 2006 Dec;44(12):4515-27.

[209] Joubert O, Voegelin J, Guillet V, Tranier S, Werner S, Colin DA, et al. Distinction between Pore Assembly by Staphylococcal alpha-Toxin versus Leukotoxins. J Biomed Biotechnol. 2007;2007:25935.

[210] Valeva A, Walev I, Pinkernell M, Walker B, Bayley H, Palmer M, et al. Transmembrane beta-barrel of staphylococcal alpha-toxin forms in sensitive but not in resistant cells. Proc Natl Acad Sci U S A. 1997 Oct 14;94(21):11607-11.

[211] Genestier AL, Michallet MC, Prevost G, Bellot G, Chalabreysse L, Peyrol S, et al. Staphylococcus aureus Panton-Valentine leukocidin directly targets mitochondria and induces Bax-independent apoptosis of human neutrophils. J Clin Invest. 2005 Nov;115(11):3117-27.

[212] Konig B, Koller M, Prevost G, Piemont Y, Alouf JE, Schreiner A, et al. Activation of human effector cells by different bacterial toxins (leukocidin, alveolysin, and erythrogenic toxin A): generation of interleukin-8. Infect Immun. 1994 Nov;62(11):4831-7.

[213] Konig B, Prevost G, Piemont Y, Konig W. Effects of Staphylococcus aureus leukocidins on inflammatory mediator release from human granulocytes. J Infect Dis. 1995 Mar;171(3):607-13.

[214] Colin DA, Monteil H. Control of the oxidative burst of human neutrophils by staphylococcal leukotoxins. Infect Immun. 2003 Jul;71(7):3724-9.

[215] de Bentzmann S, Tristan A, Etienne J, Brousse N, Vandenesch F, Lina G. Staphylococcus aureus isolates associated with necrotizing pneumonia bind to basement membrane type I and IV collagens and laminin. J Infect Dis. 2004 Oct 15;190(8):1506-15.
[216] Grojec PL, Jeljaszewicz J. Effect of staphylococcal leukocidin on mouse leukocyte system. Zentralbl Bakteriol Mikrobiol Hyg [A]. 1981 Nov;250(4):446-55.
[217] Szmigielski S, Jeljaszewicz J, Wilczynski J, Korbecki M. Reaction of rabbit leucocytes to staphylococcal (Panton-Valentine) leucocidin in vivo. J Pathol Bacteriol. 1966 Apr;91(2):599-604.
[218] Ward PD, Turner WH. Identification of staphylococcal Panton-Valentine leukocidin as a potent dermonecrotic toxin. Infect Immun. 1980 May;28(2):393-7.
[219] Labandeira-Rey M, Couzon F, Boisset S, Brown EL, Bes M, Benito Y, et al. Staphylococcus aureus Panton-Valentine leukocidin causes necrotizing pneumonia. Science. 2007 Feb 23;315(5815):1130-3.
[220] adiou C, Dumitrescu O, Croze M, Gillet Y, Dohin B, Slayman DH, et al. Panton-Valentine leukocidin is expressed at toxic levels in human skin abscesses. Clin Microbiol Infect. 2008 Dec;14(12):1180-3.
[221] Young MF. Bone matrix proteins: their function, regulation, and relationship to osteoporosis. Osteoporos Int. 2003;14 Suppl 3:S35-42.
[222] Clark BR, Keating A. Biology of bone marrow stroma. Ann N Y Acad Sci. 1995 Dec 29;770:70-8.
[223] Tristan A, Benito Y, Montserret R, Boisset S, Dusserre E, Penin F, et al. The signal peptide of Staphylococcus aureus panton valentine leukocidin LukS component mediates increased adhesion to heparan sulfates. PLoS ONE. 2009;4(4):e5042.
[224] Gomez MI, Lee A, Reddy B, Muir A, Soong G, Pitt A, et al. Staphylococcus aureus protein A induces airway epithelial inflammatory responses by activating TNFR1. Nat Med. 2004 Aug;10(8):842-8.
[225] Diep BA, Palazzolo-Ballance AM, Tattevin P, Basuino L, Braughton KR, Whitney AR, et al. Contribution of Panton-Valentine leukocidin in community-associated methicillin-resistant Staphylococcus aureus pathogenesis. PLoS ONE. 2008;3(9):e3198.
[226] Gonzalez BE, Martinez-Aguilar G, Hulten KG, Hammerman WA, Coss-Bu J, Avalos-Mishaan A, et al. Severe Staphylococcal sepsis in adolescents in the era of community-acquired methicillin-resistant Staphylococcus aureus. Pediatrics. 2005 Mar;115(3):642-8.
[227] Deurenberg RH, Stobberingh EE. The evolution of Staphylococcus aureus. Infect Genet Evol. 2008 Dec;8(6):747-63.
[228] Lee SM, Ender M, Adhikari R, Smith JM, Berger-Bachi B, Cook GM. Fitness cost of staphylococcal cassette chromosome mec in methicillin-resistant Staphylococcus aureus by way of continuous culture. Antimicrob Agents Chemother. 2007 Apr;51(4):1497-9.
[229] Diep BA, Stone GG, Basuino L, Graber CJ, Miller A, des Etages SA, et al. The arginine catabolic mobile element and staphylococcal chromosomal cassette mec linkage: convergence of virulence and resistance in the USA300 clone of methicillin-resistant Staphylococcus aureus. J Infect Dis. 2008 Jun 1;197(11):1523-30.
[230] Diep BA, Otto M. The role of virulence determinants in community-associated MRSA pathogenesis. Trends Microbiol. 2008 Aug;16(8):361-9.

[231] Kaito C, Omae Y, Matsumoto Y, Nagata M, Yamaguchi H, Aoto T, et al. A novel gene, fudoh, in the SCCmec region suppresses the colony spreading ability and virulence of Staphylococcus aureus. PLoS ONE. 2008;3(12):e3921.

[232] Klevens RM, Morrison MA, Nadle J, Petit S, Gershman K, Ray S, et al. Invasive methicillin-resistant Staphylococcus aureus infections in the United States. Jama. 2007 Oct 17;298(15):1763-71.

[233] Moncada S, Higgs A. The L-arginine-nitric oxide pathway. N Engl J Med. 1993 Dec 30;329(27):2002-12.

[234] Bubeck Wardenburg J, Bae T, Otto M, Deleo FR, Schneewind O. Poring over pores: alpha-hemolysin and Panton-Valentine leukocidin in Staphylococcus aureus pneumonia. Nat Med. 2007 Dec;13(12):1405-6.

[235] Kaplan SL. Community-acquired methicillin-resistant Staphylococcus aureus infections in children. Semin Pediatr Infect Dis. 2006 Jul;17(3):113-9.

[236] Browne LP, Mason EO, Kaplan SL, Cassady CI, Krishnamurthy R, Guillerman RP. Optimal imaging strategy for community-acquired Staphylococcus aureus musculoskeletal infections in children. Pediatr Radiol. 2008 Aug;38(8):841-7.

[237] Ferry T, Etienne J. Community acquired MRSA in Europe. Bmj. 2007 Nov 10;335(7627):947-8.

[238] Naimi TS, LeDell KH, Como-Sabetti K, Borchardt SM, Boxrud DJ, Etienne J, et al. Comparison of community- and health care-associated methicillin-resistant Staphylococcus aureus infection. Jama. 2003 Dec 10;290(22):2976-84.

[239] Betz RR, Cooperman DR, Wopperer JM, Sutherland RD, White JJ, Jr., Schaaf HW, et al. Late sequelae of septic arthritis of the hip in infancy and childhood. J Pediatr Orthop. 1990 May-Jun;10(3):365-72.

[240] Blickman JG, van Die CE, de Rooy JW. Current imaging concepts in pediatric osteomyelitis. Eur Radiol. 2004 Mar;14 Suppl 4:L55-64.

[241] Arnold SR, Elias D, Buckingham SC, Thomas ED, Novais E, Arkader A, et al. Changing patterns of acute hematogenous osteomyelitis and septic arthritis: emergence of community-associated methicillin-resistant Staphylococcus aureus. J Pediatr Orthop. 2006 Nov-Dec;26(6):703-8.

[242] Pannaraj PS, Hulten KG, Gonzalez BE, Mason EO, Jr., Kaplan SL. Infective pyomyositis and myositis in children in the era of community-acquired, methicillin-resistant Staphylococcus aureus infection. Clin Infect Dis. 2006 Oct 15;43(8):953-60.

[243] Gonzalez BE, Teruya J, Mahoney DH, Jr., Hulten KG, Edwards R, Lamberth LB, et al. Venous thrombosis associated with staphylococcal osteomyelitis in children. Pediatrics. 2006 May;117(5):1673-9.

[244] Kaplan SL, Hulten KG, Gonzalez BE, Hammerman WA, Lamberth L, Versalovic J, et al. Three-year surveillance of community-acquired Staphylococcus aureus infections in children. Clin Infect Dis. 2005 Jun 15;40(12):1785-91.

[245] El-Maghraby TA, Moustafa HM, Pauwels EK. Nuclear medicine methods for evaluation of skeletal infection among other diagnostic modalities. Q J Nucl Med Mol Imaging. 2006 Sep;50(3):167-92.

[246] Loffler CA, Macdougall C. Update on prevalence and treatment of methicillin-resistant Staphylococcus aureus infections. Expert Rev Anti Infect Ther. 2007 Dec;5(6):961-81.

[247] Cole WG, Dalziel RE, Leitl S. Treatment of acute osteomyelitis in childhood. J Bone Joint Surg Br. 1982;64(2):218-23.

[248] Meller I, Manor Y, Bar-Ziv J, Torok G. Pediatric update #8. Acute hematogenous osteomyelitis in children. Long-term results of surgical treatment. Orthop Rev. 1989 Jul;18(7):824-31.

[249] Infection of the musculoskeletal system, in Tachdjian's pediatric orthopaedics. JA Herring ed, Saunders.Philadelphia, 2008, 4th edition.:2089-177.

[250] Morrissy RT. Bone and joint sepsis. In Lovell and Winter's pediatric orthopaedics.ed, Lippincott, Philadelphia, 2001.(5th edition):459-505.

[251] El-Sayed AM. Treatment of early septic arthritis of the hip in children: comparison of results of open arthrotomy versus arthroscopic drainage. J Child Orthop. 2008 Jun;2(3):229-37.

[252] Givon U, Liberman B, Schindler A, Blankstein A, Ganel A. Treatment of septic arthritis of the hip joint by repeated ultrasound-guided aspirations. J Pediatr Orthop. 2004 May-Jun;24(3):266-70.

[253] Kaplan SL. Osteomyelitis in children. Infect Dis Clin North Am. 2005 Dec;19(4):787-97, vii.

[254] Hawkshead JJ, 3rd, Patel NB, Steele RW, Heinrich SD. Comparative severity of pediatric osteomyelitis attributable to methicillin-resistant versus methicillin-sensitive Staphylococcus aureus. J Pediatr Orthop. 2009 Jan-Feb;29(1):85-90.

[255] Dombrowski JC, Winston LG. Clinical failures of appropriately-treated methicillin-resistant Staphylococcus aureus infections. J Infect. 2008 Aug;57(2):110-5.

[256] Bocchini CE, Hulten KG, Mason EO, Jr., Gonzalez BE, Hammerman WA, Kaplan SL. Panton-Valentine leukocidin genes are associated with enhanced inflammatory response and local disease in acute hematogenous Staphylococcus aureus osteomyelitis in children. Pediatrics. 2006 Feb;117(2):433-40.

[257] Martinez-Aguilar G, Hammerman WA, Mason EO, Jr., Kaplan SL. Clindamycin treatment of invasive infections caused by community-acquired, methicillin-resistant and methicillin-susceptible Staphylococcus aureus in children. Pediatr Infect Dis J. 2003 Jul;22(7):593-8.

[258] Frank AL, Marcinak JF, Mangat PD, Tjhio JT, Kelkar S, Schreckenberger PC, et al. Clindamycin treatment of methicillin-resistant Staphylococcus aureus infections in children. Pediatr Infect Dis J. 2002 Jun;21(6):530-4.

[259] Sattler CA, Mason EO, Jr., Kaplan SL. Prospective comparison of risk factors and demographic and clinical characteristics of community-acquired, methicillin-resistant versus methicillin-susceptible Staphylococcus aureus infection in children. Pediatr Infect Dis J. 2002 Oct;21(10):910-7.

[260] Wu KC, Chiu HH, Wang JH, Lee NS, Lin HC, Hsieh CC, et al. Characteristics of community-acquired methicillin-resistant Staphylococcus aureus in infants and children without known risk factors. J Microbiol Immunol Infect. 2002 Mar;35(1):53-6.

[261] Lamp KC, Friedrich LV, Mendez-Vigo L, Russo R. Clinical experience with daptomycin for the treatment of patients with osteomyelitis. Am J Med. 2007 Oct;120(10 Suppl 1):S13-20.

[262] Daum RS. Clinical practice. Skin and soft-tissue infections caused by methicillin-resistant Staphylococcus aureus. N Engl J Med. 2007 Jul 26;357(4):380-90.

[263] Popovich K, Hota B, Rice T, Aroutcheva A, Weinstein RA. Phenotypic prediction rule for community-associated methicillin-resistant Staphylococcus aureus. J Clin Microbiol. 2007 Jul;45(7):2293-5.

[264] Hageman JC, Patel J, Franklin P, Miscavish K, McDougal L, Lonsway D, et al. Occurrence of a USA300 vancomycin-intermediate Staphylococcus aureus. Diagn Microbiol Infect Dis. 2008 Dec;62(4):440-2.

[265] Ohlsen K, Ziebuhr W, Koller KP, Hell W, Wichelhaus TA, Hacker J. Effects of subinhibitory concentrations of antibiotics on alpha-toxin (hla) gene expression of methicillin-sensitive and methicillin-resistant Staphylococcus aureus isolates. Antimicrob Agents Chemother. 1998 Nov;42(11):2817-23.

[266] Dumitrescu O, Boisset S, Badiou C, Bes M, Benito Y, Reverdy ME, et al. Effect of antibiotics on Staphylococcus aureus producing Panton-Valentine leukocidin. Antimicrob Agents Chemother. 2007 Apr;51(4):1515-9.

[267] Dumitrescu ON, Protti DA, Majumdar S, Zeilhofer HU, Wassle H. Ionotropic glutamate receptors of amacrine cells of the mouse retina. Vis Neurosci. 2006 Jan-Feb;23(1):79-90.

[268] Mukhtar TA, Wright GD. Streptogramins, oxazolidinones, and other inhibitors of bacterial protein synthesis. Chem Rev. 2005 Feb;105(2):529-42.

[269] Sabharwal V, Marchant CD. Fluoroquinolone use in children. Pediatr Infect Dis J. 2006 Mar;25(3):257-8.

In: Methicillin-Resistant Staphylococcus Aureus... ISBN: 978-1-60741-398-1
Editor: Charles L. Kolendi

Chapter 5

MRSA: THE VIRULENCE, EPIDEMIOLOGY AND PERSPECTIVE DIAGNOSTICS AND THERAPY

Julianna Kurlenda[1*] and Mariusz Grinholc[2]

[1]Department of Clinical Bacteriology in Hospital, Monte Cassino 13, 75-340 Koszalin, Poland

[2]Intercollegiate Faculty of Biotechnology University of Gdansk and Medical University of Gdansk, Department of Biotechnology, Laboratory of Molecular Diagnostics, Kladki 24, 80-822 Gdansk, Poland

ABSTRACT

Methicillin resistant *Staphylococcus aureus* is one of the most common pathogens responsible for health-care associated infections as well as community-acquired ones. It is the etiological factor of the wide spectrum of infections. Pathogenesis of those infections is multilevel, thus vaguely described and understood. Virulence factors include components facilitating microbial colonization as well as invasion with severe clinical symptoms, and are often responsible for the chronic state of the disease. Lax hygiene procedures and inappropriate antibiotic policy facilitate the wide spread of this pathogen within a hospital environment. *S. aureus* is also responsible for epidemic infections within human communities. Antibiotic-resistant strains become a serious therapeutic problem, since in such cases a limited spectrum of expensive antibiotics may be used and because of a prolonged hospitalization period resulting in economical losses. Hence, the development of diagnostics enabling faster and proper identification of the etiological factor and its drug-resistance mechanisms, followed by more appropriate clinical treatment, seems to be important. There is also an urgent need for the development of alternative therapeutics, as well as alternative therapeutic methods. The following work is a recapitulative paper on current knowledge concerning staphylococcal diseases, virulence factors, epidemiology, and prospective diagnostics and treatment.

Keywords: alternative therapies, methicillin resistant *Staphylococcus aureus*, virulence, epidemiology, diagnostics, alternative therapy

---

*Corresponding author: Tel/fax (48-58) 523 64 25, E-mail: bakteriologia@op.pl

Introduction

Methicillin resistant *Staphylococcus aureus* (MRSA) is one of the most common pathogens responsible for hospital infections, and recently it has also been discovered that for community-acquired infections as well. It can cause a broad spectrum of infections through local invasion, toxin-mediated diseases to generalized infections (James 2001; Healy et al. 2004; Whyte et al. 2005; Fowler, Jr. et al. 2005a; Browatzki et al. 2006; Collins 2007; Benfield et al. 2007) and Staphylococcal Foodborn Diseases (SFD) (Murray 2005). This pathogen possesses a wide spectrum of factors responsible for colonization and then invasion, manifesting as acute infections, which can lead to chronic disease (von Eiff et al. 2001a; Spanu et al. 2005; Ehrlich et al. 2005; Neut et al. 2007; Weigel et al. 2007). *S. aureus* is a bacterium commonly present in the human population and constant or part-time carrier frequency in the nasal vestibule is estimated at 30–60% (Kalmeijer et al. 2000; Pan et al. 2005; Wertheim et al. 2005; Graham et al. 2006). Therapeutic problems are caused by infections with strains that are resistant not only to methicillin (methicillin-resistant *S. aureus*, MRSA) (Ito et al. 2001; Ito and Hiramatsu 2003) and in consequence to all β-lactam antibiotics, but also to many other group of antimicrobial therapeutics. The presence of this pathogen enables its local distribution, e.g., within a hospital ward and/or between hospital wards (hospital or healthcare acquired) HA-MRSA (Sola et al. 2002; Savas et al. 2005; Conceicao et al. 2007; Kuint et al. 2007; Kurlenda et al. 2007a). In different types of human communities it can cause infections of an epidemic character (community acquired) CA-MRSA (Naimi et al. 2003; Ellis et al. 2004; Faria et al. 2005; Graham et al. 2006; Lorber 2006; Maree et al. 2007). CA-MRSA and HA-MRSA strains also spread globally between countries and continents both in the hospital, as well as in the non-hospital environment (HA and CA) (Gosbell 2005; Deurenberg et al. 2007). If it is the case, the clone is either highly virulent, and/or carries genes responsible for the resistance to many antibiotics, as in the in case of HA-MRSA strains (Sola et al. 2002; Savas et al. 2005; Conceicao et al. 2007; Kuint et al. 2007; Huang et al. 2007; Kurlenda et al. 2007a); then it becomes a great problem both, therapeutically—due to the limited number of antibiotics available—as well as economically—due to the necessity of expensive drugs and the prolonged time spent in the hospital (Karchmer et al. 2002; Rayner and Munckhof 2005; Trampuz and Zimmerli 2006; Livermore 2006; Karchmer 2006; Finch 2006; Dzwonkowska et al. 2007). The increase of costs is also a result of special procedure initiations needed for controlling the wide spread of the pathogen, such as hygienic and isolation procedures, identifying the carrier and, next, its eradication (Marshall et al. 2004; Coia et al. 2006; Kurlenda et al. 2007a). This indicates the importance of the improvement of diagnostic methods which enable rapid identification of the etiological factor and earlier and more precise detection of mechanisms responsible for drug resistance, especially low-level resistance, which allows us to avoid diagnostic mistakes (Donlan et al. 2001; Tan et al. 2001; Kipp et al. 2003; Tarkin et al. 2003; Struelens and Denis 2006; Fenollar et al. 2006). Due to the wide presence of multiresistant strains, research regarding the development of new therapeutic strategies, including antibiotic and alternative therapies as well as prophylaxis strategies which are important for patients before invasive medical procedures, is constantly being conducted (Hiramatsu et al. 1997; Kampf et al. 2003; Kaiser et al. 2004; Rayner and Munckhof 2005; Kowalski et al. 2005; Trampuz and Zimmerli

2006; Livermore 2006; Tacconelli 2006; Karchmer 2006; Finch 2006; Goldstein 2007; Witte et al. 2007).

STAPHYLOCOCCUS AUREUS INFECTIONS

S. aureus can cause local infections which can in later stages develop into generalized, such as bacteremia and sepsis, both in adults (Gottlieb et al. 2000; Priest and Peacock, Jr. 2005; Whyte et al. 2005; Collins 2007; Benfield et al. 2007; Desachy et al. 2007) and in children (Healy et al. 2004; Regev-Yochay et al. 2005; Kuint et al. 2007). Systemic infections develop when the protective barriers of the host are broken and the pathogen is spread into the bloodstream with the formation of secondary infection sources as endocarditis (Fowler, Jr. et al. 2005b; Chesi et al. 2006), pneumonia (Le et al. 2001; Francis et al. 2005; Elizur et al. 2007; Cunnington et al. 2009), osteomyelitis (Lew and Waldvogel 2004; Priest and Peacock, Jr. 2005) and deep abscess formation (Moran et al. 2006). Practically every local infection can lead to a systemic disease. The cause can be a foreign body, as happens in the case of device-related diseases such as: intravascular catheters (Priest and Peacock, Jr. 2005; Collins and Hampton 2005; Fowler, Jr. et al. 2005a), propylene nets (Abele-Horn et al. 2000), ventriculoperitoneal shunts (Spanu et al. 2005), pacemakers (Seifert et al. 2003), and different kinds of implants used in orthopedics (Hudson et al. 1999; Jarraud et al. 2002; Lew and Waldvogel 2004; Davis 2005) or postoperative and traumatic wounds (Abele-Horn et al. 2000), skin and soft tissue infections (Ladhani and Garbash 2005; Vayalumkal and Jadavji 2006).

The second group of infections are toxin-mediated diseases (Ladhani and Garbash 2005; Murray 2005; Vayalumkal and Jadavji 2006; Chi et al. 2006) such as Toxic Shock Syndrome (TSS) (Murray 2005; Chi et al. 2006), Bullous impetigo, Staphylococcal Scalded Skin Syndrome (SSSS) (Gravet et al. 2001; Ladhani and Garbash 2005; Vayalumkal and Jadavji 2006; Chi et al. 2006; Kurlenda et al. 2009) and Staphylococcal Foodborne Diseases (SFD) (Omoe et al. 2002; Sandhu and Kanwar 2004; Murray 2005; Naffa et al. 2006; Nema et al. 2007).

VIRULENCE FACTORS

MRSA strains are characterized with almost the same spectrum of virulence factors as MSSA (methicillin-sensitive *S. aureus*) strains. The slight differences between MRSA and MSSA virulence factors would be discussed along this paragraph.

The molecular background of infection caused by *S. aureus* is multifactoral, depending on the expression of a large group of specific surface proteins bound to the bacterial wall and proteins secreted by the bacteria as well. The expression of most virulence factors is controlled by the complex global regulation locus *agr (*ang. *accessory gene regulator*) (Morfeldt et al. 1988; Peng et al. 1988) and two additional loci, *sar* (ang. *Staphylococcus accessory regulator*) and *xpr (*ang. *extracellular* protein *sar*) (Cheung et al. 1992; Cheung and Projan 1994). Recent researches suggest that the *agr* locus is a positive regulator of genes encoding secreted factors and a negative regulator of those associated with the production of surface proteins. The *sar* locus on the other hand is a positive regulator of both virulence factors (Figure 1). It seems likely that the participance of these systems in the regulation of virulence in *S. aureus* is more complex (Shenkman et al. 2001; Jarraud et al. 2002).

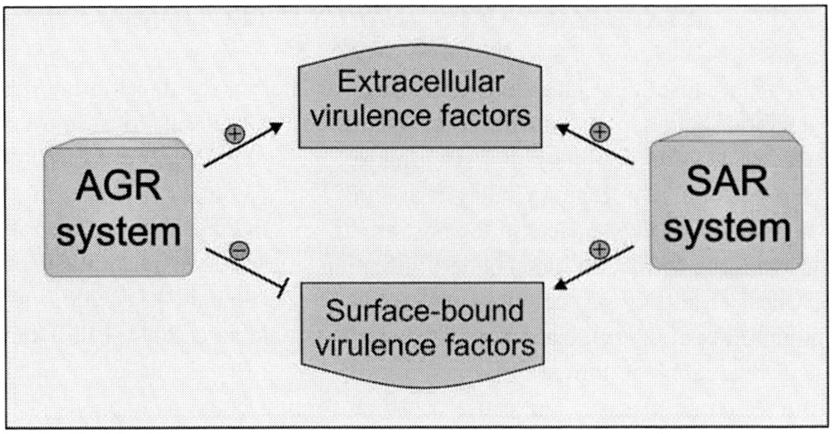

Figure 1. **The regulatory systems of the expression of most virulence factors.** The expression of virulence factors is controlled by the complex global regulation locus *agr* (accessory gene regulator) and two additional loci, sar (Staphylococcus accessory regulator) and xpr (extracellular protein sar); *agr* locus is a positive regulator of genes encoding secreted factors, and a negative regulator of those associated with the production of surface proteins. The sar locus is a positive regulator of both groups of virulence factors.

The importance of both regulatory systems is indicated by the fact that mutants, lacking the activity of both the *agr* and the *sar* genes, are much less virulent as shown in a mouse model of pneumonia. Moreover, mutations in both genes significantly decrease the ability of *S. aureus* to cause invasive infection, and only strains comprising the *agr* and *sar* loci induce bacteremia and cause pneumonia in mice (Heyer et al. 2002). It is worth to emphasize that *agr* regulon polymorphism could correlate with the types of infections. This could give as the molecular marker of the adjusted prophylaxis (Table 1) (Jarraud et al. 2002).

Table 1. *Agr* group according to the disease type

agr group	Disease type		
	Exfoliative toxin mediated disease	Enterotoxin mediated disease	TSST-1 mediated disease
I		++	
II	+	+	
III			++
IV	++		

Virulence factors like peptydoglican or protein A are bound to the bacterial outer cell wall (Lee et al. 2004; Voyich et al. 2005). A factor playing an important role in the process of infection is the electric charge of the bacterial outer cell wall. Non-virulent strains are characterized by hydrophilic outer layers (negative charge), whereas the hydrophobic character of outer layers (positive charge) is a factor which can facilitate bacterial growth (Peschel and Collins 2001; Hiemstra et al. 2004). These strains are resistant to CAMP (Cationic Antimicrobial Peptides) which are important elements in the defense system of the host (Figure 2) (Peschel 2002).

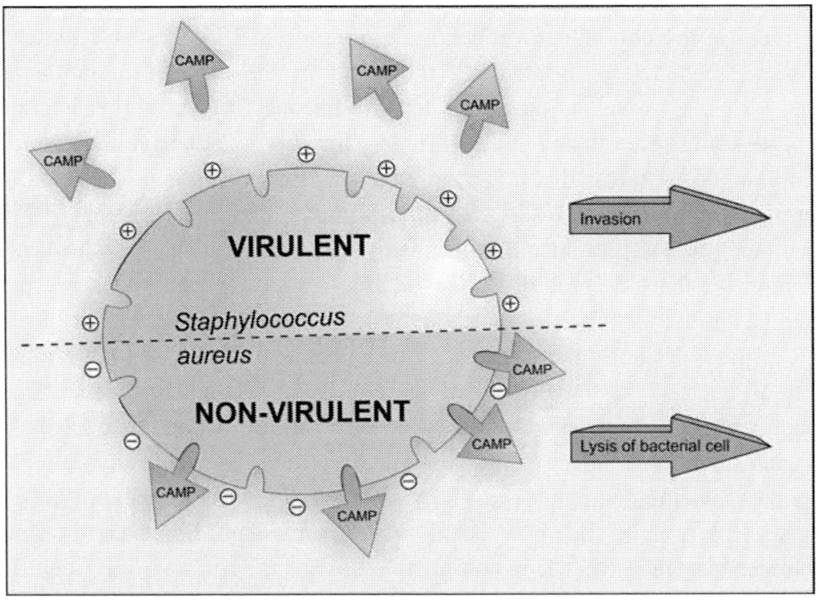

Figure 2. **Electric charge of the bacterial outer cell wall as the virulence factor.** Non-virulent strains are characterized by hydrophilic outer layers (negative charge), whereas the hydrophobic character of outer layers (positive charge) is a factor which can facilitate bacterial growth. These strains are resistant to CAMP (*Cationic Antimicrobial Peptides*) which are important elements in the defense system of the host.

In order to initiate invasive infections, *S. aureus* must adhere to the extracellular matrix (ECM) substrates of the host cells. In this process proteins bound to the surface of the bacterial cell wall play an important role. They enable the colonization of the host tissues due to their adhesive properties associated with the presence of ligand-binding domains. In this way *S. aureus* initiates colonization and adhesins participate in this binding. They belong to the family of proteins described as MSCRAMM (Microbial Surface Components Recognizing Adhesive Matrix Molecules), which comprises proteins mostly covalently bound to the peptidoglycan of the bacterial cell wall (Josefsson et al. 1998; Mazmanian et al. 2003; Roche et al. 2004; Clarke et al. 2004; Grundmeier et al. 2004; Walsh et al. 2004; Que et al. 2005; Marraffini and Schneewind 2005). These proteins include mainly the Collagen-Binding Protein (Cna), Clumping factor A (ClfA) and Fibronectin-Binding Protein A (FnBPA) (Foster and Hook 1998) but also Elastin-bindig protein (Ebps) and Bone sialoprotein-binding protein (Bbp) (Hudson et al. 1999).

As previously mentioned, the adhesion of bacteria to cellular components and extracellular matrix is a prerequisite for the colonization of host tissue. After a certain period of time implants become coated with host proteins, enabling the pathogen to adhere to its surface and initiate infection. *S. aureus* produces many surface proteins which mediate this process. It has been examined that some MRSA strains displayed weak adhesion to the surface of coated plasma proteins or proteins of the extracellular matrix *in vitro*. This has been associated with the fact that these strains produce the plasmin sensitive (PLs) protein. The *pls* gene, which encodes this protein, lies within the *mecDNA* region, which is characteristic solely for the MRSA strains and within which the *mecA* gene, responsible for methicillin resistance, is located (Savolainen et al. 2001). Researchers underline the role of the Pls protein in the non-adhesive phenotype of some MRSA strains. It has been observed that adhesion, mediated by protein A, increased in the case of mutant strains with an inactivated *pls* gene, as well as the ability to bind fibronectin, was higher in comparison with the wild type strains (Savolainen et al. 2001). As for now no adhesive function of this protein has been found, and it seems that it is a antiadhesin. However, it is possible that Pls possesses an unidentified role indicated by its structure. Some of the repeat regions of Pls can also be found in other proteins present in Gram positive bacteria, which are adhesins. Moreover, it has been reported that this protein is cleaved as a result of plasminogen receptor activation. This cleavage can be associated with the adhesive properties of Pls which have not yet been known. It is possible that during one stage of infection Pls prevents adhesion, enabling the expansion of bacteria in the organism, and on another stage promotes the adhesion of staphylococci to host tissue (Savolainen et al. 2001). Little is known about the role of Pls in the pathogenesis of staphylococci infections, and there is little literature data on this subject. A few papers available do not show any correlation between the presence of the gene encoding the Pls protein and the type of infection induced, however, the obtained results indicate lower virulence of strains comprising this gene as it was only identified in strains isolated from local infections and from carriers (Kurlenda et al. 2007b). On the basis of clinical observations we can assume that presence of that gene is a significant factor of staphylococcal pathogenesis and induce less virulent phenotype. No relevant differences between mortality in MRSA and MSSA infected patients were observed (not published data). We think that further investigations concerning above mentioned task are necessary.

The extracellular matrix protein-binding (Emp) protein has not been thoroughly examined yet. Mutants with an inactivated *emp* gene show a substantial reduction in the ability to adhere to fibrinogen and fibronectin, although all of the examined *S. aureus* strains comprise this gene. However, as in the case of the *pls* gene, no correlation was found between the presence of this gene and the type of infection (Hussain et al. 2001; Kurlenda et al. 2007b).

A serious clinical problem in the case of *S. aureus* is the frequent conversion of the acute infection to a chronic one. It seems that this is due to bacterial plurality, which is characterized as both phenotypic and genotypic diversity. It is estimated that more than 90% of bacteria in the natural population form a biofilm which is a collection of many ecological niche and different phenotypes. Due to the fact that antibiotics can only effect planktonic bacteria responsible for acute infections and not those forming a biofilm, what makes eradicating chronic infections even harder a task. Microbes inside this community are very well protected from of the host's immunological system, antibiotics and antibacterial agents what enables the survival in a biofilm (Ehrlich et al. 2005; Weigel et al. 2007). The production of the biofilm allows bacteria to adhere through extracellular polysaccharides to smooth surfaces of different medical devices (Donlan et al. 2001; Weigel et al. 2007). With the increase of frequency in implant use, the risk of developing chronic staphylococci infection rises. Infection of a prosthetic biomedical implant can be the cause of bone marrow infection and, in consequence, osteomyelitis and acute sepsis (Cramton et al. 1999). The biofilm production is regulated by Autoinducer Protein (AIP) which can influence through RNAIII on activation or inhibition of the *agr* regulon and modulate *quorum sensing* system. Microorganisms through the production and reaction to AIP can evaluate the number of their own and others's cells. This *quorum sensing* is available when AIP reach appropriate concentration in the growth environment.

The formation of a biofilm is associated with the production of a polysaccharide intercellular adhesin (PIA), which is composed of linear β-1,6-linked glucosaminylglycans. It has been discovered that the intercellular adhesion (*ica*) locus, *icaADB* and *C*, is essential for the production of adhesins, and its components *icaA* and *icaD* are both involved in its synthesis *in vitro*. However, it has also been reported that despite the presence of the *ica* gene in some species, in conditions that favor biofilm production, they do not undertake its synthesis. It is thought that this could be due to point mutations in the *ica* locus or the presence of so far unidentified factors influencing its production (Cramton et al. 1999).

However, among all of the strains examined with both methods: spectrophotometric and culture on the Congo Reg Agar plates, only 47% of MRSA strains and 69% of MSSA strains produced biofilms *in vitro*, therefore not all strains possessing the *ica* operon can form biofilm (Grinholc et al. 2007b). Moreover, the morphology of the biofilm producing strains' colonies differed between MRSA and MSSA strains. The results of the CRA assay obtained for MSSA strains were less equivocal than those for MRSA strains. MSSA colonies of slime-producing strains were black and only about 1% of the results could not be precisely interpreted. MRSA colony color on CRA varied from black, violet, brown, bordeaux, dark red to red and the percentage of such ambiguous results was higher (13.9%). 96% compatibility was achieved when research was carried out using both methods to determine the ability to produce biofilm (Grinholc et al. 2007b).

After the invasion of the host tissue, bacteria can encounter changes of temperature, pH, O_2 level and Mg^{+2}, Ca^{+2}, Fe^{+3} concentration, which could modify bacterial activity. These factors can induce bacteria in the postexponential phase to produce divergent virulence factors, such as hydrolytic enzymes e.g. serine proteases, lipases, hialuronidases, nucleases, staphylokinases, coagulases, that main function is to digest the host tissues supplying bacteria with appropriate nutritional compounds (Ruzin and Novick 2000; Kaneko and Kamio 2004; Makris et al. 2004). Much of the scientists' attention is given to the *S. aureus* proteases, virulence factors enabling bacteria the colonization and invasion of the host organisms (Redpath et al. 1991; Karlsson and Arvidson 2002; Michel et al. 2006; Sabat et al. 2008). However, one of the main mediators of infection are numerous toxins produced by *S. aureus* on various stages of infection, these include cytotoxins (hemolysins α, β, γ, δ) with hemolytic and proinflammatory properties (Dinges et al. 2000) and PV-leukocidin (Panton-Valentine leukocidin) (Issartel et al. 2005). Staphylococci produce various secreted proteins such as enterotoxins A, B, C, D, E, G, H, I, J, K, L, M, N, and O (Dinges et al. 2000; Murray 2005; Desachy et al. 2007), exfoliative toxins A, B, D (Yamasaki et al. 2005; Yamasaki et al. 2006) and toxic shock syndrome toxin-1 (TSST-1), which aid the bacteria in adapting to new conditions (Dinges et al. 2000; Holtfreter et al. 2004; Yamasaki et al. 2005; Desachy et al. 2007).

In recent years community-acquired MRSA (CA-MRSA) strains have become a problem (Larsen et al. 2009). These strains possess the co-transcribing genes *lukS-PV* and *lukF-PV* responsible for the production of a two-component, pore forming cytolytic toxin with high specificity against human polymorphonuclear cells and macrophages, termed Panton-Valentine leukocidin LukF-PV/LukS-PV (PVL) (Kaneko and Kamio 2004). It can induce tissue necrosis and has been associated with lethal necrotizing pneumonia, cellulitis, abscesses and furuncles (Francis et al. 2005; Shukla 2005; Wargo and Eiland 2005; Lorber 2006; McCaig et al. 2006; Elizur et al. 2007; Cunnington et al. 2009). *S. aureus* acquires those virulence genes from specific temperate phages: 0PVL, a defective phage; 0Sa2mw, a MW2 lysogenised phage; and 0SLT (Staphylococcal Leucocytolytic Toxin), a functional phage. Only the prophage 0SLT has been able to convert the PVL negative strain into PVL producing *S. aureus* (Narita et al. 2001; Fey et al. 2003). It is estimated that around 2% of the population of clinical *S. aureus* strains posses in their genome a gene responsible for the production of PVL, whereas among MRSA strains it is more frequent (Shukla 2005; Rossney et al. 2005). Recently Monecke et al. (Monecke et al. 2007) presented a hypothesis that points to a different origin of CA-MRSA than that involving transfer though the phage PVL gene. 100 *S. aureus* strains isolated from skin and soft tissue infections were examined and the presence of the PVL gene was determined in 30 isolates. Only 3 of them were MRSA strains. MSSA PVL-positive strains belonged to various multilocus sequence types mainly *agr*I/ST22, *agr*IICC5, *agr*III/CC30 and *agr*IV/ST121. However, some of them (*agr*I/ST152, *agr*III/ST80 and *agr*III/ST96) were genetically related to CA-MRSA. These results can indicate that the wide spreading of CA-MRSA strains can be a result of transferring a new cassette SCC*mec* between already existing MSSA PVL-positive strains (Monecke et al. 2007). The similar results were obtained by Boyle-Vavra et al. (Boyle-Vavra and Daum 2007). Anyway, the importance of PVL in virulrnce of CA-MRSA is contradictory. Wang et al. (2007) described a group of cytolytic peptide related to the phenol-solublemodulin (PMS) of *Staphylococcus epidermidis* but their role in virulence is still poorly understood (Wang et al. 2007).

S. aureus is regarded as a typical extracellular pathogen often causing pyogenic infections. Local infections as well as pyogenic skin and soft tissue infections, and also systemic infections e.g. endocarditis, occur due to the ability of the bacteria to colonize host tissue and produce extracellular enzymes and toxins which destroy the structures of the host tissue and also protect the bacteria from the host immune response (Lowy 2000).

The presence of Small Colony Variant (SCV) in a population can be the reason for the intracellular survival of bacteria, which leads to chronic and/or recurring infections. Obviously, its occurrence may have a significant meaning in the recurring infections of *S. aureus* etiology in patients with cystic fibrosis, osteomyelitis, chronic skin infections and device-related infections (Kahl et al. 1998; von Eiff et al. 2000b; von Eiff et al. 2001b; Seifert et al. 2003; Spanu et al. 2005).

SCV is a naturally present subpopulation whose growth on routine media takes longer than that of colonies from parental cells. Most of them do not have pigmentation or hemolysis and the size of a colony is around 10 times smaller (Looney 2000; von Eiff et al. 2000a). They can be auxotrophic for thiamine or hemin, less frequently for thymidine and menadione, which is a result of a deletion within genes responsible for the synthesis of these compounds (Schaaff et al. 2003). If the growth medium is supplemented with these substances they grow as fast as parental cells. Similarly to natural auxotrophes behave as *hemB* and *hemD* mutants. Their phenotype is the same as natural SCVs. As reported by Vesga et al. (Vesga et al. 1996), the intracellular milieu of endothelial cells selects SCV and they are alfa-toxin deficient, which leads to the inhibition of host cell lysis and the intracellular environment allows them to survive protecting them from the activity of the immune system and antibiotics (Vesga et al. 1996; Kahl et al. 1998; Baumert et al. 2002; Bates et al. 2003).

EPIDEMIOLOGY

S. aureus by means of its wide spreading and constant presence in the human environment can be the cause of infection and disease of epidemic character. This specifically applies to closed environments connected to health care such as hospitals and long-term care homes, where people are prone to infections, particularly immunocompromised patients in severe conditions or patients in vulnerable age groups (children and elderly). In the case of the youngest age group, that is new-borns and infants, epidemic staphylococcal infections with various clinical presentations have been described (Le, I et al. 2001; Regev-Yochay et al. 2005; Kuint et al. 2007).

Due to the usually short stay in newborn care units infections in this age group are usually caused by MSSA strains, however, hospital infections caused by MRSA have also been reported (Chi et al. 2006). Kuint et al. published an 11-year observation of newborn hospitalized in intensive care units where various clinical types of *S. aureus* infections were observed in 43 children. In 20 cases, that is in nearly half of the patients, MRSA MDR (multidrug resistant) strains were isolated (Kuint et al. 2007).

The problem of hospital infections caused by methicillin-resistant *S. aureus* is known and associated with hospital treatment from the 1960's (Hryniewicz 1999). The frequency of bacteremia of this etiology, according to research carried out in the years 1999-2000 in Europe, depending on the country ranged from 44.4% (Greece) to 0.6% (Denmark and Holland) (Tiemersma et al. 2004). According to the research of SENTRY carried out in the years 1997-1999 in USA, Canada, Latin America, and Western Pacific, the percentage of infection in these areas was 25.3; 19.2; 20.6 and 21.6, respectively (Diekema et al. 2001). The *mecA* gene, responsible for the mechanism of methicillin resistance, is of 2.1 kb size and is present on the mobile elements of the bacterial chromosome termed the Staphylococcal chromosome cassette *mec* (SCC*mec*). This genetic element described by Katayama et al. is responsible for the stability of the *mecA* gene, chromosome recombinase genes and the SCC*mec* complex (Katayama et al. 2000; Ito and Hiramatsu 2003; Ito et al. 2003). Numerals from I to VII were assigned to it depending on its size, which is in the range from 20.9 to 66.9 kb (Ito et al. 2003; Higuchi et al. 2008; Kurt et al. 2009; Nulens et al. 2009). Recently, the cassette number VIII found in Canadian epidemic strain was described (Zhang et al. 2009). SCC*mec* cassettes I, II, and III contain the determinants of resistance towards many antibiotics (Ito et al. 2003; Higuchi et al. 2008; Kurt et al. 2009) and are present mainly in HA-MRSA (hospital acquired MRSA) strains (Ito et al. 2003; Conceicao et al. 2007; Huang et al. 2007). Long term prevalence of the same multidrug resistant MRSA clones were described in hospital environments, even for a couple of years (Mulvey et al. 2005; Diep et al. 2006; Conceicao et al. 2007; Kurlenda et al. 2007a) At the same time the clonal expansion over large distances between countries and continents was also reported (Sola et al. 2002; Gosbell 2005; Deurenberg et al. 2007). To determine strain differentiation the method previously used was PFGE (Pulsed Field Gel Electrophoresis). During PFGE bacterial DNA is immobilized in agarose blocks, in which it is digested with restriction enzymes. Next during electrophoresis DNA is separated to fragments of different length. Electrophoresis is run in an alternating electrical fields, which allows the separation of DNA fragments in the range of 10 to 800 kb (Olive and Bean 1999). The whole genetic material in the cell is analyzed. The additional enzyme digestion of the genome with enzymes such as *Sma*I, results in the formation of DNA fragments of different length and number (Murchan et al. 2003). Another technique used in epidemiological studies is RFLP (Restriction Fragments Length Polymorphism) in which differentiation is based on the analysis of restriction patterns obtained through the digestion of the genetic material with restriction enzymes. In the case of *S. aureus* the amplified product of the coagulase gene (*coa*) or the polymorphic X region of the A protein (*spa*) are digested with restriction enzymes *Alu*I or *Rsa*I, respectively (Frenay et al. 1996; Hookey et al. 1998; Grinholc et al. 2008b). Currently, the *spa*-typing is thought to be one of the most usefull and powerfull typing methods (Karynski et al. 2008b). One of the most simple PCR-based methods used for differentiation is RAPD (Randomly Amplified Polymorphic DNA). This technique is based on the use of primers consisting of 9 to 11 nucleotides which begin the amplification of DNA fragments in many regions of the genome simultaneously. The products are separated in an agarose gel. This method shows sufficient level of differentiation for epidemic investigations (Figure 3) (van Belkum A. et al. 1994; Andrasevic et al. 1999; Kurlenda et al. 2007a).

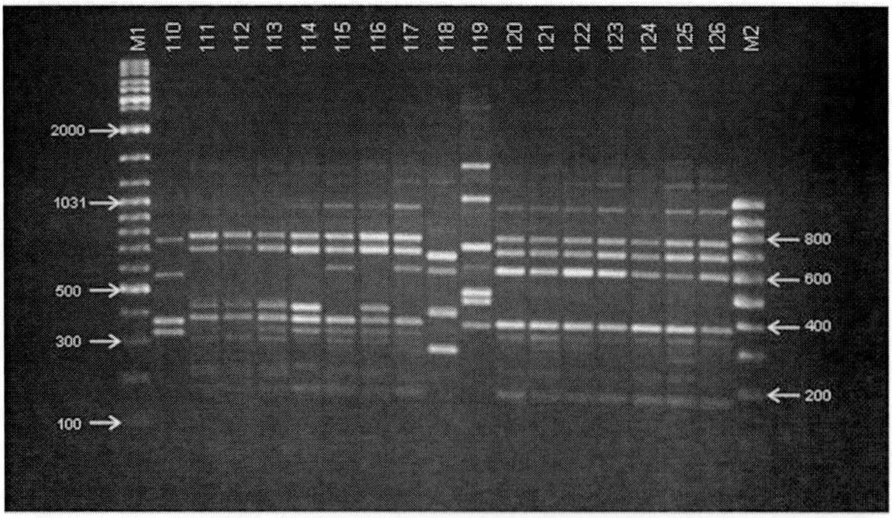

Figure 3. **RAPD PCR patterns distinguished among isolated MRSA strains.** M1, (GeneRuler DNA Ladder Mix) and M2 (GeneRuler 100bp DNA Ladder) - molecular weight DNA markers. Lines 110 – 126 – various RAPD patterns of separate clinical MRSA strains isolated from different patients.

Another described and used typing method is multiplex PCR called originally Multiple-Locus Variable-Number of Tandem Repeats Analysis (MLVA), and presently Multiple-Locus VNTR Fingerprinting (MLVF). It analyses the variation in number of repeats in seven individual genes (*sspA*, *spa*, *clfA*, *clfB*, *sdrC*, *sdrD* and *sdrE*). The discriminatory power of

MLVF method is comparable with that obtained with PFGE analysis (Sabat et al. 2006; Luczak-Kadlubowska et al. 2008; Karynski et al. 2008a). Another method used for strain typing is MLST (Multilocus Sequence Typing). In the year 2000 Enright et al. described the use of the MLST technique in epidemiological studies regarding both MRSA and methicillin-sensitive MSSA strains for the first time (Enright et al. 2000). MLST is the genotyping version of MLEE (Multilocus Enzyme Electrophoresis), an electrophoretic analysis of sequenced variable region, fragments of 0.5 kb in size and of seven housekeeping genes of *S. aureus*: *arcC*, *aroE*, *glif*, *gmk*, *pta*, *tpi* and *yqiL* encoding bacterial reference enzymes (Deurenberg et al. 2007). The accepted nomenclature is based on defining clones with the letters ST combined with the numbers and numbers of the SCC*mec* cassettes for e.g. the Brazilian/Hungarian clone which was present in countries of Europe, in the USA and South America is defined as ST239-III (Deurenberg et al. 2007). At present PFGE and MLST are regarded as the most sensitive methods for strain typing (Oliveira et al. 2002); however, as they are technically demanding and laborious, for routine purposes, regarding limited territorial ranges, more simple and quick techniques such as RAPD are used (van Leeuwen W. et al. 1996; Kurlenda et al. 2007a). Currently the most widely used method for differentiation is clone determination with the use of MLST. It allows to determine the spreading of strains, maintenance on a particular region and the appearance of new strains (Gosbell 2005; Deurenberg et al. 2007). Introducing universal nomenclature for the description of clones allows to compare centers and also evaluate maintenance of clones in time and their exchange. Conceicao et al. presented the results of a 10 years observation where during the years 1994-1998 in Hungary the type ST239-III clone dominated and then totally vanished in the year 2003 replaced by a Southern German clone ST228-I (Conceicao et al. 2007) which was present only in Europe (Deurenberg et al. 2007). In USA in the years 1996-2004, in the region of San Francisco, the clone ST5-II was present among hospital strains. This clone was present also in Europe, Australia, Japan and South America (Diep et al. 2006). In 2005 in the USA among dialyzed patients, 92% of the isolated MRSA strains from infections belonged to the clone ST5-II (Collins 2007). Two of the biggest clones, Iberian ST247-I and Brazilian/Hungarian ST-239-III, have spread into countries of Europe and the USA, Australia, Asia and South America (Gosbell 2005; Deurenberg et al. 2007).

The MRSA and the MSSA strains can cause all types of infections in adults as well as in children (Sola et al. 2002; Collins 2007; Kuint et al. 2007; Huang et al. 2007; Kurlenda et al. 2007a; Kurlenda et al. 2009). The higher virulence of HA-MRSA strains has not been proven and the existing research presents often ambiguous results. Melzer et al. in research carried out in the years 1995-2000 compared 382 patients with HA-MRSA bacteremia and 433 with MSSA. During the first stage of analysis higher mortality was observed in the HA-MRSA, 11.8% in comparison with MSSA, 5.1%. However, after verification of these results, it was revealed that patients from the first group had higher risk factors and after this correction the mortality was 7.1% and 6.2%, respectively (Melzer et al. 2003). The analysis of other virulence factors such as the ability to form biofilm, the presence of genes responsible for adhesion production (Grinholc et al. 2007b), the amount of protein A produced, ability to form capsules and the extent of adherence to the human epithelial cell line Hep-2 (Aathithan et al. 2001), no differences were determined between MRSA and MSSA strains (Aathithan et al. 2001). The obvious difference between MRSA and MSSA strains occur in the case of the Panton-Valentine leukocidin gene which presence is increased among CA-MRSA strains than MSSA ones (Gosbell 2005; Deurenberg et al. 2007).

It should be noted that problems associated with HA-MRSA infections are related to their resistance not only to beta-lactam antibiotics but also their lack of sensitivity towards other antibacterial drugs. According to many authors resistance to fluoroquinolones ranges from 75% to 100%; gentamicin 59% - 100%, clindamycin 60% - 100%, erythromycin 71% - 100%, trimetoprim/sulfamethoxazole 30%-97% (Figure 4) (Sola et al. 2002; Savas et al. 2005; Conceicao et al. 2007; Kuint et al. 2007; Huang et al. 2007; Kurlenda et al. 2007a). According to the first report in 1997 regarding the therapeutic failure of vancomycin in the treatment of MRSA with reduced sensitivity towards this antibiotic, the development of drug resistance towards glycopeptides, which are an important group of antibacterial drugs in the case of HA-MRSA infections, has become a threat (Hiramatsu et al. 1997). In Scotland in 2002 the death of a patient treated with vancomycin for a MRSA infection was reported. The GISA (glycopeptide intermediate resistant *S. aureus*) strain was isolated from this patient (SCIEH 2002). In the same year in USA a VRSA (vancomycin resistant *S. aureus*) strain, with MIC for vancomycin 120 mg/l, was isolated from a swab obtained from a catheter exit site. It is regarded that the risk of the development of vancomycin resistance is associated with group II polymorphism at the accessory gene regulator (*agr*) locus (Soukos et al. 2003; Sakoulas et al. 2003; Moise-Broder et al. 2004). Sakoulas et al. reported that among 148 MRSA strains examined, 56% belonged to *agr* group II whereas among 33 MSSA strains only 24%. The ability to develop heteroresistance towards glycopeptides for *agr* group I, II and IV were also examined with genetically engineered *agr* null mutants. Mutants and their parent wild type strains were grown on medium supplemented with 1 mg/l vancomycin. A strain from group II only developed heteroresistance (Sakoulas et al. 2003). Moise-Broder et al. determined that 69% of patients with infections caused by strains from group II did not respond to vancomycin therapy (Moise-Broder et al. 2004).

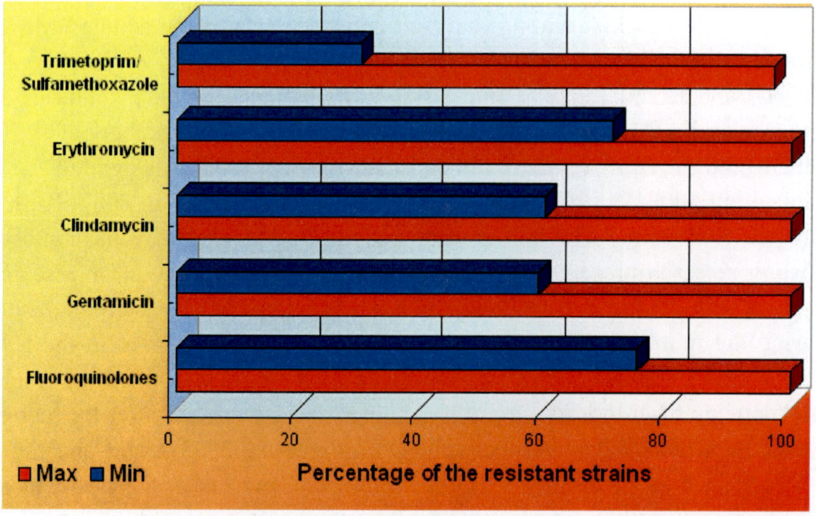

Figure 4. **The percentage distribution of *S. aureus* strains resistant to selected antimicrobial drugs.** The figure indicates the minimum (min) and maximum (max) values present in published data according to different authors.

According to various authors healthcare associated MRSA infections are more frequent among high-risk patients (Gosbell 2005; Kowalski et al. 2005; Tacconelli 2006). Bagger et al. during a five year observation of patients at a cardiosurgery ward reported that the frequency of postoperative MRSA infections depended upon the socio-economical status of the patient. Patients from lower socio-economic backgrounds acquired infections 7 times more frequently and mortality among these patients was 6 times higher. The most deprived areas 2.2%, the least deprived 0.3% $p = 0.0040$ (Bagger et al. 2004). Among other risk factors the authors mention: recent hospitalization, close contact with people which were recently hospitalized, a recent surgical procedure, intubation, hemodialysis, chronic illness, indwelling medical device, recent antibiotic therapy (Kaiser et al. 2004; Gosbell 2005; Kowalski et al. 2005; Tacconelli 2006; Seybold et al. 2006; Collins 2007). Tacconelli et al. proposed a point scale based on six variables to determine carrier probability and the development of MRSA infection: recent hospitalization (3 p.); age > 60 years (2 p.); the use of more than 2 antibiotics in the recent 30 days (3 p.); transfer from another hospital or long term hospitalization (3 p.); the need for hemodialysis (2 p.); recent MRSA infection (4 p.). Obtaining >= 10 points is considered a predisposition for MRSA infection in 98% (Table 2) (Tacconelli 2006).

From the 1990s in many countries (Fey et al. 2003; Faria et al. 2005), in the out-patients infections caused by MRSA strains defined as CA-MRSA (community-acquired MRSA) has been observed. Those strains caused severe necrotic pneumonia in children (Le et al. 2001; Lorber 2006; Cunnington et al. 2009) and adults (Dufour et al. 2002; Francis et al. 2005; Kowalski et al. 2005; Elizur et al. 2007), as well as endocarditis (Fowler, Jr. et al. 2005b), prosthetic join infections (Kourbatova et al. 2005) and brain abscess (Enany et al. 2007). However, most of the infections of this etiology concern the skin and soft tissues (Dufour et al. 2002; Fey et al. 2003; Kowalski et al. 2005; Mulvey et al. 2005; Manzur et al. 2007). According to different authors it can can reach 75% (Naimi et al. 2003; Wannet et al. 2004), 77% (Lorber 2006), and 84% (Faria et al. 2005).

Table 2. Point scale determining carrier probability and the development of MRSA infection (according to Tacconelli et al. 2006)

Variables	Points
Recent hospitalization	3
Age > 60 years	2
Use of more than 2 antibiotics in the recent 30 days	3
Transfer from another hospital long term hospitalization	3
Need for hemodialysis	2
Recent MRSA infection	4
Obtaining ≥ 10	98 % predisposition for infection

CA-MRSA strains often possess *lukF-lukS* genes in their genome encoding the ability to produce a bicomponent cytolitic toxin leukocidin Panton-Valentine (PVL) (Gosbell 2005; Deurenberg et al. 2007) and are characterized by the presence of the SCC*mec* type IV cassette (Kowalski et al. 2005; Diep et al. 2006; Deurenberg et al. 2007) or V (Garnier et al. 2006; Enany et al. 2007). However, as some authors report, not all strains with the type IV cassette possess the PVL gene (Diep et al. 2006; Huang et al. 2007). Huang et al. reported that among the examined CA-MRSA strains the PVL gene was present in 50% of strains (Huang et al. 2007), according to other authors in 16% (Graham et al. 2006), 56% (Hedin and Fang 2007), 58% (Ellis et al. 2004), 77% (Naimi et al. 2003). Cuevas et al. did not find a strain with the PVL gene among the examined CA-MRSA strains with the type IV cassette (Cuevas et al. 2007). CA-MRSA, in comparison with HA-MRSA strains, apart from their resistance to beta-lactams, are characterized by high sensitivity towards the remaining antibiotics, mostly above 90%. Whereas their resistance towards lincosamides and macrolides is observed at a various degree (Dufour et al. 2002; Naimi et al. 2003; Ellis et al. 2004; Kowalski et al. 2005; Lorber 2006). According to different authors the sensitivity towards erythromycin and clindamycin was 44% and 83% (Kowalski et al. 2005), 18% and 87% (Lorber 2006), 44% and 83% (Naimi et al. 2003), or 20% and 95.6%, respectively. As reported by Fey et al. among CA-MRSA type IV isolates, 81% were resistant only to beta-lactams (Fey et al. 2003). In other study drug resistance was clone-dependent, where clone ST30-IV(USA 1100) displayed high sensitivity towards antibiotics and clone ST8-IVa (USA 300) displayed the highest resistance and 89.4% were resistant to erythromycin, 11.7% towards clindamycin and 62.8% and 24.5% towards ciprofloxacin and tetracycline, respectively (Diep et al. 2006). However, apart from the more frequent presence of the PVL gene no differences, comparing with HA-MRSA in the distribution of genes encoding virulence factors such as adhesins or toxins, were found (Fey et al. 2003; Graham et al. 2006; Diep et al. 2006; Huang et al. 2007).

CA-MRSA quickly spread in the out-patient population with the distinct domination of some clones in different periods, countries or continents. The most frequently determined clone ST8-IV(USA 300) was identified both in European countries and in the USA, South Africa and Asia (Pan et al. 2005; Maree et al. 2007; Deurenberg et al. 2007; Krziwanek et al. 2007; Manzur et al. 2007; Hedin and Fang 2007), whereas in Europe the ST80-IV clone was mainly present (Dufour et al. 2002; Wannet et al. 2004; Denis et al. 2004; Witte et al. 2004; Faria et al. 2005; Tristan et al. 2007). As some authors point out, in recent years a polyclonal tendency has appeared and so have new clones (Tristan et al. 2007; Yu et al. 2007; Krziwanek et al. 2007; Hedin and Fang 2007). These strains have wide spread so much that they are more frequently isolated from hospitalized patients (Saiman et al. 2003; Kaiser et al. 2004; Regev-Yochay et al. 2005; Kuint et al. 2007; Maree et al. 2007). This has caused an increase in MRSA infections in hospitals, in Scandinavian countries and Holland (Kaiser et al. 2004; Skov and Westh 2005; Nulens et al. 2009), where in the 80s strict MRSA monitoring was introduced, which allowed to maintain infections on a low level <1%. Since 2001 the situation has changed due to two factors, either the enrolment of patients with CA-MRSA infections (Kaiser et al. 2004; Francis et al. 2005; Faria et al. 2005; Lorber 2006) or the infections which developed in a hospital have been introduced by previously enrolled carriers of CA-MRSA (Saiman et al. 2003; Collins and Hampton 2005; Kourbatova et al. 2005; Maree et al. 2007; Bauer et al. 2009). A new source of CA-MRSA wide spreading can in the future be pig farms and people employed there (Duijkeren et al. 2008). During an international conference in Denmark 272 participants associated with the agriculture industry, among which 12.5% were CA-MRSA carriers from different countries, were analyzed (Wagenaar et al. 2007). 24 cases of MRSA infections in horses hospitalized in the years 2003-2005 in veterinary clinics were also reported where clone CA-MRSA ST254-IV was found in five animals (Cuny et al. 2006). Different types of MRSA were found among dogs and their owners (Boost et al. 2007).

A risk factor for CA-MRSA acquisition is considered to be the age group of a patient. Younger patients such as children, athletes, soldiers are in the higher risk group (Naimi et al. 2003; Ellis et al. 2004; Faria et al. 2005; Graham et al. 2006; Lorber 2006; Maree et al. 2007). Also contact sports (Kazakova et al. 2005), intravenous drug intake (IVDU) (Pan et al. 2005; Fowler, Jr. et al. 2005a) and military service (Ellis et al. 2004) are regarded as risk factors.

DIAGNOSTICS AND DIFFICULTIES

An important issue, from the therapeutic and epidemiologic point of view, is the identification of infection and carriers of HA-MRSA and CA-MRSA. Identifying carriers is critical for the control of the wide spread of multiresistant strains and reducing the risk of infection development because *S. aureus* nasal carriage occurs quite frequently, approximately in 20% to 60% of a normal population (Toshkova et al. 2001; Pan et al. 2005; Wertheim et al. 2005; Graham et al. 2006). Kampf et al. (Kampf et al. 2003) reported that *S. aureus* carriage among hospital staff was approximately 33.8% of which 1.6% were MRSA strains (Kampf et al. 2003). The example of Holland and Scandinavian countries shows the importance of carriage monitoring where after introducing strict monitoring of enrolled patients the percentage of MRSA incidents was low and maintained on a level <1% (Kaiser et al. 2004; Skov and Westh 2005). Rapid evaluation of the etiology of infection is also important in order to apply the proper treatment, as in the case of MRSA it is critical in empiric treatment not to begin treatment with beta-lactam antibiotics. Introducing molecular biology methods into diagnostics reduces the time required to obtain information regarding the etiology of the infection and improves the identification of carriers (Nilsson et al. 2005; Huletsky et al. 2005; Holfelder et al. 2006; Cunningham et al. 2007; Ornskov et al. 2008). With the use of a method combining a 16 h preincubation of analyzed samples in a selective enrichment broth with the addition of methicillin and real-time PCR for the genes *nuc* and *mecA* the highest efficiency in detecting MRSA was obtained for nostril and throat samples. In the case of samples of different origin, detectability increased only slightly which could be associated with low inoculum. Samples containing less than 10^4 CFU/ml gave negative results in PCR tests. The advantage of the described method is the possibility to identify MRSA in the examined sample in 24 h and reduce screening costs (Nilsson et al. 2005; Ornskov et al. 2008). Holfelder et al. used the method where after amplification the hybridization with an oligonucletide probe specific for SCC*mec* of MRSA was performed. In comparison to culture screening methods the sensitivity was 94.59% and specificity 98.73% (positive predictive value PPV was 85.37%, negative predictive value NPV was 99.57%) (Holfelder et al. 2006). Huletsky et al. used real-time PCR to analyze nasal specimens from 162 patients and showed that results can be obtained in less than 1 h. In this method primer specific for the SCC*mec* cassette were also used. The specificity was 98.4% and sensitivity 100% (PPV 95.3%, NPV 100%) (Huletsky et al. 2005). Cunningham et al. introducing the PCR method obtained the results in 1 day for 613 patients reducing the transmission risk from 13.89/1000 patient/days to 4.9/1000 patient/days ($P<0.05$) (Cunningham et al. 2007). An alternative to conventional methods, which require time, and to molecular methods, which are expensive and require appropriate equipment, can be commercial methods such as BacLite Rapid MRSA (Acolyte Biomedica, UK). This method enables the identification of MRSA in

5 h directly from clinical samples and is based on determining the activity of adenylate kinase (AK), which catalyzes the conversion of ATP to ADP with the release of energy. In the assay, AK detection is combined with selective broth enrichment and magnetic microparticle extraction coupled with monoclonal mouse antibodies against *S. aureus* is performed to capture MRSA. In the next step bounded *S. aureus* cells are lysed and AK is released. After adding luciferin and luciferase in the presence of ATP light is emitted. In research carried out for 1382 nasal swab samples the obtained specificity was 95.7% (NPV 98.7%) and sensitivity 90.4% (Johnson et al. 2006).

The molecular methods enable determining the etiology of infections, with strains forming biofilm and Small Variant Colonies (SCV) where conventional methods are insufficient.

Kipp et al. (2003) used a 16S rRNA-directed *in situ* hybridization technique in tissue samples from a patient with a brain abscess. An EUB 338 fluorescently labelled oligonucleotide probe specific to eubacteria and an oligonucleotide probe SA-P1 specific to *S. aureus* were used. The obtained results were verified with a conventional bacterial culture and the *S. aureus* strains isolated from the patient was identified as the SCV hemin auxotrophic (Kipp et al., 2003).

Blood culture in systemic, life-threatening infections are very valuable diagnostic material. However, the time required to obtain a result in the case of positive cultures in automatic systems is around 48 h. Taking into consideration that a infection can be caused by MRSA it is essential to obtain the result as quickly as possible, as the therapeutic possibilities in this case are limited. Rapid and simple direct identification bacteria from positive blood culture is achieved by fluorescence *in situ* hybridisation (FISH), using DNA oligonucleotyde probe or peptide nucleic acid probe (PNP), labelled with fluorescein target 16S RNA region specific for *S. aureus* (Oliveira *et al.*, 2002; Gonzalez *et al.*, 2004; Gescher *et al.*, 2008). Comparing to other molecular methods described above, FISH is relatively cheap, not demanding any expensive equipment apart from fluorescence microscope. The sensitivity ranged 97-100%, specificity 99.4-100% and results were comparable in all three papers (Oliveira *et al.*, 2002; Gonzalez *et al.*, 2004; Gescher *et al.*, 2008).

Tan et al. carried out research for 141 positive blood cultures based on real-time fluorescence PCR. Cultures used for research were classified based on Gram staining as Gram positive cocci. *S. aureus* was identified in the case of 52 cultures, of which 19 were classified as MRSA. Amplification was carried out with the use of a primer for the Sa442 and *mecA* fragment. *S. aureus* identification showed 96% sensitivity and 100% specificity. Detection of methicillin resistance showed 100% specificity and sensitivity and the time required to obtain the end result was about 2 h (Tan et al. 2001).

A challenge for a laboratory is the identification of resistance mechanisms in a situation when they are on a low level or have heterogeneous nature. Low-level resistance can be caused by antibiotic pressure but also by other factors such as disinfecting agents. Not identifying low-level resistance can have both therapeutic and epidemiological consequences (Kampf et al. 2003; Goldstein 2007). In the case of *S. aureus* due to therapeutic options this problem concerns determining the resistance towards oxacillin and vancomycin. The mechanism of resistance towards methicillin, based on the ability of *S. aureus* strains to produce a changed PBP 2a protein (penicillin binding protein) which has lost its ability to bind beta-lactam antibiotics, results in the lack of sensitivity towards a big therapeutically relevant group of antibiotics (Ito et al. 2001). Witte et al. presented the results of their research regarding 74 *mecA*-positive strains for which the MIC for oxacillin was 0.5 – 1.0 mg/l. The authors compared different phenotypic methods. In the case of microbroth dilution the obtained results indicated that in compliance with the CLSI criteria (MIC break point for oxacillin >= 4 mg/l) 25% of *mecA*-positive strains were classified as sensitive. In the disc diffusion method the sensitivity for oxacillin was 92% and for 30 μg of cefoxitin 96%. Oxacillin-salt agar screening test, oxacillin-salt-sulbactam tube test and chromagar MRSA identified the correct low level of resistance towards oxacillin. For all of the 74 strains the agglutination test identifying the PBP2 protein was positive (Witte et al. 2007). Kampf et al. reported that among medical staff carriers (doctors and nurses) 1.6% of the strains were *mecA*-positive but methicillin sensitive. Only after the induction with beta-lactam the expresssion of resistance mechansims towards methicillin were shown noticeable (Kampf et al. 2003).

In the case of vancomycin resistance the first *S. aureus* strain with low-level vancomycin resistance with MIC 8-16 mg/l was reported in 1997 by Hiramatsu et al. (Hiramatsu et al. 1997). Howden et al. (Howden et al. 2004) described therapeutic failure in 19 patients from which *S. aureus* with reduced vancomycin susceptibility (MICs 2-4 mg/l) was isolated. In agreement with the CLSI standards strains displaying a MIC value of 8 mg/l are characterized as vancomycin intermediate *S. aureus* (VISA) (Hiramatsu et al. 1997; Hiramatsu 2001; Krzyszton-Russjan et al. 2002). The heterogenic nature of drug resistance (h-VISA) can be the cause of therapeutic failure as mechanism determination requires maintaining specific conditions of analysis (Hiramatsu 2001). It has been reported that some clinical MRSA strains isolated in Japan classified as sensitive towards vancomycin MIC <8 mg/l generate resistance with a frequency of 10^{-6} or higher. After an overnight culture on medium without antibiotics to an inoculum density of 10^7 CFU/ml and a culture on agar medium with vancomycin in the concentration of 4 mg/l the growth of several hundreds of cells, for which MIC was equal to or above 8 mg/l, were detected. They are regarded as precursors for VRSA represented by strain Mu3, which was isolated from pneumonia. The patient from which this strain was isolated at the beginning responded to therapy, however, after 9 days of treatment the infection resumed (Hiramatsu 2001). The environment for resistance gene transfer to glycopeptides can *in vivo* be a polymicrobial biofilm. Weigel et al. reports a case of a patient with the indwelling nephrostomy tube from which a few species of bacteria were isolated, among others MRSA and *Enterococcus faecium* VRE. The authors reported that the VRE strain transferred the *vanA* gene localized on the 100 kb plasmid as well as resistance genes for tetracycline *tet(U)* and *tet(S)* to the MRSA strain (Weigel et al. 2007).

Alternative Therapies

S. aureus is a pathogen, which displays the ability to omit all barriers of the host defense system as it possesses a wide spectrum of virulence factors. Despite the increasing knowledge on this subject it is still difficult in many cases to prevent or effectively treat staphylococcal infections.

As a result of the increasing percentage of infections caused by MRSA strains, particularly the presence of MRSA strains displaying intermediate sensitivity or resistance to vancomycin, antibiotic therapy is becoming more difficult and is burdened with a high failure risk (Rayner and Munckhof 2005; Levy 2005; Trampuz and Zimmerli 2006; Livermore 2006; Karchmer 2006; Finch 2006). This is the reason for the continuing research regarding the pathogenesis of staphylococcal infections as a basis for developing alternative methods of therapy as opposed to antibiotic therapy.

I. Antibacterial Natural Compounds

Research is carried out involving the use of natural substance with antibacterial activity obtained from both bacteria (Severina et al. 1998; Giacometti et al. 2000; Kokai-Kun et al. 2003; Nascimento et al. 2006; Oh et al. 2006), plants (Filipowicz et al. 2003; Yin et al. 2004; Yang et al. 2006; Adesokan et al. 2007) and animals (Giacometti et al. 2000; Bexfield et al. 2004; Giacometti et al. 2006). Bacteriocins, antimicrobial peptides produced by Staphylococcus, which inhibit the growth of *S. aureus* are termed staphylococcins. They can be produced by both *S. aureus* and *S. epidermidis*. Research carried out regarding the activity of 7 staphylococcins against MRSA strains indicated that 2 bacteriocins auricin A53 and Pep5 produced by *S. epidermidis* displayed the widest inhibition spectrum, inhibiting the growth of 87.5% of examined strains (Nascimento et al. 2006). Gram-positive bacteria from the *Lactococcus spp.* genus produce bacteriocins with a polycationic peptide (lantibiotics) character and a spectrum of activity against both MRSA and MSSA strains (Severina et al. 1998; Giacometti et al. 2000; Oh et al. 2006) similarly as cationic 34-residue peptides produced by *Lactococcus lactis* (Giacometti et al. 2000) and nisin (Severina et al. 1998). Another bacteria-derived compound which is being examined is lisostaphin. Originally isolated from *Staphylococcus simulans* it is a glicyliglicine endopeptidase capable of specifically cleaving the cross-linking pentaglicyne bridges in the cell wall of *S. aureus*. Used in the form of a cream in a 0.5% concentration to eradicate carriers in cotton rats, it was found more effective in comparison with 2% mupirocin and 5% nisin (Kokai-Kun et al. 2003). Good therapeutic effects were obtained in experimental rabbit keratitis using lysostaphin in the form of eye drops in the concentration of 0.3% and in the case of endophtalamitis 0.1% (Dajcs et al. 2002). Most of the experiments were performed *in vitro* but the obtained results allow to presume that in the future these preparations could find use in the treatment of topic infections, particularly because bacteriocins do not affect the host tissues (Kokai-Kun et al. 2003; Oh et al. 2006). It has been reported that the LC_{50} for bacteriocin isolated from *Lactococcus spp.* HY 44 in the concentration of around 50 mg/ml did not show a proliferative effect on human fibroblasts and no allergic reaction in the human patch test (Oh et al. 2006). Filipowicz et al. examined the activity of three components of Juniper Berry Oil A, B and C. The highest activity against MRSA strains (MIC 1.2 μg/ml) in comparison with a control strain ATCC 25923 was determined for compound A comprising the highest concentration of (-)ά-pinene, p-cymene and β-pinene (Filipowicz et al. 2003). Among animal-derived preparations, a Renalexin 20-residue peptide isolated from the skin of a bullfrog *Rana catesbeiana* displayed very high activity against clinical MRSA strains (MIC_{90} 16 mg/ml) (Giacometti et al. 2000). A native, thermostabile antibacterial factor isolated from third-instar larvae of *Lucilia sericata maggots*, composed of two fractions <500 Da and 0.5 – 3 kDa demonstrated the highest activity against MRSA and MSSA strains (Bexfield et al. 2004).

Another strategy is the use of a compound, namely Polyoxotungstates (POT), which sensitizes MRSA strains to beta-lactam antibiotics through the reduction of PBP2 expression.

The mechanism responsible for this occurrence is probably a stress response of MRSA to POT. The effect of sensitization to β-lactams was achieved also in the case of other substances such as fosfomycin, triazine dye, Triton X-100, tannin and some flavonoids (Tajima 2005).

II. Cationic Antimicrobial Peptides

Cationic antimicrobial peptides (CAMPs) such as defensins, cathelicidin and thrombocidins are positively charged (+2 to +7) amphipathic molecules with antibacterial properties (Peschel and Collins 2001; Zanetti et al. 2002; Hiemstra et al. 2004). Produced by the innate immune system they protect the skin and epithelia from the invasion of bacteria (Stryjewski et al. 2007). They play a key role in the innate immune response and help in the functioning of neutrophils and platelets, therefore play a dual role both through their antibacterial activity and activation of components of the immune system (Zanetti et al. 2002; Pei 2004). The mechanism of action is based on their dual hydrophobic and hydrophilic nature and is important for initial interactions with the bacterial membrane which leads to the disruption of the ion homeostasis (Peschel and Collins 2001; Peschel 2002; Pei 2004). Research is carried out regarding their use in therapy of natural peptides and their synthetic analogues (Benincasa et al. 2003; Giacometti et al. 2004; Cirioni et al. 2006a; Ghiselli et al. 2007) and other compounds with similar characters and modes of action (Jin et al. 2005; Chin et al. 2007).

The research was conducted *in vivo* by Cirioni et al. who pre-treated central venous catheters (CVC) with 27-residue cathelicidin BMAP-28 and subsequently with vancomycin or linezolid a reduction in the biofilm bacterial load 10^3 to 10^1 CFU/ml was obtained and bacteremia became undetectable. The *in vitro* MIC for linezolid in the presence of BMAP-28 decreased 4-fold in comparison with a control system where BMAP-28 was not used (Cirioni et al. 2006a). In experimental peritonitis acuta in mice caused by MRSA, applying BMAP-28 a similar survival rate was obtained as in a control, non-infected group (Benincasa et al. 2003). BMAP-28 not only acted synergistically with antibiotics reducing MIC but also protected mice from the development of sepsis. When alive *S. aureus* cells were administered to mice and antibiotic and BMAP-28 therapy was applied, all the factors induced a similar reduction of mortality in infected mice. When dead cells were administered only BMAP induced a reduction in mortality through decreasing the level of tumor necrosis factor-α and interleukin-6 in the serum indicating its ability to inhibit the development of septic shock (Giacometti et al. 2004).

Caregnins CSA-8 and CSA-13 are synthetic cationic steroid molecules, which resemble in their mode of action endogenous antimicrobial peptides. Tests towards 50 clinical hGISA and GISA strains, and 4 VRSA CSA-13 showed high activity with a 4-fold lower MIC (1 μg/ml) for VRSA in comparison with CSA-8. MIC_{90} for hGISA and GISA was 1 μg/ml and for CSA-8 MIC_{90}, 8 μg/ml (Chin et al. 2007).

Endogenous peptides in contrary to the previously described compounds seem to possess potential in parenteral therapy. However, similarly to most compounds with antibacterial activity there is a risk of a lack of efficacy towards strains inducing infections. The negatively net charge of staphylococcal teichoic acid and phospholipids play a key role in the susceptibility to CAMPs (Peschel and Collins 2001; Peschel 2002; Hiemstra et al. 2004). Due to the fact that most endogenous peptides have a positive charge, interionic forces lead to the accumulation of peptides in the cell wall, interactions with the surface of the cell membrane, integration with the lipid layer and pore formation (Peschel and Collins 2001; Peschel 2002; Hiemstra et al. 2004). It is regarded that acquiring resistance to cationic peptides is based on the modulation of the electrical charge of the wall and cell membrane, which leads to the reduction in the interactions with peptide antibiotics. This can take place through the modification of teichoic acid in the cell wall by incorporation of D-alanine or modification of phosphatidilglycerol by L-lysine as well as in the efflux pump (gacA) mechanism (Peschel 2002).

III. The Protection of Biofilm Formation

The protection strategy for biofilm formation is based on the prevention of staphylococci adhesion to the host cell or to the surface of synthetic materials used in various types of catheters and implants (Balaban et al. 2003; Dell'Acqua et al. 2004; Cirioni et al. 2006b; Balaban et al. 2007). The introduction of new less hydrophobic biomaterials (Campoccia et al. 2006) and also covering their surface with antibiotics and antiseptics can prevent the adhesion of bacteria-forming biofilm (Richards et al. 2006; Ghiselli et al. 2007; Bahna et al. 2007). In the case of biofilm production disruption of *quorum sensing* through blocking the formation of biofilm and the expression of virulence factors with RNAIII-Inhibiting Peptide (RIP) reduced the growth of bacteria and allows access to an antibiotic (Balaban et al. 2001; Balaban et al. 2003; Korem et al. 2003; Cirioni et al. 2006b). RIP is a hektapeptide, which in *S. aureus* inhibits the synthesis of toxins through interference with the RNAIII Activating Protein (RAP) (Korem et al. 2003). RAP is a protein, which activates the production of many toxins through phosphorylation of the RAP target – TRAP. If its expression or phosphorylation is disrupted by RIP then the expression of the virulence regulator, *agr* gene, is inhibited and *quorum sensing* is blocked (Korem et al. 2003; Cirioni et al. 2006b). Balaban et al. reported that RIP inhibited the adhesion to HaCat (human skin keratinocytes) and Hep-2 (human epithelial cells) and also reduced the adhesion to dialysis catheters both those made of polyurethane and silicone (Balaban et al. 2003). Moreover RIP was found more effective against strains with high adhesion affinity (Balaban et al. 2003). *In vitro* studies (Cirioni et al. 2006b) revealed that the treatment of *S. aureus* cells in a biofilm with RIP and a subsequent addition of an antibiotic induced a significant decrease in MICs and MBCs comparable to values obtained for planktonic cells. It was shown that RIP did not solely display any bacteriostatic or bactericidal activity (Cirioni et al. 2006b). The results of this research were confirmed in an *in vivo* model of rats with staphylococci CVC (central venous catheter) infection. The application of RIP and antibiotics greatly reduced the bacterial load in the biofilm by 6 $\log_{10}$ (Cirioni et al. 2006b). In research carried out by Balaban et al. bacteria preincubated with RIP were used as a factor inducing infection *in vivo*. It was reported that strains which were treated with RIP caused less frequent infections in comparison with wild strains. In the case of keratitis and osteomyelitis in rabbits by 40% and 58%, respectively, septic arthritis in mice by 60% and mastitis in cows from 70 – 100% (Balaban et al. 2000).

IV. Bacteriophages

Up till now bacteriophages were used in therapy mainly in countries of Eastern Europe such as Poland (Slopek et al. 1987) and the former Soviet Union (Keilman 1979; Sulakvelidze et al. 2001; Parfitt 2005). There are no formal regulations and standards for biocteriophage therapy despite their use in large number of patients (Keilman 1979; Slopek et al. 1987; Sulakvelidze et al. 2001; Parfitt 2005). There is still a lack of well documented research with controls, therefore conclusions as to the usefulness and dangers of bacteriophage therapy cannot be made (Sulakvelidze et al. 2001). Despite this very fact, phage therapy seems to be worth further investigation due to its concepts. The phage kills only pathogenic bacterial cells and in the case of *S. aureus* infections both MSSA and MRSA strains (Matsuzaki et al. 2003; Capparelli et al. 2007). Whereas the antimicrobial activity of an antibiotic includes both pathogenic bacteria as well as the physiological flora. A key problem is the selection of a phage so that it will be specific not only towards certain species but also towards different strains (Sulakvelidze et al. 2001). Sulakvelidze et al. mention other advantages of phage therapy. Phages replicate in the site of infection and do not cause side effects and bacteria resistant to some phages can be sensitive towards others. A drawback of this sort of therapy can be the narrow spectrum of sensitive strains, which can require the use of a mixture of phages. It is essential also to select lysogenic phages due to the risk of transferring genes encoding toxins and resistance to antibiotics (Sulakvelidze et al. 2001). Parfitt et al. describes the clinical use of phage therapy in Georgia, where this therapy helped to reach a clinical state enabling plastic surgery in three patients with purulent staphylococci infected burns caused by a radioactive agent (Parfitt 2005). This therapy is based on using synthetic skin soaked in phages from which they are slowly released to the environment of the wound as a consequence of the slow biodegradation of the artificial skin (Parfitt 2005). Keilman et al. presented results regarding the local treatment of purulent bone inflammation. Polyvalent phages were used in combination with locally applied antiseptics parallel to surgical intervention and antibiotic treatment. As a result a reduction in purulent complications from 23.2% to 4.8% was observed (Keilman 1979). New research regarding this subject has also been presented in several publications (Matsuzaki et al. 2003; Wills et al. 2005; Capparelli et al. 2007). In the experimental treatment of abscesses in rabbits with LS2a phages a significant eradication of infection was obtained (Wills et al. 2005) and the use of MR11 phage to treat peritonitis in mice reduced the mortality in comparison with a non-treated control group, where the mortality was > 90% (Matsuzaki et al. 2003). The MR11 phage was selected due the fact that it has the widest spectrum towards various *S. aureus* strains and also becouse it does not transfer toxin and resistance genes (Matsuzaki et al. 2003). Capparelli et al. with the use of a suspension of M^{Sau} phages obtained total infection eradication in mice within 4 days. This phage also has a wide spectrum of activity towards many *S. aureus* strains including MRSA. In comparison with a control group the survival rate was 97% (Capparelli et al. 2007). In a model of catheter-related infection phage λ 80, a reduction of 79.4% in the optical density of the 24 h biofilm was observed (Del Pozo et al. 2007). Phages in the form of vectors have also been used to increase the efficacy of therapy (Embleton et al. 2005; Yacoby et al. 2007). The bacteriophage 75 are used to transfer a photosensitizer Tin (IV) chlorine e6 and light-activated antimicrobial agents to *S. aureus* cells. While being exposed to light, this complex killed different cells both MSSA, MRSA as well as VISA. The use of a control consisting of only a phage allowed to evaluate that the killing effect was not a consequence of bacteria lysis by the phage (Embleton et al. 2005).

V. Photodynamic Therapy

Photodynamic Therapy (PDT) is based on the use of chemicals termed photosensitizers. Photosensitizers accumulate in specifically selected tissues or cells and their activation is a result of the activity light of adequate wavelength. This leads to the generation of singlet oxygen and free radicals responsible for the cytotoxic effect towards specific cells (Hamblin et al. 2002; Romanova et al. 2003; Mohr and Redecker-Klein 2003; Jori and Brown 2004). Until now, some mechanisms responsible for the cytotoxic effect have been examined and characterized. Depending upon the photosensitizer used the mechanism can be associated with disrupting the cell wall, inactivating enzymes or the indirect damage to the genetic material of the bacteria (genomic and plasmid DNA) (Romanova et al. 2003; Sharma et al. 2008).

Until now research concerned mainly preliminary investigations such as analyzing the survival rate of bacteria *in vitro*, depending on the type and dose of the applied photosensitizer, wavelength and time of illumination (Maisch et al. 2005; Grinholc et al. 2007a; Sharma et al. 2008; Grinholc et al. 2008a; Grinholc et al. 2008b).

An important issue, from the clinical point of view, is that multi-resistant MRSA strains are as easily killed in response to photodynamic therapy as wild MSSA strains, and the development of resistance mechanisms to this therapy seems unlikely in comparison with the quick gaining of resistance towards different antibiotics and other antibacterial agents (Hamblin and Hasan 2004; Grinholc et al. 2007a). Grinholc et al. reported that the response to photodynamic inactivation with the use of protoporphiryn diarginate ($PPArg_2$) was different among various clinical MSSA and MRSA strains and ranged from 0 to 3 $\log_{10}$ unit reduction in viable counts. Assuming a reduction of 1-2 $\log_{10}$ units as a criteria of sensitivity and reduction of 2-3 $\log_{10}$ units high sensitivity, this indicates that 62.5% of MSSA and 40% of MRSA were sensitive and shows that multiresistance correlated with the resistance to photodynamic therapy ($p<0.05$) (Grinholc et al. 2008b). Comparing *in vitro* biofilm-forming strains with nonbiofilm-forming strains, a higher killing efficacy was observed in the latter. Among the highly sensitive strains (>2 $\log_{10}$) 75% did not produce biofilm (Grinholc et al. 2008b). It is worth to emphasize that strain-dependance in photodynamic inactivation was not studied so far, thus it is necessary to continue the investigations till this phenomenon would not be explained. Determination of the molecular marker indicating strain response to photoinactivation would lead to more effective treatment (Grinholc et al. 2008b). Sharma et al. reported high *in vitro* efficacy of toluidine blue (TBO) used simultaneously with laser light in the reduction of MRSA biofilm where the reduction effect was dependent upon the light dose (Sharma et al. 2008).

So far there are very few papers regarding *in vivo* studies (Soukos et al. 1998; Komerik et al. 2003; Wilson 2004). On the basis of the results obtained by Bisland et al. on animal models it can be assumed that photodynamic therapy is potentially one of the best therapeutic options for treating osteomyelitis of *S. aureus* etiology, especially that these infections are associated with biofilm production (Bisland et al. 2006). An undoubtable advantage of photodynamic therapy is based on mechanisms like singlet oxygen generation making it impossible to induce effective resistance mechanisms (Wilson 2004). Importantly, photodynamic therapy in comparison with antibiotic treatment does not influence the physiological bacterial flora (Romanov et al. 2001; Wilson 2004) due to the possibility of the specific introduction of the photosensitizer to selected bacterial cells and also to the restricted illumination area (Hamblin and Hasan 2004).

An additional advantage of this method is the low cytotoxic effect on host cells along with high efficacy towards pathogens (Soncin et al. 2002; Grinholc et al. 2008a) and also it does not cause any direct or delayed damage to the genetic material (Zeina et al. 2002; Zeina et al. 2003).

VI. Staphylococcal Vaccines

Literature regarding staphylococcal vaccines is very comprehensive, however there is still a problem with the introduction of a efficient vaccine in the prevention of infections caused by *S. aureus*. This can be associated with the fact that a human organism throughout its whole life is in contact with this bacterium as many people are constant or part time carriers of *S. aureus* on the mucus of the nasal and oral cavity. The selection of patients for this preventative therapy should be based on recognizing the risk factors and should be proceeded by research, which would allow a individualized approach (Fattom et al. 2004). In earlier years an antigen was regarded as the whole bacterial cell, however, recently, research is aiming at obtaining antibodies against specific virulence factors such as adhesins, toxins or invasins (Otto 2008). Despite the multicomponential regulation of pathogenesis, interference in any component can reduce virulence and symptoms of disease (Balaban et al. 2000). Nonetheless the only vaccine StaphVAX^R, for which research is most advanced, is still not available (Deresinski 2006). StaphVAX^R is a bivalent vaccine composed of *S. aureus* capsular polysaccharides (CPS) types 5 and 8 conjugated with a non-toxic *Pseudomonas aeruginosa* exotoxin (Fattom et al. 2004). However, research which reached phase II and III of clinical studies indicate that the vaccine offers partial protection from bacteremia lasting for a few weeks in 60% of immunized patients (Shinefield 2006). Due to the planned expansion of the activity spectrum with *S. aureus* 336, *S. epidermidis* PS-1, *S. epidermidis* GP-1 and PVL antigens, Nabi Biopharmaceuticals suspended the research (Deresinski 2006). Other preparations comprising immunoglobulins against ClfA, ABC transporter GrfA or lipoteichoic acid are in the II and III stages of clinical trials (Deresinski 2006). A target in the bacterial cell which is of interest to researchers as potential antigens for vaccines are the adhesion proteins MSCRAMM. As these proteins are located on the surface of the cell, antibodies directed against them could directly block the binding of the microorganism to the host tissue preventing colonization and infection initiation (Otto 2008). The research is carried out in two directions, immunization through the application of surface proteins and in consequence inducing a immune response (Rennermalm et al. 2001; Josefsson et al. 2001; Shannon et al. 2006; Kuklin et al. 2006; Zhou et al. 2006), but also through DNA vaccines comprising genes encoding specific surface proteins. Genes in eukaryotic cells induce the expression of encoded proteins generating an immune response in the form of specific antibodies. As shown these proteins possess the ability to modify also the cell-mediated immune response (Brouillette et al. 2002; Kerro-Dego et al. 2006; Gaudreau et al. 2007; Therrien et al. 2007). Parallel research is carried out regarding the possibility of passive immunization with the use of antibodies obtained through the immunization of animals with specific antigens (Rennermalm et al. 2001; Josefsson et al. 2001; Vernachio et al. 2003; Hall et al. 2003; Shannon et al. 2006). Both single surface proteins as well as conjugated proteins can serve as antigens. Extracellular Fibrynogen-Binding Protein (Efb) used for the immunization of mice with foreign-body-associated wound infection induced a strong immune response, which reduced the severity of the infection (Shannon et al. 2006). Josefsson et al. reported that in experimental septic arthritis in mice, immunization with recombinant ClfA induced a less-severe course of disease and passive immunization with rat and rabbit anti-ClfA antibodies protected against severe arthritis and sepsis-induced death (Josefsson et al. 2001). A similar effect was obtained with the use of a human polyclonal immunoglobulin (SA-IGIV), specific for ClfA as complementary therapy to vancomycin treatment in experimental *S. aureus* endocarditis in mice. Combination therapy showed to be more effective in both early and well-established endocarditis (Vernachio et al. 2003). ClfB, a protein binding cytokeratin-10 plays an important role in the colonization of the nasal carriage, a known risk factor for *S. aureus* infection. Mice immunized systemically or intranasally with a vaccine containing the A domain of ClfB displayed a lower level of colonization than the control group. Passive immunization with monoclonal ClfB-Mab antibodies also reduced colonization (Schaffer et al. 2006). Iron surface determinant B (IsdB) is an iron-sequestering protein, which is conserved in diverse clinical MSSA and MRSA strains. Its application with an adjuvant induced a strong response in mice protecting them from infection and sepsis and in *Rhesus macaques* after a single dose, a five-fold increase in the production of antibodies was observed (Kuklin et al. 2006). Due to the potentially wider protection spectrum and stronger immune response research regarding conjugated vaccines containing two or more antigens are being carried out. A recombinant vaccine containing Cna-FnBP induced a strong humoral response, of which the highest was around the 5 and 6 week after immunization. This resulted in a higher survival rate in infected mice in comparison with a group immunized only with Cna or FnB (Zhou et al. 2006). In a mouse model of abscess formation immunization with a vaccine containing four proteins IsdA, IsdB, SdrD, SdrE induced complete protection in comparison with a group immunized with a single antigen. These four antigens were selected as they induced the highest level of antibodies among fifteen tested antigens. A functional correlation between the level of antibodies and opsonophagocytic properties were observed (Stranger-Jones et al. 2006). Conjugated vaccines were also examined in the context of DNA-mediated immunization and expression of the products of genes. The immune response in mice was evaluated after DNA vaccination with plasmids encoding an individual antigen and a chimeric protein composed of the surface-located GapB and GapC proteins. The recombinant vaccine induced a stronger humoral response than the vaccine bearing the single gene (Kerro-Dego et al. 2006). In the case of a polyvalent vaccine containing genes encoding the Sortase enzyme (ClfA-FnBPA-Srt) it was reported that the combination of genes induced a better response than each plasmid alone. Moreover, the conjugated vaccine induced a T1 and T2 response has a protective effect from the complex pathogenesis of *S. aureus* (Gaudreau et al. 2007). However, not all of the obtained results regarding vaccines containing surface proteins are clear. Therrien et al. reported that despite immunization with a DNA vaccine containing the *cna* gene and a humoral and cell-based immune response, the vaccine did not protect the mice from letal-induced infection. Moreover, the only mice which died in the experiment where the ones which were immunized (Therrien et al. 2007). A surface polysaccharide poly-N-acetyl glucosamine (PNAG), present in *S. aureus* and *S. epidermidis*, participates in biofilm production (Maira-Litran et al. 2004). Its production is encoded by a gene present within the *ica* locus, which is present in most clinical *S. aureus* isolates. Research indicates that this compound displays immunogenic properties and therefore could be a good vaccine candidate (McKenney et al. 1999; Maira-Litran et al. 2004; Maira-Litran et al. 2005). In a mouse renal infection model the application of PNAG induced a high level of IgG antibodies (> 500). After 5 days of immunization mice were challenged with *S. aureus* CP5 and CP8. In comparison with the control group, in immunized mice a reduction in CFU per gram of kidney was observed. Rabbits immunized with PNAG produced higher levels of antibodies (> 2500), which were held at a constant level for at least 8 months. The passive immunization of mice with these antibodies significantly reduced or totally eliminated bacteria from the kidneys (McKenney et al. 1999). In an experimental endocarditis model in a rat with intra-aortic catheters a protective effect was obtained through passive immunization with purified PNAG (Maira-Litran et al. 2004). Maira-Litran et al. compared the efficacy of PNAG and its deacylated (dPNAG) form conjugated with diphteria toxoid (DT) carrier protein used to immunize animals. In three of four tested *S. aureus* strains a marked increase in the level of antibodies was obtained with dPNAG as opposed to native PNAG. Passive immunization of mice with anti-dPNAG-DT rabbit sera induced higher elimination of bacteremia from 54% to 91%, while anti-PNAG were ineffective (Maira-Litran et al. 2005). Due to the fact that the efficacy of antibodies against a carbohydrate vaccine may not include all *S. aureus* strains as not all express biofilm production, despite the presence of the *ica* locus (Grinholc et al. 2007b), new candidates for the production of an efficient vaccine in chronic infection therapy and prevention are being sought for. Proteins, which are immunogenic in MRSA-induced chronic osteomyelitis with biofilm formation were examined in a rabbit model of tibial infection. Among 22 identified antigens in the cell fraction, antibodies for some of them such as lipase, autolysin or lipoprotein were produced at an early stage on the 14th day of infection. In the case of other antigens namely, transketolase or elongation factor, antibodies were produced during late stages of infection on the 42nd day. The authors suggest that antigens, which are recognized earlier are good vaccine candidates and those recognized in late stages can be used in adjuvant therapy (Brady et al. 2006). Another approach involves the use of vaccines against extracellularly secreted proteins with superantigen (exotoxin) properties such as TSST-1, enterotoxin B (LeClaire et al. 2002), enterotoxin C (Hu et al. 2006; Chang et al. 2008) and α-hemolysin (Hla) (Bubeck and Schneewind 2008). The intranasal application of conjugated vaccines containing non-toxic mutant toxic shock syndrome toxin-1 (TSST-1) and non-toxic mutant heat labile toxin (mLT) of *Escherichia coli* effectively induced an increase in the level of anty-TSST-1 antibodies in the serum and in the nasal secretion. They protected mice from both colonization of the nasal cavity and also from the systemic infection (Narita et al. 2008). Two high immunogenic fragments of enterotoxin B (SEB) generated the production of specific antibodies in chickens. Passive immunization of rhesus monkeys with these antibodies protected them from death after intranasal application of a lethal dose of SEB (LeClaire et al. 2002). In other experiments, intranasal immunization with a non-toxic double SEC mutant and cholera toxin (CT) as an adjuvant also protected mice from infection (Hu et al. 2006). In turn Bubeck et al. with the use of a mutant of α-hemolysin (Hla H35L), which does not possess the ability to form pores, obtained specific antibodies in mice, which protected them from *S. aureus*-induced pneumonia. Moreover, the obtained results indicate that the transfer of anti-Hla antibodies not only protected naive animals from the development of infection but also human lung epithelial cells from damage during infection (Bubeck and Schneewind 2008). In other study mice with pneumonia that additionally developed skin infection challenged with epidemic community acquired MRSA (CA-MRSA) strain USA300 after immunization with PVL revealed a high anti-PVL response. Depending on the way of immunization, the protection against one of the infection could be obtained. Intranasally vaccination protected mice from a pulmonary infection but not from intradermal one, however, a subcutanely vaccination protected mice against an intradermal infection but not a pulmonary one (Brown 2009). A vaccine directed against the PBP2a protein could improve therapeutic strategies against MRSA and long-term prognosis in a significant way. Senna et al. cloned the internal region from the transpeptidase domain from PBP2a into a mammalian expression vector, which was used as a DNA vaccine in the immunization of mice. After three immunizations a significant increase in the level of specific anti-PBP2a antibodies was obtained.

Conclusion

The treatment of high risk patients with MRSA infection without applying appropriate antibiotic policy within hospitals generates the development of various antibiotic-resistance mechanisms among microorganisms and lack of clinical efficacy. The existence of many resistance mechanisms limits the usage of commonly available antibiotics and antimicrobial drugs. Moreover, the spread of MRSA strains over the world generates the risk of an increased number of difficult-to-treat infections. It should lead us to develop an alternative and effective therapeutic option, as well as prophylaxis of microbial infections. It demands to continue investigations concerning the biology, virulence mechanisms and emidemiology of the MRSA strains.

Acknowledgments

We are grateful to Prof. Grzegorz Wegrzyn (University of Gdansk, Poland) for scientific consultations, Grzegorz Fila (University of Gdansk, Poland) for providing the graphic figures, and Anna Kawiak and Agnieszka Wojtkowiak for linguistic consultations.

References

Aathithan, S., Dybowski, R. & French, G. L. (2001). Highly epidemic strains of methicillin-resistant *Staphylococcus aureus* not distinguished by capsule formation, protein A content or adherence to HEp-2 cells. *Eur J Clin Microbiol Infect Dis., 20*, 27-32.

Abele-Horn, M., Schupfner, B., Emmerling, P., Waldner, H. & Goring, H. (2000). Persistent wound infection after herniotomy associated with small-colony variants of *Staphylococcus aureus*. *Infection, 28*, 53-54.

Adesokan, A., Akanji, M. A. & Yakubu, M. T. (2007). Antibacterial potentials of aqueous extracts of Enatia chlorantha stem bark. *Afr J Biotech*, *6*, 2502-2505.

Andrasevic, A. T., Power, E. G., Anthony R. M., Kalenic S. & French, G. L. (1999). Failure of bacteriophage typing to detect an inter-hospital outbreak of methicillin-resistant *Staphylococcus aureus* (MRSA) in *Zagr*eb subsequently identified by random amplification of polymorphic DNA (RAPD) and pulsed-field gel electrophoresis (PFGE). *Clin Microbiol Infect*, *5*, 634-642.

Bagger, J. P., Zindrou, D. & Taylor, K. M. (2004). Postoperative infection with methicillin-resistant *Staphylococcus aureus* and socioeconomic backgroud. *Lancet*, *363*, 706-708.

Bahna, P., Dvorak, T., Hanna, H., Yasko, A. W., Hachem, R. & Raad, I. (2007). Orthopaedic metal devices coated with a novel antiseptic dye for the prevention of bacterial infections. *Int J Antimicrob Agents*, *29*, 593-596.

Balaban, N., Cirioni, O., Giacometti, A., Ghiselli, R., Braunstein, J. B., Silvestri, C., Mocchegiani, F., Saba, V. & Scalise, G. (2007). Treatment of *Staphylococcus aureus* biofilm infection by the quorum-sensing inhibitor RIP. *Antimicrob Agents Chemother*, *51*, 2226-2229.

Balaban, N., Collins, L. V., Cullor, J. S., Hume, E. B., Medina-Acosta, E., Vieira, Da MO., O'Callaghan R., Rossitto, P. V., Shirtliff, M. E., Serafim, Da, S. L., Tarkowski, A. & Torres, J. V. (2000). Prevention of diseases caused by *Staphylococcus aureus* using the peptide RIP. *Peptides*, *21*, 1301-1311.

Balaban, N., Goldkorn, T., Gov, Y., Hirshberg, M., Koyfman, N., Matthews, H. R., Nhan, R. T., Singh, B. & Uziel, O. (2001). Regulation of *Staphylococcus aureus* pathogenesis via target of RNAIII-activating Protein (TRAP). *J Biol Chem.*, *276*, 2658-2667.

Balaban, N., Goldkorn, T., Nhan, R. T., Dang, L. B., Scott, S., Ridgley, R. M., Rasooly, A., Wright, S. C., Larrick, J. W., Rasooly, R. & Carlson, J. R. (1998). Autoinducer of virulence as a target for vaccine and therapy against *Staphylococcus aureus*. *Science*, *280*, 438-440.

Balaban, N., Gov, Y., Bitler, A. & Boelaert, J. R (2003). Prevention of *Staphylococcus aureus* biofilm on dialysis catheters and adherence to human cells. *Kidney Int.*, *63*, 340-345.

Bates, D. M., Von, E. C., McNamara, P. J., Peters, G., Yeaman, M. R., Bayer, A. S. & Proctor, R. A. (2003). *Staphylococcus aureus* menD and hemB mutants are as infective as the parent strains, but the menadione biosynthetic mutant persists within the kidney. *J Infect Dis.*, *187*, 1654-1661.

Bauer, C. C., Apfalter, P., Daxboeck, F., Bachhofner, N., Stadler, M., Blacky, A., ab-Elschahawi, M. & Assadian, O. (2009). Prevalence of Panton-Valentine leukocidin genes in methicillin-resistant *Staphylococcus aureus* isolates phetotypically consistent with community-acquired MRSA, 1999-2007, Vienna General Hospital. *Eur J Clin Microbiol Infect Dis*, *28*, 909-912.

Baumert, N., Von, E. C., Schaaff, F., Peters, G., Proctor, R. A. & Sahl, H. G. (2002). Physiology and antibiotic susceptibility of *Staphylococcus aureus* small colony variants. *Microb Drug Resist*, *8*, 253-260.

Benfield, T., Espersen, F., Frimodt-Moller, N., Jensen, A. G., Larsen, A. R., Pallesen, L. V., Skov, R., Westh, H. & Skinhoj, P. (2007). Increasing incidence but decreasing in-hospital mortality of adult *Staphylococcus aureus* bacteraemia between 1981 and 2000. *Clin Microbiol Infect*, *13*, 257-263.

Benincasa, M., Skerlavaj, B., Gennaro, R., Pellegrini, A. & Zanetti, M. (2003). In vitro and in vivo antimicrobial activity of two alpha-helical cathelicidin peptides and of their synthetic analogs. *Peptides, 24*, 1723-1731.

Bexfield, A., Nigam, Y., Thomas, S. & Ratcliffe, N. A. (2004). Detection and partial characterisation of two antibacterial factors from the excretions/secretions of the medicinal maggot Lucilia sericata and their activity against methicillin-resistant *Staphylococcus aureus* (MRSA). *Microbes Infect, 6*, 1297-1304.

Bisland, S. K., Chien, C., Wilson, B. C. & Burch, S. (2006). Pre-clinical in vitro and in vivo studies to examine the potential use of photodynamic therapy in the treatment of osteomyelitis. *Photochem Photobiol Sci., 5*, 31-38.

Boost, M. V., O'Donoghue, M. M. & Siu, K. H. (2007). Characterisation of methicillin-resistant *Staphylococcus aureus* isolates from dogs and their owners. *Clin Microbiol Infect, 13,* 731-733.

Boyle-Vavra, S. & Daum, R. S. (2007). Community-acquired methicillin-resistant *Staphylococcus aureus*: the role of Panton-Valentine leukocidin. *Lab Invest, 87*, 3-9.

Brady, R. A., Leid, J. G., Camper, A. K., Costerton, J. W. & Shirtliff, M. E. (2006). Identification of *Staphylococcus aureus* proteins recognized by the antibody-mediated immune response to a biofilm infection. *Infect Immun, 74*, 3415-3426.

Brouillette, E., Lacasse, P., Shkreta, L., Belanger, J., Grondin, G., Diarra, M. S., Fournier, S., & Talbot, B. G. (2002). DNA immunization against the clumping factor A (ClfA) of *Staphylococcus aureus*. *Vaccine, 20*, 2348-2357.

Browatzki, M., Borst, M. M., Katus, H. A. & Kranzhofer, R. (2006). Purulent pericarditis and pleural empyema due to *Staphylococcus aureus* septicemia. *Int J Cardiol, 107*, 117-118

Bubeck, W. J. & Schneewind, O. (2008). Vaccine protection against *Staphylococcus aureus* pneumonia. *J Exp Med., 205*, 287-294.

Campoccia, D., Montanaro, L. & Arciola, C. R. (2006). The significance of infection related to orthopedic devices and issues of antibiotic resistance. *Biomaterials, 27*, 2331-2339.

Capparelli, R., Parlato, M., Borriello, G., Salvatore, P. & Iannelli, D. (2007). Experimental phage therapy against *Staphylococcus aureus* in mice. *Antimicrob Agents Chemother*, 51, 2765-2773.

Chang, B. S., Moon, J. S., Kang, H. M., Kim, Y. I., Lee, H. K., Kim, J. D., Lee, B. S., Koo, H. C. & Park, Y. H. (2008). Protective effects of recombinant staphylococcal enterotoxin type C mutant vaccine against experimental bovine infection by a strain of *Staphylococcus aureus* isolated from subclinical mastitis in dairy cattle. *Vaccine, 26*, 2081-2091.

Chesi, G., Colli, A., Mestres, C. A., Gambarati, G., Boni, F. & Gherli, T. (2006). Multiresistant-MRSA tricuspid valve infective endocarditis with ancient osteomyelitis locus. *BMC Infect Dis., 6*, 124.

Chi, C. Y., Wang, S. M., Lin, H. C. & Liu, C. C. (2006). A clinical and microbiological comparison of *Staphylococcus aureus* toxic shock and scalded skin syndromes in children. *Clin Infect Dis., 42*, 181-185.

Chin, J. N., Rybak, M. J., Cheung, C. M., Savage, P. B.(2007).Antimicrobial activities of ceragenins against clinical isolates of resistant *Staphylococcus aureus*. *Antimicrob Agents Chemother, 51*, 1268-1273.

Cirioni, O., Giacometti, A., Ghiselli, R., Bergnach, C., Orlando, F., Mocchegiani, F., Silvestri, C., Licci, A., Skerlavaj, B., Zanetti, M., Saba, V. & Scalise, G. (2006a). Pretreatment of central venous catheters with the cathelicidin BMAP-28 enhances the efficacy of antistaphylococcal agents in the treatment of experimental catheter-related infection. *Peptides, 27*, 2104-2110.

Cirioni, O., Giacometti, A., Ghiselli, R., Dell'Acqua, G., Orlando, F., Mocchegiani, F., Silvestri, C., Licci, A., Saba, V., Scalise, G. & Balaban, N. (2006b). RNAIII-inhibiting peptide significantly reduces bacterial load and enhances the effect of antibiotics in the treatment of central venous catheter-associated *Staphylococcus aureus* infections. *J Infect Dis., 193*, 180-186.

Clarke, S. R., Wiltshire, M. D. & Foster, S. J. (2004). IsdA of *Staphylococcus aureus* is a broad spectrum, iron-regulated adhesin. *Mol Microbiol, 51*, 1509-1519.

Coia, J. E., Duckworth, G. J., Edwards, D. I., Farrington, M., Fry, C., Humphreys, H., Mallaghan, C. & Tucker, D. R. (2006). Guidelines for the control and prevention of meticillin-resistant *Staphylococcus aureus* (MRSA) in healthcare facilities. *J Hosp Infect,*, 63 Suppl 1, S1-44

Collins, F. & Hampton, S. (2005). Hand-washing and methicillin-resistant *Staphylococcus aureus*. *Br J Nurs., 14*, 703-707.

Collins, R. J. (2007). Community-acquired methicillin-resistant *Staphylococcus aureus* in a group home setting. *Consult Pharm., 22*, 763-767.

Conceicao, T., Ires-De-Sousa, M., Fuzi, M., Toth, A., Paszti, J., Ungvari, E., Van Leeuwen W. B., Van B. A., Grundmann, H. & De, L. H. (2007). Replacement of methicillin-resistant *Staphylococcus aureus* clones in Hungary over time: a 10-year surveillance study. *Clin Microbiol Infect, 13*, 971-979.

Cramton, S. E., Gerke, C., Schnell, N. F., Nichols, W. W. & Gotz, F. (1999). The intercellular adhesion (ica) locus is present in *Staphylococcus aureus* and is required for biofilm formation. *Infect Immun, 67*, 5427-5433.

Cuevas, O., Cercenado, E., Bouza, E., Castellares, C., Trincado, P., Cabrera, R. & Vindel, A. (2007). Molecular epidemiology of methicillin-resistant *Staphylococcus aureus* in Spain: a multicentre prevalence study (2002). *Clin Microbiol Infect, 13*, 250-256.

Cunningham, R., Jenks, P., Northwood, J., Wallis, M., Ferguson, S. & Hunt, S. (2007). Effect on MRSA transmission of rapid PCR testing of patient admitted to critical care. *J Hosp Infect., 65*, 24-28.

Cunnington, A., Brick, T., Cooper, M., Danin, J., Hunt, D., Jeanes, A., Kearns, A. M., Nolan, M. & Lyall, H. (2009) Severe invasive Panton-Valentine Leucocidin positive *Staphylococcus aureus* infections in children in London, UK. *J Infect, 59,* 28-36.

Cuny, C., Kuemmerle, J., Stanek, C., Willey, B., Strommenger, B. & Witte, W. (2006). Emergence of MRSA infections in horses in a veterinary hospital: strain characterisation and comparison with MRSA from humans. *Euro Surveill, 11*, 44-47.

Dajcs, J. J., Thibodeaux, B. A., Girgis, D. O., Shaffer, M. D., Delvisco, S. M. & O'Callaghan, R. J. (2002). Immunity to lysostaphin and its therapeutic value for ocular MRSA infections in the rabbit. *Invest Ophthalmol Vis Sci., 43*, 3712-3716.

Davis, J. S. (2005). Management of bone and joint infections due to *Staphylococcus aureus*. *Intern Med J.,* 35 Suppl 2, S79-S96.

Del Pozo, J. L., Alonso, M., Arciola, C. R., Gonzalez, R., Leiva, J., Lasa, I. & Penades, J. (2007). Biotechnological war against biofilms. Could phages mean the end of device-related infections? *Int J Artif Organs, 30*, 805-812.

Dell'Acqua, G., Giacometti, A., Cirioni, O., Ghiselli, R., Saba, V., Scalise, G., Gov, Y., Balaban, N. (2004). Suppression of drug-resistant Staphylococcal Infections by the quorum-sensing inhibitor RNAIII-inhibiting peptide. *J Infect Dis., 190*, 318-320.

Denis, O., Deplano, A., Nonhoff, C., De, R. R., De, M. R., Rottiers, S., Vanhoof, R. & Struelens, M. J. (2004). National surveillance of methicillin-resistant *Staphylococcus aureus* in Belgian hospitals indicates rapid diersification of epidemic clones. *Antimicrob Agents Chemother, 48*, 3625-3629.

Deresinski, S. (2006). Antistaphylococcal vaccines and immunoglobulins: current status and future prospects. *Drugs, 66*, 1797-1806.

Desachy, A., Lina, G., Vignon, P., Hashemzadeh, A., Denis, F., Etienne, J., Francois, B. & Ploy, M. C. (2007). Role of superantigenic strains in the prognosis of community-acquired methicillin-susceptible *Staphylococcus aureus* bacteraemia. *Clin Microbiol Infect., 13*, 1131-1133.

Deurenberg, R. H., Vink, C., Kalenic, S., Friedrich, A. W., Bruggeman, C. A., Stobberingh, E. E. (2007). The molecular evolution of methicillin-resistant *Staphylococcus aureus. Clin Microbiol Infect, 13*, 222-235.

Diekema, D. J., Pfaller, M. A., Schmitz, F. J., Smayevsky, J., Bell, J., Jones, R. N. & Beach, M. (2001). Survey of infections due to Staphylococcus species: frequency of occurrence and antimicrobial susceptibility of isolates collected in the United States., Canada, Latin America, Europe, and the Western Pacific region for the SENTRY Antimicrobial Surveillance Program, 1997-1999. *Clin Infect Dis.,* 32 Suppl 2, S114-S132.

Diep, B. A., Carleton, H. A., Chang, R. F., Sensabaugh, G. F. & Perdreau-Remington, F (2006). Roles of 34 virulence genes in the evolution of hospital- and community-associated strains of methicillin-resistant *Staphylococcus aureus. J Infect Dis.,* 193, 1495-1503.

Dinges, M. M., Orwin, P. M. & Schlievert, P. M. (2000). Exotoxins of *Staphylococcus aureus. Clin Microbiol Rev., 13*, 16-34, table.

Donlan, R. M., Murga, R., Bell, M., Toscano, C. M., Carr, J. H., Novicki, T. J., Zuckerman, C., Corey, LvC. & Miller, J. M. (2001). Protocol for detection of biofilms on needleless connectors attached to central venous catheters. *J Clin Microbiol, 39*, 750-753.

Dufour, P., Gillet, Y., Bes, M., Lina, G., Vandenesch, F., Floret, D., Etienne, J. & Richet, H. (2002). Community-acquired methicillin-resistant *Staphylococcus aureus* infections in France: emergence of a single clone that produces Panton-Valentine leukocidin. *Clin Infect Dis., 35*, 819-824.

Duijkeren, E., Ikawaty, R., Broekhuizen-Stins, M. J., Jansen, M. D., Spalburg, E. C., de Neeling, A. J., Allart, J. G., van Nes, A., Wagenaar, J. A. & Fluit, A. C. (2008). Transmission of methicillin-resistant *Staphylococcus aureus* strains between different kinds of pig farms. *Vet Microbiol, 126,* 383-389.

Dzwonkowska, J., Kurlenda, J., Baczkowski, B., Mazurkiewicz, S., Uzunov, I., Ziolkowski, W. & Markowicz, A. (2007). The effect of antibiotic therapy on the incidence of *Staphylococcus aureus* infections in orthopaedic patients. *Ortop Traumatol Rehabil, 9*, 532-547.

Ehrlich, G. D., Hu, F. Z., Shen, K., Stoodley, P. & Post, J. C. (2005). Bacterial plurality as a general *mec*hanism driving persistence in chronic infections. *Clin Orthop Relat Res.*, 20-24.

Elizur, A., Orscheln, R. C., Ferkol, T. W., Atkinson, J. J., Dunne, W. M. Jr., Buller, R. S., Armstrong, J. R., Mardis, E. R., Storch, G. A. & Cannon, C. L. (2007). Panton-Valentine Lekocidin-positive methicillin-resistant *Staphylococcus aureus* lung infection in patients with cystic fibrosis. *Chest, 131,* 1718-1725.

Ellis, M. W., Hospenthal, D. R., Dooley, D. P., Gray, P. J. & Murray, C. K. (2004). Natural history of community-acquired methicillin-resistant *Staphylococcus aureus* colonization and infection in soldiers. *Clin Infect Dis., 39*, 971-979.

Embleton, M. L., Nair, S. P., Heywood, W., Menon, D. C., Cookson, B. D. & Wilson, M. (2005). Development of a novel targeting system for lethal photosensitization of antibiotic-resistant strains of *Staphylococcus aureus. Antimicrob Agents Chemother, 49*, 3690-3696.

Enany, S., Higuchi, W., Okubo, T., Takano, T., Enany, M. & Yamamoto, T. (2007). Brain abscess caused by Panton-Valentine leukocidin-positive community-acquired methicillin-resistant *Staphylococcus aureus* in Egypt., April 2007. *Euro Surveill, 12*, E070927.

Enright, M. C., Day, N. P., Davies, C. E., Peacock, S. J. & Spratt, B. G. (2000). Multilocus sequence typing for characterization of methicillin-resistant and methicillin-susceptible clones of *Staphylococcus aureus. J Clin Microbiol, 38*, 1008-1015.

Faria, N. A., Oliveira, D. C., Westh, H., Monnet, D. L., Larsen, A. R., Skov, R. & De, L. H. (2005). Epidemiology of emerging methicillin-resistant *Staphylococcus aureus* (MRSA) in Denmark: a nationwide study in a country with low prevalence of MRSA infection. *J Clin Microbiol, 43*, 1836-1842.

Fattom, A., Fuller, S., Propst, M., Winston, S., Muenz, L., He, D., Naso, R., Horwith, G (2004). Safety and immunogenicity of a booster dose of *Staphylococcus aureus* types 5 and 8 capsular polysaccharide conjugate vaccine (StaphVAX) in hemodialysis patients. *Vaccine*, *23*, 656-663.

Fenollar, F., Roux, V., Stein, A., Drancourt, M. & Raoult, D. (2006). Analysis of 525 samples to determine the usefulness of PCR amplification and sequencing of the 16S rRNA gene for diagnosis of bone and joint infections. *J Clin Microbiol*, *44*, 1018-1028.

Fey, P. D., Said-Salim, B., Rupp, M. E., Hinrichs, S. H., Boxrud, D. J., Davis, C. C., Kreiswirth, B. N. & Schlievert, P. M. (2003). Comparative molecular analysis of community- or hospital-acquired methicillin-resistant *Staphylococcus aureus. Antimicrob Agents Chemother*, *47*, 196-203.

Filipowicz, N., Kaminski, M., Kurlenda, J., Asztemborska, M. & Ochocka, J. R. (2003). Antibacterial and antifungal activity of juniper berry oil and its selected components. *Phytother Res*., *17*, 227-231.

Finch, R (2006). Gram-positive infections: lessons learnt and novel solutions. *Clin Microbiol Infect, 12*, 3-8.

Foster, T. J. & Hook, M. (1998). Surface protein adhesins of *Staphylococcus aureus. Trends Microbiol, 6*, 484-488.

Fowler, V. G., Jr.., Justice, A., Moore, C., Benjamin, D. K., Jr., Woods, C. W., Campbell, S., Reller, L. B., Corey, G. R., Day, N. P. & Peacock, S. J (2005a). Risk factors for hematogenous complications of intravascular catheter-associated *Staphylococcus aureus* bacteremia. *Clin Infect Dis*., *40*, 695-703.

Fowler, V. G., Jr., Miro, J. M., Hoen, B., Cabell, C. H., Abrutyn, E., Rubinstein, E., Corey, G. R., Spelman, D., Bradley, S. F., Barsic, B., Pappas, P. A., Anstrom, K. J., Wray, D., Fortes, C. Q., Anguera, I., Athan, E., Jones, P., Van Der Meer, J. T., Elliott, T. S., Levine,

D. P. & Bayer, A. S. (2005b). *Staphylococcus aureus* endocarditis: a consequence of medical progress. *JAMA*, *293*, 3012-3021.

Francis, J. S., Doherty, M. C., Lopatin, U., Johnston, C. P., Sinha, G., Ross, T., Cai, M., Hansel, N. N., Perl, T., Ticehurst, J. R., Carroll, K., Thomas, D. L., Nuermberger, E. & Bartlett, J. G. (2005). Severe community-onset pneumonia in healthy adults caused by methicillin-resistant *Staphylococcus aureus* carrying the Panton-Valentine leukocidin genes. *Clin Infect Dis.*, *40*, 100-107.

Frenay, H. M., Bunschoten, A. E., Schouls, L. M., Van Leeuwen, W. J., Vandenbroucke-Grauls, C. M., Verhoef, J., Mooi, F. R. (1996). Molecular typing of methicillin-resistant *Staphylococcus aureus* on the basis of protein A gene polymorphism. *Eur J Clin Microbiol Infect Dis.*, *15*, 60-64.

Garnier, F., Tristan, A., Francois, B., Etienne, J., Age-Corre, M., Martin, C., Liassine, N., Wannet, W., Denis, F., Ploy, M. C. (2006). Pneumonia and new methicillin-resistant *Staphylococcus aureus* clone. *Emerg Infect Dis.*, *12*, 498-500.

Gaudreau, M. C., Lacasse, P. & Talbot, B. G. (2007). Protective immune responses to a multi-gene DNA vaccine against *Staphylococcus aureus*. *Vaccine*, *25*, 814-824.

Gescher, D. M., Kovacevic, D., Schmiedel, D., Siemoneit, S., Mallmann, C., Halle, E., Gobel, U. B. & Moter, A. (2008). Fluorescence in situ hybridisation (FISH) accelerates identification of Gram-positive cocci in positive blood cultures. *Int J Antimicrob Agents, 32 Suppl 1,* S51-S59.

Ghiselli, R., Giacometti, A., Cirioni, O., Mocchegiani, F., Silvestri, C., Orlando, F., Kamysz, W., Licci, A., Nadolski, P., Della, V. A., Lukasiak, J., Scalise, G. & Saba, V. (2007). Pretreatment with the protegrin IB-367 affects Gram-positive biofilm and enhances the therapeutic efficacy of linezolid in animal models of central venous catheter infection. *JPEN J Parenter Enteral Nutr.*, *31*, 463-468.

Giacometti, A., Cirioni, O., Barchiesi, F., Scalise, G. (2000). In-vitro activity and killing effect of polycationic peptides on methicillin-resistant *Staphylococcus aureus* and interactions with clinically used antibiotics. *Diagn Microbiol Infect Dis.*, *38*, 115-118.

Giacometti, A., Cirioni, O., Ghiselli, R., Bergnach, C., Orlando, F., D'Amato, G., Mocchegiani, F., Silvestri, C., Del Prete, M. S., Skerlavaj, B., Saba, V., Zanetti, M., & Scalise, G (2004). The antimicrobial peptide BMAP-28 reduces lethality in mouse models of staphylococcal sepsis. *Crit Care Med.*, *32*, 2485-2490.

Giacometti, A., Cirioni, O., Ghiselli, R., Orlando, F., Silvestri, C., Ranzone, G., Testa, I., Mocchegiani, F., Della, V. A., Saba, V., Scaloni, A., Scalise, G. (2006). Distinctin improves the efficacies of glycopeptides and betalactams against staphylococcal biofilm in an experimental model of central venous catheter infection. *J Biomed Mater Res A, 81A*, 233-239.

Goldstein, F. (2007). The potential clinical impact of low-level antibiotic resistance in *Staphylococcus aureus*. *J Antimicrob Chemother, 59*, 1-4.

Gonzalez, V., Padilla, E., Gimenez, M., Vilaplana, C., Perez, A., Fernandez, G., Quesada, M. D., Pallares, M. A. & Ausina, V. (2004). Rapid diagnosis of *Staphylococcus aureus* bacteremia using *S. aureus* PNA FISH. *Eur J Clin Microbiol Infect Dis, 23,* 396-398.

Gosbell, I. B. (2005). Epidemiology., clinical features and management of infections due to community methicillin-resistant *Staphylococcus aureus* (cMRSA). *Intern Med J., 35* Suppl 2, S120-S135.

Gottlieb, G. S., Fowler, V. G., Jr., Kong, L. K., McClelland, R. S., Gopal, A. K., Marr, K. A., Li, J., Sexton, D. J., Glower, D. & Corey, G. R (2000). *Staphylococcus aureus* bacteremia in the surgical patient: a prospective analysis of 73 postoperative patients who developed *Staphylococcus aureus* bacteremia at a tertiary care facility. *J Am Coll Surg., 190*, 50-57.

Graham, PL., III., Lin, S. X. & Larson, E. L (2006). A U.S. population-based survey of *Staphylococcus aureus* colonization. *Ann Intern Med., 144*, 318-325.

Gravet, A., Couppie, P., Meunier, O., Clyti, E., Moreau, B., Pradinaud, R., Monteil, H. & Prevost, G. (2001). *Staphylococcus aureus* isolated in cases of impetigo produces both epidermolysin A or B and LukE-LukD in 78% of 131 retrospective and prospective cases. *J Clin Microbiol., 39*, 4349-4356.

Grinholc, M., Kawiak, A., Kurlenda, J., Graczyk, A. & Bielawski, K. P. (2008a). Photodynamic effect of protoporphyrin diarginate (PPArg2) on methicillin-resistant *Staphylococcus aureus* and human dermal fibroblasts. *Acta Biochim Pol., 55*, 85-90.

Grinholc, M., Szramka, B., Kurlenda, J., Graczyk, A. & Bielawski, K. P. (2008b). Bactericidal effect of photodynamic inactivation against methicillin-resistant and methicillin-susceptible *Staphylococcus aureus* is strain-dependent. *J Photochem Photobiol B, 90*, 57-63.

Grinholc, M., Szramka, B., Olender, K. & Graczyk, A. (2007a). Bactericidal effect of photodynamic therapy against methicillin-resistant *Staphylococcus aureus* strain with the use of various porphyrin photosensitizers. *Acta Biochim Pol., 54*, 665-670.

Grinholc, M., Wegrzyn, G. & Kurlenda, J. (2007b). Evaluation of biofilm production and prevalence of the icaD gene in methicillin-resistant and methicillin-susceptible *Staphylococcus aureus* strains isolated from patients with nosocomial infections and carriers. *FEMS Immunol Med Microbiol., 50*, 375-379.

Grundmeier, M., Hussain, M., Becker, P., Heilmann, C., Peters, G. & Sinha, B. (2004). Truncation of fibronectin-binding proteins in *Staphylococcus aureus* strain Newman leads to deficient adherence and host cell invasion due to loss of the cell wall anchor function. *Infect Immun, 72*, 7155-7163.

Hall, A. E., Domanski, P. J., Patel, P. R., Vernachio, J. H., Syribeys, P. J., Gorovits, E. L., Johnson, M. A., Ross, J. M., Hutchins, J. T. & Patti, J. M. (2003). Characterization of a protective monoclonal antibody recognizing *Staphylococcus aureus* MSCRAMM protein clumping factor A. *Infect Immun, 71*, 6864-6870.

Hamblin, M. R. & Hasan, T. (2004). Photodynamic therapy: a new antimicrobial approach to infectious disease? *Photochem Photobiol Sci., 3*, 436-450.

Hamblin, M. R., O'Donnell, D. A., Murthy, N., Rajagopalan, K., Michaud, N., Sherwood, M. E. & Hasan, T. (2002). Polycationic photosensitizer conjugates: effects of chain length and Gram classification on the photodynamic inactivation of bacteria. *J Antimicrob Chemother, 49*, 941-951.

Healy, C. M., Palazzi, D. L., Edwards, M. S., Campbell, J. R. & Baker, C. J (2004). Features of invasive staphylococcal disease in neonates. *Pediatrics, 114*, 953-961.

Hedin, G. & Fang, H. (2007). Epidemiology of methicillin-resistant *Staphylococcus aureus* in Southern Stockholm., 2000-2003. *Microb Drug Resist., 13*, 241-250.

Hiemstra, P. S., Fernie-King, B. A., McMichael, J., Lachmann, P. J. & Sallenave, J. M. (2004). Antimicrobial peptides: mediators of innate immunity as templates for the development of novel anti-infective and immune therapeutics. *Curr Pharm Des., 10*, 2891-2905.

Higuchi, W., Takano, T., Teng, L. J. & Yamamoto, T. (2008). Structure and specific detection of staphylococcal cassette chromosome *mec* type VII. *Biochem Biophys Res Commun*, *377*, 752-756.

Hiramatsu, K. (2001). Vancomycin-resistant *Staphylococcus aureus*: a new model of antibiotic resistance. *Lancet Infect Dis.*, *1*, 147-155.

Hiramatsu, K., Hanaki, H., Ino, T., Yabuta, K., Oguri, T. & Tenover, F. C. (1997). Methicillin-resistant *Staphylococcus aureus* clinical strain with reduced vancomycin susceptibility. *J Antimicrob Chemother*, *40*, 135-136.

Holfelder, M., Eigner, U., Turnwald, A. M., Witte, W., Weizenegger, M. & Fahr, A. (2006). Direct detection of methicillin-resistant *Staphylococcus aureus* in clinical specimens by a nucleic acid-based hybridisation assay. *Clin Microbiol Infect*, *12*, 1163-1167.

Holtfreter, S., Bauer, K., Thomas, D., Feig, C., Lorenz, V., Roschack, K., Friebe, E., Selleng, K., Lovenich, S., Greve, T., Greinacher, A., Panzig, B., Engelmann, S., Lina, G. & Broker, B. M. (2004). egc-Encoded superantigens from *Staphylococcus aureus* are neutralized by human sera much less efficiently than are classical staphylococcal enterotoxins or toxic shock syndrome toxin. *Infect Immun*, *72*, 4061-4071.

Hookey, J. V., Richardson, J. F. & Cookson, B. D. (1998). Molecular typing of *Staphylococcus aureus* based on PCR restriction fragment length polymorphism and DNA sequence analysis of the coagulase gene. *J Clin Microbiol*, *36*, 1083-1089.

Howden, B. P., Ward, P. B., Charles, P. G., Korman, T. M., Fuller, A., Du, C. P., Grabsch, E. A., Roberts, S. A., Robson, J., Read, K., Bak, N., Hurley, J., Johnson, P. D., Morris, A. J., Mayall, B. C. & Grayson, M. L. (2004). Treatment outcomes for serious infections caused by methicillin-resistant *Staphylococcus aureus* with reduced vancomycin susceptibility. *Clin Infect Dis.*, *38*, 521-528.

Hryniewicz, W. (1999). Epidemiology of MRSA. *Infection*, *27* Suppl 2, S13-S16.

Hu, D. L., Omoe, K., Narita, K., Cui, J. C., Shinagawa, K. & Nakane, A. (2006). Intranasal vaccination with a double mutant of staphylococcal enterotoxin C provides protection against *Staphylococcus aureus* infection. *Microbes Infect*, *8*, 2841-2848.

Huang, Y. H., Tseng, S. P., Hu, J. M., Tsai, J. C., Hsueh, P. R. & Teng, L. J. (2007). Clonal spread of SCC*mec* type IV methicillin-resistant *Staphylococcus aureus* between community and hospital. *Clin Microbiol Infect*, *13*, 717-724.

Hudson, M. C., Ramp, W. K. & Frankenburg, K. P (1999). *Staphylococcus aureus* adhesion to bone matrix and bone-associated biomaterials. *FEMS Microbiol Lett.*, *173*, 279-284.

Huletsky, A., Lebel, P., Picard, F. J., Bernier, M., Gagnon, M., Boucher, N. & Bergeron, M. G. (2005). Identification of methicillin-resistant *Staphylococcus aureus* carriage in less than 1 hour during a hospital surveillance program. *Clin Infect Dis.*, *40*, 976-981.

Hussain, M., Becker, K., Von, E. C., Schrenzel, J., Peters, G. & Herrmann, M. (2001). Identification and characterization of a novel 38.5-kilodalton cell surface protein of *Staphylococcus aureus* with extended-spectrum binding activity for extracellular matrix and plasma proteins. *J Bacteriol.*, *183*, 6778-6786.

Issartel, B., Tristan, A., Lechevallier, S., Bruyere, F., Lina, G., Garin, B., Lacassin, F., Bes, M., Vandenesch, F. & Etienne, J. (2005). Frequent carriage of Panton-Valentine leucocidin genes by *Staphylococcus aureus* isolates from surgically drained abscesses. *J Clin Microbiol.*, *43*, 3203-3207.

Ito, T. & Hiramatsu, K. (2003). [MRSA (methicillin-resistant *Staphylococcus aureus*)]. *Nippon Rinsho*, *61* Suppl 3, 164-170.

Ito, T., Katayama, Y., Asada, K., Mori, N., Tsutsumimoto, K., Tiensasitorn, C. & Hiramatsu, K (2001). Structural comparison of three types of staphylococcal cassette chromosome *mec* integrated in the chromosome in methicillin-resistant *Staphylococcus aureus*. *Antimicrob Agents Chemother*, *45*, 1323-1336.

Ito, T., Okuma, K., Ma, X. X., Yuzawa, H. & Hiramatsu, K. (2003). Insights on antibiotic resistance of *Staphylococcus aureus* from its whole genome: genomic island SCC. *Drug Resist Updat, 6*, 41-52.

James, P. R. (2001). Isolated pericardial empyema secondary to *Staphylococcus aureus*. *Eur J Intern Med., 12,* 377-379.

Jarraud, S., Mougel, C., Thioulouse, J., Lina, G., Meugnier, H., Forey, F., Nesme, X., Etienne, J. & Vandenesch, F. (2002). Relationships between *Staphylococcus aureus* genetic background, virulence factors, *agr* groups (alleles), and human disease. *Infect Immun*, *70*, 631-641.

Jin, T., Bokarewa, M. & Tarkowski, A. (2005). Urokinase-type plasminogen activator, an endogenous antibiotic. *J Infect Dis*., *192*, 429-437.

Johnson, G., Millar, M. R., Matthews, S., Skyrme, M., Marsh, P., Barringer, E., O'Hara, S., & Wilks, M. (2006). *Evaluation of BacLite Rapid MRSA,* a rapid culture based screening test for the detection of ciprofloxacin and methicillin resistant *S. aureus* (MRSA) from screening swabs, p 83.

Jori, G. & Brown, S. B. (2004). Photosensitized inactivation of microorganisms. *Photochem Photobiol Sci*., *3*, 403-405.

Josefsson, E., Hartford, O., O'Brien, L., Patti, J. M., Foster, T. (2001). Protection against experimental *Staphylococcus aureus* arthritis by vaccination with clumping factor A, a novel virulence determinant. *J Infect Dis*., *184*, 1572-1580.

Josefsson, E., McCrea, K. W., Ni, E. D., O'Connell, D., Cox, J., Hook, M. & Foster, T. J. (1998). Three new members of the serine-aspartate repeat protein multigene family of *Staphylococcus aureus*. *Microbiology*, *144* (Pt 12), 3387-3395.

Kahl, B., Herrmann, M., Everding, A. S., Koch, H. G., Becker, K., Harms, E., Proctor, R. A., Peters, G. (1998). Persistent infection with small colony variant strains of *Staphylococcus aureus* in patients with cystic fibrosis. *J Infect Dis*., *177*, 1023-1029.

Kaiser, A. M., Schultsz, C., Kruithof, G. J., bets-Ossenkopp, Y., Vandenbroucke-Grauls, C (2004). Carriage of resistant microorganisms in repatriates from foreign hospitals to The Netherlands. *Clin Microbiol Infect. 10*, 972-979.

Kalmeijer, M. D., Van Nieuwland-Bollen, E., Bogaers-Hofman, D., De Baere, G. A. (2000). Nasal carriage of *Staphylococcus aureus* is a major risk factor for surgical-site infections in orthopedic surgery. *Infect Control Hosp Epidemiol. 21*, 319-323.

Kampf, G., Meyer, B. & Goroncy-Bermes, P. (2003). Comparison of two test methods for the determination of sufficient antimicrobial activity of three commonly used alcohol-based hand rubs for hygienic hand disinfection. *J Hosp Infect*, *55*, 220-225.

Kaneko, J. & Kamio, Y. (2004). Bacterial two-component and hetero-heptameric pore-forming cytolytic toxins: structures, pore-forming *mec*hanism, and organization of the genes. *Biosci Biotechnol Biochem, 68*, 981-1003.

Karchmer, A. W. (2006). From theory to practice: resistance in *Staphylococcus aureus* and new treatments. *Clin Microbiol Infect*, *12*, 15-21.

Karchmer, T. B., Durbin, L. J., Simonton, B. M. & Farr, B. M. (2002). Cost-effectiveness of active surveillance cultures and contact/droplet precautions for control of methicillin-resistant *Staphylococcus aureus*. *J Hosp Infect*, *51*, 126-132.

Karlsson, A. & Arvidson, S. (2002). Variation in extracellular protease production among clinical isolates of *Staphylococcus aureus* due to different levels of expression of the protease repressor sarA. *Infect Immun*, *70*, 4239-4246.

Karynski, M., Sabat, A. J., Empel, J. & Hryniewicz, W. (2008b). Molecular surveillance of methicillin-resistant *Staphylococcus aureus* by multiple-locus variable number tandem repeat fingerprinting (formerly multiple-locus variable number tandem repeat analysis) and spa typing in a hierarchic approach. *Diagn Microbiol Infect Dis.*, *62*, 255-262.

Karynski, M., Sabat, A. J., Empel, J. & Hryniewicz, W. (2008a). Molecular surveillance of methicillin-resistant *Staphylococcus aureus* by multiple-locus variable number tandem repeat fingerprinting (formerly multiple-locus variable number tandem repeat analysis) and spa typing in a hierarchic approach. *Diagn Microbiol Infect Dis.*, *62*, 255-262.

Katayama, Y., Ito, T. & Hiramatsu, K. (2000). A new class of genetic element, staphylococcus cassette chromosome *mec*, encodes methicillin resistance in *Staphylococcus aureus*. *Antimicrob Agents Chemother*, *44*, 1549-1555.

Kazakova, S. V., Hageman, J. C., Matava, M., Srinivasan, A., Phelan, L., Garfinkel, B., Boo, T, McAllister, S., Anderson, J., Jensen, B., Dodson, D., Lonsway, D., McDougal, LK., Arduino, M., Fraser, V. J., Killgore, G., Tenover, F. C., Cody, S. & Jernigan, D. B (2005). A clone of methicillin-resistant *Staphylococcus aureus* among professional football players. *N Engl J Med.*, *352*, 468-475.

Keilman, V. K. (1979). Prevention of purulent complications in the combined treatment of open fractures of crus bones. *Sov Med.*, 52-55.

Kerro-Dego, O., Prysliak, T., Potter, A. A. & Perez-Casal, J. (2006). DNA-protein immunization against the GapB and GapC proteins of a mastitis isolate of *Staphylococcus aureus*. *Vet Immunol Immunopathol*, *113*, 125-138.

Kipp, F., Ziebuhr, W., Becker, K., Krimmer, V., Hobeta, N., Peters, G. & Von E. C. (2003). Detection of *Staphylococcus aureus* by 16S rRNA directed in situ hybridisation in a patient with a brain abscess caused by small colony variants. *J Neurol Neurosurg Psychiatry, 74*, 1000-1002.

Kokai-Kun, J. F., Walsh, S. M., Chanturiya, T. & Mond, J. J. (2003). Lysostaphin cream eradicates *Staphylococcus aureus* nasal colonization in a cotton rat model. *Antimicrob Agents Chemother*, *47*, 1589-1597.

Komerik, N., Nakanishi, H., MacRobert, A. J., Henderson, B., Speight, P. & Wilson, M. (2003). In vivo killing of Porphyromonas gingivalis by toluidine blue-mediated photosensitization in an animal model. *Antimicrob Agents Chemother*, *47*, 932-940.

Korem, M., Sheoran, A. S., Gov, Y., Tzipori, S., Borovok, I. & Balaban, N. (2003). Characterization of RAP., a quorum sensing activator of *Staphylococcus aureus*. *FEMS Microbiol Lett.*, *223,* 167-175.

Kourbatova, E. V., Halvosa, J. S., King, M. D., Ray, S. M., White, N. & Blumberg, H. M. (2005). Emergence of community-associated methicillin-resistant *Staphylococcus aureus* USA 300 clone as a cause of health care-associated infections among patients with prosthetic joint infections. *Am J Infect Control*, *33*, 385-391.

Kowalski, T. J., Berbari, E. F. & Osmon, D. R. (2005). Epidemiology., treatment, and prevention of community-acquired methicillin-resistant *Staphylococcus aureus* infections. *Mayo Clin Proc., 80,* 1201-1207.

Krziwanek, K., Luger, C., Sammer, B., Stumvoll, S., Stammler, M., Metz-Gercek, S. & Mittermayer, H. (2007). PVL-positive MRSA in Austria. *Eur J Clin Microbiol Infect Dis., 26,* 931-935.

Krzyszton-Russjan, J., Gniadkowski, M., Polowniak-Pracka, H., Hagmajer, E., & Hryniewicz, W. (2002). The first *Staphylococcus aureus* isolates with reduced susceptibility to vancomycin in Poland. *J Antimicrob Chemother, 50,* 1065-1069.

Kuint, J., Barzilai, A., Regev-Yochay, G., Rubinstein, E., Keller, N. & Maayan-Metzger, A. (2007). Comparison of community-acquired methicillin-resistant *Staphylococcus aureus* bacteremia to other staphylococcal species in a neonatal intensive care unit. *Eur J Pediatr, 166,* 319-325.

Kuklin, N. A., Clark, D. J., Secore, S., Cook, J., Cope, L. D., McNeely, T., Noble, L., Brown, M. J., Zorman, J. K., Wang, X. M., Pancari, G., Fan, H., Isett, K., Burgess, B., Bryan, J., Brownlow, M., George, H., Meinz, M., Liddell, M. E., Kelly, R., Schultz, L., Montgomery, D., Onishi, J., Losada, M., Martin, M., Ebert, T., Tan, C. Y., Schofield, T. L., Nagy, E., Meineke, A., Joyce, J. G., Kurtz, M. B., Caulfield, M. J., Jansen, K. U., McClements, W. & Anderson, A. S. (2006). A novel *Staphylococcus aureus* vaccine: iron surface determinant B induces rapid antibody responses in rhesus macaques and specific increased survival in a murine *S. aureus* sepsis model. *Infect Immun, 74,* 2215-2223.

Kurlenda, J., Grinholc, M., Jasek, K. & Wegrzyn, G. (2007a). RAPD typing of methicillin-resistant *Staphylococcus aureus*: a 7-year experience in a Polish hospital. *Med Sci Monit, 13,* MT13-MT18.

Kurlenda, J., Grinholc, M. & Wegrzyn, G. (2007b). Presence of *cna, emp* and *pls* genes and pathogenicity of methicillin-resistant *Staphylococcus aureus* strains. *World J Microbiol Biotech, 24,* 591-594.

Kurlenda, J., Grinholc, M., Krzyszton-Russjan, J. & Wisniewska, K. (2009). Epidemiological investigation of nosocomial outbreak of staphylococcal skin diseases in neonatal ward. *Antonie Van Leeuwenhoek, 95,* 387-394.

Kurt, K., Alderborn, A., Nilsson, M., Strommenger, B., Witte, W. & Nubel, U. (2009). Multiplexed genotyping of methicillin-resistant *Staphylococcus aureus* isolates by use of padlock probes and tag microarrays. *J Clin Microbiol, 47,* 577-585.

Ladhani, S. & Garbash, M. (2005). Staphylococcal skin infections in children: rational drug therapy recommendations. *Paediatr Drugs., 7,* 77-102.

Larsen, A. R., Stegger, M., Bocher, S., Sorum, M., Monnet, D. L. & Skov, R. L. (2009). Emergence and characterization of community-associated methicillin-resistant Staphyloccocus aureus infections in Denmark., 1999 to 2006. *J Clin Microbiol, 47,* 73-78.

Le, T., I., Mariani-Kurkdjian, P., Collignon, A., Gravet, A., Clermont, O., Brahimi, N., Gaudelus, J., Aujard, Y., Navarro, J., Beaufils, F. & Bingen, E. (2001). Breast milk transmission of a Panton-Valentine leukocidin-producing *Staphylococcus aureus* strain causing infantile pneumonia. *J Clin Microbiol, 39,* 728-729.

LeClaire, R. D., Hunt, R. E. & Bavari, S. (2002). Protection against bacterial superantigen staphylococcal enterotoxin B by passive vaccination. *Infect Immun, 70,* 2278-2281.

Lee, L. Y., Liang, X., Hook, M. & Brown, E. L. (2004). Identification and characterization of the C3 binding domain of the *Staphylococcus aureus* extracellular fibrinogen-binding protein (Efb). *J Biol Chem., 279*, 50710-50716.

Levy, S. B. (2005). Antibiotic resistance-the problem intensifies. *Adv Drug Deliv Rev., 57*, 1446-1450.

Lew, D. P. & Waldvogel, F. A. (2004). Osteomyelitis. *Lancet, 364*, 369-379.

Livermore, D. M. (2006). Can beta-lactams be re-engineered to beat MRSA? *Clin Microbiol Infect, 12* Suppl 2, 11-16.

Looney, W. J. (2000). Small-colony variants of *Staphylococcus aureus. Br J Biomed Sci., 57*, 317-322.

Lorber, B. (2006). Update in infectious diseases. *Ann Intern Med., 145*, 354-360.

Lowy, F. D. (2000). Is *Staphylococcus aureus* an intracellular pathogen? *Trends Microbiol, 8*, 341-343.

Luczak-Kadlubowska, A., Sabat, A., Tambic-Andrasevic, A., Payerl-Pal, M., Krzyszton-Russjan, J. & Hryniewicz, W. (2008). Usefulness of multiple-locus VNTR fingerprinting in detection of clonality of community- and hospital-acquired *Staphylococcus aureus* isolates. *Antonie Van Leeuwenhoek, 94*, 543-553.

Maira-Litran, T., Kropec, A., Goldmann, D. & Pier, G. B. (2004). Biologic properties and vaccine potential of the staphylococcal poly-N-acetyl glucosamine surface polysaccharide. *Vaccine, 22*, 872-879.

Maira-Litran, T., Kropec, A., Goldmann, D. A. & Pier, G. B. (2005). Comparative opsonic and protective activities of *Staphylococcus aureus* conjugate vaccines containing native or deacetylated Staphylococcal Poly-N-acetyl-beta-(1-6)-glucosamine. *Infect Immun, 73*, 6752-6762.

Maisch, T., Bosl, C., Szeimies, R. M., Lehn, N. & Abels, C. (2005). Photodynamic effects of novel XF porphyrin derivatives on prokaryotic and eukaryotic cells. *Antimicrob Agents Chemother, 49*, 1542-1552.

Makris, G., Wright, J. D., Ingham, E. & Holland, K. T. (2004). The hyaluronate lyase of *Staphylococcus aureus* - a virulence factor? *Microbiology, 150*, 2005-2013.

Manzur, A., Vidal, M., Pujol, M., Cisnal, M., Hornero, A., Masuet, C., Pena, C., Gudiol, F. & Ariza, J. (2007). Predictive factors of meticillin resistance among patients with *Staphylococcus aureus* bloodstream infections at hospital admission. *J Hosp Infect, 66*, 135-141.

Maree, C. L., Daum, R. S., Boyle-Vavra, S., Matayoshi, K. & Miller, L. G. (2007). Community-associated methicillin-resistant *Staphylococcus aureus* isolates causing healthcare-associated infections. *Emerg Infect Dis., 13*, 236-242.

Marraffini, L. A. & Schneewind, O. (2005). Anchor structure of staphylococcal surface proteins. V. Anchor structure of the sortase B substrate IsdC. *J Biol Chem., 280*, 16263-16271.

Marshall, C., Wolfe, R., Kossmann, T., Wesselingh, S., Harrington, G. & Spelman, D. (2004). Risk factors for acquisition of methicillin-resistant *Staphylococcus aureus* (MRSA) by trauma patients in the intensive care unit. *J Hosp Infect, 57*, 245-252.

Matsuzaki, S., Yasuda, M., Nishikawa, H., Kuroda, M., Ujihara, T., Shuin, T., Shen, Y., Jin, Z., Fujimoto, S., Nasimuzzaman, MD., Wakiguchi, H., Sugihara, S., Sugiura, T., Koda, S., Muraoka, A. & Imai, S. (2003). Experimental protection of mice against lethal

Staphylococcus aureus infection by novel bacteriophage phi MR11. *J Infect Dis.*, *187*, 613-624.

Mazmanian, S. K., Skaar, E. P., Gaspar, A. H., Humayun, M., Gornicki, P., Jelenska, J., Joachmiak, A., Missiakas, D. M. & Schneewind, O. (2003). Passage of heme-iron across the envelope of *Staphylococcus aureus*. *Science*, *299*, 906-909.

McCaig, L. F., McDonald, L. C., Mandal, S. & Jernigan, D. B. (2006). *Staphylococcus aureus*-associated skin and soft tissue infections in ambulatory care. *Emerg Infect Dis.*, *12*, 1715-1723.

McKenney, D., Pouliot, K. L., Wang, Y., Murthy, V., Ulrich, M., Doring, G., Lee, J. C., Goldmann, D. A. & Pier, G. B. (1999). Broadly protective vaccine for *Staphylococcus aureus* based on an in vivo-expressed antigen. *Science*, *284*, 1523-1527.

Melzer, M., Eykyn, S. J., Gransden, W. R., & Chinn, S. (2003). Is methicillin-resistant *Staphylococcus aureus* more virulent than methicillin-susceptible *S. aureus*? A comparative cohort study of British patients with nosocomial infection and bacteremia. *Clin Infect Dis.*, *37*, 1453-1460.

Michel, A., Agerer, F., Hauck, C. R., Herrmann, M., Ullrich, J., Hacker, J. & Ohlsen, K (2006). Global regulatory impact of ClpP protease of *Staphylococcus aureus* on regulons involved in virulence, oxidative stress response, autolysis, and DNA repair. *J Bacteriol*, *188*, 5783-5796.

Mohr, H. & Redecker-Klein, A. (2003). Inactivation of pathogens in platelet concentrates by using a two-step procedure. *Vox Sang.*, *84,* 96-104.

Moise-Broder, P. A., Sakoulas, G., Eliopoulos, G. M., Schentag, J. J., Forrest, A. & Moellering, R. C., Jr. (2004). Accessory gene regulator group II polymorphism in methicillin-resistant *Staphylococcus aureus* is predictive of failure of vancomycin therapy. *Clin Infect Dis.*, *38*, 1700-1705.

Monecke, S., Slickers, P., Ellington, M. J., Kearns, A. M. & Ehricht, R. (2007). High diversity of Panton-Valentine leukocidin-positive, methicillin-susceptible isolates of *Staphylococcus aureus* and implications for the evolution of community-associated methicillin-resistant *S. aureus*. *Clin Microbiol Infect*, *13*, 1157-1164.

Moran, G. J., Krishnadasan, A., Gorwitz, R. J., Fosheim, G. E., McDougal, L. K., Carey, RB., & Talan, D. A. (2006). Methicillin-resistant *S. aureus* infections among patients in the emergency department. *N Engl J Med.*, *355*, 666-674.

Mulvey, M. R., MacDougall, L., Cholin, B., Horsman, G., Fidyk, M. & Woods, S. (2005). Community-associated methicillin-resistant *Staphylococcus aureus*, Canada. *Emerg Infect Dis.*, *11*, 844-850.

Murchan, S., Kaufmann, M. E., Deplano, A., De, R. R., Struelens, M., Zinn, C. E., Fussing, V., Salmenlinna, S., Vuopio-Varkila, J., El, S. N, Cuny, C., Witte, W., Tassios, P. T., Legakis, N., Van, L. W., Van, B. A., Vindel, A., Laconcha, I., Garaizar, J., Haeggman, S., Olsson-Liljequist, B., Ransjo, U., Coombes, G. & Cookson, B. (2003). Harmonization of pulsed-field gel electrophoresis protocols for epidemiological typing of strains of methicillin-resistant *Staphylococcus aureus*: a single approach developed by consensus in 10 European laboratories and its application for tracing the spread of related strains. *J Clin Microbiol, 41*, 1574-1585.

Murray, R. J. (2005). Recognition and management of *Staphylococcus aureus* toxin-mediated disease. *Intern Med J., 35* Suppl 2, S106-S119.

Naffa, R. G., Bdour, S. M., Migdadi, H. M. & Shehabi, A. A. (2006). Enterotoxicity and genetic variation among clinical *Staphylococcus aureus* isolates in Jordan. *J Med Microbiol*, *55*, 183-187.

Naimi, T. S., LeDell, K. H., Como-Sabetti, K., Borchardt, S. M., Boxrud, D. J., Etienne, J., Johnson, S. K., Vandenesch, F., Fridkin, S., O'Boyle, C., Danila, R. N. & Lynfield, R (2003). Comparison of community- and health care-associated methicillin-resistant *Staphylococcus aureus* infection. *JAMA*, *290*, 2976-2984.

Narita, K., Hu, D. L., Tsuji, T. & Nakane, A. (2008). Intranasal immunization of mutant toxic shock syndrome toxin 1 elicits systemic and mucosal immune response against *Staphylococcus aureus* infection. *FEMS Immunol Med Microbiol*, *52*, 389-396.

Narita, S., Kaneko, J., Chiba, J., Piemont, Y., Jarraud, S., Etienne, J. & Kamio, Y. (2001). Phage conversion of Panton-Valentine leukocidin in *Staphylococcus aureus*: molecular analysis of a PVL-converting phage, phiSLT. *Gene*, *268*, 195-206.

Nascimento, J. S., Ceotto, H., Nascimento, S. B., Giambiagi-Demarval, M., Santos, K. R., & Bastos, M. C. (2006). Bacteriocins as alternative agents for control of multiresistant staphylococcal strains. *Lett Appl Microbiol*, *42*, 215-221.

Nema, V., *Ag*rawal, R., Kamboj, D. V., Goel, A. K. & Singh L. (2007). Isolation and characterization of heat resistant enterotoxigenic *Staphylococcus aureus* from a food poisoning outbreak in Indian subcontinent. *Int J Food Microbiol*, *117*, 29-35.

Neut, D., Van Der Mei, H. C., Bulstra, S. K. & Busscher, H. J. (2007). The role of small-colony variants in failure to diagnose and treat biofilm infections in orthopedics. *Acta Orthop*, *78*, 299-308.

Nilsson, P., Alexandersson, H. & Ripa, T. (2005). Use of broth enrichment and real-time PCR to exclude the presence of methicillin-resistant *Staphylococcus aureus* in clinical samples: a sensitive screening approach. *Clin Microbiol Infect*, *11*, 1027-1034.

Nulens, E., Stobberingh, E. E., Smeets, E., van Dessel, H., Welling, M. A., Sebastian, S., van Tiel, F. H., Beisser, P. S. & Deurenberg, R. H. (2009). Genetic diversity of methicillin-resistant *Staphylococcus aureus* in a tertiary hospital in The Netherlands between 2002 and 2006. *Eur J Clin Microbiol Infect Dis, 28,* 631-639.

Oh, S., Kim, S. H., Ko, Y., Sim, J. H., Kim, K. S., Lee, S. H., Park, S. & Kim, Y. J. (2006). Effect of bacteriocin produced by Lactococcus sp. HY 449 on skin-inflammatory bacteria. *Food Chem Toxicol*, *44*, 1184-1190.

Olive, D. M. & Bean, P. (1999). Principles and applications of methods for DNA-based typing of microbial organisms. *J Clin Microbiol*, *37*, 1661-1669.

Oliveira, D. C., Tomasz, A. & De, L. H. (2002). Secrets of success of a human pathogen: molecular evolution of pandemic clones of meticillin-resistant *Staphylococcus aureus*. *Lancet Infect Dis*, *2*, 180-189.

Omoe K., Ishikawa, M., Shimoda, Y., Hu D. L., Ueda, S. & Shinagawa, K, (2002). Detection of seg, seh, and sei genes in *Staphylococcus aureus* isolates and determination of the enterotoxin productivities of *S. aureus* isolates Harboring seg, seh, or sei genes. *J Clin Microbiol*, 40, 857-862.

Ornskov, D., Kolmos, B., Bendix, H. P., Nederby, N. J., Brandslund, I. & Schouenborg, P. (2008). Screening for methicillin-resistant *Staphylococcus aureus* in clinical swabs using a high-throughput real-time PCR-based method. *Clin Microbiol Infect*, *14*, 22-28.

Otto, M. (2008). Targeted immunotherapy for staphylococcal infections : focus on anti-MSCRAMM antibodies. *BioDrugs*, *22*, 27-36.

Pan, E. S., Diep, B. A., Charlebois, E. D., Auerswald, C., Carleton, H. A., Sensabaugh, G. F. & Perdreau-Remington, F. (2005). Population dynamics of nasal strains of methicillin-resistant *Staphylococcus aureus*--and their relation to community-associated disease activity. *J Infect Dis.*, *192*, 811-818.

Parfitt, T. (2005). Georgia: an unlikely stronghold for bacteriophage therapy. *Lancet 365*, 2166-2167.

Pei, L. (2004). New approaches to develop anti-staphylococcal agents. *Curr Med Chem -Anti Inf Agents,* 267-278.

Peschel, A. (2002). How do bacteria resist human antimicrobial peptides? *Trends Microbiol*, *10*, 179-186

Peschel, A. & Collins, L. V. (2001). Staphylococcal resistance to antimicrobial peptides of mammalian and bacterial origin. *Peptides*, *22*, 1651-1659.

Priest, D. H. & Peacock, J. E., Jr. (2005). Hematogenous vertebral osteomyelitis due to *Staphylococcus aureus* in the adult: clinical features and therapeutic outcomes. *South Med J.*, *98*, 854-862.

Que, Y. A., Haefliger, J. A., Piroth, L., Francois, P., Widmer, E., Entenza, J. M., Sinha, B., Herrmann, M., Francioli, P., Vaudaux, P. & Moreillon, P. (2005). Fibrinogen and fibronectin binding cooperate for valve infection and invasion in *Staphylococcus aureus* experimental endocarditis. *J Exp Med.*, *201*, 1627-1635.

Rayner, C. & Munckhof, W. J. (2005). Antibiotics currently used in the treatment of infections caused by *Staphylococcus aureus*. *Intern Med J.*, *35* Suppl 2, S3-16.

Redpath, M. B., Foster T. J. & Bailey, C. J. (1991). The role of the serine protease active site in the mode of action of epidermolytic toxin of *Staphylococcus aureus*. *FEMS Microbiol Lett.*, *65*, 151-155.

Regev-Yochay, G., Rubinstein, E., Barzilai, A., Carmeli, Y., Kuint, J., Etienne, J., Blech, M., Smollen, G., Maayan-Metzger, A., Leavitt, A., Rahav, G. & Keller, N. (2005). Methicillin-resistant *Staphylococcus aureus* in neonatal intensive care unit. *Emerg Infect Dis.*, *11*, 453-456.

Rennermalm, A., Li, Y. H., Bohaufs, L., Jarstrand, C., Brauner, A., Brennan, F. R. & Flock, J. I. (2001). Antibodies against a truncated *Staphylococcus aureus* fibronectin-binding protein protect against dissemination of infection in the rat. *Vaccine*, *19*, 3376-3383.

Richards, R. G., Harris, L. G., Schneider, E. & Haas, N. (2006). Antiseptics and antibiotics on implants. *Int J Care Injured.*, *37*, 113-116.

Roche, F. M., Downer, R., Keane, F., Speziale, P., Park P. W. & Foster, T. J. (2004). The N-terminal A domain of fibronectin-binding proteins A and B promotes adhesion of *Staphylococcus aureus* to elastin. *J Biol Chem.*, *279*, 38433-38440.

Romanov, V. P., Lebedev, L. R., Tereshchenko, T. A., Nazarikova, N. I. & Masycheva, V. I. (2001). [Photoinactivation of microorganisms in the presence of photosensitizers]. *Zh Mikrobiol Epidemiol Immunobiol*, 24-26.

Romanova, N. A., Brovko, L. Y., Moore, L., Pometun, E., Savitsky, A. P., Ugarova, N. N. & Griffiths, M. W. (2003). Assessment of photodynamic destruction of Escherichia coli O157:H7 and Listeria monocytogenes by using ATP bioluminescence. *Appl Environ Microbiol*, *69*, 6393-6398.

Rossney, A., Morgan, P. & O'Connell, B. (2005). Community-acquired PVL+ MRSA in Ireland: a preliminary report. *Euro Surveill, 10*, E050421.

Ruzin, A. & Novick, R. P. (2000). Equivalence of lauric acid and glycerol monolaurate as inhibitors of signal transduction in *Staphylococcus aureus*. *J Bacteriol*, *182*, 2668-2671.

Sabat, A., Malachowa, N., Miedzobrodzki, J. & Hryniewicz, W. (2006). Comparison of PCR-based methods for typing *Staphylococcus aureus* isolates. *J Clin Microbiol*, *44*, 3804-3807.

Sabat A. J., Wladyka B., Kosowska-Shick K., Grundmann H., Van Dijl J. M., Kowal J., Appelbaum P. C., Dubin A. & Hryniewicz, W. (2008). Polymorphism, genetic exchange and intragenic recombination of the aureolysin gene among *Staphylococcus aureus* strains. *BMC Microbiol*, *8*, 129.

Saiman, L., Cronquist, A., Wu, F., Zhou, J., Rubenstein, D., Eisner, W., Kreiswirth, B. N., & La-Latta, P. (2003). An outbreak of methicillin-resistant *Staphylococcus aureus* in a neonatal intensive care unit. *Infect Control Hosp Epidemiol*, *24*, 317-321.

Sakoulas, G., Eliopoulos, G. M., Moellering, R. C., Jr., Novick, R. P., Venkataraman L., Wennersten, C., DeGirolami, P. C., Schwaber, M. J. & Gold, H. S (2003). *Staphylococcus aureus* accessory gene regulator (*agr*) group II: is there a relationship to the development of intermediate-level glycopeptide resistance? *J Infect Dis.*, *187*, 929-938.

Sandhu K. & Kanwar, A. J. (2004). Generalized bullous impetigo in a neonate. *Pediatr Dermatol*, *21*, 667-669.

Savas, L., Duran, N., Onlen, Y., Savas, N. & Erayman, M. (2005). Prospective analysis of antibiotic susceptibility patterns of MRSA in a Turkish University Hospital. *Turk J Med Sci.*, *35*, 323-327.

Savolainen, K., Paulin, L., Westerlund-Wikstrom, B., Foster, T. J., Korhonen, T. K. & Kuusela P. (2001). Expression of pls, a gene closely associated with the *mecA* gene of methicillin-resistant *Staphylococcus aureus*, prevents bacterial adhesion in vitro. *Infect Immun*, *69*, 3013-3020.

Schaaff, F., Bierbaum, G., Baumert, N., Bartmann, P. & Sahl, H. G. (2003). Mutations are involved in emergence of aminoglycoside-induced small colony variants of *Staphylococcus aureus*. *Int J Med Microbiol*, 293, 427-435.

Schaffer, A. C., Solinga, R. M., Cocchiaro, J., Portoles, M., Kiser, K. B., Risley, A., Randall, S. M., Valtulina, V., Speziale, P., Walsh, E., Foster, T. & Lee, J. C. (2006). Immunization with *Staphylococcus aureus* clumping factor B, a major determinant in nasal carriage, reduces nasal colonization in a murine model. *Infect Immun*, *74*, 2145-2153.

SCIEH, (2002). Glycopeptide-Intermediate *Staphylococcus aureus,* (GISA). p 321.

Seifert, H., Wisplinghoff, H., Schnabel, P. & Von E. C. (2003). Small colony variants of *Staphylococcus aureus* and pacemaker-related infection. *Emerg Infect Dis., 9*, 1316-1318.

Senna, J. P., Roth, D. M., Oliveira, J. S., Machado, D. C. & Santos, D. S. (2003). Protective immune response against methicillin resistant *Staphylococcus aureus* in a murine model using a DNA vaccine approach. *Vaccine*, *21*, 2661-2666.

Severina, E., Severin, A. & Tomasz, A. (1998). Antibacterial efficacy of nisin against multidrug-resistant Gram-positive pathogens. *J Antimicrob Chemother*, *41*, 341-347.

Seybold, U., Kourbatova, E. V., Johnson, J. G., Halvosa, S. J., Wang, Y. F., King, M. D., Ray, S. M. & Blumberg, H. M. (2006). Emergence of community-associated methicillin-resistant *Staphylococcus aureus* USA300 genotype as a major cause of health care-associated blood stream infections. *Clin Infect Dis.*, *42,* 647-656.

Shannon, O., Uekotter, A. & Flock, J. I. (2006). The neutralizing effects of hyperimmune antibodies against extracellular fibrinogen-binding protein, Efb, from *Staphylococcus aureus*. *Scand J Immunol, 63*, 184-190.

Sharma, M., Visai, L., Bragheri, F., Cristiani, I., Gupta, P. K. & Speziale, P. (2008). Toluidine blue-mediated photodynamic effects on staphylococcal biofilms. *Antimicrob Agents Chemother, 52*, 299-305.

Shinefield, H. R. (2006). Use of a conjugate polysaccharide vaccine in the prevention of invasive staphylococcal disease: is an additional vaccine needed or possible? *Vaccine, 24*, Suppl 2, S2-S9.

Shukla, S. K. (2005). Community-associated methicillin-resistant *Staphylococcus aureus* and its emerging virulence. *Clin Med Res., 3*, 57-60.

Skov, R. & Westh, H. (2005). [Epidemic increase of methicillin resistant *Staphylococcus aureus* (MRSA) in Denmark--interventions required now]. *Ugeskr Laeger, 167*, 1396.

Slopek, S., Weber-Dabrowska, B., Dabrowski, M. & Kucharewicz-Krukowska, A. (1987). Results of bacteriophage treatment of suppurative bacterial infections in the years 1981-1986. *Arch Immunol Ther Exp (Warsz), 35*, 569-583.

Sola, C., Gribaudo, G., Vindel, A., Patrito, L. & Bocco, J. L. (2002). Identification of a novel methicillin-resistant *Staphylococcus aureus* epidemic clone in Cordoba, Argentina, involved in nosocomial infections. *J Clin Microbiol, 40*, 1427-1435.

Soncin, M., Fabris, C., Busetti, A., Dei, D., Nistri, D., Roncucci, G. & Jori, G. (2002). Approaches to selectivity in the Zn(II)-phthalocyanine-photosensitized inactivation of wild-type and antibiotic-resistant *Staphylococcus aureus*. *Photochem Photobiol Sci., 1*, 815-819.

Soukos, N. S., Mulholland, S. E., Socransky, S. S. & Doukas, A. G. (2003). Photodestruction of human dental plaque bacteria: enhancement of the photodynamic effect by photo*mec*hanical waves in an oral biofilm model. *Lasers Surg Med., 33*, 161-168.

Soukos, N. S., Ximenez-Fyvie, L. A., Hamblin, M. R., Socransky, S. S. & Hasan, T. (1998). Targeted antimicrobial photochemotherapy. *Antimicrob Agents Chemother, 42*, 2595-2601.

Spanu, T., Romano, L., D'Inzeo, T., Masucci, L., Albanese, A., Papacci, F., Marchese, E., Sanguinetti, M. & Fadda, G, (2005). Recurrent ventriculoperitoneal shunt infection caused by small-colony variants of *Staphylococcus aureus*. *Clin Infect Dis., 41*, e48-e52.

Stranger-Jones, Y. K., Bae, T. & Schneewind, O. (2006). Vaccine assembly from surface proteins of *Staphylococcus aureus*. *Proc Natl Acad Sci U S A, 103*, 16942-16947.

Struelens, M. J. & Denis, O. (2006). Rapid molecular detection of methicillin-resistant *Staphylococcus aureus*: a cost-effective tool for infection control in critical care? *Crit Care, 10*, 128.

Stryjewski, M. E., Hall, R. P., Chu, V. H., Kanafani, Z. A., O'Riordan, W. D., Weinstock, M. S., Stienecker, R. S., Streilein, R., Dorschner, R. A., Fowler, V. G., Jr., Corey, G. R., & Gallo, R. L. (2007). Expression of antimicrobial peptides in the normal and involved skin of patients with infective cellulitis. *J Infect Dis., 196*, 1425-1430.

Sulakvelidze, A., Alavidze, Z. & Morris, J. G., Jr. (2001). Bacteriophage therapy. *Antimicrob Agents Chemother, 45*, 649-659.

Tacconelli, E. (2006). New strategies to identify patients harbouring antibiotic-resistant bacteria at hospital admission. *Clin Microbiol Infect, 12*, 102-109.

Tajima, Y. (2005). Polyoxotungstates reduce the beta-lactam resistance of methicillin-resistant *Staphylococcus aureus*. *Mini Rev Med Chem.*, *5*, 255-268.

Tan, T. Y., Corden, S., Barnes, R. & Cookson, B. (2001). Rapid identification of methicillin-resistant *Staphylococcus aureus* from positive blood cultures by real-time fluorescence PCR. *J Clin Microbiol*, *39*, 4529-4531.

Tarkin, I. S., Henry, T. J., Fey, P. I., Iwen, P. C., Hinrichs, S. H. & Garvin, K. L. (2003). PCR rapidly detects methicillin-resistant staphylococci periprosthetic infection. *Clin Orthop Relat Res.*, 89-94.

Therrien, R., Lacasse, P., Grondin, G. & Talbot, B. G. (2007). Lack of protection of mice against *Staphylococcus aureus* despite a significant immune response to immunization with a DNA vaccine encoding collagen-binding protein. *Vaccine*, *25*, 5053-5061.

Tiemersma, E. W., Bronzwaer, S. L., Lyytikainen, O., Degener, J. E., Schrijnemakers, P., Bruinsma, N., Monen, J., Witte, W. & Grundman, H. (2004). Methicillin-resistant *Staphylococcus aureus* in Europe., 1999-2002. *Emerg Infect Dis.*, *10*, 1627-1634.

Toshkova, K., Annemuller, C., Akineden, O. & Lammler, C. (2001). The significance of nasal carriage of *Staphylococcus aureus* as risk factor for human skin infection. *FEMS Microbiol Lett.*, *202*, 17-24.

Trampuz, A. & Zimmerli, W. (2006). Antimicrobial agents in orthopaedic surgery: Prophylaxis and treatment. *Drugs*, *66*, 1089-1105.

Tristan, A., Bes, M., Meugnier, H., Lina, G., Bozdogan, B., Courvalin, P., Reverdy, M. E., Enright, M. C., Vandenesch, F. & Etienne, J. (2007). Global distribution of Panton-Valentine leukocidin--positive methicillin-resistant *Staphylococcus aureus*, 2006. *Emerg Infect Dis.*, *13*, 594-600.

Van Belkum, A., Bax, R. & Prevost, G. (1994). Comparison of four genotyping assays for epidemiological study of methicillin-resistant *Staphylococcus aureus*. *Eur J Clin Microbiol Infect Dis.*, *13*, 420-424.

Van Leeuwen, W., Sijmons, M., Sluijs, J., Verbrugh, H. & Van, B. A. (1996). On the nature and use of randomly amplified DNA from *Staphylococcus aureus*. *J Clin Microbiol*, *34*, 2770-2777.

Vayalumkal, J. V. & Jadavji, T. (2006). Children hospitalized with skin and soft tissue infections: a guide to antibacterial selection and treatment. *Paediatr Drugs*, *8*, 99-111.

Vernachio, J., Bayer, A. S., Le, T., Chai, Y. L., Prater, B., Schneider, A., Ames, B., Syribeys, P., Robbins, J. & Patti, J. M. (2003). Anti-clumping factor A immunoglobulin reduces the duration of methicillin-resistant *Staphylococcus aureus* bacteremia in an experimental model of infective endocarditis. *Antimicrob Agents Chemother*, *47*, 3400-3406.

Vesga, O., Groeschel, M. C., Otten, M. F., Brar, D. W., Vann, J. M. & Proctor, R. A. (1996). *Staphylococcus aureus* small colony variants are induced by the endothelial cell intracellular milieu. *J Infect Dis.*, *173*, 739-742.

Von Eiff, C., Becker, K., Metze, D., Lubritz, G., Hockmann, J., Schwarz, T. & Peters, G (2001b). Intracellular persistence of *Staphylococcus aureus* small-colony variants within keratinocytes: a cause for antibiotic treatment failure in a patient with darier's disease. *Clin Infect Dis.*, *32*, 1643-1647.

Von Eiff, C., Becker, K., Metze, D., Lubritz, G., Hockmann, J., Schwarz, T. & Peters, G (2001a). Intracellular persistence of *Staphylococcus aureus* small-colony variants within keratinocytes: a cause for antibiotic treatment failure in a patient with darier's disease. *Clin Infect Dis.*, *32*, 1643-1647.

Von Eiff, C., Proctor, R. A. & Peters, G. (2000a). *Staphylococcus aureus* small colony variants: formation and clinical impact. *Int J Clin Pract Suppl*, 44-49.

Von Eiff, C., Proctor, R. A. & Peters, G. (2000b). *Staphylococcus aureus* small colony variants: formation and clinical impact. *Int J Clin Pract Suppl*, 44-49.

Voyich, J. M., Braughton, K. R., Sturdevant, D. E., Whitney, A. R., Said-Salim, B., Porcella, S. F., Long, R. D., Dorward, D. W., Gardner, D. J., Kreiswirth, B. N., Musser, J. M. & DeLeo, F. R, (2005). Insights into *mec*hanisms used by *Staphylococcus aureus* to avoid destruction by human neutrophils. *J Immunol, 175*, 3907-3919.

Wagenaar, J., Van, D. E., Troelstra, A., Van De, G. A., Kluytmans, J., Mevius, D., Tiemersma, E. W. & Van Der, W. P. (2007). [Questions and answers about MRSA in farm animals]. *Tijdschr Diergeneeskd*, *132*, 558-560.

Walsh, E. J., O'Brien, L. M., Liang, X., Hook, M. & Foster, T. J (2004). Clumping factor B., a fibrinogen-binding MSCRAMM (microbial surface components recognizing adhesive matrix molecules) adhesin of *Staphylococcus aureus*, also binds to the tail region of type I cytokeratin 10. *J Biol Chem*., *279*, 50691-50699.

Wang, R., Braughton, K. R., Kretschmer, D., Bach, T. H., Queck, S. Y., Li, M., Kennedy, A. D., Dorward, D. W., Klebanoff, S. J., Peschel, A., DeLeo, F. R. & Otto, M. (2007). Identification of novel cytolytic peptides as key virulence determinants for community-associated MRSA. *Nat Med, 13,* 1510-1514.

Wannet, W. J., Spalburg, E., Heck, M. E., Pluister, G. N., Willems, R. J. & De Neeling, A. J, (2004). Widespread dissemination in The Netherlands of the epidemic berlin methicillin-resistant *Staphylococcus aureus* clone with low-level resistance to oxacillin. *J Clin Microbiol, 42*, 3077-3082.

Wargo, K. A. & Eiland, E. H, III (2005). Appropriate antimicrobial therapy for community-acquired methicillin-resistant *Staphylococcus aureus* carrying the Panton-Valentine leukocidin genes. *Clin Infect Dis*., *40*, 1376-1378.

Weigel, L. M., Donlan, R. M., Shin, D. H., Jensen, B., Clark, N. C., McDougal, L. K., Zhu, W., Musser, K. A., Thompson, J., Kohlerschmidt, D., Dumas, N., Limberger, R. J. & Patel J. B. (2007). High-level vancomycin-resistant *Staphylococcus aureus* isolates associated with a polymicrobial biofilm. *Antimicrob Agents Chemother*, *51*, 231-238.

Wertheim, H. F., Melles, D. C., Vos, M. C., Van, L. W., Van, B. A., Verbrugh, H. A., & Nouwen, J. L. (2005). The role of nasal carriage in *Staphylococcus aureus* infections. *Lancet Infect Dis*., *5*, 751-762.

Whyte, D., Monahan, R., Boyle, L., Slevin, B., FitzGerald, R., Barron, D., De, F. J. & Kelleher, K, (2005). The incidence of *S. aureus* bacteraemia in acute hospitals of the Mid-Western Area, Ireland, 2002-2004. *Euro Surveill*, *10*, 75-77.

Wills, Q. F., Kerrigan, C. & Soothill, J. S. (2005). Experimental bacteriophage protection against *Staphylococcus aureus* abscesses in a rabbit model. *Antimicrob Agents Chemother*, *49*, 1220-1221.

Wilson, M. (2004). Lethal photosensitisation of oral bacteria and its potential application in the photodynamic therapy of oral infections. *Photochem Photobiol Sci*., *3*, 412-418.

Witte, W., Cuny, C., Strommenger, B., Braulke, C. & Heuck, D. (2004). Emergence of a new community-acquired MRSA strain in Germany. *Euro Surveill, 9*, 16-18.

Witte, W., Pasemann, B. & Cuny, C. (2007). Detection of low-level oxacillin resistance in *mecA*-positive *Staphylococcus aureus*. *Clin Microbiol Infect*, *13*, 408-412.

Yacoby, I., Bar, H. & Benhar, I. (2007). Targeted drug-carrying bacteriophages as antibacterial nanomedicines. *Antimicrob Agents Chemother*, *51*, 2156-2163.

Yamasaki, O., Tristan, A., Yamaguchi, T., Sugai, M., Lina, G., Bes, M., Vandenesch, F. & Etienne, J. (2006). Distribution of the exfoliative toxin D gene in clinical *Staphylococcus aureus* isolates in France. *Clin Microbiol Infect, 12*, 585-588.

Yamasaki, O., Yamaguchi, T., Sugai, M., Chapuis-Cellier, C., Arnaud, F., Vandenesch, F., Etienne, J. & Lina, G. (2005). Clinical manifestations of staphylococcal scalded-skin syndrome depend on serotypes of exfoliative toxins. *J Clin Microbiol*, *43*, 1890-1893.

Yang, X. P., Li, E. W., Zhang, Q., Yuan, C. S. & Jia, Z. J. (2006). Five new iridoids from Patrinia rupestris. *Chem Biodivers, 3*, 762-770.

Yin, S., Fan, C. Q., Wang, Y., Dong, L. & Yue, J. M. (2004). Antibacterial prenylflavone derivatives from Psoralea corylifolia, and their structure-activity relationship study. *Bioorg Med Chem.*, *12*, 4387-4392.

Yu, C., Sun, J., Zheng, L. & Ji, Y. (2007). Genomic analysis of gene expression of *Staphylococcus aureus*. *Methods Mol Biol., 391*, 169-178.

Zanetti, M., Gennaro, R., Skerlavaj, B., Tomasinsig, L. & Circo, R. (2002). Cathelicidin peptides as candidates for a novel class of antimicrobials. *Curr Pharm Des.*, *8*, 779-793.

Zeina, B., Greenman, J., Corry, D. & Purcell, W. M (2002). Cytotoxic effects of antimicrobial photodynamic therapy on keratinocytes in vitro. *Br J Dermatol*, *146*, 568-573.

Zeina, B., Greenman, J., Corry, D. & Purcell, W. M. (2003). Antimicrobial photodynamic therapy: assessment of genotoxic effects on keratinocytes in vitro. *Br J Dermatol*, *148*, 229-232.

Zhang, K., McClure, J. A., Elsayed, S. & Conly, J. M. (2009). Novel staphylococcal cassette chromosome *mec* type, tentatively designated type VIII, harboring class A *mec* and type 4 ccr gene complexes in a Canadian epidemic strain of methicillin-resistant Staphylococcus aureus. *Antimicrob Agents Chemother,* 53, 531-540.

Zhou, H., Xiong, Z. Y., Li, H. P., Zheng, Y. L. & Jiang, Y. Q. (2006). An immunogenicity study of a newly fusion protein Cna-FnBP vaccinated against *Staphylococcus aureus* infections in a mice model. *Vaccine*, *24*, 4830-4837.

In: Methicillin-Resistant Staphylococcus Aureus...
Editor: Charles L. Kolendi

ISBN: 978-1-60741-398-1

Chapter 6

RECENT ADVANCES IN MEDICINAL CHEMISTRY OF NEW QUINOLONES WITH POTENTIAL ANTI-MRSA ACTIVITY

Saeed Emami
Department of Medicinal Chemistry, Faculty of Pharmacy, Mazandaran University of Medical Sciences, Sari, Iran

ABSTRACT

Quinolone antibacterial agents are among the most attractive drugs in the field of anti-infective chemotherapy. Quinolones target the essential bacterial type II topoisomerases, which are involved in DNA replication and metabolism. Despite a large number of quinolones approved for the treatment of bacterial infections, there have been progressive efforts for the discovery of new quinolones to overcome the problem of growing bacterial resistance. Over the years, there has been growing concern surrounding the increased prevalence of drug resistance in *Staphylococcus aureus*, especially methicillin-resistant *Staphylococcus aureus* (MRSA). MRSA used to be susceptible to the fluoroquinolones when they were introduced in the early 1980s but a rapid emergence of resistance to these drugs was observed soon after the increasing use of fluoroquinolones from the mid 1980s. Progressive efforts have been made to optimize the biological activity of the quinolones against *S. aureus* and drug-resistant strains including MRSA. Much of the improved potency of modern quinolones against MRSA has been achieved by modification of the N-1, C-7, and C-8 substituents on the quinolone ring system. Several newly developed quinolones and related compounds exhibit enhanced activity against *S. aureus* and have proved useful against MRSA strains and are under rapid preclinical and clinical development.

This chapter will discuss the structural requirements of quinolones for anti-MRSA activity and introduce the newly developed quinolones to manage MRSA infections, their in-vitro effectiveness based on recent studies and structure-activity relationships.

Introduction

Methicillin-resistant *Staphylococcus aureus* (MRSA) is a multidrug-resistant pathogen that has historically associated with hospitals and healthcare facilities. Over the years, there has been growing concern surrounding the increased prevalence of drug resistance in *S. aureus*, especially hospital-associated MRSA (HA-MRSA), which limits drug options [1]. Until late in 1997, vancomycin was the only agent available with reliable activity against serious MRSA infections. Other agents, including rifampin, novobiocin, daptomycin, teicoplanin, tigecycline, streptogramins, oxazolidinones and quinolones have been used to treat patients infected with MRSA in an ongoing effort to expand treatment options [2].

The quinolones are synthetic antibacterial agents, first used clinically in the 1960s with the introduction of nalidixic acid. The utility of nalidixic acid was restricted to Gram-negative bacteria. In the 1960s, MRSA had only recently been described; quinolone compounds with activity against staphylococci were not available until the introduction of fluoroquinolones, such as ciprofloxacin and ofloxacin, two decades later. MRSA used to be susceptible to the fluoroquinolones when they were introduced in the early 1980s but a rapid emergence of resistance to these drugs was observed soon after the increasing use of fluoroquinolones from the mid 1980s. Improvement to the chemical structure of early quinolones, based on the addition of a fluorine atom at the C-6 position and modification of the C-7 position, dramatically enhanced antibacterial spectrum and improved pharmacokinetics. Much of the improved potency of modern quinolones against staphylococci has been achieved by tinkering with the N-1, C-7, and C-8 substituents on the quinolone ring system.

The widespread emergence of antibiotic resistance Gram-positive bacteria has stimulated the search for effective new quinolone antibacterial agents in recent years. Generally, the newer quinolones retain the broad spectrum activity of the currently used quinolones, such as levofloxacin and moxifloxacin, and also display improved activity against *S. aureus*, including some MRSA isolates. Several new agents have proved useful against MRSA strains and are under rapid preclinical and clinical development.

This chapter will discuss the pharmacophoric requirements of quinolones for anti-staphylococcal activity and the newly developed quinolones to manage MRSA infections, their in-vitro effectiveness based on recent studies and structure-activity relationships.

History of Quinolones Development

Quinolones are synthetic drugs developed through structural modification of the 'quinoline' skeleton. In 1962, during the process of synthesis of chloroquine (an antimalarial agent), a napththyridone (8-azaquinolone) derivative, nalidixic acid (Figure 1), was discovered which possessed bactericidal activity [3]. The use of nalidixic acid was limited because of its narrow spectrum, low serum levels, and toxicity issues, but it regained attention in the 1980s for the treatment of diarrhea and urinary tract infections following the development of resistance in *Shigella* and *Escherichia coli* to other classes of antibiotics used at that time [4]. Thereafter, novel compounds of this family, such as oxolinic acid and pipemidic acid (Figure 1), were synthesized and introduced into clinical practice, although the clinical indication for these quinolones still remained only for urinary tract infections.

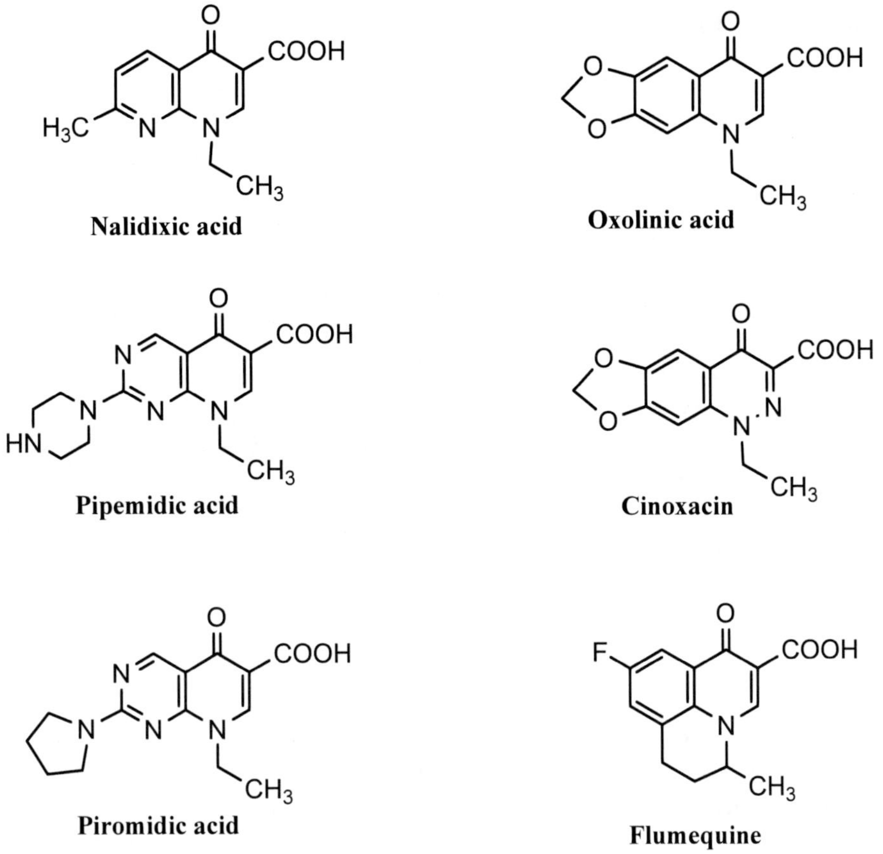

Figure 1. Older quinolones.

The introduction of a piperazin-1-yl moiety at 7-position (to provide pipemidic acid) did improve the activity against Gram-negatives, broadening the spectrum to include *Pseudomonas* spp. Pipemidic acid also had some activity against Gram-positive species. This piperazine ring has subsequently been shown to increase the ability of the quinolones to penetrate the bacterial cell wall, thus enhancing activity. The first compound with a fluorine atom at position 6 was flumequine (patented in 1973) and this compound gave the first indications that activity against Gram-positive organisms (e.g. *S. aureus*) could be improved in this class [5]. Following the introduction of flumequine, the second generation of quinolones had the major feature of a fluorine substituent at position 6, which increased activity markedly. These early compounds were most potent against Gram-negative organisms; thus their activity against *Streptococcus pneumoniae* was too marginal to warrant clear indications for use in the treatment of respiratory tract infections, and the emergence of resistance soon reduced their potential against *S. aureus* [6, 7]. The real breakthrough came with the combination of these two features in norfloxacin (Figure 2), a 6-fluoroquinolone with a piperazine ring at position 7. Although norfloxacin does have activity against a broad spectrum of Gram-negatives and has some Gram-positive activity, the combination of its pharmacokinetic profile and activity were still not adequate for systemic use [8-10].

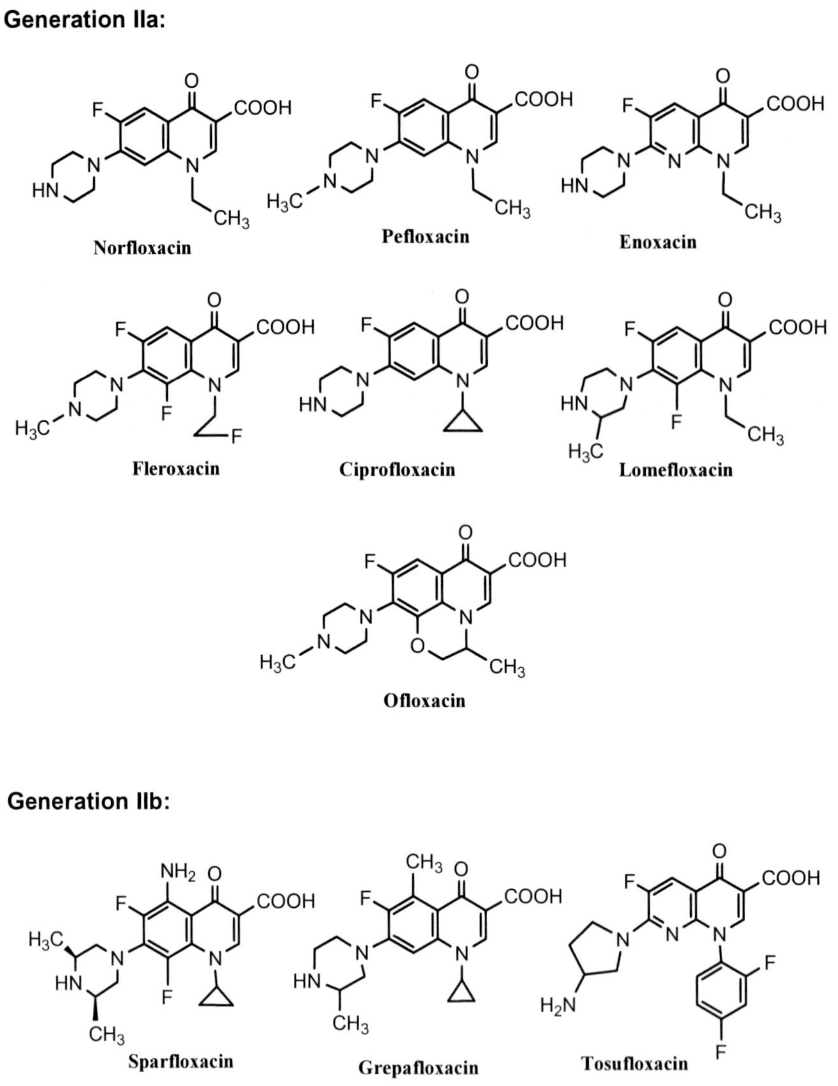

Figure 2. Second-generation of quinolones (6-fluoroquinolones).

Following the introduction of norfloxacin and several related compounds including pefloxacin, enoxacin, fleroxacin and lomefloxacin (Figure 2), the substitution of a cyclopropyl group at position N-1, as in ciprofloxacin, gave a marked improvement in activity. Ofloxacin exemplify another structural modification in the form of a bridging ring between the N-1 and position 8, which also gives improved activity [5]. The advantages of these drugs are that their spectrum includes Gram-positive organisms as well as Gram-negatives, and that they are well absorbed from the gastrointestinal tract, providing adequate blood levels to allow their use for systemic infections. However, their activity against Gram-positive cocci is inadequate for the treatment of respiratory tract infections. Similarly, although the development of resistance to these agents is clearly less of a problem than that with nalidixic acid, in some species, particularly *S. aureus* and *Pseudomonas aeruginosa*, resistance to this group has become a problem [11].

The quinolones were further optimized by the addition of different functional groups to the N-1, C-5, C-7 and C-8 positions of their respective basic molecules. A group of compounds showing more dramatic improvements are sparfloxacin, grepafloxacin, tosufloxacin (Figure 2), levofloxacin, gatifloxacin, clinafloxacin, sitafloxacin, moxifloxacin, trovafloxacin and gemifloxacin (Figure 3). They all show good activity against most Gram-positives, with clinafloxacin, gemifloxacin and sitafloxacin having the best activity [12, 13].

Optimising all other substituents has permitted the removal of the fluorine atom at position 6 (which has been claimed to be involved in genotoxicity and central nervous system defects possibly involved in genotoxicity), giving rise to the new generation of quinolones; termed des-fluoroquinolones, with garenoxacin as its first representative (Figure 4). The extensive research efforts have enabled a better definition of the structural moieties or elements around the basic pharmacophore that offer the best combination of clinical efficacy, reduced resistance selection, and safety.

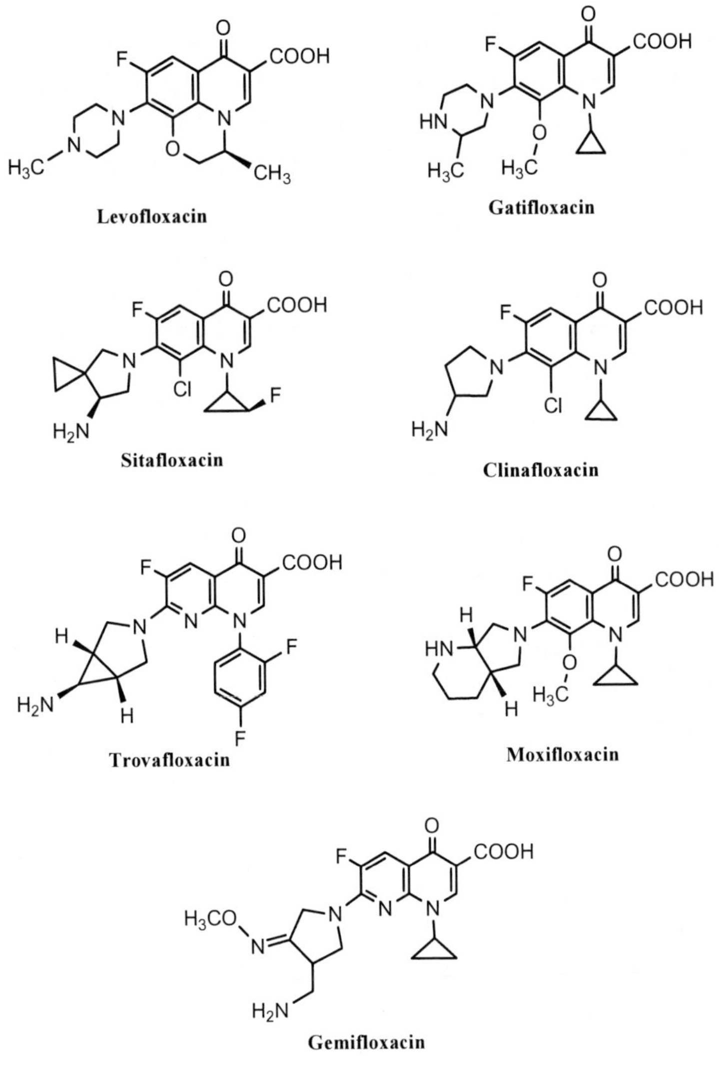

Figure 3. Third-generation quinolones with improved anti-staphylococcal activity.

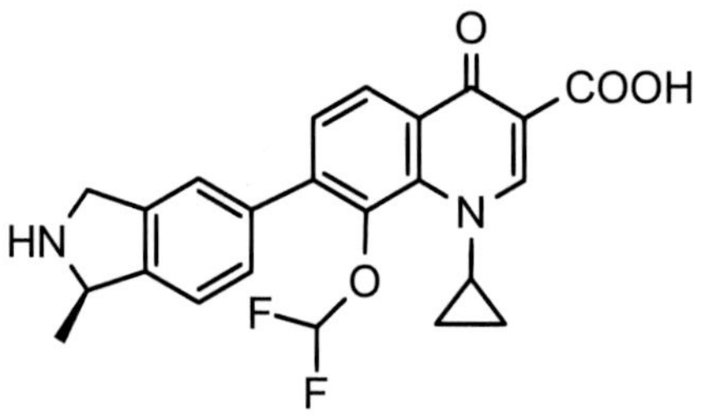

Garenoxacin

Figure 4. Structure of garenoxacin; a representative 6-desfluoroquinolones.

Mechanism of Action

Quinolones exert their antibacterial action via inhibition of homologous type II topoisomerases, DNA gyrase and DNA topoisomerase IV. Type II topoisomerases are able to cut both strands of the DNA and pass one strand through this transient gap before resealing the DNA. Both topoisomerase enzymes are essential for bacterial growth, but they cannot complement one another [14, 15].

During transcription process or DNA synthesis, the double-stranded DNA is unzipped to allow either mRNA or a new DNA strand to be synthesized. This unzipping of the DNA causes topological stress upstream of the RNA polymerase or replication fork and induces the formation of positive supercoils that need to be removed. To relieve this stress and to remove the positive supercoils, a topoisomerase known as DNA gyrase makes double-stranded breaks in the DNA. With DNA gyrase, the cut DNA strand is passed through itself to propagate what is known as negative supercoiling. After DNA synthesis, the daughter chromosomes are unlinked by another topoisomerase, topoisomerase IV, in a process called decatenation. Both of these enzymes belong to the class of type II topoisomerases, use a double-strand-passage mode of action, are essential to bacteria and are the targets of the quinolones [16].

DNA gyrase is composed of two GyrA and two GyrB monomeric subunits, which are encoded by the *gyrA* and *gyrB* genes, respectively. Topoisomerase IV is composed of two ParC and two ParE (in *S. aureus* named GrlA and GrlB, respectively) subunits. The two subunits of this tetrameric enzyme are encoded by the *parC* and *parE* genes (referred to as *grlA* and *grlB* in *S. aureus*). GyrA is similar in structure to ParC, and GyrB is similar in structure to ParE. GyrA and GrlA are the subunits responsible for the physical cutting of the DNA, whereas GyrB and GrlB provide the energy for supercoiling of decatenation by the breakdown of ATP [17, 18].

The activity of quinolones stems primarily from the formation of ternary complexes between DNA and type II topoisomerases. Complex formation reversibly inhibits DNA synthesis and cell growth, which is probably responsible for the bacteriostatic action of quinolones. Their lethal action is thought to be a separate event from complex formation and probably relates to the release of free DNA ends from the DNA gyrase–quinolone complexes [19-23]. According to this model, four quinolone molecules are envisaged to bind to the single-stranded DNA regions opened up in the gyrase-DNA complex by covalent attachment of the two GyrA subunits to each complementary DNA strand. Two quinolones are envisaged to lie above the other pair of drug molecules, making hydrophobic interactions with each other through substituents on N-1 and C-8 positions. Binding to DNA strands is suggested to involve a hydrogen bonding domain on the drug comprising the C-3 carboxyl group and the ketone at C-4. Furthermore, Shen et al. postulated that the substituent on C-7 is involved in drug-enzyme interactions [24].

A key to understanding quinolone action in *S. aureus* is the observation that the drugs can act preferentially through gyrase, through topoisomerase IV, or through both targets in a manner dependent on quinolone structure. In general, several studies have highlighted substantial variations in the in-vitro inhibitory concentrations for DNA gyrase and topoisomerase IV, depending on both the bacterial species and the molecule being studied. These data, which are roughly consistent with MIC values and data obtained from analysis of resistant mutants, confirm that DNA gyrase is the preferred target of quinolones in Gram-negative bacteria. The situation is more complex in Gram-positive bacteria [25]. It was proposed that effects of quinolones are more severe on topoisomerase IV in Gram-positive bacteria (e.g., *S. aureus*) than its counterparts in Gram-negative bacteria (e.g., *E. coli*). This scenario is consistent with a species-specific phenomenon in which topoisomerase IV would be the primary target for all quinolones in *S. aureus*, for example. However, it was also recognized that sparfloxacin was a notable exception to the rule in that mutations in GrlA caused minimal loss of potency compared to other quinolones, such as pefloxacin, ciprofloxacin, levofloxacin and trovafloxacin against *S. aureus* [26, 27]. In addition, first-step mutations were found to map in *gyrA* when selected for reduced susceptibility to sparfloxacin, suggesting that DNA gyrase is the primary target for sparfloxacin in *S. aureus* [28, 29]. Furthermore, both DNA gyrase and topoisomerase IV appeared to act as dual targets for clinafloxacin and gemifloxacin [30, 31].

In a study by Nilius et al., trovafloxacin and ciprofloxacin showed 8- and 19-fold greater potencies, respectively, against *S. aureus* topoisomerase IV than against *S. aureus* DNA gyrase. In contrast, the interaction of ABT-492 with both *S. aureus* topoisomerases was nearly equivalent, with three-fold greater activity against DNA gyrase than against topoisomerase IV. By direct comparison, ABT-492 was nearly three-fold more active than trovafloxacin and was about six-fold more active than ciprofloxacin against *S. aureus* DNA gyrase [32]. In addition, the effective target inhibition data suggest that non-fluorinated quinolones have a dual-target mechanism of action against *S. aureus* and their potency against *S. aureus* could be attributable to partial inhibition of both the targets [33].

Concisely, quinolones can be classified into 3 categories (types I, II and III) based on their ability to inhibit purified *S. aureus* DNA gyrase and topoisomerase IV and also the effect of *grlA* and *gyrA* mutations on MIC [34, 35]. Most quinolones fall within the type I category (norfloxacin, enoxacin, fleroxacin, ciprofloxacin, lomefloxacin, trovafloxacin, grepafloxacin, ofloxacin, levofloxacin and gemifloxacin) because they are more influenced by topoisomerase IV inhibition than DNA gyrase inhibition. The type II category including sparfloxacin and nadifloxacin show a greater decrease in activity against a *gyrA* mutant strain than a *grlA* mutant strain – indicating a preference for DNA gyrase [34]. It should be noted however, that the results of some studies do not concur with this analysis for sparfloxacin [36, 37]. The type III quinolones include clinafloxacin, gatifloxacin, moxifloxacin and pazufloxacin, which have dual-targeting properties where either enzyme is inhibited at approximately the same extent and *gyrA* or *grlA* mutant strains produce only a minimal 2-fold increase in MIC [34]. Other data suggest that sitafloxacin [37] and garenoxacin [38] are also type III quinolones.

QUINOLONE RESISTANCE IN *S. AUREUS*

Quinolone resistance in *S. aureus* emerged rapidly in several hospitals after the introduction of ciprofloxacin and strikingly so in MRSA, rising to prevalence of over 80% in some cases [39, 40]. The early finding with the older quinolone ciprofloxacin indicated in vitro activity against both MSSA and MRSA, although marginal with MIC_{90}s of 0.5μg/mL [41]. Unfortunately, after few years almost all MRSA had developed resistance to quinolones. Blumberg et al. reported that prior to the introduction of ciprofloxacin in 1988, 159 MSSA and 131 MRSA were found to be susceptible to ciprofloxacin. One year after introduction, 79% of MRSA and 13.6% of MSSA were ciprofloxacin resistant [39]. This situation is also illustrated well by a study in Hong Kong where resistance to ciprofloxacin in MRSA had reached 82% in 1993 from a relatively low level of 9% in 1988 [42].

The decreased susceptibility of MRSA to quinolones is not the result of increased mutation rates; MSSA have been shown to have the same mutation rates as MRSA [43]. The early findings demonstrated that although initial cases of colonization or infection with quinolone-resistance MRSA can be related to quinolone use, the spread of MRSA is more likely due to hospital transmission of existing resistant MRSA. Thereby prevalence of MRSA over MSSA due to greater opportunities for transmission of MRSA between hospitalized patients, many of whom have received antibiotics. Because MRSA are commonly multidrug-resistant, selective advantage of these strains over MSSA strains may occur with a broad range of antibiotic exposures, not just quinolones [44].

There does appear to be a relation between quinolone resistance and methicillin resistance. It seems likely that an association exists between the action of quinolones on mec(A)-positive *S. aureus* and the increase in the proportion of CFU/mL expressing methicillin resistance. Growth in the presence of various quinolones increased the proportion of the mec(A)-positive heteroresistant *S. aureus* that developed methicillin resistance (MIC of oxacillin >128 μg/mL). The resultant MRSA strains also showed a 1.5- to 3-fold increase in quinolone MICs [43].

Mechanisms of Quinolone Resistance in *S. aureus*

The resistance of *S. aureus* to quinolones involves at least three different mechanisms, which are often combined in highly resistant organisms. One mechanism is the active efflux of the drugs by the NorA transporter [45]. The two other mechanisms result from modifications of the quinolone molecular targets of the bacterium, i.e., the DNA gyrase (*gyrA* mutants) [46] and/or the topoisomerase IV (*grlA* mutants) [47]. As with other bacterial species, resistance to quinolones in *S. aureus* develops in a step-wise fashion by the accumulation of mutations. Resistance occurs due to changes in aminoacids (Table 1), particularly those in certain regions of each enzyme subunit called the quinolone-resistance-determining-region (QRDR), which make the enzyme less sensitive to inhibition by quinolones. For some of the more common QRDR mutations it seems that the aminoacid changes reduce the affinity of the enzyme-DNA complex for the quinolone [48]. Level of resistance of a target enzyme can increase as additional aminoacids are altered by additional mutations [49].

Experimental studies have shown that mutations in *grlA* occur before *gyrA* with *S. aureus* selected on ciprofloxacin and these latter mutations are silent in the absence of additional mutations in *grlA* [50]. This is in general agreement with ciprofloxacin being a type I quinolone based on target preference. With sparfloxacin, the reverse is true with mutations in *gyrA* occurring before *grlA* [28] again in keeping with the target preference of a type II quinolone. The same classical quinolone resistance alteration in GrlA (Ser80 to Phe or Tyr and Glu84 to Lys) and GyrA (Ser84 to Leu or Lys and Glu88 to Lys or Val) have been found in laboratory mutants and clinical isolates [23, 51]. Other mutations such as Asp73 to Gly in GyrA have been fond clinically but are considerably less common [52].

Table 1. Common mutations in type II topoisomerases of *S. aureus* and MICs of ciprofloxacin against clinical isolates of *S. aureus*.

Alteration of amino acid		MIC (μg/mL)	References
GyrA	GrlA		
–	–	0.25	[47]
–	Ser80→Phe/Tyr	2–16	[51], [52]
Ser84→Leu	–	16	[53]
Ser84→Leu	Ser80→Tyr	16	[51]
Ser84→Leu	Ser80→ Phe/Tyr	12.5–800	[52]
Ser84→Leu	Glu84→ Lys	100–800	[52]
Ser84→Leu	Ser80→ Phe/Tyr + Glu84Lys	100–>800	[52]
Glu88→Lys	Ser80→ Phe	8–32	[51]
Glu88→Lys	Ser80→ Phe + Glu84Val	32–256	[51]

Resistance mutations that alter the structure of the GyrB and ParE subunits are less common than those altering the structure of GyrA and ParC subunits. Single changes in ParC seem to be sufficient to cause clinical resistance to ciprofloxacin, but not necessarily to other more potent quinolones [51]. For more potent quinolones in staphylococci, single ParC mutations often generate reduced susceptibility without full resistance as defined in the clinical laboratory [51, 54, 55].

A mutation in the more sensitive enzyme results in an increase in the MIC of a quinolone, whereas a mutation in the less sensitive enzyme generally causes resistance only in the presence of resistance mutations in the primary target. A quinolone with similar affinities for both targets is little affected by a mutation in one of the enzymes, and concurrent mutations in both enzymes are necessary for resistance to develop [30, 56-58]. Thus, for dual-targeting quinolones with similar potency against both target enzymes, there may be little reduction in susceptibility with any single mutation in either topoisomerase IV or DNA gyrase [30].

Several studies pay attention to identify the structural features of quinolones and provide a rationale for dissociated propensity for development of resistance amongst *S. aureus* and other Gram-positive bacteria. Concerning on the C-6 position, the primary objective of the non-fluorinated quinolones design effort was achieved with a series of quinolones that are relatively unaffected by the serine mutation in DNA gyrase. However, the data reviewed by Roychoudhury et al. showed that those non-fluorinated quinolones also have highly desirable properties against the second target of quinolones, topoisomerase IV, especially in *S. aureus* [33, 59]. Hartman-Neumann et al. found that the non-fluorinated quinolone, garenoxacin able to select mutants with the known changes in GyrA and ParC [60].

Fukuda et al. demonstrated the importance of an 8-methoxy substituent by comparing gatifloxacin and AM-1147 (an 8-methoxy quinolone) and their respective 8-unsubstituted counterparts. They showed that the 8-methoxy derivatives to select mutants at a lower frequency than their 8-unsubstituted counterparts and to prefer DNA gyrase [61]. Furthermore, Ince and Hooper [62] underlined the importance of the 8-methoxy substituent by comparing gatifloxacin, its 8-desmethoxy derivative and ciprofloxacin. Gatifloxacin was found to be most active against mutants of *S. aureus*, and a novel substitution outside the QRDR of GrlA was detected: Lys23→Asn. Two novel mutations in *grlB* were also found: Pro25→His and Pro451→Gln. Allelic exchange experiments confirmed the role of the novel mutations in resistance, suggesting that the QRDR in GrlA should be expanded to include these mutations. Zhao et al. showed that 8-methoxy ciprofloxacin derivatives are more lethal than 8-bromo , 8-ethoxy, and 8-H derivatives for *S. aureus*, especially when topoisomerase IV is resistant. The methoxy group also increases lethality against wild-type cells when protein synthesis was inhibited [63].

Role of Efflux Pumps

The second resistance mechanism in *S. aureus* is mediated through active efflux by the over-expression of certain efflux pumps. The efflux pump system is a mechanism that allows immediate survival of bacteria in the presence of an antimicrobial agent by actively expelling that agent across the cell membrane, thereby reducing the intracellular concentrations to sub-lethal levels [21, 64].

Low level of resistance to quinolones can occur by over-expression of these efflux pumps, usually due to mutations that increase transcription of the structural gene for the pump either from effects on the gene promoter or from alterations in other regulatory proteins that affect pump expression. In the case of *S. aureus* including MRSA, low level of resistance to ciprofloxacin can result from the over expression of a pump called NorA [21]. Mutations causing this increase and low-level resistance have been seen in the *norA* gene promoter causing increased stability of *norA* mRNA and in other regulatory proteins [21]. Because NorA is a multi-drug efflux pump, its increased expression also reduces susceptibility to chloramphenicol and some dyes, such as ethidium bromide. NorA also contributes indirectly to other types of resistance because exposure of bacteria to sub-therapeutic drug concentrations promotes the selection and expression of higher-level adaptive resistance mechanisms, such at target mutations [65, 66].

Sulavik and Barg [67] examined mutants of clinical isolates of MRSA and MSSA with low-level resistance to ciprofloxacin. The comparison of generated first-step mutants and mutants with low-level ciprofloxacin resistance showed that in vitro selection did not alter the distribution into different classes of first-step mutants, with about 10% of the first-step mutants showing increased efflux.

The efflux pump action is dependent on the ability of quinolone molecule to bind to the bacterial efflux protein and be exported [68]. Norfloxacin, ciprofloxacin and ofloxacin seem to be affected more than sparfloxacin, gatifloxacin, sitafloxacin, gemifloxacin, moxifloxacin, trovafloxacin and garenoxacin, suggesting that these last quinolones may be poorer substrates for these pumps [21, 64, 69-71]. Thus, efflux-mediated resistance mechanisms seem to affect quinolone agents to different extents, depending on the physicochemical properties and structural characteristics of the individual quinolones. Previously, only hydrophilic character of the fluoroquinolones was thought to be an important factor for NorA-mediated transport. Aeschlimann et al. [70] examined the in vitro antibacterial activities and post-antibiotic effects of ciprofloxacin, norfloxacin and levofloxacin, in genetically related strains of *S. aureus*. They used the wild-type and two NorA-mediated resistant strains for susceptibility testing and determination of time-kill curves and post-antibiotic effect. It was found that the NorA inhibitors, reserpine and omeprazole dramatically improve the activities of the more hydrophilic quinolones (ciprofloxacin and norfloxacin). In addition, hydrophilic quinolones, such as norfloxacin and ciprofloxacin, appear more prone to efflux than more hydrophobic molecules like grepafloxacin and gatifloxacin in Gram-positive organisms [72, 73]. However, Takenouchi et al. demonstrated that the bulkiness of the C-7 substituent and the hydrophobicity and bulkiness of the C-8 substituent, not the molecular hydrophobicity, was correlated with the activity of quinolones [74]. Of the quinolones tested by Beyer et al. and Madras-Kelly et al., levofloxacin has the least bulky C-7 substituent (*N*-methylpiperazine), sparfloxacin a slightly larger derivative (bearing a 3,5-dimethylpiperazine) and moxifloxacin, the bulkiest C-7 substituent (diazabicyclo moiety). On the other hand, sparfloxacin is a rather hydrophobic drug, while levofloxacin and moxifloxacin are much more hydrophilic. Sparfloxacin and moxifloxacin were relatively unaffected by NorA-mediated efflux, whereas levofloxacin was highly affected [69, 75].

In order to find quinolones that are refractory to efflux-mediated resistance in Gram-positive bacteria, Kerns et al. discovered a novel series of piperazinyl-linked quinolone dimers, which exhibited potent antibacterial activity against drug resistant strains of *S. aureus*, including strains possessing resistance due to the NorA multidrug efflux pump and a mutation in the QRDR of topoisomerase IV. They suggested that a bulky C-7 side chain might account for the lack of effective dimer efflux through the *norA* pump [76, 77]. These data demonstrate that the bulk at C-7 appears to be the key structural characteristic for avoidance of efflux. Gemifloxacin, moxifloxacin, trovafloxacin, and garenoxacin are the currently available agents with the greatest bulk at C-7 position, and they appear least affected by efflux-mediated resistance [25, 64].

General Structural Features (Structure-Activity Relationships)

Progressive efforts have been made to optimize the biological activity of the quinolones against Gram-positive bacteria including *S. aureus*. The minimum pharmacophore required for significant antibacterial activity consists of the 4-pyridone ring with a 3-carboxylic acid group (Figure 5). The nitrogen at N-1 is almost indispensable for activity and the carboxylic acid at position 3 and the carbonyl group at position 4 are considered critical for binding to cleaved or perturbed DNA. Therefore, the 3-carboxylate and 4-carbonyl groups are considered essential for antimicrobial activity [78]. Thus, the 1,4-dihydro-4-oxopyridine-3-carboxylic acid associated with a 5,6-fused aromatic ring is the common chemical feature of quinolones and 1-, 2-, 5-, 6-, 7-, and 8-positions are the major targets of chemical variation (Figure 5). The antibacterial activity of quinolones depends on the nature of peripheral substituents and their spatial arrangements [57]. The purpose of our brief discussion about structure-activity relationships (SAR) of quinolones is to demonstrate the changes or effects occurred due to modifications of different substituents in particular position of the pharmacophore, and influenced the antibacterial activity especially against *S. aureus*.

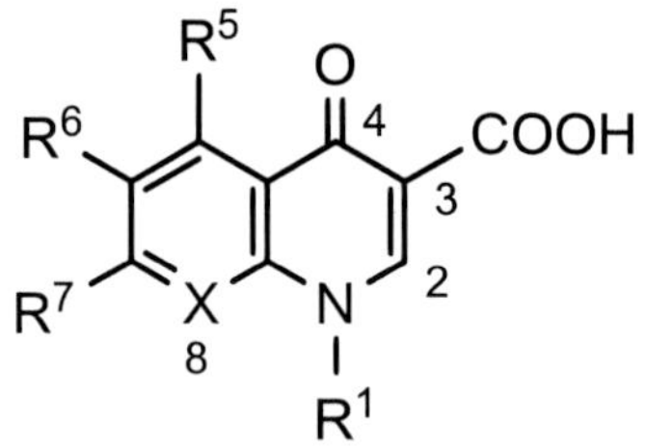

Figure 5. 4-Quinolone-3-carboxylic acid; common chemical feature of quinolone antibacterials and 1-, 5-, 6-, 7-, and 8-positions as the major targets of chemical variation.

Position 1

Antibacterial activity is greatly influenced by the hydrophobic and steric bulk of N-1 substituent [79]. A cyclopropyl substituent is now considered as a particularly potent modification at this position; most of the successful compounds against *S. aureus* contain this substituent including ciprofloxacin, sparfloxacin, grepafloxacin, gatifloxacin, clinafloxacin, gemifloxacin, moxifloxacin, and garenoxacin. Sitafloxacin contains a fluorinated cyclopropyl group [*cis*-oriented (1*R*,2*S*)-2-fluorocyclopropyl] at N-1 [80, 81]. The addition of a 2,4-difluorophenyl group at position 1 [e.g. tosufloxacin and trovafloxacin] also improved activity against *S. aureus* as well as other Gram-positive bacteria [82]. It has been observed that in N-1 aryl substituted quinolones, the N-1 aryl ring should be oriented out of the plane of quinolone ring for its antibacterial activity [83]. Another distinct structure at this position is found in ofloxacin, levofloxacin, pazufloxacin and nadifloxacin (Figure 6) which have a fused ring between positions 1 and 8. For instance, ofloxacin has a tricyclic ring structure with a methyl group attached to the asymmetric C-3 position on the oxazine ring, thus connecting positions 1 and 8 with a fused ring. Although this has been a useful alternative to the cyclopropyl substituent, it has been also shown that the *S*- isomer (levofloxacin) is 2-4 times more active than ofloxacin against both methicillin-susceptible and resistant isolate of quinolone-susceptible staphylococci [84].

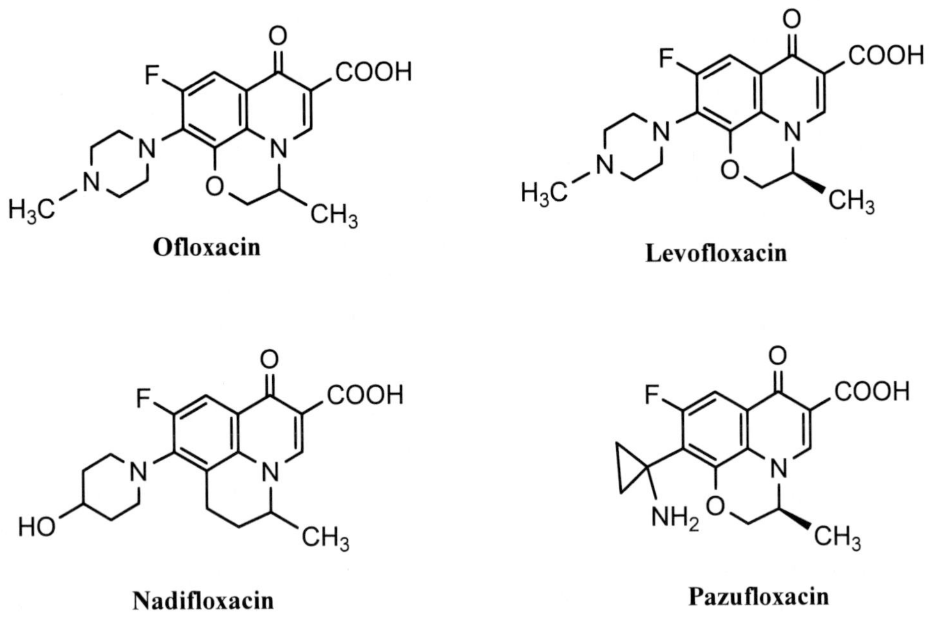

Figure 6. Quinolones with a fused ring between positions 1 and 8.

Position 2

In general, no analog modified at C-2 position remains on the market bearing other than a CH substituent. Very little is known about the SAR of quinolone having substituents at C-2 position; as loss of bioactivity has been found with methyl, hydroxyl or methylthio substituents. Moreover, significant decrease in activity was found with 2-aza-quinolones derivatives of norfloxacin. However a ring between C-1 and C-2 position was shown to have biological activity. The C-2 position is left unsubstituted because of its proximity to the enzyme binding site [7, 9, 10, 85]. In addition, It is seems that the COOH at C-3 needs to be coplanar with the C-4 carbonyl group so that it can effectively hydrogen bond with DNA bases made available by strand separation catalyzed by type II topoisomerases. The steric deficit that substitution at C-2 presumably causes can be partly overcome by rigidifying a C-2 substituent in a ring connecting it with N-1, especially when the group attached to C-2 can accept a hydrogen bond and form a virtual ring with the C-3 substituent [87]. In agreement with this idea is the outstanding potency seen with formation of a thiazolone ring fused to the carbonyl ring [88]. With thiazolone ring formation, the NH moiety is strongly acidified by resonance stemming from the aromaticity of the ring when enolization takes place. This produces a coplanar carboxyl surrogate and leads to an enhancement of antibacterial activity.

Position 5

Introduction of some substituents such as halogen, nitro, amino, hydroxy and alkyl groups at C-5 were initially thought to reduce antibacterial activity of the quinolones. Subsequently, it has been shown that introduction of bulky substituents, halogen and methoxy at this position reduce activity markedly; this is probably a consequence of interference with the active binding site at positions 3 and 4 [10, 78]. However, modestly sized substituents, such as an amino (as in sparfloxacin) or a methyl group (as in grepafloxacin) can markedly increase in vitro activity against *S. aureus* [10, 89]. Thus, substitutions at this position are thought to contribute to potency against *S. aureus* and other Gram-positive microorganisms [90-92]. It is also notable that the influence of C-5 substituent markedly depends on the substitution pattern at N-1, C-7 and C-8. A large number of current quinolones including ciprofloxacin, ofloxacin, levofloxacin, gatifloxacin, clinafloxacin, sitafloxacin, gemifloxacin, moxifloxacin, trovafloxacin and garenoxacin have only hydrogen at this position.

Position 6

Early investigations on the C-6 position were generally limited to the terminus of a CH (nalidixic acid), methylenedioxy bridge to C-7 (oxolinic acid and cinoxacin), and N (pipemidic acid and piromidic acid) (Figure 1). These early compounds had moderate activity against Gram-negatives and no activity against Gram-positives. The addition of fluorine substituent at C-6 markedly improved antimicrobial activity [93, 94]. Flumequine was the first compound to be developed with a fluoro- group at C-6, and gave the first indications that modifications of the basic chemical structure could improve Gram-positive activity [5]. The influence of fluorine at this position is essential for high activity as evidenced by its enhanced gyrase inhibition and cell penetration which has become the basis for generic name fluoroquinolones [57, 94]. After introduction of norfloxacin and attractive properties of 6-fluoro- substituent, a large number of 6-fluoroquinolones were marketed. Subsequently, it has been shown in a series of quinolones that the impact of 6-fluoro- substituent is diminished when the molecule contains spatial arrangements with other helpful substituents. For example, studies with garenoxacin (a 6-desfluorinated-8-difluoromethoxyquinolone, Figure 4) and its 6-fluoro-analog revealed no difference in potency against *S. aureus* whether the C-6 substituent is fluorine or hydrogen. The objective of the 6-desfluoro-quinolones optimization effort was to design compounds with potent activity against important Gram-positive pathogens *S. aureus* and *S. pneumoniae* and adequate activity against MRSA strains, while maintaining broad-spectrum antibacterial activity [33, 59, 95, 96]. During the initial investigation of the 6-desfluoro-quinolones, it was noticed that removal of the fluorine at the position C-6 systematically lowered the level of genotoxicity, based on a comparison of the non-fluorinated quinolones and the corresponding 6-fluoroquinolones [33]. It then became clear that the 6-desfluoro-quinolones series was a useful platform for reoptimization of the quinolone backbone toward improved potency against *S. aureus* and *S. pneumoniae*. The 6-desfluoro-quinolones, for example garenoxacin, show greater potency than the newer fluoroquinolone moxifloxacin against both sensitive and resistant strains of these organisms [56].

Moreover, recent studies revealed that good antibacterial activity was retained by replacing the C-6 fluorine atom with an amino- group and reoptimizing the other substituents.

For the 6-amino compounds as well, the overall potency is highly dependent on substituents at positions C-7 and C-8. Thus, two 6-aminoquinolones I and II (illustrated in Figure 7), displayed the best activity against Gram-positive bacteria. Further structural modification of 6-aminoquinolones led to the discovery of 6-amino-8-methyl-quinolone (III), which is about 50-times more potent than ciprofloxacin against *S. aureus* [97-99]. The fluorine at C-6 is also replaced by nitro- group to get highly potent nitroquinolones [100]. They are potent inhibitor of *Streptococcus* and *Staphylococcus*. Other C-6 substituents that have been investigated, with less satisfactory results, are Cl, Br, acetyl, CN, and methyl groups.

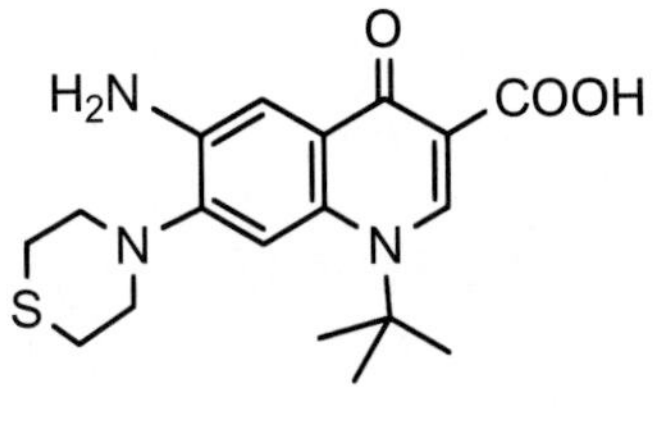

6-Aminoquinolone I

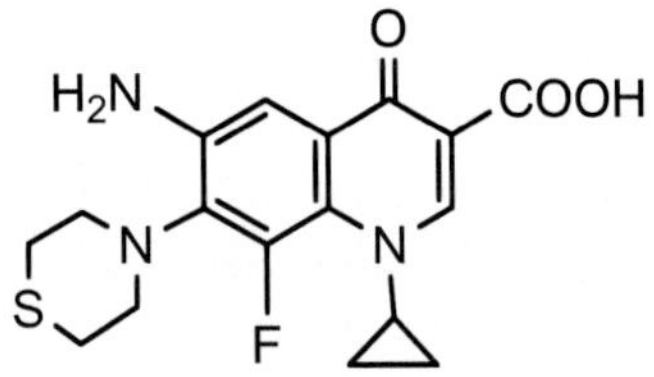

6-Aminoquinolone II

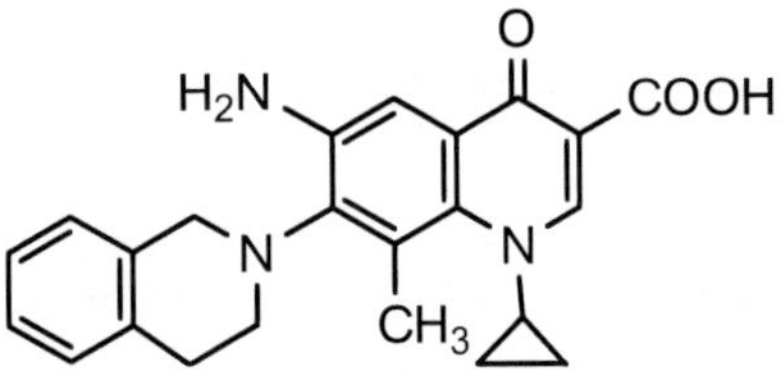

6-Aminoquinolone III

Figure 7. 6-Aminoquinolones.

Position 7

Many thousands of analogs have been prepared employing various substituents at C-7 position, leading to the conclusion that a cyclic amino- moiety is usually best. Furthermore, this 7-cyclic amine moiety of quinolones possesses enough structural flexibility to allow product optimization. The substituents at position C-7 are associated with a number of key attributes, such as antibacterial spectrum, bioavailability and adverse effects. The nature of the C-7 substituent also strongly affects the target preferences (DNA gyrase and/or topoisomerase IV) of quinolones. A five- or six-membered cycloamino moiety (such as, pyrrolidine or piperazine rings, respectively) is the most commonly found substitution at C-7 position. After the discovery of norfloxacin the piperazinyl moiety or its *N*-methyl analogue became a standard feature among quinolones. Unsubstituted piperazine rings [for example in norfloxacin, enoxacin or ciprofloxacin] confer potency against Gram-negative bacteria, while the addition of methyl groups can improve both oral absorption and anti-staphylococcal activity (generally, anti-Gram-positive activity) [5, 29, 101]. Pefloxacin, fleroxacin, ofloxacin and levofloxacin have a *N*-methyl piperazine, lomefloxacin, grepafloxacin and gatifloxacin have a 3-methyl piperazine, and sparfloxacin has a 3,5-dimethyl piperazine. These substituted piperazine-containing compounds have greater activity against Gram-positives, and are believed to have enhanced penetration into the bacterial cell [10, 102].

Pyrrolidinyl substituents with a pendant amine are also common and suitable substitution at position 7, and are associated with enhanced activity against Gram-positives. Clinafloxacin and tosufloxacin both have a 3-amino pyrrolidine substituent and sitafloxacin has a spiro-aminopyrrolidine group at C-7 [78, 80]. Similar to piperazine ring, introduction of methyl groups on the pyrrolidine ring also enhances activity against Gram-positive pathogens and helps to overcome some physicochemical and pharmacokinetics disadvantages [86, 101].

The introduction of bicyclic amino- groups onto position 7 has resulted in agents (moxifloxacin and trovafloxacin) with significant anti-Gram-positive activity and marked lipophilicity [82, 103]. Moxifloxacin has a diazabicyclic ring, while trovafloxacin has an amino-azabicyclic ring [82, 103, 104].

A number of quinolones (e.g., garenoxacin) have a certain aryl moiety instead of cyclic amine at C-7. Garenoxacin has (isoindolin-5-yl) moiety at C-7 with outstanding Gram-positive coverage, especially against *S. aureus* and *S. pneumoniae* [56, 105].

The pyrrolidinyl or piperazinyl moiety attached to C-7 position of quinolones possesses enough structural flexibility to allow modification and product optimization. In addition, a position on the quinolone molecule, where substitutions of bulky groups are permitted, is C-7 position [24, 106]. In addition, it has been proposed that for *S. aureus* and other Gram-positive organisms, increasing molecular mass and bulkiness of substituent at C-7 position is not a barrier to penetration [107]. Staphylococci do not possess an outer membrane, and so lack outer membrane proteins and lipopolysaccharide. Accumulation of quinolone drugs by *S. aureus* is thought to take place by simple diffusion across the cytoplasmic membrane [108, 109]. Accordingly, comprehensive structure-activity relationship studies on moieties attached to the C-7 cyclic amine of quinolones have been explored to identify the optimum structural requirements for designing the new anti-staphylococcal agents. We described different series of *N*-substituted piperazinyl quinolones containing certain bulky substituent connected to the piperazine unit of 7-piperazinyl quinolones (Figure 8). Some of these derivatives exhibit high activity against staphylococci more potent than their parent *N*-piperazinyl quinolones [106, 110-123].

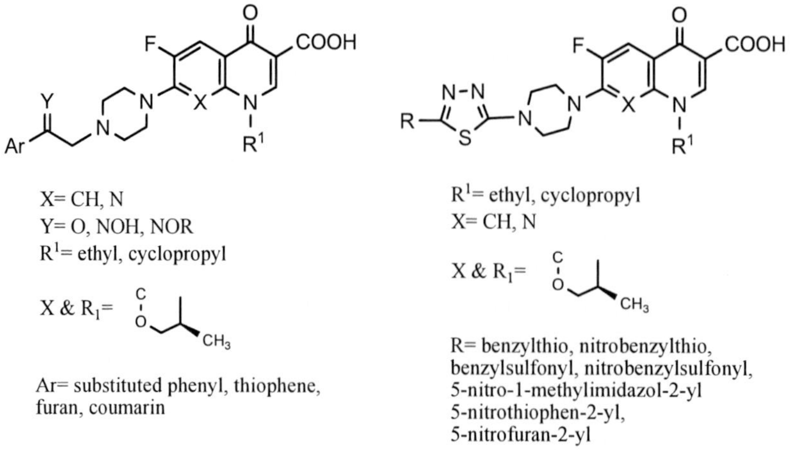

Figure 8. *N*-substituted piperazinyl quinolones.

Because of high tolerance for structural variation at the C-7 position of the quinolone ring, several quinolone-containing hybrids via C-7 connection, including quinolone-nitrothiophene, quinolone-nitrofuran and quinolone-nitroimidazole hybrids have been synthesized and evaluated [119-123]. The comparison of MIC values of these hybrids and their parent quinolones revealed that some of them have a more potent antibacterial activity against *S. aureus* (Figure 9).

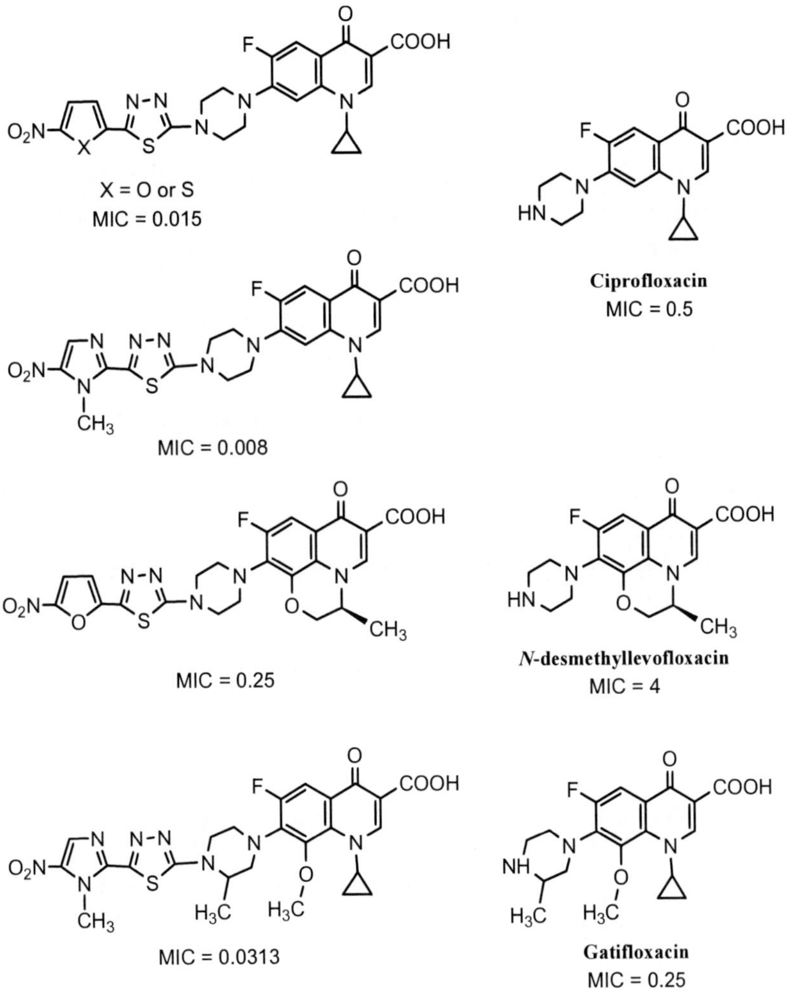

Figure 9. Quinolone-nitrofuran, quinolone-nitrothiophene and quinolone-nitroimidazole hybrids with more potent antibacterial activity against *S. aureus* relative to their parent quinolones (compared by MICs, μg/mL).

In order to find new C-7 modified quinolones, Kerns et al. described symmetric and asymmetric C-7-piperazinyl-linked dimers of quinolones (Figure 10) possessing potent antibacterial activity against drug-resistant strains of *S. aureus* [76, 77].

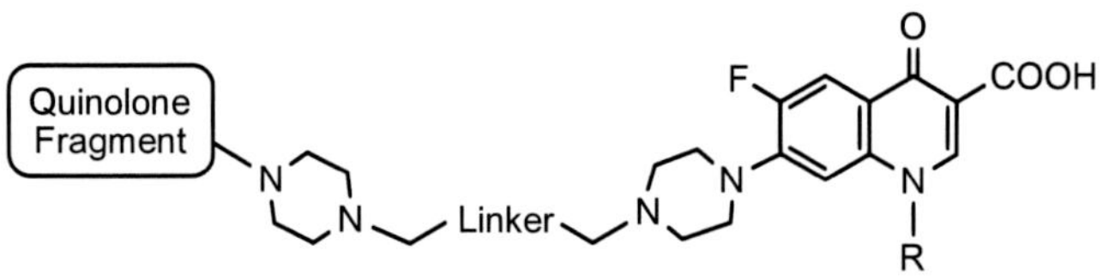

Figure 10. C-7-piperazinyl-linked dimers of quinolones, possessing potent antibacterial activity against drug-resistant strains of *S. aureus*.

Position 8

Substituents at C-8 position play an important role in altering oral pharmacokinetics, broadening the spectrum of activity and reducing the selection of mutants [63, 98, 124, 125]. Although, a number of quinolones such as grepafloxacin have no substituent at the position 8, but have a good activity against *S. aureus* and other Gram-positive pathogens. In addition, some naphthyridines (tosufloxacin, gemifloxacin, and trovafloxacin) in which C-8 of quinolone is replaced by a nitrogen atom, have excellent activity against *S. aureus*.

Halogen substituents, as well as a methyl or methoxy also increase the in vitro activity against *S. aureus*, even in those bacteria resistant to older fluoroquinolones. Early work showed that the 8-fluoro group (as in lomefloxacin and sparfloxacin) and 8-chloro group (as in clinafloxacin and sitafloxacin) were the most favorable substituents in terms of antibacterial activity; however, quinolones containing C-8 halogens tend to exhibit phototoxicity and other unacceptable side-effects. In contrast, newer quinolones such as gatifloxacin and moxifloxacin, with a methoxy group at the C-8 position exhibit enhanced activity against Gram-positives but does not seen to carry any risk of phototoxicity and cytotoxicity [11, 12, 85, 86, 103, 124, 125].

Lu et al. examined bacteriostatic and bactericidal activities of 12 fluoroquinolones against a gyrase-topoisomerase IV double mutant of *S. aureus*. They indicated that C-8 halogen and C-8 methoxy groups enhanced activity against *S. aureus* and the MIC_{99} was reduced seven- to eightfold for the *S. aureus* mutant by C-8 bromine, chlorine, and methoxy groups. Also, they showed that mutant *S. aureus* was affected more than the wild-type by the addition of a C-8 substituent. C-8 halogen and methoxy groups also improved the ability to kill staphylococci [63, 126].

Newer Quinolones with Potential Anti-MRSA Activity

Several newly developed quinolones and related compounds including 6-fluoroquinolones (Figure 11), 6-fluoronaphthyridones (Figure 12), 6-desfluoroquinolones (Figure 13), quinazolinediones and 4-quinolizinones (2-pyridones) (Figure 14) exhibit enhanced activity against *S. aureus* and have proved useful against MRSA. The structures and activities of these categories of quinolones will be discussed in this section.

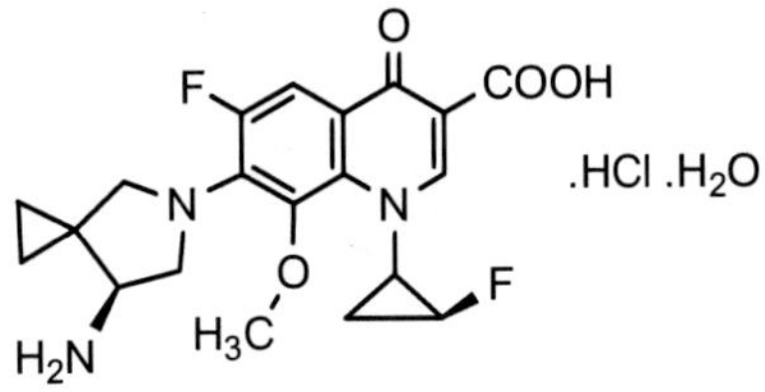

DK-507k

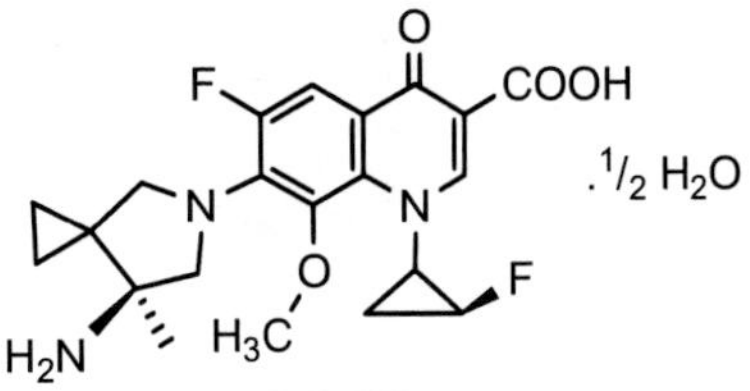

DC-159a

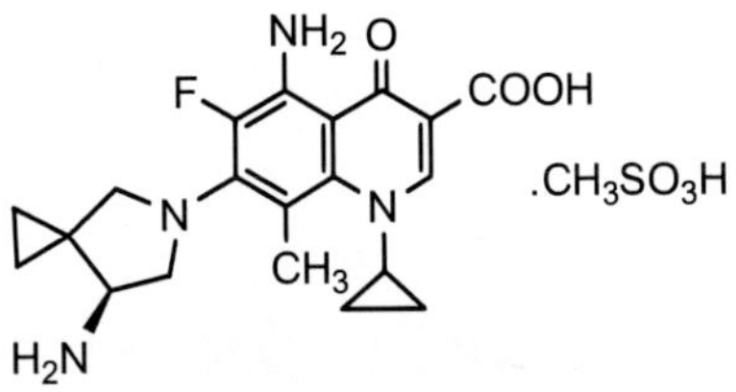

Olamufloxacin (HSR-903)

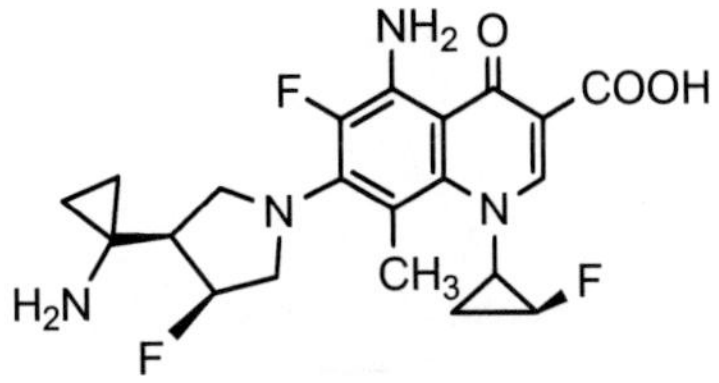

DQ-113

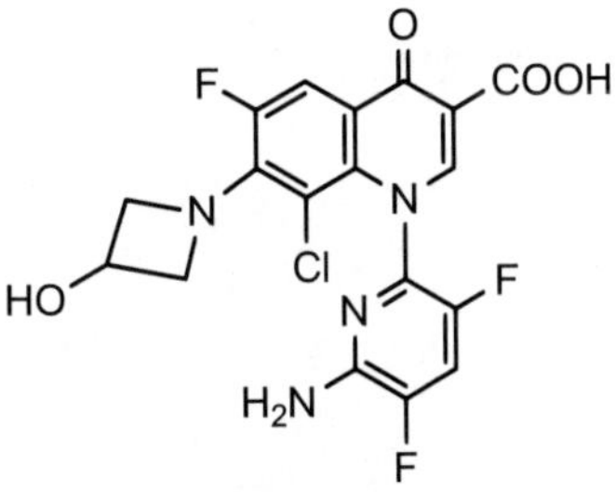

Delafloxacin (ABT-492, RX-3341)

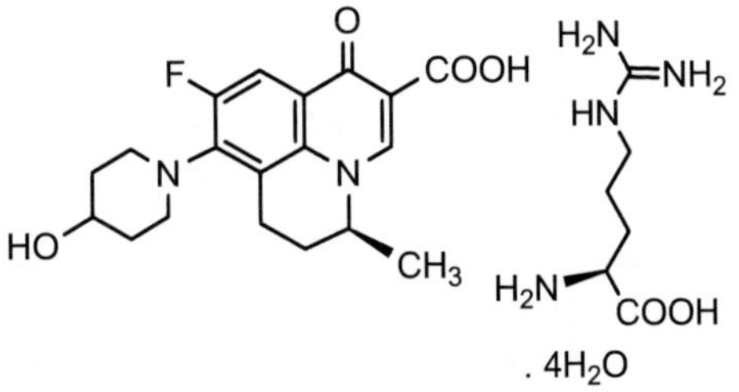

WCK 771

Figure 11. Newer 6-fluoroquinolones possessing potent anti-MRSA activity.

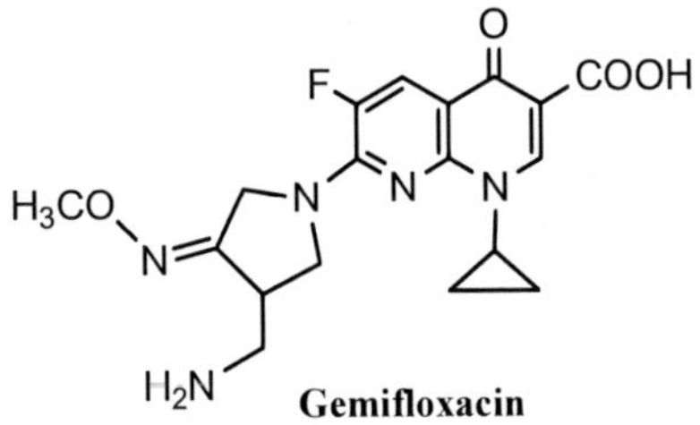

Gemifloxacin

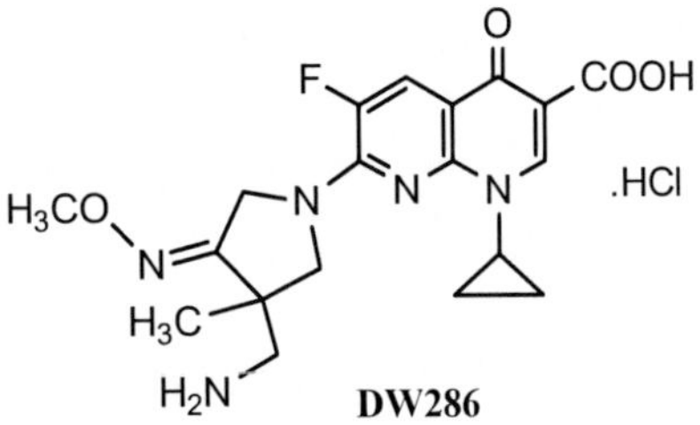

DW286

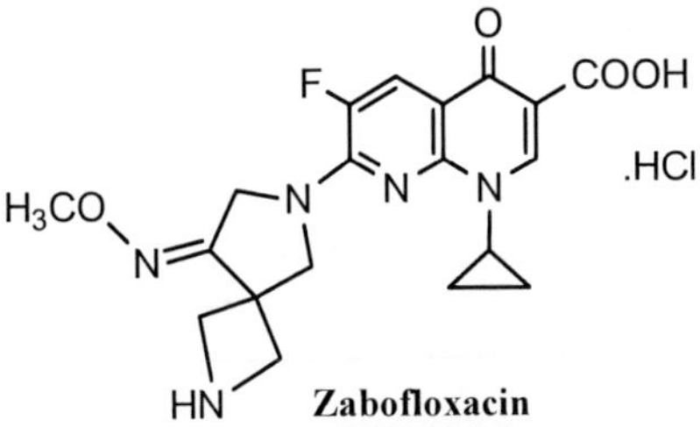

Zabofloxacin (DW-224a)

Figure 12. Newer 6-fluoronaphthyridones exhibiting potent anti-MRSA activity.

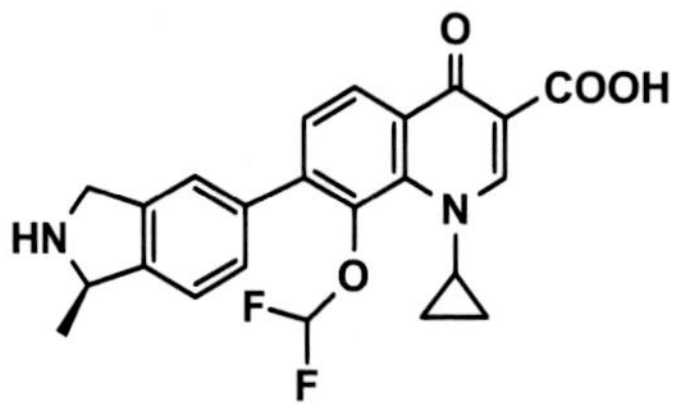

Garenoxacin

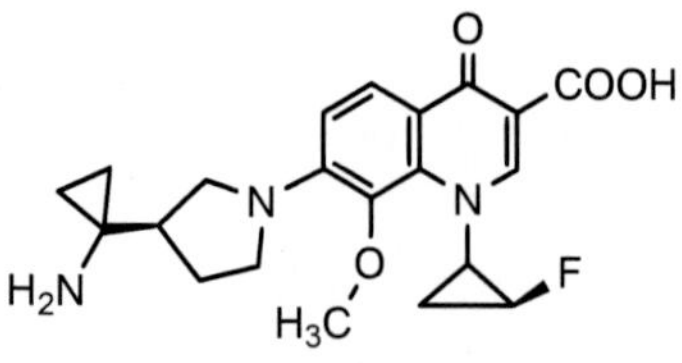

DX-619

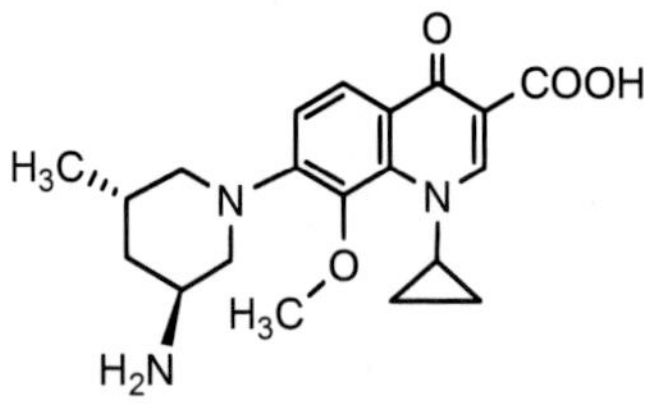

Nemonoxacin (TG-873870)

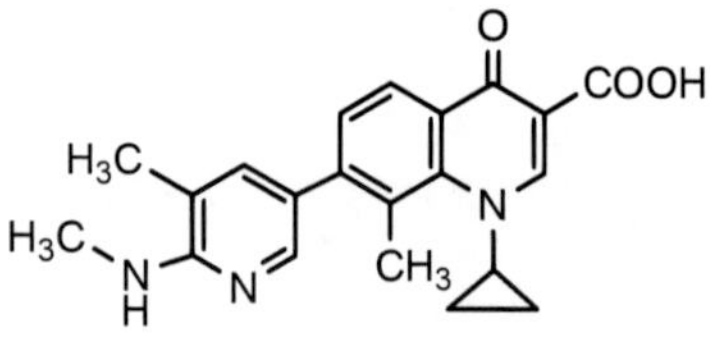

Ozenoxacin (T-3912)

Figure 13. Newer 6-desfluoroquinolones having potent anti-MRSA activity.

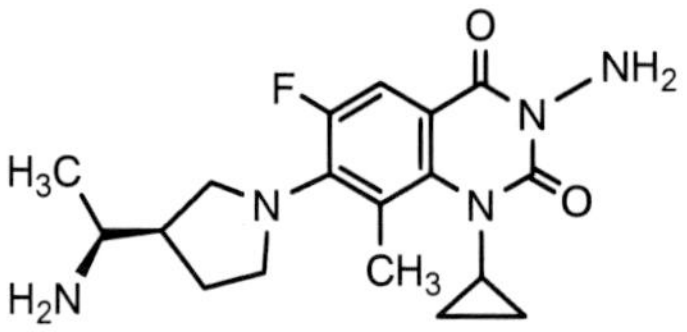

PD 0305970

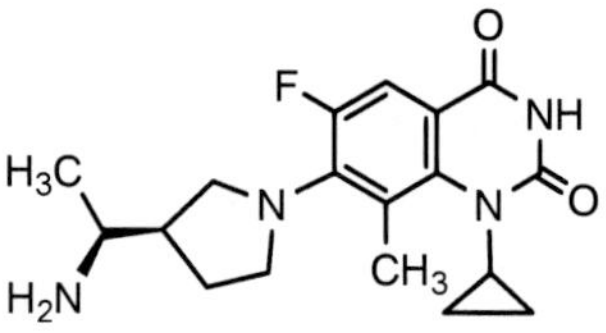

PD 0326448

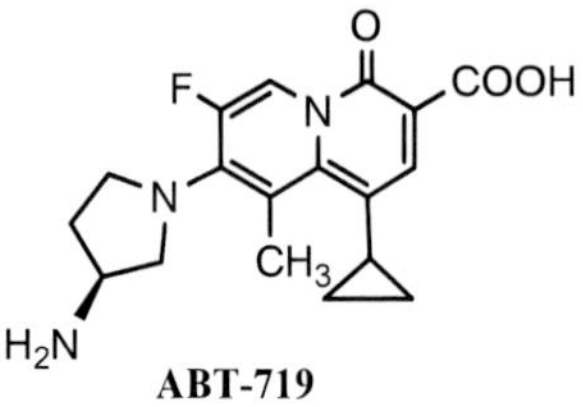

ABT-719

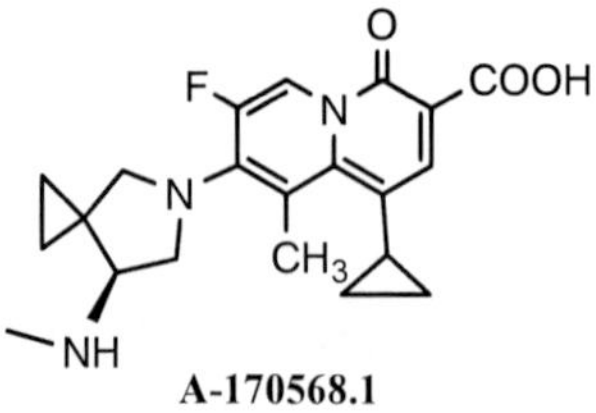

A-170568.1

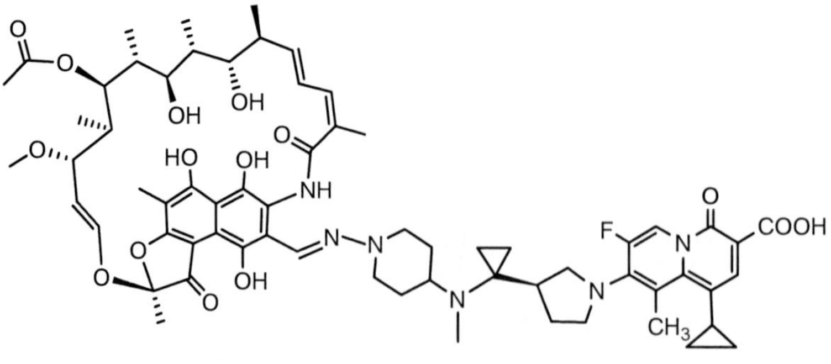

CBR-2092

Figure 14. Quinolone related compounds with potential anti-MRSA activity.

6-FLUOROQUINOLONES

DK-507k

DK-507k, 7-[(7*S*)-7-amino-5-azaspiro [2.4]heptan-5-yl]-6-fluoro-1-[(1*R*,2*S*)-2-fluoro-1-cyclopropyl]-1,4- dihydro-8-methoxy-4-oxo-3-quinolinecarboxylic acid hydrochloride hydrate (Daiichi, Japan) is an 8-methoxyquinolone bearing a modified 3-aminopyrrolidine.

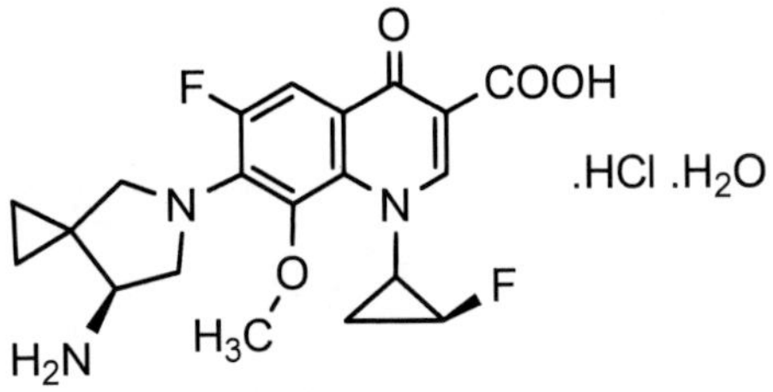

DK-507k

It is structurally related to sitafloxacin. DK-507k possesses an 8-methoxy group instead of 8-chloro substituent of sitafloxacin. Chlorine substituent, as well as a methoxy at C-8 position increases the in vitro activity against *S. aureus*, even in those bacteria resistant to older fluoroquinolones. Several studies showed that the 8-chloro group (as in clinafloxacin and sitafloxacin) was the most favorable substituents in terms of antibacterial activity [12, 103]; however, quinolones containing C-8 halogens tend to exhibit phototoxicity and other unacceptable side-effects. In contrast, newer quinolones such as DK-507k, with a methoxy group at the C-8 position presumably exhibit enhanced activity against Gram-positives and lesser risk of phototoxicity.

DK-507k had a broad antibacterial spectrum against both Gram-positive and Gram-negative bacteria [127]. Otani et al. compared the antibacterial activities of DK-507k with those of other quinolones including ciprofloxacin, gatifloxacin, levofloxacin, moxifloxacin, sitafloxacin, and garenoxacin. Against MSSA strains, DK-507k showed activities comparable to those of sitafloxacin and moxifloxacin, twofold lower than those of garenoxacin, and two- to fourfold higher than those of the other quinolones tested at the MIC_{90}s (Table 2). Against ofloxacin-susceptible MRSA strains, DK-507k showed activities comparable to those of sitafloxacin; 2-fold lower than those of garenoxacin; and 2- to 16-fold higher than those of levofloxacin, ciprofloxacin, gatifloxacin, and moxifloxacin (Table 3). Furthermore, against ofloxacin-resistant MRSA strains the activities of DK-507k and sitafloxacin were the highest among those of the quinolones tested [128].

Table 2. Comparative MICs (μg/mL) of DK-507k and other quinolones against 25 strains of MSSA.

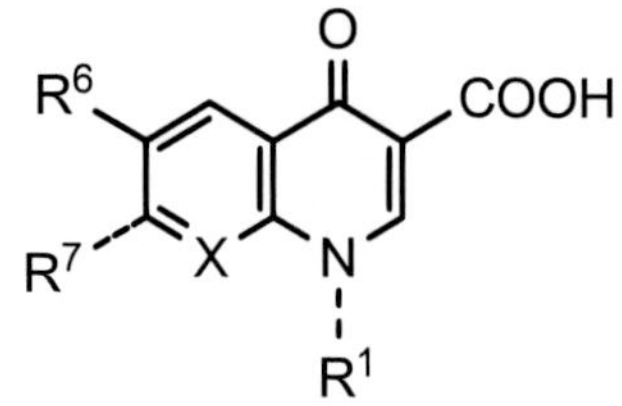

Compound	X	R^1	R^6	R^7	Range	MIC_{50}	MIC_{90}
DK-507k	$C{-}OCH_3$		F	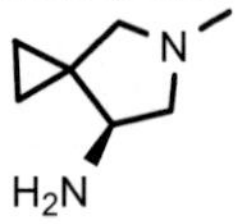	0.03–0.25	0.06	0.12
Sitafloxacin	$C{-}Cl$		F	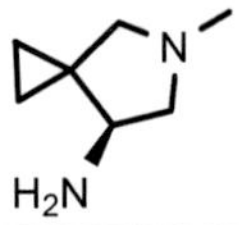	0.008–0.12	0.03	0.12
Ciprofloxacin	$C{-}H$		F	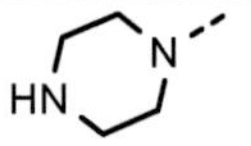	0.25–4	0.5	1
Gatifloxacin	$C{-}OCH_3$		F	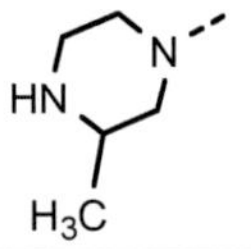	0.06–0.5	0.12	0.25
Moxifloxacin	$C{-}OCH_3$		F	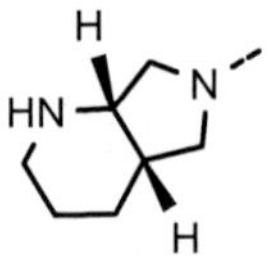	0.06–0.25	0.06	0.12
Garenoxacin	$C{-}OCHF_2$		H	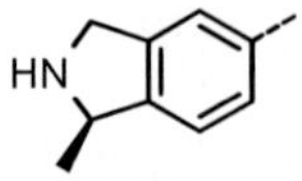	≤0.004–0.12	0.03	0.06
Levofloxacin	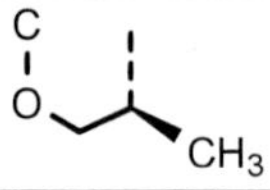		F	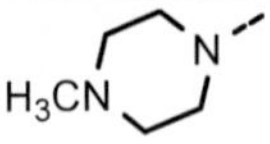	0.12–1	0.25	0.5

Table 3. Comparative MICs (µg/mL) of DK-507k and other quinolones against MRSA.

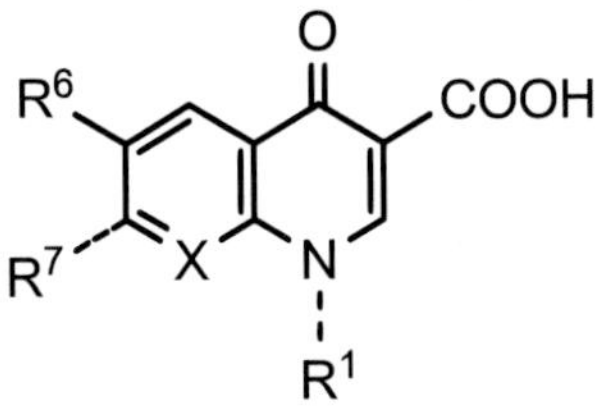

Compound	X	R[1]	R[6]	R[7]	Organism (number)					
					MRSA (24) , ofloxacin susceptible			MRSA (25), ofloxacin resistant		
					Range	MIC_{50}	MIC_{90}	Range	MIC_{50}	MIC_{90}
DK-507k	C–OCH_3	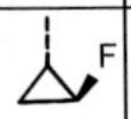	F	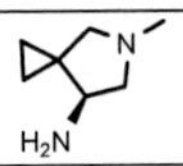	0.015–0.06	0.03	0.06	0.25–32	1	4
Sitafloxacin	C–Cl	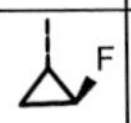	F	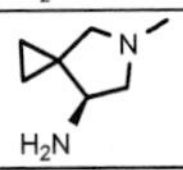	0.008–0.06	0.03	0.06	0.25–32	1	4
Ciproflox-acin	C–H		F	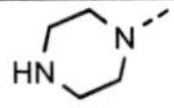	0.25–2	0.5	1	16–>128	128	>128
Gatifloxacin	C–OCH_3		F	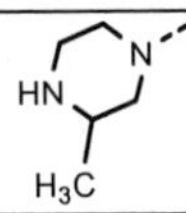	0.06–0.25	0.12	0.25	2–>128	8	16
Moxifloxacin	C–OCH_3		F	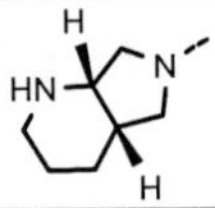	0.03–0.25	0.12	0.12	1–64	4	8
Garenoxacin	C–$OCHF_2$		H	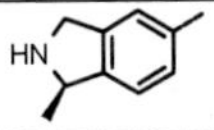	0.008–0.06	0.03	0.03	0.25–64	2	8
Levofloxacin	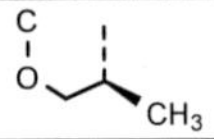		F	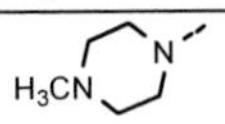	0.12–1	0.25	0.5	4–>128	16	64

The efficacy of DK-507k in comparison with gatifloxacin, moxifloxacin and levofloxacin has been also evaluated against septicemia models with MSSA and MRSA in mice (Table 4). DK-507k was also the most active quinolone against MSSA and MRSA (ED_{50} = 1.07–9.23 mg/kg) [128].

Pharmacokinetics studies with radiolabeled DK-507k in rats and monkeys showed it to be rapidly absorbed after oral administration and to be excreted predominantly by the biliary route [129]. Urinary excretion in rats and monkeys was approximately 19.4 and 41.0% of the administrated dose, respectively. Plasma protein binding in rats, monkeys and humans was about 40%.

Table 4. Efficacy of DK-507k in comparison with gatifloxacin, moxifloxacin and levofloxacin on septicemia models in mice

Compound	Organism					
	S. aureus 037114 (MSSA)			*S. aureus* 037004 (MRSA)		
	MIC (μg/mL)	ED_{50} (mg/kg)	95% confidence interval (mg/kg)	MIC (μg/mL)	ED_{50} (mg/kg)	95% confidence interval (mg/kg)
DK-507k	0.06	1.07	0.74–1.45	1	9.23	5.14–16.94
Gatifloxacin	0.12	4.66	3.84–5.43	4	67.36	46.01–142.96
Moxifloxacin	0.06	5.48	3.34–9.82	2	56.70	45.39–73.08
Levofloxacin	0.25	16.20	12.02–19.57	16	>100.00	

DC-159a

DC-159a, (+)-7-[(7*S*)-7-amino-7-methyl-5-azaspiro [2.4]heptan-5-yl]-6-fluoro-1-[(1*R*,2*S*)-2-fluoro-1-cyclopropyl]-1,4- dihydro-8-methoxy-4-oxo-3-quinolinecarboxylic acid hemihydrate is a novel orally administered fluoroquinolone developed by Daiichi Pharmaceutical Co., Ltd. (Tokyo, Japan).

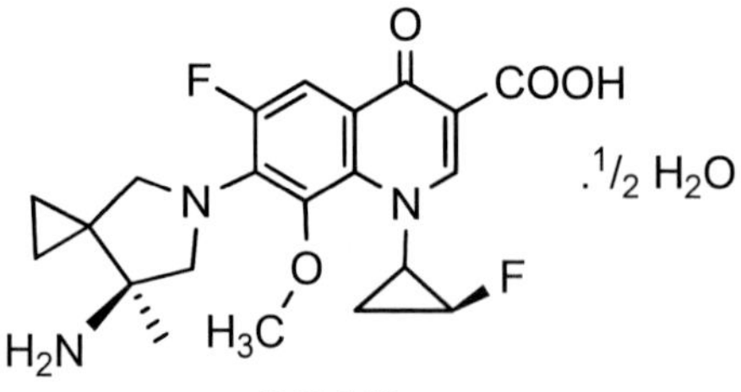

DC-159a

The distinctive substituents of DC-159a are a (1*R*,2*S*)-2-fluoro-1-cyclopropyl at N-1 position, a 7-[(7*S*)-7-amino-7-methyl-5-azaspiro [2.4]heptan-5-yl]- residue and an 8-methoxy group. The 2-fluoro-1-cyclopropyl group is also found in the structures of other new quinolones including sitafloxacin, DQ-113, DK-507k, and DX-619. The 7-amino-5-azaspiro [2.4]heptan-5-yl-residue is a new functionality for the C-7 position of quinolones which is found in sitafloxacin, DK-507k, A-170568.1 and HSR-903. The 8-methoxy group is also found in earlier quinolones such as gatifloxacin and moxifloxacin, and in new investigational quinolones for example DK-507k, DX-619, WCK 1152, WCK 1153 and nemonoxacin (TG-873870). Indeed, DC-159a is a methyl analog of DK-507k.

DC-159a is an 8-methoxy fluoroquinolone that possesses a broad spectrum of antibacterial activity, with extended activity against Gram-positive pathogens, especially streptococci and staphylococci from patients with community-acquired infections.

The antibacterial activities of DC-159a and the comparator drugs ciprofloxacin, moxifloxacin, garenoxacin, gemifloxacin and levofloxacin were evaluated by Hoshino et al. against Gram-positive and Gram -negative bacteria [130]. The MIC_{90} of DC-159a against MSSA and quinolone-susceptible MRSA was 0.06 μg/mL (Table 5). Against MSSA, DC-159a was 32-fold more active than ciprofloxacin, 8-fold more active than levofloxacin, 2-fold more active than moxifloxacin and gemifloxacin, and 2-fold less active than garenoxacin. Against quinolone-resistant MRSA, DC-159a (MIC_{90} = 8 μg/mL) showed the highest activity compared with the activities of the other quinolones tested. The MIC_{90} of DC-159a against methicillin-susceptible coagulase-negative staphylococci (MSCNS) and methicillin-resistant coagulase-negative staphylococci (MRCNS) was 0.5 μg/mL. Against MRCNS, DC-159a showed the highest activity compared with the activities of the other quinolones tested (Table 6).

In a separate study, Jones et al. evaluated the activity of DC-159a against clinical isolates of staphylococci (Table 7). DC-159a exhibited potent activity against MSSA strains, with MIC_{50}s and MIC_{90}s of 0.03 μg/mL [131]. This level of activity was 16-fold greater than that of ciprofloxacin but 2-fold less than that of gemifloxacin (MIC_{90} = 0.015 μg/mL). MRSA strains derived from patients with nosocomial infections showed the greatest susceptibility to DC-159a (MIC_{90} = 2 μg/mL). However, the range of DC-159a MICs was 32- to 64-fold greater than that for MSSA strains. The activity of DC-159a against 30 well-characterized isolates from patients with CA-MRSA infection was tested, and the MIC_{50} and MIC_{90} results for these strains were identical to those for the MSSA strains. Overall, among the five groups of staphylococci tested, DC-159a at ≤0.5 μg/mL inhibited all isolates in four of the groups (the exception was the nosocomial MRSA group). DC-159a showed a potency most similar to that of gemifloxacin (MIC_{90} range, 0.016 to 2 μg/mL) and markedly greater than the potencies of the other fluoroquinolones tested.

Olamufloxacin (HSR-903)

Olamufloxacin (HSR-903) is a newly synthesized oral fluoroquinolone with a broad spectrum of antibacterial activity. It was first synthesized by Hokuriku Pharmaceutical Co. Ltd., Fukui, Japan [132, 133], and its chemical name is 5-amino-7-[(7*S*)-7-amino-5-azaspiro [2.4]heptan-5-yl]-1-cyclopropyl-6-fluoro-1,4-dihydro-8-methyl-4-oxoquinoline-3-carboxylic acid methanesulfonate.

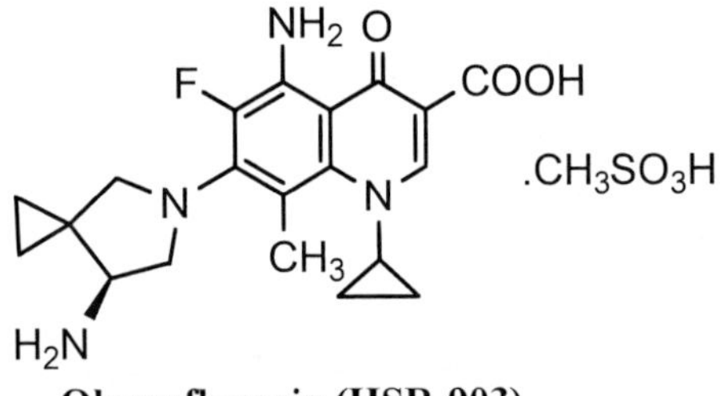

Olamufloxacin (HSR-903)

The distinctive structural features of olamufloxacin are an amino moiety at the C-5 position, a 7-amino-5-azaspiro [2.4]heptan-5-yl ring at the C-7 position, and a methyl group at the C-8 position. In olamufloxacin, the 8-methyl group would contribute to the high hydrophobicity. In general, a 5- amino moiety, as found in sparfloxacin and DQ-113, enhances potency against Gram-positive bacteria. Few quinolones such as DQ-113, T-3912, have a methyl group at the C-8 position. The 7-amino-5-azaspiro [2.4]heptan-5-yl ring, as found in sitafloxacin, DK-507k, A-170568.1 and DC-159a is a cycloproyl-fused aminopyrrolidine ring. The aminopyrrolidine scaffold as found in tosufloxacin, clinafloxacin, and ABT-719 also enhances potency against Gram-positive bacteria [10]. In olamufloxacin, these substituents presumably contribute to the strong activity against Gram-positive bacteria.

The preliminary antibacterial data of olamufloxacin against MRSA were published by Takahashi et al. [134] and Watanabe et al. [135], respectively. In the comparative study of Takahashi et al., olamufloxacin inhibited 90% of MSSA and MRSA clinical isolates at 0.78 and 1.56 mg/mL, respectively, and its activity against MRSA was 16-fold higher than those of sparfloxacin and levofloxacin and 64-fold higher than that of ciprofloxacin (Table 8).

The in vitro activity of olamufloxacin against clinical isolates of respiratory pathogens was evaluated by Watanabe et al. [135] in comparison with ciprofloxacin, sparfloxacin and ofloxacin (Table 9). Olamufloxacin was 2 to 32 times more active than ofloxacin, ciprofloxacin, and sparfloxacin against Gram-positive bacteria, including MSSA and MRSA.

One of the reasons why the antibacterial activity ratio of olamufloxacin to ciprofloxacin is greater for MRSA and ciprofloxacin-resistant *S. aureus* than that for MSSA is considered to be that olamufloxacin, with its high hydrophobicity, is little influenced by NorA. Quinolones with high hydrophobicity, such as sparfloxacin, are also reported to be active against a *norA*-mediated resistant strain [45, 136].

After oral administration to animals, olamufloxacin is well absorbed and is distributed into various tissues, including lung and kidney, but not the central nervous system [137]. It was demonstrated that olamufloxacin efficiently distributes in both epithelial lining fluid and alveolar macrophage, two potential and representative sites for pulmonary infections [138]. In the view of kinetic analysis, there existed the higher influx than efflux clearances across the alveolar barrier in favor of olamufloxacin penetration into epithelial lining fluid. Moreover, both rapid permeability across the alveolar macrophage cell membranes and avid binding to the membrane phospholipids were referred to account for the high accumulation of olamufloxacin within alveolar macrophage. In a toxicological study, olamufloxacin (30 mg/kg, intravenously) did not cause phototoxicity in guinea pigs [132].

Table 5. Chemical structures and comparative MICs (µg/mL) of DC-159a and other quinolones against MSSA and MRSA.

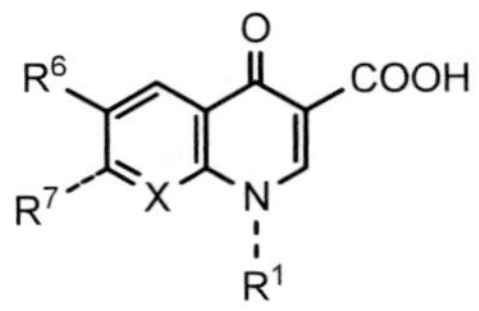

Compound	X	R[1]	R[6]	R[7]	Organism (number)					
					MSSA (48)			Quinolone-susceptible MRSA (46) [Quinolone-resistant MRSA (49)]		
					Range	MIC_{50}	MIC_{90}	Range	MIC_{50}	MIC_{90}
DC-159a	C–OCH_3		F	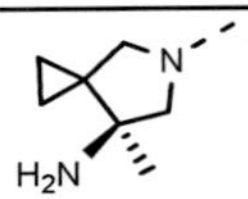	0.015–1	0.03	0.06	0.015–0.06 [0.25–16]	0.03 [1]	0.06 [8]
Ciprofloxacin	C–H		F	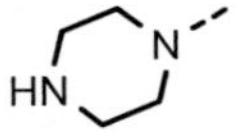	0.12–32	0.25	2	0.12–1 [8–>64]	0.5 [32]	0.5 [>64]
Moxifloxacin	C–OCH_3		F	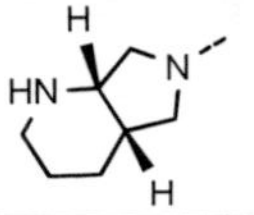	0.015–2	0.06	0.12	0.015–0.06 [0.5–64]	0.03 [2]	0.06 [32]
Garenoxacin	C–$OCHF_2$		H	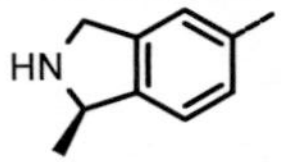	0.008–1	0.03	0.03	0.008–0.03 [0.25–32]	0.015 [1]	0.03 [32]
Gemifloxacin	N		F	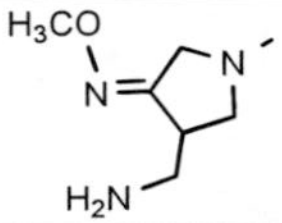	0.008–1	0.03	0.12	0.008–0.06 [0.25–128]	0.03 [1]	0.03 [32]
Levofloxacin	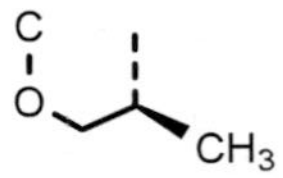		F	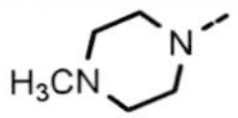	0.06–8	0.25	0.5	0.12–0.5 [4–>128]	0.25 [8]	0.25 [>128]

Table 6. Comparative MICs (µg/mL) of DC-159a and other quinolones against coagulase-negative staphylococci.

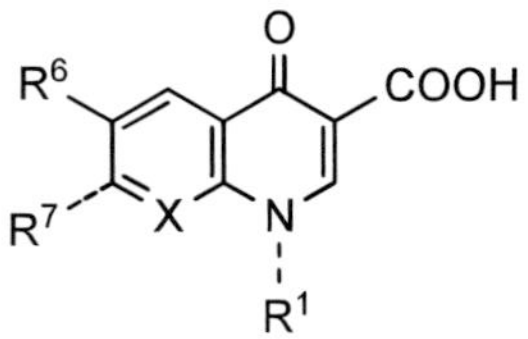

Compound	X	R[1]	R[6]	R[7]	Organism (number)					
					MSCNS (48)			MRCNS (46)		
					Range	MIC_{50}	MIC_{90}	Range	MIC_{50}	MIC_{90}
DC-159a	C–OCH_3		F	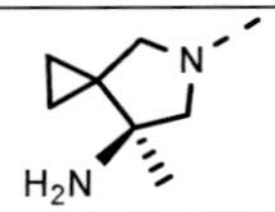	0.015–1	0.03	0.5	0.03–8	0.25	0.5
Ciprofloxacin	C–H		F	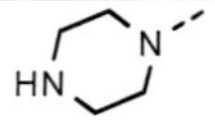	0.06–32	0.25	4	0.06–>64	4	32
Moxifloxacin	C–OCH_3		F	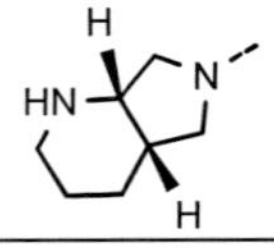	0.03–8	0.06	1	0.03–64	1	4
Garenoxacin	C–$OCHF_2$		H	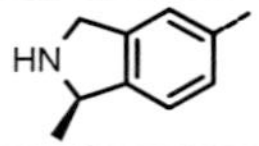	0.015–8	0.03	1	0.015–32	1	2
Gemifloxacin	N		F	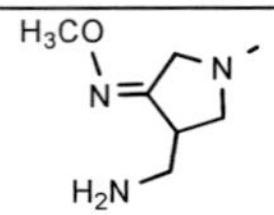	≤0.004–4	0.015	0.5	0.008–32	0.5	1
Levofloxacin	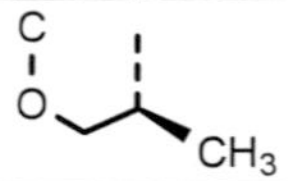		F	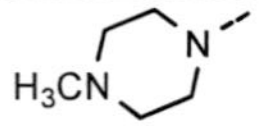	0.06–32	0.25	4	0.12–>128	4	16

Table 7. Comparative MICs (μg/mL) of DC-159a and other quinolones against MSSA and MRSA (nosocomial and community acquired isolates).

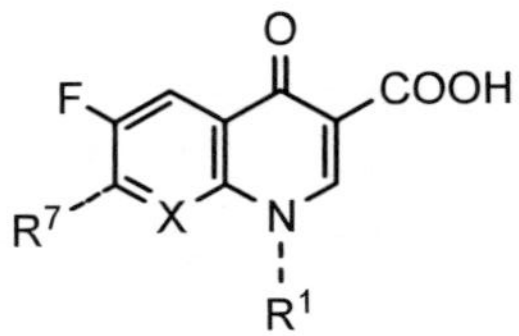

Compound	X	R^1	R^7	Organism (number)					
				MSSA (30)			MRSA, nosocomial (30) [community acquired (30)]		
				Range	MIC_{50}	MIC_{90}	Range	MIC_{50}	MIC_{90}
DC-159a	C–OCH_3		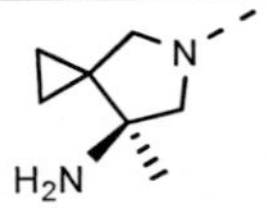	0.015–0.06	0.03	0.03	0.5–4 [0.008–0.5]	1 [0.03]	2 [0.03]
Ciprofloxacin	C–H		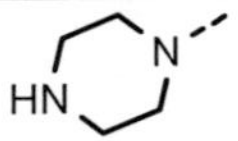	0.06–0.5	0.25	0.5	>4 [≤0.25–>4]	>4 [≤0.25]	>4 [0.5]
Moxifloxacin	C–OCH_3		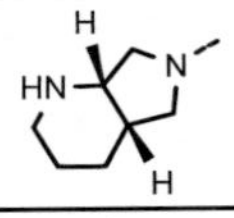	≤0.03–0.12	≤0.03	0.06	1–>4 [≤0.03–1]	2 [≤0.03]	4 [0.06]
Gatifloxacin	C–OCH_3		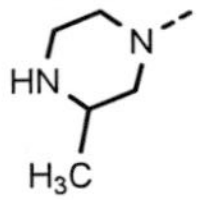	≤0.03–0.12	0.06	0.12	1–>4 [≤0.03–2]	4 [0.06]	>4 [0.12]
Levofloxacin	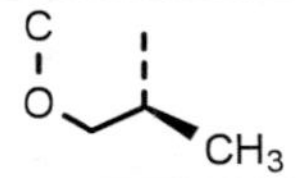		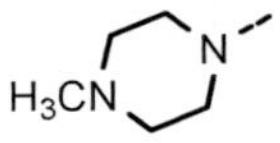	0.06–0.25	0.12	0.25	4–>8 [0.12–4]	>8 [0.12]	>8 [0.25]

Table 8. Comparative in vitro activities (MICs, µg/mL) of olamufloxacin (HSR-903) and other quinolones against clinical isolates of staphylococci.

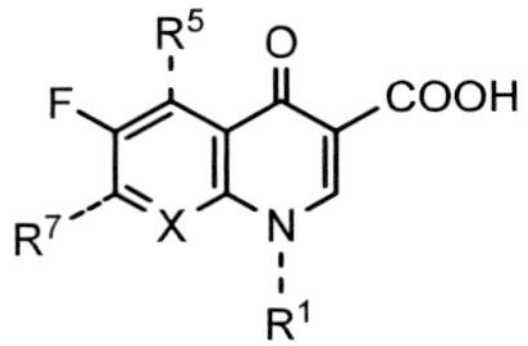

Compound	X	R[1]	R[5]	R[7]	Organism (number)					
					MSSA (38)			MRSA (33)		
					Range	MIC_{50}	MIC_{90}	Range	MIC_{50}	MIC_{90}
Olamufloxacin (HSR-903)	C–CH_3		NH_2	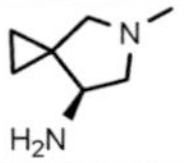	0.006–6.25	0.025	0.78	0.006–3.13	0.39	1.56
Ciprofloxacin	C–H		H	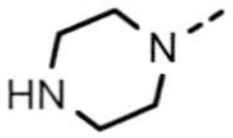	0.20–100	0.78	6.25	0.20– >100	12.5	100
Lomefloxacin	C–F		H	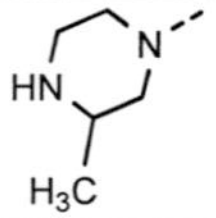	0.39–>100	0.78	25	0.20– >100	25	>100
Sparfloxacin	C–F		NH_2	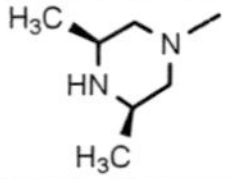	0.025–25	0.10	12.5	0.012–50	3.13	25
Levofloxacin	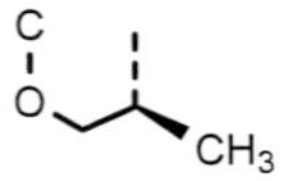		H	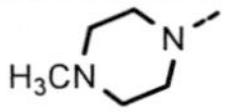	0.10–12.5	0.20	12.5	0.10–50	6.25	25

Table 9. Comparative in vitro activities (MICs, µg/mL) of olamufloxacin (HSR-903), ciprofloxacin, sparfloxacin and ofloxacin against clinical isolates of staphylococci.

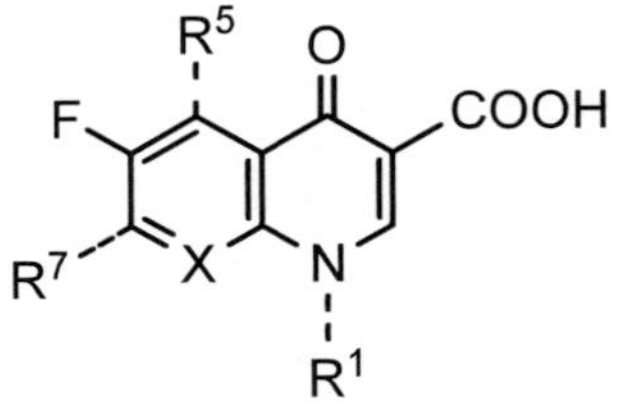

Compound	X	R^1	R^5	R^7	Organism (number)					
					MSSA (20)			MRSA (20)		
					Range	MIC_{50}	MIC_{90}	Range	MIC_{50}	MIC_{90}
HSR-903	C–CH_3		NH_2	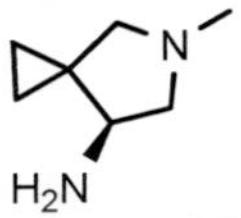	≤0.03–0.06	0.06	0.06	≤0.03–2	0.5	2
Ciprofloxacin	C–H		H	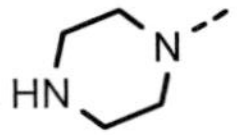	0.12–1	0.25	0.5	2–>64	16	>64
Sparfloxacin	C–F		NH_2	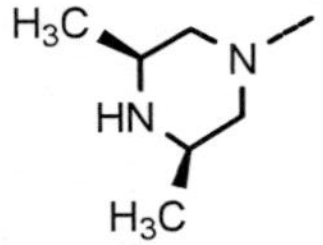	0.06–0.12	0.06	0.12	0.06–16	8	16
Ofloxacin	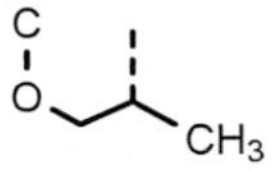		H	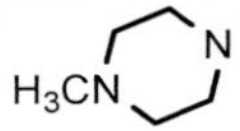	0.12–0.5	0.25	0.5	0.5–32	16	32

DQ-113

DQ-113 (formerly D61-1113), 5-amino-7-[(3*R*,4*S*)-3-(1-aminocyclopropyl)-4-fluoropyrrolidin-1-yl]-6-fluoro-1-[(1*R*,2*S*)-2-fluorocyclopropyl]-1,4-dihydro-8-methyl-4-oxoquinoline-3-carboxylic acid is a highly functionalized fluoroquinolone.

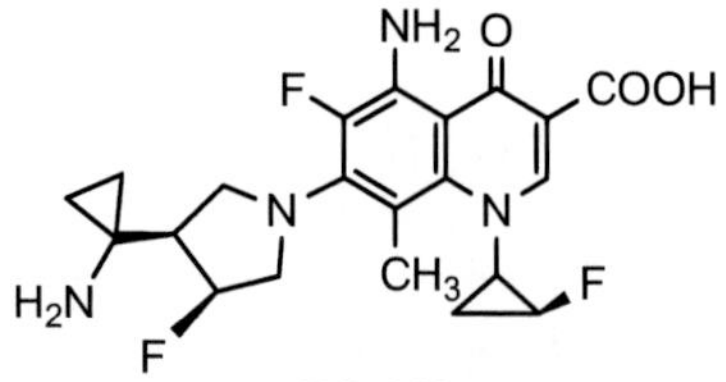

DQ-113

This poly-fluorinated compound characterized by having three fluoro- substituents located on quinolone core, cyclopropyl ring and pyrrolidine moiety. Its N-1 substituent is similar to sitafloxacin. The 5-amino group (as found in sparfloxacin) and the 8-methyl substituent (as found in ABT-719, olamufloxacin and T-3912) presumably contribute to enhance potency against Gram-positive pathogens. DQ-113 shares structural similarity with olamufloxacin based on possessing both 5-amino and 8-methyl groups.

Several in vitro and in vivo studies were reported with DQ-113. Tanaka *et al.* [139] published data on the antibacterial activity of DQ-113, in comparison with those of currently available quinolones, ciprofloxacin, gatifloxacin, gemifloxacin, levofloxacin, moxifloxacin, sitafloxacin, sparfloxacin, tosufloxacin, and garenoxacin (Tables 10 and 11). The MIC_{90}s of DQ-113 against MSSA and MSCNS were both 0.03 μg/mL. MIC_{90}s against ofloxacin-susceptible MRSA, ofloxacin-resistant (MIC of ofloxacin, ≥8 μg/mL) MRSA, and MRCNS were 0.008, 0.25, and 0.06 μg/mL, respectively. The antibacterial activity against MSSA was 2-fold higher than those of garenoxacin and gemifloxacin; 4-fold higher than those of sitafloxacin, tosufloxacin and moxifloxacin. Against ofloxacin-susceptible MRSA, DQ-113 activity was 4-fold higher than that of garenoxacin and at least 8-fold higher than those of the other reference compounds at MIC_{90} levels. Furthermore, DQ-113 showed the highest activity against ofloxacin-resistant MRSA among the compounds tested. Against MRCNS, the MIC_{90} of DQ-113 was at least 8-fold lower than those of the reference drugs.

Hong *et al.* compared DQ-113 respect to sitafloxacin, moxifloxacin, levofloxacin, and ciprofloxacin for potential to select mutational resistance in multi-resistant Gram-positive cocci. DQ-113 was the most effective agent for preventing the emergence of single-step mutants of MSSA and MRSA, with only one mutant selected from the 10 strains compared to mutants being selected from 4 strains by moxifloxacin, 5 strains by sitafloxacin, 7 strains by levofloxacin, and 8 strains by ciprofloxacin. In this study, all single-step mutants were susceptible to 0.03 μg/mL of DQ-113, 0.25 μg/mL of sitafloxacin, 0.5 μg/mL of moxifloxacin, 2 μg/mL of levofloxacin, and 8 μg/mL of ciprofloxacin, with frequencies of 10^{-6} and 10^{-10}, with no apparent differences between the quinolones [140].

Kaneko *et al.* [141] compared the effects of DQ-113 to those of vancomycin and teicoplanin in murine models of hematogenous pulmonary infections caused by MRSA and vancomycin-insensitive *S. aureus* (VISA). The MICs of DQ-113, vancomycin and teicoplanin for MRSA were 0.125, 1.0, and 0.5 μg/mL, respectively; and those for VISA were 0.25, 8.0, and 8.0 μg/mL, respectively. Kaneko *et al.* also concluded that DQ-113 effectively can reduce the number of bacteria in MRSA and VISA hematogenous infection models and significantly improves the rates of survival of immunocompromised mice infected with VISA compared with the rates achieved with vancomycin and teicoplanin. Thus, DQ-113 is potentially effective for the treatment of hematogenous pulmonary infections caused by MRSA and VISA strains.

Delafloxacin (ABT-492, RX-3341)

Delafloxacin (formerly A-319492, WQ-3034, ABT-492 and RX-3341) is a new fluoroquinolone that differs from other members of the class by two structural features: the 6-amino-3,5-difluoropyridine-2-yl moiety at the N-1 position and the 3-hydroxyazetidine-1-yl substituent at the C-7 position of the quinolone core. Delafloxacin is 1-(6-amino-3,5-difluoropyridine-2-yl)-8-chloro-6-fluoro-7-(3-hydroxyazetidine-1-yl)-4-oxo-1,4-dihydroquinoline-3-carboxylic acid, currently being developed by Rib-x Pharmaceutical Inc.

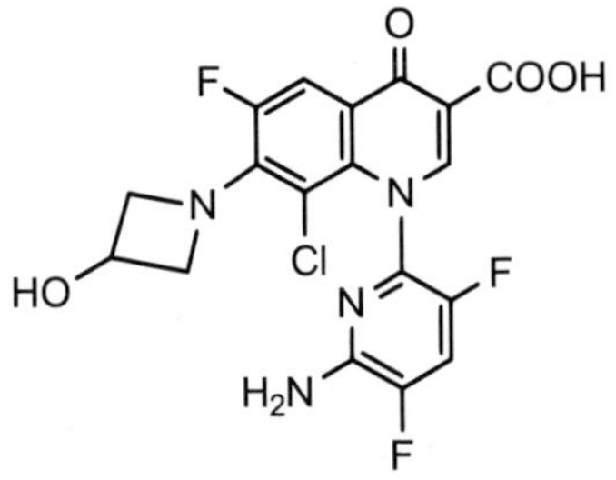

Delafloxacin
(ABT-492, RX-3341)

Preliminary data reported by Abbott Laboratories indicate that delafloxacin has a broad spectrum of activity with increased potency against multidrug-resistant bacteria [32]. Comparative study of the *in vitro* activity of delafloxacin published by Harnett et al. indicate that, this compound is significantly more active than other quinolones against staphylococci, but a higher MIC_{90} was observed for MRSA (0.5 μg/mL) compared with MSSA (0.008 μg/mL). Anti-staphylococcal activity of delafloxacin was superior for all species tested by Harnett et al., with good activity displayed against MRSA strains (MIC_{90} = 0.5 μg/mL) [142]. Delafloxacin has been shown to be at least 8-fold more potent than ciprofloxacin, gatifloxacin, moxifloxacin, gemifloxacin and levofloxacin (MIC_{90} of ≤ 0.5 μg/mL for delafloxacin versus > 4 μg/mL for all other quinolones) against ciprofloxacin resistant MRSA (Table 12).

In 2006, Rib-x Pharmaceutical Inc. licensed this antibacterial agent from a Japanese company, Wakunaga Pharmaceuticals, opening up a new antibiotic niche for Rib-x with invaluable clinical data. Based on activity against MRSA and other highly pathogenic microorganisms, delafloxacin has successfully undergone Phase II studies for oral treatment of respiratory infections, and Phase I studies as an intravenous formulation. These studies have shown delafloxacin to be both clinically efficacious and safe. The oral administration of delafloxacin has been shown to have no evidence of either phototoxicity or QTc prolongation in Phase I studies and has demonstrated efficacy in two Phase II clinical trials (community acquired pneumonia and bronchitis when compared to levofloxacin) [143].

Table 10. Comparative MICs (μg/mL) of DQ-113 and important quinolones against MSSA and MRSA.

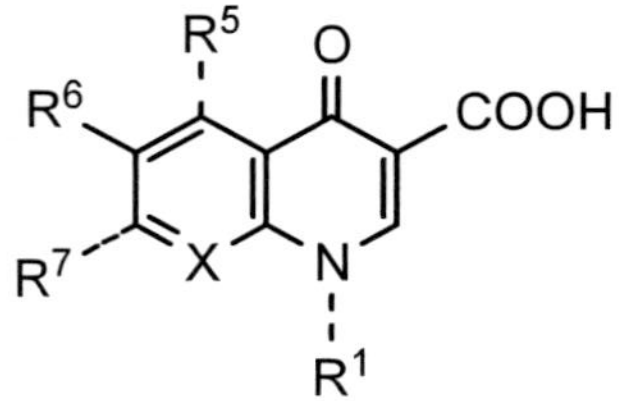

Compound	X	R^1	R^5	R^6	R^7	Organism (number)					
						MSSA (25)			MRSA, ofloxacin susceptible (24) [MRSA (25), ofloxacin resistant]		
						Range	MIC_{50}	MIC_{90}	Range	MIC_{50}	MIC_{90}
DQ-113	C–CH_3		NH_2	F	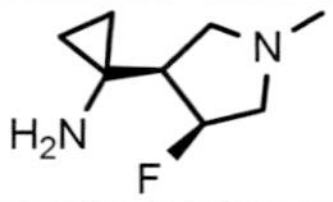	≤0.004–0.06	0.008	0.03	≤0.004–0.015 [0.03–2]	≤0.004 [0.06]	0.008 [0.25]
Sitafloxacin	C–Cl		H	F	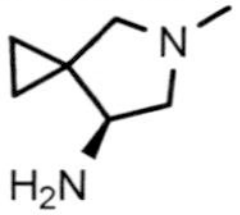	0.008–0.12	0.03	0.12	0.008–0.06 [0.25–32]	0.03 [1]	0.06 [4]
Ciprofloxacin	C–H		H	F	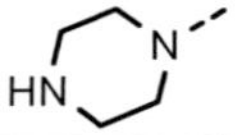	0.25–4	0.5	1	0.25–2 [16–>128]	0.5 [128]	1 [>128]
Gatifloxacin	C–OCH_3		H	F	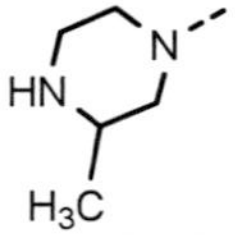	0.06–0.5	0.12	0.25	0.06–0.25 [2–>128]	0.12 [8]	0.25 [16]

Table 10. (Continued)

Moxifloxacin	C–OCH_3		H	F	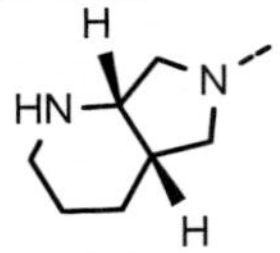	0.06–0.25	0.06	0.12	0.03–0.25 [1–64]	0.12 [4]	0.12 [8]
Garenoxacin	C–$OCHF_2$		H	H	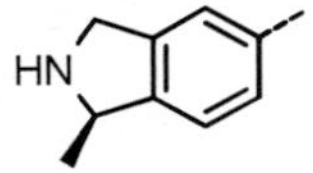	≤0.004–0.12	0.03	0.06	0.008–0.06 [0.25–64]	0.03 [2]	0.03 [8]
Gemifloxacin	N		H	F	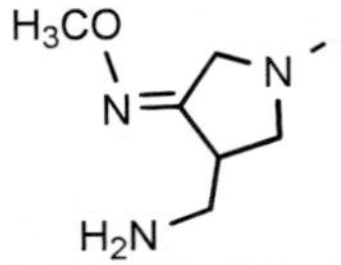	0.03–0.25	0.06	0.06	0.015–0.06 [-]	0.06 [-]	0.06 [-]
Sparfloxacin	C–F		NH_2	F	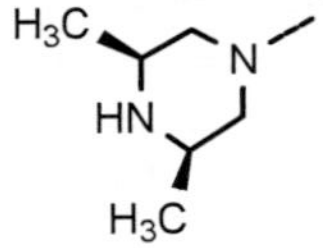	0.06–0.25	0.12	0.25	0.03–0.12 [2–>128]	0.06 [8]	0.12 [32]
Tosufloxacin	N	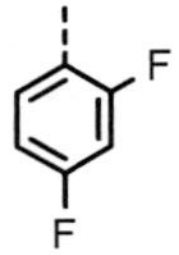	H	F	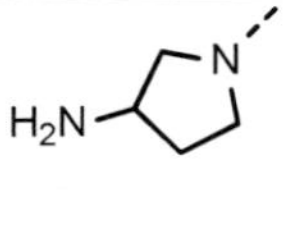	0.015–0.25	0.06	0.12	0.015–0.12 [1–64]	0.06 [32]	0.06 [32]
Levofloxacin	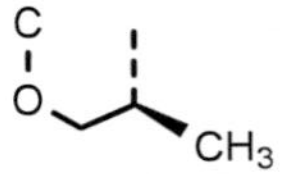		H	F	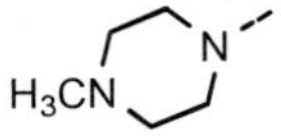	0.12–1	0.25	0.5	0.12–1 [4–>128]	0.25 [16]	0.5 [64]

Table 11. Comparative MICs (μg/mL) of DQ-113 and important quinolones against coagulase-negative staphylococci.

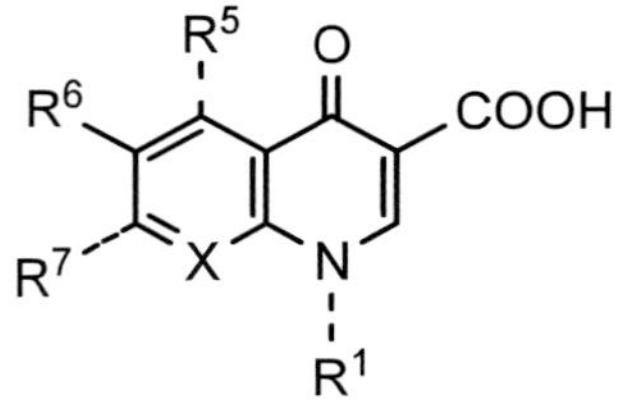

Compound	X	R^1	R^5	R^6	R^7	Organism (number)					
						MSCNS (27)			MRCNS (36)		
						Range	MIC_{50}	MIC_{90}	Range	MIC_{50}	MIC_{90}
DQ-113	C–CH_3		NH_2	F	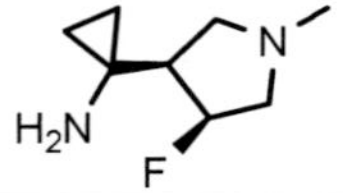	≤0.004–0.06	0.008	0.03	≤0.004–0.12	0.03	0.06
Sitafloxacin	C–Cl		H	F	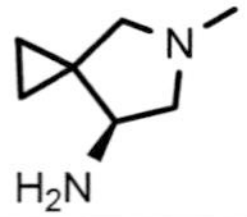	0.008–0.25	0.03	0.25	0.008–0.5	0.12	0.5
Ciprofloxacin	C–H		H	F	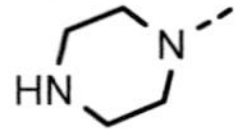	0.06–64	0.25	8	0.12–128	8	64
Gatifloxacin	C–OCH_3		H	F	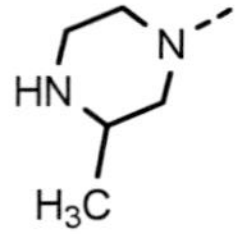	0.06–4	0.25	2	0.12–4	2	4

Table 11. (Continued)

Moxifloxacin	C–OCH_3		H	F	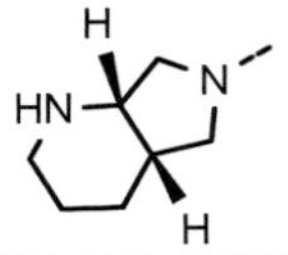	0.03–4	0.12	1	0.06–8	1	4
Garenoxacin	C–$OCHF_2$		H	H	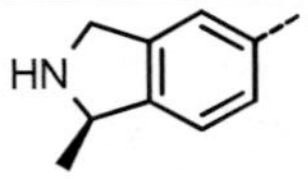	0.015–4	0.06	1	0.03–4	1	4
Gemifloxacin	N		H	F	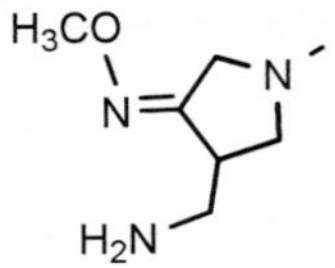	0.008–2	0.06	0.25	0.015–4	0.5	2
Sparfloxacin	C–F		NH_2	F	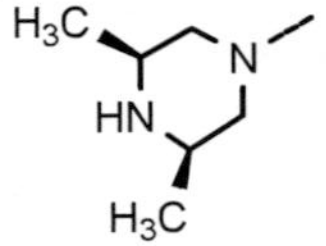	0.03–8	0.12	4	0.06–16	4	8
Tosufloxacin	N	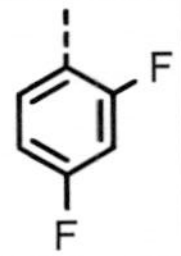	H	F	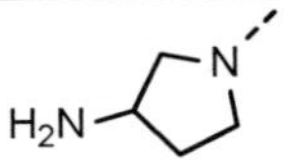	0.015–16	0.06	4	0.03–16	4	16
Levofloxacin	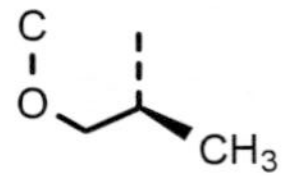		H	F	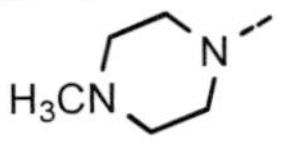	0.12–8	0.25	4	0.12–32	4	16

Table 12. Comparative MICs (μg/mL) of delafloxacin and other quinolones against MSSA and MRSA.

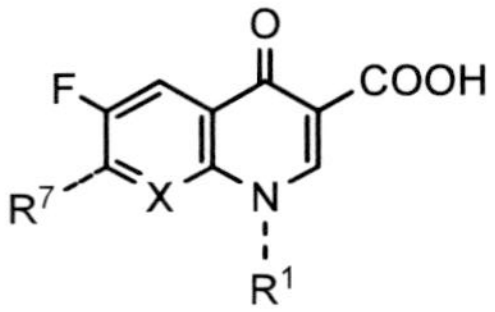

Compound	X	R[1]	R[7]	Organism (number)					
				MSSA (50)			MRSA (25)		
				Range	MIC_{50}	MIC_{90}	Range	MIC_{50}	MIC_{90}
Delafloxacin	C–Cl	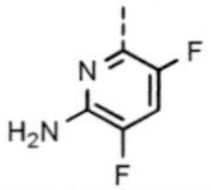	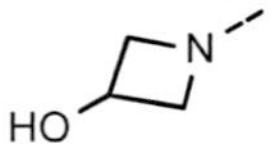	0.001–0.12	0.004	0.008	0.008–1	0.5	0.5
Ciprofloxacin	C–H		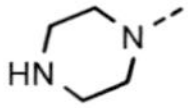	0.12–128	0.5	1	0.5– >128	64	128
Gatifloxacin	C–OCH_3		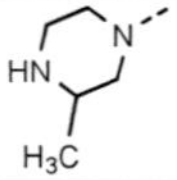	0.06–8	0.12	0.25	0.25–32	4	8
Moxifloxacin	C–OCH_3		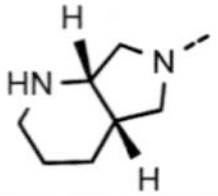	0.03–2	0.06	0.12	0.12–8	2	4
Gemifloxacin	N		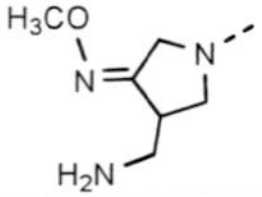	0.015–8	0.03	0.06	0.03–16	8	8
Levofloxacin	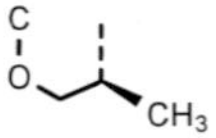		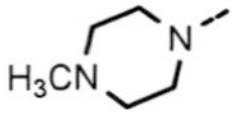	0.12–16	0.25	0.5	0.5–128	16	16

WCK 771

WCK 771, the L-arginine salt of *S*-(–)-nadifloxacin [*S*-(–)-9-fluoro-6,7-dihydro-8-(4-hydroxypiperidin-1-yl)-5-methyl-1-oxo-1*H*,5*H*-benzo[*i,j*] quinolizine-2-carboxylic acid L-arginine salt] (Figure 15) is a broad-spectrum tricyclic quinolone that is active against quinolone-resistant staphylococci and MRSA and is being studied in phase II clinical trials. *RS*-(±)-nadifloxacin was introduced only for topical use as a liniment against *Propionobacterium acnes* [144, 145]. Structurally, nadifloxacin has a lipophilic tricyclic benzoquinolizine nucleus, with a 4-hydroxypiperidinyl moiety at C-8 position. This 4-hydroxypiperidine moiety is a particular moiety without distal basic functionality, which is unusual for a side chain of quinolone. Most marketed quinolones bear side chains with basic nitrogen functionality, but this compound along with delafloxacin have a hydroxylcyclic amine without distal amine.

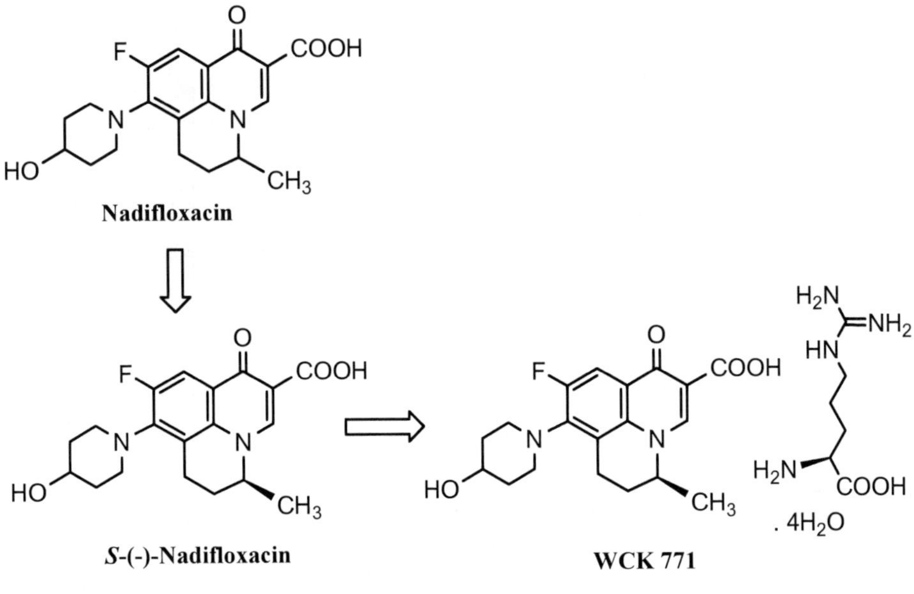

Figure 15. Development of WCK 771.

Treatment with *RS*-(±)-nadifloxacin has been reported to result in an unusually lower incidence of the emergence of resistant mutants. A Japanese study by Nishijima et al. against *S. aureus* isolates, including MRSA strains, collected during the period from 1994 to 2000 did not show significant increase in the MIC_{90} of nadifloxacin [146, 147].

Similar to the case of levofloxacin [(*S*)-isomer] and ofloxacin (racemic), the levorotatory (*S*)-isomer of nadifloxacin is 64–256 times more potent than its *R*-(+)-isomer and approximately twice as active as the *RS*-(±)-nadifloxacin against Gram-positive and -negative bacteria [144, 148]. Because the *S*-(–)-isomer is primarily responsible for antibacterial activity, the potency of WCK 771 [arginine salt of the *S*-(–) isomer] is two to four times higher than that of racemic nadifloxacin [149].

Table 13. Comparative MICs (μg/mL) of WCK 771 and representative quinolones against clinical isolates of staphylococci.

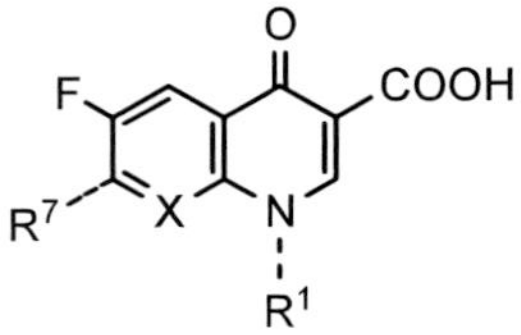

Compound	X	R^1	R^7	Organism (number)					
				MSSA (244)			MRSA (176)		
				Range	MIC_{50}	MIC_{90}	Range	MIC_{50}	MIC_{90}
WCK 771	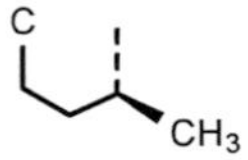		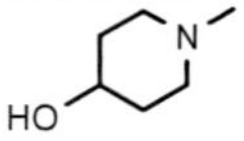	0.015–0.25	0.03	0.03	0.015–4.0	0.5	1.0
Levofloxacin	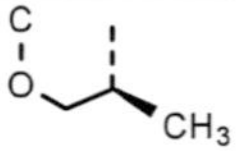		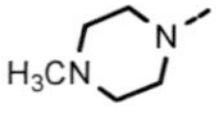	0.12–1.0	0.25	0.5	0.125–64	4.0	16.0
Ciprofloxacin	C–H		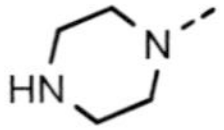	0.12–2.0	0.5	1.0	0.25–128	4.0	64
Clinafloxacin	C–Cl		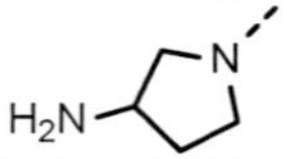	0.015–0.25	0.03	0.06	0.015–4.0	0.25	1.0
Moxifloxacin	C–OCH_3		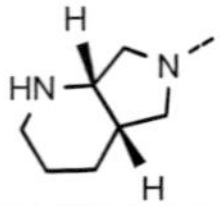	0.03–0.5	0.03	0.12	0.03–8.0	1.0	4.0
Trovafloxacin	N	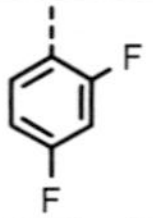	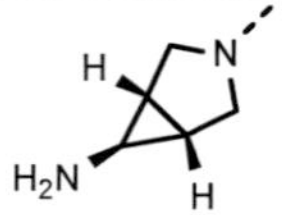	0.015–0.5	0.06	0.12	0.015–16	1.0	4.0

The unusual physicochemical properties of *S*-(–)-nadifloxacin, such as high partition coefficients, low aqueous solubility and having a single p*K*a value distinct it from representative quinolones. However, it is envisaged that the less aqueous solubility may be compensated by its high potency coupled with the lipophilicity-driven permeability, which makes *S*-(–)-nadifloaxcin optimizable for an oral administration. Alternatively, it is also envisaged that a water-soluble salt form of *S*-(–)-nadifloaxcin can be considered for injection preparation. Various prodrugs and salts of *S*-(–)-nadifloxacin were studied by de Souza *et al.* and the L-arginine salt of *S*-(–)-nadifloxacin (WCK 771) was selected as an entity for further development as a potential drug candidate [150].

The comparative study of WCK 771 with other quinolones demonstrates that this new agent is a highly potent anti-staphylococcal quinolone with improved potency against even quinolone-resistant strains of *S. aureus* and coagulase-negative staphylococci. In Patel *et al.* study, the potency of WCK 771 against 176 strains of MRSA (MIC_{90} = 1.0 μg/mL) was considerably higher than that of levofloxacin (MIC_{90} = 16 μg/mL) or ciprofloxacin (MIC_{90} = 64 μg/mL). It was two- to fourfold more potent than moxifloxacin and trovafloxacin (Table 13). Additionally, WCK 771 has potency comparable to that of clinafloxacin [151].

Also, Patel *et al.* evaluated the potency of WCK 771 in animal models of staphylococcal infection. WCK 771 was effective by both the oral and the subcutaneous routes in mice infected intraperitoneally with quinolone-susceptible MSSA strains. For infections caused by quinolone-resistant MRSA strains, the activity of WCK 771 administered subcutaneously was superior to those of trovafloxacin and sparfloxacin, with an ED_{50} range of 27.8 to 46.8 mg/kg. The activity of WCK 771 was superior to those of moxifloxacin, vancomycin, and linezolid in a mouse cellulitis model of infection caused by one strain of MSSA and two strains of MRSA, with effective doses of 2.5 and 5 mg/kg for the MSSA strain and 10-fold higher effective doses for MRSA strains. WCK 771, like vancomycin and linezolid, eradicated MRSA from mouse liver, spleen, kidney, and lung when it was administered subcutaneously at a dose of 50 mg/kg for four doses [151].

In assessing the impact of the NorA efflux pump on WCK 771 activity by Patel et al., it was observed that the activity of the drug against a NorA-hyperexpressing *S. aureus* strain was not affected by the presence of reserpine and that there was little difference between the MICs for $NorA^-$ strains and those for $NorA^+$ [151].

The strong potency of WCK 771 against MSSA and MRSA strains is possibly due to its unique mechanism of action and its resistance to efflux-mediated resistance, resulting in the combination of high-affinity targeting of staphylococcal DNA gyrase, coupled with efflux resistance, creating high intracellular drug concentrations.

To estimate the range of target mutations that may affect the activity of WCK 771 and to determine its primary and secondary targeting properties, Bhagwat *et al.* characterized sequentially selected staphylococcal mutants and determined their quinolone susceptibilities. WCK 771 was compared with other quinolones possessing improved anti-staphylococcal activities and diverse targeting properties, such as moxifloxacin, trovafloxacin, and garenoxacin [152]. There was a 2-fold increase in the MICs of WCK 771 for single *gyrA* mutants, indicating that DNA gyrase is its primary target. All first- and second-step mutants selected by WCK 771 revealed *gyrA* and *grlA* mutations, respectively. The MICs of WCK 771 and clinafloxacin were found to be superior to those of other quinolones against strains with double and triple mutations. This study demonstrated that *gyrA* is the primary target of WCK 771 and its target preference is remarkably different from those of quinolones with dual targets (garenoxacin and moxifloxacin) and topoisomerase IV-specific quinolones (trovafloxacin). WCK 771 displays an activity profile comparable to that of clinafloxacin, a dual-acting quinolone with a high affinity to DNA gyrase.

Based on the pharmacological profile of WCK 771, it could be considered as a drug candidate for use against nosocomial multidrug-resistant Gram-positive infections, in particular MRSA and VRSA infections.

6-FLUORONAPHTHYRIDONES

Gemifloxacin

Gemifloxacin (SB-265805, LB20304a) possessing excellent activity against streptococci and staphylococci, is a fluoronaphthyridone derivative with a 4-(aminomethyl)-3-(methoxyimino)pyrrolidine substituent at the C-7 position. Its N-1 side chain includes a cyclopropyl group characteristic of safer fluoroquinolones in contrast to the difluorophenyl substituent of other naphthyridones, tosufloxacin and trovafloxacin. Gemifloxacin carries a methoxyimino- group with (*Z*)-configuration on C-7 pyrrolidine ring. This compound is a racemic (*R*, *S*) mixture.

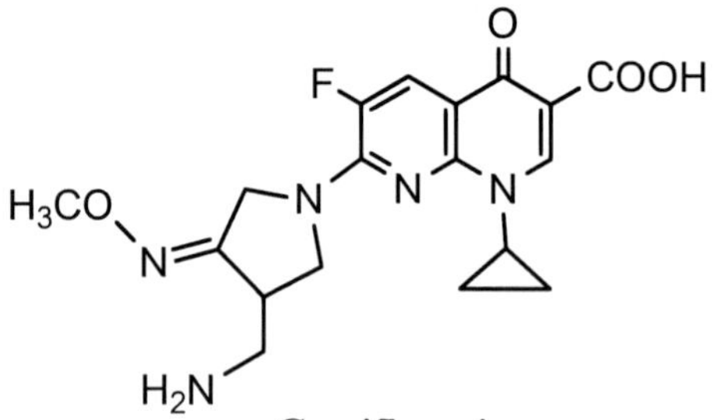

Gemifloxacin

Increased bulkiness of C-7 position appears to confer protection from efflux exporter proteins and diminishes the likelihood of resistance in wild-type strains [69]. Replacement of the carbon at the C-8 position with nitrogen enhances antibacterial potency, apparently due to the increased activity against both DNA gyrase and topoisomerase IV so that ≥2 mutations in the QRDRs are usually required for drug resistance to develop.

Preliminary evaluation of gemifloxacin against MSSA indicated that gemifloxacin was equiactive to trovafloxacin (MIC_{90}s = 0.125 μg/mL) and more active than comparator agents (MIC_{90} = 1 μg/mL for ofloxacin; MIC_{90} = 2 μg/mL for ciprofloxacin). The activity of gemifloxacin against MRSA was equal to trovafloxacin (MIC_{90}s = 4 μg/mL). However, gemifloxacin was more potent against MRSA than other quinolones (MIC_{90}s = 32 μg/mL for grepafloxacin, ofloxacin, and ciprofloxacin) [153].

In another study, gemifloxacin demonstrated better antibacterial activity against 39 strains of *S. aureus* isolated from patients with sinusitis than levofloxacin, gatifloxacin, and moxifloxacin, with an MIC_{90} <0.03 μg/mL compared with 0.25, 0.125, and 0.125 μg/mL of comparator drugs, respectively [154]. Against 346 contemporary clinical isolates of methicillin and ciprofloxacin-sensitive *S. aureus* in the US and Canada, gemifloxacin demonstrated 0.03 μg/mL compared with 1–16 μg/mL for amoxicillin/clavulanate and azithromycin. Gemifloxacin, however, failed to show efficient antibacterial activity against 41 methicillin-resistant ATCC strains of *S. aureus* and 160 methicillin- and ciprofloxacin-resistant clinical isolates from 11 North American medical centers; MIC_{90} values against the 2 groups of MRSA were 2 and 16 μg/mL, respectively [155, 156].

DW286

DW286, 7-[3-(aminomethyl)-4-(methoxyimino)-3-methylpyrrolidin-1-yl]-1-cyclopropyl-6-fluoro-4-oxo-1,4-dihydro[1,8]naphthyridine-3-carboxylic acid hydrochloric acid salt, is a fluoronaphthyridone developed by Dong Wha (South Korea). DW286 is methyl analog of gemifloxacin. In fact, the pyrrolidine ring of gemifloxacin has been substituted by a methyl group at position 3, resulted in DW286. Structurally, gemifloxacin and DW286 are distinguished from other pyrrolidin-1-ylquinolones by attaching a methoxyimino group on pyrrolidine ring.

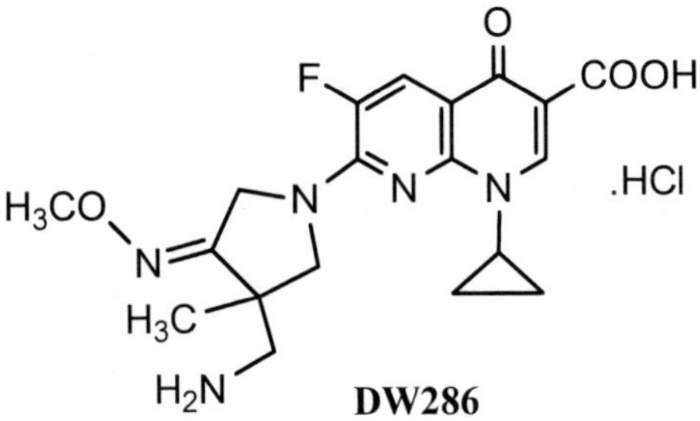

DW286

The in vitro and in vivo activities of DW286 have been compared with those of ciprofloxacin, gemifloxacin, sparfloxacin, and trovafloxacin (Table 14). DW286 has potent antibacterial activity against staphylococci, including MRSA and quinolone-resistant *S. aureus* (QRSA). The in vitro activity of DW286 was stronger than that of mentioned quinolones. In an experimental mouse model of systemic infection caused by three *S. aureus* strains, including MRSA and QRSA, DW286 demonstrated the most potent activity, as found in vitro. DW286 was >8-fold more active against QRSA than the other quinolones [157]. It also exhibited better in vitro activity than several quinolones against mutant strains that developed quinolone resistance in vitro and was a poor substrate for efflux pumps [158].

Zabofloxacin (DW-224a)

Zabofloxacin (formerly DW-224a) is a new fluoroquinolone with the following chemical formula: 1-cyclopropyl-6-fluoro-7-[8-(methoxyimino)-2,6-diazospiro[3,4]oct-6-yl]-4-oxo-1,4-dihydro[1,8]naphthyridine-3-carboxylic acid hydrochloride [159, 160]. Zabofloxacin synthesized by the research laboratory of Dong Wha Pharmaceutical Industry, Ltd. (Anyang, Republic of Korea).

Table 14. Efficacy of DW286 in comparison with ciprofloxacin, gemifloxacin, sparfloxacin and trovafloxacin on systemic infections in mice

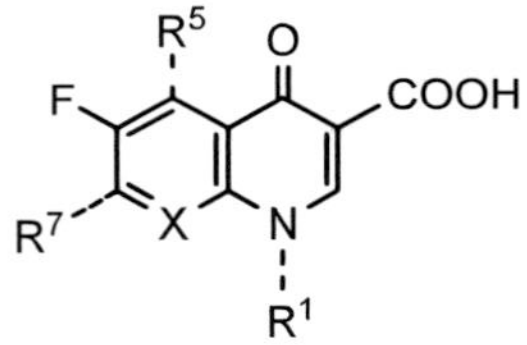

Compound	X	R^1	R^5	R^7	Microorganism [challenge dose (CFU/ml)]					
					S. aureus Smith [1.2×10^7]		MRSA MB4-19 [3.1×10^8]		QRSA MB4-20 [5.0×10^8]	
					MIC (μg/ml)	PD_{50} (mg/kg)	MIC (μg/ml)	PD_{50} (mg/kg)	MIC (μg/ml)	PD_{50} (mg/kg)
DW286	N		H	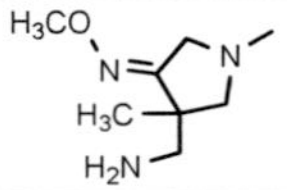	0.004	0.12	0.004	0.12	0.25	5.83
Gemifloxacin	N		H	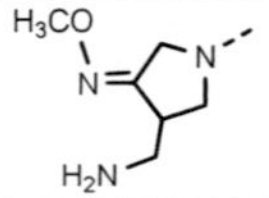	0.031	0.77	0.016	0.70	2	>40
Ciprofloxacin	C–H		H	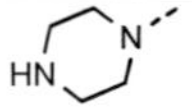	0.5	3.43	0.25	3.99	32	>40
Sparfloxacin	C–F		NH_2	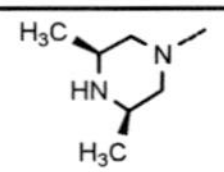	0.063	0.54	0.031	0.22	32	>40
Trovafloxacin	N	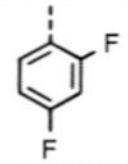	H	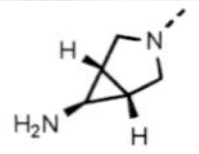	0.016	0.32	0.016	0.17	8	>40

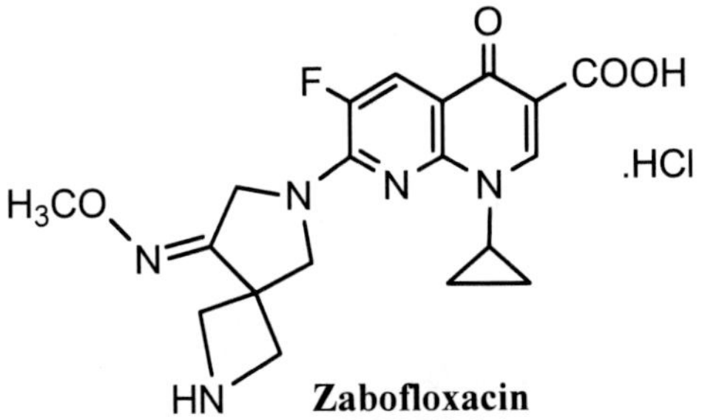

Zabofloxacin (DW-224a)

Different studies demonstrated that zabofloxacin has a potent in vitro and in vivo antibacterial activity against both Gram-positive and Gram-negative bacteria.

The in vitro efficacy of zabofloxacin in comparison with those of ciprofloxacin, moxifloxacin, and gemifloxacin has been demonstrated by Park et al. (Table 15) [161]. Zabofloxacin was slightly superior to gemifloxacin and much more active than moxifloxacin and ciprofloxacin against clinical isolates of Gram-positive bacteria. Against MSSA and MRSA, the antibacterial activity of zabofloxacin was comparable to that of gemifloxacin but 2- to 16-fold more potent than those of moxifloxacin and ciprofloxacin (Table 15). The MIC_{90}s of zabofloxacin against MSSA and MRSA were 0.03 μg/mL and 4 μg/mL, respectively.

Against coagulase-negative staphylococci (both MSCNS and MRCNS), the activity of zabofloxacin was also 2- to 16-fold more potent than those of gemifloxacin, ciprofloxacin and moxifloxacin (Table 16).

In the systemic infection model caused by *S. aureus* giorgio, the ED_{50}s of zabofloxacin, ciprofloxacin, moxifloxacin and gemifloxacin were 1.11, 16.52, 1.78, and 1.37 mg/kg of body weight, respectively. Especially against quinolone-resistant *S. aureus* P197 (an MRSA), zabofloxacin was more effective than ciprofloxacin, moxifloxacin, and gemifloxacin [161].

The safety of zabofloxacin after oral administration has been studied in mice, rats and dogs [162]. Zabofloxacin appears to have exerted no adverse effects on the central nervous, cardiovascular and respiratory systems with the exception of the effect on the QT interval prolongation [163].

6-DESFLUOROQUINOLONES

Garenoxacin (T-3811, BMS-284756)

Garenoxacin (T-3811, BMS-284756) is a new 6-desfluoroquinolone that was developed by Bristol-Myers Squibb (New York) under license from Toyama Chemical Co. (Toyama, Japan). Bristol-Myers Squibb returned all rights for the drug to Toyama in 2003 and an agreement was signed in 2004 [164]. Its structural feature is the absence of fluorine at C-6 but has fluorine incorporated through a C-8 difluoromethoxy group. In addition, garenoxacin has a certain aryl moiety (isoindolin-5-yl) instead of cyclic amine at C-7, which connected to the quinolone core via carbon-carbon bond [165, 166].

Table 15. Comparative MICs (μg/mL) of zabofloxacin (DW-224a), gemifloxacin, ciprofloxacin and moxifloxacin against MSSA and MRSA.

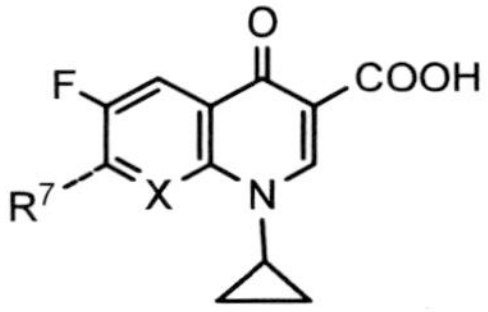

Compound	X	R^7	Organism (number)					
			MSSA (62)			MRSA (120)		
			Range	MIC_{50}	MIC_{90}	Range	MIC_{50}	MIC_{90}
Zabofloxacin (DW-224a)	N	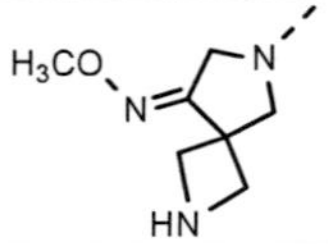	0.008–0.06	0.015	0.03	0.008–32	2	4
Gemifloxacin	N	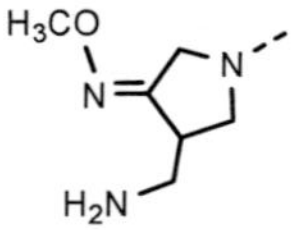	0.008–0.06	0.015	0.03	0.008–64	2	4
Ciprofloxacin	C-H	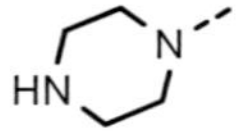	0.06–0.5	0.25	0.5	0.125–>64	32	>64
Moxifloxacin	$C-OCH_3$	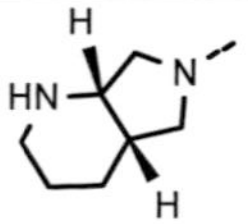	0.015–0.25	0.06	0.06	0.03–>64	4	16

Table 16. Comparative MICs (μg/mL) of zabofloxacin (DW-224a), gemifloxacin, ciprofloxacin and moxifloxacin against coagulase-negative staphylococci.

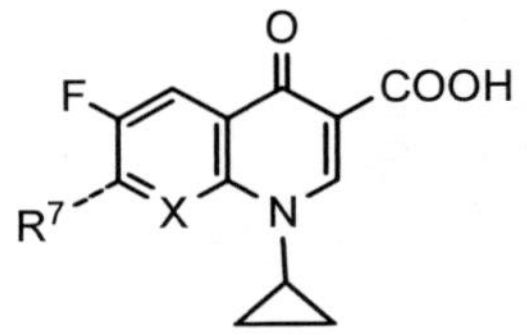

Compound	X	R^7	Organism (number)					
			MSCNS (14)			MRCNS (38)		
			Range	MIC_{50}	MIC_{90}	Range	MIC_{50}	MIC_{90}
Zabofloxacin (DW-224a)	N	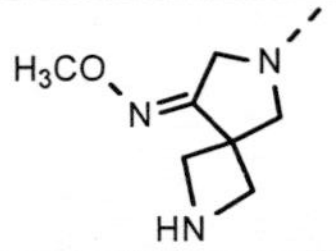	0.008–0.25	0.015	0.125	0.008–4	0.25	2
Gemifloxacin	N	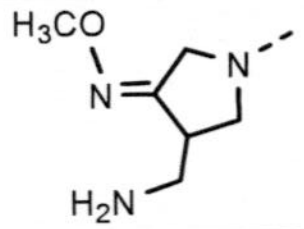	0.008–0.25	0.015	0.25	0.008–8	0.25	4
Ciprofloxacin	C-H	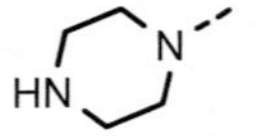	0.125–2	0.125	2	0.06–64	8	32
Moxifloxacin	$C\text{-}OCH_3$	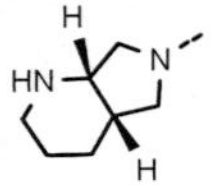	0.03–2	0.125	0.25	0.06–16	2	8

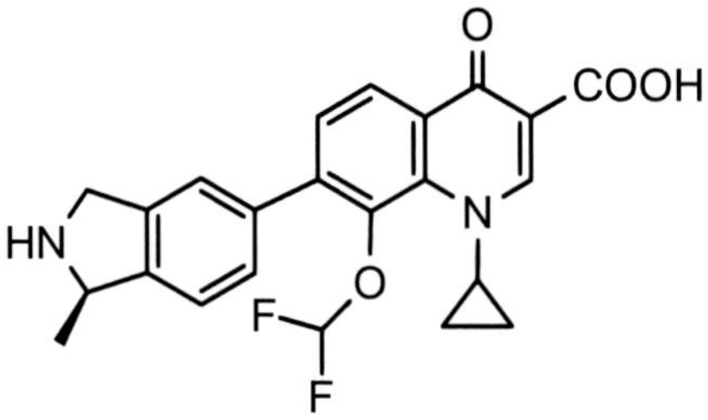

Garenoxacin

Garenoxacin has been evaluated for treatment of respiratory, urinary tract, and skin and soft tissue infections caused by susceptible and resistant pathogens [164]. This non-fluorinated quinolone is reported to be active against MRSA strains that are resistant to other fluoroquinolones and selects fluoroquinolone-resistant mutants at a lower frequency than older quinolones [157, 167]. Antibacterial studies have indicated that MSSA and MRSA were susceptible to garenoxacin at MIC breakpoint of 0.5 μg/mL and 1 μg/mL, respectively. Based on studies with reserpine as an efflux pump inhibitor, the presence of reserpine did not result in a decrease in garenoxacin MICs for any of the isolates tested. Thus, garenoxacin is not a substrate for efflux pump activity via NorA [38, 168]. However, a recent study of *S. aureus* clinical isolates in the Asia-Pacific region indicated that 1% and 9% of MSSA and MRSA isolates resistant to ciprofloxacin were also resistant to 4 μg/mL garenoxacin, respectively [165].

DX-619

DX-619 is a new 6-desfluoroquinolone which was developed as a drug used for the treatment of infections caused by multidrug-resistant Gram-positive organisms. Its chemical name is 7-[(3*R*)-3-(1-aminocyclopropyl)- pyrrolidin-1-yl]-1-[(1*R*,2*S*)-2-fluorocyclopropyl]-8-methoxy-1,4-dihydro-4-oxoquinoline-3-carboxylic acid.

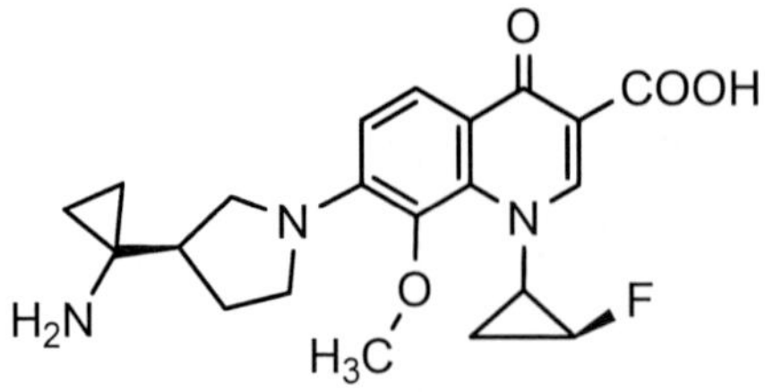

DX-619

DX-619 shares structural similarity with DK-507k, DC-159a based on possessing both 2-fluorocyclopropyl at N-1 and methoxy group at C-8. A certain pyrrolidine ring bearing aminocyclopropyl pendent at C-7 position, differentiates DX-619 from earlier quinolones and 6-desfluoroquinolones. Indeed, the aminomethyl residue which is found in gemifloxacin and DW286, fused with cylopropane in DX-619 resulted in aminocyclopropyl pendent. Development of 6-desfluoroquinolones such as DX-619 demonstrated that the impact of 6-

fluoro- substituent is diminished when the molecule contains spatial arrangements with other appropriate substituents.

Watanabe et al. [169] tested susceptibilities of MRSA isolates from both hospitals and the community to DX-619 in comparison with sitafloxacin, levofloxacin, and other anti-MRSA antibiotics. They used different categories of strains: healthcare-associated MRSA (HA-MRSA) strains, community-associated MRSA (CA-MRSA)/non-multiresistant oxacillin-resistant *S. aureus* (NORSA) strains; healthcare-associated methicillin-resistant coagulase-negative staphylococci (HA-MRCNS) strains and community-associated methicillin-resistant coagulase-negative staphylococci (CA-MRCNS) strains. DX-619 showed the lowest MIC_{90} values for all categories of strains tested, irrespective of the degree of glycopeptide resistance. The six strains with MIC values of >128 μg/mL of levofloxacin commonly carried two mutations in *gyrA* and two mutations in *grlA*. DX-619 showed potent activity against strains with MIC values of 2 μg/mL.

In a Japanese study, Yamamoto et al. [170] surveyed and characterized highly virulent, community-acquired MRSA strains with Panton-Valentine leucocidin (PVL) genes (PVL^+ strains of MRSA) isolated from patients in hospitals. They also examined the in vitro susceptibilities of PVL^+ MRSA strains to 31 antibacterial agents, including DX-619 and sitafloxacin. DX-619 showed the greatest activity against PVL^+ MRSA among the antibacterial agents tested (MIC_{50} and MIC_{90} = 0.008 μg/mL). This activity was 8-fold greater than that of sitafloxacin (MIC_{90} = 0.063 μg/mL), 32-fold greater than that of levofloxacin (MIC_{90} = 0.25 μg/mL), and 125-fold greater than that of vancomycin (MIC_{90} = 1 μg/mL).

Bogdanovich et al. [171] examined the activity of DX-619 against 259 methicillin- and quinolone-susceptible and –resistant staphylococci compared to those of ciprofloxacin, gatifloxacin, moxifloxacin, sitafloxacin and levofloxacin (Tables 17 and 18). DX-619 had the lowest MICs against 131 *S. aureus* strains (MIC_{90} = 0.5 μg/mL) and 128 coagulase-negative staphylococci (MIC_{90} = 0.125 μg/mL). Among strains tested, 76 *S. aureus* strains and 51 coagulase-negative staphylococci were resistant to ciprofloxacin. DX-619 had the lowest MIC_{90} values against 127 quinolone-resistant staphylococci (0.5 μg/mL), followed by sitafloxacin (4 μg/mL), moxifloxacin (8 μg/mL), gatifloxacin (16 μg/mL), levofloxacin (32 μg/mL), and ciprofloxacin (>32 μg/mL). DX-619 and sitafloxacin were also more active than other tested drugs against selected mutants and had the lowest mutation frequencies in single-step resistance selection.

According to Strahilevitz et al. studies against wild-type *S. aureus* [172], DX-619 was 16- to 32-fold, twofold, and four- to eightfold more potent than ciprofloxacin, gemifloxacin, and garenoxacin, respectively. DX-619 manifested equal fourfold increases in MIC against a common *parC* mutant and a common *gyrA* mutant and selected for mutants at up to two- to fourfold its MIC, consistent with dual-targeting properties. Studies with purified topoisomerase IV and gyrase from *S. aureus* also showed that DX-619 had similar activity against topoisomerase IV and gyrase. Susceptibility studies with DX-619 and an array of efflux pump substrates with and without reserpine, an inhibitor of efflux pumps, suggested that resistance in DX-619-selected mutants is affected by mechanisms other than mutations in topoisomerases or known reserpine-inhibitable pumps in *S. aureus* and thus are likely novel [172].

Table 17. Comparative MICs (µg/mL) of DX-619 and important quinolones against MSSA and MRSA.

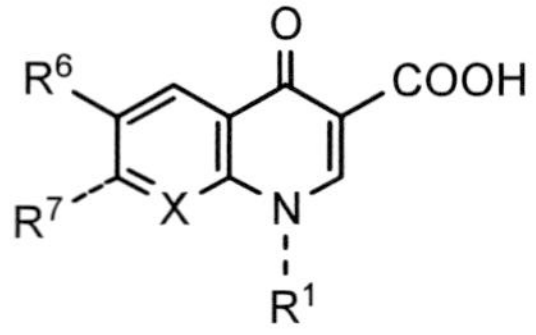

Compound	X	R[1]	R[6]	R[7]	Organism (number)					
					MSSA (69)			MRSA (62)		
					Range	MIC_{50}	MIC_{90}	Range	MIC_{50}	MIC_{90}
DX-619	C–OCH_3		H	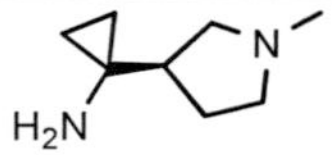	≤0.002–0.5	0.008	0.125	0.008–2.0	0.125	1.0
Ciprofloxacin	C–H		F	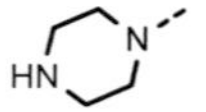	≤0.06–>32.0	0.5	>32.0	0.5–>32.0	>32.0	>32.0
Gatifloxacin	C–OCH_3		F	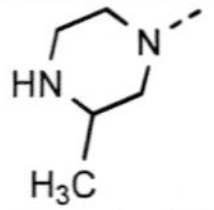	0.03–16.0	0.125	8.0	0.125–>32.0	8.0	16.0
Moxifloxacin	C–OCH_3		F	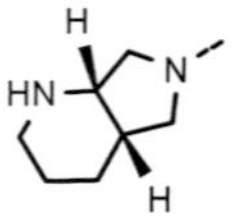	≤0.016–8.0	0.06	4.0	0.06–>32.0	2.0	16.0
Sitafloxacin	C–Cl		F	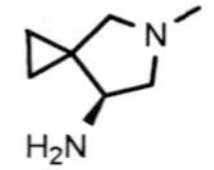	≤0.008–8.0	0.06	1.0	0.03–32.0	1.0	8.0
Levofloxacin	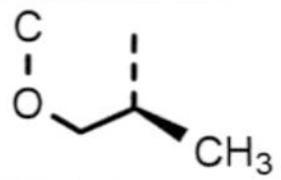		F	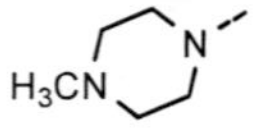	0.06–>32.0	0.25	16.0	0.25–>32.0	16.0	>32.0

Table 18. Comparative MICs (μg/mL) of DX-619 and important quinolones against coagulase-negative staphylococci.

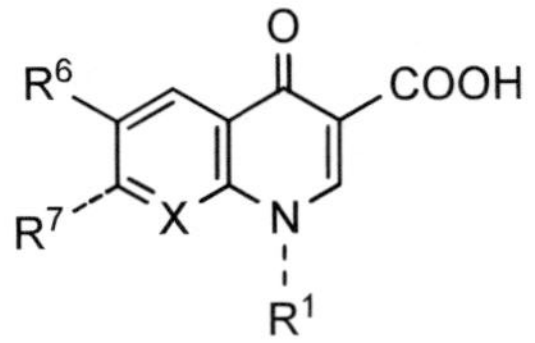

Compound	X	R^1	R^6	R^7	Coagulase-negative staphylococci (number)					
					MSSA (67)			MRSA (61)		
					Range	MIC_{50}	MIC_{90}	Range	MIC_{50}	MIC_{90}
DX-619	C–OCH_3		H	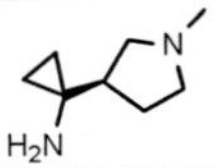	0.004–0.25	0.016	0.06	0.004–0.25	0.06	0.125
Ciprofloxacin	C–H		F	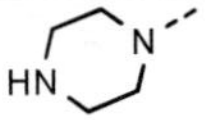	≤0.06–>32.0	0.25	8.0	0.25–>32.0	16.0	>32.0
Gatifloxacin	C–OCH_3		F	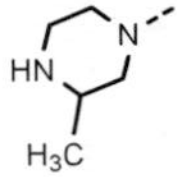	0.06–4.0	0.125	2.0	0.06–32.0	2.0	4.0
Moxifloxacin	C–OCH_3		F	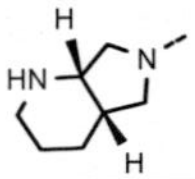	0.03–4.0	0.125	1.0	0.03–8.0	1.0	4.0
Sitafloxacin	C–Cl		F	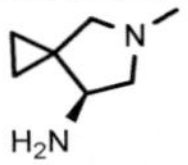	0.016–0.25	0.03	0.25	0.016–2.0	0.25	0.5
Levofloxacin	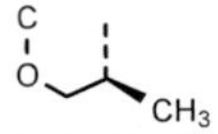		F	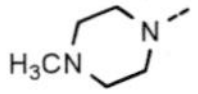	0.125–16.0	0.25	8.0	0.125–>32	8.0	16.0

In general, DX-619 is a potent agent against both MSSA and MRSA, and coagulase-negative staphylococci, suggesting that it would be a promising candidate for the treatment of MRSA infection if sufficient in vivo concentrations were safely attained. Preliminary studies by Daiichi Sankyo Co. Ltd. have suggested that DX-619 has a promising safety profile [169].

Nemonoxacin (TG-873870)

Nemonoxacin (TG-873870) is a 6-desfluorinated quinolone, developed by TaiGen Biotechnology Co. (Taipei, Taiwan). Its chemical name is 7-[(3*S*,5*S*)-3-amino-5-methylpiperidin-1-yl]-1-cyclopropyl-8-methoxy-4-oxo-1,4-dihydroquinoline-3-carboxylic acid [173].

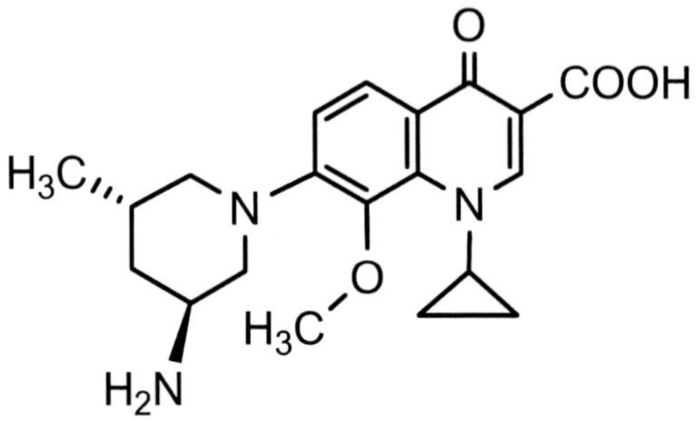

Nemonoxacin (TG-873870)

Beside the absence of fluorine atom at C-6, the distinctive structural feature of nemonoxacin is a (3*S*,5*S*)-3-amino-5-methylpiperidin-1-yl moiety at C-7. The most frequent substitution of quinolones at C-7 is piperazine or pyrrolidine moiety. Thus quinolones with cyclic amine at C-7 can categorize as piperazin-1-yl- and pyrrolidin-1-yl- quinolones. The third category is piperidin-1-ylquinolones. Nemonoxacin along with WCK 771, WCK 1152 and WCK 1153 are piperidin-1-ylquinolone. Also, the N-1-cyclopropyl and C-8-methoxy substituents are preserved in this molecule similar to gatifloxacin, moxifloxacin, WCK 1152 and WCK 1153.

Nemonoxacin has good activity against MRSA, particularly toward community-acquired MRSA. Also, This agent has an excellent activity toward vancomycin-intermediate or resistant *S. aureus*. In a comparative in vitro study with other quinolones in clinical use, nemonoxacin is less prone to the development of resistance and is active against quinolone-resistant clinical strains [174].

Ozenoxacin (T-3912)

Ozenoxacin (T-3912) is a new non-fluorinated quinolone (similar to garenoxacin), developed by Toyama Chemical Co. Its chemical name is 1-cyclopropyl-8-methyl-7-[5-methyl-6-(methylamino)-3-pyridinyl]-4-oxo-1,4-dihydro-3-quinolinecarboxylic acid [175].

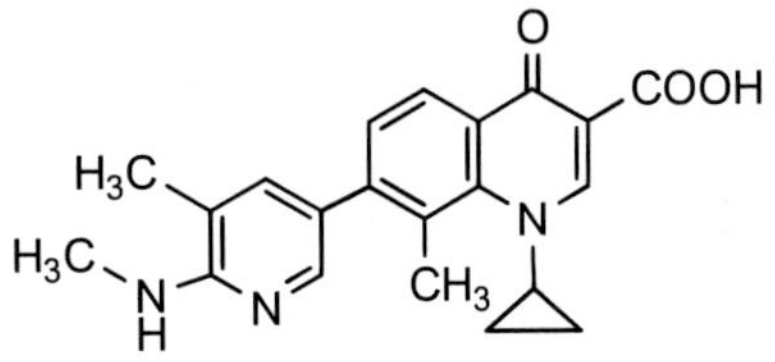

Ozenoxacin (T-3912)

Its structural modifications respect to typical quinolones is the absence of 6-fluoro, the presence of 8-methyl and the attachment of certain aminoaryl moiety instead of aminocyclic amine to the C-7 position. Among the quinolones, there are only a limited number of molecules in which substituent attached to C-7 through carbon-carbon linkage. In ozenoxacin similar to garenoxacin, the C-7 aryl ring is connected to the quinolone core via carbon-carbon bond.

The antibacterial activity of ozenoxacin has been compared with that of nadifloxacin, ofloxacin, levofloxacin, clindamycin, erythromycin and gentamicin (Table 19) [176]. The *in vitro* activity of ozenoxacin against MSSA, ofloxacin-resistant MRSA was four-fold to 16000-fold greater than that of other agents at the MIC_{90} for the clinical isolates. Yamakawa et al. compared the inhibitory activity of ozenoxacin with other quinolones for DNA gyrase and topoisomerase IV of *S. aureus* SA113. Ozenoxacin showed the greatest inhibitory activity for both enzymes among the quinolones tested. This study showed that the activity of ozenoxacin was not influenced by *grlA* mutation in *S. aureus*, and the degree of MIC increase of ozenoxacin for *grlA-gyrA* double and triple mutants was lowest among the quinolones tested (nadifloxacin, levofloxacin and ofloxacin). Yamakawa et al. concluded that ozenoxacin is potentially a useful quinolone for the treatment of skin and soft-tissue infections and that its potent bactericidal activity might be able to shorten the treatment period.

Table 19. Antibacterial activity (MIC, µg/mL) of ozenoxacin (T-3912) and other antibacterial agents against clinical isolates of staphylococci.

Compound	Organism (number)					
	MSSA (25)			MRSA, ofloxacin-resistant (23)		
	Range	MIC_{50}	MIC_{90}	Range	MIC_{50}	MIC_{90}
Ozenoxacin (T-3912)	0.00313 –0.00625	0.00625	0.00625	0.025–0.2	0.2	0.2
Nadifloxacin	0.0125–0.05	0.025	0.05	0.78–1.56	1.56	1.56
Ofloxacin	0.2–1.56	0.39	0.78	12.5–100	50	50
Levofloxacin	0.1–0.39	0.2	0.39	3.13–50	25	25
Clindamycin	0.1–>100	0.1	0.2	>100	>100	>100
Erythromycin	0.2–>100	0.2	>100	>100	>100	>100
Gentamicin	0.05–100	0.39	100	0.2–>100	25	100

QUINAZOLINEDIONES

PD 0305970 and PD 0326448

Quinazoline-2,4-diones such as PD 0305970 and PD 0326448 are new bacterial type II topoisomerase inhibitors structurally related to the quinolones. The chemical name of PD 0305970 is 3-amino-7-{(*R*)-3-[(*S*)-1-amino-ethyl]-pyrrolidin-1-yl}-1-cyclopropyl-6-fluoro-8-methyl-1*H*-quinazoline-2,4-dione. PD 0326448 is the 3-desamino analog of PD 0305970.

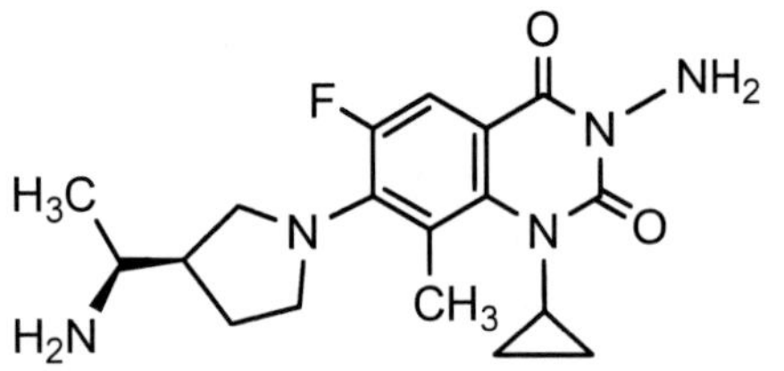

PD 0305970

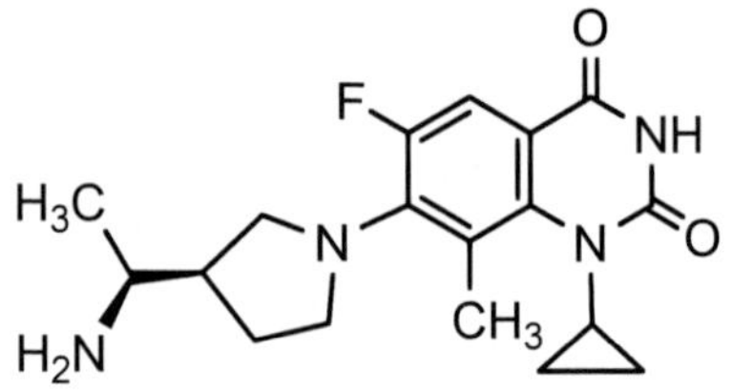

PD 0326448

Historically, quinazolinediones as antibacterial appeared in the literature as early as 1961, but it was not until 2004 that researchers at Pfizer reported similar antibacterial agents [177]. PD 0305970 and PD 0326448 were developed to introduce an orally active quinazolinedione, displaying antibacterial activity against fastidious Gram-negative and Gram-positive organisms, including multidrug- and fluoroquinolone-resistant organisms. In Huband et al. study, the respective MIC_{90} values against staphylococci for PD 0305970 were from 0.125 to 0.5 μg/mL. The MIC_{90}s of PD 0326448 were generally twofold higher than that of PD 0305970 (Table 20) [178, 179].

Table 20. Comparative antibacterial activities (MICs, μg/mL) of PD 0305970, PD 0326448, garenoxacin, gatifloxacin and levofloxacin against clinical isolates of staphylococci.

Compound	*Staphylococcus aureus* (number)								
	Oxacillin susceptible (20)			Oxacillin resistant (47)			Oxacillin resistant, levofloxacin resistant (35)		
	Range	MIC_{50}	MIC_{90}	Range	MIC_{50}	MIC_{90}	Range	MIC_{50}	MIC_{90}
PD 0305970	0.06–0.125	0.125	0.125	0.06–0.5	0.125	0.25	0.25–1	0.25	0.5
PD 0326448	0.125–0.5	0.25	0.25	0.125–1	0.25	0.5	0.25–4	1	4
Garenoxacin	0.015–2	0.06	0.06	0.03–4	2	2	0.25–32	2	4
Gatifloxacin	0.03–2	0.125	0.125	0.06–8	4	8	1–64	8	16
Levofloxacin	0.125–4	0.25	0.25	0.125–8	8	8	16–>64	32	64

4-QUINOLIZINONES (2-PYRIDONES)

A-170568.1

A-170568.1 belongs to the 2-pyridones class of type II topoisomerase inhibitors, with following chemical name: 8-(7-methylamino-5-azaspiro[2.4]hept-5-yl)-1-cyclopropyl-7-fluoro-9-methyl-4*H*-4-oxo-quinolizine-3-carboxylic acid.

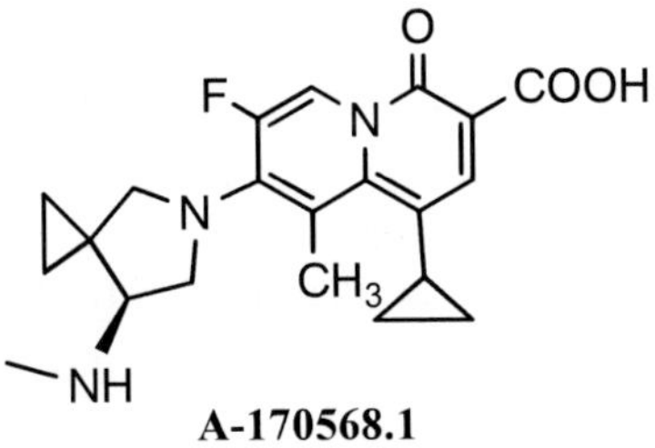

A-170568.1

It is structurally distinguished from ABT-719 by a certain bicyclic amine, 7-methylamino-5-azaspiro[2.4]hept-5-yl, replacing the 3-aminopyrrolidin-1-yl moiety located at C-8 position of ABT-719. The 7-amino-5-azaspiro[2.4]hept-5-yl residue is found in several new quinolones such as sitafloxacin, DC-159a, DK-507k and HSR-903. A-170568.1 is more hydrophobic than ABT-719 due to the *N*-methyl and bicyclic amine functionalities. In data presented by Nilius et al. [180] A-170568.1 showed superior activity as compared to ciprofloxacin, sparfloxacin, trovafloxacin, and clinafloxacin.

The efficacy of A-170568.1 against experimental infections was reported by Meulbroek, et al. A-170568.1 was less efficacious than ciprofloxacin and trovafloxacin against Gram-negative organisms in the mouse protection test [181]. However, against Gram-positive organisms, A-170568.1 displayed significantly better efficacy than both ciprofloxacin and vancomycin. The ED_{50} against ciprofloxacin-resistant MRSA in rat was at 27.1 mg/kg/day, 2-fold higher than that of vancomycin.

ABT-719

ABT-719, 8-[(3*S*)-aminopyrrolidin-1-yl]-1-cyclopropyl-7-fluoro-9-methyl-4*H*-4-oxo-quinolizine-3-carboxylic acid hydrochloride monohydrate, belong to the 2-pyridones class of type II topoisomerase inhibitors.

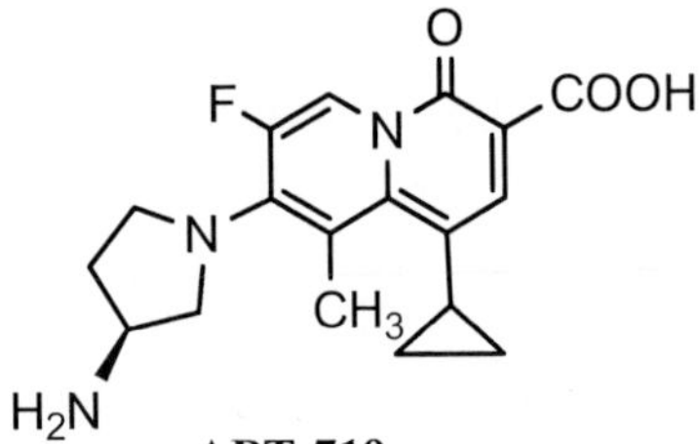

ABT-719

ABT-719 was also referred to as A-86719.1 for the hydrochloride salt and as A-86719.0 for the free base [182]. The hydrochloride salts of ABT-719 was found to be extremely soluble in water (>16 mg/mL). Beside the 2-pyridones structure, the distinctive structural features of ABT-719 are a 3-aminopyrrolidinyl moiety at the C-8 position, as found in tosufloxacin and clinafloxacin, and a methyl group at the C-9 position. Also, a cyclopropyl group has been preserved at position 1 similar to ciprofloxacin.

Antibacterial activity of ABT-719 in comparison to the activities of ciprofloxacin, sparfloxacin and clinafloxacin, were reported by Flamm et al. [183]. In general, against Gram-positive bacteria, ABT-719 showed superior activity as compared to ciprofloxacin, sparfloxacin, trovafloxacin, and clinafloxacin. ABT-719 was found to be particularly active against resistant organisms. MIC_{90} values of ABT-719 against ciprofloxacin-resistant *S. aureus* were 0.25 and 0.5 μg/mL, respectively, which were 2- to 4-fold better than that of trovafloxacin and up to 512-fold better than that of ciprofloxacin. MIC_{90} of ABT-719 against MRSA (1 μg/mL) was identical to that of clinafloxacin, and significantly lower than those of ciprofloxacin and sparfloxacin (Table 21).

Table 21. Comparative antibacterial activities (MICs, μg/mL) of ABT-719, ciprofloxacin, clinafloxacin, and sparfloxacin against *S. aureus*.

Compound	Organism (number)					
	Ciprofloxacin-susceptible MSSA (21) [Ciprofloxacin-resistant MSSA (25)]			MRSA (18)		
	Range	MIC_{50}	MIC_{90}	Range	MIC_{50}	MIC_{90}
ABT-719	0.008–0.015 [≤0.03–2]	0.008 [0.25]	0.015 [0.25]	0.06–1	0.25	1.0
Ciprofloxacin	0.12–0.5 [8–>128]	0.25 [16]	0.5 [64]	8–>128	16	>128
Clinafloxacin	0.015–0.03 [0.25–2]	0.03 [0.5]	0.03 [1]	0.25–2.0	0.25	1.0
Sparfloxacin	0.03–0.06 [0.12–16]	0.03 [8]	0.06 [16]	0.5–16	4	8

Chin et al. have also published a study of ABT-719 against 369 Gram-positive clinical isolates including MSSA and MRSA [184]. The in vitro activity of ABT-719 was compared with that of ciprofloxacin, sparfloxacin, and sitafloxacin. ABT-719 was found to be superior to ciprofloxacin and sparfloxacin, and comparable to or more potent than sitafloxacin. ABT-719 exhibited MIC_{90}s of ≤0.12 μg/mL against both ciprofloxacin-susceptible MSSA and MRSA, and ≤1 μg/mL against ciprofloxacin-resistant MRSA.

Alder et al. published several study of ABT-719 in systemic infections caused by important pathogens including staphylococci. ABT-719 displayed superior efficacy over both ciprofloxacin and vancomycin against ciprofloxacin-sensitive *S. aureus* with ED_{50}s <3.1 mg/kg/day, compared to 8-18 mg/kg/day for ciprofloxacin and vancomycin. ABT-719 was effective against ciprofloxacin-resistant MRSA with an ED_{50} value of 50.0 mg/kg/day, which was at least 10 times lower than that of ciprofloxacin. However, ABT-719 was 7-fold less efficacious than vancomycin against MRSA in these in vivo studies, despite being nearly 8-fold lower in MIC [182].

CBR-2092

In a program pursued by Cumbre Pharmaceuticals Inc. toward the rifamycin-based hybrids, a series of compounds was prepared in which rifamycin and a quinolone pharmacophore were covalently connected. Among several hybrids, CBR-2092 has been developed to Phase I clinical trials for infections involving biofilms with indwelling medical devices (catheters, prosthetic devices).

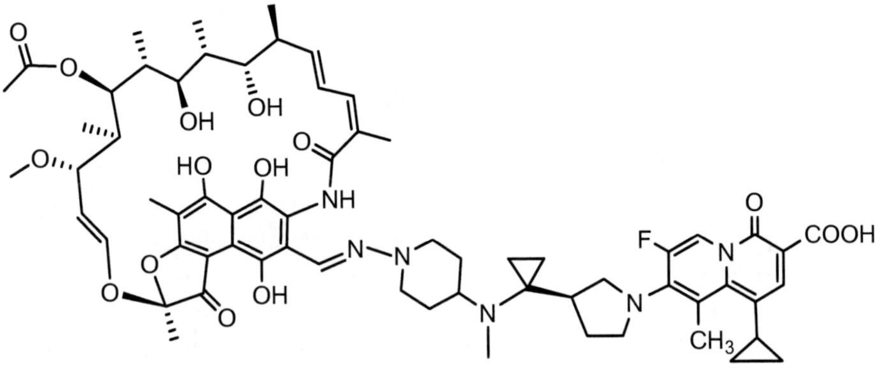

CBR-2092

In the CBR-2092 molecule, the rifamycin SV pharmacophore is combined with a quinolone pharmacophore derived from ABT-719 [185]. ABT-719 is the most extensively characterized member of the 2-pyridone series and exhibits equipotent inhibitory activity against ciprofloxacin-resistant isolates of *S. aureus* and MRSA [186]. Rifamycins are a RNA polymerase inhibitor that have proven efficacy in the treatment of persistent bacterial infections [187, 188]. As all of the pharmacophoric features of the rifamycin identified as critical elements in the RNA polymerase interaction are preserved in CBR-2092 and the quinolone scaffold is appended to the C-3 position via hydrazone functionality, it is perhaps not surprising that CBR-2092 exhibits in vitro potency as an RNA polymerase inhibitor that is nearly equivalent to that of rifampin. Thus, in biochemical studies, CBR-2092 exhibited rifampin-like potency as an inhibitor of RNA polymerase and was an equipotent inhibitor of DNA gyrase and DNA topoisomerase IV [185].

CBR-2092 exhibits potent activity against prevalent staphylococci and streptococci; especially, against the MRSA. For the MRSA isolates, the MIC ranges determined for CBR-2092 and rifampin were $\leq$0.004 to 2 and $\leq$0.004 to >4 μg/ml, respectively (Table 22).

In studies of the intracellular killing of *S. aureus*, CBR-2092 exhibited prolonged bactericidal activity that was superior to the activities of moxifloxacin, rifampin, and a combination of moxifloxacin and rifampin [189].

Table 22. Comparative antibacterial activities (MICs, μg/mL) of CBR-2092, rifampin and ciprofloxacin against *S. aureus*.

Compound	Organism (number)					
	MSSA (51)			MRSA (54)		
	Range	MIC_{50}	MIC_{90}	Range	MIC_{50}	MIC_{90}
CBR-2092	≤0.004–0.03	0.008	0.015	≤0.004–2	0.015	0.015
Rifampin	≤0.008–0.06	0.015	0.015	≤ 0.008–>4	0.015	0.015
Ciprofloxacin	0.06–>4	0.25	1	0.12–>4	>4	>4

CONCLUSION

Progressive efforts have been made to optimize the biological activity of the quinolones against *S. aureus* and drug-resistant strains including MRSA [190-194]. Much of the improved potency of modern quinolones against staphylococci has been achieved by tinkering with the N-1, C-7, and C-8 substituents on the quinolone ring system. Even by development of 6-desfluoroquinolones, it demonstrated that the spatial arrangements with appropriate substituents at N-1, C-7, and C-8 could diminish the impact of 6-fluoro-substituent.

As mentioned above, the N-1 cyclopropyl group, which was originally described for ciprofloxacin, remains one of the most effective functionalities for providing anti-MRSA quinolones. Combination of the N-1 cyclopropyl or 2-fluorocyclopropyl with a C-8 substituent (halogen, methyl, methoxy or difluoromethoxy) can improve activity of quinolones against MRSA. Unusual SAR was found in the case of gemifloxacin, zabofloxacin and DW286, where the naphthyridone nucleus provided MICs improved over those of the corresponding C-8 quinolones. Introduction of an aminopyrrolidinyl substituent at C-7 generally results in increased anti-MRSA activity compared to the piperazine derivative. Increasing the lipophilicity of a quinolone, however, as has been effected in many new agents, generally tends to increase potency against MRSA while somewhat attenuating Gram-negative potency.

Several newly developed quinolones and related compounds including 6-fluoroquinolones (delafloxacin, olamufloxacin, Dk-507k, DC-159a, DQ-113 and WCK 771), 6-fluoronaphthyridones (gemifloxacin, zabofloxacin and DW286), 6-desfluoroquinolones (garenoxacin, nemonoxacin, ozenoxacin and DX-619), quinazolinediones (PD 0305970 and PD 0326448) and 4-quinolizinones (ATB-719 and CBR-2092) exhibit enhanced activity against *S. aureus* and have proved useful against MRSA strains and are under rapid preclinical and clinical development.

REFERENCES

[1] Deresinski, S. Methicillin-resistant *Staphylococcus aureus:* an evolutionary, epidemiologic, and therapeutic odyssey. *Clin Infect Dis*, 2005, 40, 562–573.

[2] Shittu, A; Lin, J. Newer antistaphylococcal agents: in-vitro studies and emerging trends in *Staphylococcus aureus* resistance. *Wounds*, 2006, 18, 129–146.

[3] Lescher, GY; Froelich, EJ; Gruett, MD; Bailey, JH; Brundage, RP. 1,8-Naphthyridine derivatives: a new class of chemotherapy agents. *J Med Pharm Chem*, 1962, 5, 1063–1068.

[4] Wise, R. Norfloxacin – a review of pharmacology and tissue penetration. *J Antimicrob Chemother*, 1984, 13 (Suppl. B), 59–64.

[5] Appelbaum, PC; Hunter, PA. The fluoroquinolone antibacterials: past, present and future perspectives. *Int J Antimicrob Agents*, 2000, 16, 5–15.

[6] Koga, H; Itoh, A; Murayama, S; Suzue, S; Irikura, T. Structure-activity relationships of antibacterial 6,7- and 7,8-disubstituted 1-alkyl-1,4-dihydro-4-oxoquinoline-3-carboxylic acids. *J Med Chem*, 1980, 23, 1358–63.

[7] Asahina, Y; Ishizaki, T; Suzue, S. Recent advances in structure activity relationships in new quinolones. *Prog Drug Res*, 1992, 38, 57–106.

[8] Andriole, VT. *The Quinolones*, Academic Press: London, 1988.

[9] Wolfson, JS; Hooper, DC. *Quinolone Antimicrobial Agents*, American Society for Microbilogy: Washington DC, 1989.

[10] Domagala, JM. Structure-activity and structure side effect relationships for the quinolone antibacterials. *J Antimicrob Chemother*, 1994, 33, 685–706.

[11] Ball, P; Fernald, A; Tillotson, G. Therapeutic advances of new fluoroquinolones. *Expert Opin Inv Drugs*, 1998, 7, 761–783.

[12] Zhanel, GG; Ennis, K; Vercaigne, L; Walkty, A; Gin, AS; Embil, J; Smith, H; Hoban, DJ. A critical review of the fluoroquinolones: focus on respiratory infections. *Drugs*, 2002, 62, 13–59.

[13] De Sarro, A; De Sarro, G. Adverse reactions to fluoroquinolones. an overview on mechanistic aspects. *Curr Med Chem*, 2001, 8, 371–384.

[14] Hooper, DC. Mechanisms of fluoroquinolone resistance. *Drug Resis Updates*, 1999, 2, 38–55.

[15] Hooper, DC. Bacterial topoisomerases, anti-topoisomerases, and antitopoisomerase resistancc. *Clin Infect Dis*, 1998, 27(Suppl 1), S54–63.

[16] Schmitz, FJ; Higgins, PG; Mayer, S; Fluit, AC; Dalhoff, A. Activity of quinolones against Gram-positive cocci: mechanisms of drug action and bacterial resistance. *Eur J Clin Microbiol Infect Dis*, 2002, 21, 647–659.

[17] Wang, JC; Lynch, AS. Transcription and DNA supercoiling. *Curr Opin Genet Dev*, 1993, 3, 764–768.

[18] Levine, C; Hiasa, H; Marians, KJ. DNA gyrase and topoisomerase IV: Biochemical activities, physiological roles during chromosome replication, and drug sensitivities. *Biochim Biophys Acta Gene Struct Expression*, 1998, 1400: 29–43.

[19] Hawkey, PM. Mechanisms of quinolone action and microbial response. *J Antimicrob Chemother*, 2003, 51(Suppl. S1), 29–35.

[20] Drlica, K. Control of bacterial DNA supercoiling. *Mol Microbiol*, 1992, 6, 425–433.

[21] Hooper, DC. Fluoroquinolone resistance among Gram-positive cocci. *Lancet Infect Dis*, 2002, 2, 530–538.

[22] Wigley, DB. Structure and mechanism of DNA topoisomerases. *Ann Rev Biophys Biomol Struct*, 1995, 24, 185–208.

[23] Drlica, K; Zhao, X. DNA gyrase, topoisomerase IV, and the 4-quinolones. *Microbiol Mol Biol Rev*, 1997, 61, 377–392.

[24] Shen, LL; Mitscher, LA; Sharma, PN; O'Donnell, TJ; Chu, WT; Cooper, CS; Rosen, T; Pernet, AG. Mechanism of inhibition of DNA gyrase by quinolone antibacterials: a co-operative drug-DNA binding model. *Biochemistry*, 1989, 28, 3886–3894.

[25] Van Bambeke, F; Michot, JM; Van Eldere, J; Tulkens, PM. Quinolones in 2005: an update. *Clin Microbiol Infect*, 2005, 11, 256–280.

[26] Anderson, VE; Zaniewski, RP; Kaczmarek, FS; Gootz, TD; Osheroff, N. Quinolones inhibit DNA religation mediated by *Staphylococcus aureus* topoisomerase IV. Changes in drug mechanism across evolutionary boundaries. *J Biol Chem*, 1999, 274, 35927–35932.

[27] Gootz, TD; Zaniewski, RP; Haskell, SL; Kaczmarek, FS; Maurice, AE. Activities of trovafloxacin compared with those of other fluoroquinolones against purified topoisomerases and *gyrA* and *grlA* mutants of *Staphylococcus aureus*. *Antimiocrob Agents Chemother*, 1999, 43, 1845–1855.

[28] Ruiz, J; Sierra, JM; De Anta, MT; Vila, J. Characterization of sparfloxacin-resistant mutants of *Staphylococcus aureus* obtained in vitro. *Int J Antimicrob Agents*, 2001, 18, 107–112.

[29] Hooper, DC. Mode of action of fluoroquinolones. *Drugs*, 1999, 58 (Suppl. 2), 6–10.

[30] Pan, XS; Fisher, LM. DNA gyrase and topoisomerase IV are dual targets of clinafloxacin action in *Streptococcus pneumoniae*. *Antimicrob Agents Chemother*, 1998, 42, 2810–2816.

[31] Heaton, VJ; Ambler, JE; Fisher, LM. Potent antipneumococcal activity of gemifloxacin is associated with dual targeting of gyrase and topoisomerase IV, an in vivo target preference for gyrase, and enhanced stabilization of cleavable complexes in vitro. *Antimicrob Agents Chemother*, 2000, 44, 3112–3117.

[32] Nilius, AM; Shen, LL; Hensey-Rudloff, D; Almer, LS; Beyer, JM; Balli, DJ; Cai, Y; Flamm, RK. In vitro antibacterial potency and spectrum of ABT-492, a new fluoroquinolone. *Antimicrob Agents Chemother*, 2003, 47, 3260–3269.

[33] Roychoudhury, S; Ledoussal, B. Non-fluorinated quinolones (NFQs): new antibacterials with unique properties against quinolone-resistant Gram-positive pathogens. *Curr Drug Targets - Infectious Disorders*, 2002, 2, 51–65.

[34] Takei, M; Fukuda, H; Kishii, R; Hosaka, M. Target preference of 15 quinolones against *Staphylococcus aureus*, based on antibacterial activities and target inhibition. *Antimicrob Agents Chemother,* 2001, 45, 3544–3547.

[35] Schulte, A; Heisig, P. In vitro activity of gemifloxacin and five other fluoroquinolones against defined isogenic mutant of *Escherichia coli*, *Pseudomonas aeruginosa* and *Staphylococcus aureus*. *J Antimicrob Chemother*, 2000, 46, 1037–1046.

[36] Blanche, F; Cameron, B; Bernard, FX; Maton, L; Manse, B; Ferrero, L; Ratet, N; Lecoq, C; Goniot, A; Bisch, D; Crouzet, J. Differential behaviors of *Staphylococcus aureus* and *Escherichia coli* type II DNA topoisomerases. *Antimicrob Agents Chemother*, 1996, 40, 2714–2720.

[37] Tanaka, M; Onodera, Y; Uchida, Y; Sato, K; Hayakawa, I. Inhibitory activities of quinolones against DNA gyrase and topoisomerase IV purified from *Staphylococcus aureus.* *Antimicrob Agents Chemother*, 1997, 41, 2362–2366.

[38] Schmitz, FJ; Boos, M; Mayer, S; Jagusch, H; Fluit, AC. Increased in vitro activity of the novel des-fluoro(6) quinolone BMS-284756 against genetically defined clinical isolates of *Staphylococcus aureus*. *J Antimicrob Chemother*, 2002, 49, 283–287.

[39] Blumberg, HM; Rimland, D; Carroll, DJ; Terry, P; Wachsmuth, IK. Rapid development of ciprofloxacin resistance in methicillin-susceptible and -resistant *Staphylococcus aureus*. *J Infect Dis*, 1991, 163, 1279–1285.

[40] Harnett, N; Brown, S; Krishnan, C. Emergence of quinolone resistance among clinical isolates of methicillin-resistant *Staphylococcus aureus* in Ontario, Canada. *Antimicrob Agents Chemother*, 1991, 35, 1911–1913.

[41] Barry, AL; Jones, RN. In vitro activity of ciprofloxacin against gram-positive cocci. *Am J Med*, 1987, 82, 27–32.

[42] Scheel, O; Lyon, DJ; Rosdahl, VT; Adeyemi-Doro, FA; Ling, TK; Cheng, AF. In-vitro susceptibility of isolates of methicillin-resistant *Staphylococcus aureus* 1988-1993. *J Antimicrob Chemother*, 1996, 37, 243–251.

[43] Schmitz, FJ; Fluit, AC; Hafner, D; Beeck, A, Perdikouli, M; Boos, M; Scheuring, S; Verhoef, J; Köhrer, K; Von Eiff, C. Development of resistance to ciprofloxacin, rifampin, and mupirocin in methicillin-susceptible and –resistant *Staphylococcus aureus* isolates. *Antimicrob Agents Chemother*, 2000, 44, 3229–3231.

[44] Lipsitch, M; Samore, MH. Antimicrobial use and antimicrobial resistance: a population perspective. *Emerg Infect Dis*, 2002, 8, 347–354.

[45] Yoshida, H; Bogaki, M; Nakamura, S; Ubukata, K; Konno, M. Nucleotide sequence and characterization of the *Staphylococcus aureus norA* gene, which confers resistance to quinolones. *J Bacteriol*, 1990, 172, 6942–6949.

[46] Goswitz, JJ; Willard, KE; Fasching, CE; Peterson, LR. Detection of *gyrA* mutations associated with ciprofloxacin resistance in methicillin- resistant *Staphylococcus aureus*: analysis by polymerase chain reaction and automated direct DNA sequencing. *Antimicrob Agents Chemother*, 1992, 36, 1166–1169.

[47] Ferrero, L; Cameron, B; Manse, B; Lagneaux, D; Crouzet, J; Famechon, A; Blanche, F. Cloning and primary structure of *Staphylococcus aureus* DNA topoisomerase IV: a primary target of fluoroquinolones. *Mol Microbiol*, 1994, 13, 641–653.

[48] Willmott, CJ; Maxwell, A. A single point mutation in the DNA gyrase A protein greatly reduces binding of fluoroquinolones to the gyrase-DNA complex. *Antimicrob Agents Chemother*, 1993, 37, 126–127.

[49] Fournier, B; Hooper, DC. Mutations in topoisomerase IV and DNA gyrase of *Staphylococcus aureus*: Novel pleiotropic effects on quinolone and coumarin activity. *Antimicrob Agents Chemother*, 1998, 42, 121–128.

[50] Ng, EY; Trucksis, M; Hooper, DC. Quinolone resistance mutations in topoisomerase IV: relationship of the *flqA* locus and genetic evidence that topoisomerase IV is the primary target and DNA gyrase the secondary target of fluoroquinolones in *Staphylococcus aureus*. *Antimicrob Agents Chemother*, 1996, 40, 1881–1888.

[51] Schmitz, FJ; Hofmann, B; Hansen, B; Scheuring, S; Lückefahr, M; Klootwijk, M; Verhoef, J; Fluit, A; Heinz, HP; Köhrer, K; Jones, ME. Relationship between ciprofloxacin, ofloxacin, levofloxacin, sparfloxacin and moxifloxacin (BAY 12-8039)

MICs and mutations in *grlA*, *grlB*, *gyrA* and *gyrB* in 116 unrelated clinical isolates of *Staphylococcus aureus*. *J Antimicrob Chemother*, 1998, 41, 481–484.

[52] Wang, T; Tanaka, M; Sato, K. Detection of *grlA* and *gyrA* mutations in 344 *Staphylococcus aureus* strains. *Antimicrob Agents Chemother*, 1998, 42, 236–240.

[53] Shreedhan, S; Oram, M; Jensen, B; Peterson, LR; Fisher, LM. DNA gyrase *gyr*A mutations in ciprofloxacin-resistant strains of *Staphylococcus aureus*: close similarities with quinolone resistance mutations in *Escherichia coli*. *J Bacteriol*, 1990, 172, 7260–7262.

[54] Bast, DJ; Low, DE; Duncan, CL; Kilburn, L; Mandell, LA; Davidson, RJ; De Azavedo, JC. Fluoroquinolone resistance in clinical isolates of *Streptococcus pneumoniae*: Contributions of type II topoisomerase mutations and efflux to levels of resistance. *Antimicrob Agents Chemother*, 2000, 44, 3049–3054.

[55] Piddock, LJV; Johnson, MM; Simjee, S; Pumbwe, L. Expression of efflux pump gene *pmrA* in fluoroquinolone-resistant and -susceptible clinical isolates of *Streptococcus pneumoniae*. *Antimicrob Agents Chemother*, 2002, 46, 808–812.

[56] Ince, D; Zhang, X; Silver, LC; Hooper, DC. Dual targeting of DNA gyrase and topoisomerase IV: target interactions of garenoxacin (BMS-284756, T-3811ME), a new desfluoroquinolone. *Antimicrob Agents Chemother*, 2002, 46, 3370–3380.

[57] Bryskier, A; Chantot, JF. Classification and structure-activity relationships of fluoroquinolones. *Drugs*, 1995, 49 (Suppl. 2), 16–28.

[58] Peterson, LR. Quinolone molecular structure-activity relationships: what we have learned about improving antimicrobial activity. *Clin Infect Dis*, 2001, 33 (Suppl 3), S180–S186.

[59] Roychoudhury, S; Twinem, TL; Makin, KM; McIntosh, EJ; Ledoussal, B; Catrenich, CE. Activity of non-fluorinated quinolones (NFQs) against quinolone-resistant *Escherichia coli* and *Streptococcus pneumoniae*. *J Antimicrob Chemother*, 2001, 48, 29–36.

[60] Hartman-Neumann, S; DenBleyker, K; Pelose, LA; Lawrence, LE; Barett, JF; Douherty, TJ. Selection and genetic characterization of *Streptococcus pneumoniae* mutants resistant to the des-F(6) quinolone BMS-284756. *Antimicrob Agents Chemother*, 2001, 45, 2865–2870.

[61] Fukuda, H; Kishii, R; Takei, M; Hosaka, M. Contributions of the 8-methoxy group of gatifloxacin to resistance selectivity, target preference, and antibacterial activity against *Streptococcus pneumoniae*. *Antimicrob Agents Chemother*, 2001, 45, 1649–1653.

[62] Ince, D; Hooper, DC. Mechanisms and frequency of resistance to gatifloxacin in comparison to AM-1121 and ciprofloxacin in *Staphylococcus aureus*. *Antimicrob Agents Chemother*, 2001, 45, 2755–2764.

[63] Zhao, X; Wang, JY; Xu, C; Dong, Y; Zhou, J; Domagala, J; Drlica, K. Killing of *Staphylococcus aureus* by C-8-methoxy fluoroquinolones. *Antimicrob Agents Chemother*, 1998, 42, 956–958.

[64] Zhanel, GG; Hoban, DJ; Schurek, K; Karlowsky, JA. Role of efflux mechanisms on fluoroquinolone resistance in *Streptococcus pneumoniae* and *Pseudomonas aeruginosa*. *Int J Antimicrob Agents*, 2004, 24, 529–535.

[65] Markham, PN; Neyfakh, AA. Inhibition of the multidrug transporter NorA prevents emergence of norfloxacin resistance in *Staphylococcus aureus*. *Antimicrob Agents Chemother*, 1996, 40, 2673–2674.

[66] Markham, PN. Inhibition of the emergence of ciprofloxacin resistance in *Streptococcus pneumoniae* by the multidrug efflux inhibitor reserpine. *Antimicrob Agents Chemother*, 1999, 43, 988–989.
[67] Sulavik, CM; Barg, NL. Examination of methicillin-resistant and methicillin-susceptible *Staphylococcus aureus* mutants with low-level fluoroquinolone resistance. *Antimicrob Agents Chemother*, 1998, 42, 3317–3319.
[68] Scheld, WM. Maintaining fluoroquinolone class efficacy: review of influencing factors. *Emerg Infect Dis*, 2003, 9, 1–9.
[69] Beyer, R; Pestova, E; Millichap, JJ; Stosor, V; Noskin, GA; Peterson, LR. A convenient assay for estimating the possible involvement of efflux of fluoroquinolones by *Streptococcus pneumoniae* and *Staphylocccus aureus*: evidence for diminished moxifloxacin, sparfloxacin, and trovafloxacin efflux. *Antimicrob Agents Chemother*, 2000, 44, 798–801.
[70] Aeschlimann, JR; Dresser, LD; Kaatz, GW; Rybak, MJ. Effects of NorA inhibitors on *in vitro* antibacterial activities and postantibiotic effects of levofloxacin, ciprofloxacin, and norfloxacin in genetically related strains of *Staphylococcus aureus*. *Antimicrob Agents Chemother*, 1999, 43, 335–340.
[71] Piddock, LJ; Johnson, MM. Accumulation of 10 fluoroquinolones by wild-type or efflux mutant *Streptococcus pneumoniae*. *Antimicrob Agents Chemother*, 2002, 46, 813–820.
[72] Brenwald, NP; Gill, MJ; Wise, R. Prevalence of a putative efflux mechanism among fluoroquinolone-resistant clinical isolates of *Streptococcus pneumoniae*. *Antimicrob Agents Chemother,* 1998, 42, 2032–2035.
[73] Zeller, V; Janoir, C; Kitzis, MD; Gutmann, L; Moreau, NJ. Active efflux as a mechanism of resistance to ciprofloxacin in *Streptococcus pneumoniae*. *Antimicrob Agents Chemother,* 1997, 41, 1973–1978.
[74] Takenouchi, T; Tabata, F; Iwata, Y; Hanzawa, H; Sugawara, M; Ohya, S. Hydrophilicity of quinolones is not an exclusive factor for decreased activity in efflux-mediated resistant mutants of *Staphylococcus aureus*. *Antimicrob Agents Chemother*, 1996, 40, 1835–1842.
[75] Madras-Kelly, KJ; Daniels, C; Haegbloom, M; Thompson, M. Pharmacodynamic characterization of efflux and topoisomerase IV-mediated fluoroquinolone resistance in *Streptococcus pneumoniae*. *J Antimicrob Chemother*, 2002, 50, 211–218.
[76] Kerns, RJ; Rybak, MJ; Kaatz, GW; Vaka, F; Cha, R; Grucz, RG; Diwadkar, VU; Ward, TD. Piperazinyl-linked fluoroquinolone dimers possessing potent antibacterial activity against drug-resistant strains of *Staphylococcus aureus*. *Bioorg Med Chem Lett*, 2003, 13, 1745–1749.
[77] Kerns, RJ; Rybak, MJ; Kaatz, GW; Vaka, F; Cha, R; Grucz, RG; Diwadkar, VU. Structural features of piperazinyl-linked ciprofloxacin dimers required for activity against drug-resistant strains of *Staphylococcus aureus*. *Bioorg Med Chem Lett*, 2003, 13, 2109–2112.
[78] Bhanot, SK; Singh, M; Chatterjee, NR. The chemical and biological aspects of fluoroquinolones: reality and dreams. *Curr Pharm Design*, 2001, 7, 311–335.
[79] De Sarro, A; De Sarro, G. Adverse reactions to fluoroquinolones. an overview on mechanistic aspects. *Curr Med Chem*, 2001, 8, 371–384.

[80] Takemura, M; Hayakawa, I. Recent advances in the field of quinolones. *Il Farmaco*, 2001, 56, 37–40.

[81] Kimura, Y; Atarashi, S; Kawakami, K; Sato, K; Hayakawa, I. (Fluorocyclopropyl)quinolones. 2. Synthesis and Stereochemical structure-activity relationships of chiral 7-(7-amino-5-azaspiro[2.4]heptan-5-yl)-1-(2-fluorocyclopropyl)quinolone antibacterial agents. *J Med Chem*, 1994, 37, 3344–3352.

[82] Brighty, KE; Gootz, TD. The chemistry and biological profile of trovafloxacin. *J Antimicrob Chemother*, 1997, 39 (Suppl. B), 1–14.

[83] Chung, SJ; Kim, DH. Synthesis and evaluation of 3-fluoro-2-piperazinyl-5,8,13-trihydro-5-oxoquino[1,2-a][3,1]benzoxazine-6-carboxylic acids as potential antibacterial agents. *Arch Pharm*, 1997, 330, 63–66.

[84] Morrissey, I; Hoshino, K; Sato, K; Yoshida, A; Hayakawa, I; Bures, MG; Shen, LL. Mechanism of differential activities of ofloxacin enantiomers. *Antimicrob Agents Chemother*, 1996, 40, 1775–1784.

[85] Tillotson, GS; Blondeau, JM. Structure–activity–function evaluation of the fluoroquinolones. In: Adam D, Finch RG and Hunter PA (eds.) *Moxifloxacin in Practice,* Maxim Medical (1999) 91–101.

[86] Tillotson, GS. Quinolones: structure–activity relationships and future predictions. *J Med Microbiol,* 1996, 44, 320–324.

[87] Segawa, J; Kitano, M; Kazuno, K; Matsuoka, M; Shirahase, I; Ozaki, M; Matsuda, M; Tomii, Y; Kise, M. Studies on pyridonecarboxylic acids. 1. Synthesis and antibacterial evaluation of 7-substituted-6-halo-4-oxo-4H-[1,3]thiazeto[3,2-a]quinoline-3- carboxylic acids. *J Med Chem*, 992, 35, 4727–4738.

[88] Chu, DT; Hallas, R; Clement, JJ; Alder, J; McDonald, E; Plattner, JJ. Synthesis and antitumour activities of quinolone antineoplastic agents. *Drugs Exp Clin Res*, 1992, 18, 275–282.

[89] Yoshida, T; Yamamoto, Y; Orita, H; Kakiuchi, M; Takahashi, Y; Itakura, M; Kado, N; Mitani, K; Yasuda, S; Kato, H; Itoh, Y. Studies on quinolone antibacterials. IV. Structure-activity relationships of antibacterial activity and side effects for 5- or 8-substituted and 5,8-disubstituted-7-(3-amino-1-pyrrolidinyl)-1-cyclopropyl-1, 4-dihydro-4-oxoquinoline-3-carboxylic acids. *Chem Pharm Bull (Tokyo)*, 1996, 44, 1074–1085.

[90] Jaillon, P; Morganroth, J; Brumpt, I; Talbot, G. Overview of electrocardiographic and cardiovascular safety data for sparfloxacin. Sparfloxacin Safety Group. *J Antimicrob Chemother*, 1996, 37 (Suppl. A), 161–167.

[91] Schentag, JJ. Sparfloxacin: a review. *Clin Ther*, 2000, 22, 372–378.

[92] Wiedemann, B; Heisig, P. Antibacterial activity of grepafloxacin. *J Antimicrob Chemother*, 1997, 40 (Suppl. A), 19–25.

[93] Bryskier, A; Chantot, JF. Classification and structure-activity relationships of fluoroquinolones. *Drugs*, 1995, 49 (Suppl. 2), 16–28.

[94] Gootz, TD; Brighty, KE. Fluoroquinolone antibacterials: SAR mechanism of action, resistance, and clinical aspects. *Med Res Rev*, 1996, 16, 433–486.

[95] Lawrence, LE; Wu, P; Fan, L; Gouveia, KE; Card, A; Casperson, M; Denbleyker, K; Barrett, JF. The inhibition and selectivity of bacterial topoisomerases by BMS-284756 and its analogues. *J Antimicrob Chemother*, 2001, 48, 195–201.

[96] Barry, AL; Fuchs, PC; Brown, SD. In vitro activities of three nonfluorinated quinolones against representative bacterial isolates. *Antimicrob Agents Chemother*, 2001, 45, 1923–1927.

[97] Cecchetti, V; Clementi, S; Cruciani, G; Fravolini, A; Pagella, PG; Savino, A; Tabarrini, O. 6-Aminoquinolones: A new class of quinolone antibacterials? *J Med Chem*, 1995, 38, 973–982.

[98] Cecchetti, V; Fravolini, A; Lorenzini, MC; Tabarrini, O; Terni, P; Xin, T. Studies on 6-aminoquinolones: Synthesis and antibacterial evaluation of 6-amino-8-methyl quinolones. *J Med Chem*, 1996, 39, 436–445.

[99] Cecchetti, V; Fravolini, A; Palumbo, M; Sissi, C; Tabarrini, O; Terni, P; Xin, T. Potent 6-desfluoro-8-methylquinolones as new lead compounds in antibacterial chemotherapy. *J Med Chem*, 1996, 39, 4952–4957.

[100] Artico, M; Mai, A; Spardella, G; Massa, S; Musiu, C; Lostia, S; Demontis, F; La Colla; F. Nitroquinolones with broad-spectrum antimycobacterial activity in vitro. *Bioorg Med Chem Lett*, 1999, 9, 1651–1656.

[101] Gootz, TD; Brighty, KE. *The Quinolones*. 2nd ed., Academic Press, 1998.

[102] Sun, J; Sakai, S; Tauchi, Y; Deguchi, Y; Chen, J; Zhang, R; Morimoto, K. Determination of lipophilicity of two quinolone antibacterials, ciprofloxacin and grepafloxacin, in the protonation equilibrium. *Eur J Pharm Biopharm*, 2002, 54, 51–58.

[103] Ball, P. Moxifloxacin (Avelox): an 8-methoxyquinolone antibacterial with enhanced potency. *Int J Clin Pract*, 2000, 54, 329–332.

[104] Kim, JH; Kang, JA; Kim, YG; Kim, JW; Lee, JH; Choi, EC; Kim, BK. In vitro and in vivo antibacterial efficacies of CFC-222, a new fluoroquinolone. *Antimicrob Agents Chemother*, 1997, 41, 2209–2213.

[105] Weller, TMA; Andrews, JM; Jevons, G; Wise, R. The in vitro activity of BMS-284756, a new des-fluorinated quinolone. *J Antimicrob Chemother*, 2002, 49, 177–184.

[106] Foroumadi, A; Emami, S; Davood, A; Moshafi, MH; Sharifian, A; Tabatabaiee, M; Tarhimi Farimani, H; Sepehri, G; Shafiee, A. Synthesis and in-vitro antibacterial activities of N-substituted piperazinyl quinolones. *Pharm Sci*, 1997, 3, 559.

[107] Piddock, LJV; Jin, YF; Griggs, DJ. Effect of hydrophobicity and molecular mass on the accumulation of fluoroquinolones by *Staphylococcus aureus*. *J Antimicrob Chemother*, 2001, 47, 261–270.

[108] McCaffrey, C; Bertasso, A; Pace, J; Georgopapadakou, NH. Quinolone accumulation in *Escherichia coli*, *Pseudomonas aeruginosa*, and *Staphylococcus aureus*. *Antimicrob Agents Chemother*, 1992, 36, 1601–1605.

[109] Piddock, LJV; Hall, MC; Wise, R. Mechanism of action of lomefloxacin. *Antimicrob Agents Chemother*, 1990, 34, 1088–1093.

[110] Foroumadi, A; Emami, S; Mehni, M; Moshafi, MH; Shafiee, A. Synthesis and antibacterial activity of N-[2-(5-bromothiophen-2-yl)-2-oxoethyl] and N-[(2-5-bromothiophen-2-yl)-2-oximinoethyl] derivatives of piperazinyl quinolones. *Bioorg Med Chem Lett*, 2005, 15, 4536–4539.

[111] Foroumadi, A; Oboudiat, M; Emami, S; Karimollah, A; Saghaee, L; Moshafi, MH; Shafiee, A. Synthesis and antibacterial activity of *N*-[2-[5-(methylthio)thiophen-2-yl]-2-oxoethyl] and *N*-[2-[5-(methylthio)thiophen-2-yl]-2-(oxyimino)ethyl] piperazinylquinolone derivatives. *Bioorg Med Chem*, 2006, 14, 3421–3427.

[112] Foroumadi, A; Ghodsi, S; Emami, S; Najjari, S; Samadi, N; Faramarzi, MA; Beikmohammadi, L; Shirazi, FH; Shafiee, A. Synthesis and antibacterial activity of new fluoroquinolones containing a substituted *N*-(phenethyl)piperazine moiety. *Bioorg Med Chem Lett*, 2006; 16, 3499–3503.

[113] Foroumadi, A; Mohammadhosseini, N; Emami, S; Letafat, B; Faramarzi, MA; Samadi, N; Shafiee, A. Synthesis and antibacterial activity of new 7-piperazinylquinolones containing functionalized 2-(furan-3-yl)ethyl moiety. *Arch Pharm*, 2007, 340, 47–52.

[114] Letafat, B; Emami, S; Mohammadhosseini, N; Faramarzi, MA; Samadi, N; Shafiee, A; Foroumadi, A. Synthesis and antibacterial activity of new *N*-[2-(thiophen-3-yl)ethyl] piperazinyl quinolones. *Chem Pharm Bull,* 2007, 55, 894–898.

[115] Foroumadi, A; Emami, S; Mansouri, S; Javidnia, A; Saeid-Adeli, N; Shirazi, FH; Shafiee, A. Synthesis and antibacterial activity of levofloxacin derivatives with certain bulky residues on piperazine ring. *Eur J Med Chem*, 2007, 42, 985–992.

[116] Emami, S; Foroumadi, A; Faramarzi, MA; Samadi, N. Synthesis and antibacterial activity of quinolone-based compounds containing a coumarin moiety. *Arch Pharm Chem Life Sci*, 2008, 341, 42–48.

[117] Foroumadi, A; Emami, S; Hassanzadeh, A; Rajaee, M; Sokhanvar, K; Moshafi, MH; Shafiee, A. Synthesis and antibacterial activity of N-(5-benzylthio-1,3,4-thiadiazol-2-yl) and N-(5-benzylsulfonyl-1,3,4-thiadiazol-2-yl)piperazinyl quinolone derivatives. *Bioorg Med Chem Lett*, 2005, 15, 4488–4492.

[118] Foroumadi, A; Firoozpour, L; Emami, S; Mansouri, S; Ebrahimabadi, AH; Asadipour, A; Amini, M; Saeid-Adeli, N; Shafiee, A. Synthesis and antibacterial activity of *N*-[5-(chlorobenzylthio)-1,3,4-thiadiazol-2-yl] piperazinyl quinolone derivatives. *Arch Pharm Res*, 2007, 30, 138–145.

[119] Foroumadi, A; Soltani, F; Moshafi, MH; Ashraf-Askari, R. Synthesis and in vitro antibacterial activity of some *N*-(5-aryl-1,3,4-thiadiazole-2-yl)piperazinyl quinolone derivatives. *Il Farmaco*, 2003, 58, 1023–1028.

[120] Foroumadi, A; Mansouri, S; Emami, S; Mirzaei, J; Sorkhi, M; Saeid-Adeli, N; Shafiee, A. Synthesis and antibacterial activity of nitroaryl thiadiazole-levofloxacin hybrids. *Arch Pharm*, 2006, 339, 621–624.

[121] Jazayeri, S; Moshafi, MH; Firoozpour, L; Emami, S; Rajabalian, S; Haddad , M; Pahlavanzadeh, F; Esnaashari, M; Shafiee, A; Foroumadi; A. Synthesis and antibacterial activity of nitroaryl thiadiazole-gatifloxacin hybrids. *Eur J Med Chem*, 2009, 44, 1205–1209.

[122] Foroumadi, A; Mansouri, S; Kiani, Z; Rahmani, A. Synthesis and in vitro antibacterial evaluation of N-[5-(5-nitro-2-thienyl)-1,3,4-thiadiazole-2-yl] piperazinyl quinolones. *Eur J Med Chem*, 2003, 38, 851–854.

[123] Foroumadi, A; Ashraf-Askari, R; Moshafi, MH; Emami, S; Zeynali, A. Synthesis and in vitro antibacterial activity of N-[5-(5-nitro-2-furyl)-1,3,4-thiadiazol-2-yl]piperazinyl quinolone derivatives. *Pharmazie*, 2003, 58, 432–433.

[124] Dong, Y; Xu, C; Zhao, X; Domagala, J; Drlica, K. Fluoroquinolone action against mycobacteria: effects of C-8 substituents on growth, survival, and resistance. *Antimicrob Agents Chemother*, 1998, 42, 2978–2984.

[125] Lu, T; Zhao, X; Drlica, K. Gatifloxacin activity against quinolone-resistant gyrase: allele-specific enhancement of bacteriostatic and bactericidal activities by the C-8-methoxy group. *Antimicrob Agents Chemother*, 1999, 43, 2969–2974.

[126] Lu, T; Zhao, X; Li, X; Drlica-Wagner, A; Wang, J-Y; Domagala, J; Drlica, K. Enhancement of fluoroquinolone activity by C-8 halogen and methoxy moieties: action against a gyrase resistance mutant of *Mycobacterium smegmatis* and a gyrase-topoisomerase IV double mutant of *Staphylococcus aureus*. *Antimicrob Agents Chemother*, 2001, 45, 2703–2709.

[127] Kawakami, K, Ohki, H, Kimura, K; et al., DK-507k, a new 8-methoxyquinolone: synthesis and biological evaluation of 7-[(3-amino-4-substituted)pyrrolidin-1-yl] derivatives. In: Abstracts of the *41st Interscience Conference on Antimicrobial Agents and Chemotherapy*, 2001, Abstract F-546, p. 218.

[128] Otani, T; Tanaka, M; Ito, E; Kurosaka, Y; Murakami, Y; Onodera, K; Akasaka, T; Sato, K. In vitro and in vivo antibacterial activities of DK-507k, a novel fluoroquinolone. *Antimicrob Agents Chemother*, 2003, 47, 3750–3759.

[129] Tanaka, M, Tachibana, M, Seki, H, Ohzone, Y, DK-507k, a new 8-methoxyquinolone: pharmacokinetics in rats and monkeys. In: Abstracts of the *41st Interscience Conference on Antimicrobial Agents and Chemotherapy*, 2001; Abstract F-548, p. 218.

[130] Hoshino, K; Inoue, K; Murakami, Y; Kurosaka, Y; Namba, K; Kashimoto, Y; Uoyama, S; Okumura, R; Higuchi, S; Otani, T. In vitro and in vivo antibacterial activities of DC-159a, a new fluoroquinolone. *Antimicrob Agents Chemother*, 2008, 52, 65–76.

[131] Jones, RN; Fritsche, TR; Sader, HS. Antimicrobial activity of DC-159a, a new fluoroquinolone, against 1,149 recently collected clinical isolates. *Antimicrob Agents Chemother*, 2008, 52, 3763–3775.

[132] Okezaki, E; Watanabe, Y; Hirose, T; Yoshida, T; Aoki, Y; Kato, H. Antibacterial activity of HSR-903, a new novel quinolone, abstr. F202, p. 148. In: Program and abstracts of the *35th Interscience Conference on Antimicrobial Agents and Chemotherapy*. American Society for Microbiology, 1995; Washington, D.C.

[133] Yoshida, T; Kado, N; Okezaki, E; Yasuda, S; Kato, H. Synthesis and antibacterial activity of HSR-903 and related compounds, abstr. F58, p. 110. In: Program and abstracts of the *36th Interscience Conference on Antimicrobial Agents and Chemotherapy*. American Society for Microbiology, 1996, Washington, D.C.

[134] Takahashi, Y; Masuda, N; Otsuki, M; Miki, M; Nishino, T. In vitro activity of HSR-903, a new quinolone. *Antimicrob Agents Chemother*, 1997, 41, 1326–1330.

[135] Watanabe, A; Tokue, Y; Takahashi, H; Kikuchi, T; Kobayashi, T; Gomi, K; Fujimura, S; Nukiwa, T. In vitro activity of HSR-903, a new oral quinolone, against bacteria causing respiratory infections. *Antimicrob Agents Chemother*, 1999, 43, 1767–1768.

[136] Ng, EYW; Trucksis, M; Hooper, DC. Quinolone resistance mediated by *norA*: physiologic characterization and relationship to *flqB*, a quinolone resistance locus on the *Staphylococcus aureus* chromosome. *Antimicrob Agents Chemother*, 1994, 37, 1345–1355.

[137] Murata, M; Takahara, E; Nagata, O; Kato, H; Tamai, I; Tsuji, A. Carrier-mediated tissue distribution and pharmacokinetics of HSR-903, a new quinolone, abstr. F203, p. 148. In: Program and abstracts of the *35th Interscience Conference on Antimicrobial Agents and Chemotherapy*. American Society for Microbiology, 1995; Washington, D.C.

[138] Sun, J; Deguchi, Y; Tauchi, Y; He, Z; Cheng, G; Morimoto, K. Distribution characteristics of orally administered olamufloxacin, a newly synthesized

fluoroquinolone antibacterial, in lung epithelial lining fluid and alveolar macrophage in rats. *Eur J Pharm Biopharm*, 2006, 64, 238–245.

[139] Tanaka, M; Yamazaki, E; Chiba, M; Yoshihara, K; Akasaka, T; Takemura, M; Sato, K. In vitro antibacterial activities of DQ-113, a potent quinolone, against clinical isolates. *Antimicrob Agents Chemother*, 2002; 46, 904–908.

[140] Hong, SG; Moland, ES; Wickman, PA; Black, JA; Hossain, A; Hanson, ND; Thomson, KS. In vitro studies with DQ-113 and comparison fluoroquinolones to determine propensities to select resistance in gram-positive cocci. *Antimicrob Agents Chemother*, 2007, 51, 1512–1514.

[141] Kaneko, Y; Yanagihara, K; Miyazaki, Y; Tsukamoto, K; Hirakata, Y; Tomono, K; Kadota, J; Tashiro, T; Murata, I; Kohno, S. Effects of DQ-113, a new quinolone, against methicillin- and vancomycin-resistant *Staphylococcus aureus*-caused hematogenous pulmonary infections in mice. *Antimicrob Agents Chemother*, 2003, 47, 3694–3698.

[142] Harnett, SJ; Fraise, AP; Andrews, JM; Jevons, G; Brenwald, NP; Wise, R. Comparative study of the in vitro activity of a new fluoroquinolone, ABT-492. *J Antimicrob Chemother*, 2004, 53, 783–792.

[143] Cheng, Y. Rib-X Pharmaceuticals repaints the antibioticscape. *Tech Impact @ Yale*, 2008, 2, 4.

[144] Ishikawa, H; Tabusa, F; Miyamoto, H; Kano, M; Ueda, H; Tamaoka, H; Nakagawa, K. Studies on antibacterial agents. I. Synthesis of substituted 6,7-dihydro-1-oxo-1*H*,5*H*-benzo[*i,j*]-quinolizine-2-carboxylic acids. *Chem Pharm Bull*, 1989, 37, 2103–2108.

[145] Kurokawa, I; Akamatsu, H; Nishigima, S; Asada, Y; Kawabata, S. Clinical and bacteriologic evaluation of OPC-7251 in patients with acne: a double blind group comparison study vs cream base. *J M Acad Dermatol*, 1991, 25, 674–681.

[146] Nishijima, S, Kurokawa, I; Nakaya, H. Susceptibility change to antibiotics of *Staphylococcus aureus* strains isolated from skin infections between July 1994 and November 2000. *J Infect Chemother*, 2002, 8, 187–189.

[147] Nishijima, S; Nakagawa, M; Sugiyama, T; Akamatsu, H; Horio, T; Kawabata, S; Fujita, M. Sensitivity of *Staphylococcus aureus*, isolated from skin infections in 1994 to 19 antimicrobial agents. *J Int Med Res*, 1995, 23, 328–334.

[148] Morita, S; Otsubo, K; Matsubara, J; Ohtnai, T; Uchida, M. An efficient synthesis of a key intermediate towards (*S*)-(-)-nadifloxacin. *Tetrahedron: Asymmetry*, 1995, 6, 245–254.

[149] Nishijima, S; Namura, S; Akamatsu, H; Kawai, S; Asada, Y; Kawabata, S; Fujita, M. In vitro activity of nadifloxacin against both methicillin-susceptible and -resistant clinical isolates of *Staphylococcus aureus* from patients with skin infections. *Drugs*, 1995, 49(Suppl. 2), 230–232.

[150] De Souza, NJ; Gupte, SV; Deshpande, PK; Desai, VN; Bhawsar, SB; Yeole, RD; Shukla, MC; Strahilevitz, J; Hooper, DC; Bozdogan, B; Appelbaum, PC; Jacobs, MR; Shetty, N; Patel, MV; Jha, R; Khorakiwala, HF. A chiral benzoquinolizine-2-carboxylic acid arginine salt active against vancomycin-resistant *Staphylococcus aureus*. *J Med Chem*, 2005, 48, 5232–5242.

[151] Patel, MV; De Souza, NJ; Gupte, SV; Jafri, MA; Bhagwat, SS; Chugh, Y; Khorakiwala, HF; Jacobs, MR; Appelbaum, PC. Antistaphylococcal activity of WCK

771, a tricyclic fluoroquinolone, in animal infection models. *Antimicrob Agents Chemother*, 2004, 48, 4754–4761.

[152] Bhagwat, SS; Mundkur, LA; Gupte, SV; Patel, MV; Khorakiwala, HF. The anti-methicillin-resistant *Staphylococcus aureus* quinolone WCK 771 has potent activity against sequentially selected mutants, has a narrow mutant selection window against quinolone-resistant *Staphylococcus aureus*, and preferentially targets DNA gyrase. *Antimicrob Agents Chemother*, 2006, 50, 3568–3579.

[153] Greka, P; Souli, M; Athanasiou, K; Mandaraka, A; Katsala, D; Bourousi, M; Giamarellou, H. *39th Interscience Conference on Antimicrobial Agents and Chemotherapy*, 1999, Abstract No. E-1495.

[154] Goldstein, EJ; Conrads, G; Citron, DM; Merriam, V; Warren, Y; Tyrrell, K. In vitro activity of gemifloxacin compared to seven other oral antimicrobial agents against aerobic and anaerobic pathogens isolated from antral sinus puncture specimens from patients with sinusitis. *Diagn Microbiol Infect Dis*, 2002, 42, 113–118.

[155] Fuchs, PC; Barry, AL; Brown, SD. In vitro activity of gemifloxacin against contemporary clinical bacterial isolates from eleven North American medical centers, and assessment of disk diffusion test interpretive criteria. *Diagn Microbiol Infect Dis*, 2000, 38, 243–253.

[156] Blondeau, JM; Hansen, G; Metzler, KL; Borsos, S; Irvine, LB; Blanco, L. In vitro susceptibility of 4903 bacterial isolates to gemifloxacin—an advanced fluoroquinolone. *Int J Antimicrob Agents*, 2003, 22, 147–154.

[157] Yun, HJ; Min, YH; Lim, JA; Kang, JW; Kim, SY; Kim, MJ; Jeong, JH; Choi, YJ; Kwon, HJ; Jung, YH; Shim, MJ; Choi, EC. In vitro and in vivo antibacterial activities of DW286, a new fluoronaphthyridone antibiotic. *Antimicrob Agents Chemother*, 2002, 46, 3071–3074.

[158] Kim, MJ; Yun, HJ; Kang, JW; Kim, S; Kwak, JH; Choi, EC. In vitro development of resistance to a novel fluoroquinolone, DW286, in methicillin-resistant *Staphylococcus aureus* clinical isolates. *J Antimicrob Chemother*, 2003, 51, 1011–1016.

[159] Kim, HJ; Seol, MJ; Park, HS; Choi, DR; Seong, SK; Shin, HK; Kwak, JH. Antimicrobial activity of DW-224a, a new fluoroquinolone, against *Streptococcus pneumoniae*. *J Antimicrob Chemother*, 2006, 57, 1256–1258.

[160] Kwon, AR; Min, YH; Ryu, JM; Choi, DR; Shim, MJ; Choi, EC. In vitro and in vivo activities of DW-224a, a novel fluoroquinolone antibiotic agent. *J Antimicrob Chemother*, 2006, 58, 684–688.

[161] Park, HS; Kim, HJ; Seol, MJ; Choi, DR; Choi, EC; Kwak, JH. In vitro and in vivo antibacterial activities of DW-224a, a new fluoronaphthyridone. *Antimicrob Agents Chemother*, 2006, 50, 2261–2264.

[162] Kim, EJ; Shin, WH; Kim, KS; et al. Safety pharmacology of DW-224a, a novel fluoroquinolone antibiotic agent. *Drug Chem Toxicol,* 2004, 27, 295–307.

[163] Han, J; Kim, JC; Chung, MK; Kim, B; Choi, DR. Subacute toxicity and toxicokinetics of a new antibiotic, DW-224a, after single and 4-week repeated oral administration in dogs. *Biol Pharm Bull*, 2003, 26, 832–839.

[164] Bush, K; Macielag, M; Weidner-Wells, M. Taking inventory: antibacterial agents currently at or beyond phase 1. *Curr Opin Microbiol*, 2004, 7, 466–476.

[165] Christiansen, KJ; Bell, JM; Turnidge, JD; Jones, RN. Antimicrobial activities of garenoxacin (BMS 284756) against Asia-Pacific region clinical isolates from the SENTRY program, 1999 to 2001. *Antimicrob Agents Chemother*, 2004, 48, 2049–2055.
[166] Fung-Tomc, JC; Minassian, B; Kolek, B; Huczko, E; Aleksunes, L; Stickle, T; Washo, T; Gradelski, E; Valera, L; Bonner, DP. Antibacterial spectrum of a novel des-fluoro(6) quinolone, BMS-284756. *Antimicrob Agents Chemother*, 2000, 44, 3351–3356.
[167] Firsov, AA; Vostrov, SN; Lubenko, IY; Arzamastsev, AP; Portnoy, YA; Zinner, SH. ABT492 and levofloxacin: comparison of their pharmacodynamics and their abilities to prevent the selection of resistant *Staphylococcus aureus* in an in vitro dynamic model. *J Antimicrob Chemother*, 2004, 54, 178–186.
[168] Pfaller, MA; Beach, ML; Gordon, KA; Jones, RN. Comparative antimicrobial spectrum and activity of BMS284756 (T-3811; a desfluoroquinolone) tested against an international collection of staphylococci and enterococci, including in vitro test development and international comparisons. *J Chemother*, 2002, 14, 13–18.
[169] Watanabe, S; Ito, T; Hiramatsu, K. Susceptibilities of healthcare- and community-associated methicillin-resistant staphylococci to the novel des-F(6)-quinolone DX-619. *J Antimicrob Chemother*, 2007, 60, 1384–1387.
[170] Yamamoto, T; Dohmae, S; Saito, K; Otsuka, T; Takano, T; Chiba, M; Fujikawa, K; Tanaka, M. Molecular characteristics and in vitro susceptibility to antimicrobial agents, including the des-fluoro(6) quinolone DX-619, of Panton-Valentine leucocidin-positive methicillin-resistant *Staphylococcus aureus* isolates from the community and hospitals. *Antimicrob Agents Chemother,* 2006, 50, 4077–4086.
[171] Bogdanovich, T; Esel, D; Kelly, LM; Bozdogan, B; Credito, K; Lin, G; Smith, K; Ednie, LM; Hoellman, DB; Appelbaum, PC. Antistaphylococcal activity of DX-619, a new des-F(6)-quinolone, compared to those of other agents. *Antimicrob Agents Chemother*, 2005, 49, 3325–3333.
[172] Strahilevitz, J; Truong-Bolduc, QC; Hooper, DC. DX-619, a novel des-fluoro(6) quinolone manifesting low frequency of selection of resistant *Staphylococcus aureus* mutants: quinolone resistance beyond modification of type II topoisomerases. *Antimicrob Agents Chemother*, 2005, 49, 5051–5057.
[173] Recommended International Nonproprietary Names: List 58, *WHO Drug Information*, 2007, 21, 256.
[174] 48th Annual Interscience Conference on Antimicrobial Agents and Chemotherapy / *46th Annual Meeting of Infectious Diseases Society of America*, 2008; Washington, D.C.
[175] Recommended International Nonproprietary Names: List 58, *WHO Drug Information*, 2007, 21, 257.
[176] Yamakawa, T; Mitsuyama, J; Hayashi, K. In vitro and in vivo antibacterial activity of T-3912, a novel non-fluorinated topical quinolone. *J Antimicrob Chemother*, 2002, 49, 455–465.
[177] Tran, TP; Ellsworth, EL; Stier, MA; Domagala, JM; Showalter, HDH; Gracheck, SJ; Shapiro, MA; Joannides, TE; Singh, R. Synthesis and structural–activity relationships of 3-hydroxyquinazoline-2,4-dione antibacterial agents. *Bioorg Med Chem Lett*, 2004, 14, 4405–4409.
[178] Huband, MD; Cohen, MA; Zurack, M; Hanna, DL; Skerlos, LA; Sulavik, MC; Gibson, GW; Gage, JW; Ellsworth, E; Stier, MA; Gracheck, SJ. In vitro and in vivo activities of

PD 0305970 and PD 0326448, new bacterial gyrase/topoisomerase inhibitors with potent antibacterial activities versus multidrug-resistant gram-positive and fastidious organism groups. *Antimicrob Agents Chemother*, 2007, 51, 1191–1201.

[179] Ellsworth, EL; Tran, TP; Showalter, HD; Sanchez, JP; Watson, BM; Stier, MA; Domagala, JM; Gracheck, SJ; Joannides, ET; Shapiro, MA; Dunham, SA; Hanna, DL; Huband, MD; Gage, JW; Bronstein, JC; Liu, JY; Nguyen, DQ; Singh, R. 3-aminoquinazolinediones as a new class of antibacterial agents demonstrating excellent antibacterial activity against wild-type and multidrug resistant organisms. *J Med Chem*, 2006, 49, 6435–6438.

[180] Nilius, AM; Hensey-Rudloff, DM; Almer, LS; Fung, A; Arminger, YL; Chu, D; Flamm, RK. Comparative in vitro antibacterial activity of A-170568.1, a novel 2-pyridone antibacterial agent. Abstract F79. *38th Interscience Conference on Antimicrobial Agents and Chemotherapy*; 1998.

[181] Armiger, YL; Chu, DTW; Fung, AKL; Li, Q; Wang, W; Nilius, A; Alder, J; Ewing, P; Stone, G; Meulbroek, J; Bui, M; Shen, LL; Paige, L; Or, YS; Plattner, JJ. The discovery of A-165753 and A-170568, two potent broad spectrum antimicrobial agents. Abstract F86. *38th Interscience Conference on Antimicrobial Agents and Chemotherapy*; 1998.

[182] Alder, J; Clement, J; Meulbroek, J; Shipkowitz, N; Mitten, M; Jarvis, K; Oleksijew, A; Hutch, T; Sr Paige, L; Flamm, R; Chu, D; Tanaka, K. Efficacies of ABT-719 and related 2-pyridones, members of a new class of antibacterial agents, against experimental bacterial infections. *Antimicrob Agents Chemother*, 1995, 39, 971–975.

[183] Flamm, RK; Vojtko, C; Chu, DT; Li, Q; Beyer, J; Hensey, D; Ramer, N; Clement, JJ; Tanaka, SK. In vitro evaluation of ABT-719, a novel DNA gyrase inhibitor. *Antimicrob Agents Chemother*, 1995, 39, 964–970.

[184] Chin, NX; Chu, D; Nesu, HC. In vitro antibacterial activity of A-86719.1, a new class of DNA gyrase inhibitor. Abstract F49. *34th Interscience Conference on Antimicrobial Agents and Chemotherapy*, 1994.

[185] Robertson, GT; Bonventre, EJ; Doyle, TB; Du, Q; Duncan, L; Morris, TW; Roche, ED; Yan, D; Lynch, AS. In vitro evaluation of CBR-2092, a novel rifamycin-quinolone hybrid antibiotic: studies of the mode of action in *Staphylococcus aureus*. *Antimicrob Agents Chemother*, 2008, 52, 2313–2323.

[186] Li, Q; Mitscher, LA; Shen, LL. The 2-pyridone antibacterial agents: bacterial topoisomerase inhibitors. *Med Res Rev*, 2000, 20, 231–293.

[187] Artsimovitch, I; Vassylyeva, MN; Svetlov, D; Svetlov, V; Perederina, A; Igarashi, N; Matsugaki, N; Wakatsuki, S; Tahirov, TH; Vassylyev, DG. Allosteric modulation of the RNA polymerase catalytic reaction is an essential component of transcription control by rifamycins. *Cell*, 2005, 122, 351–363.

[188] Campbell, EA; Korzheva, N; Mustaev, A; Murakami, K; Nair, S; Goldfarb, A; Darst, SA. Structural mechanism for rifampicin inhibition of bacterial RNA polymerase. *Cell Microbiol*, 2001, 104, 901–912.

[189] Robertson, GT; Bonventre, EJ; Doyle, TB; Du, Q; Duncan, L; Morris, TW; Roche, ED; Yan, D; Lynch, AS. In vitro evaluation of CBR-2092, a novel rifamycin-quinolone hybrid antibiotic: microbiology profiling studies with staphylococci and streptococci. *Antimicrob Agents Chemother*, 2008, 52, 2324–34.

[190] Emami, S; Shafiee, A; Foroumadi, A. Structural features of new quinolones and relationship to antibacterial activity against Gram-positive bacteria. *Mini-Rev Med Chem*, 2006, 6, 375–386.

[191] Nseir, S; Ader, F; Marquette, CH; Durocher, A. Impact of fluoroquinolone use on multidrug-resistant bacteria emergence. *Pathol Biol (Paris)*, 2005, 53, 470–475.

[192] John, JF; Harvin, AM. History and evolution of antibiotic resistance in coagulase-negative staphylococci: Susceptibility profiles of new anti-staphylococcal agents. *Ther Clin Risk Manag*, 2007, 3, 1143–1152.

[193] Chide, OE; Orisakwe, OE. Structural development, haematological immunological and pharmacological effects of quinolones. *Recent Pat Antiinfect Drug Discov*, 2007, 2, 157–168.

[194] De Souza, MV. New fluoroquinolones: a class of potent antibiotics. *Mini Rev Med Chem*, 2005, 5, 1009–1017.

In: Methicillin-Resistant Staphylococcus Aureus... ISBN: 978-1-60741-398-1
Editor: Charles L. Kolendi

Chapter 7

METHICILLIN-RESISTANT *STAPHYLOCOCCUS AUREUS*: ETIOLOGY, AT-RISK POPULATIONS AND TREATMENT

Mine Hosgor-Limoncu, Safak Ermertcan, Huseyin Tasli, Bayrı Erac and Ayse Nur Yurtman

Ege University Faculty of Pharmacy, Department of Pharmaceutical Microbiology, 35100, Izmir,Turkey

ABSTRACT

After beta-lactamase resistant antibiotics, including methicillin, nafcillin and cephalosporins had been introduced to the clinical use; the first methicillin-resistant *Staphylococcus aureus* (MRSA) isolates were described in 1961. Since then, MRSA clones have become a problem in health-care institutions throughout the world. Community acquired MRSA (CA-MRSA) isolates have also threatened human health for the last two decades. In multicentral surveillance studies, MRSA prevalence was determined as 1-2 % in North European countries but it may reach up to 80 % in some Far East countries.

Many risk factors have been found to be associated with MRSA infections. Recent hospitalization or surgery, residence in long-term care facilities, dialysis, broad-spectrum antibiotic usage and indwelling percutaneous medical devices or catheter are the most common risk factors for nosocomial MRSA infections. The risk factors for CA-MRSA infections have not been well established but several important factors have been described in the literature: HIV infection, recent antibiotic usage, intravenous and intranasal drug usage, living in closed populations, lower socioeconomic status, homelessness, non-hygienic tattoo and health-care workers.

The infection or colonization of health-care workers with MRSA is highly important since it is a cause of transmission and an increase in its incidence. Cellulites, boils and furuncles are among the most common clinical presentations of skin and soft tissue infections (SSTIs) caused by MRSA. These organisms are also responsible for severe infections such as endocarditis, pneumonia and bacteremia. In addition to invasive infections, MRSA produces a variety of toxins that lead to some clinical manifestations such as staphylococcal scalded skin syndrome and toxic shock syndrome. Although CA-

MRSA has been traditionally associated with SSTI, there are increasing reports of pneumonia, bacteremia and infective endocarditis caused by CA-MRSA. Hospital acquired MRSA (HA-MRSA) produces more severe clinical conditions associated with increased mortality and morbidity among patients.

Treatment options for both mild SSTI and severe nosocomial infections are limited due to increasing antimicrobial resistance. After MRSA isolates emerged, glycopeptide antibiotics (vancomycin and teicoplanin) have been used more commonly and this resulted in the occurrence of isolates with reduced susceptibility or resistance to vancomycin. Trimetoprim-sulfamethoxazole, rifampin, linezolid, doxycycline, clindamycin and occasionally fluoroquinolones have been used in treating mild SSTIs. In more severe invasive infections, besides vancomycin, other parenteral agents such as linezolid, daptomycin, and tigecycline are commonly used. This review will focus on etiology, at-risk populations and treatment options of MRSA infections.

Risk Factors and at Risk Populations for MRSA Infections

Methicillin resistant *Staphylococcus aureus* (MRSA) infections, which are associated with risk factors including length of hospitalization (particularly in the intensive care unit), surgical intervention, receiving care in nursing home settings, long-term or repeated antibiotic treatment, presence of catheters and percutaneous medical devices, have been reported since 1960's [1-3]. The report of the first community acquired (CA)-MRSA case in USA in 1980 weakened the common belief that MRSA causes infections, which emerge only in hospital settings and in patients carrying certain risk factors [4]. Since then, the diagnosis of CA-MRSA and the related infections has become more important than ever. Categorizing all MRSA infections as community-acquired without taking into consideration whether the patient receives treatment in an ambulatory care setting or a nursing home proved to be a wrong approach [5]. In order to contribute to a new categorization, Friedman et al. conducted a study including three hospitals in North Carolina, USA [6]. In their study, a positive culture obtained 48 hours or longer after hospitalization was defined as a “hospital acquired” (nosocomial) infection. If the patient had been transferred from another hospital, the evaluation was based on the date on which the patient was first admitted to a hospital. An infection was considered as “health center or hospital associated” (HA) if one or more of the following factors were present: a) receiving in-home IV treatment or receiving wound care service at home in the last one month b) applying to a hospital or hemodialysis center or receiving IV chemotherapy in the last one month c) a hospital stay of at least two or three days in the last three months d) dwelling in a nursing home or rehabilitation center. Patients who did not carry one of the above mentioned factors and had positive blood culture were considered as having “community acquired” infections. Nevertheless, staphylococcus cassette chromosome *mec* (SCC*mec*) type, genes encoding virulence factors (Panton-Valentine-Leukocidin (PVL) and other toxins), Pulsed Field Gel Electrophresis (PFGE) and Multi Locus Sequence Typing (MLST) profiles should be identified for a reliable differentiation of HA-MRSA and CA-MRSA strains [1, 7].

The risk factors for CA-MRSA infection include receiving IV or intranasal drugs, recent use of antibiotics, cardiovascular or pulmonary diseases, diabetes, malignancy, underlying diseases such as HIV infection and chronic skin diseases, living in crowded dwellings, low socio-economical level, nasal MRSA carriage, and working as a healthcare personnel. The reporting of CA-MRSA infections in some patients not carrying the aforementioned risk factors and the reporting of hospital outbreaks resulting from these strains have suggested that the epidemiology of the CA-MRSA infections has undergone a change [1, 2, 8-10]. The communities where the infections occur as outbreaks include prisoners, individuals receiving IV drugs, athletes, military trainess and homosexuals. Although CA-MRSA mostly causes skin and soft tissue infections, it may also cause clinical manifestations such as sepsis and necrotizing pneumonia [11].

MRSA infections have also been reported amongst individuals involved in sports such as soccer, basketball, voleyball, football, rugby, wrestling, weightlifting and fencing. Kazakova et al. [12] reported skin infections due to MRSA in the players of a football team. The authors stated that the use of antibiotics, impaired skin integrity, the skin-to-skin contact between the players, contaminated environment, the violation of hygiene rules by the players and their trainers were all contributors to the outbreak. It has been suggested that the direct skin-to-skin contact, which is a part of the sports based on physical struggle with the opponents -such as football-, is a factor contributing to the spread of MRSA. The position of a player during a game, injured skin and the collective use of the sports equipment in sports branches, where there is no direct contact with the rival, constitute risk factors for MRSA infection [13]. A similar risk is also valid for populations such as prisoners and military troops where communal life is practised. Collecting nasal culture samples from the athletes, preventing colonization, informing the athletes and the trainers on MRSA infections and drawing attention on hygiene rules are some of the precautions that may be useful in reducing MRSA infections.

CA-MRSA strains are responsible for most of the skin and soft tissue infections in patients using IV drugs. This may result from the fact that *S. aureus* nasal and skin colonization rates are higher in individuals using IV drugs. It has been suggested that contaminated drug injection and using drugs by inhalation -due to its consequent damage to mucosa- increase the probability of *S. aureus* colonization [14, 15]. It is known that *S. aureus* nasal carriage increases the infection risk associated with this bacterium. Huang et al. [16], investigated whether IV drug usage constitute a risk factor for CA-MRSA infections by using the matched case-control method. 127 CA-MRSA patients, 127 CA-methicillin susceptible *S. aureus* (MSSA) and a control group consisting of 381 individuals with no MRSA infection history who were considered as representative of the general population were included in the study. The demographic data, medical histories, length of hospital stay and social characteristics (cigarette and alcohol consumption, IV drug use, etc.) were gathered from the medical records retrospectively. Although IV drug history was present in about half of the CA-MRSA patients (49%), this rate was 17% in the CA-MSSA group and 9% in the non-infected patients. However, a significant difference was not determined between the patient group and the control group in terms of ethnicity and underlying diseases such as viral hepatitis, cancer and diabetes. The authors concluded that monitoring CA-MRSA infections in individuals using IV drugs may be effective in reducing the prevalence of these infections in the general population.

It is known that *S. aureus* carriage is an important risk factor for nosocomial and community acquired infections in numerous populations. In a study conducted by Lu et al. [17], samples received from a total of 2,231 subjects consisting of 1,838 community residents (including children from all age groups as well) and 393 subjects from health care facility-related settings were tested for *S. aureus* colonization and an inquiry form was filled. It was determined that 24.1% and 4.1% of those included in the study were colonized with *S. aureus* and MRSA, respectively. 3.5% of the community subjects had nasal MRSA colonization while the rate of healthcare acquired nasal MRSA rate was 7.63%. It was determined that *S. aureus* colonization rates were significantly higher in the individuals younger than twenty years and in those aged between 71 and 80, as compared to the remaining age groups. It was seen that the presence of a gastrointestinal disease and recent hospital admission were important risk factors for CA-MRSA colonization while recent medication was closely associated with CA-MRSA colonization. The individuals included in the study were assigned into two groups as those who had predisposing factors such as receiving medical service within the previous 3 months and being a health care worker and those who did not. MRSA colonization rate was 3.4% in the individuals who did not carry such risk factors and 4.6% in those who did. Taking the information obtained by molecular methods into consideration, the authors concluded that MRSAs are able to spread in community independently of hospital settings.

The prevalance of *S. aureus* carriage among healthy children in the age group of 4 to 6 years in Afyonkarahisar, Turkey and the related risk factors were investigated in a study. Of the 1134 healthy children, *S. aureus* was detected in 322 (28.4%) while MRSA was isolated in only three (0.3%) of them. No differences were observed in *S. aureus* colonization between the girls and boys. The carriage rates were found to be higher among the children whose parents were low-educated, whose families were large and whose mothers were health care workers. It was also determined that the parents of all the children in whom MRSA was isolated, were health care workers [18]. In another study, conducted in another region of Turkey, a total of 2015 students including 1012 high school and 1003 primary school students were examined for CA-MRSA carriage rate and the related risk factors. Nasal MRSA carriage was not observed in any of the students while MSSA colonization rate was determined as 14.7% (296/2015). It was observed that MSSA carriage was higher among the students from higher socioeconomic backgrounds. A significant correlation was not determined between the following risk factors: admission to a hospital or undergoing surgery in the last one year, antibiotic use or skin/soft tissue lesion in the last six months, the presence of one or more family members who are aged under 15 years or who are healthcare workers, and living in a crowded house [19].

MRSA colonization in dialysis patients increases the infection risk and plays an important role in its transmission to healthcare workers and their families. Lu et al. [15] investigated *S. aureus* nasal colonisation in a dialysis unit of a regional hospital three times in one year and gathered information regarding the age and sex of the patients and their health records from the previous year. The authors also performed the molecular typing of the isolated strains. Although nasal carriage of *S. aureus* was higher in the peritenoal dialysis patients (43.4%) than in the hemodialysis patients (21.8%), it was determined that the MRSA carriage rates were similar between the two groups. A significant difference was not observed between the patients and the health care workers in *S. aureus* and MRSA carriage whereas it was observed that nasal MRSA carriage was higher in the families of the health care workers than in the families of the patients. The presence of a lung disease, recent hospital admission and use of antibiotics were found to be significantly correlated with MRSA carriage. However, it was determined that factors including sex, diabetes, hypertension, kidney diseases, gastrointestinal system disorders, nasal diseases and undergoing a medical examination were not significantly related to MRSA carriage. In another study aimed at determining the *S. aureus* carriage rates in diabetic and non-diabetic hemodialysis patients and the antibiotic sensitivity profile of these strains, *S. aureus* was detected in 148 (56.7%) of the 261 hemodialysis patients and 26 (16.2%) of the 160 individuals comprising the control group. MRSA positivity was 20.6% in the hemodialysis patients and no MRSA strains were isolated in the control group. Additionally, a significant correlation was observed between the time elapsed during the hemodialysis and *S. aureus* isolation rate. It was determined that age, sex and diabetes were not significantly correlated with *S. aureus* nasal carriage. The authors concluded that the nasal carriage rate of *S. aureus* was higher in the hemodialysis patients than in the control group and diabetes did not constitute a risk factor [20].

Although only nasal carriage has been taken into consideration in the numerous studies on *S. aureus* carriage, it has been reported that about 25% of the *S. aureus* carriers are throat carriers. In one study conducted by Mertz et al. [21] it was determined that, of the 3464 individuals involved with the health care system at different levels, 428 (12.4%) were exclusive throat carriers and 1260 (36.4%) were exclusive nasal carriers or both nasal and throat carriers. It was concluded that the most important independent factor for exclusive throat carriage was being younger than 30. The authors also stated that not being in contact with the health care system and being young may be considered as risk factors for exclusive throat carriage. The data obtained from the study indicates that nasal screening alone is insufficient in investigating *S. aureus* carriage and most probably, the exact carriage rate is much higher.

Premature or low birth weight, underlying chronic disease, long-term use of antibiotics, invasive and surgical interventions, the presence of a catheter and long-term hospitalization are considered as risk factors for pediatric populations [9, 22]. Wang et al. [8] evaluated the clinical data of the patients of a Taiwanese hospital who had been infected with CA-MRSA in the last five years retrospectively, along with the associated risk factors, and categorized the isolated MRSA strains by using molecular methods. Among the patients younger than 18 years, those in whom MRSA had been isolated in the first 72 hours after hospitalization, were considered as having CA-MRSA and were included in the study. The 19 patients meeting the inclusion criteria were evaluated and it was observed that their mean age was 2 years and 8 months, 12 (68%) of them were male and none of them were carrying the typical risk factors of MRSA infection. According to the study, the different antibiogram patterns of the HA-MRSA and CA-MRSAs isolated from the same hospital and their clustering in different groups during the molecular typing studies indicate that CA-MRSAs emerge independently of the hospital settings and this finding reinforces the idea that they most probably emerge under the pressure of the antibiotics, which are widely used in the community. Moreover, it was determined that the CA-MRSAs isolated in the study were different from the CA-MRSA clones isolated in the USA, Europe and South Pacific. Some studies performed in the USA and Europe support the findings of this study. Bratu et al. [9] reported an outbreak involving seven patients in whom skin and soft tissue infections had developed due to CA-MRSA in the gynecology and obstetrics departments and the newborn unit. It was determined that the outbreak strain isolated from six newborns and a mother was SCC*mec* type IV and carried the genes responsible for PVL and staphylococcal enterotoxin K production. Another finding was that the MRSA strains isolated from the hospital personnel during the same period were unrelated to the outbreak strain and MRSA was not isolated from the samples collected from the environment. In the study of David et al. [1] which reported a CA-MRSA infection involving four infants and a hospital employee, it was determined that, during the outbreak, the number of the patients in the neonatal and delivery units was above the units' capacity, hygiene conditions were not satisfactory in the neonatal unit and the liquid soap dispensers and disinfectant cleaners were broken. In a study retrospectively examining MRSA colonization and infection that had been observed in the infants hospitalized in a neonatal intensive care unit of a hospital in Atlanta, USA, between 1993 and 2006; CA-MRSA colonization was determined in 130 (3.5%) of the 3707 newborns. Multivariate analyses showed that smoking and normal delivery in mothers who receive antibiotic treatment prior to delivery were independent variables for CA-MRSA colonization in newborns. It was determined that systematic antibiotic use prior delivery and probably endotracheal intubation was associated with HA-MRSA isolation in the mothers who had undergone cesarean section and were non-smokers. In the study, which put forth that the leading cause of MRSA colonization in newborns in the high risk group is CA-MRSAs, the authors concluded that CA-MRSA colonization generally develops with birth and HA-MRSA strains mostly spread nosocomially [23].

The healthcare workers, who serve as a bridge between the society and healthcare centers, have an important role in the epidemiology of the MRSA infections. It has been shown in many studies that MRSA strains are transmitted from the healthcare employees to their families and to the patients [15, 24-26]. It is important to fully adhere to the infection control precautions and inform the healthcare personnel about MRSA infections. Wolf et al. [27] evaluated the thoughts of nurses employed in long-term nursing homes about MRSA and infection control and determined that only 59% of the participants were thinking that MRSA constituted a threat for patients. MRSA carriage rates vary among the health care workers (0% to 59%) and its average prevalence is 4.6%. The leading risk factors for MRSA colonization and infection in healthcare workers are various skin infections, urinary system infections, sinusitis, otitis, rhinitis; recent use of antibiotics; working in a region or department where MRSA prevalence is high; being in close contact with the patients, not adhering strictly to infection control precautions and heavy work load. The eradication of MRSA carriage among healthcare workers is important for both patient and community health [28].

There are numerous studies in the literature focusing on the incidence and risk factors of bloodstream infections resulting from *S. aureus*. Some of these studies only mention the causative agent as MRSA or MSSA while others tend to detect bloodstream infections resulting from CA-MRSA and HA-MRSA by using molecular methods. Manzur et al. [29] evaluated a series of basic clinical and epidemiologic data on two groups of patients: those diagnosed with bloodstream infections due to MRSA in the first 48 hours following their admission to hospital and those who had bloodstream infections due to MSSA. The collected data included information on age, sex, underlying disease (diabetes; rheumatoid arthritis; chronic renal insufficiency and dialysis; IV drug use; and immunosuppression); previous hospitalization; previous positive MRSA culture result; long-term stay in a nursing home; bloodstream infection origin, empirical antibiotic treatment and its clinical outcome. The comparison of the patients with bloodstream infections due to MRSA with the control group revealed that the mean age was higher in the patient group (69.0 / 58.3), which contained more women than men (54% / 34.7%). It was determined that both groups had generally similar characteristics in terms of diabetes, hemodialysis and underlying disease. In this study, which used only the patient group with positive MSSA bloodstream infection as the control group, the multivariate analysis of the risk factors for the bloodstream infection due to

MRSA showed that previous positive MRSA culture result and long-term stay in a nursing home were independent variables. According to the results of a study conducted in the Calgary Health Region (Canada) where 1.2 million residents live, which was aimed investigating the epidemiologic profile and incidence of *S. aureus* bloodstream infections and the antibiotic resistance profile of the causative strains between 2000 and 2006, it was observed that, although rates of health care-associated and nosocomial MSSA bacteremia did not change significantly throughout the study, rates of community-acquired MSSA bacteremia gradually decreased as the rates of MRSA bacteremia dramatically increased. Dialysis, organ transplantation, HIV infection, cancer and diabetes were found to be important risk factors for bloodstream infections both due to MSSA and MRSA [30]. In their study investigating the risk factors associated with nosocomial bloodstream infections due to SCC*mec* type IV MRSA, Vidal et al. [31] used the patients with infection due to SCC*mec* type III MRSA diagnosed at least 48 hours after hospital admission as the controls. The authors determined that the leading risk groups in nosocomial bloodstream infections due to MRSA SCC*mec* type IV were children younger than one year, women and the patients in whom central venous catheter had been less frequently used. In their second analysis excluding the patients in the newborn unit, they found that a low "acute physiology and chronic health evaluation score" and undergoing less surgical interventions, as well as less frequent use of venous catheter and being a woman, were associated with the infection. The authors stressed that traditional risk factors are not valid for patients infected with SCC*mec* type IV MRSA and the patients infected with this strain have better health and undergo a smaller number of invasive procedures, as compared to the patients infected with SCC*mec* type III MRSA [31].

Although it is important to determine the presence of risk factors in a patient, they alone are not enough to start an empirical treatment. Additionally, due to the discrepancies in the antibiotic resistance profiles of CA-MRSA and HA-MRSA, molecular methods should absolutely be used along with epidemiologic data in the differentiation of these strains.

Antibiotic Resistance of MRSA

Methicillin-resistant *S. aureus* is known as one of the most important causes of nosocomial infections throughout the world. The gradual increase in methicillin resistance and the consequent multidrug resistance complicates the treatment of MRSA infections considerably.

HA-MRSAs and CA-MRSAs are different from each other in terms of resistance profile. Multidrug resistance is seen in HA-MRSA strains. It should be considered that these strains can be resistant not only to beta-lactam antibiotics but also to macrolids, lincosamides, tetracyclines, chinolones, TMP/SMX and aminoglycosides. In contrast, CA-MRSAs are susceptible to clindamycin, TMP/SMX, tetracycline and fluoroquinolones [32]. Due to the the multidrug resistance of MRSA, the antibiotics that can be used against MRSA infections are few in number. Soon after the introduction of each new antibiotic, particularly glycopeptides, new resistant strains have emerged.

With the onset of the antibiotics era, the selective pressure due to the use of antibiotics resulted in the development of resistance in staphylococci in a short period of time. This process began with the resistance due to beta-lactamase production that appeared with the use of penicilin G as treatment and continued with the introduction of each novel antibiotic into clinical use. No later than 1950s, epidemic clones became resistant against penicillin G, streptomycin, chloramphenycol, oxytetracycline in succession and eventually against macrolides [33]. Back then, the most important multiresistant representatives were the isolates named as "80/81 phage type". According to the present MLST technique, these isolates belong to the ST30 group. It has been shown that, in addition to antibiotic resistance, a 80/81 subclone also produces PVL. The 80/81 clone spreaded to all hospitals and communal settings around the world and caused skin and soft tissue infections in particular. With the introduction of methicillin into clinical use, a decrease was observed in the prevalence of the 80/81 clone. In the recent years, this clone reappeared in European countries, in the form of CA-MRSA [33, 34].

Although the problem of resistance due to beta-lactamase production was overcome with the introduction of methicillin in 1959, soon after this development, in 1961, a *S. aureus* strain resistant to this antibiotic was isolated in England. Since then, MRSA strains have gradually become widespread throughout the world [35].

The resistance mechanism against methicillin was elucidated by Hartman et al. [36] in 1981. These authors determined a different penicillin-binding protein that is involved in cell wall synthesis. They showed that this protein, which is named as PBP2a or PBP2', has a low affinity for all beta-lactam antibiotics and is produced more abundantly than PBP2. Since beta lactam antibiotics can not bind to PBP2a in MRSA strains, resistance to all beta-lactam agents is observed in these strains.

The cause of the methicillin resistance in *S. aureus* is the presence of the *mecA* gene. PBP2a is encoded by *mecA*. The *mecA* gene is regulated by repressor *mecI* gene and *mecR1*, a beta-lactam sensing signal-transducer integral membrane protein. In the absence of beta-lactams, *mecI* interferes with the transcription of both *mecA* and *mecR1-mecI* genes. However, in the presence of beta-lactams, *mecR1* is blocked through an autocatalytic process and its cytoplasmic metalloprotease domain becomes activated. Metalloprotease blocks the *mecI* gene that is bound to the promoter region of *mecA*. Consequently, *mecA* transcription and PBP2a synthesis occur [37].

The *mecA* gene is located on SCC*mec*, a mobile genetic element. Until today, at least five SCC*mec* types (I-V) and many variants of these *SSCmec* types have been described [37-39]. The size of the various types of SCC*mec* elements change between 20.9 kb and 66.9 kb. As SCC*mec* type I (34.3 kb), IV (20.9-24.3 kb) and V (28 kb) encode only the resistance to beta-lactam antibiotics; SCC*mec* type II (53.0 kb) and III (66.9 kb) cause multi-resistance. They achieve this resistance by plasmids integrated into the gene cassette (pUB110, pI258 ve pT181) and a transposon (Tn 554). Plasmid pUB110 carries the *ant(4')* gene; which is responsible for the resistances against canamycin, tobramycin, bleomycin; pI258 encodes resistance to penicillin and resistance to heavy metals; pT181 encodes tetracyclin resistance, and Tn554 carries the *ermA* gene that causes resistance to macrolides, lincosamides and streptogramins (MLS) [37]. Aside from SCC*mec*, *S. aureus* carry resistance elements located in the various parts of the chromosome or on the plasmids.

It is possible to differentiate CA-MRSA and HA-MRSA according to the SCC*mec* types. SCC*mec* IV, V and PVL exotoxin presence is observed in CA-MRSAs whereas; HA-MRSAs harbor SCC*mec* I, II, III types. As a result, HA-MRSA strains show multidrug resistance [39, 40]. SCC*mec* type IV and V, which are found in CA-MRSAs, lack multidrug resistance and are considered to be easily transferred to other *S. aureus* strains [41]. Type IV is considerably susceptible to non-beta-lactam antimicrobials. However, it is also highly virulent due to its high level of PVL production [41, 42].

Typically, SCC*mec* elements contain *mecA*, which acts as the methicillin resistance determinant, along with its regulator genes (*mec* gene complex); and *ccr* (cassette chromosome recombinases) gene complexes, which mediate the mobility of these elements, i.e., their integration into and excision from a region [43]. SCC*mec* elements are differentiated according to the different combinations of *mec* and *ccr* allelles [39]. SCC*mec* also contains insertion sequences (e.g. IS431) and genes regulating *mecA* transcription [e.g. ΔmecR1 (SCC*mec* type I, IV ve V) or *mecR1* and *mecI* (type II and type III)]. These genes are included in the *mec* gene complex. To date, 5 main classes of *mec* gene complexes (A-E) have been identified [44-46].

SCC*mec* integrates into *attBscc*, a specific site, in the *S. aureus* chromosome. This site is located at the 3' end of an open reading frame named as orfX; of which the function remains unknown. The identified *ccr* genes are *ccrA1* and *ccrB1* in SCC*mec* type I; *ccrA2* and *ccrB2* in SCC*mec* types II and IV; *ccrA3* and *ccrB3* in SCC*mec* type III; *ccrA4* and *ccrB4* in the HDE288 strain carrying type IV SCC*mec*; and *ccrC* in SCC*mec* type V. The regions, which lie outside the *ccr* and *mec* complex, are known as the J (junkyard) regions [37]. All SCC*mec* elements include the following three regions: the J1 region extending from the right side of the chromosome to the *ccr* genes, the J2 region extending from the *ccr* gene to the *mec* complex and the J3 region that lies between the *mec* complex and the left end of SCC*mec* [47, 48].

Although the exact origin of SCC*mec* is still unknown, it is believed that this cassette originates from non-*S. aureus* staphylococci. The 87.8% amino-acid sequence identity between a PBP in *Staphylococcus sciuri* and a PBP2a in MRSA suggests that *S. sciuri* might harbour the ancestor of PBP2a [49]. Moreover, in-vivo horizontal transfer of the *mecA* gene between the two staphylococci has been shown [50]. Other studies have concluded that the SCC*mec* types or SCC elements that do not contain *mecA* may act as reservoirs for the antibiotic resistance of *S. aureus* [47, 51-53].

Since the early 1990's, numerous molecular methods have been developed to identify the epidemic MRSA strains throughout the world. These methods mainly aim at determining the genetical and evolutionary relations between the strains and the widespread clones and subclones. Such methods include PFGE, SmaI-macrorestriction pattern analysis [54], MLST [55-57], SCC*mec* typing [58, 59] and typing of the variable tandem repeat of protein A (*spa* typing) [60-62].

Viewed from a historical perspective, HA-MRSA and CA-MRSA strains are different from each other in terms of both epidemiology and evolutionary relations. Since the late 1980's, an increase in HA-MRSA rate has been observed in nearly all countries except Holland and some North European countries [63]. The majority of the HA-MRSA strains identified throughout the world belong to the 11 major clones of the five clonal complexes. Some of these epidemic MRSAs (ST239, ST247, ST254, ST22, ST45) have shown continental and intercontinental spread. It has been suggested that this spread may be associated with a common SCC*mec* shared by these MRSA strains. However, it has also been alleged that that clonal lineages carrying different SCC*mec* elements, such as ST5 and ST8, might have evolved by combining with different SCC*mec* elements in different places and at different times [33, 56].

Community acquired MRSA was first detected in 1998, in the USA [64] and then in Australia, in 2000 [65]. Since 2002, CA-MRSA has become a growing problem throughout the world. These strains, which are more virulent than HA-MRSAs, have begun to supersede HA-MRSAs in the healthcare setting [10, 66]. According to the PFGE analyses, CA-MRSA clone types are not associated with hospital clones [67]. On the other hand, MLST analyses showed that the clonal diversity of CA-MRSA is greater than that of HA-MRSA [56]. Although CA-MRSA strains usually carry SCC*mec* types IV and V [41, 46, 68]; it was shown in two studies that some CA-MRSA strains contain SCC*mec* types I, II or III [69, 70]. On the other hand, in another study, SCC*mec* type IV was reported in HA-MRSA isolates [71]. These discrepant findings complicate the differentiation of CA-MRSA and HA-MRSA solely based on the SCC*mec* type.

In the United States, two CA-MRSA clones, ST1 (USA 400) ve ST8 (USA 300) are very prevalent and they carry the SCC*mec* type IV element. These clones are responsible for the CA-MRSA outbreaks in crowded populations. In some regions, mostly these strains are isolated from the deep-seated infections of the skin [10, 72]. Similarly to the hospital acquired MRSA, some community acquired isolates can spread between countries and continents. For example, it was reported that community acquired ST80 strain spread to many European countries and ST8 (USA 300) clone began to be reported in European countries [73].

Studies have proved that MRSA strains have increased dramatically throughout the world over the years. For instance, the rate of nosocomial MRSA strains was 2.4% in the USA in 1973, whereas it rose to 35% in 1993. Similarly, the same rate, which was 1.7% in 1990 in Germany rose to 8.4 in 1995 [74]. According to the data of SENTRY antimicrobial surveillance program, MRSA prevalence rates in Australia, Japan, Latin America countries, the USA and European countries between 1997 and 1999 were 22.4%, 66.8%, 34.9%, 32.4% and 26.5%, respectively [75, 76].

In Europe, MRSA prevalence varies from one country to another. The prevalence in the North European countries and Holland changes between 1 to 2% while it has been shown that this rate increases in the Eastern and Southern European countries such as Italy, Greece and France where a prevalence of up to 80% has been reported [63]. The low prevalence in the Scandinavian countries and Holland is associated by the low antibiotic selective pressure and the screening of the patients at high risk for MRSA prior to hospitalization [77]. According to the ARMed surveillance report including Mediterranean countries, mean MRSA prevalance rate between 2003 and 2005 was 39% [78]. The USA Surveillance Network Database, which collects data from 300 clinical microbiology laboratories in the USA, reported the MRSA rate in 2005 as 59% in non-intensive care patients, 55% in intensive care patients and 48% in ambulatory patients [79].

Klevens et al. [80] conducted a study aimed to describe the incidence and distribution of invasive MRSA infections in 9 US communities in 2005. Throughout the surveillance period, 8987 invasive MRSA cases were reported. According to the report, most of the MRSA infections were health care-associated: 5250 (58.4%) were community-onset infections, 2389 (26.6%) were hospital-onset infections; 1234 (13.7%) were community-associated infections and 114 (1.3%) could not be classified. Although CA-MRSA prevalence is low throughout the world today, it keeps on increasing. Salgado et al. [81] conducted a meta-analysis of 57 studies on MRSA prevalence and determined that most of the individuals with CA-MRSA possessed at least one risk factor. The same study determined a CA-MRSA prevalence rate of

0.24% in the individuals who were not carrying any risk factors. In another study, CA-MRSA prevalence in Europe was reported as 0.03-1.5% [77].

Vancomycin Resistance

Due to the spread of MRSA around the world in the early 1980s and its multiresistance, glycopeptides remained as the first choice of treatment during the many years, in which novel antibiotics were not yet introduced into clinical use. Vancomycin resistance dates back to 1997 when a vancomycin intermediate *S. aureus* (VISA) strain was reported by Hiramatsu et al. [82]. In the following years, VISA strains were reported in Europe, America and Asia [83, 84]. In 2002, a more important development took place: following the transferring of a *vanA* operon from a vancomycin resistant enterococcus strain to an MRSA strain, the first clinical vancomycin-resistant *Staphylococcus aureus* (VRSA) strain was isolated in the USA. Since 2002, high-level VRSA (MIC ≥32 ug/mL) has been detected in seven clinical samples [83, 84]. However, it has been determined that the resistance in the VISA isolates is not limited solely to vancomycin but to teicoplanin as well. These isolates were defined as glycopeptide intermediate *S. aureus* (GISA). Treatment failures with vancomycin resulted in the detection of heterogeneous glycopeptide-resistant VISA (hVISA) and heterogeneous glycopeptide - resistant GISA (hGISA) [84].

Resistance to vancomycin occurs via two mechanisms. The thicker than normal peptidoglycan layer in the cell wall of VISA strains prevents the antibiotic from reaching its site of activity and consequently, resistance develops. As for VRSA strains, the acquistion of transferable genetic elements, *vanA* and *vanB*, results in resistance and subsequent to the change in the target site of the drug, d-alanin-d-alanin peptide transforms into d-alanin-d-lactate peptide. This peptide change brings out high-level resistance to vancomycin [32].

Linezolid Resistance

Linezolid, of which the antibacterial spectrum resembles that of vancomycin, is an antibiotic that is effective against to all gram positive bacteria including beta-lactam and vancomycin resistant bacteria [32]. The first linezolid resistant MRSA strain was reported in the USA in 2000 from a patient with MRSA associated peritonitis, after a 4-week-long linezolid treatment [85]. Similarly, Wilson et al. [86] determined that the MRSA strains isolated from a patient with empyema, in whom tracheotomy and drainage was performed, were sensitive to linezolid. At the end of a three-week-long linezolid treatment, the MRSA strains became resistant to linezolid. Molecular studies have shown that linezolid resistance occurs as a result of the mutations in 23 rRNA [85].

Quinupristin/Dalfopristin Resistance

The first *S. aureus* isolate resistant to quinupristin/dalfopristin, a streptogramin derivate, was reported in France, in 1975. The quinopristin/dalfopristin resistance rates in clinical isolates change between 0% to 44% [87]. The resistance to quinupristin/dalfopristin develops through several mechanisms including enzymatic modification, active transport of efflux mediated by an ATP-binding protein and the alterations in the target site of the drug [88].

The acetyltranspherase enzyme, which dismantles the antibiotic, and *vatA*, *vatB*, and *vatC* genes, which are responsible for the production of this enzyme, are responsible for the resistance to quinupristin. The resistance mechanism to the drug also involves the *vgaA*, *vgaB* ve *vgaAv* genes that encode the ATP-binding proteins, which are considered as having a role in the active transport of quinupristin [87].

Erm, *vgb* and *msr* genes are responsible for the resistance to dalfopristin. The *ermA, ermB* and *ermC* genes are responsible for enzyme formation, which causes the methylation of adenine located in the 23S rRNA binding site. The *vgbA* and *vgbB* genes are plasmid-mediated resistance genes responsible for the production of the enzyme that causes the inactivation of quinupristin/ dalfopristin. The *msrA* and *msrB* genes, on the other hand, result in dalfopristin resistance mediated by active transport after the induction of erythromycin [87].

Quinolone Resistance

Fluoroquinolones were introduced into clinical use in the 1980's. However, resistance to this group of antiobitics began to develop as *S. aureus* isolates became widespread. This caused a decrease in the use of fluoroquinolones. When fluoroquinolones are used in the the treatment of infections due to other pathogens, *S. aureus* which is in a colonized state may become resistant as a result of its exposure to the subtherapeutic concentrations of the antibiotic [89].

Quinolones are effective against DNA girase and topoisomerase IV in bacteria. The amino acid alterations that occur in the quinolone resistance determining region (QDRD), which encodes the enzyme, decrease the affinity of the drug for its target site. ParC (referred to as GrlA in *S. aureus*), a subunit of topoisomerase IV and GyrA, a subunit of DNA girase, are the two sites where mutation is most commonly observed in the development of resistance against quinolones [90]. Clinical studies have shown that ParC mutation is sufficient alone for the development of resistance against ciprofloxacine but not for the resistance against new generation fluoroquinolones [91]. Another type of resistance against quinolones is the efflux pump-mediated resistance [89].

Daptomycin Resistance

Although the mechanism of the resistance against daptomycin has not been clarified completely yet, Patel et al. [92] stressed that a cross resistance may develop between vancomycin and daptomycin. Vancomycin resistance should be taken into consideration during treatment with daptomycin since the thickened cell wall in vancomycin-resistant strains may prevent daptomycin from reaching the cell membrane.

In a prospective study including 120 patients with bacteremia and endocarditis, daptomycin treatment was performed and during the treatment and an increase was observed in MIC of daptomycin in 5% of the isolates 93]. Similarly, in some studies, daptomycin resistance development has been observed in the clinical MRSA isolates obtained from patients treated with daptomycin [94,95]. Still, daptomycin resistance is low in the clinical setting [96].

Virulence Factors and Pathogenesis

Staphylococcus aureus expresses many potential virulence factors and both structural and secreted products play a role in the pathogenesis of its infections. It has been shown that, except one, all pandemic MRSA clones contain at least one gene encoding a superantigenic toxin. Superantigenic toxins are mostly present in HA-MRSA clones while PVL genes exist in CA-MRSA clones [97, 98].

Nasal and Non-Nasal Colonization

Anterior nares is the primary reservoirs for *S. aureus*. From that site, bacteria may disseminate to the skin. It has been observed that when *S. aureus* nasal carriage is eliminated by the use of intranasal mupirocin, colonization decreases in the other body sites. The decrease in the risk of clinical infection, which occurs when colonization is eradicated, indicates the role of nasal colonization in the pathogenesis of *S. aureus* infections [97, 99, 100]. The most important mechanism, particularly in the pathogenesis of HA-MRSA infections, is the colonization of MRSAs that is followed by infection. However, the contact with patients or with the objects contaminated with bacteria is more important than colonization in the development of *S. aureus* infections. Studies have shown that, among patients with CA-MRSA infections, colonization is more commonly seen in the other body sites (axillae, inguinal, groin and gastrointestinal tract) than in nares [97, 99].

Environmental Sources

In athletes, skin–to-skin contact, especially contact with broken skin facilitates the spread of infection. Additionally, the sharing of contaminated fomites (razors, towels, soap bars, etc.) plays an important role in the spread of CA-MRSA infections. Contamination from the environment is not specific to CA-MRSAs. In hospitals, HA-MRSA strains have been isolated from stethoscopes, pagers, bed spaces, and workstations. Healthcare workers are the leading actors of MRSA spread. The control of healthcare workers and the treatment of carriers are prerequisites for treatment [99, 101].

Microbial Surface Components Recognizing Adhesive Matrix Molecules

The surface proteins, which play a role in the adhesion of *S. aureus* to host tissues are called "microbial surface components recognizing adhesive matrix molecules" (MSCRAMMs). These molecules, which consist of many proteins, bind to the molecules of the host including collagen, fibronectin, and fibrinogen. It has been suggested that

MSCRAMMs play a key role in the development of endovascular, bone and joint, and prosthetic-device infections [102]. After *S. aureus* adheres to host tissues and prosthetic materials, it increases in number and can form biofilms (slime) to protect itself from host defenses and antimicrobials. Another defense mechanism of *S. aureus* is the production of an antiphagocytic microcapsule. The zwitterionic capsule (both positively and negatively charged) can also induce abscess formation. The MSCRAMM protein A may prevent opsonization by binding to the Fc portion of immunoglobulins. *Staphylococcus aureus* also secretes chemotaxis inhibitory protein of staphylococci and the extracellular adherence protein, which prevent neutrophil extravasation and chemotaxis. Leukocidins produced by *S. aureus* destroys leukocytes by forming pores in the cell membrane [97, 103, 104]. By the enzymes, such as proteases, lipases, and elastases, *S. aureus* invades host tissues and metastasizes to other organs [97].

The regulation of expression of staphylococcal virulence genes is of great importance in pathogenesis. Expression of the MSCRAMMs generally occurs during the logarithmic growth phase whereas secretion of proteins, such as toxins, occurs during the stationary phase. During infection, the early expression of the MSCRAMM proteins facilitates initial colonization of *S.aureus* in the tissue sites, whereas the later elaboration of toxins facilitates spread. The accessory gene regulator (agr) is a quorum-sensing system that is of great importance in the regulation of staphylococcal virulence [105].

Panton Valentine Leukocidin

Panton Valentine leukocidin is the most important virulence factor detected in CA-MRSA strains. Along with, γ-haemolysin and other leukocidins, it is a member of the bicomponent synergohymenotropic toxin family. PVL is encoded by *lukF-PV* and *lukS-PV,* two genes that are borne by different phages. The S and F proteins, produced by these genes, destroy host cell membranes through a synergistic effect. PVL is active on human polymorphonuclear neutrophils (PMNs), monocytes and macrophages. By opening membrane calcium channels, it brings out an increase in the calcium level in the cell and liberates inflammatory mediators such as leukotriene B4, IL-8 and histamine [97, 106-108]. Experimental studies on rabbits have shown that, when injected intadermally, PVL induces severe inflammatory lesions and necrosis on the skin with capillary dilation, chemotaxis, PMN infiltration and PMN karyorrhexis [109]. On the other hand, no reactions have been observed in the rabbits immunized against LukS-PV and LukF-PV [110, 111]. Labandeira et al. [112] used isogenic PVL-positive and PVL-negative strains in their mouse pneumonia models. They detemined that PVL-positive strains cause necrotizing pneumonia whereas PVL-negative strains only cause leukocytic invasion. These findings indicate that PVL plays a role in the occurrence of community-onset necrotizing pneumonia by causing inflammation and necrosis in respiratory epithelium and parenchyma [108, 113]. It is highly possible that PVL is responsible for peripheral leucopenia observed in patients with community-onset pneumonia [98].

There is a strong epidemiologic association between PVL and CA-MRSA infections. In a study conducted in France, 593 *S. aureus* strains were screened for PVL and it was determined that all CA-MRSA strains and none of the HA-MRSA strains were positive for PVL [114]. Skin and soft tissue infections (SSTIs) and necrotizing pneumonia mostly result from PVL-positive strains. Lina et al. [106] detected PVL in 85% of the strains associated with CA-pneumonia and in 93% of the strains associated with invasive skin infections such as furunculosis.

Exfoliative Toxins

To date, three exfoliative toxins have been described: A, B and D. These toxins behave like epidermolytic protease and cleave desmoglein 1, a protein that attaches epidermal cells in the granular layer of epidermis. Consequently, they cause the exfoliation of epidermis without inflammation and necrolysis of the skin [115]. In various parts of the world, MRSA strains carrying exfoliative toxin genes (*eta* gene and *etb* gene) and bullous impetigo caused by these strains have been reported [116].

Toxic Shock Syndrome Toxin-1

Toxic shock syndrome toxin-1 (TSST-1) is the leading virulence factor that plays a role in toxic shock syndrome, staphylococcal scarlet fever and various suppurative infections. This toxin, which is a superantigen, is released by staphylococci, which carry the *tst-gene*. The *tst*-positive MRSA clones have been mostly isolated from children (median age 3 years) and hospital-acquired infections. It has been shown that SCCmecIV cassette is present in most of these MRSA clones [98, 117]. It has been determined that a TSST-1-secreting MRSA clone, which predominates in Japan, is responsible for a nosocomial disease called neonatal toxic shock syndrome-like exanthematous disease. This clone was also isolated in France and it was determined that it continues to spread through Europe [98, 118].

Phenol-Soluble Modulins

Phenol-soluble modulins (PSMSs) are cytolytic peptides encoded by an operon that exists in the *S. aureus* genome. In addition to their proinflammatory and chemotactic effects, they also have in vitro and in vivo leukocidal activity. HA-MRSA produces a small number of or no PSMs at all, whereas CA-MRSA produces a great number of PSMs. It has been reported that the difference in lytic activity between these two MRSAs resulted from PSMs and this finding has been reinforced by the experiments conducted on infected mouse models [99, 119].

MRSA Infections

Skin and Soft Tissue Infections

Before the 90's, the causative agents of SSTIs were MSSA or *Streptococcus pyogenes*, however, in the early 90's, MRSA strains began to be isolated in these infections. Although

HA-MRSA strains may often be isolated in SSTIs, the causative agent is usually CA-MRSA. Formerly, such infections were observed in specific groups such as individuals with health care contact and intravenous drug users. However, today they are detected even in people who are young and do not carry any known risk factors. SSTIs have been reported in healthy newborns, healthy children, healthy adults, pregnant women, postpartum women, intravenous drug users, prisoners, homeless persons, men who have sex with men, athletes, tattoo recipients and soldiers. Some important factors in SSTI transmissions are direct contact with infected patients, colonized subjects, contaminated environment and sharing of personal items [120-122].

The most commonly seen SSTIs are cellulitis and abscess while fasciitis and pyomyositis are rare. Abcesses have been reported in about 50-75% and cellulitis has been reported in about 25-50% of the patients. These lesions are generally single, localized in the extremities and usually do not cause fever and leukocytosis. Furuncles (boils), on the other hand, are mostly multiple and appear as outbreaks. They can progress to abscesses and cellulitis and recurrences may occur. Folliculitis is a less common clinical manifestation. Although impetigo and scalded-skin sydrome are rare manifestations of SSTIs, they can be seen in children. The other rare infections are pyomyositis and myositis, which usually involve the lower extremities or pelvis. All affected patients present with pain and fever. Unlike viral myositis, an increase is observed in the white blood cell count of patients with myositis due to MRSA and their creatine kinase levels usually remain within the normal range. Thus, the increase in white blood cell count is important in differentiating myositis due to MRSA from viral myositis. Bacteremia and septic shock may develop in patients with pyomyositis and myositis [123-126]. A subacute form of necrotizing fasciitis is generally seen among middle aged intravenous drug users and middle aged patients with diseases such as hepatitis C or diabetes. Staphylococcal toxic shock syndrome, Waterhouse-Friderichsen syndrome and purpura fulminans are rare MRSA-induced skin syndromes [127]. The prognosis of SSTIs due to MRSA is generally good and the hospitalization rate of adults hovers between 16% and 44%. However, recurrence is common in these infections [120].

CDC introduced the "Five C's of CA-MRSA Transmission" for CA-MRSA SSTI prevention. These are contact, cleanliness, compromised skin, contaminated objects, and crowded conditions. Skin-to-skin contact with the patients should be avoided, attention should be paid to personal hygiene, personal items should not be shared with others, activities that may cause skin abrasions should be refrained from, attention should be paid to environmental hygiene and overcrowding should be avoided [121].

Pneumonia

Staphylococcus aureus is not a pathogen that is commonly encountered in community acquired-pneumonia (CAP). CAP cases are generally associated with influenza epidemics. However, *S. aureus* is one of the foremost causes of CAP requiring hospitalization and intensive care and necrotizing pneumonia cases. *Staphylococcus aureus* is responsible for 25.5% of the hospitalizations due to CAP. PVL-producing strains, in particular, have been found to be associated with the occurrence of necrotizing pneumonia in healthy adults and children. High fever, leucopenia, haemoptysis, pleural effusion, shock, and respiratory failure following influenza-like illness are seen in these patients. MRSA infection should be kept in mind in case of recent use of broad-spectrum antibiotics (fluoroquinolones), intravenous drug use, recent MRSA infection, cavitary lesions on chest radiograph, and multilobar involvement [128].

Staphylococcus aureus is responsible for about 30-50% of the hospital-acquired pneumonia (HAP), ventilator-associated pneumonia (VAP) and health care-associated pneumonia (HCAP) cases. Diagnosis of healthcare-associated pneumonia can be made if a patient with pneumonia presents with at least one of the following risk factors: hospitalization in an acute care hospital for two or more days within the last 90 days, residence in a nursing home or long-term care facility; receiving intravenous antibiotic therapy, chemotherapy or wound care within the past 30 days; attending a hospital or hemodialysis. Other risk factors in MRSA-associated HAP/VAP include steroid use, chronic obstructive pulmonary disease (COPD), antimicrobial use in the preceding 90 days, and high frequency of antibiotic resistance in the hospital or community [129].

Bone and Joint Infections

Osteomyelitis is a bone infection caused by pyogenic microorganisms. It may present with an acute, subacute or chronic course. *Staphylococcus aureus* is the causative agent in more than 90% of the cases of acute hematogenous osteomyelitis, which is commonly seen in children and mostly in the metaphysis of long bones. In addition to local infection findings (local tenderness and movement restriction in adjacent joint), the patients have fever, irritability and lethargy. The subacute and chronic forms of these infections generally develop in adults and secondary to open wounds. MRSA is one of the most common pathogens encountered in these forms of this infection. Patients with these infections present with localized pain in the infected site, erythema, movement restriction, draining sinus tracts, deformity and instability. [130, 131]

Bloodstream Infections

In the recent years, there has been an increase in the incidence of *S. aureus* BSIs. It is evident that this increase results from the increase in BSIs associated with MRSA, since the incidence rates of BSIs due to MSSA are relatively constant. In a study conducted in 2008, the incidence rate of *S. aureus* BSIs was reported as 19.7 cases/100,000 population/year [30]. Similar studies have reported this rate as varying between 8.4-31.4 cases/100,000 population/year [132-134]. Bloodstream infections are divided into three groups as community-acquired, nosocomial and health care-associated. MRSA is a one of the pathogens, which are most frequently encountered in all three groups of bloodstream infections. Laupland et al. [30] identified bone and joint, soft tissue, lower respiratory, and endovascular infections as the most common focal sources of primary bacteremia. MRSA bacteremia risk factors include old age, previous MRSA infection, prior antibiotic exposure, diabetes, dialysis dependence, organ transplantation, hepatitis C, HIV infection and malignancies (especially hematologic malignancies) [30, 135]. Studies have shown that the rates of duration of hospitalization and mortality are much higher in MRSA cases than in MSSA cases [30, 136]. As a result of the spread of CA-MRSA strains to hospitals, they have begun to be isolated more commonly as pathogens causing BSIs. In one study, CA-MRSA was found to be responsible for 20% of nosocomial BSIs and 28% of health care associated BSIs. In the presence of at least one of the factors including concomitant injection drug use, necrotizing pneumonia, and skin and soft tissue infections, CA-MRSA BSIs should be borne in mind [10].

In about one third of the patients with bacteremia, metastatic foci of infections may develop in joints, kidney, central nervous system, lungs, liver, spleen, bone, and heart valves. The most dangerous of all complications resulting from bacteremia is endocarditis [128].

A study conducted in a USA hospital reported the standardized mortality rate of MRSA infections as 6.3 deaths/100.000 population. The same study found a case-fatality rate of 18% [80]. On the other hand, the case-fatality and standardized mortality rates reported in a study conducted in Canada were 25% and 4.7 deaths/100,000 population/year, respectively. As for a study conducted in Denmark, the same rates were 21% and 5.1 deaths/100,000 population/year [30, 133].

Treatment of MRSA Infections

A short while after the use of penicillins in the treatment of *S. aureus* infections, resistance was developed against these antibiotics and consequently, penicillins (methicillin, nafcillin) and cephalosporins resistant to beta-lactamases were developed in the early 1960's. In 1961, MRSA strains resistant to all beta-lactamases emerged [40, 137, 138]. Since the time MRSA strains were reported, an increase has been observed in HA-MRSA infections, of which mortality and morbidity rates are too high. On the other hand, in the recent years, significant increases have been observed in the CA-MRSA infections, which demonstrate marked differences in antibiotic sensitivity profile [139, 140]. The appearance of MRSA strains and their rapid spread has caused an increase in the use of glycopeptide antibiotics (vancomycin and teicoplanin) [83, 88, 141]. In addition to this, various glycopeptide resistant MRSA strains have begun to be reported. The emergence of GRSA, homogeneous and heterogeneous GISA and glycopeptide tolerance has caused the production of novel antimicrobials (streptogramins (quinupristin/dalfopristin) (Q/D), oxazolidinones (linezolid) and cyclic lipopeptides (daptomycin) [83, 141, 142]. Unlike HA-MRSAs, CA-MRSA strains are sensitive to many antibiotics other than beta-lactamases. However, the effect of most of these antimicrobials on CA-MRSA infections has not been thoroughly investigated and serious ambiguities regarding resistance development are present. The studies about this subject mostly focus on agents such as TMP/SMX, clindamycin, and tetracyclines. However, rifampin is another agent that is still commonly used but shown to be increasingly subject to bacterial resistance. The data regarding effectiveness of rifampin against CA-MRSA strains are limited [40, 142-144].

Due to increasing resistance, treatment options are limited in MRSA infections. The choice of the appropriate antimicrobial therapy varies according to the severity of the disease and the MRSA strain. Vancomycin is the first choice in the treatment of serious MRSA infections. However; although rare, treatment failures have been reported with this antimicrobial. In localized soft tissue infections, oral antimicrobials are recommended in addition to surgical incision and drainage. On the other hand, hospitalization and intravenous treatment are required in severe infections. TMP/SMX, clindamycin, tetracyclines, rifampin, linezolid and rarely fluoroquinolones are used for mild CA-MRSA infections. However, the treatment for severe infections requires novel parenteral antimicrobial agents including linezolid, daptomycin and tigecycline, as well as vancomycin [137, 141, 142]. Although Daptomycin is a lipopeptide approved by the Food and Drug Administration (FDA), British Society for Antimicrobial Chemotherapy (BSAC) does not recommend it. BSAC recommends glycopeptides and linezolid as the first choice for the treatment and prophylaxis of MRSAs infections. Although quinupristin/dalfopristin combination is also used for treatment, its use is limited due to high resistance profile, tolerance and problems of drug interaction [83].

Trimethoprim-Sulfamethoxazole

TMP/SMX is one of the most commonly used bactericidal agents against CA-MRSA infections. In-vitro studies have shown that TMP/SMX has a more rapid bactericidal activity than linezolid, rifampin, clindamycin, minocycline and moxifloxacin in HA-MRSA and CA-MRSA strains [40, 137, 141, 145]. In their study conducted in 2005, Fridkin et al. [123] reported that 97% of the CA-MRSA isolates they examined were TMP/SMX susceptible. However, they stressed that the treatment should be conducted according to the sensitivity results because of the regional differences in resistance. On the other hand, in another study conducted in 2007, TMP/SMX resistance was reported as 33.7% [146]. Since TMP/SMX has good bioavailability and is cost-efficient, it constitutes an alternative in case of resistance to vancomycin. Thus, it is a reasonable choice in the treatment of MRSA infections. However, its comparison with vancomycin and linezolid in terms of effectiveness, cost and reliability has not been satisfactorily made [137, 147]. Rates of treatment success of MRSA infections with TMP/SMX were reported to be high in a retrospective study conducted in 2008 [141].

Tetracyclines

Although resistance to minocycline and doxycycline has been observed in MRSA strains recently, they are still preferred in the treatment of mild CA-MRSA infections due to their perfect oral bioavailability, tissue penetration and quality of being easily tolerated. They may sometimes be used in combination with rifampin, even though sufficient data on this combination have not been collected [137, 148]. In a study on 480 MRSA strains, which was performed in Asia-Pacific and South Africa, tetracycline resistance rate was found as 64.6% [146]. In a retrospective study, Ruhe et al. [149] showed that tetracyclines (doxycycline and minocycline) were effective on CA-MRSA strains. Another study on 100 nosocomial strains conducted in 2005, reported a tetracycline resistance rate of 85% [150].

Fluoroquinolones

An increasing resistance has been observed in *S. aureus* against fluoroquinolones and this resistance is common. Thus, they are not recommended in the empirical treatment of skin and soft-tissue lesions resulting from CA-MRSAs. Some studies have reported that resistance develops rapidly during the use of fluoroquinolones in the treatment of MRSA infections [151].

Clindamycin

Clindamycin inhibits the protein synthesis in the bacterium by affecting the ribosomes. It is effective against most of the CA-MRSA strains and is one of the first choices of treatment. However, this antibiotic is not recommended due to the high-level resistance developed in serious MRSA infections. There are regional differences in clindamycin resistance. Inducible macrolide- lincosamide- streptogramin (MLS)B resistance may emerge during clindamycin treatment [137, 142]. Therefore, inducible resistance should be investigated before its use in the treatment. In a study conducted on children between 2001 and 2004, an increase in the resistance of CA-MRSA strains against clindamycin was shown. Inducible clindamycin resistance was reported in the clindamycin-sensitive and erythromycin-resistant strains [152]. Various studies have shown that clindamycin resistance rate is 13.3% in CA-MRSA strains and 36-56% in HA-MRSA strains [153, 154]. The activity of clindamycin; as well as that of daptomycin, doxycycline, linezolid, TMP-SMX and vancomycin; against CA-MRSA was shown both in vivo and in vitro by a study conducted in 2008. According to that study, the effectiveness order of antibiotics in the strains where inducible clindamycin resistance was not seen and in the strains with inducible resistance was as follows, respectively: daptomycin > clindamycin > doxycycline > vancomycin = linezolid = TMP-SMX and daptomycin > vancomycin = linezolid > clindamycin [142].

Vancomycin

Vancomycin is a glycopeptide antibiotic with bactericidial effect. It is the first choice of treatment in serious MRSA infections. Since it became available, it has been commonly used due to its partial reliability, the low probability of resistance development and the fact that it had no alternatives until recently. However, along with the increase in MIC values, clinical insufficiency of the antibiotic has begun to be reported recently [128, 155]. Since 1996, strains that are intermediate resistant, and since 2002, strains that are completely resistant to vancomycine have been reported [137, 143, 155-157]. Although varying from one country to another, the hVISA rates of Europe was reported as 0.21-20%, and this rate was found to change between 0.54% and 1.93% for Asia [158, 159]. In a study conducted between 2000 and 2006, an increase in MIC values for vancomycin was reported in MRSA strains [160]. However, there are also studies in which vancomycin resistance is not determined. In a study conducted in 2007, vancomycin and teicoplanin resistances were not seen in 480 MRSA strains [146]. A study conducted in Taiwan showed that the rate of nosocomial MRSA infections rose from 39% in 1991 to 78% in 2002 and the use of glycopeptides increased along with other antibiotics. In the study where 100 MRSA strains were found to be completely susceptible to vancomycin, only one strain was found to be intermediate to teicoplanin [150]. In the study of Manzur et al. [29] conducted in 2007, resistance to glycopeptides and TMP-SMX was not determined in the 50 MRSA strains isolated from patients with bloodstream infections while the resistance rates for ciprofloxacin, clindamycin, tetracycline and rifampin were found as 100%, 50% and 30%, respectively.

Linezolid

Linezolid, a member of oxazolidinones, has a bacteriostatic effect. It has both intravenous and oral formulations. Its oral bioavailability is 100%. Following vancomycin, it is the second antibiotic, of which the effect on MRSA strains has been most commonly studied. Since it can be used orally, it is the best alternative to vancomycin. Linezolid's hematologic side effect and the resistance development of *S. aureus* strains are the most important disadvantages of this antibiotic. One year after the antibiotic's introduction into clinical use in 2000, the first linezolid-resistant MRSA strain was reported. However, the reported resistant strains are few in number and linezolid's effectiveness against MRSA is still so high [137, 143, 157]. Its distribution to the body fluids and tissues is very good. Compared to vancomycin, the treatment of nosocomial pneumonia due to MRSA and VAP is more successful with linezolid [156, 161]. In addition to linezolid's protein synthesis inhibition, it also inhibits exotoxin production and accordingly, it is the most preferred agent in the treatment of pneumonias resulting from PVL-producing MRSAs [29]. The superiority of linezolid to vancomycin in the treatment of complicated SSTI (cSSTI) and surgical infections has been shown in various studies. Nevertheless, the prospective controlled studies about linezolid are not sufficient [88, 137, 143, 156, 162]. In a study conducted in 2006, it was seen that 5000 MRSA strains obtained from 50 medical centers were susceptible to linezolid [163]. On the other hand, two studies conducted in 2005 and 2007 did not detect any MRSA strains resistant to linezolid [146, 150]. In some patients, lactic acidose has been reported during linezolid treatment. Combined with serotonergic agents, hyperpyrexia and serotonin syndrome involving mental changes may occur. If the drug is used over three months, peripheral and optic neuropathy may develop [88, 143].

Quinupristin/Dalfopristin

They belong to the streptogramin family. The two antibiotics inhibit the protein synthesis in bacteria through a synergistic effect. Dalfopristin inhibitis the early stage of protein synthesis while quinupristin inhibits its late stage. It is effective against MRSAs. Compared to other agents (71%), its success rate in the treatment of skin and soft tissue infections due to MRSAs is 68% [88, 161]. Its success rate in the treatment of hospital acquired MRSA pneumonias is similar to that of vancomycin. Structural or inducible MLSB resistance against Q/D has been shown in *S. aureus* strains [88, 164]. In two studies conducted in 2005 and 2007, the rate of resistance to Q/D was 8% and 0.8%, respectively [146, 150]. Dose adjustment may be necessary in patients with hepatic insufficiency. Its most prominent side effects are arthralgia and myalgia. It has more side effects than the other agents [88, 156]. Thus, quinupristin/dalfopristin combination is commonly preferred when standard agents prove insufficient or intolerance occurs [88].

Daptomycin

Daptomycin is a novel lipopeptide antibiotic, which is effective against MRSAs. It ensures depolarization of the cell membrane, disfunction of macromolecule synthesis and collapse of the organism without causing disintegration. It has a bactericidial effect. The greatest advantage of daptomycin is that it destroys the microorganism without disintegrating it [165]. Its post-antibiotic effect lasts between 1.5 and 6 hours [83]. Its use in the treatment of cSSTI infections due to MRSA was approved by the FDA in 2003. Various in vitro and in vivo studies demonstrating the bactericidial effects of daptomycin have been reported [142, 150, 166]. Since only one dose of the antibiotic is used per day and its side effect profile is low, it is considered as a drug of choice in the treatment of MRSAs. Its effectiveness in the treatment of the blood stream infections, endocarditis and SSTI infections due to MRSAs is considered as equivalent to that of vancomycin. There are some studies in the literature, which report that daptomycin is highly successful in the treatment of osteomyelitis due to MRSA. It is not recommended for the treatment of MRSA penumonia since it is inhibited by pulmonary surfactant [128]. About two years after daptomycin was introduced to clinical use, resistance in MRSA strains to the antibiotic was reported. MRSA strains exhibiting reduced susceptibility to vancomycin may also exhibit this phenomenon with daptomycin. Resistance to the antibiotic has been rarely shown in clinical studies and the mechanism of this resistance has not been clarified yet [88]. During long-term use of the drug, serum creatine kinase levels should be monitored [143].

Tigecycline

Tigecycline is a member of the glyclycycline family. Structurally, it is a semisynthetic derivative of minocycline and has a bacteriostatic effect. It exerts its activity by inhibiting protein synthesis. Tigecycline's outstanding quality is its effectiveness against Gram-positive bacteria, which develop resistance against tetracycline through ribosomal protection and efflux mechanisms. Its use against cSSTI due to MRSAs and complicated intraabdominal infections is approved by the FDA. It is highly effective against MRSAs. Phase 3 studies determined that it has an effect equivalent to that of vancomycin in the treatment of cSSTI. Studies have shown that its most common side effects are nausea and vomittting [137, 143, 167].

Antibiotics That Are under Investigation

Novel Glycopeptides / Oritavancin – Telavancin – Dalbavancin: These are semi-synthetic lipoglycopeptides. They exert bactericidial effect by inhibiting the cell wall synthesis. They are effective against MRSA infections [168]. The molecular structure of oritavancin includes 4-epi-vancomycin, so its activity spectrum resembles that of vancomycin. It has been shown that it is effective on the rabbit endocarditis model [169]. A phase 2 study demonstrated that oritavancin is as effective as vancomycin against *S. aureus* bloodstream infections [170]. In vitro studies have shown that it accumulates slowly in lysosomes and is very slowly released [171]. Currently, there are no clinical data on the effect of oritavancin against MRSA infections [172]. Telavancin is obtained by the alkylation of vancomycin. Unlike vancomycin, it is effective on biofilm models [173]. A phase 3 study showed that the effect of telavancin is equivalent to that of vancomycin in the treatment of cSSTI due to MRSA [174]. Dalbavancin, a second generation glycopeptide, is a teicoplanin analogue. It has been shown that it is highly active against MRSAs in animal models. It has also been determined that its effectiveness and tolerance is better than vancomycin in cathether-related bloodstream infections. It is used once a week due to its long half life. It has been shown that it is as effective as linezolid against cSSTIs [88, 175, 176].

Cephalosporins

Ceftobiprole: It is a cephalosporin with an extended spectrum of activity. It has not been approved by FDA yet. Due to its quality of binding to PBP2a, it is effective against MRSAs [177]. Ceftobiprole is a weak substrate for AmpC beta-lactamases and is a weak inducer like cefepim. It is the first beta-lactam antibiotic that is effective against MRSAs. Phase 3 studies have shown that the clinical success rate of ceftobiprole in the treatment of cSSTI and nosocomial pneumonias due to MRSA is higher than that of vancomycin. Phase 3 studies investigating the effect of the antibiotic on community acquired pneumonias still continue. Studies conducted until now have indicated that ceftobiprole is a reliable and tolerable agent. Due to its extended spectrum, it can be used alone against infections requiring combined therapy. Still, further studies on its effectiveness and reliability are needed [155, 172, 178].

Ceftaroline: It is another cephalasporin in the study phase. It has been shown by phase 2 studies that ceftaroline is a reliable and effective agent against MRSA infections [143].

Conclusion

In the recent years, MRSA strains have become important pathogens due to the resistance they have developed against various antibiotics. Compared to HA-MRSA strains, CA-MRSA strains are susceptible to most antibiotics. Agents such as ciprofloxacin, clindamycin and TMP/SMX are effective against CA-MRSAs. As far as HA-MRSAs are concerned, the treatment options are limited. Glycopeptides constitute the first choice of treatment agianst serious MRSA infections. Due to increasing resistance and tolerance, clinical insufficiency of these agents has been reported. Linezolid, daptomycin and tigecycline are novel alternative agents. Additionally, it seems likely that novel lipoglycopeptides (oritavancin – telavancin – dalbavancin), two wide-spectrum cephalosporins (ceftobiprole and ceftaroline), iclaprim – a novel diaminopyrimidine – and agents such as CS-023/RO-4908463 and monoclonal antibody tefibazumab will be introduced as new treatment alternatives in the next few years.

REFERENCES

[1] David, MD; Kearns, AM; Gossain, S; Ganner, M; Holmes, A. Community-associated meticillin-resistant *Staphylococcus aureus*: nosocomial transmission in a neonatal unit. *J Hosp Infect*, 2006, 64, 244-250.

[2] Viallon, A; Marjollet, O; Berthelot P; Carricajo, A; Guyomarc'h, S; Robert, F; Zeni, F; Bertrand, JC. Risk factors associated with methicillin-resistant *Staphylococcus aureus* infection in patients admitted to the ED. *Am J Emerg Medicine*, 2007, 25, 880–886.

[3] Tambyah, PA; Kumarasinghe, G. Methicillin-resistant *Staphylococcus aureus* control at the national university hospital, Singapore: a historical perspective. *Ann Acad Med Singapore*, 2008, 37, 855-60.

[4] Centers for Disease Control. Community-acquired methicillin-resistant *Staphylococcus aureus* infections-Michigan. *MMWR Morb Mortal Wkyl Rep.*, 1981, 30, 185–7.

[5] Morin, CA; Hadler JL. Population-based ıncidence and characteristics of community-onset *Staphylococcus aureus* infections with bacteremia in 4 metropolitan Connecticut areas, 1998. *J Infect Dis.*, 2001, 184, 1029–34.

[6] Friedman, ND; Kaye, KS; Stout, JE; McGarry, SA; Trivette, SL; Briggs, JP; Lamm, W; Clark, C; MacFarquhar, J; Walton, AL; Reler, LB; Sexton, DJ. Health care–associated bloodstream infections in adults: a reason to change the accepted definition of community-acquired infections. *Ann Intern Med.* 2002, 137, 791-7.

[7] Yang, JA; Kim, JY; Yoon, YK; Kim, S; Park, DW; Sohn, JW; Sim, HS; Kim, MJ. Epidemiological and genetic characterization of methicillin-resistant *Staphylococcus aureus* ısolates from the ear discharge of outpatients with chronic otitis media. *J Korean Med Sci.*, 2008, 23, 762-6.

[8] Wang, CC; Lo,WT; Chu, ML; Siu, LK. Epidemiological typing of community-acquired methicillin-resistant *Staphylococcus aureus* ısolates from children in Taiwan. *Clin Infect Dis.*, 2004, 39, 481–487.

[9] Bratu, S; Eramo, A; Kopec, R; Coughlin, E; Ghitan, M; Yost, R; Chapnick, EK; Landman, D; Quale, J. Community-associated methicillin resistant *Staphylococcus aureus* in hospital nursery and maternity units. *Emerg Infect Dis.*, 2005, 11, 808-813.

[10] Seybold, U; Kourbatova, EV; Johnson, JG; Halvosa, SJ; Wang, YF; King, MD; Ray, SM; Blumberg, HM. Emergence of community-associated methicillin-resistant *Staphylococcus aureus* USA300 genotype as a major cause of health care-associated blood stream infections. *Clin Infect Dis.*, 2006, 42, 647-56.

[11] Moran, GJ; Krishnadasan, A; Gorwitz, RJ; Fosheim, GE; McDougal, LK; Carey, RB; Talan, DA; EMERGEncy ID Net Study Group. Methicillin-resistant S. *aureus* infections among patients in the emergency department. *N Engl J Med.*, 2006, 355, 666-674.

[12] Kazakova, SV; Hageman, JC; Matava, M; Srinivasan, A; Phelan, L; Garfinkel, B; Boo, T; McAllister, S; Anderson, J; Jensen, B; Dodson, D; Lonsway, D; McDougal, LK; Arduino, M; Fraser, VJ; Killgore, G; Tenover, FC; Cody, S; Jernigan, DB. A clone of methicillin-resistant *Staphylococcus aureus* among professional football players. *N Engl J Med.*, 2005, 352, 468-475.

[13] Kirkland, EB; Adams, BB. Methicillin-resistant *Staphylococcus aureus* and athletes. *J Am Acad Dermatol*, 2008, 59, 494-502.

[14] Von Eiff, C; Becker, K; Machka, K; Stammer, H; Peters, G. Nasal carriage as a source of *Staphylococcus aureus* bacteremia. Study Group. *N Engl J Med.*, 2001, 344, 11-6.

[15] Lu, PL; Tsai, JC; Chiu, YW; Chang, FY; Chen, YW; Hsiao, CF; Siu, LK. Methicillin-resistant *Staphylococcus aureus* carriage, infection and transmission in dialysis patients, healthcare workers and their family members. *Nephrol Dial Transplant*, 2008, 23, 1659-65.

[16] Huang, H; Cohen, SH; King, JH; Monchaud, C; Nguyen, H; Flynn, NM. Injecting drug use and community-associated methicillin-resistant *Staphylococcus aureus* infection. *Diagn Microbiol Infect Dis.*, 2008, 60, 347–350.

[17] Lu, PL; Chin, LC; Peng, CF; Chiang, YH; Chen, TP; Ma, L; Siu, LK. Risk factors and molecular analysis of community methicillin-resistant S*taphylococcus aureus* carriage. *J Clin Microbiol*, 2005, 43, 132–139.

[18] Ciftci, IH; Koken, R; Bukulmez, A; Ozdemir, M; Safak, B; Cetinkaya, Z. Nasal carriage of *Staphylococcus aureus* in 4–6 age groups in healthy children in Afyonkarahisar, Turkey. *Acta Pediatr*, 2007, 96, 1043–1046.

[19] Ozgüven, A; Tünger, O; Cetin, CB; Dinç, G. Investigation of nasal carriage of community-acquired methicillin resistant *Staphylococcus aureus* in primary and high school students. *Mikrobiyol Bul.*, 2008, 42, 661-667.

[20] Duran, N; Ocak, S; Eskiocak, AF. *Staphylococcus aureus* nasal carriage among the diabetic and non-diabetic haemodialysis patients. *Int J Clin Pract.*, 2006, 60, 1204–1209.

[21] Mertz, D; Frei, R; Periat, N; Zimmerli, M; Battegay, M; Flückiger, U; Widmer, AF. Exclusive *Staphylococcus aureus* throat carriage: at-risk populations. *Arch Intern Med.*, 2009, 169, 172-178.

[22] Babazono, A; Kitajima, H; Nishimaki, S; Nakamura, T; Shiga, S; Hayakawa, M; Tanaka, T; Sato, K; Nakayama, H; Ibara, S; Une, H; Doi, H. Risk factors for nosocomial infection in the neonatal ıntensive care unit by the Japanese Nosocomial Infection Surveillance (JANIS). *Acta Med Okayama*, 2008, 62, 261-268.

[23] Seybold, U; Halvosa, JS; White, N; Voris, V; Ray, SM; Blumberg, HM. Emergence of and risk factors for methicillin-resistant *Staphylococcus aureus* of community origin in intensive care nurseries. *Pediatrics*, 2008, 122, 1039-1046.

[24] Wang, JT; Chang, SC; Ko, WJ, Chang, YY; Chen, ML; Pan, HJ; Luh, KT. A hospital-acquired outbreak of methicillin-resistant *Staphylococcus aureus* infection initiated by a surgeon carrier. *J Hosp Infect.*, 2001, 47, 104–109.

[25] Faibis, F; Laporte, C; Fiacre, A; Delisse, C; Lina, G; Demachy, MC; Botterel, F. An outbreak of methicillin-resistant *Staphylococcus aureus* surgical-site infections initiated by a healthcare worker with chronic sinusitis. *Infect Control Hosp Epidemiol* 2005; 26: 213–215.

[26] Eveillard, M; Martin, Y; Hidri, N; Boussougant, Y; Joly-Guillou ML. Carriage of methicillin-resistant *Staphylococcus aureus* among hospital employees: prevalence, duration, and transmission to households. *Infect Control Hosp Epidemiol*, 2004, 25, 114–20.

[27] Wolf, R; Lewis, D; Cochran, R; Richards, C. Nursing staff perceptions of methicillin-resistant *Staphylococcus aureus* and infection control in a long-term care facility. *J Am Med Dir Assoc.*, 2008, 9, 342-346.

[28] Albrich, WC; Harbarth, S. Health-care workers: source, vector, or victim of MRSA?. *Lancet Infect Dis*., 2008, 8, 289–301.

[29] Manzur, A; Vidal, M; Pujol, M; Cisnal, M; Hornero, A; Masuet, C; Pen˜a, C; Gudiol, F; Ariza, J. Predictive factors of meticillin resistance among patients with *Staphylococcus aureus* bloodstream infections at hospital admission. *J Hosp Infect*., 2007, 66, 135-141.

[30] Laupland, KB; Ross, T; Gregson, DB. *Staphylococcus aureus* bloodstream infections: risk factors, outcomes, and the influence of methicillin resistance in Calgary, Canada, 2000-2006. *J Infect Dis*., 2008, 198, 336-343.

[31] Vidal, PM; Trindade, PA; Garcia, TO; Pacheco, RL; Costa, SF; Reinert, C; Hiramatsu, K; Mamizuka, EM; Garcia, CP; Levin, AS. Differences between "classical" risk factors for infections caused by methicillin-resistant S*taphylococcus aureus* (MRSA) and risk factors for nosocomial bloodstream infections caused by multiple clones of the staphylococcal cassette chromosome mec type iv MRSA strain. *Infect Control Hosp Epidemiol*, 2009, 30, 139–145.

[32] Salzer, W. Antimicrobial-resistant gram-positive bacteria in PD peritonitis and the newer antibiotics used to treat them. *Perit Dial Int*., 2005, 25, 313-319.

[33] Witte, W; Cuny, C; Klare, I; Nübel, U; Strommenger, B; Werner, G. Emergence and spread of antibiotic-resistant Gram-positive bacterial pathogens. *Int J Med Microbiol*, 2008, 298, 365-77.

[34] Robinson, DA; Kearns, AM; Holmes, A; Morrison, D; Grundmann, H; Edwards, G; O'Brien, FG; Tenover, FC; McDougal, LK; Monk, AB; Enright, MC. Re-emergence of early pandemic *Staphylococcus aureus* as a community acquired methicillin-resistant clone. *Lancet*, 2005, 365, 1256–1258.

[35] Stapleton, PD; Peter, WT. Methicillin resistance in *Staphylococcus aureus*: mechanisms and modulation. *NIH Public Access*., 2002, 85, 57–72.

[36] Hartman, B; Tomasz, A. Altered Penicillin-Binding Proteins in Methicillin-Resistant Strains of *Staphylococcus aureus*. *Antimicrob Agents Chemother*, 1981, 19, 726-735.

[37] Deurenberg, RH; Vink, C; Kalenic, S; Friedrich, AW; Bruggeman, CA; Stobberingh, EE. The molecular evolution of methicillin-resistant *Staphylococcus aureus*. *Clin Microbiol Infect*., 2007, 13, 222-35

[38] Ito, T. Kuwahara, K. Hiramatsu, K. Staphylococcal cassette chromosome mec (SCC mec) analysis of MRSA. *Methods Mol Biol*., 2007, 391, 87-102.

[39] Malhotra-Kumar, S; Haccuria, K; Michiels, M; Ieven, M; Poyart, C; Hryniewicz, W; Goossens, H; MOSAR WP2 Study Team. Trends in Rapid Diagnostics for Methicillin-Resistant *Staphylococcus aureus* and Glycopeptide-Resistant *Enterococcus* Species. *J Clin Microbiol*, 2008, 46, 1577–1587.

[40] Farley, JE. Epidemiology, clinical manifestations, and treatment options for skin and soft tissue infection caused by community-acquired methicillin-resistant *Staphylococcus aureus*. *J Am Acad Nurse Pract*., 2008, 20, 85-92.

[41] Daum, RS; Ito, T; Hiramatsu, K; Hussain, F; Mongkolrattanothai, K; Jamklang, M; Boyle-Vavra, S. A novel methicillin-resistance cassette in community-acquired methicillin-resistant *Staphylococcus aureus* isolates of diverse genetic backgrounds. *J Infect Dis*., 2002, 186, 1344–1347.

[42] Baba, T; Takeuchi, F; Kuroda, M; Yuzawa, H; Aoki, K; Oguchi, A; Nagai, Y; Iwama, N; Asano, K; Naimi, T; Kuroda, H; Cui, L; Yamamoto, K, Hiramatsu, K. Genome and

virulence determinants of high virulence community-acquired MRSA. *Lancet*, 2002, 359, 1819–1827.

[43] Kondo, Y; Ito, T; Ma, XX; Watanabe, S; Kreiswirth, BN; Etienne, J; Hiramatsu, K. Combination of multiplex PCRs for staphylococcal cassette chromosome mec type assignment: rapid identification system for mec, ccr, and major differences in junkyard regions. *Antimicrob Agents Chemother*, 2007, 51, 264-74.

[44] Katayama, Y; Ito, T; Hiramatsu, K. Genetic organization of the chromosome region surrounding mecA in clinical staphylococcal strains: role of IS431-mediated mecI deletion in expression of resistance in mecA-carrying, lowlevel methicillin-resistant Staphylococcus haemolyticus. *Antimicrob Agents Chem.*, other 2001, 45, 1955–1963.

[45] Ito, T; Katayama, Y; Asada, K; Mori, N; Tsutsumimoto, K; Tiensasitorn, C; Hiramatsu, K. Structural comparison of three types of staphylococcal cassette chromosome mec integrated in the chromosome in methicillin-resistant *Staphylococcus aureus*. *Antimicrob Agents Chemother*, 2001, 45, 1323–1336.

[46] Ito, T; Okuma, K; Ma, XX; Yuzawa, H; Hiramatsu, K. Insights on antibiotic resistance of *Staphylococcus aureus* from its whole genome: genomic island SCC. *Drug Resist Update*, 2003, 6, 41–52.

[47] Shore, A; Rossney, AS; Keane, CT; Enright, MC; Coleman, DC. Seven novel variants of the staphylococcal chromosomal cassette mec in methicillin-resistant *Staphylococcus aureus* isolates from Ireland. *Antimicrob Agents Chemother*, 2005, 49, 2070–2083.

[48] Chongtrakool, P; Ito, T; Ma, XX; Kondo, Y; Trakulsomboon, S; Tiensasitorn, C; Jamklang, M; Chavalit, T; Song, JH; Hiramatsu K. Staphylococcal cassette chromosome mec (SCCmec) typing of methicillin-resistant *Staphylococcus aureus* strains isolated in 11 Asian countries: a proposal for a new nomenclature for SCCmec elements. *Antimicrob Agents Chemother*, 2006, 50, 1001–1012.

[49] Wu, SW; De Lencastre, H; Tomasz, A. Recruitment of the mecA gene homologue of Staphylococcus sciuri into a resistance determinant and expression of the resistant phenotype in *Staphylococcus aureus*. *J Bacteriol*, 2001, 183, 2417–2424.

[50] Wielders, CL; Vriens, MR; Brisse, S; de Graaf-Miltenburg, LA; Troelstra, A; Fleer, A; Schmitz, FJ; Verhoef, J; Fluit, AC. In-vivo transfer of mecA DNA to *Staphylococcus aureus*. *Lancet*, 2001, 357, 1674–1675.

[51] Hanssen, AM; Kjeldsen, G; Sollid, JU. Local variants of staphylococcal cassette chromosome mec in sporadic methicillin-resistant *Staphylococcus aureus* and methicillin-resistant coagulase-negative staphylococci: evidence of horizontal gene transfer? *Antimicrob Agents Chemother*, 2004, 48, 285–296.

[52] Corkill, JE; Anson, JJ; Griffiths, P; Anthony, Hart, C. Detection of elements of the staphylococcal cassette chromosome (SCC) in a methicillin-susceptible (mecA gene negative) homologue of a fucidin-resistant MRSA. *J Antimicrob Chemother*, 2004, 54, 229–231.

[53] Mongkolrattanothai, K; Boyle, S; Murphy, TV; Daum, RS. Novel non-mecA-containing staphylococcal chromosomal cassette composite island containing pbp4 and tagF genes in a commensal staphylococcal species: a possible reservoir for antibiotic resistance islands in *Staphylococcus aureus*. *Antimicrob Agents Chemother*, 2004, 48, 1823–1836

[54] Struelens, MJ; Bax, R; Deplano, A; Quint, WG; Van Belkum, A. Concordant clonal delineation of methicillin-resistant *Staphylococcus aureus* by macrorestriction analysis

and polymerase chain reaction genome fingerprinting. *J Clin Microbiol*, 1993, 31, 1964–1970.

[55] Enright, MC; Day, NP; Davies, CE; Peacock, SJ; Spratt, BG. Multilocus sequence typing for characterization of methicillin-resistant and methicillin-susceptible clones of *Staphylococcus aureus*. *J Clin Microbiol.*, 2000, 38, 1008–1015.

[56] Enright, MC; Robinson, DA; Randle, G; Feil, EJ; Grundmann, H; Spratt, BG. The volutionary history of methicillin-resistant *Staphylococcus aureus* (MRSA). *Proc Natl Acad Sci USA*, 2002, 99, 7687–7692.

[57] Robinson, DA; Enright, MC. Evolutionary models of the emergence of methicillin-resistant *Staphylococcus aureus*. *Antimicrob Agents Chemother*, 2003, 47, 3926–3934.

[58] Oliveira, DC; De Lencastre, H. Multiplex PCR strategy for rapid identification of structural types and variants of the mec element in methicillin-resistant *Staphylococcus aureus*. *Antimicrob Agents Chemother*, 2002, 46, 2155–2161.

[59] Okuma, K; Iwakawa, K; Turnidge, JD; Grubb, WB; Bell, JM; O'Brien, FG; Coombs, GW; Pearman, JW; Tenover, FC; Kapi, M; Tiensasitorn, C; Ito, T; Hiramatsu, K. Dissemination of new methicillin-resistant *Staphylococcus aureus* clones in the community. *J Clin Microbiol*, 2002, 40, 4289–4294.

[60] Frénay, HM; Bunschoten, AE; Schouls, LM; van Leeuwen, WJ; Vandenbroucke-Grauls, CM; Verhoef, J; Mooi, FR. Molecular typing of methicillin-resistant *Staphylococcus aureus* on the basis of protein A gene polymorphism. *Eur J Clin Microbiol Infect Dis*., 1996, 15, 60–64.

[61] Harmsen, D; Claus, H; Witte, W; Rothgänger, J; Claus, H; Turnwald, D; Vogel, U. Typing of methicillin-resistant *Staphylococcus aureus* in a university hospital setting by using novel software for spa repeat determination and database management. *J Clin Microbiol*, 2003, 41, 5442–5448.

[62] Strommenger, B; Kettlitz, C; Weniger, T; Harmsen, D; Friedrich, A; Witte, W. Assignment of *Staphylococcus aureus* isolates to groups by spa typing SmaImacrorestriction analysis and multi locus sequence typing. *J Clin Microbiol*, 2006, 44, 2533–2540.

[63] EARSS, 2005. Annual Report "http://www.eurosurveillance.org/".

[64] Herold, BC; Immergluck, LC; Maranan, MC; Lauderdale, DS; Gaskin, RE; Boyle-Vavra, S; Leitch, CD; Daum, RS. Community-acquired methicillin-resistant *Staphylococcus aureus* in children with no identified predisposing risk. *JAMA*, 1998, 279, 593–598.

[65] Turnridge, JD; Bell, JM. Methicillin-resistant Staphylococcal aureus evolution in Australia over 35 years. *Microb Drug Resist*., 2000, 6, 223–229.

[66] Chambers, HF. The changing epidemiology of *Staphylococcus aureus*? *Emerg Infect Dis*., 2001, 7, 178–182.

[67] Naimi, TS; LeDell, KH; Boxrud, DJ; Groom, AV; Steward, CD; Johnson, SK; Beser, JM; O'Boyle, C; Danila, RN; Cheek, JE; Osterholm, MT; Moore, KA; Smith, KE. Epidemiology and clonality of community-acquired methicillin-resistant *Staphylococcus aureus* in Minnesota, 1996–1998. *Clin Infect Dis*., 2001, 33, 990–996.

[68] Ito, T; Ma, XX; Takeuchi, F; Okuma, K; Yuzawa, H; Hiramatsu, K. Novel type V staphylococcal cassette chromosome mec driven by a novel cassette chromosome recombinase ccrC. *Antimicrob Agents Chemother*, 2004, 48, 2637–2651.

[69] Chung, M; Dickinson, G; De Lencastre, H; Tomasz, A. International clones of methicillin-resistant *Staphylococcus aureus* in two hospitals in Miami, Florida. *J Clin Microbiol*, 2004, 42, 542–547.

[70] Wannet, WJ; Spalburg, E; Heck, ME; Pluister, GN; Tiemersma, E; Willems, RJ; Huijsdens, XW; de Neeling, AJ; Etienne, J. Emergence of virulent methicillin-resistant *Staphylococcus aureus* strains carrying Panton–Valentine leucocidin genes in The Netherlands. *J Clin Microbiol*, 2005, 43, 3341–3345.

[71] De A Trindade, P; Pacheco, RL; Costa, SF; Rossi, F; Barone, AA; Mamizuka, EM; Levin, AS. Prevalence of SCCmec type IV in nosocomial bloodstream isolates of methicillin-resistant *Staphylococcus aureus*. *J Clin Microbiol*, 2005, 43, 3435–3437.

[72] Chambers, HF. Community-associated MRSA-resistance and virulence converge. *N Engl J Med.*, 2005, 352, 1485–1487.

[73] Witte, W; Strommenger, B; Cuny, C; Heuck, D; Nübel, U. Methicillin-resistant *Staphylococcus aureus* containing the Panton-Valentine leukocidin gene in Germany in 2005 and 2006. *J Antimicrob Chemother*, 2007, 60, 1258–1263.

[74] Martins, A; Cunha, ML. Methicillin resistance *Microbiol Immunol*, 2007, 51, 787-795.

[75] Bell, JM; Turnidge, JD. High prevalence of oxacillinresistant *Staphylococcus aureus* isolates from hospitalized patients in Asia-Pacific and South Africa: results from SENTRY antimicrobial surveillance program, 1998–1999. *Antimicrob Agents Chemother,* 2002, 46, 879–881.

[76] Diekema, DJ; Pfaller, MA; Schmitz, FJ; Smayevsky, J; Bell, J; Jones, RN; Beach, M; SENTRY Partcipants Group. Survey of infections due to Staphylococcus species: frequency of occurrence and antimicrobial susceptibility of isolates collected in the United States, Canada, Latin America, Europe, and the Western Pacific region for the SENTRY Antimicrobial Surveillance Program, 1997–1999. *Clin Infect Dis.*, 2001, 32 (suppl 2), S114–S132.

[77] Tiemersma, EW; Bronzwaer, SL; Lyytikäinen, O; Degener, JE; Schrijnemakers, P; Bruinsma, N; Monen, J; Witte, W; Grundman, H; European Antimicrobial Resistance Surveillance System Participants. Methicillin-resistant *Staphylococcus aureus* in Europe, 1999–2002. *Emerg Infect Dis.*, 2004, 10, 1627–1634.

[78] Borg, MA; De Kraker, M; Scicluna, E; van de Sande-Bruinsma, N; Tiemersma, E; Monen, J; Grundmann, H; ARMed Project Members and Collaborators. Prevalance of methicillin-resistant *Staphylococcus aureus* (MRSA) in invasive isolates from southern and eastern Mediterranean countries. *J Antimicrob Chemother*, 2007, 60, 1310-5.

[79] Tacconelli, E; Angelis, GD; Cataldo, MA; Pozzi, E; Cauda, R. Does antibiotic exposure increase the risk of methicillin-resistant *Staphylococcus aureus* (MRSA) isolation? A systematic review and meta-analysis. *J Antimicrob Chemother*, 2008, 61, 26–38.

[80] Klevens, RM; Morrison, MA; Nadle, J; Petit, S; Gershman, K; Ray, S; Harrison, LH; Lynfield, R; Dumyati, G; Townes, JM; Craig, AS; Zell, ER; Fosheim, GE; McDougal, LK; Carey, RB; Fridkin, SK; Active Bacterial Core surveillance (ABCs) MRSA Investigators. Invasive methicillin-resistant *JAMA*, 2007, 298, 1763-1771.

[81] Salgado, CD; Farr, BM; Calfee, DP. Community-acquired methicillin-resistant *Staphylococcus aureus*: a meta-analysis of prevalence and risk factors. *Clin Infect Dis.*, 2003, 36, 131–139.

[82] Hiramatsu, K; Hanaki, H; Ino, T; Yabuta, K; Oguri, T; Tenover, FC: Methicillin-resistant *Staphylococcus aureus* clinical strain with reduced vancomycin susceptibility. *J Antimicrob Chemother*, 1997, 40, 135-6.

[83] French, GL. Bactericidal agents in the treatment of MRSA infections-the potential role of daptomycin. *J Antimicrob Chemother*, 2006, 58, 1107-1117.

[84] Appelbaum, PC. Reduced glycopeptide susceptibility *Int J Antimicrob Agents*, 2007, 30, 398-408.

[85] Tsiodras, S; Gold, HS; Sakoulas, G; Eliopoulos, GM; Wennersten, C; Venkataraman, L; Moellering Jr, RC; Ferraro MJ. Linezolid resistance in a clinical isolate of *Staphylococcus aureus*. *Lancet*, 2001, 358, 207–208.

[86] Wilson, P; Andrews, JA; Charlesworth, R; Walesby, R; Singer, M; Farrell, DJ; Robbins, M. Linezolid resistance in clinical isolates of *Staphylococcus aureus*. *J Antimicrob Chemother*, 2003, 51, 186–188.

[87] Hershberger, E; Donabedian, S; Konstantinou, K; Zervos, MJ. Quinupristin-Dalfopristin Resistance in Gram-Positive Bacteria: Mechanism of Resistance and Epidemiology. *Clin Infect Dis*., 2004, 38, 92–98

[88] Lentino, JR; Narita, M; Yu, VL. New antimicrobial agents as therapy for resistant gram-positive cocci. *Eur J Clin Microbiol Infect Dis*., 2008, 27, 3-15.

[89] Lowy, FD. Antimicrobial resistance: the example of *Staphylococcus aureus*. *J Clin Invest*, 2003, 111, 1265-1273.

[90] Hooper, DC. Fluoroquinolone resistance *Lancet Infect Dis*., 2002, 2, 530-538.

[91] Schmitz, FJ; Hofmann, B; Hansen, B; Scheuring, S; Lückefahr, M; Klootwijk, M; Verhoef, J; Fluit, A; Heinz, HP; Köhrer, K; Jones, ME. Relationship between ciprofloxacin, ofloxacin, levofloxacin, sparfloxacin and moxifloxacin (BAY 12-8039) MICs and mutations in grlA, grlB, gyrA and gyrB in 116 unrelated clinical isolates of *Staphylococcus aureus*. *J Antimicrob Chemother*, 1998, 41, 481-484.

[92] Patel, JB; Jevitt, LA; Hageman, J; McDonald, LC; Tenover, FC. An association between reduced susceptibility to daptomycin and reduced susceptibility to vancomycin in *Staphylococcus aureus*. *Clin Infect Dis*., 2006, 42, 1652-1653.

[93] Fowler,VG Jr; Boucher, HW; Corey, GR; Abrutyn, E; Karchmer, AW; Rupp, ME; Levine, DP; Chambers, HF; Tally, FP; Vigliani, GA; Cabell, CH; Link, AS; DeMeyer, I; Filler, SG; Zervos, M; Cook, P; Parsonnet, J; Bernstein, JM; Price; Forrest, GN; Fätkenheuer, G; Gareca, M; Rehm, SJ; Brodt, HR; Tice; Cosgrove, SE; *S. aureus* Endocarditis and Bacteremia Study Group. Daptomycin versus standard therapy for bacteremia and endocarditis caused by *Staphylococcus aureus*. *N Engl J Med*., 2006, 355, 653–65.

[94] Marty, FM; Yeh, WW; Wennersten, CB; Venkataraman, L; Albano, E; Alyea, EP; Gold, HS; Baden, LR; Pillai, SK. Emergence of a clinical daptomycin-resistant *Staphylococcus J Clin Microbiol*, 2006, 44, 595-597.

[95] Skiest, DJ. Treatment failure resulting from resistance of *Staphylococcus aureus* to daptomycin. *J Clin Microbiol*, 2006, 44, 655-656.

[96] Cottagnoud, P. Daptomycin: a new treatment for insidious infections due to gram-positive pathogens. *Swiss Med Wkly*, 2008, 138, 93–99.

[97] Gordon, RJ; Lowy, FD. Pathogenesis of methicillin-resistant *Staphylococcus aureus* infection. *Clin Infect Dis*., 2008, 46 (Suppl 5), 350-359.

[98] Tristan, A; Ferry, T; Durand, G; Dauwalder, O; Bes, M; Lina, G; Vandenesch, F; Etienne, J. Virulence determinants in community and hospital meticillin-resistant *Staphylococcus aureus*. J Hosp Infect 2007; 65 (Suppl 2), 105-109
[99] Decker, CF. Pathogenesis of MRSA infections. *Dis Mon.*, 2008, 54, 774-779.
[100] Kluytmans, J; Van Belkum, A; Verbrugh, H. Nasal carriage of *Staphylococcus aureus*: epidemiology, underlying mechanism, and associated risks. *Clin Microbiol Rev.*, 1997, 10, 505-520.
[101] Meek, C. Isolate patients, screen staff to fight MRSA. *Can Med Assoc J.*, 2004, 171, 1158.
[102] Patti JM; Allen, BL; McGavin, MJ; Hook, M. MSCRAMM-mediated adherence of microorganisms to host tissues. *Annu Rev Microbiol*, 1994, 48, 585–617.
[103] Donlan, RM; Costerton, JW. Biofilms: survival mechanisms of clinically relevant microorganisms. *Clin Microbiol Rev.*, 2002, 15, 167–193.
[104] O'Riordan, K Lee JC. *Staphylococcus aureus* capsular polysaccharides. *Clin Microbiol Rev.*, 2004, 17, 218–234.
[105] Novick, RP. Autoinduction and signal transduction in the regulation of staphylococcal virulence. *Mol Microbiol.*, 2003, 48, 1429–1449.
[106] Lina G, Piemont Y, Godail-Gamot F, Bes, M; Peter, MO; Gauduchon, V; Vandenesh, F; Etienne, J. Involvement of Panton-Valentine leukocidin–producing *Staphylococcus aureus* in primary skin infections and pneumonia. *Clin Infect Dis.*, 1999, 29, 1128–1132.
[107] Dinges, MM; Orwin, PM; Schlievert, PM. Exotoxins of *Staphylococcus aureus. Clin Microbiol Rev.*, 2000, 13, 16-34.
[108] Diep, AD; Otto, M. The role of virulence determinants in community-associated MRSA pathogenesis. *Trends in Micobiol*, 2008, 16, 361-369.
[109] Ward, PD; Turner, WH. Identification of staphylococcal Panton-Valentine leukocidin as a potent dermonecrotic toxin. *Infect Immun*, 1980, 28, 393-397.
[110] Cribier, B; Prevost, G; Couppie, P; Finck-Barbancon, V; Grosshans, E; Piemont, Y. *Staphylococcus aureus* leukocidin: a new virulence factor in cutaneous infections? An epidemiological and experimental study. *Dermatology*, 1992, 185, 175-180.
[111] Genestier, AL; Michallet, MC; Prevost, G; Bellot, G; Chalabreysse, L; Peyrol, S; Thivolet, F; Etienne, J; Lina, G; Vallette, FM; Vandenesch, F; Genestier, F. *Staphylococcus aureus* Panton-Valentine leukocidin directly targets mitochondria and induces Bax-independent apoptosis of human neutrophils. *J Clin Invest.*, 2005, 115, 3117–3127.
[112] Labandeira-Rey, M; Couzon, F; Boisset, S, Brown, EL; Bes, M; Benito, Y; Barbu, EM; Vazquez, V; Höök, M; Etienne, J; Vandenesch, F; Bowden, MG. *Staphylococcus aureus* Panton-Valentine leukocidin causes necrotizing pneumonia. *Science*, 2007, 315, 1130–1133.
[113] Gillet, Y; Issartel, B; Vanhems, P; Fournet, JC; Lina, G; Bes, M; Vandenesch, F; Piémont, Y; Brousse, N; Floret, D; Etienne, J. Association between *Staphylococcus aureus* strains carrying gene for Panton-Valentine leukocidin and highly lethal necrotising pneumonia in young immunocompetent patients. *Lancet*, 2002, 359, 753-759.

[114] Dufour, P; Gillet, Y; Bes, M; Lina, G; Vandenesch, F; Floret, D; Etienne, J; Richet, H. Community-Acquired Methicillin-Resistant *Staphylococcus aureus* Infections in France: Emergence of a Single Clone That Produces Panton-Valentine Leukocidin. *Clin Infect Dis*., 2002, 35, 819- 824.

[115] Ladhani, S; Joannou, CL; Lochrie, DP; Evans, RW; Poston, SM. Clinical, microbial, and biochemical aspects of the exfoliative toxins causing staphylococcal scalded-skin syndrome. *Clin Microbiol Rev*., 1999, 12, 224-242.

[116] Yamaguchi T, Yokota Y, Terajima J, Hayashi, T; aepfelbacher, M; Ohara, M; Komatsuzawa, H; Watanabe, H; Sugai, M. Clonal association of *Staphylococcus aureus* causing bullous impetigo and the emergence of new methicillin-resistant clonal groups in Kansai district in Japan. *J Infect Dis*., 2002, 185, 1511-1516.

[117] Durand, G; Bes, M; Meugnier, H;. Enright, MC; Forey, F; Liassine, N; Wenger, A; Kikuchi, K; Lina, G; Vandenesch, F; Etienne, J. Detection of new methicillin-resistant *Staphylococcus aureus* clones containing the toxic shock syndrome toxin 1 gene responsible for hospital- and community-acquired infections in France. *J Clin Microbiol*, 2006, 44, 847-853.

[118] Ferry, T; Thomas, D; Genestier, AL; Bes, M; Lina, G; Vandenesch, F; Etienne, J. Comparative prevalence of superantigen genes in *Staphylococcus aureus* isolates causing sepsis with and without septic shock. *Clin Infect Dis*., 2005, 41, 771- 777.

[119] Wang, R; Braughton, KR; Kretschmer, D; Bach, THL; Queck, SY; Li, M; Kennedy, AD; Dorward, DW; Klebanoff, SJ; Peschel, A; DeLeo, FR; Otto, M. Identification of novel cytolytic peptides as key virulence determinants for community-associated MRSA. *Nat Med*., 2007, 13, 1510-1514.

[120] Stryjewski, ME; Chambers, HF. Skin and soft-tissue infections caused by community-acquired methicillin-resistant *Staphylococcus aureus*. *Clin Infect Dis*., 2008, 46(S5), 368- 377.

[121] Whitman, TJ. Community-associated methicillin-resistant *Staphylococcus aureus* skin and soft tissue infections. *Dis Mon*., 2008, 54, 780-786.

[122] Weber, JT. Community-associated methicillin-resistant *Staphylococcus aureus*. *Clin Infect Dis*., 2005, 41(Suppl), 269-272.

[123] Fridkin, SK; Hageman, J; Morrison, M; Sanza, LT; Como-Sabetti, K; Jernigan, JA; Harriman, K; Harrison, LH, Lynfield, R;. Farley, MM. Methicillin-resistant Staphylococcus aureus disease in three communities. *N Engl J Med*., 2005, 352, 1436-1444.

[124] Ruhe, JJ; Smith, N; Bradsher, RW; Menon, A. Community-onset methicillin-resistant *Staphylococcus aureus* skin and soft-tissue infections: impact of antimicrobial therapy on outcome. *Clin Infect Dis*., 2007, 44, 777-784.

[125] Buck, JM; Como-Sabetti, K; Harriman, KH; Danila, RN; Boxrud, DJ; Glennen, A; Lynfield, R. Community-associated methicillin-resistant *Staphylococcus aureus,* Minnesota, 2000–2003. *Emerg Infect Dis*., 2005, 11, 1532-1538.

[126] Pannaraj, PS; Hulten, KG; Gonzalez, BE; Mason, EO Jr; Kaplan, SL. Infective pyomyositis and myositis in children in the era of community-acquired, methicillin-resistant *Staphylococcus aureus* infection. *Clin Infect Dis*., 2006, 43, 953-960.

[127] Miller, LG; Perdreau-Remington, F; Rieg, G; Mehdi, S; Perlroth, J; Bayer, A; Tang, AW; Phung, TO; Spellberg, B. Necrotizing fasciitis caused by community-associated

methicillin-resistant *Staphylococcus aureus* in Los Angeles. *N Engl J Med.*, 2005, 352, 1445-1453.

[128] Ganesan, A. Methicillin-resistant *Staphylococcus aureus* bacteremia and pneumonia. *Dis Mon.*, 2008, 54, 787-792.

[129] Rello, J; Torres, A; Ricart, M; Valles, J; Gonzalez, J; Artigas, A; Rodriguez- Roisin, R; Ventilator-associated pneumonia by *Staphylococcus aureus*. Comparison of methicillin-resistant and methicillin-sensitive episodes. *Am J Respir Crit Care Med.*, 1994, 150, 1545-1549.

[130] Shetty, AK; Kumar, A. Osteomyelitis in adolescents. *Adolesc Med State Art Rev.*, 2007, 18, 79-94.

[131] Carek, PJ; Dickerson, LM; Sack, JL. Diagnosis and Management of Osteomyelitis. *American Family Physcian*, 2001; 63, 2413-2420.

[132] Jacobsson, G; Dashti, S; Wahlberg, T; Andersson, R. The epidemiology of and risk factors for invasive *Staphylococcus aureus* infections in western Sweden. *Scand J Infect Dis.*, 2007, 39, 6-13.

[133] Frederiksen, MS; Espersen, F; Frimodt-Møller, N; Jensen, AG; Larsen, AR; Pallesen, LV; Skov, R; Westh, H; Skinhøj, P; Benfield, T. Changing epidemiology of pediatric *Staphylococcus aureus* bacteremia in Denmark from 1971 through 2000. *Pediatr Infect Dis J.*, 2007, 26, 398-405.

[134] Benfield, T; Espersen, F; Frimodt-Møller, N; Jensen, AG; Larsen, AR; Pallesen, LV; Skov, R; Westh, H; Skinhøj, P. Increasing incidence but decreasing in-hospital mortality of adult *Staphylococcus aureus* bacteraemia between 1981 and 2000. *Clin Microbiol Infect.*, 2007, 13, 257-263.

[135] Libert, M; Elkholti, M; Massaut, J; Karmali, R; Mascart, G; Cherifi, S. Risk factors for methicillin resistance and outcome of *Staphylococcus aureus* bloodstream infection in a Belgian university hospital. *J Hosp Infect.*, 2008, 68, 17-24.

[136] Cosgrove, SE; Sakoulas, G; Perencevich, EN; Schwaber, MJ; Karchmer, AW; Carmeli, Y. Comparison of mortality associated with methicillin-resistant and methicillin-susceptible *Staphylococcus aureus* bacteremia: a meta-analysis. *Clin Infect Dis.*, 2003, 36, 53-9.

[137] Johnson, M; Decker, CF. Antimicrobial agents in treatment of MRSA infections. *Dis Mon.*, 2008, 54, 793-800.

[138] Harbarth, S; Samore, MH. Interventions to control MRSA: high time for time-series analysis? *J Antimicrob Chemother*, 2008, 62, 431–433.

[139] Gould, I. M. Who's winning the war? *Antimicrob Agents Chemother*, 2008, 52, 2156-2162.

[140] Robinson, JO; Person, JC; Christiansen, KJ; Coombs, GW; Murray, RJ. Community-associated versus healthcare-associated methicillin-resistant *Staphylococcus aureus* bacteremia: a 10-year retrospective review. *Eur J Clin Microbiol Infect Dis.*, 2008, (Epub ahead of print).

[141] Drekonja, DM; Traynor, LM; DeCarolis, DD; Crossley, KB; Johnson, JR. Treatment of non–life-threatening methicillin-resistant Staphylococcus aureus infections with alternative antimicrobial agents: a 2-year retrospective review. *Diagn Microbiol Infect Dis.*, 2008, 63, 201-207.

[142] LaPlante, KL; Leonard, SN; Andes, DR; Craig, WA; Rybak, MJ. Activities of Clindamycin, daptomycin,doxycycline, linezolid, trimethoprim-sulfamethoxazole, and vancomycin against community-associated methicillin-resistant *Staphylococcus aureus* with inducible clindamycin resistance in murine thigh infection and in vitro pharmacodynamic models. *Antimicrob Agents Chemother*, 2008, 52, 2156-2162.

[143] Napolitano, LM. Early appropriate parenteral antimicrobial treatment of complicated skin and soft tissue infections caused by methicillin-resistant *Staphylococcus aureus. Surg Infect*, 2008, 9, S-17-27.

[144] Le, J; Lieberman, JM. Management of community-associated against methicillin-resistant *Staphylococcus aureus* agents infections. *Pharmacotherapy*, 2006, 26, 1758-1770.

[145] Kaka, AS; Rueda, AM; Shelburne, SA; Hulten, K; Hamil, RJ; Musher, DM. Bactericidal activity of orally available methicillin-resistant *Staphylococcus aureus. J Antimicrob Chemother*, 2006, 58, 680-683.

[146] Biedenbach, DJ; Bell, JM; Sader, HS; Fritsche, TR; Jones, RN; Turnidge, JD. Antimicrobial susceptibility of Gram-positive bacterial isolates from the Asia–Pacific region and an in vitro evaluation of the bactericidal activity of daptomycin, vancomycin, and teicoplanin: a SENTRY Program Report (2003–2004). *Int J Antimicrob Agents*, 2007, 30, 143–149.

[147] Proctor, RA. Role of folate antagonists in the treatment of methicillin-resistant *Staphylococcus aureus* infections. *Clin Infect Dis*., 2008, 46, 584-593.

[148] Han, LL; McDougal, LK; Gorwitz, RJ; Mayer, KH; Patel, JB; Sennott, JM; Fontana, JL. High frequencies of clindamycin and tetracycline *J Clin Microbiol*, 2007, 45, 1350-1352.

[149] Ruhe, JJ; Monson, T; Bradsher, RW; Menon, A. Use of longacting tetracyclines for methicillin-resistant *Staphylococcus aureus* infections: Case series and review of the literature. *Clin Infect Dis*., 2005, 40, 1429–1434.

[150] Po-Ren, H; Wen-Huei, C; Lee-Jene, T; Kwen-Tay, L. Nosocomial infections due to methicillin-resistant *Staphylococcus aureus* and vancomycin-resistant enterococci at a university hospital in Taiwan from 1991 to 2003: resistance trends, antibiotic usage and in vitro activities of newer antimicrobial agents. *Int J Antimicrob Agents*, 2005, 26, 43–49.

[151] Daum, RS. Skin and soft-tissue infections caused by methicillin-resistant *Staphylococcus aureus. N Engl J Med*., 2007, 357, 380–390.

[152] Kaplan, SL; Hulten, KG; Gonzalez, BE; Hammerman, WA; Lamberth, L; Versalovic, J; Mason, EO. Three-year surveillance of community-acquired *Staphylococcus aureus* infections in children. *Clin Infect Dis*., 2005, 40, 1785–1791.

[153] LaPlante, KL; Rybak, MJ; Amjad, M; Kaatz, GW. Antimicrobial susceptibility and staphylococcal chromosomal cassette mec type in community- and hospital-associated methicillin-resistant *Staphylococcus aureus*. *Pharmacotherapy*, 2007, 27, 3–10.

[154] Siberry, GK; Tekle; T, Carroll, K; Dick, J. Failure of clindamycin treatment of methicillin-resistant Staphylococcus aureus expressing inducible clindamycin resistance in vitro. *Clin Infect Dis*., 2003, 37, 1257–1260.

[155] Pan, A; Lorenzotti, S; Zoncada, A. Registered and investigational drugs *Recent Pat Anti-Infect Drug Discov*., 2008, 3, 10-33.

[156] Maclayton, DO; Hall, RG. Pharmacologic treatment options for nosocomial pneumonia involving methicillin-resistant *Staphylococcus aureus*. *Ann Pharmacother*, 2007, 41, 234-244.

[157] Burkhardt, O; Pletz, MW; Mertgen, CP; Welte, T. Linezolid-the first oxazolidinone in the treatment of nosocomial MRSA pneumonia. *Recent Pat Anti-Infect Drug Discov*, 2007, 2, 123-130.

[158] Bierbaum, G; Fuchs, K; Lenz, W; Szekat, C; Sahl, HG. Presence of *Staphylococcus aureus* with reduced susceptibility to vancomycin in Germany. *Eur J Clin Microbiol Infect Dis*., 1999, 18, 691-696.

[159] Kim, MN; Hwang, SH; Pyo, YJ; Mun, HM; Pai, CH. Clonal spread of *Staphylococcus* heterogenously resistant to vancomycin in a university hospital in Korea. *J Clin Microbiol*., 2002, 40, 1376-80.

[160] Awad, SS; Elhabash, SI; Lee, L; Farrow, B; Berger, DH. Increasing incidence of methicillin-resistant *Staphylococcus aureus* skin and soft-tissue infections: Reconsideration of empiric antimicrobial therapy. *Am J Surg*., 2007, 194, 606–610.

[161] Schmidt-Ioanas, M; de Roux, A; Lode, H. New antibiotics for the treatment of severe staphylococcal infection in the critically ill patient. *Curr Opin Crit Care*, 2005, 11, 481-486.

[162] Anstead, GM; Quinones-Nazario, G; Lewis, JS. Treatment of infections caused by resistant *Staphylococcus aureus*. *Methods Mol Biol*., 2007, 391, 227-258.

[163] Jones, RN; Fritsche, TR; Sader, HS; Ross, JE. LEADER surveillance program results for 2006: An activity and spectrum analysis of linezolid using clinical isolates from the United States (50 medical centers). *Diagn Microbiol Infect Dis*., 2007, 59, 309–317.

[164] Rubinstein, E; Prokocimer, P; Talbot, GH. Safety and tolerability of quinupristin/dalfopristin: administration *J Antimicrob Chemother*, 1999, 44 Suppl A, 37-46.

[165] Livermore, DM. Future directions with daptomycin. *J Antimicrob Chemother*, 2008, 62 Suppl. 3, iii41-iii49.

[166] Leonard, SN; Rybak, MJ. Evaluation of vancomycin and daptomycin against methicillin-resistant *Staphylococcus aureus* and heterogeneously vancomycin-intermediate *S. aureus* in an in vitro pharmacokinetic/pharmacodynamic model with simulated endocardial vegetations. *J Antimicrob Chemother*, 2009, 63, 155-160.

[167] Livermore, DM. Tigecycline: what is it, and where should it be used? *J Antimicrob Chemother*, 2005, 56, 611–614.

[168] Loffler, CA; Macdougall, C. Update on prevalence and treatment of methicillin-resistant *Staphylococcus aureus* infections. *Expert Rev Anti Infect Ther*, 2007, 5, 961-981.

[169] Boylan, CJ; Campanale, K; Iverson, PW; Phillips, DL; Zeckel, ML; Parr, TR Jr. Pharmacodynamics of oritavancin (LY333328) in an neutropenic-mouse thigh model of *Staphylococcus aureus* infection. *Antimicrob Agents Chemother*, 2003, 47, 1700-1706.

[170] Mercier, RC; Hrebickova, L. Oritavancin: a new avenue for resistant Gram-positive *Expert Rev Anti Infect Ther*., 2005, 3, 325-332.

[171] Van Bambeke, F; Carryn, S; Seral, C; Chanteux, H; Tyteca, D; Mingeot-Leclercq, MP; Tulkens, PM. Cellular pharmacokinetics and pharmacodynamics of oritavancin (LY333328) glycopeptide in a model of J774 mouse macrophages, *Antimicrob Agents Chemother*, 2004, 48, 2853-2860.

[172] Cosgrove, SE; Fowler Jr, VG. Management of methicillin-resistant *Staphylococcus aureus* bacteremia. *Clin Infect Dis*., 2008, 46, S386-S393.

[173] Gander, S; Kinnaird, A; Finch, R. Telavancin: in vitro activity against staphylococci in a biofilm model, *J Antimicrob Chemother*, 2005, 56, 337-43.

[174] Stryjewski, ME; Graham, DR; Wilson, SE; O'Riordan, W; Young, D; Lentnek, A; Ross; DP; Fowler, VG; Hopkins, A; Friedland, HD; Barriere, SL; Kitt, MM; Corey, GR. Telavancin versus vancomycin Assessment of Telavancin in Complicated Skin and Skin-Structure infections Study. *Clin Infect Dis*., 2008, 46, 1683-1693.

[175] Raghavan, M; Linden, PK. Newer treatment options for skin and soft tissue infections. *Drugs*, 2004, 64, 1621-1642.

[176] Malabarba, A; Goldstein, BP. Origin, structure and activity in vitro and in vivo of dalbavancin. *J Antimicrob Chemother*, 2005, 55 (Supp 2):S15-20.

[177] Jones, ME. In-vitro profile of a new beta-lactam, ceftobiprole, with activity against methicillin-resistant *Staphylococcus aureus. Clin Micobiol Infect*, 2007, 13 Suppl 2, 17-24.

[178] Zhanel, GG; Lam, A; Schweizer, F; Thomson, K; Walkty, A; Rubinstein, E; Gin, AS; Hoban, DJ; Noreddin, AM; Karlowsky, JA. Ceftobiprole: a review of a broad-spectrum *Am J Clin Dermatol*, 2008, 9, 245-254.

In: Methicillin-Resistant Staphylococcus Aureus...
Editor: Charles L. Kolendi
ISBN: 978-1-60741-398-1

Chapter 8

MULTIDRUG-RESISTANT *ACINETOBACTER BAUMANNII* INFECTIONS: CHARACTERISTICS AND TREATMENT OPTIONS

***A. Rodríguez-Guardado*[1]*, A. Blanco*[2]*, F. Pérez*[3]*, M. Lantero*[3]*, M. Álvarez Vega*[4]*, J. M. Torres*[4] *and J. A. Cartón*[1]**

[1]Infectious Diseases Unit
[2]Intensive Care Unit
[3]Microbiology Unit
[4]Neurosurgery Unit
Hospital Universitario Central de Asturias. Spain

INTRODUCTION

Interest in *Acinetobacter spp.* has been growing for the past 30 years. One of the main reasons for the present increased interest in this genus is the emergence of multiresistant strains—some of which are pan-resistant to antibiotics—that suddenly cause an outbreak of infection [1].

Multidrug-resistant *Acinetobacter baumannii* is a rapidly emerging pathogen in the health-care setting, where it causes infections that include bacteraemia, pneumonia, meningitis, urinary tract infection, and wound infection. The mortality associated with these infections is high. The crude mortality rate associated with bacteremia is approximately 52% and that associated with pneumonia ranges from 23% to 73% [2,3]. It is recognized to be among the most difficult antimicrobial-resistant gram negative bacilli to control and treat due to the organism's ability to survive under a wide range of environmental conditions and to persist for extended periods of time on surfaces—which make it a frequent cause of outbreaks of infection and an endemic, health-care-associated pathogen. Recently, an increased rate of infections caused by *A.baumanii* strains resistant to the antibiotics traditionally used in this therapy has been reported [3-5]. This is an especially severe event in some infections like post-surgical meningitis because the election of the antibiotic depends not only on the sensitivity of *A.baumanii* but also on the penetrability of the chosen antibiotic through the blood-brain barrier. This multidrug-resistance complicates the treatment of these infections and makes imperative the search for new agents and the return of old drugs as we consider the optimal treatment of these multidrug-resistant organisms.

The focus of this review is to summarize the current state of knowledge regarding *Acinetobacter baumannii* in relation to this role, including their taxonomy, identification, microbiology, epidemiology, infections, and resistance to antibiotics and the potential therapeutic approaches.

Microbiological Characteristics

The genus *Acinetobacter* consists of ubiquitous gram-negative bacilli that were originally identified in the 1930s [3,6]. Subsequently, several changes were introduced into the taxonomic classification of the *Acinetobacter* species. The genus *Acinetobacter* initially encompassed a heterogeneous collection of non-pigmented, oxidase-positive and oxidase-negative gram-negative rods [3,6-8]. The genus *Acinetobacter* is now defined as including gram-negative (but sometimes difficult to distain) coccobacilli, with a DNA G1C content of 39 to 47 mol%, that are strictly aerobic, nonmotile, nonfermentative, catalase-positive, and oxidase-negative. The good growth occurs on 40° C without growth factor requirements, while nitrates are reduced only rarely [3]. In the rapid growth phase, it appears as bacilli and in the stationary phase as cocobacilli [9].

Over 20 species belonging to the genus *Acinetobacter* have been identified [10]. *Acinetobacter baumannii* (genomic species 2) is the species of greatest clinical importance, and is typically associated with outbreaks in the hospital setting [11]. Other species that have been associated with disease in humans belong to the *A. calcoaceticus-A. baumannii* complex that includes *Acinetobacter* genomic species 3, *A. johnsonii*, *A. lwoffii*, and *A. calcoaceticus subsp anitratus* [9].

Epidemiology

Increasing isolation of multidrug-resistant *Acinetobacter baumannii* (MDR-ab) has been reported worldwide, and it is now one of the most difficult nosocomially acquired gram-negative pathogens to control and treat [12]. *Acinetobacter baumannii* has the ability to use various sources of nutrition which accounts for its growth on routine laboratory media. This explains its survival as an environmental pathogen. Some strains of *Acinetobacter* can survive environmental desiccation for weeks, a characteristic that promotes transmission through fomite contamination in hospitals [13]. The capacity of *Acinetobacter* species to survive on most environmental surfaces for long periods of time suggests that all animate and inanimate entities should be considered reservoirs.

In health-care settings, colonized and infected patients are often the sources of *A. baumannii* infections; however, the ability of the organism to survive for prolonged periods on environmental surfaces also has contributed to protracted outbreaks in these settings [12]. *Acinetobacter* has been isolated from pasteurized milk, frozen foods, chilled poultry, foundry and hospital air, vaporizer mist, tap water faucets, peritoneal dyalisate baths, bedside urinals, washcloths, door handles, keyboards, angiography catheters, ventilators, contaminated gloves, duodenoscopes, laryngoscope blades, plasma protein fraction, and hospital pillows [14-16]. This ability to grown on medical equipment and in a hospital environment emphasizes the need for special attention to disinfection of these places as they are in a later chapter [17-20].

Certain strains (clones) have been shown to be distributed between hospitals in wide geographical areas [21]. However, the origin of a suddenly appearing epidemic A. *baumannii* strain in a hospital is often unknown, and it remains possible that A. *baumannii* has a reservoir in the nonhospitalised human population, from which strains can be introduced into a hospital. In humans, *Acinetobacter* can colonize skin, wounds, and the respiratory and gastrointestinal tracts. Up to 25% of healthy ambulatory adults exhibit cutaneous colonization and 7% of adults and infants have transient pharyngeal colonization. It is the most common gram-negative organism persistently carried on the skin of hospital personnel and it has been found to colonizate 45% of inpatient tracheostomy sites and 41% of fecal samples in intensive care units [22-23].

Acinetobacter baumannii can cause community-acquired infections although less frequently than nosocomially infections. Multidrug-resistant *Acinetobacter* infection has been reported among patients residing in rehabilitation and long-term care facilities, as well as in acute care hospitals [24-26]. Several factors maintain the presence of multidrug-resistant *Acinetobacter* species in the health-care setting including the presence of susceptible patients, the presence of patients already colonized or infected with the organism, selective pressure from antimicrobial use and incomplete compliance with infection-control procedures [27].

In addition to transmission, emergence of resistance occurs in the context of selective pressure from broad-spectrum antimicrobial therapy, such as therapy involving carbapenems or third-generation cephalosporins. The relative contributions of antimicrobial selective pressure and transmission between patients on the emergence of multidrug-resistance *Acinetobacter* species are not known [28-30].

Recently multidrug-resistant *Acinetobacter* deep wound infections, osteomyelitis and bacteraemia have been reported among military personnel with traumatic injuries during the war in Iraq and Afghanistan [31-34]. Theories that previously colonized soldiers are autoinoculated or that *Acinetobacter* species from local soil or water are introduced during traumatic injury have not been supported by cultures of specimens obtained from healthy soldiers, soil samples, water samples, or samples from fresh wounds. Current literature suggests that these infections are associated with health care and are acquired by soldiers in medical facilities during the processes of stabilization, emergency treatment and evacuation through the military medical system [35-39]. There is a possibility of transmission of these strains to the hospital destination of these patients so the Public Health Agency of Canada, in conjunction with the Canadian Forces, have developed an alert to inform hospitals of the potential for importation of *Acinetobacter baumannii*, and the appropriate precautionary measures that should be taken to prevent secondary spread within hospitals [34,40-42].

ANTIMICROBIAL RESISTANCE

Acinetobacter spp. display mechanisms of resistance to all existing antibiotic classes as well as a prodigious capacity to acquire new determinants of resistance [43]. The capacity of the Acinetobacter species for extensive antimicrobial resistance may be due in part to the organism's relatively impermeable outer membrane and its environmental exposure to a large reservoir of genes.

Various terms have been used (sometimes interchangeably) to denote antimicrobial-resistant phenotypes of *Acinetobacter* spp. The most common definitions of multidrug resitance are carbapenem resistance or resistance to $\geq$ 3 classes of antimicrobials [44].

The frequently observed multiresistance of *Acinetobacter* sp. isolates complicates antibiotic therapy and varies greatly between geographical regions, institutions, and even hospital wards [45]. Some studies have reported an average rate of carbapenem resistance of 27% among *Acinetobacter* spp. in European countries [46].

Mechanisms of Resistance

Resistance mechanisms for the *Acinetobacter* species generally fall into 3 categories: a) antimicrobial-inactiving enzymes; b) reduced access to bacterial targets; or c) mutations that change targets or cellular functions. Because the *Acinetobacter* spp are gram-negative organisms, they possess an additional outer membrane that acts as a permeation barrier [2].

Resistance mechanisms that are expressed frequently in nosocomial strains of acinetobacter include β-lactamases, alterations in cell-wall channels (porins), and efflux pumps. *A. baumannii* can become resistant to quinolones through mutations in the genes *gyrA* and *parC* and can become resistant to aminoglycosides by expressing aminoglycoside-modifying enzymes [13,44].

Resistance to beta-lactams

The mechanism of resistance to ß-lactams involve a) their hydrolysis by beta-lactamases, b) changes in penicillin-binding proteins (PBPs) that prevent their action, c) alterations in the structure and number of porin proteins that result in decreased permeability to antibiotics through the outer membrane of the bacterial cell, and (iv) the activity of efflux pumps that further decrease the concentration of antibiotic within the bacterial cell [47].

AmpC β-lactamases are chromosomally encoded cephalosporinases intrinsic to all *A. baumannii*. Usually, such β-lactamases have a low level of expression that does not cause clinically appreciable resistance; however, the addition of a promoterinsertion sequence, IS*Aba*1, next to the *ampC* gene increases β-lactamase production, causing treatment- limiting resistance to cephalosporins [48] Although porin channels in *A. baumannii* are poorly characterized, it is known that reduced expression or mutations of bacterial porin proteins can hinder passage of β-lactam antibiotics into the periplasmic space, leading to antibiotic resistance.

In addition to the class C cephalosporinase discussed above, other b-lactamases have been reported in *A. baumannii* [44]. These include the TEM-1 type [49,50], SHV type [3,51], CTX-M type [52], PER-1 [53,54], and VEB-1 [55,56] b-lactamases. Although they are important, it is difficult to assess their impact on resistance in the presence of the AmpC cephalosporinase.

The most important problem in multidrug-resistant *Acinetobacter* is the resistance to carbapenems. This resistance is mediated for various mechanisms:

- metallo-β-lactamase (VIM, IMP)
- OXA carbapenemases (class D).
- cell permeability changes.
- target (PBPs) change.

The metallo-β-lactamases and OXA carbapenemases are the most important mechanism of resistant in *Acinetobacter baumannii* because confer resistance to carbapenems. These enzymes belong to either Ambler Class B (metalo-beta-lactamases) or class D (oxacillinases). Metalo-B-Lactamases are efficient carbapenemases and three groups of this enzyme class have been found in Acinetobacter spp the IMP-like MBL, the VIM-like MBL or SIM-1. IMP and VIM confer high-level resistance to carbapenems and most other B-lactams with the exception of aztreonam. MBL are located in class-1 integrons and may be transferred and expressed along with resistance genes to other antimicrobials such as aminoglycosides [45,47,57-59].

The other group of enzymes (carbapenem-hydrolysing oxacillinases) consists of oxacillinases with intrinsic carbapenemase activity. The significant contribution of these enzymes to carbapenem resistance in *A. baumannii* has been emphasized, particularly when they are accompanied by IS*Aba1* and IS*Aba3* in the naturally occurring plasmid [60-62].

A third mechanism of carbapenem resistance involves porins, which are outer-membrane proteins that allow antimicrobials, such as B-lactams, to permeate into the bacterial cell. Loss or modification of porin proteins has been shown to confer carbapenem resistance and high-level resistance is observed in the presence of both loss of porin function and expression and production of carbapenemases [47,63].

Additionally a mechanism that contributes to carbapenem resistance is the overexpression of bacterial efflux pumps, which decreases the concentration of β-lactam antibiotics in the periplasmic space [64]. To cause clinical resistance in *Acinetobacter spp*, efflux pumps usually act in association with overexpression of AmpC β-lactamases or carbapenemases. With less b-lactam entering the periplasmic space, the weak enzymatic activity of the b-lactamase is amplified In addition to removing β-lactam antibiotics, efflux pumps can actively expel quinolones, tetracyclines, chloramphenicol, disinfectants, and tigecycline [65].

The resistance of *A. baumannii* to carbapenems is also explained by reduced expression of PBP-2 [62]. Establishing the relative contribution of the action of beta-lactamases, beta-lactam penetration through OMPs, and interaction with other mechanisms of resistance, and the control of their expression presents important questions for the future [47].

Resistance to aminoglycosides

Resistance to aminoglycosides is mediated by 3 mechanisms: alteration of the ribosomal target, reduction of uptake, and enzymatic modification of the drug. Alteration of the ribosomal target only affects streptomycin. The second mechanism is fairly common in *Acinetobacter spp*, but resistance to aminoglycosides in *A. baumannii* is mediated principally by aminoglycoside-modifying enzymes. These include aminoglycoside o-phosphotransferases, aminoglycoside N-acetyltransferases, and aminoglycoside o-nucleotidyltransferases and are mediated primarily by plasmids or transposons that can play a role in the spread of resistance. These enzymes may also be chromosomally located [3,66,67].

Resistance to quinolones

Resistance of *A. baumannii* to quinolones is often caused by modifications in the structure of DNA gyrase secondary to mutations in the quinolone resistance-determining regions of the *gyrA* and *parC* genes [68,69]. These changes result in a lower affinity for the binding of the quinolone to the enzyme-DNA complex. As mentioned above, a second mechanism of resistance to the quinolones is mediated by efflux systems that decrease intracellular drug accumulation [47].

Resistance to tetracyclines

Two different mechanisms of resistance to tetracyclines have been described in *A. baumannii*. TetA and TetB are specific transposon-mediated efflux pumps; TetB determines the efflux of both tetracycline and minocycline, whereas TetA drives only the efflux of tetracycline. The second mechanism is the ribosomal protection protein, which shields the ribosome from the action of tetracycline. The *tet*(M) gene encodes this protein, which serves to protect the ribosome from tetracycline, doxycycline, and minocycline [3,47].

Tygecycline is the first antibiotic of a new class called glycylcyclines that may be useful for multidrug-resistant *Acinetobacter* [70]. Tigecycline overcomes the two major mechanisms of resistance to tetracyclines (ribosomal protection and efflux), but tigecycline resistance emerging during therapy has been reported [71]. This resistance appears to be attributable to the overexpression of AdeABC multidrug efflux pump [72]. Probably this mechanism can be upregulated by the previous use of other antibiotics (eg, fluoroquinolones, aminoglycosides) that are administered frequently in patients with serious infections [73].

Resistance to polymyxins

There are several suggested mechanisms of resistance to colistin/polymyxin B in Gram-negative bacteria, most of which involve changes in the outer membrane. The mechanism of resistance to colistin likely resides in modifications in the lipopolysaccharide of *A. baumannii* (acidification, acylation, or presence of antigens that interfere with binding of the antibiotic to the cell membrane) [3]. It had been reported a strong association between the use of colistin and the development of resistance in A.baumannii clinical isolates [74].

VIRULENCE AND PATHOGENICITY

A. baumannii is regarded as a low-grade pathogen, with limited virulence. However, this organism has certain features that allow it to increase the virulence of those strains implicated in infections. The invasiveness of the bacteria may be related to substances that protect the surface of phagocytosis, as is the polysaccharide capsule. The presence of fimbriae along with the dish allows the adhesion to human epithelial cells. It has studied the production of enzymes as butyrate esterase, caprylate esterase, leucine and aryl amidasa which may damage tissue lipids and the potentially toxic role of the lipopolysaccharide component of the cell wall and the presence of lipid A. Apart from owning factors Acinetobacter spp in common with other gram-negative bacteria produce a lipopolysaccharide that can take effect endotoxigénico through the lipid A and that would highlight the ability of these microorganisms to capture the iron through the secretion of siderophores, which allows them to survive in the body human. On the other hand, some strains of *Acinetobacter spp* can producing slime which was considered to be the main factor responsible for the enhancement of virulence in mixed infections [1,3, 75].

All this factors act together in the pathogenicity of *Acinetobacter spp*. The capsular polysaccharides and lipopolysaccharides (LPSs) act in synergy by blocking the access of the human complement to the bacterial cell wall, thus preventing its lytic activity on bacterial membranes. Furthermore, the hydrophilicity conferred by the presence of capsular polysaccharides, together with other nonspecific adherence factors (i.e., fimbriae), prompts the colonization of human epithelial cells by pathogen strains [76]. Enzymes able to damage tissue lipids are also produced during infectious stages of the colonization. Interestingly, the action of LPSs in the pathogenicity of *Acinetobacter* is not limited to the protection of bacterial cells from host defenses; these molecules are in fact provided with a powerful endotoxic activity, and their production in vivo is thought to be related to *Acinetobacter* septicemia symptoms, given the ability of *Acinetobacter* LPSs to induce lethal toxicity in mice and pyrogenicity in rabbits [77].

LPSs are characteristic of the Gram-negative bacteria outer membrane. Their chemical nature affects the physical properties of the cell envelope, thus determining the interactions with the surrounding environment as well as the host-pathogen recognition mechanism. From a structural point of view can be divided in two categories: smooth-type LPSs, composed of the glycolipid portion of lipid A, the core oligosaccharide, and the *O*-polysaccharide or *O*-specific chain; and rough-type LPSs (R-LPSs), in which the *O*-polysaccharide is missing. The structural variability that characterizes the polysaccharide moiety decreases in the core oligosaccharide and primarily in lipid A, whose architecture is usually conserved within the genus. All virulent strains of *Acinetobacter* have been found to possess smooth-type LPSs. Nevertheless, the most powerful immunostimulator center of the LPS is the lipid A glycolipid moiety, which is able to activate the innate immune system, triggering cytokine production that, when uncontrolled, can lead to septic shock [78].

ANTIBIOTIC USEFUL FOR ACINETOBACTER INFECTIONS

There is a distinct class of antimicrobial agents that are considered as potentially effective for the treatment of *A baumannii* infections that include: Sulbactam, Anti-pseudomonal penicillins or cephalosporins, carbapenems, monobactams, aminoglycosides, fluoroquinolones, glycylcyclines and colistin [79]. However the apparition of strains with high resistance to carbapenems, aminoglycosides and fluoroquinolones [80,81] presents a challenge to the clinician so sulbactam, tigecycline and colistin represent the current therapeutic approaches, which are associated with satisfactory efficacy.

Carbapenems

Imipenem and meropenem, have been regarded as the treatment of choice for severe *A baumannii* infections [3],, although the increase of resistance rates of *A baumannii* isolates to carbapenems, especially in Europe and North America has diminished its usefulness. In a recent surveillance study [82] about 2,902 strains of *A. baumannii* the susceptibility of imipenem varied across the regions, from fairly good in North America and Europe (74%) to poor or moderate in Latin America (60.6%) and Asia/Pacific Rim (69.2% and 59.6%, respectively). On the other hand carbapenem-resistant *A baumannii* strains might only rarely be susceptible to other antipseudomonal [79] and the treatment became very difficult.

As described in previous chapters, Carbapenem resistance in *A baumannii* is mediated by several mechanism including plasmid or chromosomally encoded carbapenemases (mainly OXA-23-like, OXA-24-like, or OXA-58-like class D β-lactamases), as well as metallo-β lactamases (class B β lactamases), efflux pump mechanisms, penicillin-binding protein alterations, and modifications or loss of outer membrane proteins (porins) [48,79]. More than one of these resistance determinants could be present in the same strain conferring high-level resistance.

Recently, a new carbapenem has been developed. Doripenem is a novel forthcoming carbapenem that possesses a broad spectrum of activity against Gram-negative bacteria, similar to that of meropenem, while retaining the spectrum of imipenem against Gram-positive pathogens [83]. It has been approved for the treatment of complicated urinary-tract and complicated intra-abdominal infections. An indication for hospital-acquired pneumonia including ventilator-associated pneumonia is pending. Its principal features are: a) bactericidal action against most species; b) B-lactamase stability to commonly occurring enzymes, including the emerging extended spectrum B-lactamases (ESBLs); (c) pharmacokinetic and pharmacodynamic qualities similar to those of meropenem (half-life of 1 h) with minimal risk of convulsive adverse reactions; d) postantibiotic effects of nearly 2 h in vitro for *P.aeruginosa*; and e) low serum protein binding (8.9%). Doripenem is stable enough in the presence of renal dehydropeptidase I that it need not be coadministered with a dehydropeptidase I inhibitor such as cilastatin. Pharmacokinetic studies have supported a 3-times-daily dosing schedule [84].

The most frequent adverse reactions are: nausea, diarrhea, headache, phlebitis, hypersensitivity reaction and *C. difficile*-associated colitis. In contrast to imipenem, doripenem did not cause EEG changes and seizures in animal studies [85]

Doripenem may have an important role in the treatment of *Acinetobacter* infection because it has a lower MIC than ertapenem or imipenem when tested against *Acinetobacter spp.* [86,87] and some isolates that are intermediate or resistant to other carbapenems may be susceptible to doripenem and will require additional susceptibility testing [88,89]. A recent study [87] realized about 12,581 isolates collected between 2005 and 2006 and tested by doripenem, imipenem and meropenem. The MIC(90) of doripenem (0.12) was comparable to that of meropenem (0.12) and superior to that of imipenem (2), though susceptibility of isolates exceeded 99% for all evaluated carbapenems. Against imipenem-susceptible *Acinetobacter spp.*, doripenem (MIC(90) = 2, 89.1%S) was twice as active by MIC(90) as were imipenem and meropenem. However none of the carbapenems was active against *A. baumannii* strains expressing plasmid-mediated carbapenemases [88].

Colistin

Colistin belongs to the Polymyxins group and is available in two forms, colistin sulfate and colistimethate sodium. Colistimethate sodium is a non-active prodrug that is used for parenteral administration because of its lower toxicity [90]. Polymyxins show bactericidal activity against A baumannii that acts by interacting with anionic lipopolysaccharide moieties on the bacterial cell membrane, thereby leading to increased membrane permeability [91,92], and resistance rates against these agents have remained low despite their more frequent use and the presence of carbapenem-resistant strains [93,94,95].

The dose of colistin is 2.5–5 mg/Kg/day of colistin base in two-to-four divided doses. In persons with > 60 Kg bodyweight, the doses more frequently used are 80–160 mg/8 hours. The bactericidal activity of colistin is concentration dependent, therefore administering large doses in less frequent intervals may be a favourable approach [96-99]. Isolated case reports described the cure of Acinetobacter baumannii bacteraemia with continuous intravenous colistin (2,000,000 units per 24 h) in patients with allergic reactions to other treatments. In this context colistin may be a useful salvage regimen [100]. Dose adjustment is required in the presence of renal failure.

Colistin has demonstrated alone or in combination with other antimicrobials like rifampicin, its usefulness in several infections due to *A. baumannii* as ventilator- associated pneumonia, meningitis, bacteraemia, etc. [100-104].

The most common toxic effects of colistin are nephrotoxicity and the emergence of polyneuropathy. For this reason, the systemic administration of polymyxins had been abandoned for about 20 years in most areas of the world. However, the problem of infections due to multidrug-resistant (MDR) Gram-negative bacteria such us *Pseudomonas aeruginosa* and *Acinetobacter baumanniii* has led to the re-use of polymyxins. Recent studies showed that the rate of development of nephrotoxicity in patients treated with colistin for Gramnegative bacterial infections is similar to that related to other antimicrobial agents [105-109]. Most non-comparative studies have reported a rate of renal toxicity associated with polymyxin therapy of between 8% and 21%. Patients with pre-existing renal dysfunction seem to be more susceptible to the renal adverse effects of polymyxins [108-109]. Other side effects are the apparition of neurotoxicity, bronchoconstriction, chest tightness or cough following administration via the respiratory tract, and chemical meningitis with intraventricular or intrathecal use but has been infrequently reported [110-111].

Resistance to colistin is much lower but it has already been reported. In recent reports, risk factors significantly associated with the isolation of colistin-resistant isolates were age, duration of intensive-care-unit stay, surgical procedures, use of colistin, use of monobactams, duration of use of colistin—although in the multivariable model, the use of colistin was identified as the only independent risk factor [112]. To prevent this phenomenon, colistin should be used judiciously, given that treatment options for colistin-resistant gram-negative bacteria are limited.

Sulbactam

Sulbactam was introduced in the eighties as a beta-lactamase inhibitor in combination with beta-lactamic antibiotics [4,5]. Sulbactam have bacteriostatic activity against

Acinetobacter spp through binding to PBP2. The most frequently used combination is ampicillin/sulbactam (ratio 2:1) although the two agents are not synergetic [103-105]. The combination of ampicillin-sulbactam has been used by some authors in doses of 2 gr to 6–8 hours [45,113].

The use of ampicillim/sulbactam had been described in the literature in a retrospective series. The combination therapy has shown similar effectiveness to imipenem therapy in *A baumannii* ventilator-associated pneumonia or bacteraemia caused by multidrug-resistance [113-118]. Favourable clinical outcomes have also been reported with sulbactam or combination ampicillin/sulbactam therapy in patients with other types of nosocomial infections like meningitis [119]. Recent studies compared the efficacy of ampicillin/sulbactam to polymyxins therapy and it appears to be more efficacious than polymyxins, which was an independent factor associated with mortality during treatment [120].

Therefore, ampicillin-sulbactam is a sensible option for the treatment of life-threatening *Acinetobacter* infections. However the role of sulbactam in the treatment of *Acinetobacter* infections is limited due to the apparition of multidrug-resistant strains.

Glycylcyclines

Glycylcyclines are a novel class of antimicrobial agents related to the tetracyclines. Tigecycline, a semisynthetic derivative of minocycline, is the first commercially available agent of this class. It has a similar mechanism of action to that of the tetracyclines, and shows bacteriostatic activity against *A baumannii* because it interfere with bacterial protein synthesis through ribosomal binding and thus exhibits time-dependent bactericidal activity [121-122].

Tigecycline is extensively distributed into many tissues, resulting in a prolonged half-life that justifies twice-daily dosing. Fifty-nine percent of a tigecycline dose is excreted through the liver, and 33% is excreted through the kidney. In patients with severe hepatic impairment (Child's class C), the normal dose of tigecycline (100-mg initial dose, then 50 mg twice daily) should be reduced to 25 mg twice daily after the normal loading dose. No adjustments are required for any level of renal impairment [123].

Tigecycline is able to evade the most common mechanisms of resistance to tetracyclines in *A baumannii*, including efflux pumps encoded by the *tet*(A) and *tet*(B) determinants, and ribosomal protection mechanisms. Nevertheless, several unique multidrug efflux pumps have been shown to reduce organism susceptibility to tygecycline and the rate of resistance is nearly 6% in strains in Europe and America [124]. In-vitro studies have shown that resistance to tigecycline is mediated by the expression of resistance-nodulation division-type multidrug efflux pumps in *A. baumannii* suggesting that resistance may emerge during therapy [125].

Clinical experience regarding the use of tigecycline for the treatment of patients with multidrug-resistant *A baumannii* infections is accumulating. In the available relevant reports, tigecycline has been used in a small number of critically ill patients—mainly as part of combination antibiotic regimens—for the treatment of various types of infections, including ventilator-associated pneumonia and primary or secondary bacteraemia [126-128]. In first case series, most patients had a good clinical outcome. However, failure of tigecycline to clear *A baumannii* bacteraemia was reported in a few cases [129-130]. On the other hand pharmacokinetic and pharmacodynamic data indicate that tigecycline blood concentrations seem to be suboptimal for maximal antibacterial activity to be exerted at this compartment.

The emergence of resistance in *A baumannii* during tigecycline therapy is an additional point for consideration. There is a crescent number of reports which indicate that resistance to tigecycline developed during the treatment. In recent review reporting retrospective data regarding 42 severely ill patients, among whom 31 had respiratory tract infection (in 4 cases with secondary bacteraemia) and 4 had bacteraemia tigecycline therapy (in combination with other antibiotics in 28 patients) was effective in 32/42 cases but in three cases the resistance development during the treatment [129]

Clinicians should be aware that tigecycline MICs for A. *baumannii* isolates may increase during therapy with tigecycline after only brief exposure to the drug. Patients receiving tigecycline for *Acinetobacter* should be monitored for the development of clinical resistance, and isolates should be monitored for evidence of microbiologic resistance.Aminoglycosides

Aminoglycosides have shown moderate rates of antimicrobial activity against *A baumannii.* In worldwide collections of *Acinetobacter* spp isolates, susceptibility rates to amikacin were approximately 60% [92] and the activity of aminoglycosides is lower for multidrug-resistant isolates of *A baumannii* compared with non-multidrugresistant ones [131].

Reports about the clinical use of aminoglycosides in human beings are scarce, and refer to cases of bacteraemia [132] or meningitis [104] in which aminoglycosides have been used in combination with other classes of antimicrobial agents.

Fluoroquinolones

Fluoroquinolones have moderate antimicrobial activity against *A baumannii.* Althought levofloxacin has been shown to yield a lower MIC compared with ciprofloxacin and ofloxacin against *Acinetobacter spp o*verall resistance rates to ciprofloxacin and levofloxacin in clinical isolates are around 50%. [133]. On the other hand the activity of fluoroquinolones against recent multidrug-resistant or imipenem-resistant isolates has been reported to be low [92,132].

The clinical dates are based on in vitro studies. In this studies levofloxacin has shown similar effectiveness to imipenem in a mouse pneumonia model against a strain susceptible to both agents [134]. However, no comprehensive evidence exists regarding the effectiveness of fluoroquinolones for the treatment of human infections caused by *A baumannii.*

Combination Therapy

Against infections multidrug- resistant *Acinetobacter baumannii* isolates several novel antibiotic combinations demonstrate increased activity in vitro compared with that of any single agent. Whether these combinations yield outcomes that are improved over those seen with a polymyxin or other agent alone remains to be determined. However, against infections with species resistant to all antibiotics, including polymyxins, novel combinations are the only remaining therapeutic option.Several drugs combinations havee been tested mostly involving colisitin , carbapenems, rifampicin, fluorquinolones and sulbactam [135].

Although most of the studies were done *in vitro* and in animal models of pneumonia, they show that the combined therapy, especially the combination with rifampin, obtained better results than monotherapy with carbapenems or colistin [136,137]. The colistin and rifampicin combination (colistin sulphomethate sodium 2 million IU three times a day and intravenous rifampicin 10 mg/kg every 12 h). has been probed in the treatment of ventilator-associated pneumonia or bacteremia caused by carbapenem-resistant Acinetobacter baumannii with clinical response of 76% and mortality of 21% [138,139]. Some studies [140] has compared the efficacy of intravenous and aerosolized colistin combined with rifampicin in the treatment of critically patients with nosocomial infections caused by multiresistant *A. baumannii* with good results.

The imipenem and ripampicin combination had been studied in some clinical papers.The combination of imipenem with rifampicin, although synergistic in vitro,was not associated with clinical benefits in a small study. [141] In a small retrospective cohort study of patients with infections caused by multidrug-resistant, including carbapenem-resistant, Gram-negative bacteria, the combination of colistin with meropenem was found to be inferior in terms of patient survival compared with colistin monotherapy, although infection-related outcomes did not differ. The combination of imipenem and tigecycline has not been shown to be synergistic [45]. The combination of imipenem with sulbactam has showed synergistic activity in vitro although the clinical usefulness of this combination in patients infected with carbapenem resistant *A baumannii* is not well established [142].

The assessment of in-vitro synergy of the combination of imipenem with a polymyxin against carbapenem-resistant *A baumannii* strains has provided mixed findings. Among other polymyxin-based combination regimens for multidrug-resistant *A baumannii*, the combination of colistin with a carbapenem has shown in vitro either synergistic or indiff rent activity 143,144 but clinical effectiveness has not been substantiated 145. By contrast, the combination of polymyxin B and tigecycline has resulted in indiff erent activity146.

Clinical data are currently insufficient to favour the use of specific combination regimens for the treatment of multidrug-resistant *A baumannii* infections.

Specific Infections Due to A. Baummanii

1. Nosocomial Infections

Acinetobacter baumannii is an important cause of nosocomial infections. Risk factors specific for nosocomial infection include length of hospital stay, surgery, wounds, underlying severity of illness, previous infection, fecal colonization with *Acinetobacter*, treatment with broad-spectrum antibiotics, parenteral nutrition, indwelling central intravenous or urinary catheters or other invasive procedures, admission to a burn unit or intensive care unit, parenteral nutrition, and mechanical ventilation [28,147].

Nosocomial meningitis

A. baumannii meningitis is an infrequent infection mostly associated with intraventricular catheters, CSF fistula or head trauma [148,149].

A. baummanii meningitis is more frequent in patients with intraventricular hemorrhages, head trauma, longer hospital stays and admitted to the intensive care unit [104,150,151]. Others risk factors are the presence of intraventricular catheters longer than 7 days, like in all post-surgical meningitis [152], and the previous use of antimicrobials.

The clinical manifestations of *A. baumannii* post-surgical meningitis are unremarkable, mostly fever and progressive loss of consciousness. Therefore these manifestations are attributed to the underlying disease, delaying this way the final diagnosis of meningeal infection , and representing a prognostic factor of mortality [104,153].

Acinetobacter spp is a genus with tendence to quickly develop resistances to multiple antimicrobials. This complicates the treatment of these infections makes imperative the search of new agents and the return of old drugs. Until the apparition of resistances carbapenems alone or combined with amikacin was the gold standard in the treatment of this infection. However the resistance to the new antibiotics in the last years has determined the use of alternative agents such as ampicillin-sulbactam or colistin [154]. Sulbactam was introduced in the eighties as a beta-lactamase inhibitor in combination with beta-lactamic antibiotics [5] with "in vitro" activity against *Acinetobacter spp*, enhanced by its affinity for penicillin-binding proteins. Intravenous sulbactam penetrates only 1% the blood-brain barrier but increased up to 32% with meningeal inflammation [119]. The combination of ampicillin-sulbactam has been used by some authors in doses of 2 gr to 6-8 hours with mortalities of 20-25% [104,119].

The presence of multiresistance and the poor penetration of many drugs through the blood-brain barrier has forced the use of intrathecal therapies initially with aminoglycosides and most recently with colistin. The use of intrathecal and intravenous amikacin has been reported in the treatment of meningitis refractary to standard therapy in a small number of cases con rates of cure by 50-60%, [155,156]. The exact dose of intrathecal amikacin has not been established. In recent study a 20mg/day dose was used with cure rates of 83.4% and without side-effects [104,116].

The role of colistin in the treatment of meningitis was initially limited due to its poor CSF penetration (25% of serum levels) without modification in the presence of meningeal inflammation [157]. However it has been used in isolated clinical cases by intrathecal and IV routes with good results [104,158]. The dose of intrathecal colistin is variable, most of the authors use 5mg/12 hours. In recent study the doses were 10 mg/12 hours without increasing side-effects [104]. In this paper all the patients treated with intravenous and intrathecal colistin survived. Although the small number of cases limit the statistical meaning of this findings the data show that colistin can be an effective and safe alternative for this infection [104].

New antibiotics have recently appeared such as tigecycline, member of a new class of antibiotics, the glycines. In animal studies of meningitis, tigecycline in a unique dose of 20 mg/kg/ day has shown CSF concentrations higher than 1μg/ml 3 hours after the infusion. However enough data regarding its ability to pass the blood-brain barrier and its future use in the treatment of meningitis are lacking [159-160]. Polymyxins have also been used for the treatment of nosocomial meningitis caused by multidrug-resistant *A baumannii.* [161]. Clinical effectiveness has been reported for treatment via the intravenous route alone, although studies have only included a small number of patients [157,158,162]. Because of suboptimal penetration through the blood–brain barrier [157] colistin has been used via the intraventricular or the intrathecal route in addition to systemically administered colistin [104,110,163,164] or other antibacterial agents, such as amino glycosides [104,110,163,164].

In the series the lack of removal of the intraventricular catheters is associated with higher mortality. Therefore the early removal of the intraventricular devices is imperative to obtain the cure of the infection. Other factors conditioning a higher mortality were the absence of an adequate treatment and the delay in starting therapy emphasizing the value of an adequate empirical treatment [104].

Table 1. Main clinical studies on *A. baumannii meningitis* treatment

Reference	Design	Patients	Treatment	Dosing	Cure
Levin et al (1999)[162]	Prospective nonrandomized	5 CNS infections	colistin	2.5-5 mg/kg/day iv divided into 3 doses	80%
Jiménez-Mejias et al (2000)[158]	Case report	1 meningits	colistin	5 mg/Kg/day iv in 3 dosis	Cure
Jiménez-Mejias et al (2002)[157]	Case report	1 meningitis	colistin	5 mg/Kg/day iv in 4 dosis	Cure
Levin et al (2003) [117]	Prospective nonrandomized	2 CNS infections	Ampicillin-sulbactam	6 gr/ day in 3 dosis	None
Jimenez-Mejias et al (1997)[119]	Retrospective	8 pos-tsurgical meningitis	Ampicillin-sulbactam	2 gr/iv every 6-8 hours	75% cure
Cawley et al (2002)[166]	Case reprort	Meningitis	Ampicillin-sulbactam	2 gr/iv every 3 hours	Cure
Siegman-Igra et al(1993) [148]	Retrospective	25 nosocomial meningitis	Various	Not reported	3 death
Nguyen et al (1994) [155]	Retrospective	7 post-surrgical meningitis	2 imipenem+ amikacin 3 imipenem+ amikacin + intrathecal aminoglicosides 2 others treatments	-Not reported	86%
Benifla M et al (2004) [166]	Case report	1 Surgical meningitis	intrathecal colistin	40.000 U/day	Cure
Gleeson T et al (2005) [167]	Case report	Surgical meningitis	Rifampin Meropenem	600 mg/24 hours Dose not reported	cure
Vasen W et al (2000) [168]	Case report	Post-surgical meningitis	Intrathecal Colistin	10 mg/24 hours 21 days	Cure
Karakitsos et al (2006) [169]	Retrospective	Postsurgical meningitis	Ripampin Colistin	10 mg/Kg/12h 10-20 mg day	Cure

Hospital-acquired/ventilator-associated pneumonia (VAP).

Acinetobacter spp are a major cause of VAP and are associated with mortality rates up to 50–70% [170] in some cases associated to inappropriate initial antibiotic therapy [171]. The incidence of this microorganism varies from site to site, but it is the second commonest aetiological agent among the gram-negative bacteria. Longer periods of hospitalization, longer time on mechanical ventilation and prior use of antibiotics are the recognized factors increasing the risk of VAP due to *Acinetobacter spp* [172]. The presentation of VAP due to *Acinetobacter baumannii* only sensitive to colistin is associated with previous treatment with carbapenems [173], previous episode of VAP, prior antimicrobial therapy for >10 days [174,175]. Treatment needs to clearly differentiate infection from colonization, and the agents with the most antimicrobial activity are imipenem/cilastatin, amikacin, colistin, ampicillin/sulbactam and tigecycline.

The treatment with imipenen is considerated the treatment of choice in carbapenems-susceptible strains. However a experimental study about the useful of combination treatment with amikacina showed that the combination of imipenem plus amikacin was inferior to imipenem alone with the imipenem-susceptible strain , despite their in-vitro synergy, and was inferior to amikacin alone with the imipenem-resistant strain. Combined use of imipenem with amikacin was less efficacious than monotherapy, probably because of a drug-drug interaction that resulted in decreased pharmacokinetic and pharmacodynamic parameters for both antimicrobial agents [176]. Further studies assessed prospectively or retrospectively treatment with colistin for patients with ventilator-associated pneumonia caused by *A baumannii* [101,106,107]. Specifically, no difference in clinical response was reported between patients treated with intravenous colistin for infections caused by organisms susceptible only to polymyxins and those treated with other agents, mainly carbapenems, for infections caused by carbapenem-susceptible organisms.

Colistin may also be administered by inhalation. The dose of aerosolized colistin may range between 500,000 UI every 12 hours and 2,000,000 IU every 8 hours. Several studies have reported favourable clinical outcomes with the use of aerosolised colistin along with intravenous colistin or other parenteral antimicrobial agents for nosocomial pneumonia caused by multidrug-resistant *A baumannii* [102,177]. In a recent study about sixty critically ill patients that received aerosolized colistin (2,000,000 million international units) during 16 in concomitant intravenous treatment with colistin or other antimicrobial agents the bacteriological and clinical response of VAP was observed in 83.3% patients. No adverse effects related to inhaled colistin were recorded. Aerosolized colistin may be considered as adjunctive to intravenous treatment in patients with VAP due to MDR Gram-negative bacteria susceptible to colistin in critically ill patients [178].

Reports of patients with polymyxin-only susceptible gram-negative nosocomial pneumonia treated with inhaled, but without concurrent intravenous treatment are rare. Notably, in some of the few relevant cases reported in the literature, administration of nebulised colistin as the sole microbiologically active treatment against *A baumannii* nosocomial pneumonia has shown clinical effectiveness nearly to 60% [179-180].

The efficacy and safety of ampicillin-sulbactam for MDR *Acinetobacter baumannii* VAP has been evaluated in a few reports with clinical improvement of 67% [181] and high-dose ampicillin/sulbactam were comparably safe and effective treatments with respect to colistin for critically ill patients with MDR *A. baumannii* VAP [182].

The role of tigecycline alone or in combination with other antimicrobials for treatment of ventilator-associated pneumonia (VAP) had shown best clinical response (84%) in a small retrospective case series. However four patients had clinical failure: three with VAP and one with VAP plus bacteremia that developed resistance to tigecycline during therapy, and three patients with VAP had a recurrence of infection. Tigecycline may be effective when used alone or in combination with other antimicrobials for VAP caused by multidrug-resistant A. baumannii. However the emergence of a resistant strain while one patient was receiving therapy concerns us [126]. A very important item regarding hospital-associated pneumonia is prevention. There is convincing evidence to suggest that specific interventions can be employed to prevent HAP/VAP. The evidence-based interventions focus on the prevention of aerodigestive tract colonization (avoidance of unnecessary antibiotics and stress ulcer prophylaxis, use of sucralfate for stress ulcer prophylaxis, chlorhexidine oral rinse, selective digestive decontamination, short-course parenteral prophylactic antibiotics in high-risk patients) and the prevention of aspiration of contaminated secretions (preferred oral intubation, appropriate intensive care unit staffing, avoidance of tracheal intubation with the use of mask ventilation, application of weaning protocols and optimal use of sedation to shorten the duration of mechanical ventilation, semirecumbent positioning, minimization of gastric distension, subglottic suctioning, avoidance of ventilator circuit changes/manipulation, routine drainage of ventilator circuit condensate) [183].

Bloodstream infections

Bacteremia, followed by respiratory tract and surgical wound infections, is the most significant infection caused by *A. baumannii* [184]. The known risk factors for A. baumannii bacteremia are previous colonization with *A. baumannii*, invasive procedures and the use of broad-spectrum antimicrobials [185,186]. In other series found risk factors are immunosupression and mechanical ventilation [187-188]. Consequently, episodes of bacteremia due to *A. baumannii* occur most frequently in critically-ill patients admitted to an intensive care unit [189]. The clinical manifestations of bacteremia by *A. baumannii* are not specific. The most common sources of bacteremia are intravascular catheters and the respiratory tract [189-191]. *A. baumannii* bacteremia is associated with a high crude mortality rate, but it is difficult to distinguish morbidity and mortality attributable to *A. baumannii* from that attributable to the common and severe co-morbidity in these patients. The risk factor associated with higher mortality are: presence of inappropriate antimicrobial treatment, age > 65 years, development of septic shock and the severity of the underlying disease [192-196].

Generally, imipenem is the most active agent against *A. baumannii*. However, the description of imipenem-resistant *A. baumannii* strains is becoming increasingly common. Among the different antimicrobial regimens prescribed, the combination of a carbapenem and ampicillin-sulbactam was associated with a better outcome than the combination of a carbapenem and amikacin, or a carbapenem alone [190,192,197,198].The imipenem resistance has a significant impact on the mortality rate among patients with *Acinetobacter* bacteraemia [*1*99].

2. Community-Acquired Infections

Although *Acinetobacter* infections have been attracting increasing attention during recent years because they cause nosocomial infections, the available literature suggests that the pathogen can cause community-acquired infections [200]. The most frequent community-acquired *Acinetobacter baumannii* infections are pneumonia and bacteremia [201,202] and occasionally described are meningitis, empyema or endocarditis [203,204]. These infections have high mortality rates (56%). Most community-acquired *Acinetobacter* infections have been reported from countries with tropical or subtropical climate [201]. Comorbidity was commonly present in all patients mainly chronic obstructive pulmonary disease, renal disease, liver cirrhosis, diabetes mellitus, heavy smoking and excess alcohol consumption [205]. The most important infection is pneumonia because of its high mortality. This is a distinct syndrome with a high incidence of bacteremia—nearly 100%. It is most frequent during the wet season. Ninety-two percent of patients had a fulminant course presenting with septic shock and respiratory failure. *A. baumannii* should be considered as a possible etiologic agent in community-acquired lobar pneumonia when (1) patients with a fulminant course present during the warmer and more humid months of the year, and (2) patients are younger alcoholics. Early antibiotherapy is required because of the fulminant clinical course, with approximatively 50% fatality rate [206].

Infection Control and Prevention

Several studies have indicated that colonization with multidrug-resistant gram-negative bacilli is a frequent precursor of true infection. Specific sites of colonization may be amenable to eradication strategies. Hand colonization contributes significantly to *Acinetobacter* transmission and can be controlled by proper hand washing, glove use, and use of antiseptic- or alcohol-based soaps [207].

The digestive tract has only rarely been implicated as a major site for *Acinetobacter* colonization [208,209]. If this is documented during an outbreak, selective decolonization with polymyxin B, or aminoglycosides could be attempted.

The strategies for infection control and prevention of outbreaks caused by multiresistant *A.baumanii.* need to be simultaneous and multiples to be useful. Identification of colonized and infected patients, combined with implementation of infection-control measures such as hand-hygiene and contact-isolation precautions, might help prevent transmission of this organism within medical facilities These strategies have included [210-215]:

- Institution of active surveillance of groin, axillary, and/or wound cultures for *A. baumannii* for all patients.
- Use of contact precautions for colonized or infected patients
- Increased availability and use of alcohol-based hand rubs, daily review of antibiotic susceptibility of all clinical isolates
- Select cefepime or carbapenem-resistant isolates
- Review patient records prospectively to determine geographic location, nosocomial acquisition, and presence of clinical infection
- Emphasize contact precautions to prevent transmission of isolates
- Assign a dedicated infection-control nurse to affected area or areas
- Use polymyxin B to selectively decontaminate colonized wounds
- Consider inhaled polymyxin B for pulmonary colonization or infection (to supplement parenteral therapy)
- Use surveillance techniques to monitor personnel, equipment, and environmental contamination
- Instruct housekeeping to decontaminate the inanimate environment on a regular schedule
- Close contaminated unit if necessary
- Use molecular epidemiologic techniques to determine clonality of isolates.

Bibliography

[1] Joly-Guillou ML. Clinical impact and pathogenicity of *Acinetobacter*. *Clin Microbiol Infect.,* 2005, 11, 868-73.

[2] Jain, R; Danziger, LH. Multidrug-resistant *Acinetobacter* infections: an emerging challenge to clinicians. *Ann Pharmacother,* 2004, 38, 1449-59.

[3] Bergogne-Berezin, E; Towner, KJ. *Acinetobacter spp.* As nosocomial pathogens: microbiological, clinical, and epidemiological features. *Clin Microbiol Rev*., 1996, 9, 148-165.

[4] Levin, AS. Multiresistant *Acinetobacter* infections. A role for sulbactam combinations in overcoming an emerging worldwide problem. *Clinical and Microbiological Infection,* 2002, 8, 144-153

[5] Fernandez-Viladrich, P; Corbella, X; Corral, L; Tubau, F; Mateu, A .Successful treatment of ventriculitis due to carbapenem-resistant *Acinetobacter baumannii* with intraventricular colistin sulfomethate sodium. *Clin Infect Dis*., 1999, 28, 916-917.

[6] Cowan, ST. Unusual infections following cerebral operations: with a description of Diplococcus mucosus (von Lingelsheim). *Lancet,* 1938, 2, 1052.

[7] Brisou, J; Prevot, AR. Studies on bacterial taxonomy. X. The revision of species under Acromobacter group. *Ann Inst Pasteur*, (Paris) 1954, 86, 722.

[8] Ingran, M; Shewan, JW. Introductory reflections on the *Pseudomonas-Acinetobacter* group. *J Appl Bacteriol,* 1960, 23, 373-378.

[9] Juni, E. Genetics and physiology of *Acinetobacter. Annu Rev Microbiol*, 1978, 32, 349

[10] Dijkshoorn, L, Van Der, Toorn J. *Acinetobacter* species: which do we mean? *Clin Infect Dis*., 1992, 15, 748.

[11] Seifert, H, Baginski, R, Schulze, A, Pulverer, G. The distribution of *Acinetobacter* species in clinical culture materials. *Zentralbl Bakteriol*, 1993, 279, 544.

[12] Urban, C; Segal-Maurer, S; Rahal, JJ. Considerations in control and treatment of nosocomial infections due to multidrug-resistant *Acinetobacter baumannii. Clin Infect Dis*., 2003, 36, 1268-74;

[13] Muñoz-Price, SL; Weinstein, RA. *Acinetobacter* Infection. *N Engl J Med*., 2008, 358, 1271-81.

[14] Allen, DM; Hartman, BJ. *Acinetobacter* species. In: Mandell, Douglas and Bennett's.*Principles and Practice of Infectious Diseases,* 5th Edition.;

[15] Cordes, LG; Brink, EW; Checo, PJ; et al. A cluster of *Acinetobacter* pneumonia in foundry workers. *Ann Intern Med*., 1981, 95, 688-693

[16] Allen, KD; Green, HT. Hospital outbreak of multi-resistant *Acinetobacter anitratus:* An airborne mode of spread? *J Hosp Infect*, 1987, 9, 110-119.

[17] Bernards, AT; Harinck, HI; Dijkshoorn, L; Van Der Reijden, TJ; Van Den Broek, PJ. Persistent *Acinetobacter baumannii*? Look inside your medical equipment. *Infect Control Hosp Epidemiol,* 2004, 25, 1002-4.

[18] Maragakis, LL; Cosgrove, SE; Song, X; Kim, D; Rosenbaum, P; Ciesla, N; Srinivasan, A; Ross, T; Carroll, K; Perl, TM. An outbreak of multidrug-resistant *Acinetobacter baumannii* associated with pulsatile lavage wound treatment. *JAMA*, 2004, 292, 3006-11.

[19] Das, I; Lambert, P; Hill, D; Noy, M; Bion, J; Elliott, T Carbapenem-resistant *Acinetobacter* and role of curtains in an outbreak in intensive care units. *J Hosp Infect,* 2002, 50, 110-4.

[20] Wilks, M; Wilson, A; Warwick, S; et al. Control of an outbreak of multidrug-resistant *Acinetobacter baumannii-calcoaceticus* colonization and infection in an intensive care unit (ICU) without closing the ICU or placing patients in isolation. *Infect Control Hosp Epidemiol*, 2006, 27, 654-658.

[21] Dijkshoorn, L; Van Aken, E; Shunburne, L; Van Der Reijden, TJ; Bernards, AT; Nemec, A; Towner, KJ. Prevalence of *Acinetobacter baumannii* and other *Acinetobacter* spp. in faecal samples from non-hospitalised individuals. *Clin Microbiol Infect,* 2005, 11, 329-32.

[22] Larson, EL. Persistent carriage of gram-negative bacteria on hands. *Am J Infect Control*, 1981, 9, 112-119.

[23] Corbella, X; Pujol, M; Ayats, J; et al. Relevance of digestive tract colonization in the epidemiology of nosocomial infections due to multiresistant *Acinetobacter baumannii. Clin Infect Dis*., 1996, 23, 329–334.

[24] Furuno, JP; Hebden, JN; Standiford, HC; Perencevich, EN; Miller, RR; Moore, AC; Strauss, SM; Harris, AD. Prevalence of methicillin-resistant *Staphylococcus aureus* and *Acinetobacter baumannii* in a long-term acute care facility. *Am J Infect Control,* 2008, 36, 468-71.

[25] Apisarnthanarak, A; Pinitchai, U; Thongphubeth, K; Yuekyen, C; Warren, DK; Fraser, VJ; Thammasat University Pandrug-Resistant *Acinetobacter baumannii* Control Group. A multifaceted intervention to reduce pandrug-resistant *Acinetobacter baumannii* colonization and infection in 3 intensive care units in a Thai tertiary care center: a 3-year study. *Clin Infect Dis*., 2008, 47, 760-7.

[26] Stephens, C; Francis, SJ; Abell, V; DiPersio, JR; Wells, P. Emergence of resistant *Acinetobacter baumannii* in critically ill patients within an acute care teaching hospital and a long-term acute care hospital. *Am J Infect Control*, 2007, 35, 212-5.

[27] Siegel, JD; Rhinehart, E; Jackson, M; Chiarello, L; Health Care Infection Control Practices Advisory Committee. 2007 Guideline for Isolation Precautions: Preventing Transmission of Infectious Agents in Health Care Settings. *Am J Infect Control*, 2007, 35(10 Suppl 2), S65-164.

[28] Maragakis, LL; Perl, TM. *Acinetobacter baumannii*: Epidemiology, antimicrobial resistance, and treatment options. *Clin Infect Dis*., 2008, 46, 1254-63.

[29] Lee, SO; Kim, NJ; Choi, SH; Hyong, Kim, T; Chung, JW; Woo, JH; Ryu, J; Kim, YS. Risk factors for acquisition of imipenem-resistant *Acinetobacter baumannii*: a case-control study. *Antimicrob Agents Chemother*, 2004, 48, 224-8.

[30] Kim, YA; Choi, JY; Kim, CK; Kim, CO; Kim, MS; Choi, SH; Chin, BS; Han, SH; Lee, HS; Choi, HK; Jeoung, SJ; Song, YG; Yum, JH; Yong, D; Lee, K; Kim, JM. Risk factors and outcomes of bloodstream infections with metallo-beta-lactamase-producing *Acinetobacter*. *Scand J Infect Dis*., 2008, 40, 234-40.

[31] Sebeny, PJ; Riddle, MS; Petersen, K. *Acinetobacter baumannii* skin and soft-tissue infection associated with war trauma. *Clin Infect Dis*., 2008, 47, 444-9.

[32] Yun, HC; Branstetter, JG; Murray, CK. Osteomyelitis in military personnel wounded in Iraq and Afghanistan. *J Trauma*, 2008, 64(2 Suppl), S163-8

[33] Calhoun, JH; Murray, CK; Manring, MM. Multidrug-resistant organisms in military wounds from Iraq and Afghanistan. *Clin Orthop Relat Res.*, 2008, 466, 1356-62.

[34] Stuart, TL; Mulvey, M; Simor, AE; Tien, HC; Battad, A; Taylor, G; Vayalumkal, JV; Weir, C; Ofner, M; Gravel, D; Paton, S. *Acinetobacter baumannii* in casualties returning from Afghanistan. *Can J Infect Control*, 2007, 22(3), 152-4.

[35] Griffith, ME; Lazarus, DR; Mann, PB; Boger, JA; Hospenthal, DR; Murray, CK. *Acinetobacter* skin carriage among US army soldiers deployed in Iraq. *Infect Control Hosp Epidemiol*, 2007, 28, 720-2.

[36] Turton, JF; Kaufmann, ME; Gill, MJ; Pike, R; Scott, PT; Fishbain, J; Craft, D; Deye, G; Riddell, S; Lindler, LE; Pitt, TL. Comparison of *Acinetobacter baumannii* isolates from the United Kingdom and the United States that were associated with repatriated casualties of the Iraq conflict. *J Clin Microbiol*, 2006, 44, 2630-4.

[37] Griffith, ME; Ceremuga, JM; Ellis, MW; Guymon, CH; Hospenthal, DR; Murray, CK. *Acinetobacter* skin colonization of US Army Soldiers. *Infect Control Hosp Epidemiol,* 2006, 27, 659-61.

[38] Centers for Disease Control and Prevention (CDC). *Acinetobacter baumannii* infections among patients at military medical facilities treating injured U.S. service members, 2002-2004, *MMWR Morb Mortal Wkly Rep.*, 2004, 53, 1063-6.

[39] Griffith, ME; Gonzalez, RS; Holcomb, JB; Hospenthal, DR; Wortmann, GW; Murray, CK. Factors associated with recovery of *Acinetobacter baumannii* in a combat support hospital. *Infect Control Hosp Epidemiol*, 2008, 29, 664-6.

[40] Whitman, TJ; Qasba, SS; Timpone, JG; Babel, BS; Kasper, MR; English, JF; Sanders JW; Hujer, KM; Hujer, AM; Endimiani, A; Eshoo, MW; Bonomo, RA. Occupational transmission of *Acinetobacter baumannii* from a United States serviceman wounded in Iraq to a health care worker. *Clin Infect Dis.*, 2008, 47, 439-43.

[41] Wortmann, G; Weintrob, A; Barber, M; Scott, P; Zoll, ST; Eshoo, MW; Sampath, R; Ecker, DJ; Massire, C. Genotypic evolution of *Acinetobacter baumannii* strains in an outbreak associated with war trauma. *Infect Control Hosp Epidemiol*, 2008, 29, 553-5.

[42] Whitman, TJ. Infection control challenges related to war wound infections in the ICU setting. *J Trauma*, 2007, 62(6 Suppl), S53.

[43] Falagas, ME; Koletsi, PK; Bliziotis, IA. The diversity of definitions of multidrug-resistant (MDR) and pandrug-resistant (PDR) *Acinetobacter baumannii* and *Pseudomonas aeruginosa*. *J Med Microbiol*, 2006, 55, 1619-29

[44] Bonomo, RA; Szabo, D. Mechanisms of multidrug resistance in Acinetobacter species and Pseudomonas aeruginosa. *Clin Infect Dis.*, 2006, 43, Suppl 2, S49-S56.

[45] Gilad, J; Carmeli, Y. Treatment Options for Multidrug-Resistant Acinetobacter Species . *Drugs.*, 2008, 68, 165-189

[46] Van Looveren, M; Goossens, H. ARPAC Steering Group. Antimicrobial resistance of Acinetobacter spp. in Europe. *Clin Microbiol Infect*, 2004, 10, 684-704.

[47] Perez, F; Hujer, AM; Hujer, KM; Decker, BK; Rather, PN; Bonomo, RA. Global challenge of multidrug-resistant *Acinetobacter baumannii. Antimicrob Agents Chemother*, 2007t, 51, 3471-84.

[48] Poirel, L; Nordmann, P. Carbapenem resistance in Acinetobacter baumannii; mechanisms and epidemiology. *Clin Microbiol Infect.*, 2006, 12, 826-836

[49] Vila, J; Marcos, A; Marco, F; et al. In vitro antimicrobial production of b-lactamases, aminoglycoside-modifying enzymes, and chloramphenicol acetyltransferase by and

susceptibility of clinical isolates of *Acinetobacter baumannii. Antimicrob Agents Chemother*, 1993, 37, 138–41.

[50] Bou, G; Cervero, G; Dominguez, MA; Quereda, C; Martinez-Beltran, J. Characterization of a nosocomial outbreak caused by a multiresistant *Acinetobacter baumannii* strain with a carbapenem-hydrolyzing enzyme: high-level carbapenem resistance in *A. baumannii* is not due solely to the presence of b-lactamases. *J Clin Microbiol*, 2000, 38, 3299–305.

[51] Huang, ZM; Mao, PH; Chen, Y; Wu, L; Wu, J. Study on the molecular epidemiology of SHV type b-lactamase-encoding genes of multiple drug-resistant *Acinetobacter baumannii. Zhonghua Liu Xing Bing Xue Za Zhi.*, 2004, 25, 425–7.

[52] Vahaboglu, H; Ozturk, R; Aygun, G; et al. Widespread detection of PER-1-type extended-spectrum b-lactamases among nosocomial *Acinetobacter* and *Pseudomonas aeruginosa* isolates in Turkey: a nationwide multicenter study. *Antimicrob Agents Chemother*, 1997, 41, 2265–9.

[53] Yong, D; Shin, JH; Kim, S; et al. High prevalence of PER-1 extended spectrum b-lactamase-producing *Acinetobacter* spp. in Korea. *Antimicrob Agents Chemother*, 2003, 47, 1749–51.

[54] Poirel, L; Karim, A; Mercat, A; et al. Extended-spectrum b-lactamase producing strain of *Acinetobacter baumannii* isolated from a patient in France. *J Antimicrob Chemother*, 1999, 43, 157–8.

[55] Poirel, L; Menuteau, O; Agoli, N; Cattoen, C; Nordmann, P. Outbreak of extended-spectrum b-lactamase VEB-1-producing isolates of *Acinetobacter baumannii* in a French hospital. *J Clin Microbiol*, 2003, 41, 3542–7.

[56] Carbonne, A; Naas, T; Blanckaert, K; et al. Investigation of a nosocomial outbreak of extended-spectrum b-lactamase VEB-1-producing isolates of *Acinetobacter baumannii* in a hospital setting. *J Hosp Infect*, 2005, 60, 14–8.

[57] Poirel, L; Nordmann, P. Carbapenem resistance in *Acinetobacter baumannii*: mechanisms and epidemiology. *Clin Microbiol Infect.*, 2006, 12, 826-36.

[58] Walsh, TR. The emergence and implications of metallo-beta-lactamases in Gram-negative bacteria. *Clin. Microbiol. Infect*, 2005, 11(Suppl. 6), 2–9.

[59] Walsh, TR; Toleman, MA; Poirel, L; Nordmann, P. Metallo-ß- lactamases: the quiet before the storm? *Clin. Microbiol. Rev.*, 2005, 18, 306–325.

[60] Walther-Rasmussen, Hoiby, JN. OXA-type carbapenemases. *J. Antimicrob Chemother,* 2006, 57, 373–383.

[61] Coelho, J; Woodford, N; Afzal-Shah, M; et al. Occurrence of OXA-58-like carbapenemases in *Acinetobacter* spp. Collected over 10 years in three continents. *Antimicrob Agents Chemother*, 2006, 50, 756-8

[62] Heritier, C; Poirel, L; Lambert, T; et al. Contribution of acquired -hydrolyzing oxacillinases to carbapenem resistance in *Acinetobacter baumannii. Antimicrob Agents Chemother*, 2005 Aug, 49 (8), 3198-202

[63] Bou, G; Cervero, G; Dominguez, MA; et al. Characterization of a nosocomial outbreak caused by a multiresistant *Acinetobacter baumannii* strain with a carbapenem-hydrolyzing enzyme: high-level carbapenem resistance in *A. baumannii* is not due solely to the presence of beta-lactamases. *J Clin Microbiol*, 2000, 38, 3299-305

[64] Fernandez-Cuenca, F; Martinez-Martinez, L; Conejo, MC; et al. Relationship between beta-lactamase production, outer membrane protein and penicillin-binding protein

profiles on the activity of carbapenems against clinical isolates of *Acinetobacter baumannii. J Antimicrob Chemother*, 2003, 51, 565-74

[65] Peleg, AY; Adams, J; Paterson, DL. Tigecycline efflux as a mechanism for nonsusceptibility in Acinetobacter baumannii. Antimicrob *Agents Chemother*, 2007, 51, 2065-9.

[66] Seward, RJ; Lambert, T; Towner, KJ. Molecular epidemiology of aminoglycoside resistance in *Acinetobacter* spp. *J Med Microbiol*, 1998, 47, 455–62.

[67] Nemec, A; Dolzani, L; Brisse, S; Van Den Broek, P; Dijkshoorn, L. Diversity of aminoglycoside-resistance genes and their association with class 1 integrons among strains of pan-European *Acinetobacter baumannii* clones. *J Med Microbiol*, 2004, 53, 1233–40.

[68] Vila, J; Ruiz, J; Goni, P; Marcos, A; Jimenez, De Anta, T. Mutation in the *gyrA* gene of quinolone-resistant clinical isolates of *Acinetobacter baumannii. Antimicrob Agents Chemother*, 1995, 39, 1201–3.

[69] Vila, J; Ruiz, J; Goni, P; Jimenez De Anta, T. Quinolone-resistance mutations in the topoisomerase IV *parC* gene of *Acinetobacter baumannii. J Antimicrob Chemother,* 1997, 39, 757–62.

[70] Karageorgopoulos, DE; Kelesidis, T; Kelesidis, I; Falagas, ME. Tigecycline for the treatment of multidrug-resistant (including carbapenem-resistant) *Acinetobacter* infections: a review of the scientific evidence. *J Antimicrob Chemother,* 2008, 62, 45-55.

[71] Reid, GE; Grim, SA; Aldeza, CA; Janda, WN; Clark, NM. Rapid development of Acinetobacter baumannii resistance to tigecycline. *Pharmacotherapy*, 2007, 27, 1198-1201.

[72] Bosó-Ribelles, V; Romá-Sánchez, E; Carmena, J; Cáceres, C; Bautista, D. Tigecycline: a new treatment choice against *Acinetobacter baumannii. Recent Patents Anti-Infect Drug Disc.,* 2008, 3, 117-22.

[73] Gallagher, JC; Rouse, HM. Tigecycline for the treatment of *Acinetobacter* infections: a case series. *Ann Pharmacother*, 2008, 42, 1188-94.

[74] Gilad, J; Eskira, S; Riesenberg, K; Schlaeffer, F; Hyam, E; Borer, A. Emergence of nosocomial colistin-resistant *Acinetobacter baumannii. 45th Interscience Conference on Antimicrobial Agents and Chemotherapy*; Washington, DC, USA; Dec, 16–19, 2005.

[75] Goel, VK; Kapil, A. Monoclonal antibodies against the iron regulated outer membrane proteins of *Acinetobacter baumannii* are bactericidal. *BMC Microbiol*, 2001, 1, 16–24.

[76] E. Bergogne-Bérézin, ML; Joly-Guillou, KJ. Towner. *Acinetobacter: Microbiology, Epidemiology, Infections, Management.* Editors CRC Press, 1996.

[77] Obana, Y. Pathogenic significance of *Acinetobacter calcoaceticus*: analysis of experimental infection in mice. *Microbiol. Immunol,* 1986, 30, 645–657.

[78] Leone, S; Sturiale, L; Pessione, E; Mazzoli, R; Giunta, C; Lanzetta, R; Garozzo, D; Molinaro, A; Parrilli, M. Detailed characterization of the lipid A fraction from the nonpathogen *Acinetobacter radioresistens* strain S13. *Journal of Lipid Research*, 2007, 48, 1045-1051.

[79] Karageorgopoulos, DE; Falagas, ME. Current control and treatment of multidrug-resistant *Acinetobacter baumannii* infections. *Lancet Infect Dis*., 2008, 8, 751–762.

[80] Dijkshoorn, L; Nemec, A; Seifert, H. An increasing threat in hospitals: multidrug-resistant *Acinetobacter baumannii. Nat Rev Microbiol*, 2007, 5, 939–51.

[81] Van Looveren, M; Goossens, H. Antimicrobial resistance of *Acinetobacter* spp in Europe. *Clin Microbiol Infect*, 2004, 10, 684–704.

[82] Reinert, RR; Low, DE; Rossi, F; Zhang, X; Wattal, C; Dowzicky, MJ. Antimicrobial susceptibility among organisms from the Asia/ Pacific Rim, Europe and Latin and North America collected as part of TEST and the in vitro activity of tigecycline. *J Antimicrob Chemother*, 2007, 60, 1018–29.

[83] Poulakou, G; Giamarellou, H. Doripenem: an expected arrival in the treatment of infections caused by multidrug-resistant Gram-negative pathogens. *Expert Opin Investig Drugs,* 2008, 17, 749-71.

[84] Rice, LB. Challenges in identifying new antimicrobial agents effective for treating infections with *Acinetobacter baumannii* and *Pseudomonas aeruginosa. Clin Infect Dis*., 2006, 43 Suppl 2, S100-5.

[85] Horiuchi, M; Kimura, M; Tokumura, M; Hasebe, N; Arai, T; Abe, K. Absence of convulsive liability of doripenem, a new carbapenem antibiotic, in comparison with beta-lactam antibiotics. *Toxicology*, 2006, 222, 114-24

[86] Jones, RN; Sader, HS; Fritsche, TR. Comparative activity of doripenem and three other carbapenems tested against Gram-negative bacilli with various beta-lactamase resistance mechanisms. *Diagn Microbiol Infect Dis*., 2005, 52, 71-4

[87] Pillar, CM; Torres, MK; Brown, NP; Shah, D; Sahm, DF. In vitro activity of doripenem, a carbapenem for the treatment of challenging infections caused by gram-negative bacteria, against recent clinical isolates from the United States. *Antimicrob Agents Chemother*, 2008, 52, 4388-99.

[88] Jones, RN; Huynh, HK; Biedenbach, DJ. Activities of doripenem (S-4661) against drug-resistant clinical pathogens. *Antimicrob Agents Chemother*, 2004, 48, 3136-40.

[89] Zhanel, GG; Wiebe, R; Dilay, L; Thomson, K; Rubinstein, E; Hoban, DJ; Noreddin, AM; Karlowsky, JA. Comparative review of the carbapenems. *Drugs*, 2007, 67, 1027-52.

[90] Falagas, ME; Kasiakou, SK. Colistin: the revival of polymyxins for the management of multidrug-resistant Gram-negative bacterial infections. *Clin Infect Dis*., 2005, 40, 1333–41.

[91] Giamarellou, H. Treatment options for multidrug-resistant bacteria. *Expert rev Anti Infect Ther*., 2006, 4, 601-18

[92] Gales, AC; Jones, RN; Sader, HS. Global assessment of the antimicrobial activity of polymyxin B against 54,731 clinical isolates of Gram-negative bacilli: report from the SENTRY antimicrobial surveillance programme (2001–2004). *Clin Microbiol Infect*, 2006, 12, 315–21.

[93] Henwood, CJ; Gatward, T; Warner, M; et al. Antibiotic resistance among clinical isolates of *Acinetobacter* in the UK, and in vitro evaluation of tigecycline (GAR-936). *J Antimicrob Chemother*, 2002, 49, 479–87.

[94] Song, JY; Kee, SY; Hwang, IS; et al. In vitro activities of carbapenem/sulbactam combination, colistin, colistin/rifampicin combination and tigecycline against carbapenem-resistant *Acinetobacter baumannii. J Antimicrob Chemother*, 2007, 60, 317–22.

[95] Scheetz, MH; Qi, C; Warren, JR; et al. In vitro activities of various antimicrobials alone and in combination with tigecycline against carbapenem-intermediate or -resistant *Acinetobacter baumannii. Antimicrob Agents Chemother*, 2007, 51, 1621–26.

[96] Falagas, ME; Kasiakou, SK. Colistin: the revival of polymyxins for the management of multidrug-resistant gram-negative bacterial infections. *Clin Infect Dis*., 2005, 40, 1333–41.

[97] Li, J; Nation, RL; Turnidge, JD; et al. Colistin: the re-emerging antibiotic for multidrug-resistant Gram-negative bacterial infections. *Lancet Infect Dis*., 2006, 6, 589–601.

[98] Falagas, ME; Kasiakou, SK; Tsiodras, S; et al. The use of intravenous and aerosolized polymyxins for the treatment of infections in critically ill patients: a review of the recent literature. *Clin Med Res*., 2006, 4, 138–46.

[99] Li, J; Nation, RL; Milne, RW; et al. Evaluation of colistin as an agent against multi-resistant Gram-negative bacteria. *Int J Antimicrob Agents*, 2005, 25, 11–25.

[100] Michalopoulos, A; Kasiakou, SK; Rosmarakis, ES; Falagas, ME. Cure of multidrug-resistant *Acinetobacter baumannii* bacteraemia with continuous intravenous infusion of colistin. *Scand J Infect Dis*., 2005, 37, 142-5.

[101] Kallel, H; Hergafi, L; Bahloul, M; et al. Safety and efficacy of colistin compared with imipenem in the treatment of ventilator-associated pneumonia: a matched case-control study. *Intensive Care Med*., 2007, 33, 1162–67.

[102] Falagas, ME; Kasiakou, SK; Kofteridis, DP; Roditakis, G; Samonis, G. Effectiveness and nephrotoxicity of intravenous colistin for treatment of patients with infections due to polymyxin-onlysusceptible (POS) Gram-negative bacteria. *Eur J Clin Microbiol Infect Dis*., 2006, 25, 596–99.

[103] Michalopoulos, A; Kasiakou, SK; Mastora, Z; Rellos, K; Kapaskelis, AM; Falagas, ME. Aerosolized colistin for the treatment of nosocomial pneumonia due to multidrug-resistant Gram-negative bacteria in patients without cystic fibrosis. *Crit Care*, 2005, 9, R53–59.

[104] Rodriguez Guardado, A; Blanco, A; Asensi, V; et al. Multidrugresistant *Acinetobacter* meningitis in neurosurgical patients with intraventricular catheters: assessment of different treatments. *J Antimicrob Chemother*, 2008, 61, 908–13.

[105] Falagas, ME; Rizos, M; Bliziotis, IA; Rellos, K; Kasiakou, SK; Michalopoulos, A. Toxicity after prolonged (more than four weeks) administration of intravenous colistin. *BMC Infect Dis*., 2005, 5, 1.

[106] Garnacho-Montero, J; Ortiz-Leyba, C; Jimenez-Jimenez, FJ; et al. Treatment of multidrug-resistant *Acinetobacter baumannii* ventilator associated pneumonia (VAP) with intravenous colistin: a comparison with imipenem-susceptible VAP. *Clin Infect Dis*., 2003, 36, 1111–18.

[107] Reina, R; Estenssoro, E; Saenz, G; et al. Safety and efficacy of colistin in Acinetobacter and Pseudomonas infections: a prospective cohort study. *Intensive Care Med*., 2005, 31, 1058–65.

[108] Falagas, ME; Kasiakou, SK. Toxicity of polymyxins: a systematic review of the evidence from old and recent studies. *Crit Care*, 2006, 10, R27.

[109] Falagas, ME; Fragoulis, KN; Kasiakou, SK; Sermaidis, GJ; Michalopoulos, A. Nephrotoxicity of intravenous colistin: a prospective evaluation. *Int J Antimicrob Agents*, 2005, 26, 504–07.

[110] Ng, J; Gosbell, IB; Kelly, JA; Boyle, MJ; Ferguson, JK. Cure of multiresistant *Acinetobacter baumannii* central nervous system infections with intraventricular or

intrathecal colistin: case series and literature review. *J Antimicrob Chemother*, 2006, 58, 1078–81.

[111] Evagelopoulou, P; Katsaros, A; Myrianthefs, P; et al. Colistin and rhabdomyolysis: a causative agent or an innocent bystander? *Intensive Care Med.*, 2007, 33, 556–7.

[112] Matthaiou, DK; Michalopoulos, A; Rafailidis, PI; Karageorgopoulos, DE; Papaioannou, V; Ntani, G; Samonis; G; Falagas, ME. Risk factors associated with the isolation of colistin-resistant gram-negative bacteria: a matched case-control study. *Crit Care Med.*, 2008, 36, 807-11.

[113] Rafailidis, PI; Ioannidou, EN; Falagas, ME. Ampicillin/sulbactam: current status in severe bacterial infections. *Drugs*, 2007, 67, 1829–49.

[114] Ko, WC; Lee, HC; Chiang, SR; et al. In vitro and in vivo activity of meropenem and sulbactam against a multidrug-resistant *Acinetobacter baumannii* strain. *J Antimicrob Chemother*, 2004, 53, 393–95.

[115] Corbella, X; Ariza, J; Ardanuy, C; et al. Efficacy of sulbactam alone and in combination with ampicillin in nosocomial infections caused by multiresistant *Acinetobacter baumannii*. *J Antimicrob Chemother*, 1998, 42, 793–802.

[116] Wood, GC; Hanes, SD; Croce, MA; Fabian, TC; Boucher, BA. Comparison of ampicillin-sulbactam and imipenem-cilastatin for the treatment of *Acinetobacter* ventilator-associated pneumonia. *Clin Infect Dis.*, 2002, 34, 1425–30.

[117] Levin, AS; Levy, CE; Manrique, AE; Medeiros, EA; Costa, SF. Severe nosocomial infections with imipenem-resistant *Acinetobacter baumannii* treated with ampicillin/sulbactam. *Int J Antimicrob Agents*, 2003, 21, 58–62.

[118] Betrosian, AP; Frantzeskaki, F; Xanthaki, A; Georgiadis, G. High-dose ampicillin-sulbactam as an alternative treatment of late-onset VAP from multidrug-resistant *Acinetobacter baumannii*. *Scand J Infect Dis.*, 2007, 39, 38–43.

[119] Jimenez-Mejias, ME; Pachon, J; Becerril, B; Palomino-Nicas, J; Rodriguez-Corbacho, A; Revuelta, M. Treatment of multidrug-resistant *Acinetobacter baumannii* meningitis with ampicillin/sulbactam. *Clin Infect Dis.*, 1997, 24, 932–35.

[120] Oliveira, MS; Prado, GV; Costa, SF; Grinbaum, RS; Levin, AS. Ampicillin/sulbactam compared with polymyxins for the treatment of infections caused by carbapenem-resistant *Acinetobacter spp*. *J Antimicrob Chemother*, 2008, 61, 1369-75.

[121] Fraise, AP. Tigecycline: the answer to beta-lactam and fluorquinolone resistance? *J Infect*, 2006, 53, 293-300;

[122] Nathwani, D. Tigecycline: clinical evidence and formulary positioning. *Int J Antimicrobial Agents*, 2005, 25, 185-92.

[123] Rice, LB. Challenges in identifying new antimicrobial agents effective for treating infections with *Acinetobacter baumannii* and *Pseudomonas aeruginosa*. *Clin Infect Dis.*, 2006, 43 Suppl 2, S100-5.

[124] Seifert, H; Stefanik, D; Wisplinghoff, H. Comparative in vitro activities of tygecycline and 11 other antimicrobial agents against 215 epidemiologically defined multidrug – resistant *Acinetobacter baumannii* isolates. *J Antimicrob Chemother*, 2006, 58, 1099-100.

[125] Peleg, AY; Adams, J; Paterson, DL. Tigecycline efflux as a mechanism for nonsusceptibility in *Acinetobacter baumannii*. *Antimicrob Agents Chemother*, 2007, 51, 2065–69.

[126] Schafer, JJ; Goff, DA; Stevenson, KB; Mangino, JE. Early experience with tigecycline for ventilator-associated pneumonia and bacteremia caused by multidrug-resistant *Acinetobacter baumannii. Pharmacotherapy*, 2007, 27, 980–87.
[127] Karageorgopoulos, DE; Kelesidis, T; Kelesidis, I; Falagas, ME. Tigecycline for the treatment of multidrug-resistant (including carbapenem-resistant) acinetobacter infections: a review of the scientific evidence. *J Antimicrob Chemother*, 2008, 62, 45–55.
[128] Anthony, KB; Fishman, NO; Linkin, DR; Gasink, LB; Edelstein, PH; Lautenbach, E. Clinical and microbiological outcomes of serious infections with multidrug-resistant Gram-negative organisms treated with tigecycline. *Clin Infect Dis.,* 2008, 46, 567–70.
[129] Gallagher, JC; Rouse, HM. Tigecycline for the treatment of Acinetobacter infections: a case series. *Ann Pharmacother*, 2008, 42, 1188-94.
[130] Reid, GE; Grim, SA; Aldeza, CA; Janda, WM; Clark, NM. Rapid development of *Acinetobacter baumannii* resistance to tigecycline. *Pharmacotherapy*, 2007, 27, 1198-201.
[131] Halstead, DC; Abid, J; Dowzicky, MJ. Antimicrobial susceptibility among *Acinetobacter calcoaceticus-baumannii* complex and Enterobacteriaceae collected as part of the Tigecycline Evaluation and Surveillance Trial. *J Infect*, 2007, 55, 49–57
[132] Kuo, LC; Lai, CC; Liao, CH; et al. Multidrug-resistant *Acinetobacter baumannii* bacteraemia: clinical features, antimicrobial therapy and outcome. *Clin Microbiol Infect*, 2007, 13, 196–98
[133] Sahm, DF; Critchley, IA; Kelly, LJ; et al. Evaluation of current activities of fluorquinolones against gram-negative bacilli using centralized in vitro testing and electronic surveillance. *Antimicrob Agents Chemother*, 2001, 45, 267-74.
[134] Joly-Guillou, ML; Wolff, M; Farinotti, R; Bryskier, A; Carbon, C. In vivo activity of levofl oxacin alone or in combination with imipenem or amikacin in a mouse model of *Acinetobacter baumannii* pneumonia. *J Antimicrob Chemother*, 2000, 46, 827–30.
[135] Rahal, JJ. Novel antibiotic combinations against infections with almost completely resistant *Pseudomonas aeruginosa* and *Acinetobacter species*. *Clin Infect Dis*., 2006, 43 Suppl 2, S95-9.
[136] Song, JY; Cheong, HJ; Lee, J; Sung, AK; Kim, WJ. Efficacy of monotherapy and combined antibiotic therapy for carbapenem-resistant Acinetobacter baumannii pneumonia in an immunosuppressed mouse model. *Int J Antimicrob Agents*, 2009, 33, 33-9.
[137] Montero, A; Ariza, J; Corbella, X; Doménech, A; Cabellos, C; Ayats, J; Tubau, F; Borraz, C; Gudiol, F. Antibiotic combinations for serious infections caused by carbapenem-resistant *Acinetobacter baumannii* in a mouse pneumonia model. *J Antimicrob Chemother,* 2004, 54, 1085-91.
[138] Song, JY; Lee, J; Heo, JY; Noh, JY; Kim, WJ; Cheong, HJ; Hwang, IS. Colistin and rifampicin combination in the treatment of ventilator-associated pneumonia caused by carbapenem-resistant *Acinetobacter baumannii. Int J Antimicrob Agents*, 2008, 32, 281-4.
[139] Bassetti, M; Repetto, E; Righi, E; Boni, S; Diverio, M; Molinari, MP; Mussap, M; Artioli, S; Ansaldi, F; Durando, P; Orengo, G; Bobbio Pallavicini, F; Viscoli, C. Colistin and rifampicin in the treatment of multidrug-resistant *Acinetobacter baumannii* infections. *J Antimicrob Chemother*, 2008, 61, 417-20.

[140] Motaouakkil, S; Charra, B; Hachimi, A; Nejmi, H; Benslama, A; Elmdaghri, N; Belabbes, H; Benbachir, M. Colistin and rifampicin in the treatment of nosocomial infections from multiresistant *Acinetobacter baumannii*. *J Infect.*, 2006, 53, 274-8.

[141] Saballs, M; Pujol, M; Tubau, F; Peña, C; Montero, A; Domínguez, MA; Gudiol, F; Ariza, J. Rifampicin/imipenem combination in the treatment of carbapenem-resistant *Acinetobacter baumannii* infections. *J Antimicrob Chemother*, 2006, 58, 697-700.

[142] Lee, NY; Wang, CL; Chuang, YC; et al. Combination carbapenemsulbactam therapy for critically ill patients with multidrug-resistant *Acinetobacter baumannii* bacteremia: four case reports and an in vitro combination synergy study. *Pharmacotherapy*, 2007, 27, 1506–11.

[143] Timurkaynak, F; Can, F; Azap, OK; Demirbilek, M; Arslan, H; Karaman, SO. In vitro activities of non-traditional antimicrobials alone or in combination against multidrug-resistant strains of *Pseudomonas aeruginosa* and *Acinetobacter baumannii* isolated from intensive care units. *Int J Antimicrob Agents*, 2006, 27, 224–28.

[144] Yoon, J; Urban, C; Terzian, C; Mariano, N; Rahal, JJ. In vitro double and triple synergistic activities of polymyxin B, imipenem, and rifampin against multidrug-resistant *Acinetobacter baumannii*. *Antimicrob Agents Chemother*, 2004, 48, 753–57.

[145] Falagas, ME; Rafailidis, PI; Kasiakou, SK; Hatzopoulou, P; Michalopoulos, A. Eff ectiveness and nephrotoxicity of colistin monotherapy *vs* colistin-meropenem combination therapy for multidrug-resistant Gram-negative bacterial infections. *Clin Microbiol Infect*, 2006, 12, 1227–30.

[146] Scheetz, MH; Qi, C; Warren, JR; et al. In vitro activities of various antimicrobials alone and in combination with tigecycline against carbapenem-intermediate or -resistant *Acinetobacter baumannii*. *Antimicrob Agents Chemother*, 2007, 51, 1621–26.

[147] Lortholary, O; Fagon, JY; Hoi, AB; et al. Nosocomial acquisition of multi-resistant *Acinetobacter baumannii*: Risk factors and prognosis.*Clin Infect Dis.*, 1995, 20, 790-796.

[148] Siegman-Igra, Y; Bar, Yosef; S; Gorea, A; Avram, J. Nosocomial A*cinetobacter* meningitis secondary to invasive procedures: report of 25 cases and review. *Clin Infect Dis.*, 1993, 17, 843-849.

[149] Seifert, H; Ritcher, W; Pulverer, G. Clinical and bacteriological features of relapsing shunt-associated meningitis due to *Acinetobacter baumannii*. *Eur J Clin Microbiol Infect Dis.*, 1995, 14, 130-134.

[150] Chen, HP; Lai, Ch; Chan, Y; Chen, T; Liu, CH; Fung, CH; Liu, CH. Clinical significance of *Acinetobacter* species isolated from cerebrospinal fluid. *Scand J Infect Dis.*, 2005, 37, 669-675

[151] Rodríguez-Guardado, A; Maradona, JA; Asensi, V; Cartón, JA; Pérez, F; Blanco, A; Arribas, JM. Meningitis postquirúrgica por *Acinetobacter baumannii*: estudio de 22 casos y revisión de la literatura. *Rev Clin Esp.*, 2001, 201, 497-500

[152] Mayhall, CG; Archer, NH; Archer Lamb, V; Spadora, AC; Baggett, JW; Ward, JD; Narayan, RK. Ventriculostomy-related infections. A prospective epidemiologic study. *N Engl J Med.*, 1984, 310, 553-559.

[153] Wang, KW; Chang, WN; Huang, CR; Tsai, NW; Tsui, HW; Wang, HC; Su, TM; Rau, CS; Cheng, BC; et al. Post-neurosurgical nosocomial bacterial meningitis in adults: microbiology, clinical features, and outcomes. *J Clin NeuroSci.*, 2005, 12, 647-650

[154] Mahgoub, S; Ahmed, J; Glatt, AE. Completely resistant *Acinetobacter baumannii* strains. *Infect control Hosp Epidemiol,*, 2002, 23, 477-79.

[155] Nguyen, MH; Harris, SP; Muder, RR; Pasculle, AW. Antibiotic- resistant *Acinetobacter* meningitis in neurosurgical patients. *Neurosurgery*, 1994, 35, 851-855.

[156] Gilbert, VE; Velas, Jr JD; Natelson, SE; Tyler, WA. Treatment of cerebrospinal fluid leaks and gram-negative bacillary meningitis with large doses of intratecal amikacin and systemic antibiotics. *Neurosurgery*, 1986, 18, 402-406.

[157] Jiménez-Mejías, ME; Pichardo-Guerrero, C; Marquez-Rivas, FJ; Martin-Lozano, D; Prados, T; Pachón, J. Cerebrospinal fluid penetration and pharmacokinetic /pharmacodynamic parameters of intravenously administered colistin in a case of multidrug-resistant *Acinetobacter baumannii* meningitis. *Eur J Clin Microbiol Infect Dis.*, 2002, 21, 212-214

[158] Jiménez-Mejías, ME; Becerril, B; Marquez-Rivas, FJ; Pichardo, C; Cuberos, L; Pachón, J. Sucessfull treatment of multidrug- resistant *Acinetobacter baumannii* meningitis with intravenous colisitn sulfomethate sodium. *Eur J Clin Microbiol Infect Dis.,* 2000, 19, 970-1

[159] Pankey, GA. Tigecycline. *J Antimicrob Chemother*, 2005, 56, 470-480

[160] Noskin, GA. Tigecycline: A new glycylcycline for treatment of serious infections. *Clin Infect Dis.*, 2005, 41, S303-14

[161] Falagas, ME; Bliziotis, IA; Tam, VH. Intraventricular or intrathecal use of polymyxins in patients with Gram-negative meningitis: a systematic review of the available evidence. *Int J Antimicrob Agents*, 2007, 29, 9–25

[162] Levin, AS; Barone, AA; Penco, J; et al. Intravenous colistin as therapy for nosocomial infections caused by multidrug-resistant *Pseudomonas aeruginosa* and *Acinetobacter baumannii*. *Clin Infect Dis.*, 1999, 28, 1008–11

[163] Katragkou, A; Roilides, E. Successful treatment of multidrugresistant *Acinetobacter baumannii* central nervous system infections with colistin. *J Clin Microbiol*, 2005, 43, 4916–17.

[164] Paramythiotou, E; Karakitsos, D; Aggelopoulou, H; Sioutos, P; Samonis, G; Karabinis, A. Post-surgical meningitis due to multiresistant *Acinetobacter baumannii*. Effective treatment with intravenous and/or intraventricular colistin and therapeutic dilemmas. *Med Mal Infect*, 2007, 37, 124–25

[165] Cawley, MJ; Suh, C; Lee, S; Ackerman, BH. Nontraditional dosing of ampicillin-sulbactam for multidrug-resistant *Acinetobacter baumannii* meningitis. *Pharmacotherapy*, 2002, 22, 527-532

[166] Benifla, M; Zucker, G; Cohen, A; Alkan, M. Successful treatment of *Acinetobacter* meningitis with intrathecal polymixyn E. *J Antimicrob Chemother*, 2004, 54, 290-292

[167] Gleeson, T; Petersen, K; Mascola, J. Successful treatment of *Acinetobacter* meningitis with meropenm and rifampicin. *J Antimicrob chemother*, 2005, 56, 602-603.

[168] Vasen, W; Desmery, P; Ilutovich, S; Di Martino, A. Intrathecal use of colistin. *Journal of Clinical Microbiology*, 2000, 38, 3523.

[169] Karakitsos, D; Paramythiotou, E; Samonis, G; et al. Is intraventricular colistin an effective and safe treatment for post-surgical ventriculitis in the intensive care unit? *Acta Anaesthesiol Scand*, 2006, 50, 1309-1310.

[170] Montero, A; Corbella, X; Ariza, J. Clinical relevance of *Acinetobacter baumannii* ventilator-associated pneumonia. *Crit Care Med.*, 2003, 31, 2557-9

[171] Kollef, KE; Schramm, GE; Wills, AR; Reichley, RM; Micek, ST; Kollef, MH. Predictors of 30-day mortality and hospital costs in patients with ventilator-associated pneumonia attributed to potentially antibiotic-resistant gram-negative bacteria. *Chest*, 2008, 134, 281-7.

[172] Luna, CM; Aruj, PK. Nosocomial *Acinetobacter* pneumonia. *Respirology*, 2007, 12, 787-91.

[173] Garnacho-Montero, J; Ortiz-Leyba, C; Fernández-Hinojosa, E; Aldabó-Pallás, T; Cayuela, A; Marquez-Vácaro, JA; Garcia-Curiel, A; Jiménez-Jiménez, FJ. *Acinetobacter baumannii* ventilator-associated pneumonia: epidemiological and clinical findings. *Intensive Care Med.*, 2005, 31, 649-55.

[174] Medina, J; Formento, C; Pontet, J; Curbelo, A; Bazet, C; Gerez, J; Larrañaga, E. Prospective study of risk factors for ventilator-associated pneumonia caused by *Acinetobacter species*. *J Crit Care*, 2007, 22, 18-26.

[175] Rios, FG; Luna, CM; Maskin, B; Saenz Valiente, A; Lloria, M; Gando, S; Sosa, C; Baquero, S; Llerena, C; Petrati, C; Apezteguia, C. Ventilator-associated pneumonia due to colistin susceptible-only microorganisms. *Eur Respir J.*, 2007, 30, 307-13.

[176] Bernabeu-Wittel, M; Pichardo, C; García-Curiel, A; Pachón-Ibáñez, ME; Ibáñez-Martínez, J; Jiménez-Mejías, ME; Pachón, J. Pharmacokinetic/pharmacodynamic assessment of the in-vivo efficacy of imipenem alone or in combination with amikacin for the treatment of experimental multiresistant *Acinetobacter baumannii* pneumonia. *Clin Microbiol Infect*, 2005, 11, 319-25.

[177] Berlana, D; Llop, JM; Fort, E; Badia, MB; Jodar, R. Use of colistin in the treatment of multiple-drug-resistant Gram-negative infections. *Am J Health Syst Pharm.*, 2005, 62, 39–47.

[178] Michalopoulos, A; Fotakis, D; Virtzili, S; et al. Aerosolized colistin as adjunctive treatment of ventilator-associated pneumonia due to multidrug-resistant Gram-negative bacteria: a prospective study. *Respir Med.*, 2008, 102, 407–12.

[179] Kwa, AL; Loh, C; Low, JG; Kurup, A; Tam, VH. Nebulized colistin in the treatment of pneumonia due to multidrug-resistant *Acinetobacter baumannii* and *Pseudomonas aeruginosa*. *Clin Infect Dis.*, 2005, 41, 754–57.

[180] Michalopoulos, A; Kasiakou, SK; Mastora, Z; et al. Aerosolized colistin for the treatment of nosocomial pneumoniae due to multidrug-resistant Gram-negative bacteria in patients without cystic fibrosis. *Crit Care*, 2005, 9, R53-59.

[181] Betrosian, AP; Frantzeskaki, F; Xanthaki, A; Georgiadis, G. High-dose ampicillin-sulbactam as an alternative treatment of late-onset VAP from multidrug-resistant *Acinetobacter baumannii*. *Scand J Infect Dis*, 2007, 39, 38-43.

[182] Betrosian, AP; Frantzeskaki, F; Xanthaki, A; Douzinas, EE. Efficacy and safety of high-dose ampicillin/sulbactam vs. colistin as monotherapy for the treatment of multidrug resistant *Acinetobacter baumannii* ventilator-associated pneumonia. *J Infect*, 2008, 56, 432-6.

[183] Kollef, MH. Prevention of hospital-associated pneumonia and ventilator-associated pneumonia. *Crit Care Med.*, 2004, 32, 1396-405.

[184] Cisneros, JM. Rodríguez-Baño JNosocomial bacteremia due to Acinetobacter baumannii: epidemiology, clinical features and treatment. *Clin Microbiol Infect,* 2002, 8, 687-93.

[185] Shih, MJ; Lee, NY; Lee, HC; Chang, CM; Wu, CJ; Chen, PL; Ko, NY; Ko, WC. Risk factors of multidrug resistance in nosocomial bacteremia due to *Acinetobacter baumannii*: a case-control study. *J Microbiol Immunol Infect*, 2008, 41, 118-23.

[186] Wareham, DW; Bean, DC; Khanna, P; Hennessy, EM; Krahe, D; Ely, A; Millar, M. Bloodstream infection due to *Acinetobacter spp*: epidemiology, risk factors and impact of multi-drug resistance. *Eur J Clin Microbiol Infect Dis.*, 2008, 27, 607-12.

[187] Wisplinghoff, H; Edmond, MB; Pfaller, MA; Jones, RN; Wenzel, RP; Seifert, H. Nosocomial bloodstream infections caused by *Acinetobacter* species in United States hospitals: clinical features, molecular epidemiology, and antimicrobial susceptibility. *Clin Infect Dis.*, 2000, 31, 690-7.

[188] García-Garmendia, JL; Ortiz-Leyba, C; Garnacho-Montero, J; Jiménez-Jiménez, FJ; Pérez-Paredes, C; Barrero-Almodóvar, AE; Gili-Miner, M. Risk factors for *Acinetobacter baumannii* nosocomial bacteremia in critically ill patients: a cohort study. *Clin Infect Dis.*, 2001, 33, 939-46.

[189] Kuo, LC; Lai, CC; Liao, CH; Hsu, CK; Chang, YL; Chang, CY; Hsueh, PR. Multidrug-resistant Acinetobacter baumannii bacteraemia: clinical features, antimicrobial therapy and outcome. *Clin Microbiol Infect.*, 2007, 13, 196-8.

[190] Wisplinghoff, H; Edmond, MB; Pfaller, MA; Jones, RN; Wenzel, RP; Seifert, H. Nosocomial bloodstream infections caused by *Acinetobacter species* in United States hospitals: clinical features, molecular epidemiology, and antimicrobial susceptibility. *Clin Infect Dis.*, 2000, 31, 690-7.

[191] Tseng, YC; Wang, JT; Wu, FL; Chen, YC; Chie, WC; Chang, SC. Prognosis of adult patients with bacteremia caused by extensively resistant *Acinetobacter baumannii. Diagn Microbiol Infect Dis.*, 2007, 59, 181-90.

[192] Cisneros, JM; Reyes, MJ; Pachón, J; Becerril, B; Caballero, FJ; García-Garmendía, JL; Ortiz, C; Cobacho, AR. Bacteremia due to *Acinetobacter baumannii:* epidemiology, clinical findings, and prognostic features. *Clin Infect Dis.*, 1996, 22, 1026-32.

[193] Poutanen, SM; Louie, M; Simor, AE. Risk factors, clinical features and outcome of *Acinetobacter* bacteremia in adults. *Eur J Clin Microbiol Infect Dis.*, 1997, 16, 737-40.

[194] Valero, C; García Palomo, JD; Matorras, P; Fernández-Mazarrasa, C; González, Fernández; C; Fariñas, MC. *Acinetobacter* bacteraemia in a teaching hospital, 1989-1998. *Eur J Intern Med.*, 2001, 12, 425-429.

[195] Choi, JY; Park, YS; Kim, CO; Park, YS; Yoon, HJ; Shin, SY; Kim, YA; Song, YG; Yong, D; Lee, K; Kim, JM. Mortality risk factors of *Acinetobacter baumannii* bacteraemia. *Intern Med J.,* 2005, 35, 599-603.

[196] Grupper, M; Sprecher, H; Mashiach, T; Finkelstein, R. Attributable mortality of nosocomial *Acinetobacter* bacteremia. *Infect Control Hosp Epidemiol*, 2007, 28, 293-8.

[197] Kuo, LC; Lai, CC; Liao, CH; Hsu, CK; Chang, YL; Chang, CY; Hsueh, PR. Multidrug-resistant *Acinetobacter baumannii* bacteraemia: clinical features, antimicrobial therapy and outcome. *Clin Microbiol Infect*, 2007, 13, 196-8

[198] Jellison, TK; Mckinnon, PS; Rybak, MJ. Epidemiology, resistance, and outcomes of Acinetobacter baumannii bacteremia treated with imipenem-cilastatin or ampicillin-sulbactam. *Pharmacotherapy*, 2001, 21, 142-8.

[199] Kwon, KT; Oh, WS; Song, JH; Chang, HH; Jung, SI; Kim, SW; Ryu, SY; Heo, ST, Jung, DS; Rhee, JY; Shin, SY; Ko, KS; Peck, KR; Lee, NY. Impact of imipenem

resistance on mortality in patients with *Acinetobacter* bacteraemia. *J Antimicrob Chemother*, 2007, 59, 525-30.

[200] Falagas, ME; Karveli, EA; Kelesidis, I; Kelesidis, T. Community-acquired *Acinetobacter* infections. *Eur J Clin Microbiol Infect Dis*., 2007, 26, 857-68.

[201] Wang, JT; McDonald, LC; Chang, SC; Ho, M. Community-acquired *Acinetobacter baumannii* bacteremia in adult patients in Taiwan. *J Clin Microbiol*., 2002, 40, 1526-9.

[202] Chen, MZ; Hsueh, PR; Lee, LN; Yu, CJ; Yang, PC; Luh KT. Severe community-acquired pneumonia due to *Acinetobacter baumannii*. *Chest*, 2001, 120, 1072-7.

[203] Chang, WN; Lu, CH; Huang, CR; Chuang, YC. Community-acquired *Acinetobacter* meningitis in adults. *Infection*, 2000, 28, 395-7.

[204] Alonso Fernández, A; Alvarez-Sala Walter, R; Prados Sánchez, C; Mayoralas Alises, S; Villamor León, J. Community-acquired empyema caused by *Acinetobacter baumannii*. *An Med Interna*, 2001, 18, 529-30.

[205] Leung, WS; Chu, CM; Tsang, KY; Lo, FH; Lo, KF; Ho, PL. Fulminant community-acquired *Acinetobacter baumannii* pneumonia as a distinct clinical syndrome. *Chest*, 2006, 129, 102-9.

[206] Drault, JN; Herbland, A; Kaidomar, S; Mehdaoui, H; Olive, C; Jouanelle, J. Community-acquired Acinetobacter baumannii pneumonia. *Ann Fr Anesth Reanim*, 2001, 20, 795-8.

[207] Kollef, MH; Fraser, VJ. Antibiotic resistance in the intensive care unit. *Ann Intern Med*, 2001, 134, 298–314.

[208] Timsit, JF; Garrait, V; Misset, B; et al. The digestive tract is a major site for *Acinetobacter baumannii* colonization in intensive care unit patients. *J Infect Dis*., 1993, 168, 1336–7.

[209] Agusti, C; Pujol, M; Argerich, MJ; et al. Short-term effect of the application of selective decontamination of the digestive tract on different body site reservoir ICU patients colonized by multi-resistant *Acinetobacter baumannii*. *J Antimicrob Chemother*, 2002, 49, 205–8.

[210] Podnos, YD; Cinat, ME; Wilson, SE; Cooke, J; Gornick, W; Thrupp, LD. Eradication of multi-drug resistant *Acinetobacter* from an intensive care unit. *Surg Infect (Larchmt)*, 2001, 2, 297-301.

[211] Aygun, G; Demirkiran, O; Utku, T; et al. Environmental contamination during a carbapenem-resistant *Acinetobacter baumannii* outbreak in an intensive care unit. *J Hosp Infect*., 2002, 52, 259-62.

[212] Zanetti, G; Blanc, DS; Federli, I; et al. Importation of *Acinetobacter baumannii* into a burn unit: a recurrent outbreak of infection associated with widespread environmental contamination. *Infect Control Hosp Epidemiol*, 2007, 28, 723-725.

[213] Wang, SH; Sheng, WH; Chang, YY; et al. Healthcare-associated outbreak due to pan-drug resistant *Acinetobacter baumannii* in a surgical intensive care unit. *J Hosp Infect*, 2003, 53, 97-102

[214] Urban, C; Segal-Maurer, S; Rahal, JJ. Considerations in control and treatment of nosocomial infections due to multidrug-resistant *Acinetobacter baumannii*. *Clin Infect Dis*., 2003, 36, 1268-74

[215] Siegel, JD; Rhinehart, E; Jackson, M; Chiarello, L. Healthcare Infection Control Practices Advisory Committee. Management of multidrug-resistant organisms in healthcare settings. Atlanta: *Centers for Disease Control and Prevention*, 2006.

Available at: http://www.cdc.gov/ncidod/dhqp/pdf/ar/mdroguideline2006.pdf. Accessed 9 March 2007.

In: Methicillin-Resistant Staphylococcus Aureus...
Editor: Charles L. Kolendi

ISBN: 978-1-60741-398-1

Chapter 9

The Anatomy of an Outbreak: Lessons Learned from an Outbreak of Methicillin-Resistant *Staphylococcus aureus* among Newborns and a Review of the Literature

Nevio Cimolai*

Department of Pathology and Laboratory Medicine, Faculty of Medicine, The University of British Columbia.

Abstract

As methicillin-resistant *Staphylococcus aureus* (MRSA) has become established in acute care hospitals, high risk or routine newborn care has not been spared. What was once considered community-acquired MRSA has also become established nosocomially and will inevitably affect newborns. An example of a neonatal intensive care unit outbreak with MRSA is highlighted, and the lessons gleaned from the same are detailed. A multi-directional approach to outbreak control is imperative.

The spectrum of MRSA infection in the newborn essentially parallels that due to methicillin-susceptible *Staphylococcus aureus*. Spread to the newborn can occur from maternal, other family, caregiver, and environmental sources. Risk factors for MRSA acquisition include low birth weight, decreased gestational age, multiple gestation, and intensive care. Once established as an outbreak, or for endemic settings otherwise, containment efforts have been quite variable, but some newborn centres have not been successful. MRSA poses several logistical problems for active treatment of infection or for attempted eradication of the carrier state. Prevention strategies for the neonatal intensive care unit are critical in the current context of a global MRSA pandemic in which hypervirulent clones are becoming established in many settings.

With the ability to control various factors given the immobility of the newborn, the scenario of newborn MRSA colonization and infection lends itself to critical related research.

* Corresponding author: Department of Pathology and Laboratory Medicine Children's and Women's Centre of British Columbia Vancouver, British Columbia Canada V6H 3V4, Phone 604-271-9321, FAX 604-271-9321, e-mail: ncimolai@interchange.ubc.ca

Keywords: *Staphylococcus aureus*, methicillin-resistant, infection control, newborn, neonatal intensive care.

Introduction

As methicillin-resistant *Staphylococcus aureus* (MRSA) emerged worldwide as a significant pathogen in medical institutions, and now the general community, it was inevitable that it would not spare the venues that serve to intensively care for newborns. Such an emerging threat was clearly being established in the new millennium as was predicted from prior experience with methicillin-susceptible *Staphylococcus aureus* (MSSA).[1] MRSA is no less virulent than MSSA,[2,3] and the eventual selection of antibiotic resistance with more potent *S. aureus* virulence factors (e.g., Panton-Valentine leukocidin, e.g., toxic shock syndrome toxin (TSST-1), e.g., exfoliative toxin) has the potential to create a burdensome and formidable pathogen. One must also accept the fact that *S. aureus* is in general, whether MSSA or MRSA, commonly human associated (e.g., as normal flora) and will always likely be a frequent human pathogen as it has been in the past regardless of the aforementioned apparent virulence factors.

The dramatic impact of MRSA and its subsequent effect on newborn care have been widely chronicled in, and is ably illustrated by, the Japanese experience[4,5] where carrier rates in medical institutions have at times exceeded 75%. As the bacterium becomes institutionalized on such a gargantuan basis, the potential for significant secondary morbidity has become realized, and control efforts have become exceedingly difficult for some. The emergence of MRSA that is capable of causing neonatal toxic shock syndrome-like exanthematous disease (NTED) in Japanese nurseries is an example of how evolutionary pressure will select for more aggressive and stubborn pathogens.[4] The prospect for similar epidemic spread will be anticipated in all countries of the world unless effective strategies for containment are realized and effected.[6] The impact of MRSA on pediatric facilities in North America has also been gaining momentum.[7]

MRSA colonization and infection are preventable.[8] Whereas the medical literature is complete with examples of hospital outbreaks and attempted remedies, control efforts are not uncommonly stymied for uncertain reasons. In order to effect change, the medical community must come to grips with the complexities of MRSA. The context of the newborn intensive care unit can provide for an excellent model in the understanding of such complexity. It is with the latter realization that such a model is extensively explored in this review of a newborn intensive care outbreak. The goal herein is not to resurrect old shadows or to focus culpability, but rather to academically provide the medical community with insights that may translate into real action and real prevention.

Anatomy of an Outbreak

The venue for the outbreak was a tertiary care pediatric centre which served as the provincial reference centre for child health. The centre served both ambulatory and in-hospital patients, and included both primary and complex care. The centre was conjoined to a tertiary maternity and women's hospital as well as a small senior's auxiliary care residence. Within the pediatric centre, there resided a neonatal intensive care unit with approximately 50-60 bed spaces (including two isolation rooms and a dozen convalescent care cots), the availability of which depended on staffing and available funding. Of the total patient bed equivalents, the active patient census occupied anywhere from 30-50 spaces at any one time.

Within both the pediatric and women's centres, nosocomially acquired MRSA was not recognized over a 15 year period since the hospitals' inceptions. The first such nosocomially acquired MRSA arose in mid-1998 when a child in the pediatric surgical ward was found to have acquired MRSA several weeks after admission. Through a screening of staff and patients, the only putative linkage was medical staff who attended this patient and who also served several hospital sites including a nearby adult-oriented hospital which had widespread endemic MRSA.

Four months later in that year, the first nosocomial transmission of MRSA occurred in the neonatal intensive care unit and affected one patient. A single staff carrier was discovered, the latter who had linkage with the case previously described from the surgical ward. The staff member was decolonized successfully. Over the next three months, a further three newborns were found to carry MRSA. The second child died with MRSA sepsis, and another staff carrier linkage was established. Newborn screening was initiated and focused staff screening was included when deemed appropriate given the existing epidemiological data. In early 1999, two more MRSA-positive newborns were identified and again two carrier staff were associated with the latter. Thereafter, individual foci or small clusters of MRSA-positive newborns became apparent. At the peak of such activity, up to eight carrier infants co-existed in the unit. A variety of approaches were taken to curb the outbreak (Table 1), but the implementation and maintenance of such efforts were not uncommonly fraught with obstacles (Table 2). A number of key epidemiological findings were noted (Table 3). During a two year period, 46 infants were found carriers, several of which developed an infection, whether deep or superficial. The latter included infections of skin, eye, deep septic thrombosis, pulmonary, blood, and central nervous system foci. During the entirety, the infection control policy of the centre was to maintain strict isolation for MRSA carriers which included the use of a negative pressure isolation room, gowning, gloving, and masking. In the neonatal intensive care, only two negative pressure rooms existed. When the capacity was exceeded, carrier patients were nursed in cohort, preferably in one large room within the unit. Direct patient contact by any staff required gloves, gown, and mask. Patients were given the option of utilizing the same isolation precautions as staff or, if desiring unimpeded newborn contact, e.g., breastfeeding, of using a gown and mandating thorough handwashing. Parents who wished such unimpeded contact with a carrier newborn were only allowed access into the unit but were not allowed to use other hospital facilities. Decolonization of infants per se was not attempted although all staff carriers were excluded from work temporarily and treated for carriage successfully. Typing (coagulase gene) data revealed commonality among newborn and staff isolates. A variant of the epidemic strain was identified approximately one year into the outbreak in which previous erythromycin and clindamycin resistance was lost. The bacterium was resistant to beta-lactam agents, clindamycin, erythromycin, cotrimoxazole, and amino glycosides, but susceptible to vancomycin, rifampin, chloramphenicol, fucidic acid, and mupirocin.

Table 1. Approaches toward infection control of the outbreak.

A. *Informational*
i. Establishing an open question/answer board for all unit staff.
ii. Regular and ad hoc meetings between infection control and unit managers (nursing and medical staff).
iii. Informational sheets for staff regarding the outbreak and the justification for staff screening when required.
iv. Academic medical rounds to discuss MRSA in general and for the unit specifically.
v. Ongoing tabulation and epidemiological review by a neonatal medical fellow.
vi. Ongoing appraisal of outbreak status for hospital administration.
vii. Updating transport teams.
viii. Updating respiratory therapy support.
ix. Regular in-servicing for unit nursing staff.
x. Updating surgical and anesthesia services.
B. *Support oriented*
i. Identification of an individual neonatal intensive care unit nurse for the purpose of co-ordinating MRSA-oriented interventions and surveillance.
ii. Identification of an individual neonatologist for the purpose of co-ordinating MRSA-oriented interventions and surveillance.
iii. Twenty-four hour availability of infection control team services.
iv. Public relations plans – both internal and external.
C. *Integrational*
i. Establishing and maintaining key contacts in housekeeping and maintenance for infection control purposes.
ii. Open door policy for daily and ad hoc (hourly) interactions with infection control to discuss problems and changes, or to enhance existing infection control practices.
iii. Meetings with hospital administration and key stakeholders to mobilize resources as required.
iv. Integration with occupational health and wellness services.
v. Infection control subcommittee formed for the outbreak.
D. *Interventional*
i. Daily walk-about surveillance rounds of the unit by infection control staff.
ii. Enhanced floor cleaning protocols.
iii. Enhanced environmental cleaning.
iv. Admission flagging system for MRSA positive newborns after discharge from the unit.
v. Ongoing tabulation of the epidemiology of spread by the infection control service and associated ad hoc meetings to disseminate the information with the unit.
vi. Regularly scheduled newborn surveillance by screening cultures.
vii. Strict isolation for MRSA as possible.
viii. Monitoring of infection control technique adherence by staff, support workers, parents, and visitors.
ix. Coroner's review of an MRSA-associated death.
x. Timely discharge of culture-positive patients.
xi. Temporary gloving and gowning for all patient contact.
xii. Cohorting patients when strict isolation facilities were exceded.
xiii. Enhanced terminal cleaning of rooms after patient discharge.

Table 2. Issues arising which complicated and/or hindered the outbreak control.

A. ***General logistic***
i. Defeatism – accepting lost cause sentiment at the strata of administration, unit staff, external medical staff, and families.
ii. Flat denial by some unit staff to engage mandated infection control measures.
iii. Common infection control technique breaches at all levels.
iv. High endemicity of MRSA in a nearby large teaching hospitals.
v. Frequent visitation by students and medical staff from nearby teaching hospitals.
vi. Staff working in multiple facilities including those which were highly endemic for MRSA, e.g., respiratory therapy.
vii. Connectivity with other neonatal intensive care units in the region and the discordance for isolation and screening.
viii. Contiguity of the hospital with an adult chronic care facility which had endemic MRSA.
ix. International trainees at all levels and the potential to import MRSA.
x. Personal trauma and burn-out secondary to enhanced infection control activities – both for caregivers and families.
xi. Prevailing attitude of some to take 'harm reduction' approach rather than absolute control.
xii. An excess of care item paraphernalia in the neonatal intensive care unit setting.
xiii. An excess of other peripheral materials in the unit, e.g., quilts, covers, linens, gowns.
xiv. Maintaining food at patient bedside.
xv. Slow implementation of Coroner's report.
xvi. Denial by some to monitor typing data for MRSA isolates.
xvii. Co-emergence of an outbreak of multi-resistant *Escherichia coli* in the unit.
xviii. Other needs for isolation requirements given the multitude of other infectious agents that may be encountered in the intensive care unit setting.
xix. False information spread by news media.
xx. Inadequate design of handwashing facilities in the unit to accommodate the increasing numbers of isolation areas.
xxi. Defeatist attitude of public health representation.
B. ***Patient oriented***
i. Hospital in-patients having occasional surgery at nearby teaching hospitals where MRSA was endemic.
ii. Temporary shortage of available strict isolation beds due to high numbers of MRSA-positive newborns.
iii. Issues surrounding breast pump stations, banked milk, and donor milk.
iv. Patient profile of delivering mothers and risk for MRSA, e.g. addict mothers.
v. Rare occurrence of a new MRSA strain.
C. ***Staff oriented***
i. Expression by some infectious disease consultants that excessive efforts were being had to control the outbreak.
ii. Expression by some medical staff that infection control efforts were too stringent and that the hospital should follow the less stringent approaches then believed by their experience to have been maintained at other centres internationally.

Table 2. (Continued)

iii. Reluctance of a few infection control team members to engage in the strict control.
iv. Refusal by some staff to participate in staff screening sessions.
v. Multidisciplinary care by staff who cross over to other hospital wards and outpatient clinics.
vi. Sensitivity of some staff to particular cleaning solutions.
vii. Crossover of unit medical staff to follow-up discharged MRSA-positive patients in outpatient settings.
viii. Occasional return of MRSA-positive patients to outpatient settings where the admitting flagging system was overlooked.
D. ***Administrative***
i. Interruption of staff screening efforts by hospital administration.
ii. Fluctuating financial support from administration.
iii. Hospital impact of strike action by some workers.
iv. Intermittent suppression of occupational health and wellness information from the infection control service.
v. Political manipulation of infection control committee representation during the outbreak.
vi. Direct interference in the activities of the infection control service by administration.
vii. Direct interference in the infection control committee by administration.

Table 3. Key epidemiological findings during the outbreak.

A. Clustering of most MRSA-positive patients.
B. Common molecular fingerprint.
C. Staff carriers including physician, clerical staff, nurse, and respiratory therapist
D. MRSA found in floor samplings.
E. MRSA found in other environmental samples (door handle, computer terminal).
F. Susceptibility profile change during the outbreak for erythromycin and clindamycin.
G. Intensification of control efforts often associated with significant reduction in culture-positive patients.

The Impact of Newborn Infection

Newborn infections with MRSA are well documented and are as broad in scope as MSSA infections (Table 4) except for the fact that more experience with MSSA has been published given its long-standing recognition. Superficial eye and skin infections are among the most common as is bacteremia. The placement of intravenous catheters, central vascular access lines, endotracheal tubes, dialysis catheters, and chest tubes, along with a variety of surgical interactions, assures that MRSA can gain access to sterile sites. Infection of bone, joint, peritoneum, deep veins, central nervous system, heart, and muscle may be direct consequences of systemic invasion and metastatic spread. Japanese newborn units have published a wealth of experience with NTED in which the newborns experience a symptom complex of fever, diffuse red rash, and thombocytopenia due to TSST-1 positive MRSA.[76,78,79,81] Such infections are also being experienced elsewhere.[17,82] TSST-1 producing strains are now well documented to cause skin infections in newborn care.[62,82,90] Outbreaks have also been detailed in which strains were capable of elaborating exfoliative toxin and thus were causative of skin denudation.[41,65,74] Although most MRSA are resistant to some beta-lactam antibiotics via beta-lactamase production and yet other beta-lactam agents via an alteration in the active binding site (penicillin-binding protein) for the antibiotic, a unique outbreak of beta-lactamase negative MRSA has been reported.[41] More recent interest has been delegated to outbreak strains which are identical to those that have been frequently seen in aggressive soft tissue infections in the community and which are labeled as community-acquired MRSA.[2,19,25,28,29,45,68,69,84-86,91-95] The origin of the latter has been a subject for debate, but a pandemic of these strains is emerging. In some settings, such community-acquired MRSA are becoming resident in hospitals and eventually the source for intra-hospital spread including the newborn care areas and then back to the community; community-acquired MRSA will become a misnomer in such settings over time. Unfortunately, patient deaths are now well documented among several outbreaks, albeit for a minority of patients.[2,9,11,18,19,21,22,24,28,29,31,32,37-39,42,48-50,52,53,61,66,83,86,87,96] Some have suggested that the death frequency is greater among those with MRSA infection in contrast to MSSA infection,[83] but it is difficult to know how such comparisons may be biased by other features of such epidemic strains. Although MRSA may be present in a complex patient, the cause of death may be some other. Hospital-acquired infections in the neonatal unit, whether due to MRSA or other pathogens, are associated with prolonged hospital stay and increased costs.[21,27,97]

Table 4. Spectrum of MRSA-related infections published in the context of newborn care.

bacteremia[2,5,8-56]	
ocular infections (mainly conjunctivitis) [9,10,13,14,22,23,27,30,31,33,35,38,39,44,46,48,50,52,57-66]	
superficial skin[5,9,11,13,16,23,27,31,33,34,41,47,48,52,57,59,62,64-73]	
scalded skin syndrome[74]	exanthem/thrombocytopenia[17,70,75-82]
balanitis[62]	omphalitis[13,39,50,62,69,79] surgical wound [9,12,22,34,39,66,83]
cellulitis[49,51,66,68,69,83]	necrotizing fasciitis[84] dialysis catheter[14,22,44]
pneumonia[5,11-13,18,21,24,27,28,36,38-40,42,43,83,85-87] chest tube[11]	
endotracheal tube[9,10,25,34]	pleural fluid[36]
central nervous system[5,11,24,31-33,35,43,49,54,61,66,83,86]	
osteomyelitis[5,13,31,34,48,49,51,56,66]	septic arthritis[5,21,51,66]
abscess not defined[5,18,49,51,66,67,83]	intra-abdominal abscess[12,34]
peritonitis[31,49]	pelvic abscess[48] ilio-psoas abscess[79]
umbilical artery catheter[10,51]	thrombophlebitis[11,51,83] atrial thrombus[51]
intravenous catheter[31,34,37,49,51]	endocarditis[66,83] parotitis[21]
sialadenitis[88]	external otitis[89] urachal cyst[21]
urinary tract infection[32,34,54]	hydrocoele[48]

THE SPREAD OF MRSA

The acquisition of MRSA by the newborn may begin at birth as there is plenty of opportunity for maternal-newborn transmission with delivery. This burden will be reflected in the frequency of maternal vaginal and other site carriage in the community. It appears, however, that most MRSA acquisitions occur in the nursery – the sources include personnel, families, and the environment (including airborne). Campbell et al. have estimated that the mean interval from acquisition to actual infection is approximately three weeks, and thus mere colonization must occur at a shorter interval.[51] Given the immobility of the newborn, evidently the germ must be delivered to the patient. The frequency of subsequent colonization during outbreaks is quite variable[1] and has varied from 0.4 to 75%.[9,14,22,24,26-28,30,39,40,44,48,53,62,64,65,68,80,86,91,95,98-107] Such figures may greatly vary depending on the intensity of search for MRSA and by the possibility that the outbreak may be chronic, i.e., endemic. Likewise, the frequency of infection among those colonized is quite variable (e.g., from 0.9 to 56%)[9,10,12-14,52,57,66] and will depend on a number of factors that balance virulence and other pathogenesis factors among germ and host. More serious infections occur among low birth weight newborns,[13] but outbreaks have been detailed for full-term newborns as well.[108,109] Citations for newborn bacteremia and sepsis must be tempered with the knowledge that positive blood cultures may at times be inadvertent contaminants that are acquired from the colonized skin during venipuncture.[18,110] After acquisition in the newborn care setting, whether the individual is infected or not, a prolonged carriage may ensue even after discharge[62,65]; the latter may extend to months and possibly years although MRSA carriage is progressively lost over time in many without antibiotic treatment or other decontamination strategies. The acquisition of MRSA is thereafter associated with prolonged hospital stay; in some circumstances, MRSA infections are directly responsible for such prolongation.[21,27]

The body sites for newborn surveillance have varied, but the need to screen newborns has become evident for various reasons. In the midst of an outbreak, such screening can identify newborns who serve as carriers and hence foci for spread.[10, 13, 14, 22, 26, 27, 32, 37, 38, 40, 44, 46, 50, 52-54, 57, 66-68, 87, 92, 94, 95, 100, 104, 110-112] Screening can therefore be used for prevention. Alternatively, screening can assist the infection control and newborn services to determine the magnitude of sources or of the efforts that may be required for containment. For example, a screening study may find that the actual number of colonized newborns could overwhelm isolation facilities. Screening may further provide some measure of understanding that could be of clinical value when treatment strategies for newborn sepsis are contemplated. Fortunately, vancomycin-resistance among MRSA in nurseries is not yet a realized dilemma, but certainly screening could have tremendous value if the latter was to arise.

Apart from the desire to screen in an outbreak setting, some have used newborn screening as a mechanism to proactively determine MRSA carriers, much in the same way as such a tool has been utilized in adult settings. In the scenario where a newborn is transferred from setting to setting, and given the problem with MRSA in many nurseries, the potential usefulness of such a strategy is obvious. The timing of screening in a setting where the nursery serves a maternal facility on-site becomes more contentious. In most settings, the frequency of maternal carriage is low, and therefore most spread in a nursery will occur as an endemic phenomenon within the facility. Furthermore, the opportunity to determine who will become a carrier at birth is limited because the microbial burden expands to a detectable level well after birth and is measured in many hours to days. Whereas maternal-newborn transmission can occur at birth, it would be of less value to screen the newborns immediately after birth and at the time of immediate submission to the nursery.

Body sites for screening newborns have included a variety from respiratory, skin, and bowel sources.[9,13,14,22,26-28, 30, 32, 35, 37, 38, 40, 42, 46, 48, 50, 52, 54, 55, 62, 68, 72, 87, 91, 92, 94, 95, 101, 103, 104, 109-112] Such surveillance has constituted as minimal as a nasal swab[13,38,52,62,68,87,94] to as complex a screening as five different sites (e.g., nares, axilla, groin, umbilicus, rectal[46]). Few studies have compared the value of screening multiple sites or of one site over another.[52,101,109,113,114] In general, MRSA are found most often when more than one site is sampled, and there appears to be incremental benefit by adding more sites for sampling. There is nevertheless some discrepancy in regards to which site(s) are preferential or for that matter essential. For example, Rosenthal et al. found value for adding an umbilical culture to a nasal swab,[101] whereas Singh et al. did not find any umbilical cultures to yield MRSA.[113] Back et al.[52] found rectal swabs to add valuably to screening much like the findings of Rimland and Roberson for adults.[115] During the course of screening for whatever purpose, a carrier patient may be subsequently deemed culture-negative (i.e., non-carrier). A definition of the latter may be fraught with difficulty, although for practical purposes nonetheless desirable. Several such definitions have been published, [9,32,40,50,54,106] but there are no determinative studies to confirm the value of one over another. Generally, patients are deemed culture-negative when the accepted screening sites have been repetitively (two or three times) culture-negative over a variable period and when the sampling has been taken at different points of time. It is conceivable that a newborn may become spontaneously, or due to treatment of infection or colonization, culture-negative. Culture negativity by whatever definition is doomed to failure, however infrequent, due to the limitation of sampling and laboratory isolation methods. Some therefore may be hesitant to delist a former carrier in a newborn setting due to the consequences of spread should it arise in the context of a tertiary nursery where morbidity outcomes could be considerable. In the latter case, the patient could be isolated regardless until the time of hospital discharge.

Sources for spread to newborns have been investigated considerably. Colonized or infected patients serve as a carrier for spread but non-isolated patients may represent the subset of most concern.[44] Medical staff often inadvertently carry the bacterium on hands from patient to patient.[116] The medical staff carrier may in some circumstances be the major vector for an outbreak,[25,26,41,42,46,50,68,71,73,90,96,106,110] and hence staff screening has often been invoked.[5,9,10,13,14,22,25,27,30,32,33,35-43,46-48,50,52-54,60,62,64,66-69,71,73,74,90-92,94,96,100,102,106,110-112,114,116-118] MRSA may be spread via aerosols,[35,47,61] and therefore nearby patient care and hospital activities can affect the extent of such transmission. Environmental contamination has also been assessed frequently,[5,9,10,13,30,32,33,35,36,39,41,42,46,48,50,54,60,62,64,67,69,71,73,91,93,96,102,111,112,114,119,120] and it is apparent that numerous foci in such a complicated scenario could harbour MRSA.[121] Both staff and environmental screening have been implemented in outbreak settings. The role of spread to the nursery from other facilities may be relevant.[36,42,46,53,93,100,106]

Maternal-newborn spread has been suggested, [15, 23, 30, 35, 45, 57, 59, 61, 84, 89, 95, 114, 122-125] but it is not the usual mode of acquisition for most newborns. Some have found a very low transmission rate at birth from carrier mothers to newborns.[72] Both maternal nasal and vaginal sites may be colonized, and thus transmission can occur at the time of delivery or thereafter when maternal-newborn contact continues. There may be discordance between which site(s) in the mother yield MRSA,[99] although perineal carriage can at times be considerable.[114] The demographics of MRSA-carrier mothers have been studied in one example where such mothers were more likely to be unmarried, have lower education, and abuse drugs.[15] In the latter follow-up, however, less than 1% of positive mothers gave rise to carrier offspring. Contaminated breast milk pools have also been implicated as a source to newborns.[19,30,35,61,84,123-125] Paternal-newborn transmission has been cited,[94] as has spread among and from other family members.[126]

Several other mitigating factors have been presented as risks for MRSA spread to the newborn. Poor handwashing,[9] low staff:patient ratios,[66,104,112] and increased workload for staff who handle positive patients[10,67] are examples. Overcrowding,[9,92,105,112,127] insufficient care space,[9,67,127] and lesser than optimal environmental cleaning[9,67] have been discussed as factors too. Some have linked MRSA acquisition risk to the performance of male circumcision.[71,91] In general, there may be many logistic problems if staff, administration, and the public are less than motivated to effect necessary control measures.[128,129]

Risk Factors for MRSA

Antibiotic use is generally a risk factor for antimicrobial resistance, and this also appears to be the case for newborn MRSA acquisition.[103] This phenomenon must be tempered however with the historic finding of MRSA at a time when methicillin was not yet used.[107]

Low birth weight newborns are often overrepresented among MRSA positive infants,[5,9,38,39,83,103,130] but this is likely due to the effect of multiple co-morbidities which are inevitable in severely premature infants.[18,27] Risk factors for MRSA include not only low birth weight but also decreased gestational age, multiple gestation, gavage feeding, intubation, and respiratory ventilation.[27,39] Risk factors for infection after colonization may include low birth weight, parenteral nutrition, presence of a percutaneous central line, and endotracheal intubation. Prolonged hospital stay increases the probability of MRSA acquisition.[102]

MRSA Containment

Once confronted with MRSA in a newborn unit, containment issues become paramount, but the approaches will vary with the circumstances at hand. The identification of a primary case in the context of no endemic MRSA should prompt isolation of the patient. Newborn facilities should be endowed with isolation rooms. The identification of an outbreak involving multiple carrier patients should ideally follow with the same optimal isolation precautions. The latter may be difficult, however, if the number of such patients overburdens existing isolation facilities, staffing, or other requisite resources. Initial and apparently reasonable containment maneuvers may falter given the particular epidemiology of spread. Difficulty with control measures in achieving desired outcomes has been realized.[9]

Infection control precautions for MRSA in a newborn unit have varied,[9,59,67] but the use of barrier precautions is a reasonable minimum.[14,26,32,50,59,66,95] The latter may be modified by gloving and/or masking.[22,52,67,131] Strict isolation with negative pressure ventilation would be ideal, but not many newborn units are structured to accommodate the latter. Difficulty with control has prompted some to close the neonatal unit to further admissions.[32,37,54,66,67,114]

Staff and/or patient cohorting has been applied in order to further the segregation of the carrier pool.[9,13,26,32,36-38,42,43,46,48,49,51-54,71,87,92,95,104,111,114,132] A dedicated infection control practitioner or nursery captain may also be invoked.[1,22] Some have directed attention to enhanced handwash precautions.[9,38,39,59] For example, improvements have been proposed for using alcohol rub and gloving over routine handwashing.[133] The potential for spread from environmental sources has prompted others to enhance cleaning efforts.[9,35,121]

In some settings, the carrier health care worker has been removed from work in order to sever transmission links.[37,40,41,68,69,71,96,110] In one such context, the health care worker was to be compensated for such mandatory absence.[71] The return of the health care worker could be contingent upon having demonstrated negative screening cultures.[32] The circumstances of a staff carrier may be complicated by other carriers within that family.[90] The direct family of a carrier patient could also be investigated for carriage.[50] If such direct family are not carriers, family members may be required to meet stringent infection control precautions when entering the care unit.[23]

Scientifically Establishing Connections

There are many laboratory methods which serve as fingerprinting tools for MRSA, and a wide variety of them have been used for nursery outbreaks. Most current methods are advanced and readily available molecular techniques, and they have largely replaced the old standard of phage typing. These methods have relatively short turn-around times, and thus useful information may be available on a timely basis. Commonality among fingerprints raises the spectre of an outbreak strain in the nursery,[27,65,73,87,100] but there is considerable experience with endemic or outbreak scenarios where the presiding isolates are heterogeneous.[24,29,33,35,46,49,53,54,62,64,76,80,90,92,95,96,116,134-137]

Typing schemes may demonstrate homology, and hence probable linkage, with isolates from maternal sources,[15,59] staff carriage, or the immediate environment. Homology may be demonstrated with isolates from other hospital wards[50,106,127,134-136] or other institutions.[80,83] Some have found very little correlation between the fingerprint types of isolates from mothers and their respective newborns.[72]

Re-infection or colonization with a new strain may be identified by such differentiating tools.[43] In one circumstance, a putative bacteremia with two different MRSA genotypes was identified.[43]

Whereas typing techniques can demonstrate homology, identical 'types' do not necessarily confirm common origin. These tools are but one part of the epidemiological framework that must be taken into consideration when looking at source. Common types certainly do, however, increase the probability that two strains are from a central source.

TREATMENT OF MRSA INFECTION AND COLONIZATION

As methicillin resistance generally prevents the use of beta-lactam antibiotics, the active treatment of MRSA infection and empiric antibiotic therapy for presumed sepsis both invoke the need for alternate antibiotics. Vancomycin remains a cornerstone of such treatment, although its use in the empiric treatment of newborn sepsis in the intensive care setting has been previously practiced by most given the role of coagulase-negative staphylococci in such settings.[31,86] Although most MRSA infections respond to vancomycin, especially those of the bloodstream, there have been many documentations of ongoing sepsis and bacteremia despite vancomycin use.[66,86] In such episodes, rifampin or an alternate additional antibiotic has been used. The following case history exemplifies such circumstances.

A twin newborn was delivered at 25 weeks gestational age by Caesarean section due to preterm labour and fetal distress. The neonatal course was complicated by hyaline membrane disease, patent ductus arteriosus, intraventricular hemorrhage, jaundice, and seizures. By the sixth hospital day, oropharyngeal and umbilical colonization with MRSA was found. A presumed septic episode on the next working day led to the institution of intravenous vancomycin and cefotaxime as empiric therapy. Both blood and cerebrospinal fluid yielded MRSA. The isolates were resistant to methicillin, gentamicin, and sulphonamides, but susceptible to clindamycin, erythromycin, vancomycin, and fusidic acid. Vancomycin was administered as 18 mg/kg every 36 hours, and adequate vancomycin levels were maintained as determined by therapeutic monitoring. Despite the latter, persistently positive blood cultures were detected as late as 8 days after the onset of sepsis and vancomycin administration. Rifampin was added, and the bacteremia resolved with no further bacteremia episodes nor complications.

Therapy in newborn intensive care is surely to be complicated by numerous interactions especially including central line placement.[18] From a treatment perspective, susceptibility to vancomycin is defined by existing laboratory methods which assess a response under a given bacterial inoculum. Such methods may not truly reflect the in vivo situation when the bacterial load in blood is high or where the bacterium may be protected by biofilms or exposed to less than expected antibiotic levels (e.g., in cerebrospinal fluid or abscesses). Emerging reports of vancomycin-resistant MRSA have been published although the majority of these have found bacteria with intermediate susceptibility. Fully vancomycin-resistant MRSA pose a great concern for the future.

The treatment of carrier status in this setting has been much more controversial. The latter has mainly incorporated the use of nasal mupirocin and not systemic antibiotics. Some colonized newborns will spontaneously lose MRSA carriage as determined by screening cultures; nasal mupirocin will facilitate such decolonization for several more. Nasal mupirocin may be limited, however, when the skin is heavily colonized, and an ancillary body wash may be desirable albeit not always efficacious. Mupirocin resistance has now emerged in many settings, although most such documentations have been reported in adult care.[50,136] A lesser use of mupirocin has been associated with lesser resistance,[136] and inappropriate uses of mupirocin have been cited.[139] If mupirocin treatment is to be implemented, susceptibility of the isolate(s) *in vitro* should be determined.

It cannot be understated that screening and prevention are most beneficial when compared to the outcome of infection, colonization, and treatment.[22]

Prevention Strategies

The prevention of MRSA acquisition may take several approaches. Appropriate isolation of the colonized newborn is essential, although the extent of requisite prevention may be a subject for debate. Negative pressure ventilation with strict isolation is an aggressive stance aimed at maximal prevention.[67] Environmental decontamination should be optimized. New admissions may be routinely screened,[27,32,43,87] but the timing may be critical since screening at birth is unlikely to be rewarding. After admission to the neonatal unit, the latter may be furthered by weekly screening, particularly in outbreak settings.[40,45,48,50-54,110,112] The screening of admissions which have been resident at other medical centres is prudent.[43,50,93]

Enough experience is available to accept that the screening of staff in the context of epidemics may be justified.[116] This is more so critical when epidemiological analyses are highly suggestive of a health care worker vector. Some have chosen to screen all new health care workers especially those arriving from another institution.[42] Identified staff carriers may be subject to eradication therapy.[116,118] Others have proposed that staff screening may be pre-determined by a threshold trigger point.[137]

Antiseptic measures have been studied in various contexts. The use of alcohol gel for skin cleansing has been recommended.[87] Newborn and staff skin decontamination has been assessed with triclosan and povidone-iodine.[55,71,132] Cord and umbilical care may also have a role, and products as divergent as hexachlorophene, chlorhexidine, and triple dye have been assessed.[22,48,105,112,140,141,142]

The prophylactic use of nasal mupirocin has received attention,[106] but such an approach must be tempered with the potential for mupirocin resistance to emerge. In one example, blanket therapy with nasal mupirocin was used for staff and newborns as a preventative measure. On a theoretical basis, the concept of using natural products, i.e. bacteria, to compete with MRSA colonization at the level of the pharyngeal mucosa has created interest.[108,143]

References

[1] Cimolai, N. *Staphylococcus aureus* outbreaks among newborns: new frontiers in an old dilemma. *Am J Perinatol*, 2003, 20, 125-136.

[2] Kuint, J; Barzilai, A; Regev-Yochay, G; Rubinstein, E; Keller, N; Maayan-Metzger, A. Comparison of community-acquired methicillin-resistant *Staphylococcus aureus*

bacteremia to other staphylococcal species in a neonatal intensive care unit. *Eur J Pediatr*, 2007, 166, 319-325.

[3] Cimolai, N. Are all MRSA made equal? *Can J Microbiol*, 2002, 48, 1-7.

[4] Takahashi, N. Current status of MRSA infections in neonatal medicine in Japan. *Pediatr Int.*, 2003, 45, 221-222.

[5] Kitajima, H. Prevention of methicillin-resistant *Staphylococcus aureus* infections in neonates. *Pediatr Int.*, 2003, 45, 238-245.

[6] Peterson, LR; Hacek, DM; Robicsek, A. Case study: an MRSA intervention at Evanston Northwestern Healthcare. *Jt Comm J Qual Patient Saf.*, 2007, 33, 732-738.

[7] Jarvis, WR; Thornsberry, C; Boyce, J; Hughes, JM. Methicillin-resistant *Staphylococcus aureus* at children's hospitals in the United States. *Pediatr Infect Dis.*, 1985, 4, 651-655.

[8] Griffin, FA. Reducing methicillin-resistant *Staphylococcus aureus* (MRSA) infections. *Jt Comm J Qual Patient Saf.*, 2007, 33, 726-731.

[9] Haddad, Q; Sobayo, EI; Basit, OBA; Rotimi, VO. Outbreak of methicillin-resistant *Staphylococcus aureus* in a neonatal intensive care unit. *J Hosp Infect*, 1993, 23, 211-221.

[10] Millar, MR; Keyworth, N; Lincoln, C; King, B; Congdon, P. Methicillin-resistant *Staphylococcus aureus* in a regional neonatology unit. *J Hosp Infect*, 1987, 10, 187-197.

[11] Storch, GA; Rajagopalan, L. Methicillin-resistant *Staphylococcus aureus* bacteremia in children. *Pediatr Infect Dis.*, 1986, 5, 59-67.

[12] Ribner, BS. Endemic multiply resistant *Staphylococcus aureus* in a pediatric population. AJDC, 1987, 141, 1183-1187.

[13] Tan, KW; Tay, L; Lim, SH. An outbreak of methicillin-resistant *Staphylococcus aureus* in a neonatal intensive care unit in Singapore: a 20-month study of clinical characteristics and control. *Singapore Med J.*, 1994, 35, 277-282.

[14] Jernigan, JA; Titus, MG; Gröschel, DHM; Getchell-White, SI; Farr, BM. Effectiveness of contact isolation during a hospital outbreak of methicillin-resistant *Staphylococcus aureus*. *Am J Epidemiol*, 1996, 143, 496-504.

[15] Andrews, WW; Schelonka, R; Waites, K; Stamm, A; Cliver, SP, Moser, S. Genital tract methicillin-resistant *Staphylococcus aureus*: risk of vertical transmission in pregnant women. *Obstet Gynecol*, 2008, 111, 113-118.

[16] Overturf, GD; Sherman, MP; Scheifele, DW; Wong, LC. Neonatal necrotizing enterocolitis associated with delta toxin-producing methicillin-resistant *Staphylococcus aureus*. *Pediatr Infect Dis J.*, 1990, 9, 88-91.

[17] Durand, G; Bes, M; Meugnier, H; Enright, MC; Forey, F; Liassine, N; Wenger, A; Kikuchi, K; Lina, G; Vandenesch, F; Etienne, J. Detection of new methicillin-resistant *Staphylococcus aureus* clones containing the toxic shock syndrome toxin 1 gene responsible for hospital- and community-acquired infections in France. *J Clin Microbiol*, 2006, 44, 847-853.

[18] Denniston, S; Riordan, FAI. *Staphylococcus aureus* bacteremia in children and neonates: a 10 year retrospective review. *J Infect*, 2006, 53, 387-393.

[19] Gastelum, DT; Dassey, D; Mascola, L; Yasuda, L. Transmission of community-associated methicillin-resistant *Staphylococcus aureus* from breast milk in the neonatal intensive care unit. *Pediatr Infect Dis J.*, 2005, 24, 1122-1124.

[20] Sarkar, S; Bhagat, I; DeCristofaro, JD; Wiswell, TE; Spitzer, AR. A study of the role of multiple site blood culture in the evaluation of neonatal sepsis. *J Perinatol*, 2006, 26, 18-22.

[21] Tam, AYC; Yeung, CY. The changing pattern of severe neonatal staphylococcal infection: a 1-year study. *Aust Paediatr J*, 1988, 24, 275-278.

[22] Karchmer, TB; Durbin, LJ; Simonton, BM; Farr, BM. Cost-effectiveness of active surveillance cultures and contact/droplet precautions for control of methicillin-resistant *Staphylococcus aureus*. *J Hosp Infect.*, 2002, 51, 126-132.

[23] Morel, AS; Wu, F; Della-Latta, P; Cronquist, A; Rubenstein, D; Saiman, L. Nosocomial transmission of methicillin-resistant *Staphylococcus aureus* from a mother to her preterm quadruplet infants. *Am J Infect Control*, 2002, 30, 170-173.

[24] Loureiro, MM; De Moraes, BA; Quadra, MRR; Pinheiro, GS; Suffys, PN; Asensi, MD. Molecular epidemiology of methicillin resistant *Staphylococcus aureus* isolated from newborns in a hospital in Rio de Janeiro, Brazil. *Mem Inst Oswaldo Cruz, Rio de Janeiro* 2000, 95, 777-782.

[25] Stein, M; Novon-Venezia, S; Chmelnitsky, I; Kohelet, D; Schwartz, O; Agmon, O; Somekh, E. An outbreak of new, nonmultidrug-resistant, methicillin-resistant *Staphylococcus aureus* strain (SCC*Mec* type IIIA variant-1) in the neonatal intensive care unit transmitted by a staff member. *Pediatr Infect Dis J.*, 2006, 25, 557-559.

[26] Bertin, ML; Vinski, J; Schmitt, S; Sabella, C; Danziger-Isakov, L; McHugh, M; Procop, GW; Hall, G; Gordon, SM; Goldfarb, J. Outbreak of methicillin-resistant *Staphylococcus aureus* colonization and infection in a neonatal intensive care unit epidemiologically linked to a heathcare worker with chronic otitis. *Infect Control Hosp Epidemiol*, 2006, 27, 581-585.

[27] Khoury, J; Jones, M; Grim, A; Dunne, WM; Fraser, V. Eradication of methicillin-resistant *Staphylococcus aureus* from a neonatal intensive care unit by active surveillance and aggressive infection control measures. *Infect Control Hosp Epidemiol*, 2005, 26, 616-621.

[28] Regev-Yochay, G; Rubinstein, E; Barzilai, A; Carmeli, Y; Kuint, J; Etienne, J; Blech, M; Smollen, G; Maayan-Metzger, A; Leavitt, A; Rahav, G; Keller, N. Methicillin-resistant *Staphylococcus aureus* in neonatal intensive care unit. *Emerg Infect Dis.*, 2005, 11, 453-456.

[29] Healy, CM; Hulten, KG; Palazzi, DL; Campbell, JR; Baker, CJ. Emergence of new strains of methicillin-resistant *Staphylococcus aureus* in a neonatal intensive care unit. *Clin Infect Dis.*, 2004, 39, 1460-1466.

[30] Behari, P; Englund, J; Alcasid, G; Garcia-Houchins, S; Weber, SG. Transmission of methicillin-resistant *Staphylococcus aureus* to preterm infants through breast milk. *Infect Control Hosp Epidemiol*, 2004, 25, 778-780.

[31] Chuang, YY; Huang, TC; Lee, CY; Lin, TY; Lien, R; Chou, YH. Methicillin-resistant *Staphylococcus aureus* bacteraemia in neonatal intensive care units: an analysis of 90 episodes. *Acta Paediatr*, 2004, 93, 786-790.

[32] Liar, S; Herwaldt, LA; Singh, N. Outbreak of invasive disease caused by methicillin-resistant *Staphylococcus aureus* in neonates and prevalence in the neonatal intensive care unit. *Pediatr Crit Care Med.*, 2003, 4, 220-226.

[33] Trzcinski, K; Hryniewicz, W; Kluytmans, J; et al. Simultaneous persistence of methicillin-resistant and methicillin-susceptible clones of *Staphylococcus aureus* in a neonatal ward of a Warsaw Hospital. *J Hosp Infect*, 1997, 36, 291-303.

[34] Perez Trallero, E; Garcia Arenzana, J; Ansa Castaneda, A; Paisan Grisolia, L. Unusual multiresistant *Staphylococcus aureus* in a newborn nursery. *Am J Dis Child.*, 1981, 135, 689-692.

[35] Otter, JA; Klein, JL; Watts, TL; Kearns, AM; French, GL. Identification and control of an outbreak of ciprofloxacin-susceptible EMRSA-15 on a neonatal unit. *J Hosp Infect*, 2007, 67, 232-239.

[36] Mulhern, B; Griffin, E. An epidemic of gentamicin/cloxacillin resistant staphylococcal infection in a neonatal unit. *Ir Med J*, 1981, 74, 228-229.

[37] Price, EH; Brain, A; Dickson, JAS. An outbreak of infection with a gentamicin and methicillin-resistant *Staphylococcus aureus* in a neonatal unit. *J Hosp Infect*, 1980, 1, 221-228.

[38] Graham, DR; Correa-Villasenor, A; Anderson, RL; Vollman, JH; Baine, WB. Epidemic neonatal gentamicin-methicillin-resistant *Staphylococcus aureus* infection associated with nonspecific topical use of gentamicin. *J Pediatr*, 1980, 97, 972-978.

[39] Reboli, AC; John, Jr. JF; Levkoff, AH. Epidemic methicillin-gentamicin-resistant *Staphylococcus aureus* in a neonatal intensive care unit. *AJDC*, 1989, 143, 34-39.

[40] Davies, EA; Emmerson, AM; Hogg, GM; Patterson, MF; Shields, MD. An outbreak of infection with a methicillin-resistant *Staphylococcus aureus* in a special care baby unit: value of topical mupirocin and of traditional methods of infection control. *J Hosp Infect*, 1987, 10, 120-128.

[41] Richardson, JF; Quoraishi, AHM; Francis, BJ; Marples, RR. Beta-lactamase-negative, methicillin-resistant *Staphylococcus aureus* in a newborn nursery: report of an outbreak and laboratory investigations. *J Hosp Infect*, 1990, 16, 109-121.

[42] Coovadia, YM; Bhana, RH; Johnson, AP; Haffejee, I; Marples, RR. A laboratory-confirmed outbreak of rifampicin-methicillin resistant *Staphylococcus aureus* (RMRSA) in a newborn nursery. *J Hosp Infect*, 1989, 14, 303-312.

[43] Huang, YC; Su, LH; Wu, TL; Lin, TY. Molecular surveillance of clinical methicillin-resistant *Staphylococcus aureus* isolates in neonatal intensive care units. *Infect Control Hosp Epidemiol*, 2005, 26, 157-160.

[44] Geffers, C; Farr, BM. Risk of transmission of nosocomial methicillin-resistant *Staphylococcus aureus* (MRSA) from patients colonized with MRSA. *Infect Control Hosp Epidemiol*, 2005, 26, 114-115.

[45] Eckhardt, C; Halvosa, JS; Ray, SM; Blumberg, HM. Transmission of methicillin-resistant *Staphylococcus aureus* in the neonatal intensive care unit from a patient with community-acquired disease. *Infect Control Hosp Epidemiol*, 2003, 24, 460-461.

[46] Saiman, L; Cronquist, A; Wu, F; et al. An outbreak of methcillin-resistant *Staphylococcus aureus* in a neonatal intensive care unit. *Infect Control Hosp Epidemiol*, 2003, 24, 317-321.

[47] Aihara, M; Sakai, M; Iwasaki, M; et al. Prevention and control of nosocomial infection caused by methicillin-resistant *Staphylococcus aureus* in a premature infant ward – preventive effect of a povidone-iodine wipe of neonatal skin. *Postgrad Med J*, 1993, 69(Suppl. 3), S117-S121.

[48] Webster, J; Faoagali, JL. Endemic methicillin-resistant *Staphylococcus aureus* in a special care baby unit: a 2 year review. *J Paediatr Child Health*, 1990, 26, 160-163.
[49] Noel, GJ; Kreiswirth, BN; Edelson, PJ; et al. Multiple methicillin-resistant *Staphylococcus aureus* strains as a cause for a single outbreak of severe disease in hospitalized neonates. *Pediatr Infect Dis J*, 1992, 11, 184-188.
[50] Mean, M; Mallaret, MR; Andrini, P; Recule, C; et al. A neonatal specialist with recurrent methicillin-resistant *Staphylococcus aureus* (MRSA) carriage implicated in the transmission of MRSA to newborns. *Infect Control Hosp Epidemiol*, 2007, 28, 625-628.
[51] Campbell, JR; Zaccaria, E; Mason, EO; Baker, CJ. Epidemiological analysis defining concurrent outbreaks of *Serratia marcescens* and methicillin-resistant *Staphylococcus aureus* in a neonatal intensive-care unit. *Infect Control Hosp Epidemiol*, 1998, 19, 924-928.
[52] Back, NA; Linnemann, CC; Staneck, JL; Kotagal, UR. Control of methicillin-resistant *Staphylococcus aureus* in a neonatal intensive-care unit: use of intensive microbiologic surveillance and mupirocin. *Infect Control Hosp Epidemiol*, 1996, 17, 227-231.
[53] Fisher, A; Gavin, PJ. Control of methicillin-resistant *Staphylococcus aureus* in a neonatal intensive care unit. *Am J Infect Control*, 2004, 32, E66.
[54] Nambiar, S; Herwaldt, LA; Singh, N. Outbreak of invasive disease caused by methcillin-resistant *Staphylococcus aureus* in neonates and prevalence in the neonatal intensive care unit. *Pediatr Crit Care Med.*, 2003, 4, 220-226.
[55] Webster, J; Faoagali, JL; Cartwright, D. Elimination of methicillin-resistant *Staphylococcus aureus* from a neonatal intensive care unit after hand washing with triclosan. *J Paediatr Child Health*, 1994, 30, 59-64.
[56] Weeks, JL; Garcia-Prats, JA; Baker, CJ. Methicillin-resistant *Staphylococcus aureus* osteomyelitis in a neonate. *JAMA*, 1981, 245, 1661-1664.
[57] Hamada, T. The improvement on the nursing procedure for pregnant women and newborns and the decrease of *Staphylococcus aureus* infection in newborns and infants in our hospital. *Kansenshogaku Zasshi*, 1994, 68, 1211-1217.
[58] Anari, S; Karagama, YG; Fulton, B; Wilson, JA. Neonatal disseminated methicillin-resistant *Staphylococcus aureus* presenting as orbital cellulitis. *J Laryngol Otol*, 2005, 119, 64-67.
[59] Shiojima, T; Ohki, Y; Nako, Y; Morikawa, A; Okubo, T; Iyobe, S. Immediate control of a methicllin-resistant *Staphylococcus aureus* outbreak in a neonatal intensive care unit. *J Infect Chemother,* 2003, 9, 243-247.
[60] Ndawula, EM; Brown, L. Mattresses as reservoirs of epidemic methicillin-resistant *Staphylococcus aureus*. Lancet, 1991, 337, 488-489.
[61] Lemoine, L. Possible transmission of methicillin-resistant *Staphylococcus aureus* by expressed human breast milk. *J Hosp Infect*, 1987, 9, 95-100.
[62] Mitsuda, T; Arai, K; Fujita, S; Yokota, S. Epidemiological analysis of strains of methicillin-resistant *Staphylococcus aureus* (MRSA) infection in the nursery; prognosis of MRSA carrier infants. *J Hosp Infect*, 1995, 31, 123-134.
[63] Cimolai, N. Ocular methicillin-resistant *Staphylococcus aureus* infections in a newborn intensive care cohort. *Am J Ophthalmol*, 2006, 142, 183-184.

[64] Le Coq, M; Simon, I; Sire, C; et al. Epidemie d'infections nosocomiales a *Staphylococcus aureus* resistant a la meticilline dan une maternite. *Pathol Biol.*, 2001, 49, 16-22.

[65] Hayakawa, T; Hayashidera, T; Yoneda, K; Kagawa, S; Kusunoki, T. Unexpectedly prolonged colonization of exfoliative toxin a-producing methicillin resistant *Staphylococcus aureus* in infants. *Eur J Pediatr*, 1998, 157, 781.

[66] Gilbert, GL; Asche, V; Hewstone, AS; Mathiesen, JL. Methicillin-resistant *Staphylococcus aureus* in neonatal nurseries. *Med J Aust.*, 1982, 1, 455-459.

[67] Andersen, BM; Lindemann, R; Bergh, K; Nesheim, BI; Syversen, G; Solheim, N; Laugerud, F. Spread of methicillin-resistant *Staphylococcus aureus* in a neonatal intensive unit associated with understaffing, overcrowding and mixing of patients. *J Hosp Infect*, 2002, 50, 18-24.

[68] James, L; Gorwitz, RJ; Jones, RC; et al. Methicillin-resistant *Staphylococcus aureus* infections among healthy full-term newborns. *Arch Dis Child Fetal Neonatal Ed.* 2008, 93, F40-F44.

[69] Bratu, S; Eramo, A; Kopec, R; Coughlin, E; Ghitan, M; Yost, R; Chapnick, EK; Landman, D; Quale, J. Community-associated methicillin-resistant *Staphylococcus aureus* in hospital nursery and maternity units. *Emerg Infect Dis.*, 2005, 11, 808-813.

[70] Meguro, H. MRSA infections in the neonatal unit. *Nippon Rinsho*, 1992, 50, 1117-1121.

[71] Zafar, AB; Butler, RC; Reese, DJ; Gaydos, LA; Mennonna, PA. Use of 0.3% triclosan (Bacti-Stat*) to eradicate an outbreak of methicillin-resistant *Staphylococcus aureus* in a neonatal nursery. *Am J Infect Control*, 1995, 23, 200-208.

[72] Reusch, M; Ghosh, P; Ham, C; Klotchko, A; Singapuri, S; Everett, G. Prevalence of MRSA colonization in peripartum mothers and their newborn infants. *Scand J Infect Dis* e-pub DOI: 10.1080/00365540801946520

[73] Saunders, A; Panaro, L; McGeer, A; et al. A nosocomial outbreak of community-associated methicillin-resistant *Staphylococcus aureus* among healthy newborns and postpartum mothers. *Can J Infect Dis Med Microbiol*, 2007, 18, 128-132.

[74] Egashira, M; Tagawa, M; Hirakata, Y; Sugai, M; Moriuchi, H. Genetically confirmed nosocomial transmission of staphylococcal scalded skin syndrome between two infants in the neonatal intensive care unit. *Kansenshogaku Zasshi*, 2008, 82, 34-37.

[75] Takahashi, N; Nishida, H; Kato, H; Imanishi, K; Sakata, Y; Uchiyama, T. Exanthematous disease induced by toxic shock syndrome toxin 1 in the early neonatal period. *Lancet*, 1998, 351, 1614-1619.

[76] Nakano, M; Miyazawa, H; Kawano, Y; Kawagishi, M; Torii, K; Hasegawa, T; Iinuma, Y; Ohta, M. An outbreak of neonatal toxic shock syndrome-like exanthematous disease (NTED) caused by methicillin-resistant *Staphylococcus aureus* (MRSA) in a neonatal intensive care unit. *Microbiol Immunol*, 2002, 46, 277-284.

[77] Kato, H; Takahashi, N; Arimura, Y; Imanishi, K; Nishida, H; Uchiyama, T. The percentage of superantigen-reactive T cells in peripheral blood significantly decreases before massively increasing in patients with neonatal TSS-like exanthematous disease in the early acute phase. *J Infect Chemother*, 2002, 8, 111-114.

[78] Takahashi, N; Kato, H; Imanishi, K; Miwa, K; Yamanami, S; Nishida, H; Uchiyama, T. Immunopathophysiological aspects of an emerging neonatal infectious disease induced by a bacterial superantigen. *J Clin Invest*, 2000, 106, 1409-1415.

[79] Okada, Y; Yamataka, A; Ogasawara, Y; Matsubara, K; Watanabe, T; Lane, GJ; Miyano, T. Ilio-psoas abscess caused by methicillin-resistant *Staphylococcus aureus* (MRSA): a rare but potentially dangerous condition in neonates. *Pediatr Surg Int.*, 2004, 20, 73-74.

[80] Kikuchi, K; Takahashi, N; Piao, C; Totsuka, K; Nishida, H; Uchiyama, T. Molecular epidemiology of methicillin-resistant *Staphylococcus aureus* strains causing neonatal toxic shock syndrome-like exanthematous disease in neonatal and perinatal wards. *J Clin Microbiol*, 2003, 41, 3001-3006.

[81] Takahashi, N. Neonatal toxic shock syndrome-like exanthematous disease (NTED). *Pediatr Int*, 2003, 45, 233-237.

[82] Van Der Mee-Marquet, N; Lina, G; Quentin, R; et al. Staphylococcal exanthematous disease in a newborn due to a virulent methicillin-resistant *Staphylococcus aureus* strain containing the TSST-1 gene in Europe: an alert for neonatologists. *J Clin Microbiol*, 2003, 41, 4883-4884.

[83] Isaacs, D; Fraser, S; Hogg, G; Li, HY. *Staphylococcus aureus* infections in Australasian neonatal nurseries. *Arch Dis Child Fetal Neonatal Ed*, 2004, 89, 331-335.

[84] Dehority, W; Wang, E; Vernon, PS; Lee, C; Perdreau-Remington, F; Bradley, J. Community-associated methicillin-resistant *Staphylococcus aureus* necrotizing fasciitis in a neonate. *Pediatr Infect Dis J*, 2006, 25, 1080-1081.

[85] Yee-Guardino, S; Kumar, D; Abughali, N; Tuohy, M; Hall, GS; Kumar, ML. Recognition and treatment of neonatal community-associated MRSA pneumonia and bacteremia. *Pediatr Pulmonol*, 2008, 43, 203-205.

[86] McAdams, RM; Mazuchowski, E; Ellis, MW; Rajnik, M. Necrotizing staphylococcal pneumonia in a neonate. *J Perinatol*, 2005, 25, 677-679.

[87] Honeycutt, M; Gilliam, C; Juretschko, S; Beavers-May, T; et al. Control of methicillin-resistant *Staphylococcus aureus* in a neonatal intensive care unit by modification of 2003 Society for Healthcare Epidemiology of America guideline for preventing transmission of resistant pathogens. *Am J Infect Control*, 2004, 32, E71.

[88] McAdams, RM; Mair, EA; Rajnik, M. Neonatal suppurative submandibular sialadenitis: case report and literature review. *Int J Pediatr Otorhinolaryngol*, 2005, 69, 993-997.

[89] Santos, F; Mankarious, LA; Eavey, RD. Methicillin-resistant *Staphylococcus aureus* pediatric otitis. Arch Otolaryngol Head *Neck Surg.*, 2000, 126, 1383-1385.

[90] Mitsuda, T; Arai, K; Ibe, M; Imagawa, T; Tomono, N; Yokota, S. The influence of methicillin-resistant *Staphylococcus aureus* (MRSA) carriers in a nursery and transmission of MRSA to their households. *J Hosp Infect*, 1999, 42, 45-51.

[91] Nguyen, DM; Bancroft, E; Mascola, L; Guevara, R; Yasuda, L. Risk factors for neonatal methicillin-resistant *Staphylococcus aureus* infection in a well-infant nursery. *Infect Control Hosp Epidemiol*, 2007, 28, 406-411.

[92] David, MD; Kearns, AM; Gossain, S; Ganner, M; Holmes, A. Community-associated meticillin-resistant *Staphylococcus aureus*: nosocomial transmission in a neonatal unit. *J Hosp Infect*, 2006, 64, 244-250.

[93] Ko, KS; Park, S; Reck, KR; Shin, EJ; Oh, WS; Lee, NY; Song, JH. Molecular characterization of methicillin-resistant *Staphylococcus aureus* spread by neonates transferred from primary obstetrics to a tertiary care hospital in Korea. *Infect Control Hosp Epidemiol*, 2006, 27, 593-597.

[94] Al-Tawfiq, JA. Father-to-infant transmission of community-acquired methicillin-resistant *Staphylococcus aureus* in a neonatal intensive care unit. *Infect Control Hosp Epidemiol*, 2006, 27, 636-637.

[95] Sax, H; Posfay-Barbe, K; Harbarth, S; Francois, P; Touveneau, S; Pessoa-Silva, CL; Schrenzel, J; Dharan, S; Gervaix, A; Pittet, D. Control of a cluster of community-associated, methicillin-resistant *Staphylococcus aureus* in neonatology. *J Hosp Infect*, 2006, 63, 93-100.

[96] Farrington, M; Ling, J; Ling, T; French, GL. Outbreaks of infection with methicillin-resistant *Staphylococcus aureus* on neonatal and burns units of a new hospital. *Epidemiol Infect*, 1990, 105, 215-228.

[97] Mahieu, LM; Buitenweg, N; Beutels, P; De Dooy, JJ. Additional hospital stay and charge due to hospital-acquired infections in a neonatal intensive care unit. *J Hosp Infect*, 2001, 47, 222-229.

[98] Johnson, Z; Fitzpatrick, P; Hayes, C; Sayers, G; Pelly, H; McDonnell, B; Thornton, L; Buttimer, J. National survey of MRSA: Ireland, 1995. *J Hosp Infect*, 1997, 35, 175-184.

[99] Mitsuda, T; Arai, K; Fujita, S; Yokota, S. Demonstration of mother-to-infant transmission of *Staphylococcus aureus* by pulsed-field gel electrophoresis. *Eur J Pediatr*, 1996, 155, 194-199.

[100] Wanger, AR; Morris, SL; Ericsson, C; Singh, KV; LaRocco, MT. Latex agglutination-negative methicillin-resistant *Staphylococcus aureus* recovered from neonates: epidemiologic features and comparison of typing methods. *J Clin Microbiol*, 1992, 30, 2583-2588.

[101] Rosenthal, A; White, D; Churilla, S; Brodie, S; Katz, KC. Optimal surveillance culture sites for detection of methicillin-resistant *Staphylococcus aureus* in newborns. *J Clin Microbiol*, 2006, 44, 4234-4236.

[102] Ahoyo, AT; Baba-Moussa, L; Makoutode, M; Gbohoun, A; Bossou, R; Dramane, K; Sanni, A; Prévost, G. Incidence de *Staphylococcus aureus* résistant à la méticilline dans le service de néonatologie du centre hospitalier départemental du Zou et des Collines au Bénin. *Arch Pediatr*, 2006, 13, 1391-1396.

[103] De Almeida Silva, H; Steffen Abdallah, VO; Carneiro, CL; Gontijo Filho, PP. Infection and colonization by *Staphylococcus aureus* in a high risk nursery of a Brazilian teaching hospital. *Braz J Infect Dis*., 2003, 7, 381-386.

[104] Dunkle, LM; Naqvi, SH; McCallum, R; Lofgren, JP. Eradication of epidemic methicillin-gentamicin resistant *Staphylococcus aureus* in an intensive care nursery. *Am J Med.,* 1981, 70, 455-458.

[105] Rosenfeld, CR; Laptook, AR; Jeffery, J. Limited effectiveness of triple dye in preventing colonization with methicillin-resistant *Staphylococcus aureus* in a special care nursery. *Pediatr Infect Dis J*, 1990, 9, 290-291.

[106] Hitomi, S; Kubota, M; Mori, N; Baba, S; Yano, H; Okuzumi, K; Kimura, S. Control of a methicillin-resistant *Staphylococcus aureus* outbreak in a neonatal intensive care unit by unselective use of nasal mupirocin ointment. *J Hosp Infect*, 2000, 46, 123-129.

[107] Borowski, J; Kamienska, K; Rutecka, I. Methicillin-resistant staphylococci. *BMJ,* 1964, i,983.

[108] Uehara, Y; Kikuchi, K; Kakamura, T; Nakama, H; Agematsu, K; Kawakami, Y; Maruchi, N; Totsuka, K. H_2O_2 produced by viridans group streptococci may contribute

to inhibition of methicillin-resistant *Staphylococcus aureus* colonization of oral cavities in newborns. *Clin Infect Dis*., 2001, 32, 1408-1413.

[109] Hicks, NR; Moore, EP; Williams, EW. Carriage and community treatment of methicillin-resistant *Staphylococcus aureus*: what happens to colonized patients after discharge? *J Hosp Infect*, 1991, 19, 17-24.

[110] Berthelot, P; Grattard, F; Fascia, P; et al. Implication of a healthcare worker with chronic skin disease in the transmission of an epidemic strain of methicillin-resistant *Staphylococcus aureus* in a pediatric intensive care unit. *Infect Control Hosp Epidemiol*, 2003, 24, 299-300.

[111] Gerber, SI; Jones, RC; Scott, MV; et al. Management of outbreaks of methicillin-resistant *Staphylococcus aureus* infection in a neonatal intensive care unit: a consensus statement. *Infect Control Hosp Epidemiol*, 2006, 27, 139-145.

[112] Haley, RW; Cushion, NB; Tenover, FC; et al. Eradication of endemic methicillin-resistant *Staphylococcus aureus* infections from a neonatal intensive care unit. *J Infect Dis.,* 1995, 171, 614-624.

[113] Singh, K; Gavin, PJ; Vescio, T; Thomson, Jr RB; Deddish, RB; Fisher, A; Noskin, GA; Peterson, LR. Microbiologic surveillance using nasal cultures alone is sufficient for detection of methicillin-resistant *Staphylococcus aureus* isolates in neonates. *J Clin Microbiol*, 2003, 41, 2755-2757.

[114] Moore, EP; Williams, EW. A maternity hospital outbreak of methicillin-resistant *Staphylococcus aureus*. *J Hosp Infect*, 1991, 9, 5-16.

[115] Rimland, D; Roberson, B. Gastrointestinal carriage of methicillin-resistant *Staphylococcus aureus*. *J Clin Microbiol*, 1986, 24, 137-138.

[116] Cimolai, N. The role of healthcare personnel in the maintenance and spread of methicillin-resistant *Staphylococcus aureus*. *J Infect Public Health*, 2008, 1, 78-100.

[117] Bryan, CS; Wilson, RS; Meade, P; Sill, LG. Topical antibiotic ointments for staphylococcal nasal carriers: survey of current practices and comparison of bacitracin and vancomycin ointments. *Infect Control*, 1980, 1, 153-156.

[118] Masano, H; Fukuchi, K; Wakuta, R; Tanaka, Y. Efficacy of intranasal application of povidone-iodine cream in eradicating nasal methicillin-resistant *Staphylococcus aureus* in neonatal intensive care unit (NICU) staff. *Postgrad Med J,* 1993, 69(Suppl. 3):S122-S125.

[119] Fujimura, S; Kato, S; Hashimoto, M; Takeda, H; Maki, F; Watanabe, A. Survey of methicillin-resistant *Staphylococcus aureus* from neonates and the environment in the NICU. *J Infect Chemother*, 2004, 10, 131-132.

[120] Wright, IMR; Orr, H; Porter, C. Stethoscope contamination in the neonatal intensive care unit. *J Hosp Infect*, 1995, 29, 65-68.

[121] Cimolai, N. MRSA and the environment: implications for comprehensive control measures. *Eur J Clin Microbiol Infect Dis*., 2008, 27, 481-493.

[122] André, P; Thébaud, B; Guibert, M; Audibert, F; Lacaze-Masmonteil, T; Dehan, M. Maternal-fetal staphylococcal infections: a series report. *Am J Perinatol*, 2000, 17, 423-427.

[123] Novak, FR; Da Silva, AV; Hagler, AN; Figueiredo, AMS. Contamination of expressed human breast milk with an epidemic multiresistant *Staphylococcus aureus* clone. *J Med Microbiol*, 2000, 49, 1109-1117.

[124] Kawada, M; Okuzumi, K; Hitomi, S; Sugishita, C. Transmission of Staphylococcus aureus between healthy, lactating mothers and their infants by breastfeeding. *J Human Lact,* 2003, 19, 411-417.

[125] Cooke, RPD; Devlin, J; Robinson, MJ. Breast milk and methicillin-resistant *Staphylococcus aureus*. *J Hosp Infect*, 1987, 10, 312-315.

[126] Hollis, RJ; Barr, JL; Doebbeling, BN; Pfaller, MA; Wenzel, RP. Familial carriage of methicillin-resistant *Staphylococcus aureus* and subsequent infection in a premature neonate. *Clin Infect Dis*., 1995, 21, 328-332.

[127] Kikuchi, K. Genetic basis of neonatal methicillin-resistant *Staphylococcus aureus* in Japan. *Pediatr Int*., 2003, 45, 223-229.

[128] Cimolai, N. MRSA control and the counter-culture. *J Hosp Infect*, 2007, 66, 189-190.

[129] Cimolai, N. Methicillin-resistant *Staphylococcus aureus* public concern, and legislative mandates. *Infect Control Hosp Epidemiol*, 2007, 28, 896.

[130] Usukura, Y; Igarashi, T. Examination of severe, hospital acquired infections affecting extremely low birthweight (ELBW) infants. *Pediatr Int*., 2003, 45, 230-232.

[131] Lacey, S; Flaxman, D; Scales, J; Wilson, A. The usefulness of masks in preventing transient carriage of epidemic methicillin-resistant *Staphylococcus aureus* by healthcare workers. *J Hosp Infect*, 2001, 48, 308-311.

[132] Hasegawa, K; Kawase, S; Koshizawa, S; et al. Preventing nosocomial infection with methicillin-resistant *Staphylococcus aureus* in a neonatal intensive care unit and natural course of colonized neonates. *J Jpn Pediatr Soc*., 1993, 97, 2088-2093.

[133] Ng, PC; Wong, HL; Lyon, DJ; So, KW; Liu, F; Lam, RKY; Wong, E; Cheng, AFB; Fok, TF. Combined use of alcohol hand rub and gloves reduces the incidence of late onset infection in very low birthweight infants. *Arch Dis Child Fetal Neonatal Ed*, 2004, 89, 336-340.

[134] El-Adhami Roberts, L; Vickery, A; Inglis, B; Gibbs, A; Stewart, PR. Epidemiological analysis of a methicillin-resistant *Staphylococcus aureus* outbreak using restriction fragment length polymorphisms of genomic DNA. *J Gen Microbiol*, 1991, 137, 2713-2720.

[135] Sá-Leão, R; Santos Sanches, I; Dias, D; Peres, I; Barros, RM; De Lencastre, H. Detection of an archaic clone of *Staphylococcus aureus* with low-level resistance to methicillin in a pediatric hospital in Portugal and in international samples: relics of a formerly widely disseminated strain? *J Clin Microbiol*, 1999, 37, 1913-1920.

[136] Villari, P; Farullo, C; Torre, I; Nani, E. Molecular characterization of methicillin-resistant *Staphylococcus aureus* (MRSA) in a university hospital in Italy. *Eur J Epidemiol*, 1998, 14, 807-816.

[137] Grant, PS; Charns, LG; Rawot, BW; Benedetti, SG. Consideration to culture health care workers related to increased methicillin-resistant *Staphylococcus aureus* activity in a neonatal intensive care unit. *Am J Infect Control*, 2008 doi:10.1016/j.ajic.2008.03.006

[138] Vivoni, AM; Santos, KRN; De-Oliveira, MP; et al. Mupirocin for controlling methicillin-resistant *Staphylococcus aureus*: lessons from a decade of use at a university hospital. *Infect Control Hosp Epidemiol*, 2005, 26, 662-667.

[139] Lines, LD; Weightman, NC. Incidence and effect of incorrect prescription and administration of nasal mupirocin for methicillin-resistant *Staphylococcus aureus*. *J Hosp Infect*, 2005, 61, 170-180.

[140] Allen, KD; Ridgway, EJ; Parsons, LA. Hexachlorophane powder and neonatal staphylococcal infection. *J Hosp Infect*, 1994, 27, 29-33.

[141] Oishi, T; Iwata, S; Nonoyama, M; Tsuji, A; Sunakawa, K. Double-blind comparative study on the care of the neonatal umbilical cord using 80% ethanol with or without chlorhexidine. *J Hosp Infect*, 2004, 58, 34-37.

[142] Wilcox, MH; Hall, J; Gill, AB; Fawley, WN; Parnell, P; Verity, P. Effectiveness of topical chlorhexidine powder as an alternative to hexachlorophene for the control of *Staphylococcus aureus* in neonates. *J Hosp Infect*, 2004, 56, 156-159.

[143] Uehara, Y; Kikuchi, K; Kakamura, T; Nakama, H; Agematsu, K; Kawakami, Y; Maruchi, N; Totsuka, K. Inhibition of methicillin-resistant *Staphylococcus aureus* colonization of oral cavities in newborns by viridans group streptococci. *Clin Infect Dis*., 2001, 32, 1399-1407.

In: Methicillin-Resistant Staphylococcus Aureus...
Editor: Charles L. Kolendi

ISBN: 978-1-60741-398-1

Chapter 10

THE EMERGING AND THE RE-EMERGING OF ANTIMICROBIAL-RESISTANT GRAM-POSITIVE COCCI: EVIDENCES SUPPORTING NOVEL PHARMACEUTICAL WEAPONS

Roberto Manfredi*

Department of Internal Medicine, Aging, and Nephrologic Diseases, Division of Infectious Diseases, "Alma Mater Studiorum" University of Bologna, S. Orsola Hospital, Bologna, Italy

ABSTRACT

Introduction

Methicillin- and also vancomycin (glycopeptide)-resistant Gram-positive organisms have emerged as an increasingly problematic cause of hospital-acquired infections, also spreading into the community. Vancomycin (glycopeptide) resistance has emerged primarily among Enterococci, but the MIC values of vancomycin are also increasing for the entire *Staphylococcus* species, worldwide.

Materials and Methods

Aim of our review is to evaluate the efficacy and tolerability of newer antibiotics with activity against methicillin-resistant and glycopeptide-resistant Gram-positive cocci, on the ground of our experience at a tertiary care metropolitan Hospital, and the most recent literature evidences in this field.

* Corresponding author: Telephone: +39-051-6363355, Telefax: +39-051-343500, E-mail: Roberto. manfredi@ unibo.it

Results

Quinupristin-dalfopristin, linezolid, daptomycin, and tigecycline show an excellent *in vitro* activity, comparable of that of to vancomycin and teicoplanin for methicillin-resistant staphylococci, and superior to that of vancomycin for vancomycin-resistant isolates. Dalbavancin, televancin and oritavancin are new glycopeptide agents with excellent activity against Gram-positive cocci, and have superior pharmacodynamics properties compared to vancomycin. We review the bacterial spectrum, clinical indications and practical use, pharmacologic properties, and expected adverse events and contraindications associated with each of these novel antimicrobial agents, compared with the present standard of care.

Discussion

Quinupristin-dalfopristin is the drug of choice for vancomycin-resistant *Enterococcus faecium* infections, but it does not retain effective activity against *Enterococcus faecalis*. Linezolid acxtivity is substantially comparable to that of vancomycin in patients with methicillin-resistant *Staphylococcus aureus* (MRSA) pneumonia, although its penetration into the respiratory tract is exceptionally elevated. Tigecycline has activity against both *Enterococus* species and MRSA; it is also active against a broad spectrum of Enterobacteriaceae and anaerobes, which allows for use intraabdominal, diabetic foot, and surgical infections. Daptomycin has a rapid bactericidal activity for *Staphylococcus aureus* and it is approved in severe complications, like bacteremia and right-sided endocarditis. It cannot be used to treat pneumonia and respiratory diseases, due to its inactivation in the presence of pulmonary surfactant.

Keywords: Resistant Gram-positive cocci, Staphylococci, Enterococci, Pneumococci, Streptococci, epidemiology, clinical issues, novel antimicrobial compounds, characteristics, literature evidences.

Infections Caused by Gram-Positive Cocci : An Outline of Their Antibiotic Resistance Profile

Gram-positive cocci have re-emerged as predominant pathogens of human hosts within the past 10-15 years. After the introduction of penicillin over 60 years ago, infections by *Staphylococcus aureus*, *Streptococcus pyogenes* and *Streptococcus pneumoniae* finally became treatable. Within a short period of time, however, *S. aureus* developed resistance to penicillin. As a conseqience, penicillinase-resistant penicillins were successfully introduced with in the early 1960's. Concomitantly, resistance emerged for the penicillinase-resistant penicillins, and finally methicillin (oxacillin)-resistant *S. aureus* (MRSA) became a major hospital-acquired pathogen. Vancomycin (belonging to the class of glycopeptides) remained an active agent against MRSA and coagulase-negative Staphylococci, so that it was increasingly used durint the subsequent years, until now. From the 1990's to the present, however, the emergence of resistance to vancomycin also occurred in a significant proportion [1-3]. First among these organisms were *Enterococcus faecium* and *Enterococcus faecalis* [4].

Subsequently, vancomycin (glycopeptide)-resistant enterococci (VRE) became major hospital-acquired pathogena. In the past several years, MRSA were also spreading clonally into the community (the so-called "CA-MRSA"), leading to increased use of vancomycin-teicoplanin therapy [5]. In the late 1990's, glycopeptide resistance was reported for coagulase-negative Staphylococci [6] and then, *S. aureus* (the so-called vancomycin-intermediate *S. aureus*, or VISA, and the so-called glycopeptide-intermediate *S aureus*, or GISA). The first reported isolation of VISA occurred in Japan in 1997 [7] and more than one hundred VISA isolates have been reported in the subsequent years [8]. In the year 2002, three vancomycin-resistant *S. aureus* (VRSA) strains isolated from clinical specimens of American patients were found to have high level resistance to vancomycin (MIC >32 ug/ml) [9]. Although a number of cases of VRSA have since been described [10], these isolates fortunately have not yet become widespread.

The re-emergence of Gram-positive cocci have been well established in the setting of hospital-acquired infections, but community-acquired infections due to MRSA have become increasingly problematic during the last years [11-13]. Foreign body infections and bacteremia caused by coagulase-negative Staphylococci have also increased during time [14]. As a result, vancomycin-teicoplanin usage has increased in both inpatients and outpatiens. Although the majority of *S. aureus* strains remain susceptible *in vitro* to vancomycin, its efficacy against methicillin-sensitive *S. aureus* (MSSA) is inferior to that of penicillinase-resistant penicillins, and beta-lactam derivatives as a whole [15, 16].

Actually, MRSA is born as a multi-drug resistant pathogen. Resistance to the macrolides, lincosamides, aminoglycosides and all beta-lactam agents as a group, as well as fluoroquinolones, are also seen when MRSA is of clinical concern. Rifampin should not be used as a single agent due to rapid emergence of resistance in these microorganisms, while doxycycline and trimethoprim-sulfamethoxazole (cotrimoxazole), are bacteriostatic rather than bactericidal in their mechanisms of action [17].

S aureus is well known to be a virulent and invasive pathogen. It produces a variety of pyrogenic toxins and superantigens which contribute to its overall virulence [18]. The presence of the Panton-Valentine leukocidin may predispose to invasive skin and soft tissue infections, and also necrotizing pneumonias and other necrotizing infectious localizations. MRSA infection often has its origin from a localized skin infection, with subsequent contiguous or hematogenous spread to lungs, heart (endocarditis), central nervous system (CNS), and sometimes bones and joints and other organs and sites [19]. The prolonged duration of treatment with vancomycin or teicoplanin for severe infections, like endocarditis and osteomyelitis, may lead to more frequent and severe adverse effects (especially nephropathy, serum electrolyte imbalance, and myelotixicity). While VISA/GISA and VRSA infections have only rarely been reported, clinical hetero-resistant populations of VISA (with strains shosing MIC values > 4-16 mcg/ml) have been isolated following prolonged administration of glycopeptides. Moreover, pharmacodynamics of vancomycin may have led to unappreciated under dosing of vancomycin, thereofore predisposing to microbial resistance [20]. Coagulase-negative Staphylococci have the capability to produce a glycocalyx enabling them to attach to prosthetic materials [21]. Biofilm formation on the surfaces of medical devices (i.e. prosthetic devices, central vascular catheters), provides a protected environment for coagulase-negative Staphylococci; this biofilm formation impedes antibiotic penetration and reduces target site formation [21, 22]. As expected, catheter-related blood stream infections, CNS ventricular shunt infections, prosthetic joint infections and prosthetic valve endocarditis are commonly caused by coagulase-negative Staphylococci [23].

The large majority of these microorganisms usually are or become resistant to methicillin. Intermediate resistance to vancomycin was first reported among coagulase-negative Staphylococci, several years before it occurred among *S. aureus* strains. Unlike *S. aureus*, infections by coagulase-negative Staphylococci on prosthetic hardware tend to be insidious and more chronic. Therapy often requires a combined medical- surgical approach with removal of the infected device, and prolonged duration (usually exceeding four weeks) of antibiotic therapy thereafter.

Vancomycin-resistant Enterococci (VRE) are primarily associated with healthcare institutional acquisition in patients with co-morbid conditions. Since their peak incidence around the year 2000, several new antibiotics with excellent activity against VRE have been introduced into clinical practice in the meantime.

On the other hand, *S. pneumoniae* is the most frequent cause of community acquired pneumonia (CAP). It accounts for at least one third of patients with CAP. The crude incidence of this common pathogen rises to greater than 50%, if respiratory culture with Gram stains, and urinary antigen for *S .pneumoniae* are systematically performed. Associated bacteremia occurs in 20% of pneumococcal pneumonias and mortality is notably higher than for other respiratory pathogen. *In vitro* resistance of *S. pneumoniae* to penicillin as currently defined by Clinical Laboratory Standards Institute (CLSI) criteria, does not necessarily correlate with clinical failure. Specifically, penicillins have been favorably efficacious for pneumonia caused by penicillin-resistant pneumococci [24, 25]. These resistant isolates are often also resistant to macrolides, and *in vitro* resistance to macrolide does appear to correlate with clinical outcome [26, 27].

In adults patients, *S. pneumoniae* also represents the most common cause of meningitis. Empiric therapy for meningitis with ceftriaxone and vancomycin pending antibiotic susceptibility testing is often employed. Data from a large scale observational study of pneumococcal meningitis suggests that combination therapy may be superior to monotherapy [28].

Groups A Streptococci (whose leading organism is *Streptococcus pyogenes*), as well as other beta-hemolytic Streptococci, are often associated with life-threatening infections, especially involviong the skin and soft tissues. Group B, C, F, and G beta-hemolytic Streptococci can also cause invasive infection and becteremia. *Streptococcus agalactiae* (belonging to group B Streptococci), is a common cause of neonatal sepsis. Fortunately, susceptibility to penicillin remains stable for the majority of the above-mentioned Streptococci.

Epidemiological Experience at a Major Tertiary Care Hospital in Northern Italy

A prospective, microbiological surveillance study is ongoing since a decade at our academic Hospital (S. Orsola-Malpighi Hospital, Bologna, Italy), in order to check the epidemiological-clinical evolution of bacterial and fungal infections occurring among inpatients. All microorganisms isolated and identified from sterile sites (i.e. blood cultures, protected bronchoalveolar lavage, urine culture, and so on), are systematically tested for *in vitro* susceptibility against a consistent panel of antimicrobial compounds, and data are reported quarterly (A. Nanetti and S. Ambretti, unpublished data). Each bacterial isolate cultured from a single patient within one month, is counted only once.

Some data regarding *in vitro* sensitivity rates of *Staphylococcus aureus* and Enterococci from the year 2005 to the first nine months of the year 2008, are reported in Table 1.

With regard to *S. aureus*, we underline a progressive reduction of methicillin resistance rate (from 56.8% of strains in the year 2005, up to 44.7% of overall tested strains of the year 2008), while we confirm a maintained 100% susceptibility to all available glycopeptides. Of major interest, the introduction of novel guidelines for a correct antimicrobial use in medicine and surgery, led to a progressive reduction of the frequency of antibiotic-resistant strains, as demostrated by an increased mean *in vitro* susceptibility rate against almost all tested compounds demostrated in the year 2008, *versus* years 2005-2007, together with an apparent, progressive trend toward reduced resistances during the examined temporal span (year 2005 to year 2008). Furthermore, elevated sensitivity rates are also found for a number of “older” molecules, like cotrimoxazole (87.9% to 98.2%), chloramphenicol (83.5% to 88.6%), followed by rifampicin (60.2% to 80.4%), clindamycin (34.9% to 54.4%), and erythromycin (34.2% to 55.3%). These last compounds, which are largely available and do not imply increased costs of administration, are expected to be clinically effective, alone and/or in combination with anti-Gram-positive agents (Table 1).

When examining the temporal trend of isolation of Enterococci (as a group) in the same time period (year 2005, up to September 2008), we notice a full sensitivity to novel compounds (i.e. linezolid, daptomycin, and quinupristin/dalfopristin), and a maintained *in vitro* effectiveness of glycopeptides (94.8% to 98.1%). Vancomycin-resistant Enterococci (also called VRE) were isolated infrequently: only 9 cases in the year 2005, 6 in the year 2006, 8 in the year 2007, and only three cases in the first 9 months of the year 2009 (A. Nanetti and S. Ambretti, unpublished data). Also in the case of Enterococci, a number of “older” compounds still retain an effective activity against Enterococci, as demostrated by susceptibility rates of penicillin (49.0% to 51.6%), ampicillin (49.0% to 52.3%), and tetracyclines (38.0% to 46.6%). Also in this case, these compounds may act favorably or may be a part of a combination regimen, after *in vitro* sensitivity assays (Table 1).

Novel Antibacterial Agents with Enhanced Activity against Resistant Gram Positive Cocci

The following antibacterial agents have been approved during the last five years: quinupristin /dalfopristin [29-32], linezolid [33, 34], daptomycin [35-37], and tigecycline [38, 39]. Novel glycopeptide agents under study include dalbavancin [40], telavancin, and oritavancin [41]. Even novel cephalosporins (i.e. cefbiprole), and fluroquinolones (i.e. garenofloxacin) with enhanced activity against MRSA are in the pipeline. Some features of these novel antimicrobial molecule are summarized in Table 2 (microbiological spectrum, clinical indication, adverse events), and in Table 3 (selected pharmacological features).

Table 1. Microbiological figures from patients hospitalized at our tertiary-care Hospital (S. Orsola-Malpighi Hospital, Bologna, Italy), year 2005-2008

In vitro antimicrobial susceptibility of *Staphylococcus aureus* strains isolated from inpatients (years 2005-2008)

Antibiotic susceptibility	Year 2005 (190 strains)	Year 2006 (167 strains)	Year 2007 (131 strains)	Year 2008 (Jan. to Sep.) (103 strains)
Penicillin	7.9	7.8	9.9	7.8
Amoxicillin-clavulanate	43.2	48.5	48.9	55.3
Cefotaxime/Ceftriaxone	42.9	48.5	48.9	55.3
Methicillin/Oxacillin	43.2	48.5	49.6	55.3
Erithromycin	34.2	48.5	51.9	55.3
Clindamycin	34.9	48.5	51.9	54.4
Chloramphenicol	84.7	88.6	87.8	83.5
Rifampicin	60.2	64.0	62.1	80.4
Cotrimoxazole	87.8	98.2	93.1	95.1
Gentamicin	31.1	41.0	43.5	49.5
Vancomycin/Teicoplanin	100.0	100.0	100.0	100.0

In vitro antimicrobial susceptibility of Enterococci isolated from inpatients (years 2005-2008)

Antibiotic susceptibility	Year 2005 (206 strains)	Year 2006 (151 strains)	Year 2007 (155 strains)	Year 2008 (Jan. to Sep.) (108 strains)
Penicillin	50.0	49.0	51.6	50.9
Ampicillin	50.0	49.0	52.3	51.9
Tetracyclin	46.6	45.0	38.7	38.0
Vancomycin/Teicoplanin	98.1	95.4	94.8	96.3
Linezolid	100,0	100,0	100,0	100,0
Daptomycin	N/A	N/A	N/A	100.0
Quinupristin/Dalfopristin	100,0	100,0	100,0	100,0

Quinupristin/Dalfopristin

The so-called streptogramin antibiotic, quinupristin/dalfopristin, is a combination of two semisynthetic pristinamycin derivatives, which are represented by quinupristin and dalfopristin, in a 30:70 ratio. Resistance to quinupristin/dalfopristin can occur by several mechanisms increasing enzymatic modification, active transport of specific efflux pumps mediated by an adenosine triphosphate-binding proteinm and alteration of the target site. Resistance is rare for Streptococci and *Enterococcus faecium* [42]. This streptogramin combination acts synergistically to inhibit bacterial protein synthesis at the ribosome level. Quinupristin/dalfopristin is therefore active against *Staphylococcus aureus* (including MRSA strains), *Streptococcus pneumoniae*, and Gram-positive anaerobes such as *Clostridium* spp.,

Peptococcus spp., and *Peptostreptococcus* spp. It is effective against vancomycin-sensitive as well as vancomycin-resistant *Enterococcus faecium* (VREF), but has little *in vitro* activity against *Enterococcus faecalis*, so that it cannot be recommended until final speciation of Enterococcal organisms is concluded. Dalfopristin/quinipristin association inhibits cytochrome P450 3A4, and can inhibit agents metabolized through this pathway. Dosage adjustments may be needed in patients with hepatic dysfunction. Renal function has minimal impact on the agent's pharmacokinetics. A post-antibiotic effect is observed at 4-5 hours at 4X MIC versus Staphylococci, 7-9 hours for Streptococci, and only 4 hours for Enterococci [43].

The registered clinical indications for quinupristin/dalfopristin use include intraabdominal infections, bacteremia, urinary tract infection and skin and soft tissue infections in which Enterococci may play a relevant pathogenic role. Overall clinical success rate for patients with vancomycin-resistant *E. faecium* (VREF) proved to be 74%, while overall clinical and bacteriological success rate was 66% [44]. Patients with bacteremia, those on mechanical ventilation, and those undergoing surgery had a worse outcome as might be expected [44]. The most common and notable adverse events were arthralgias and myalgias. In a comparative clinical trial of therapy for Gram-positive skin and soft tissue infections, *S. aureus* was the most frequent pathogen isolated [45]. The clinical success rate of quinupristin/dalfopristin was comparable (68%) to the comparator agents (71%) (cefazolin, oxacillin or vancomycin). A higher incidence of drug-related adverse events occurred with quinopristin/dalfopristin as compared to other agents [46]. For those patients receiving comparator agents, the most common reason for discontinuation was treatment failure (12%) [46]. Furthermore, quinupristin/dalfopristin was compared to vancomycin in patients with hospital-acquired pneumonia [47]. Successful outcomes were similar at 56% for quinupristin/dalfopristin and 58% for vancomycin. The bacteriologic success rate was identical for both antibiotic groups, at around 54% of treated cases. Quinupristin/dalfopristin has been also used to treat patients infected by *S. aureus* intolerant of or failing standard therapies [48].

Ninety patients were treated an average of 28 days with a 71% clinical outcome of cure or improvement and bacteriologic outcome of eradication or presumed eradication. Infections included bone and joint, skin and soft tissue, bacteremia, endocarditis and respiratory tract involvement. Adverse events included arthralgias mainly (11%), myalgias (9%) and nausea (9%). However, in patients with hepatic dysfunction or liver transplantation and concurrent receipt of immunosuppressive chemotherapy, the incidence of arthralgias approached 50% of treated subjects [49, 50].

Linezolid

Linezolid is an oxazolidinone antibiotic with activity against Gram-positive pathogens including VRE, MRSA, and VISA. The unique mechanism of action of linezolid involves the inhibition of bacterial protein synthesis through binding to the domain V regions of the 23 Sr RNA gene 46. Resistance to linezolid requires mutations of multiple gene copies, and seems an infrequent phenomenon. Linezolid is 100% bioavailable when given by either oral or intravenous route. Maximal plasma levels are achieved within 1-2 hours after oral dosing. Protein binding is only around 30% with free distribution to well-perfused tissues. The drug does not require dosage alteration in the presence of renal failure, and no interaction exists for cytochrome P450 enzymes (and drugs metabolized through this last pathway). Linezolid and its two metabolites are decreased with hemodialysis, so dosing should occur postdialysis [51].

Linezolid is currently approved for skin and soft tissue infections and pneumonia due to susceptible pathogens [52]. In two controlled trials of hospital-acquired pneumonia, a trend was seen for linezolid superiority over vancomycin [53, 54]. There is little data (mainly based on observational studies), on the utility of linezolid for either bacteremia [55], or osteomyelitis [56, 57].

Based on a rabbit model, linezolid does not have sufficient CSF penetration, and should not be recommended for pneumococcal meningitis [33]. However, CNS penetration appears adequate to treat CSF shunt infections [58], and brain abscesses, too.

The myelotoxicity (especially the thrombocytopenia), is the most common serious adverse event caused by linezolid [59]; it can be ameliorated or prevented by co-administration of pyridoxine (Vitamin B6) [60-63]. Both peripheral and optic neuropathy have been reported with prolonged use of greater than four consecutive weeks [64, 65]. Lactic acidosis has also been reported and is not associated with duration of administration [65, 66]. Interaction exists between linezolid and serotonin-reuptake inhibitors (antidepressants drugs). In these last cases, a minority of patients might develop the so-called serotonin syndrome (fever, agitation with mental status changes and tremors). Due to its weak activity as a monoamine oxidase- inhibitor, linezolid should not be used concomitantly with agents such as tramadol, pethidne, duloxetine, venlafaxine, milnacipran, sibutramine, chlorpeniramine, brompheniramine, cyproheptadine, citalopram, and paroxetine [65, 67]. Drug metabolites may accumulate in the event of severe renal failure.

Daptomycin

Daptomycin is the first in a new class of antimicrobial agents: a lipopeptide antibiotic with activity against *S. aureus* (including methicillin-resistant strains), beta-hemolytic Groups A, B C and G Streptococci, and Enterococci, including ampicillin-and vancomycin- resistant strains. The mechanism of action of daptomycin is unique as le molecule causes a calcium ion dependent disruption of bacterial cell membrane potential resulting in an efflux of potassium which inhibits RNA, DNA and protein synthesis. Rare instances of resistance have occurred in clinical trials, although the mechanism of resistance has not yet been clerarly identified to date. Daptomycin was shown to have a rapidly bactericidal effect *in vitro* against Gram-positive drug-resistant pathogens. Its activity is concentration-dependent and once daily dosing is associated with significant post-antibiotic effect.

Daptomycin is highly protein bound (around 92%), with a terminal half-life of 8 hours, which allows for once daily dosing. Post-antibiotic effect proves dose dependent, and is reduced in the presence of albumin (i.e. essudates). The drug volume of distribution is low (0.1L/kg) and the Cmax (54.6mcg/ml) is unchanged at steady state is achieved by day three of administration in humans. Cmax concentrations occur at the end of a 30 minute infusion. Dosage needs to be reduced and dosing interval extended to every 48 hours in patients with reduced creatinine clearance <30mL/min; the same occurs for patients on either hemodialysis or peritoneal dialysis; the daptomycin dose is in these patients becomes 4 mg/kg every 48 hours. Daptomycin should be administered after hemodialysis as approximately 15% is cleared per 4 hour hemodialysis session. On the other hand, no dose adjustments for hepatic dysfunction are required.

In early clinical trails conducted in the years 1980's-1990's, daptomycin was given in divided daily doses of 2 mg/kg every 12 hours for skin and soft tissue infection and 3 mg/kg every 12 hours for bacteremia, achieving good clinical and bacteriological outcomes. However, rise in serum creatine phosphokinase (CPK), with myalgias and muscle weakness led to initial abandonment of this promising antibiotic. However, myopathy was reversible upon cessation of the drug. With the advent of MRSA infections, daptomycin has been re-examined and resurrected, and its dosage has been increased to 4 mg/kg daily for skin and soft-tissue infection [68], and up to 6 mg/kg daily for bacteremia and endocarditis [69]. Both indications have been approved by the United States FDA. Otherwise, daptomycin is not approved for the treatment of bacterial pneumonia; its efficacy is significantly compromised by its interaction with pulmonary surfactant [70]. Significant drug-drug interaction occurs with the statins, and patients receiving HMG-CoA reductase inhibitors; these drugs should be suspended and avoided while the patient is undergoing a daptomycin course.

Tigecycline

Tigecycline is a novel glycylcycline molecule, which is a derivative of the tetracycline minocycline. Resistance to the tetracycline class is classically mediated by ribosomal protection mechanisms or by active efflux. Tigecycline has more potent activity against tetracycline-resistant organisms, and maintains a broad antibacterial spectrum against Gran-positive and also Gram-negative pathogens. Tigecycline binds more avidly to the ribosome and either does not induce efflux proteins or is not readily exported by efflux proteins [38]. Resistant clinical isolates were associated with up-regulation of chromosomially mediated efflux pumps. Unlike original tetracyclines, tigecycline has a large volume of distribution (above 10 L/kg), the protein binding is approximately 68%, the terminal half-life of elimination is 36 hours, and less than 15% of the native drug is excreted unchanged in the urine.

Clinical trials have been conducted in patients with complicated skin and soft tissue infections and intraabdominal infections for which the drug gained its United States FDA approval.

Based on *in vitro* susceptibility data, tigecycline has a broad spectrum of activity against both Gram-positive cocci (including methicillin-resistant Staphylococci or MRSA, penicillin-resistant *Streptococcus pneumoniae*, beta-hemolytic group A and group B Streptococci, Enterococci (vancomycin-susceptible ones), and *Listeria monocytogenes.* Unlike other new agents for Gram-positive cocci, tigecycline also has extensive activity against Gram-negative pathogens including *Haemophilus influenzae, Neisseria spp* [11], Enterobacteriaceae, and non-lactose fermenters other than *Pseudomonas aeruginosa.* The MIC_{90} values for *Proteus* spp., *Providentia* spp., and *Burkholderia cepacia* is 8mcg/mL, which limits its utility in infections caused by these mentioned pathogens.

Tigecycline needs no reduction in renal impairment and it is not dialyzable. Patients with severe hepatic dysfunction (Child-Pugh stage C liver disease), should receive a lower dosage. Tigecycline activity is dependent on the time above the MIC, and the drug concentrations should be above the MIC values for at least 50% of the dosing interval.

The expected adverse effects of tigeciclyne are primarily gastrointestinal in origin, with nausea, vomiting, diarrhea and heartburn as the most frequent. As with all tetracyclines, tigecycline is contraindicated for pregnant females and for children less than 8 years of age [71]. Drug interactions of tigecycline with either digoxin or warfarin do not alter the effect of either drug. Tigecycline does not inhibit metabolism mediated by the cytochrome P450 isoforms IA2, 2C8, 2C9, 2C19, 2D6 and 3A4, so that no drug-drug interaction are expected with drugs metabolized by these cytochrome isoforms.

Dalbavancin

Dalbavancin is a true second-generation glycopeptide. Its unique pharmacokinetic profile allows once weekly dosing. It is not active against VRE, but has an excellent activity against MRSA, *S. pyogenes* and *S. pneumoniae* as well as vancomycin-susceptible Enterococci. It is bactericidal and synergistic with ampicillin against Van-A type Enterococci. The mechanism of action is the inhibition of the cell wall peptidoglycan cross-linking.

The daily dosage is 1000 mg i.v. once, followed by 500 mg i.m. 7 days later; the terminal half-life of dalbavancin is 9-12 days in humans due to protein binding of greater than 95%. Animal models of infection show excellent activity against MRSA or GISA endocarditis, penicillin-resistant *Streptococcus pneumoniae* [12] pneumonia or MRSA pouch infection, and septicemia due to Staphylococci, Streptococci or Enterococci. This antibiotic has been evaluated for catheter-related bacteremia [72] and skin and soft tissue infections [73]. Dalbavancin was effective and well tolerated in adult patients with catheter-related bacteremia caused by coagulase-negative Staphylococci, MSSA and MRSA in a comparative trial with vancomycin. In skin and soft tissue infections, a 92-94% microbiological and clinical response respectively was found in an open label phase 2 comparative dosing trial [73]. Clinical success at follow-up visits for the two dose dalbavancin group was 80% for MRSA *versus* 50% for comparator therapy (which included beta-lactams, clindamycin, vancomycin and linezolid, respectively).

Oritavancin

Oritavancin is another derivative of vancomycin: it's a chloroeremomycin with the substitution of vancosamine by epi-vancosamine. It has a similar spectrum of activity to vancomycin but with consistently lower MIC values (<1mg/L). No resistance to oritavancin has been noted among *S. aureus* strains including VISA strains, but VAN-A and VAN-B strains of Enterococci with reduced susceptibility to oritavancin have been obtained *in vitro*. The known mechanisms of resistance of oritavancin are: 1) complete elimination of D-Ala-entry precursors; 2) mutations in the VAN Sb sensor of the VAN B cluster; or 3) the expression of Van Z (the precise function of which is still unknown).

Oritavancin shows rapid, concentration dependent bactericidal activity with a concentration-dependent post-antibiotic effect exerted against both VRE and MRSA. Oritavancin activity is negatively affected by large inoculum and its activity *versus* VRE was slightly reduced in stationary phase or in acidic foci of infection. In animal models, its efficacy has been demonstrated for experimental MRSA endocarditis and *S. pneumoniae* meningitis [74, 75]. In a reliable endocarditis model, the addition of gentamicin proved to be synergistic, and able to prevent the emergence of resistant mutants. With regard to skin and soft tissue infections, oritavancin proved to be at leat equivalent to vancomycin, for both clinical and bacteriological cure (about 78% of cure rate) [76].

Telavancin

Televancin is a rapidly bactericidal lipoglycopeptide analog of vancomycin. The mechanism of action of this moleculr is by inhibition of peptidoglycan chain formation through blockage of both the transpeptidation and transglycosylation steps; and by a direct effect on the bacterial membrane dissipating membrane potential and effecting changes in cellular permeability.

The *in vitro* activity of telavancin demonstrates enhanced activity against MRSA, penicillin-resistant *S. pneumoniae,* GISA and Van-A type Enterococci. Telavancin achieves a higher volume of distribution into tissues and a prolonged half-life [77]. A high level of protein binding (93%) occurs in human plasma and repetitive dosing does not lead to accumulation. The terminal half-life is 7-9 hours at doses above 5 mg/kg [78]. Telavancin exhibit time-dependent killing activity [79].

Telavancin and its comparators of vancomycin or beta-lactam agent have been compared in a phase 2 clinical trial for skin and skin-structure infections. Clinical cure rates were similar at 92% in for telavancin *versus* 96% for comparator agents. Microbiologic rates of cure were noted to be 93% in the telavancin group and 95% among the comparator group [80]. For complicated skin and soft tissue infections, clinical cure rates were at 96% for telavancin and 90% for comparator agents. Microbiologic eradication was better with telavancin (92%) *versus* comparator agents (78%, p=0.07) [80]. Telavancin is currently under assessment in phase 3 trials of hospital-acquired pneumonia.

Adverse events associated with telavancin among evaluated patients included vomiting, paresthesias and dyspnea. Laboratory abnormalities included microalbuminemia and a decreased platelet count [81].

Clinical Indications of Novel Antibiotics with Expanded Spectrum against Resistant Gram-Positive Cocci

Skin and Soft-Tissue Infections

Skin and soft-tissue infections caused by Gram-positive cocci range from a simple cellulitis to life-threatening necrotizing fasciitis. All of the newer agents have been studied for such infections and have been found to be efficacious (Table 2). Most of the patients in these studies had less severe infections than necrotizing fasciitis as that infection requires a surgical approach as well as antibiotic therapy. All five FDA-approved agents, i.e. quinupristin/dalfopristin, linezolid, daptomycin, tigecycline and vancomycin are appropriate choices for an effective treatment of Gram-positive pathogens. Only tigecycline has activity against Gram-negative bacilli pathogens. So, tigecycline may have a major role for diabetic foot infections and infected decubitus ulcers which may be co-infected by anaerobic bacteria and aerobic Gram-negative bacilli, in addition to Gram-positive cocci.

.

Bone and Joint Infections

With regard to osteomyelitis and joint infections, Gram-positive cocci largely predominate over other microbial patogens. *S. aureus*, both MSSA and MRSA, as well as coagulase-negative Staphylococci account for greater than 50% of recovered pathogens. Unfortunately, only few studies have prospectively investigated the above-mentioned newer antibiotics in these infections [56, 57]. Aneziokoro *et al.* evaluated 20 patients who received linezolid for osteomyelitis for 6 weeks or more in a retrospective non-comparative study [82]. Fifty-five percent of cases (11 patients) achieved a cure with follow-up periods ranging from 6 to 49 months (median of 36 months). Prospective comparative studies of efficacy in bone and joint infections have not been reported to date. In two retrospective studies, 22 patients with osteomyelitis and three subjects with septic joint infections were treated with daptomycin [83, 84]. MRSA was the predominant pathogen in over 75% of patients. Daptomycin was used as salvage therapy, and its usual dose was 6 mg/kg per day. Clinical success rate was about 90%; follow up periods were one year or less. Limited data has been published with respect to bone and joint infections for dalbavancin, tigecycline or quinupristin/dalfopristin in humans. In a rabbit model of MRSA osteomyelitis, the combination of rifampin and tigecycline was compared to vancomycin with or without rifampicin, tigecycline alone, and vancomycin alone [85]. All regimens were effective (in about 90% of episodes). Untreated rabbits had spontaneous cure in 26% of cases (4/15). Tigecycline concentrations are higher in infected bone than in non-infected bone. A rabbit model of quinupristin/dalfopristin prosthetic joinT infection with MRSA was compared to vancomycin with or without rifampicin, showing an equivalent outcome [86].

Pneumonia and Lower Respiratory Tract Infections

Pneumonia due to Gram-positive cocci is common. In the community, infection is usually due to *S. pneumoniae* and occasionally *S. aureus.* Hospital-acquired pneumonia (HAP) is often caused by MRSA organisms. Linezolid was comparable to vancomycin in the therapy of MRSA-associated VAP, although a trend was seen for linezolid superiority [53, 54]. Daptomycin is not indicated for pneumonia due to its interaction with surfactant [70], while tigecycline is undergoing clinical evaluation. Quinupristin/dalfopristin has been compared to vancomycin for hospital-acquired pneumonia [47]. One hundred and 71 patients had similar clinical response rates of about 57% respectively. Drug discontinuation adverse events occurred more frequently in the quinupristin/dalfopristin group (15%), as compared to vancomycin. Only two isolated of the 87 overall strains were shown to have decreased susceptibility to quinupristin/dalfopristin during and after treatment.

Table 2. Novel antimicrobial agents for the management of resistant Gram-positive infections. Microbiological, clinical, and therapeutic features as of end of year 2008.

Stable-	Class	Microbiologically effective on				Adverse events	Clinical indications for at end-2008 (infections)						
		MRSA	MRSE	PRSP	VRE		Blood stream	Skin-soft tissue	Hospital-acquired pneumonia	Endo-carditis	Bone and joint	CNS	Intra-abdominal
Vancomycin	Glycopeptide	+	+	+	Not VISA-VRSA	Nephro- and ototoxicity Red man syndrome	+	+	+	+	+	+	+
Teicoplanin	Glycopeptide	+	+	+	Not Van-A	N/D	+	+	-	-	-	-	-
Quinupristin/ Dalfopristin	Streptogramin	+	+	+	*E. faecium*	Hepatic, phlebitis Artho-myalgias	+	+	-	-	-	-	-
Linezolid	Oxazolidinone	+	+	+	+	Neuropathy, myelotixicity, serotonin syndrome	+	+	+	-	+	-	-
Daptomycin	Lipopeptide	+	+	+	+	Arthro-myalgias, CPK rise	+	+	-	+	-	-	-
Tigecycline	Glycopeptide	+	+	+	+	Nausea, diarrhea		+	-	-	-	-	+
Dalbavancin	Glycopeptide	+	+	+	Not VAN-A	Gastrointestinal, hypokaleamia	+	+	-	-	-	-	-
Oritavancin	Glycopeptide	+	+	+	+	N/D	-	+	-	-	-	-	-
Telavancin	Lipoglycopeptid e	+	+	+	+	Altered taste, CNS, phlebitis7	-	+	-	-	-	-	-

Table 3. Novel antimicrobial agents for the management of resistant Gram-positive infections. Selected pharmacogical features, updated at the end of year 2008

Drug	Class	Pharmacodynamics	Protein binding (%)	Elimination route	Dosage adjustement	
					Renal	Hepatic
Vancomycin	Glycopeptide	AUC/MIC	10-55	Renal	+	-
Teicoplanin	Glycopeptide	AUC/MIC	90	Renal	+	-
Quinupristin/ Dalfopristin	Streptogramin	AUC/MIC	N/A	Hepatic/ feces	N/A	+
Linezolid	Oxazolidinone	AUC/MIC	31	Hepatic	N/A	N/A
Daptomycin	Lipopeptide	AUC/MIC	92	Renal	+	N/A
Tigecycline	Glycopeptide	Time above MIC	68	Biliary	N/A	+
Dalbavancin	Glycopeptide	AUC/MIC	>95	Renal	+	N/A
Oritavancin	Glycopeptide	AUC/MIC	N/A	Renal	Renal	N/A
Telavancin	Lipoglycopeptide	AUC/MIC	N/A	Renal	+	-

Intraabdominal Infections

Of the newer antibiotics, only tigecycline has been approved for intraabdominal infections. As mentioned, tigecycline's broader spectrum of activity includes Gram-negative bacilli and anaerobic bacilli. Linezolid, daptomycin, and quinupristin/dalfopristin can be used in combination with antibiotics with Gram-negative spectrum of activity such as aztreonam, and especially carbapenems, fluoroquinolones, and aminoglycosides. Of concern, quinupristin/dalfopristin has no activity against *E. faecalis*.

Bacteremia and Endocarditis

Daptomycin and quinipristin/dalfopristin have been approved by the United States FDA organisms for the treatment of Gram-positive bacteremia. In addition, daptomycin has been approved for use in *S. aureus* right-sided endocarditis [87]. Dalbavancin, linezolid, tigecycline and oritavancin have not yet been approved for bacteremia due to gram-positive cocci. Linezolid has been evaluated for Gram-positive bacteria [55, 88, 89]. Among 108 bacteremic patients receiving linezolid, eradication was seen in 91% and clinical cure was seen in 94% of episodes [55]. On the other hand, linezolid is still not approved for catheter-related becteremia and endocarditis. A randomized study of 726 patients with catheter-related bacteremia received linezolid or vancomycin; an excess number of deaths were seen for patients receiving linezolid due mainly to Gram-negative rods implicated in these infections [90]. Based on 23 case reports and three case series, a total of 63% (21/33) of patients with endocarditis were successfully cured after linezolid administration [91]. MRSA and vancomycin-intermediate *S.aureus* were the most commonly isolated cocci (24.2% and 30.3% of cases, respectively). Five cases were succesfully treated with linezolid monotherapy.

Table 4. Some experimental studies conducted *in vitro* or on animal models, regarding possible interactions between the different antimicrobial agents effective on Gram-positive cocci [89, 92-95, 98-107].

Reference quotation	Combination	Pathogens	Interaction
[97]	Daptomycin+vancomycin	GISA	Additive
[97]	Daptomycin+gentamicin	GISA	Additive
[93]	Daptomycin+gentamicin	MSSA/MRSA	Enhanced time-kill
[98]	Daptomycin+gentamicin	MSSA/MRSA	Increased killing
[99]	Daptomycin+rifampicin	MRSA	Additive
[100]	Daptomycin+gentamicin+rifampicin	MRSA	Additive
[97]	Linezolid+vancomycin	GISA	Additive
[94]	Linezolid+vancomycin	MSSA/MRSA	Antagonistic
[89, 95]	Linezolid+vancomycin	MRSA	Indifferent
[101]	Linezolid+vancomycin	MSSA/MRSA/MRSE	Increased killing
[94, 98]	Linezolid+gentamicin	MSSA/MRSA	Indifferent
[95]	Linezolid+gentamicin	MRSA	Antagonistic
[94]	Linezolid+rifampicin	MSSA/MRSA	Indifferent
[95]	Linezolid+rifampicin	MSSA/MRSA	Synergistic
[102]	Linezolid+rifampicin	MSSA	Indifferent
[101]	Linezolid+Quinupristin/dalfopristin	MRSA	Increased killing
[97]	Quinupristin/dalfopristin+vancomycin	GISA	Additive-Synergistic
[103]	Quinupristin/dalfopristin+vancomycin	MSSA/MRSA	Additive
[101]	Quinupristin/dalfopristin+vancomycin	MRSA	Increased killing
[92]	Quinupristin/dalfopristin+vancomycin	MSSA/MRSA	Synergistic
[97]	Quinupristin/dalfopristin+gentamicin	GISA	Indifferent
[104]	Quinupristin/dalfopristin+rifampicin	MSSA	Increased killing
[105]	Quinupristin/dalfopristin+rifampicin	MRSA	Synergistic
[106]	Tigecycline+vancomycin	MRSA	Indifferent
[107]	Tigecycline+gentamicin	MRSA/GISA	Increased killing
[106]	Tigecycline+rifampicin	MRSA	Indifferent

POTENTIAL SYNERGISTIC INTERACTIONS OF NEWER ANTIBIOTICS: *IN VITRO* STUDIES

In vitro interaction between the new antistaphylococcal antibiotics were virtually always indifferent, therefore leading to a possible additive effect, although a few experiments showed possible (Table 4) [89, 92-95, 98-107]. For instance, a synergistic interaction was found for quinupristin/dalfopristin and vancomycin in two independent studies [92, 93].

On the other hand, antagonistic interactions were demonstrated for the combination of linezolid plus vancomycin [94], and linezolid plus gentamicin [95]. It should be emphasized that *in vitro* interaction may not translate into clinical efficacy. Quinupristin/daflopristin in combination with vancomycin appeared to be favorable for the management of MRSA infections responding poorly to vancomycin [96]. However, we have to specify that the

MRSA isolates were of a specific genotype, accessory gene regulator (agr), which has been linked to vancomycin treatment failure [96]. Nevertheless, such information may be useful if innovative combination therapy needs to be administered to severely ill patients with invasive *S. aureus* infection unresponsive to monotherapy.

Controlled clinical trials using combinations including these new agents are indicated for patients with severe, life-threatening infections caused by gram-positive cocci, and randomized trials are strongly warranted in this somewhat unexplored field.

REFERENCES

[1] Foster, JK; Lentino, JR; Strodtman, R; DiVincenzo, C. Comparison of *in vitro* activity of quinolone antibiotics and vancomycin against gentamicin- and methicillin-resistant *Staphylococcus aureus* by time-kill kinetic studies. *Antimicrob Agents Chemother*, 1986, 30, 823-7.

[2] Tallent, SM; Bischoff, T; Climo, M; Ostrowsky, B; Wenzel, RP; Edmond, MB. Vancomycin susceptibility of oxacillin-resistant *Staphylococcus aureus* isolates causing nosocomial bloodstream infections. *J Clin Microbiol*, 2002, 40, 2249-50.

[3] Sieradzki, K; Leski, T; Dick, J; Borio, L; Tomasz, A. Evolution of a vancomycin-intermediate *Staphylococcus aureus* strain in vivo: multiple changes in the antibiotic resistance phenotypes of a single lineage of methicillin-resistant *S. aureus* under the impact of antibiotics administered for chemotherapy. *J Clin Microbiol*, 2003, 41, 1687-93.

[4] Murray, BE. Vancomycin-resistant enterococci. *Am J Med.*, 1997, 102, 284-93.

[5] Okuma, K; Iwakawa, K; Turnidge, JD; et al. Dissemination of new methicillin-resistant Staphylococcus aureus clones in the community. *J Clin Microbiol*, 2002, 40, 4289-94.

[6] Schwalbe, RS; Stapleton, JT; Gilligan, PH. Emergence of vancomycin resistance in coagulase-negative staphylococci. *N Engl J Med.*, 1987, 316, 927-31.

[7] Hiramatsu, K; Hanaki, H; Ino, T; Yabuta, K; Oguri, T; Tenover, FC. Methicillin-resistant *Staphylococcus aureus* clinical strain with reduced vancomycin susceptibility. *J Antimicrob Chemother*, 1997, 40, 135-6.

[8] Appelbaum, PC. MRSA-the tip of the iceberg. *Clin Microbiol Infect*, 2006, 12 Suppl 2, 3- 10.

[9] MMWR. *Staphylococcus aureus* resistant to vancomycin-United States 2002. Vol. 51 (26), 565.567RE, 2002.

[10] Chang, S; Sievert, DM; Hageman, JC; et al. Infection with vancomycin-resistant *Staphylococcus aureus* containing the vanA resistance gene. *N Engl J Med.*, 2003, 348, 1342-7.

[11] King, MD; Humphrey, BJ; Wang, YF; Kourbatova, EV; Ray, SM; Blumberg, HM. Emergence of community-acquired methicillin-resistant *Staphylococcus aureus* USA 300 clone as the predominant cause of skin and soft-tissue infections. *Ann Intern Med.*, 2006, 144, 309-17.

[12] Moellering, RC; Jr. The growing menace of community-acquired methicillin-resistant *Staphylococcus aureus*. *Ann Intern Med.*, 2006, 144, 368-70.

[13] Noskin, GA; Rubin, RJ; Schentag, JJ; et al. The burden of *Staphylococcus aureus* infections on hospitals in the United States: an analysis of the 2000 and 2001 Nationwide Inpatient Sample Database. *Arch Intern Med.*, 2005, 165, 1756-61.

[14] Rupp, ME; Archer, GL. Coagulase-negative Staphylococci: pathogens associated with medical progress. *Clin Infect Dis.*, 1994, 19, 231-43, quiz 244-5.

[15] Schaaff, F; Reipert, A; Bierbaum, G. An elevated mutation frequency favors development of vancomycin resistance in *Staphylococcus aureus. Antimicrob Agents Chemother*, 2002, 46, 3540-8.

[16] Chang, FY; Peacock, JE; Jr., Musher, DM; et al. *Staphylococcus aureus* bacteremia: recurrence and the impact of antibiotic treatment in a prospective multicenter study. Medicine (Baltimore) 2003, 82, 333-9.26

[17] Markowitz, N; Quinn, EL; Saravolatz, LD. Trimethoprim-sulfamethoxazole compared with vancomycin for the treatment of *Staphylococcus aureus* infection. *Ann Intern Med.*, 1992, 117, 390-8.

[18] Becker, K; Friedrich, AW; Lubritz, G; Weilert, M; Peters, G; Von, Eiff, C. Prevalence of genes encoding pyrogenic toxin superantigens and exfoliative toxins among strains of *Staphylococcus aureus* isolated from blood and nasal specimens. *J Clin Microbiol*, 2003, 41, 1434-9.

[19] Lowy, FD. *Staphylococcus aureus* infections. *N Engl J Med.,* 1998, 339, 520-32.

[20] Sakoulas, G; Moellering, RC; Jr., Eliopoulos GM. Adaptation of methicillin-resistant *Staphylococcus aureus* in the face of vancomycin therapy. *Clin Infect Dis.*, 2006, 42 Suppl 1, S40-50.

[21] Donlan, RM; Costerton, JW. Biofilms: survival mechanisms of clinically relevant microorganisms. *Clin Microbiol Rev.*, 2002, 15, 167-93.

[22] Caiazza, NC; O'Toole, GA. Alpha-toxin is required for biofilm formation by *Staphylococcus aureus*. *J Bacteriol*, 2003, 185, 3214-7.

[23] Von Eiff, C; Peters, G; Heilmann, C. Pathogenesis of infections due to coagulase-negative Staphylococci. *Lancet Infect Dis.*, 2002, 2, 677-85.

[24] Yu, VL; Chiou, CC; Feldman, C; et al. An international prospective study of pneumococcal bacteremia: correlation with *in vitro* resistance, antibiotics administered, and clinical outcome. *Clin Infect Dis.*, 2003, 37, 230-7.

[25] Peterson, LR. Penicillins for treatment of pneumococcal pneumonia: does in vitro resistance really matter? *Clin Infect Dis.*, 2006, 42, 224-33.

[26] Lonks, JR; Garau, J; Gomez, L; et al. Failure of macrolide antibiotic treatment in patients with bacteremia due to erythromycin-resistant *Streptococcus pneumoniae. Clin Infect Dis.*, 2002, 35, 556-64.

[27] Schentag, JJ; Klugman, KP; Yu, VL; et al. *Streptococcus pneumoniae* bacteremias: pharmacodynamic correlations with outcome and macrolide resistance: a controlled study. *Int J Antimicrob Agents,* 2007, 30, 264-9.

[28] Greenberg, DDR; Klugman, K; Madhi, SA; Feldman, C; Roberts, S; Morris, A; Chedid, MBF; Chiou, CC; Yu, VL. *Streptococcus pneumoniae* serotypes causing meningitis in children and adults. Proceedings of the 14th ICAAC Conference. *Washington*, DC., 2004.

[29] Carpenter, CF; Chambers, HF. Daptomycin: another novel agent for treating infections due to drug-resistant gram-positive pathogens. *Clin Infect Dis.*, 2004, 38, 994-1000.

[30] Fenton, C; Keating, GM; Curran, MP. Daptomycin. *Drugs*, 2004, 64, 445-55.

[31] Steenbergen, JN; Alder, J; Thorne, GM; Tally, FP. Daptomycin: a lipopeptide antibiotic for the treatment of serious Gram-positive infections. *J Antimicrob Chemother*, 2005, 55, 283-8.
[32] Schriever, CA; Fernandez, C; Rodvold, KA; Danziger, LH. Daptomycin: a novel cyclic lipopeptide antimicrobial. *Am J Health Syst Pharm*., 2005, 62, 1145-58.
[33] Moellering, RC. Linezolid: the first oxazolidinone antimicrobial. *Ann Intern Med.*, 2003, 138, 135-42.
[34] Birmingham, MC; Rayner, CR; Meagher, AK; Flavin, SM; Batts, DH; Schentag, JJ. Linezolid for the treatment of multidrug-resistant, gram-positive infections: experience from a compassionate-use program. *Clin Infect Dis*., 2003, 36, 159-68.27
[35] LaPlante, KL; Rybak, MJ. Daptomycin - a novel antibiotic against Gram-positive pathogens. *Expert Opin Pharmacother*, 2004, 5, 2321-31.
[36] Jeu, L; Fung, HB. Daptomycin: a cyclic lipopeptide antimicrobial agent. *Clin Ther*., 2004, 26, 1728-57.
[37] Alder, JD. Daptomycin: a new drug class for the treatment of Gram-positive infections. *Drugs Today* (Barc) 2005, 41, 81-90.
[38] Livermore, DM. Tigecycline: what is it, and where should it be used? *J Antimicrob Chemother*, 2005, 56, 611-4.
[39] Pankey, GA. Tigecycline. *J Antimicrob Chemother*, 2005, 56, 470-80.
[40] Van Bambeke, F; Van Laethem, Y; Courvalin, P; Tulkens, PM. Glycopeptide antibiotics: from conventional molecules to new derivatives. *Drugs*, 2004, 64, 913-36.
[41] Virginlar, N. MA. Glycopeptides (Dalbavancin, Oritavancin, Teicoplanin, Vancomycin). In: Yu VL, Eds. Antimicrobial Therapy and Vaccines. Vol. II: *Antimicrobial Agents*: www.antimicrobe.org, 2004.
[42] Hershberger, E; Donabedian, S; Konstantinou, K; Zervos, MJ. Quinupristin-dalfopristin resistance in gram-positive bacteria: mechanism of resistance and epidemiology. *Clin Infect Dis.*, 2004, 38, 92-8.
[43] Speciale, A; La Ferla, K; Caccamo, F; Nicoletti, G. Antimicrobial activity of quinupristin/dalfopristin, a new injectable streptogramin with a wide Gram-positive spectrum. *Int J Antimicrob Agents*, 1999, 13, 21-8.
[44] Moellering, RC; Linden, PK; Reinhardt, J; Blumberg, EA; Bompart, F; Talbot, GH. The efficacy and safety of quinupristin/dalfopristin for the treatment of infections caused by vancomycin-resistant *Enterococcus faecium*. Synercid Emergency-Use Study Group. *J Antimicrob Chemother*, 1999, 44, 251-61.
[45] Nichols, RL; Graham, DR; Barriere, SL; et al. Treatment of hospitalized patients with complicated gram-positive skin and skin structure infections: two randomized, multicentre studies of quinupristin/dalfopristin versus cefazolin, oxacillin or vancomycin. Synercid Skin and Skin Structure Infection Group. *J Antimicrob Chemother*, 1999, 44, 263-73.
[46] Meka, VG; Pillai, SK; Sakoulas, G; et al. Linezolid resistance in sequential *Staphylococcus aureus* isolates associated with a T2500A mutation in the 23S rRNA gene and loss of a single copy of rRNA. *J Infect Dis.*, 2004, 190, 311-7.
[47] Fagon, J; Patrick, H; Haas, DW; et al. Treatment of gram-positive nosocomial pneumonia. Prospective randomized comparison of quinupristin/dalfopristin versus vancomycin.Nosocomial Pneumonia Group. *Am J Respir Crit Care Med.*, 2000, 161, 753-62.

[48] Drew, RH; Perfect, JR; Srinath, L; Kurkimilis, E; Dowzicky, M; Talbot, GH. Treatment of methicillin-resistant *Staphylococcus aureus* infections with quinupristin-dalfopristin in patients intolerant of or failing prior therapy. For the Synercid Emergency-Use Study Group. *J Antimicrob Chemother*, 2000, 46, 775-84.

[49] Carver, PL; Whang, E; VandenBussche, HL; Kauffman, CA; Malani, PN. Risk factors for arthralgias or myalgias associated with quinupristin-dalfopristin therapy. *Pharmacotherapy*, 2003, 23, 159-64.

[50] Raad I, Hachem R, Hanna H. Relationship between myalgias/arthralgias occurring in patients receiving quinupristin/dalfopristin and biliary dysfunction. *J Antimicrob Chemother*, 2004, 53, 1105-8.

[51] Stalker, DJ; Jungbluth, GL. Clinical pharmacokinetics of linezolid, a novel oxazolidinone antibacterial. *Clin Pharmacokinet*, 2003, 42, 1129-40.

[52] Weigelt, J; Itani, K; Stevens, D; Lau, W; Dryden, M; Knirsch, C. Linezolid versus vancomycin in treatment of complicated skin and soft tissue infections. *Antimicrob Agents Chemother*, 2005, 49, 2260-6.

[53] Rubinstein, E; Cammarata, S; Oliphant, T; Wunderink, R. Linezolid (PNU-100766) versus vancomycin in the treatment of hospitalized patients with nosocomial pneumonia: a randomized, double-blind, multicenter study. *Clin Infect Dis.*, 2001, 32, 402-12.

[54] Wunderink, RG; Rello, J; Cammarata, SK; Croos-Dabrera, RV; Kollef, MH. Linezolid vs vancomycin: analysis of two double-blind studies of patients with methicillin-resistant *Staphylococcus aureus* nosocomial pneumonia. *Chest*, 2003, 124, 1789-97.

[55] Rayner, CR; Forrest, A; Meagher, AK; Birmingham, MC; Schentag, JJ. Clinical pharmacodynamics of linezolid in seriously ill patients treated in a compassionate use programme. *Clin Pharmacokinet*, 2003, 42, 1411-23.

[56] Rayner, CR; Baddour, LM; Birmingham, MC; Norden, C; Meagher, AK; Schentag, JJ. Linezolid in the treatment of osteomyelitis: results of compassionate use experience. *Infection*, 2004, 32, 8-14.

[57] Razonable, RR; Osmon, DR; Steckelberg, JM. Linezolid therapy for orthopedic infections. *Mayo Clin Proc.*, 2004, 79, 1137-44.

[58] Cook, AM; Ramsey, CN; Martin, CA; Pittman, T. Linezolid for the treatment of a heteroresistant *Staphylococcus aureus* shunt infection. *Pediatr Neurosurg*, 2005, 41, 102-4.

[59] Rho, JP; Sia, IG; Crum, BA; Dekutoski, MB; Trousdale, RT. Linezolid-associated peripheral neuropathy. *Mayo Clin Proc.*, 2004, 79, 927-30.

[60] Spellberg, B; Yoo, T; Bayer, AS. Reversal of linezolid-associated cytopenias, but not peripheral neuropathy, by administration of vitamin B6. *J Antimicrob Chemother*, 2004, 54, 832-5.

[61] Young, LS. Hematologic effects of linezolid *versus* vancomycin. *Clin Infect Dis.*, 2004, 38, 1065-6.

[62] Rao, N; Ziran, BH; Wagener, MM; Santa, ER; Yu, VL. Similar hematologic effects of longterm linezolid and vancomycin therapy in a prospective observational study of patients with orthopedic infections. *Clin Infect Dis.*, 2004, 38, 1058-64.

[63] Nasraway, SA; Shorr, AF; Kuter, DJ; O'Grady, N; Le, VH; Cammarata, SK. Linezolid does not increase the risk of thrombocytopenia in patients with nosocomial pneumonia:

comparative analysis of linezolid and vancomycin use. *Clin Infect Dis.*, 2003, 37, 1609-16.

[64] Kulkarni, K; Del Priore, LV. Linezolid induced toxic optic neuropathy. *Br J Ophthalmol*, 2005, 89, 1664-5.

[65] Narita, MT; B Yu, VL. Linezolid-associated peripheral and optic neuropathy, lactic acidosid and serotonin syndrome: a review. *Pharmacotherapy*, 2007, 27, 1189-97.

[66] Soriano, A; Miro, O; Mensa, J. Mitochondrial toxicity associated with linezolid. *N Engl J Med.*, 2005, 353, 2305-6.

[67] Bernard, L; Stern, R; Lew, D; Hoffmeyer, P. Serotonin syndrome after concomitant treatment with linezolid and citalopram. *Clin Infect Dis.*, 2003, 36, 1197.

[68] Arbeit, RD; Maki, D; Tally, FP; Campanaro, E; Eisenstein, BI. The safety and efficacy of daptomycin for the treatment of complicated skin and skin-structure infections. *Clin Infect Dis.*, 2004, 38, 1673-81.

[69] Fowler, VG; Cosgrove, S; Abrutyn, E; et al. Daptomycin *vs*. Standard Therapy for Staphylococcus aureus Bacteremia (SAB) and Infective Endocarditis (SAIE). 45th Annual Interscience Congress on Antimicrobial Agents and Chemotherapy. *Washington DC,* 2005.

[70] LaPlante, KL; Rybak, MJ. Daptomycin. Antimicrobial Therapy and Vaccines. Vol. II: *Antimicrobial Agents*: www.antimicrobe.org.

[71] Zhanel, GG; Homenuik, K; Nichol, K; et al. The glycylcyclines: a comparative review with the tetracyclines. *Drugs*, 2004, 64, 63-88.

[72] Raad, I; Darouiche, R; Vazquez, J; et al. Efficacy and safety of weekly dalbavancin therapy for catheter-related bloodstream infection caused by gram-positive pathogens. *Clin Infect Dis.*, 2005, 40, 374-80.

[73] Seltzer, E; Dorr, MB; Goldstein, BP; Perry, M; Dowell, JA; Henkel, T. Once-weekly dalbavancin *versus* standard-of-care antimicrobial regimens for treatment of skin and soft-tissue infections. *Clin Infect Dis.*, 2003, 37, 1298-303.

[74] Kaatz, GW; Seo, SM; Aeschlimann, JR; Houlihan, HH; Mercier, RC; Rybak, MJ. Efficacy of LY333328 against experimental methicillin-resistant Staphylococcus aureus endocarditis. *Antimicrob Agents Chemother*, 1998, 42, 981-3.

[75] Gerber, J; Smirnov, A; Wellmer, A; et al. Activity of LY333328 in experimental meningitis caused by a Streptococcus pneumoniae strain susceptible to penicillin. *Antimicrob Agents Chemother*, 2001, 45, 2169-72.

[76] Giamarellou, H ORW; Harris, H; Owen, S; Porter, S; Loutit, J. Phase 3 trial comparing 3-7 days of oritavancin vs. 10-14 days of vancomycin/cephalexin in the treatment of patients with complicated skin and skin structure infections (CSSI). In: Program and abstracts of the 43rd ICAAC Conference, Chicago, IL, September, 14-17, 2003, 2003. *American Society of Microbiology.*

[77] Barrett, JF. Recent developments in glycopeptide antibacterials. *Curr Opin Investig Drugs*, 2005, 6, 781-90.

[78] Shaw, JP; Seroogy, J; Kaniga, K; Higgins, DL; Kitt, M; Barriere, S. Pharmacokinetics, serum inhibitory and bactericidal activity, and safety of telavancin in healthy subjects. *Antimicrob Agents Chemother*, 2005, 49, 195-201.

[79] Hegde, SS; Reyes, N; Wiens, T; et al. Pharmacodynamics of telavancin (TD-6424), a novel bactericidal agent, against gram-positive bacteria. *Antimicrob Agents Chemother*, 2004, 48, 3043-50.

[80] Stryjewski, ME; Chu, VH; O'Riordan, WD; et al. Telavancin versus standard therapy for treatment of complicated skin and skin structure infections caused by gram-positive bacteria: FAST 2 study. *Antimicrob Agents Chemother*, 2006, 50, 862-7.

[81] Stryjewski, ME; O'Riordan, WD; Lau, WK; et al. Telavancin versus standard therapy for treatment of complicated skin and soft-tissue infections due to gram-positive bacteria. *Clin Infect Dis.*, 2005, 40, 1601-7.

[82] Aneziokoro, CO; Cannon, JP; Pachucki, CT; Lentino, JR. The effectiveness and safety of oral linezolid for the primary and secondary treatment of osteomyelitis. *J Chemother*, 2005, 17, 643-50.

[83] Finney, MS; Crank, CW; Segreti, J. Use of daptomycin to treat drug-resistant Gram-positive bone and joint infections. *Curr Med Res Opin.*, 2005, 21, 1923-6.30

[84] Anthony, SJ HM; Angelos, E; Stratton, CW. Clinical Experience with daptomycin in Patients with Orthopedic-Related Infections. 43rd IDSA Annual Meeting 2005. *San Francisco*, CA, 2005.

[85] Yin, LY; Lazzarini, L; Li, F; Stevens, CM; Calhoun, JH. Comparative evaluation of tigecycline and vancomycin, with and without rifampicin, in the treatment of methicillin-resistant *Staphylococcus aureus* experimental osteomyelitis in a rabbit model. *J Antimicrob Chemother*, 2005, 55, 995-1002.

[86] Howden, BP; Ward, PB; Charles, PG; et al. Treatment outcomes for serious infections caused by methicillin-resistant *Staphylococcus aureus* with reduced vancomycin susceptibility. *Clin Infect Dis.*, 2004, 38, 521-8.

[87] Fowler, VG; Jr., Boucher, HW; Corey, GR; et al. Daptomycin versus standard therapy for bacteremia and endocarditis caused by *Staphylococcus aureus*. *N Engl J Med.*, 2006, 355, 653-65.

[88] Woods, CW; Cheng, AC; Fowler, VG; Jr., et al. Endocarditis caused by *Staphylococcus aureus* with reduced susceptibility to vancomycin. *Clin Infect Dis.*, 2004, 38, 1188-91.

[89] Chiang, FY; Climo, M. Efficacy of linezolid alone or in combination with vancomycin for treatment of experimental endocarditis due to methicillin-resistant *Staphylococcus aureus*. *Antimicrob Agents Chemother*, 2003, 47, 3002-4.

[90] FDA. Information for Healthcare Professionals: Linezolid (marketed as Zyvox) March 16, 2007, 2007.

[91] Falagas, ME; Manta, KG; Ntziora, F; Vardakas, KZ. Linezolid for the treatment of patients with endocarditis: a systematic review of the published evidence. *J Antimicrob Chemother*, 2006, 58, 273-80.

[92] Kang, SL; Rybak, MJ. In-vitro bactericidal activity of quinupristin/dalfopristin alone and in combination against resistant strains of *Enterococcus* species and *Staphylococcus aureus*. *J Antimicrob Chemother,* 1997, 39 Suppl A:33-9.

[93] Tsuji, BT; Rybak, MJ. Short-course gentamicin in combination with daptomycin or vancomycin against *Staphylococcus aureus* in an *in vitro* pharmacodynamic model with simulated endocardial vegetations. *Antimicrob Agents Chemother*, 2005, 49, 2735-45.

[94] Grohs, P; Kitzis, MD; Gutmann, L. *In vitro* bactericidal activities of linezolid in combination with vancomycin, gentamicin, ciprofloxacin, fusidic acid, and rifampin against *Staphylococcus aureus*. *Antimicrob Agents Chemother*, 2003, 47, 418-20.

[95] Jacqueline, C; Caillon, J; Le Mabecque, V; et al. *In vitro* activity of linezolid alone and in combination with gentamicin, vancomycin or rifampicin against methicillin-resistant

Staphylococcus aureus by time-kill curve methods. *J Antimicrob Chemother*, 2003, 51, 857-64.

[96] Moise-Broder, PA; Sakoulas, G; Eliopoulos, GM; Schentag, JJ; Forrest, A; Moellering RC; Jr. Accessory gene regulator group II polymorphism in methicillin-resistant *Staphylococcus aureus* is predictive of failure of vancomycin therapy. *Clin Infect Dis*., 2004, 38, 1700-5.

[97] Tsuji, BT; Rybak, MJ. Etest synergy testing of clinical isolates of *Staphylococcus aureus* demonstrating heterogeneous resistance to vancomycin. *Diagn Microbiol Infect Dis*., 2006, 54, 73-7.

[98] LaPlante, KL; Rybak, MJ. Impact of high-inoculum *Staphylococcus aureus* on the activities of nafcillin, vancomycin, linezolid, and daptomycin, alone and in combination with gentamicin, in an in vitro pharmacodynamic model. *Antimicrob Agents Chemother*, 2004, 48, 4665-72.

[99] Sakoulas, G; Eliopoulos, GM; Alder, J; Eliopoulos, CT. Efficacy of daptomycin in experimental endocarditis due to methicillin-resistant *Staphylococcus aureus*. *Antimicrob Agents Chemother*, 2003, 47, 1714-8.

[100] Baltch, A RW; Bopp, L. et al. Killing of methicillin-resistant *Staphylococcus aureus* by daptomycin, gentamicin, and rifampin, singly and in combination, in broth and in human monocyte-derived macrophages, with and without GM-CSF and Interferon Activation. Proceedings of the 15th ICAAC Conference, 2005, Abstract E-1741, *American Society of Microbiology, Washington*, 2005.

[101] Allen, GP; Cha, R; Rybak, MJ. *In vitro* activities of quinupristin-dalfopristin and cefepime, alone and in combination with various antimicrobials, against multidrug-resistant Staphylococci and Enterococci in an *in vitro* pharmacodynamic model. *Antimicrob Agents Chemother*, 2002, 46, 2606-12.

[102] Dailey, CF; Pagano, PJ; Buchanan, LV; Paquette, JA; Haas, JV; Gibson, JK. Efficacy of linezolid plus rifampin in an experimental model of methicillin-susceptible *Staphylococcus aureus* endocarditis. *Antimicrob Agents Chemother*, 2003, 47, 2655-8.

[103] Kang, SL; Rybak, MJ; McGrath, BJ; Kaatz, GW; Seo SM. Pharmacodynamics of levofloxacin, ofloxacin, and ciprofloxacin, alone and in combination with rifampin, against methicillin-susceptible and -resistant *Staphylococcus aureus* in an *in vitro* infection model. *Antimicrob Agents Chemother*, 1994, 38, 2702-9.

[104] Zarrouk, V; Bozdogan, B; Leclercq, R; et al. Activities of the combination of quinupristin-dalfopristin with rifampin *in vitro* and in experimental endocarditis due to *Staphylococcus aureus* strains with various phenotypes of resistance to macrolide-lincosamide-streptogramin antibiotics. *Antimicrob Agents Chemother*, 2001, 45, 1244-8.

[105] Sambatakou, H; Giamarellos-Bourboulis, EJ; Grecka, P; Chryssouli, Z; Giamarellou, H. *In vitro* activity and killing effect of quinupristin/dalfopristin (RP59500) on nosocomial *Staphylococcus aureus* and interactions with rifampicin and ciprofloxacin against methicillin-resistant isolates. *J Antimicrob Chemother*, 1998, 41, 349-55.

[106] Petersen, PJ; Labthavikul, P; Jones, CH; Bradford, PA. *In vitro* antibacterial activities of tigecycline in combination with other antimicrobial agents determined by chequerboard and time-kill kinetic analysis. *J Antimicrob Chemother*, 2006, 57, 573-6.

[107] Mercier, RC; Kennedy, C; Meadows, C. Antimicrobial activity of tigecycline (GAR-936) against *Enterococcus faecium* and *Staphylococcus aureus* used alone and in combination. *Pharmacotherapy*, 2002, 22, 1517-23.

In: Methicillin-Resistant Staphylococcus Aureus...
Editor: Charles L. Kolendi

ISBN: 978-1-60741-398-1

Chapter 11

DHEA-Mediated Phenotypic Alterations in Staphylococcal Cell Wall Architecture

Balbina J. Plotkin[1]*and Monika I. Konaklieva[2]**

[1]Department of Microbiology[1] Midwestern University, Downers Grove, IL 60515
[2]Department of Chemistry, American University, Washington, D.C. 20016

Abstract

Clinically, 23-40% of patients with *S. aureus* infections treated with vancomycin do not respond to it (vancomycin treatment failures). These treatment failures occur in the absence of laboratory-documented vancomycin resistance. We have previously reported that resistance of methicillin-resistant *S. aureus* (MRSA) clinical isolates to vancomycin can be phenotypically increased by exposure to the androgen dehydroepiandosterone (DHEA), a steroid whose chemical structure is similar to that of cholesterol. Herein, we describe an intrinsic phenotypic response to DHEA which results in increased resistance to the host cationic defense peptide, β-1 defensin, as well as vancomycin and other antibiotics that have a positive charge. The increased resistance to vancomycin appears to correlate with a DHEA-mediated alteration in cell surface architecture. DHEA-mediated cell surface changes include alterations in: cell surface charge, surface hydrophobicity, capsule production, and carotenoid production. In addition, exposure to DHEA results in decreased resistance to lysis by Triton X-100 and lysozyme indicating alteration of murien hydrolase activity. We propose that DHEA is an interspecies quorum-like signal that triggers innate phenotypic host survival strategies in *S. aureus* including but not limited to increased carotenoid production. A side effect of this phenotypic change is increased vancomycin resistance. Furthermore this DHEA-mediated survival system may share the cholesterol-squalene pathway shown to be statin sensitive.

Keywords: Testosterone, dehydrotestosterone, DHT, dehydroepiandosterone, DHEA, *Staphylococcus aureus, Pseudomonas aeurginosa, Enterococcus faecalis, Escherichia coli,* antimicrobial susceptibility.

*Corresponding author: (630)515-6163, (630)515-7245 (FAX), bplotk@midwestern.edu

INTRODUCTION

A pathogen's survival in the host is dependent on its capability to differentially express virulence factors and survive host defenses. This fundamental ability to survive in multiple host habitats can be triggered by recognition of host substances which function as interkingdom communication signals (10, 11, 33). Niche localization, be it mucosal surface, tissue, or circulation could be informed by a microbes ability to recognize host signals e.g., insulin, catecholamines and sterols (25, 31, 35, 38, 40, 49, 60). Receipt of these chemical signals would warn the organism of potential exposure to host immune defenses including reactive oxygen intermediates; this could subsequently trigger expression of phenotypes more adapted for survival in that hostile environment. In addition, to host hormones acting as signaling compounds they could also serve as substrates for production of essential survival factors, e.g. carotenoids. An alteration in *S. aureus* that promotes its survival is its ability to alter its surface architecture and relative electronegativity (26-28, 33, 52-55). We previously reported that exposure of clinical isolates of *S. aureus* from multiple body sites to DHEA results in an increased MIC for vancomycin (60). This change was DHEA concentration specific and associated with a change in colony morphology. The presence of DHEA, a C3 hydroxylated steroid *in vivo* may contribute to treatment failure. An intrinsic alteration in the physico-chemical nature of *S. aureus* cell surfaces also appears to play a role in resistance to both drugs and host defense molecules.

The negative surface charge on staphylococci is determined by anionic phospholipids and by the teichoic acids and lipoteichoic acids, which are highly negatively-charged polymers of the cell wall and membrane complex of staphylococci (26). *S. aureus* increases its surface positive charge by incorporating D-alanine into teichoic and lipoteichoic acids and introducing L-lysine-modified phosphatidylglycerol, the only phospholipid with a positive net charge, into the cell membrane (26, 27, 52-55). Recently two surface charge mutants of *S. aureus*, *dltABCD* and *mprF*, have been described that are highly susceptible to human defensins, as well as to other highly cationic peptides. Interestingly, is that, *mprF* mutants also exhibit an increased susceptibility to vancomycin (53-55). Taken together, these findings suggest that changes in surface charge directly affect sensitivity to vancomycin. It is not known what regulates expression of this surface charge phenotype. What can be postulated is that the signal triggering this phenotype must communicate to the organism that exposure to host immune factors, e.g., highly cationic defense peptides, is imminent. An essential characteristic of such a signal would be its consistent presence at sites where exposure to harmful substances, e.g. defensins, is inevitable. From an evolutionary perspective it is also essential that these signals be present irrespective of age, circadian cycle or gender. DHEA, an androgen, fits this criterion (3, 5, 50).

DHEA is a naturally occurring steroid hormone in humans, whose levels decline with aging (21, 23, 50). This decline appears to parallel immunosenescence (7, 23, 39, 63). Experimental evidence supports a protective role for DHEA(S) in reversing immunosensecence, reducing heart disease, controlling obesity, insulin resistance, memory loss, and vascular narrowing, with its role in immunoprotection evolving (15, 29, 63). Both middle-aged adults and seniors take DHEA, which they obtain over the counter, to supplement normal endogenous androgen production in the hopes of improving their general

** Co- author: 20016-8014 (202)885-1777, FAX: (202) 885-1752 e-mail: mkonak@american.edu

health and resistance to disease (9, 15). Thus, *S. aureus* is exposed to various levels of DHEA throughout the life of the host which varies depending on host age and consumption of nutritional supplements. The focus of this study was to determine whether DHEA affected *S. aureus* cell surface architecture and how the changes may correlate to antibiotic susceptibility.

MATERIALS AND METHODS

Bacterial Stains and Growth Conditions

S. aureus ATCC 25923, a highly stable quality control strain, was grown in Mueller-Hinton medium with and without steroid as previously described unless otherwise noted (58, 60).

Chemicals

Steroids used were testosterone, estrone, estradiol, pregnenolone, cortisol dehydrotestosterone, hydrocortisone, dehydroepiandrosterone (DHEA), cholesterol, and progesterone (Sigma-Aldrich Chemical Co.). All steroids except cholesterol were prepared in 95% (v/v) ethanol then diluted at least 100 fold into the manufacturer supplied diluent (Pluronic®) or Mueller-Hinton (MH) broth (0.05% v/v highest final ethanol concentration) as described (58, 60). Cholesterol was initially dissolved in chloroform which was diluted at least 100 fold in dimethylsulfoxide (DMSO) to a final maximum DMSO concentration of 0.5% v/v. All steroids were tested at 0.1, 0.5,1.0 and 5.0 μM. Controls consisted of parallel assays done in the presence of the highest concentration of steroid diluent (0.5% ethanol or DMSO-cholesterol only). Lysostaphin and lysozyme were obtained from Sigma-Aldrich. β-1 defensin was obtained from Peptides International.

Antibiotic Sensitivity

Susceptibility in the presence of physiologically relevant concentrations of DHEA (0.1, 0.5, 1.0, 5.0 μM) was screened for using MicroScan® Gram positive Breakpoint Combo panels (Dade Behring Inc, West Sacramento, CA). The kits were used according to manufacturer's instructions (60) with results confirmed by microdilution methodology per CLSI (formally NCCLS) procedures (43, 44). All hormones (Sigma-Aldrich Chemical Co.) were prepared in 95% ethanol then diluted at least 100 fold in broth. Bacteria were suspended in the manufacturer's diluent with or without DHEA (0.1, 0.5, 1.0, 5.0 μM) prior to inoculation of the microtiter plates. Breakpoints were read by visual examination (24 and 48 hr). Samples (0.1 ml) from wells exhibiting no growth were inoculated onto Muller-Hinton agar for cidal concentration determinations. Negative controls included the highest concentration of hormone diluent, 0.05% ethanol and manufacturer's diluent alone. To confirm that the quality control strain (ATCC 29523) had not acquired a cryptic penicillinase that could be activated by DHEA, the presence of β-lactamase was tested for in all β-lactam antibiotic containing wells and growth control wells using nitrocefin disks (Cefinase; BBL Microbiology Systems) (47, 66-67). The positive control was a known β-lactamase producing *S. aureus* strain. No β-lactamase was detected. The specificity of DHEA induced alteration of antibiotic sensitivity was determined by also testing testosterone (differ from DHEA by a OH group), and dehydroxy-testosterone (DHT), at physiologically relevant concentrations (0.1, 0.5, 1.0, 5.0 μM) (58, 70, 76). In addition, the steroids hydrocortisone, dexamethasone, and cortisone were also tested at physiologically relevant concentrations (1, 20, 50 μM) (58). Antibiotic testing was done as described above for DHEA.

Binding of Radiometric DHEA

Direct evidence that DHEA interacts specificity with *S. aureus* was obtained from radiometric binding assays, done as previously described (57). Mid-logarithmic growth phase bacteria (10^8CFU/ml) were incubated (37^0 C, 30 min; 5^0 C, 18 hr) with a constant concentration of [^{3}H]DHEA (2.94 Ci/mmol; 20 nM) and increasing levels of cold DHEA (0.2 to 100μM). After incubation the cells were harvested by centrifugation, washed four times and cell-associated radioactivity determined by scintillation count. [^{3}H]DHEA associated with heat-killed cells was subtracted from that measured for live cell levels under homologous incubation conditions. All experiments were done in triplicate and repeated once. Data was analyzed and plotted using Prism/EnZ Plot (GraphPad).

QSARIS

Due to the fact that DHEA affects bacterial response to molecules from different antimicrobial classes regardless of their mechanism of drug action, QSARIS(TM) a program for quantitative structure-activity or structure-property modeling, was used to determine what if any similarities exist between the antimicrobials whose MICs were or weren't affected by the presence of DHEA. The similarity search algorithm assigns each molecule a correspondent vector in the multidimensional space formed by molecular descriptors. Similarity of molecules is expressed in terms of either the distance between these vectors in the multidimensional space, or the similarity coefficient based on the same distances. The similarity search was performed with DHEA as the pharmacophore for the 32 antimicrobials tested in the antibiotic sensitivity assays using QSARIS program Version 1.2, copyright 2001, by SciVision (www.scivision.com). The molecules similarity was based on their 3D structures using the cosine coefficient where x and y are the compared vectors, their components being $\{xi\}$ and $\{yi\}$; n is a dimension in space (the total number of descriptors) (see formula below). In this method, one starts from a molecule, such as a receptor ligand or an enzyme substrate, whose pharmacophoric groups are known for a particular target. Then a search of a database of compounds is performed to determine which ones have a similar 3D structure as pharmacophore. The top virtual hits are inspected and tested (18).

$$S(x,y) = \frac{\sum_{i=1}^{n} x_i y_i}{\left[\sum_{i=1}^{n} (x_i)^2 \sum_{i=1}^{n} (y_i)^2 \right]^{1/2}}$$

Carotenoid Production

Carotenoids were extracted from cells grown in MH broth with the steroids (0.1, 0.5, 1.0, and 5.0 µM) listed above or in medium alone essentially as decribed by Morikawa, et al. (41). After growth in media with and without steroid for 24 hr (shaking, 37 ^{0}C) the cells were harvested from 850µl of culture, washed with distilled deionized water and the pellet resuspended in 200 µl of methanol which was heated to 55^{0}C for 3 min. After the supernatant was removed, the cell pellet was re-extracted with methanol (200 µl). The extracted cell supernantant was placed in a quartz microcuvette with absorbance read at 465nm.

Surface Hydrophobicity/ Capsule Production

Plastic adherence was used to measure surface hydrophobicity as previously described (12). Bacteria were grown (O/N, Mueller Hinton broth) in flat bottom microtiter plates in the presence and absence of DHEA. After extensive washing, the plates were stained with crystal violet (50% v/v). Stain was dissolved in 300µl of 95% ethanol and absorbance read in an EIA microtiter plate reader (Dynatech Laboratories, Inc., Chantilly, Virginia) at 590 nm. Duplicate plates stained with Alcian blue (1% PBS, 30 min, room temperature) were used for capsule production determinations. Stain was dissolved in 300µl pyridine (Sigma-Aldrich). Optical density was read in an EIA reader at 450 nm.

Surface Electronegativity

Adherence to glass was used to measure surface charge (glass = negative surface charge) (16). Bacteria were grown (O/N) in Muller-Hinton broth with and without DHEA (flat bottom 24 well plates containing glass coverslips). After extensive washing the coverslips were stained with crystal violet. The stain was washed off the coverslips with 300ml of 95% ethanol and absorbance (510nm) read in a Beckman Spectrophotometer.

Autolytic Assay

Autolysis induced by Triton X-100 was measured as previously described with some modifications (8). Mid-logarithmic cells grown in Muller-Hinton broth with and without DHEA were washed (4X; cold distilled water) and resuspended in buffer (0.05M Tris-HCL 7.2 pH, 0.05% Triton X- 100). Cell lysis was measured as a decrease in O.D (600nm) over time. Each data point represents the mean of six data points +/- S.D.

Coagulase Production

Coagulase production was performed according to manufacturers' (DIFCO) directions. *S. aureus* grown in the presence and absence of DHEA was pelleted and resuspended to a starting concentration of 10^9 CFU/ml. The suspension was serial diluted then tested for coagulase production; 1:100 dilution at 4 hr, 37^0C was the lowest bacterial concentration to result in coagulation

β-Defensin

Resistance to human β-1 defensin (Peptide International) was measured in LB broth without salts essentially as described by Peschel, et al.(54) with the exception of the addition of DHEA (0.5μM).

Statistical Analysis

Unless otherwise indicated experiments were performed in triplicate and repeated at least twice. All samples were coded, and performed in a blinded fashion. A one-way analysis of variance was used to determine differences between growth conditions. If statistical significance was found, a Tukey-Kramer post-hoc analysis was applied (Instat, GraphPad Software, San Diego, CA; Table 2).

Results and Discussion

1. DHEA's Effect on Antibiotic Activity

As previously reported, exposure of *S. aureus* to DHEA (0.1μM) resulted in a 32-fold increase in vancomycin MIC (0.25 μg/ml unexposed to 8μg/ml exposed) which would now classify the organism as a vancomycin-intermediate resistant *S. aureus* or VISA (60). In addition, DHEA (0.5μM) increased vancomycin's MIC from 0.25μg/ml vancomycin to 4μg/ml; this represents a MIC that is 1μg/ml below the recommended pharmacotherapeutic trough level in patients (24, 75). DHEA's effect on *S. aureus* response to antimicrobials was not limited to vancomycin (Table 1). Furthermore, as with vancomycin, increased MICs were only measured in the presence of DHEA despite repeated serial transfers through medium with and without DHEA and antibiotic (60). This indicates that DHEA's effect on antimicrobial resistance regardless of drug class is phenotypic and not the result of mutant selection.

Table 1. Effect of DHEA on the antibiotic resistance of *Staphylococcus aureus* ATCC 25923 (MSSA)

Antibiotic[b]	Control (medium alone)	DHEA			
		0.1 μM	0.5 μM	1.0 μM	5.0 μM
Cell wall synthesis inhibitors					
Penicillin	≤ 0.03 (S)[a]	< 0.03 (S)	8 (R)	≤ 0.03 (S)	≤ 0.03 (S)
Oxacillin	0.5 (S)	< 0.5 (S)	>2 (R)	≤0.5 (S)	≤0.5 (S)
Ampicillin	≤ 0.25 (S)	< 0.25 (S)	>8 (R)	≤ 0.25 (S)	≤ 0.25 (S)
Ampicillin/sulbactam	≤8/4 (S)	16/8 (I)	16/8 (I)	≤8/4 (S)	≤8/4 (S)
Piperacilllin/tazobactam	≤1 (S)	< 1 (S)	>8 (R)	≤1 (S)	≤1 (S)
Cefepime	≤2 (S)	>16 (R)	< 2 (S)	≤2 (S)	≤2 (S)
Imipenem	≤1 (S)	< 1 (S)	>8 (R)	≤1 (S)	≤1 (S)
Meropenem	≤4 (S)	>8 (R)	>8 (R)	≤4 (S)	≤4 (S)
Amoxicillin/K Clavulanate	≤4/2 (S)	>4/2 (R)	< 4/2 (S)	≤ 4/2 (S)	≤ 4/2 (S)
Vancomycin	0.25 (S)	8 (I)	4 (S)	0.25 (S)	0.25 (S)
Protein synthesis inhibitors					
Erythromycin	0.5 (S)	1 (I)	0.5 (S)	0.5 (S)	0.5 (S)
Clindamycin	≤0.25 (S)	< 0.25 (S)	0.5 (S)	≤ 0.25 (S)	≤ 0.25 (S)
Nitrofurantoin	≤ 32 (S)	>64 (R)	>64 (R)	≤ 32 (S)	≤ 32 (S)

[a]MIC in μg/ml (S= sensitive, I = intermediate, R = resistant - Interpretive Breakpoints as indicated in NCCLS document M100-S7)

[b]Antibiotics not affected by DHEA: chloramphenicol, cefazolin, cefuroxime, ceftriaxone, cephalothin, amoxicillin/K clavulanate, levofloxacin, gentamicin, tetracycline, ciprofloxacin, amikacin, cefotaxime, rifampin, norfloxacin, trimethoprim/sulfamethoxazole

DHEA's effect on antibiotic resistance patterns was drug and hormone concentration specific with antimicrobials from multiple classes affected. The cross-class induction of multiple-drug resistance affected MIC values for members of the penicillin, cephalosporin, macrolide, aminoglycoside and nitrofurantoin classes. The cell wall active agents affected by DHEA were either zwitterionic (e.g., cefepine), or reported to have lower concentrations of bound cations than others of the same class (13, 17, 46). For the drugs that affect protein synthesis, the ability of DHEA to alter drug activity also appeared to be related to the amount of drug-associated cations which are required for the drug's transport across cytoplasmic membranes (42, 48). Additional structure-functional analysis of the drugs with QSARIS modeling indicated that drugs affected by DHEA contain functionality's which have 60% or greater similarity with respect to overall charge (amphiphilic to positive) and hydrophobicity (Log P value). Of those affected, within specific classes, the most positively charged antibiotics were affected. For example, resistance to cefipine, the most positively charged of the cephalosporins, was increased (Table 1). Additional QSARS analysis of the antibiotics using cosine coefficient as one of the similarity coefficients shows that a compound having less than 60% similarity with the structure of the DHEA was not affected by DHEA. The experimental data correlates very nicely with the similarity search performed for this group of compounds. It is interesting to note that our data correlates well with the observation that cefepime (cos. coeff 0.80) is affected by the DHEA and the rest of the cephalosporines tested are not (0.18-0.36). This DHEA-mediated alteration of response to antibiotics appeared to be specific for DHEA and not the result of nonspecific perturbation of cell wall architecture, since a chemically similar steroid, testosterone, as well as other steroids (dehydrotestosterone, dexamethasone, hydrocortisone, cortisone) had no affect on either *S. aureus* antibiotic susceptibility or growth (58, 60). Thus, antibiotic structure plays a role in DHEA's effect on drug response. However, as previously reported ESI-MS studies show that the effect of DHEA is not the result of steroid-drug interaction ((59), data not shown).

Table 2. Effect of DHEA on *S.aureus* ATCC 29523 characteristics

	Relative Hydrophobicity	Relative Electronegativity	Lysozyme MIC (μg/ml)	β–1 Defensin MIC (μg/ml)
Control	21.8 ± 0.3[a,c]	0.1363 ± 0.052 [ac]	32	25
DHEA 0.1μM	34.3 ± 0.7[b]	0.1418 ± 0.052	16	>100 [b]
DHEA 0.5μM	36.7 ± 7 [b]	0.2890 ± 0.004[b]	16	>100 [b]
DHEA 1.0μM	29.9 ± 1.6 [b]	0.2233 ± 0.011[b]	4 [b]	>100 [b]
DHEA 5.0μM	24.3 ± 2.7	0.1518 ± 0.014	8 [b]	50

[a]Mean ± S.D.
[b]Significantly different from control, $p \leq 0.01$
[c]Absorbance 560nm

2. DHEA Interaction with *S. aureus*

Direct evidence that DHEA interacts specifically with *S. aureus* was obtained from radiometric binding assays. [^{3}H]DHEA interaction with *S. aureus* at 5^{0}C, a temperature permissive for binding but inhibitory for uptake , indicates that [^{3}H]DHEA binds in a saturable manner (Figure 2) with an estimated 783 ± 30 binding sites per bacterial cell. At 37^{0}C, a permissive temperature for uptake, the data indicate that DHEA can also be taken up by *S. aureus*. Procaryotes normally contain no sterols, instead they contain polyterpenes (45, 51). An important functional structure, regardless of the phylogenetic ranking of organisms, common to all membrane sterols or sterol-like compounds is the presence of a single hydroxyl group at the C-3 position. This is a characteristic of the structure of DHEA that may relate to its ability to interact with *S. aureus*. However, the identity of a specific receptor remains unclear. BLAST amino acid sequence analysis and genomic sequence analysis appear to rule out the presence in *S. aureus* of sigma 1, the putative cell surface human receptor for DHEA, as well as the mammalian intracellular steroid nuclear receptor superfamily (1). This does leave open the possibility that DHEA does not bind to a specific receptor but intercalates into the membrane as reported for some mammalian cells (4).

3. DHEA's Effect on Cell Surface Characteristics

QSARIS analysis of DHEA's effect on antibiotic resistance indicated that DHEA affects the surface characteristics of *S. aureus* including alterations in surface logP (hydrophobicity) and relative electronegative charge. Alterations in surface characteristics mediated by DHEA included an increase in the ability to adhere to plastic, a measure of surface hydrophobicity and ability to form biofilm (6, 73) (Table 2). Plastic adherence assays revealed that growth in the presence of DHEA resulted in a significant increase in surface hydrophobicity at low to mid range DHEA concentrations, with the highest level of plastic adherence measured for cells grown in 0.5μM DHEA. Data shown in Figure 2 suggests that DHEA binds to *S. aureus*. DHEA-mediated alteration in cell membrane composition via signal transduction and/or DHEA intercalation into the membrane could account for the increase in hydrophobicity as measured by adherence to plastic.

We previously reported that in the presence of DHEA *S. aureus* exhibits colony pleomorphism with a preponderance of the small colonies similar in appearance to that described for heteroresistant *S. aureus* (56, 59, 60, 72). Heteroresistance in *S. aureus* correlates with increased vancomycin resistance (56, 72). However, DHEA-mediated colony variants were unstable; they are observed only in the presence of DHEA, regardless of medium NaCl content, which is reported to stabilize appearance of heteroresistant colonies (60). To ascertain whether DHEA-mediated alterations are analogous to changes associated with those reported for heteroresistance, a systematic comparison of DHEA-mediated changes to characteristics associated was done. This comparison tested the following heteroresistance characteristics: increased resistance to Triton-X 100, which is used to evaluate murein hydrolase autolytic activity (8, 66); increased resistance to lysozyme which cleaves the β1-4 linkage in peptidoglycan strands; increased resistance to lysostaphin, a glycylglycine endopeptidase; depressed coagulase production; decreased capsule production; and slower generation times (for review see (71)).

In contrast to phenotypes associated with heteroresistance, *S. aureus* grown in the presence of DHEA (0.5 μM) was slightly but significantly ($p < 0.05$) more sensitive to lysis by Triton X-100 and four to eight fold more sensitive to lysis by lysozyme (Figure 1 and Table 2). The effect of DHEA was hormone concentration specific for both Triton X-100 and lysozyme with the optimum concentration of DHEA at 0.5 μM. In addition, sensitivity to lysostaphin lysis, levels of capsule production and coagulase production were unaffected by the presence of DHEA. With respect to growth kinetics, although heteroresistant *S. aureus* exhibit decreased rates of growth, *S. aureus* grown in the presence of DHEA had either an increased rate of growth (1.3 fold), although at a concentration that does not correlate with levels triggering altered cell surface changes or drug MICs (5.0 μM DHEA), or were unaffected (58, 60). These results taken together indicate that the mechanism by which DHEA mediates increased resistance to vancomycin is dissimilar to that reported for heteroresistance (56, 72).

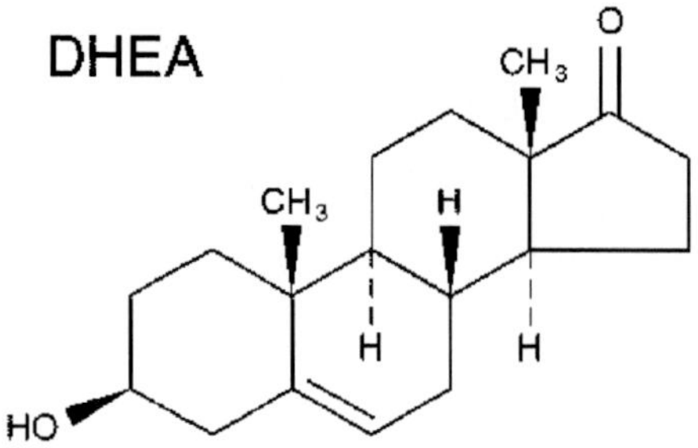

Figure 1. Chemical structure of dehydroepiandosterone (DHEA).

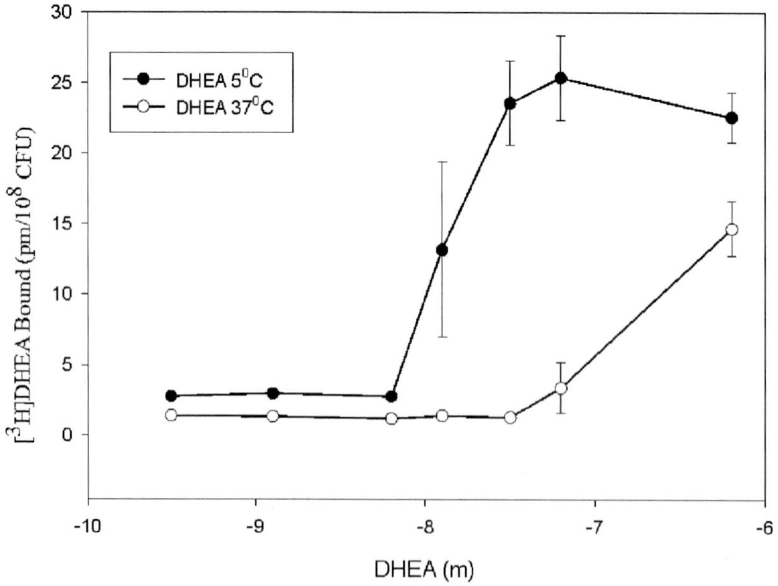

Figure 2. Binding of [^{3}H]DHEA to *S. aureus* ATCC 29523. Mid-logarithmic growth phase bacteria (10^8CFU/ml) were incubated (37^0 C, 30 min; 5^0 C, 18 hr) with a constant concentration of [^{3}H]DHEA (2.94 Ci/mmol; 20 nM) and increasing levels of cold DHEA (0.2 to 100μM in a total volume of 150μL). After incubation the cells were harvested by centrifugation, washed four times and cell-associated radioactivity determined by scintillation count. [^{3}H]DHEA associated with heat-killed cells was subtracted from that measured for live cell levels under homologous incubation conditions. All experiments were done in triplicate and repeated once. Data was analyzed and plotted using Prism/EnZ Plot (GraphPad).

An alternative to heteroresistance as a vancomycin resistance mechanism is the increase in cell surface positive charge resulting from amino acid substitutions in teichoic/lipoteichoic acids and phosphatidylglycerol (26, 53-55, 62). These alterations are also linked to increased resistance to a cationic immune defense protein human β-1 defensin (53-55). DHEA induced an increase in resistance to human β-1 defensin which was measured in LB broth without salts as described by Peschel, et al. (53) with the exception of the addition of hormone (Table 2). Exposure of *S. aureus* to DHEA (0.1, 0.5, and 1.0μM) resulted in increased resistance to β-1 defensin; a finding that supports QSARIS analysis of DHEA's effect on resistance to antibiotics (Table 2).

To directly confirm that DHEA alters *S. aureus* surface charge, adherence to glass, a negatively charged surface, was measured (Table 2) (16). DHEA (0.5 and 1.0 μM) induced a significant increase in cell surface positive charge as measured by adherence to glass. These findings may also explain the modest but significant effect of Triton X-100, a non-ionic detergent that has a preferential interaction with amphiphilic membrane proteins (20, 61). Both the increased ability to bind to glass and increased hydrophobicity also indicate that *S. aureus*' ability to form biofilms on a variety of surfaces is enhanced by the presence of DHEA including host cell surfaces which have a negative charge (2, 65). The DHEA concentrations resulting in altered surface charge and response to β-1 defensin correlate with the DHEA levels that most often resulted in alteration in drug activity (Table 1). These data support the hypothesis that the DHEA induction of resistance to cationic peptides of the innate immune system (β-1 defensin) and induction of increased resistance to weakly cationic drugs, including vancomycin, occur via a similar mechanism. A mechanism by which DHEA-may be mediating these alterations in *S. aureus* phenotype is repression of *mprF* gene transcription which has been demonstrated to result in an increase in surface charge. The negative surface charge on staphylococci is contributed to by anionic phospholipids and by the teichoic acids and lipoteichoic acids, which are highly negatively-charged polymers of the cell wall and membrane complex of staphylococci (14, 54, 55). *S. aureus* increases its surface positive charge by incorporating D-alanine into teichoic and lipoteichoic acids and introducing L-lysine-modified phosphatidylglycerol, the only phospholipid with a positive net charge into the cell membrane. *mprF* mutants exhibit increased susceptibility to vancomycin and are highly susceptible to human defensin killing. However the signaling trigger for induction of altered surface charge has not yet been described. An essential characteristic required of such a trigger would be its consistent presence in sites where exposure to harmful substances, e.g. innate immune factors, is inevitable. DHEA is a hormone present in the human host, 24hrs/day, throughout life, regardless of gender (21, 30, 50) and fits the qualifications for such a gene trigger. An alternative mechanism by which DHEA could affect overall S. *aureus* surface charge is by increasing the levels of neutrally charged molecules. Molecules which would fit this parameter and which have also been shown to correlate with increased virulence are carotenoids (32, 33, 36, 37, 69, 74). To determine if cell composition has been altered by DHEA, the effect of DHEA as well as other steroids, e.g. testosterone, dehydroxytestosterone, which lack a hydroxyl group at C3 and had no effect on staphylococcal response to vancomycin, were tested (Table 3).

Table 3. Effect of sterols on susceptibility to vancomycin and carotenoid production

	Sterol Concentration (μM)	Vancomycin MIC (μg/ml)	Carotenoid Production (Abs_{465nm})	Carotenoid Level (Test/Control)
DHEA	0.1	8.0 (I)	6.2±0.2	1.61
	0.5	4.0 (S)	14.0*±0.3	3.61*
	1.0	0.25	7.6±0.4	1.96
	5.0	0.25	6.1±0.2	1.68
Estradiol	0.1	0.25	5.4±0.3	1.91
	0.5	0.25	11.4±0.8	3.26*
	1.0	0.25	5.7±0.2	1.67
	5.0	0.25	7.0±0.2	1.8
Pregnenolone	0.1	0.25	7.6±0.3	1.16
	0.5	0.25	12.6±0.3	1.2
	1.0	0.25	6.4±0.1	1.0
	5.0	0.50	7.0±0.2	1.65
Progesterone	0.1	0.25	4.5±0.2	1.4
	0.5	0.25	4.6±0.3	2.95*
	1.0	0.25	4.0±0.2	0.89
	5.0	0.25	5.4±0.2	1.0
Cholesterol	0.1	1.0 (S)	5.1±0.2	1.3
	0.5	0.25	10.4±0.3	2.71*
	1.0	0.25	3.4±0.1	0.89
	5.0	0.25	3.9±0.2	1.0

*Significantly different from control, $p \leq 0.01$

Of the sterols tested, DHEA caused the maximal response both with respect to resistance to vancomycin and with regards to carotenoid synthesis. However elevation in carotenoid synthesis was not limited to DHEA, but was also affected by cholesterol, estradiol and progesterone. *S. aureus* produces carotenoids via pathways that have been demonstrated as being highly mutable (68, 69). Carotenoids with one ring can mimic the tetracyclic sterols. The ability to regulate carotenoid production as an intrinsic mechanism of resistance to cationic compounds would provide it with a selective advantage whether in the soil wherein vancomycin is produced, or the body to provide protection against highly cationic peptides, e.g., defensins as well as reactive oxygen intermediates (32, 33). The ability of *S. aueus* to convert sterol to carotenoid would hypothetically require a cyclase. *S. aureus* may contain such as cyclase as part of its ability to produce cholesterol ostensibly to promote its survival as an L-form (19). If such a hypothetical enzyme exists then growth in the presence of the 3-hydroxylated sterols should affect levels of carotenoid present. Growth in the presence of all 3-hydroxyl sterols tested (cholesterol, DHEA, estrone, estradiol) with the exception of pregnenolone resulted in a significantly higher ($p<0.01$) increase in carotenoid levels at sterol concentrations (0.5 μM) which overlap those affecting antibiotic response (Table 1). Growth in the presence of sterols tested which lack an OH group at the C3 position i.e., testosterone, dehydrotestosterone, progesterone, cortisol, hydrocortisone had no effect on either vancomycin MIC or carotenoid levels (data not shown).

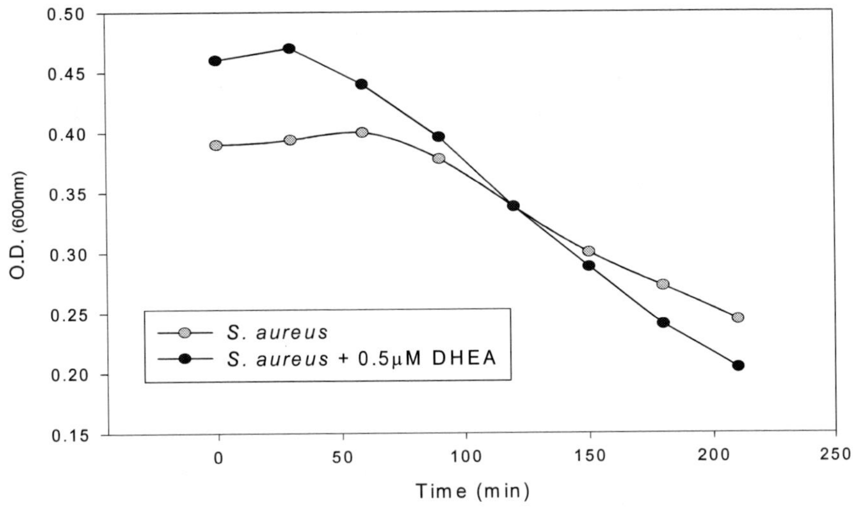

Figure 3. Effect of Triton X-100 on lysis of *S. aureus* ATCC 29523 grown in the presence (●) and absence (○) of 0.5 μM DHEA.

CONCLUSION

We propose that DHEA, and possibly other 3-hydroxylated sterols, are interkingdom signals which act as an environmental triggers for *S. aureus,* warning of the potential for exposure to highly cationic host peptides that are part of the innate immune defense system. This cell surface remodeling also, incidentally, causes an increase in resistance to those antibiotics whose associated positive charge or other chemical properties are insufficient to overcome the charge-repulsion. The staphylococcal response is at minimum increasing carotenoid levels, but may also include altered *mprF* gene expression. This alteration in cell surface charge also incidentally causes an increase in resistance to those antibiotics whose associated positive charge is insufficient to overcome the repulsive forces to allow for interaction with their cellular targets. The concentration specificity may be indicative of tight genetic regulation or membrane functional requirements. Previous studies have shown that L-forms of *S. aureus* can synthesize cholesterol, indicating the presence of a mevalonate-like pathway, the pathway which may be affected by statins (32). We reported that growth in the presence of DHEA also results in alterations of colonial morphology with the production of small to petite colony types reminiscent of persisters. Survival in the environment requires plasticity to survive the presence of antibiotics produced by soil organisms or cationic peptides of the host including lysozyme and β-1 defensin. In the soil, microbes alter their levels of carotenoids through environmentally induced changes in hopanoid synthesis. *S. aureus* while not having been demonstrated to produce hopanoids, appears instead to use the readily available 3'-hydroxy steroids present in the host to alter its membrane, increase its carotenoid production and survive the host. Interestingly, genetic changes resulting in a broad spectrum of resistance to antimicrobials, including presence of PBP2a, and the presence of efflux pumps appear to override the organism's intrinsic ability to defend itself against deleterious environmental assaults. These data indicate that *in vivo* the cell surface architecture of *S. aureus* is fundamentally different due to the presence and utilization of 3-hydroxysterols resulting in increased levels of carotenoid synthesis. The resulting alteration in cell surface charge and hydrophobic characteristics may provide a new approach to treatment management if the patient is infected with strains responsive to host-derived sterols.

REFERENCES

[1] Altschul, S; Madden, T; Shaffer, A; Zhang, J; Zhang, Z; et al. Gapped BLAST and PSI-BLAST: a new generation of protein database search programs. *Nucleic Acids Research*, 1997, 25, 3389-402.

[2] Anne FCBBFSS. Particle size and surface charge affect particle uptake by human dendritic cells in an in vitro model. *Internat J Pharmaceut*, 2005, 298, 315-22.

[3] Aoki, K; Taniguchi, H; Ito, Y; Satoh, S; Nakamura, S; et al. Dehydroepiandrosterone decreases elevated hepatic glucose production in C57BL/KsJ-db/db mice. *Life Sciences*, 2004, 74, 3075-84.

[4] Aragno, M; Cutrin, J; Mastrocola, R; Perrelli, MG; Restivo, F; et al. Oxidative stress and kidney dysfunction due to ischemia/reperfusion in rat: Attenuation by dehydroepiandrosterone. *Kidney International*, 2003, 64, 836–43.

[5] Arlt W. Dehydroepiandrosterone and ageing. Best Practice and Research in Clinical Endocrinology and Metabolism, 2004, 18, 363-80.

[6] Beenken, K; Blevins, J; Smeltzer, M. Mutation of sarA in *Staphylococcus aureus* limits biofilm formation. *Infect Immun*, 2003, 71, 4206-11.

[7] Bottasso, O; Bay, M; Besedovsky, H; Rey, Ad. Immunoendocrine alterations during human tuberculosis as an integrated view of disease pathology. *Neuroimmunomodulation*, 2009, 16, 68-77.

[8] Boyle-Vavra, S; Carey, R; Daum, R. Development of vancomycin and lysostaphin resistance in a methicillin-resistant *Staphylococcus aureus* isolate. *J Antimicrob Chemother*, 2001, 48, 617-25.

[9] Cameron, D; Braunstein, G. Androgen replacement therapy in women. *Fertility and Sterility,* 2004, 82, 273-89.

[10] Catron, D; Lange, Y; Borensztajn, J; Sylvester, M; Jones, B; Haldar, K. *Salmonella enterica* serovar Typhimurium requires nonsterol precursors of the cholesterol biosynthetic pathway for intracellular proliferation. *Infect. Immun,* 2004, 72, 1036-42

[11] Cheng, I. The effect of cholesterol on the growth of *Mycobacterium tuberculosis*, BCG and *Mycobacterium phlei*. *Scientia Sinica*, 1963, 12, 1541-52.

[12] Farzam, F; Plotkin, B. Effect of sub-MICs of antibiotics on the hydrophobicity and production of acidic polysaccharide by *Vibrio vulnificus*. *Chemotherapy*, 2001, 47, 184-93.

[13] Fietta, A; Merlini, C; Gialdroni, G; Grassi, G. Requirements for intracellular accumulation and release of clarithromycin and azithromycin by human phagocytes. *J Chemother*, 1997, 9, 23-31.

[14] Fischer, W; Koch, H; Rosel, P; Fiedler, F; Schmuck, L. Structural requirements of lipoteichoic acid carrier for recognition by the poly(ribitol phosphate) polymerase from *Staphylococcus aureus* H. A study of various lipoteichoic acids, derivatives, and related compounds. *J Biol Chem.*, 1980, 255, 4550-6.

[15] Genazzani, A; Lanzoni, C; Genazzani, A. Might DHEA be considered a beneficial replacement therapy in the elderly? *Drugs Aging,* 2007, 24, 173-85.

[16] Gross, M; Cramton, S; Gotz, F; Peschel, A. Key role of teichoic acid net charge in *Staphylococcus aureus* colonization of artificial surfaces. *Infection and Immunity*, 2001, 69, 3423-6.

[17] Hanberger, H; Nilsson, L; Maller, R; Isaksson, B. Pharmacodynamics of daptomycin and vancomycin on *Enterococcus faecalis* and *Staphylococcus aureus* demonstrated by studies of initial killing and postantibiotic effect and influence of Ca^{2+} and albumin on these drugs. *Antimicrob Agents Chemother*, 1991, 35, 1710-6.

[18] Hall, LH; Kier, L; Hall, LM. QSAR development with QSARIS. *SciVision,* 2001, 118-120.

[19] Hayami, M; Okabe, A; Sasai, K; Hayashi, H; Kanamasa, Y. Presence and synthesis of cholesterol in stable staphylococcal L-Forms. *J Bacteriol*, 1979, 140, 859-63.

[20] Helenius, A; Simons, K. Charge shift electrophoresis: Simple method for distinguishing between amphiphilic and hydrophilic proteins in detergent solution. *Proc. Natl. Acad. Sci. USA,* 1977, 74, 529-32.

[21] Herbert, J. The age of dehydroepiandrosterone. *Lancet,* 1995, 345, 1193-4.

[22] Hubert, S; Mohammed, J; Fridkin, S; Gaynes, R; McGowan, J; Tenover, F. Glycopeptide-intermediate *Staphylococcus aureus*: evaluation of a novel screening method and results of a survey of selected U.S. hospitals. *Journal of Clinical Microbiology*, 1999, 37, 3590-3.

[23] James, K; Premchand, N; Skibinska, A; Skibinska, J; Nicol, M; Mason, J. IL-6, DHEA and the aging process. *Mech Ageing Dev.*, 1997, 93, 15-24.

[24] Kahl, B; Belling, G; Reichelt, R; Herrmann, M; Karam, C; et al. Outcome assessment of minimizing vancomycin monitoring and dosing adjustments. *Pharmacotherapy,* 1999, 19, 257-66.

[25] Klosowska, K; Plotkin, B. Human insulin modulation of *Escherichia coli* adherence and chemotaxis. *American Journal of Infectious Diseases*, 2006, 2, 197-200.

[26] Koprivnjak, T; Peschel, A; Gelb, M; Liang, N; Weiss, J. Role of charge properties of bacterial envelope in bactericidal action of human group IIA phospholipase A2 against *Staphylococcus aureus*. *J Biol Chem*, 2002, 277, 47636-44.

[27] Koprivnjak, T; Weidenmaier, C; Peschel, A; Weiss, J. Wall teichoic acid deficiency in *Staphylococcus aureus* confers selective resistance to mammalian group IIA phospholipase A(2) and human beta-defensin 3. *Infect Immun,* 2008, 76, 2169-76.

[28] Kristian, S; Durr, M; Strijp, JV; Neumeister, B; Peschel, A. MprF-mediated lysinylation of phospholipids in *Staphylococcus aureus* leads to protection against oxygen-independent neutrophil killing. *Infect Immun*, 2003, 71, 546-9.

[29] Krysiak, R; Frysz-Naglak, D; Okopien, B. Current views on the role of dehydroepiandrosterone in physiology, pathology and therapy. *Pol Merkur Lekarski,* 2008, 24, 66-71.

[30] Lamberts, S; VandenBeld, A; vanderLely, A. The endocrinology of aging. *Science,* 1997, 278, 419-24.

[31] Lenard, J. Mammalian hormones in microbial cells. *Trends in Biochemical Sciences,* 1992, 17, 147-50.

[32] Liu, C; Liu, G; Song, Y; Yin, F; Hensler, M; et al. A cholesterol biosynthesis inhibitor blocks *Staphylococcus aureus* virulence. *Science*, 2008, 319, 1391-4.

[33] Liu, G; Essex, A; Buchanan, J; Datta, V; Hoffman, H; et al. *Staphylococcus aureus* golden pigment impairs neutrophil killing and promotes virulence through its antioxidant activity. *J Exp Med.*, 2005, 202, 209-15.

[34] Lowery, C; Dickerson, T; Janda, K. Interspecies and interkingdom communication mediated by bacterial quorum sensing. *Chem Soc Rev.*, 2008, 37, 1337-46.

[35] Lyte, M; Frank, C; Green, B. Production of an autoinducer of growth by norepinephrine cultured *Escherichia coli* O157:H7. *FEMS Microbiol Lett.*, 1996, 139, 155-9.

[36] Marshall, J; Wilmoth, G. Proposed pathway of trierpenoid carotenoid biosynthesis in *Staphylococcus aureus*: evidence from a study of mutants. *J Bacteriol*, 1981, 147, 914-9

[37] Martin, HD; Ruck, C; Schmidt, M; Sell, S; Beutner, S; et al. Chemistry of carotenoid oxidation and free radical reactions. *Pure Appl. Chem.,* 1999, Vol. 71, No. 12, pp. 2253±2262.

[38] Martinotti, M; Savoia, D. Effect of some steroid hormones on the growth of *Trichomonas vaginalis*. *G. Batteriol. Virol. Immunol*, 1985, 78, 52-9.

[39] Mege, J; Capo, C; Michel, B; Gastaut, J; Bongrand, P; Phagocytic cell function in aged subjects. *Neurobiol Aging*, 1988, 9, 217-20.

[40] Meier, K; Eberhard, K; Ivens, K. Effect of hormones on bacteria growth. *Naturwissenschaften*, 1954, 41, 284-5.

[41] Morikawa, K; Maruyama, A; Inose, Y; Higashide, M; Hayashi, H; Ohta, T. Overexpression of sigma factor B urges *Staphylococcus aureus* to thicken the cell wall and resist beta lactams. *Biochem Biophys Res Commun*, 2001, 288, 385-9.

[42] Mtairag, E; Abdelghaffar, H; Douhet, C; Labro, M; Role of extracellular calcium in *in vitro* uptake and intraphagocytic location of macrolides. *Antimicrob Agents Chemother*, 1995, 39, 1676-82.

[43] NCCLS. Methods for dilution antimicrobial susceptibility testing for bacteria that grew aerobically: approved standard. Wayne, Pennsylvania: National Committee for Clinical Laboratory Standards, 1997.

[44] NCCLS. 1998. Performance standards for antimicrobial susceptibility testing: eighth informational supplement. Villanova, PA: National Committee for Clinical Laboratory Standards

[45] Nes, W. Role of sterols in membranes. *Lipids*, 1974, 9, 596-612.

[46] Niebergall, P; Hussar, D; Cressman, W; Sugita, E; Doluisio, J. Metal binding tendencies of various antibiotics. *J Pharm Pharmacol*, 1966, 18, 729-38.

[47] Novick, R; Morse, S. *In vivo* transmission of drug resistance factors between strains of *Staphylococcus aureus*. *J Exp Med.*, 1967, 125, 45-59.

[48] Obaseiki-Ebor, E. Enhanced *Escherichia coli* susceptibility to nitrofurantoin by EDTA and multiple aminoglycoside antibiotics resistance mutation. *Chemother*, 1984, 30, 88-91.

[49] O'Connor, C; Essmann, M; Larsen, B. 17-ß-Estradiol upregulates the stress response in *Candida albicans*: implications for microbial virulence. *Infect. Dis. Obstet. Gynecol*, 1998, 6, 176-81.

[50] Orentreich, N; Brind, J; Vogelman, J; Andres, R; Baldwin, H. Longterm longitudinal measurements of plasma dehydroepiandrosterone sulfate in normal men. *Journal of Clinical Endocrinology and Metabolism*, 1992, 75, 1002-4.

[51] Ourisson, G; Rohmer, M. Procaryotic polyterpenes: phylogenetic precursors of sterols. *Current Topics in membranes and transport*, 1982, 17, 153-82.

[52] Peschel, A; Jack, R; Otto, M; Collins, L; Staubitz, P; et al. Interaction of tachyplesin 1 and gallidermin with cacterial cells: *Staphylococcus aureus* resistance to human defensins and evasion of neutrophil killing via the novel virulence factor MprF is based on modification of membrane lipids with l-lysine. *J Exp Med.*, 2001, 193, 1067-76.

[53] Peschel, A; Jack, R; Otto, M; Collins, L; Staubitz, P; et al. *Staphylococcus aureus* resistance to human defensins and evasion of neutrophil killing via the novel virulence factor MprF is based on modification of membrane lipids with l-lysine. *J Exp Med.*, 2001, 193, 1067-76.

[54] Peschel, A; Otto, M; Jack, R; Kalbacher, H; Jung, G; Gotz, F. Inactivation of the dlt operon in *Staphylococcus aureus* confers sensitivity to defensins, protegrins, and other antimicrobial peptides. *J Biol Chem.*, 1999, 274, 8405-10.

[55] Peschel, A; Vuong, C; Otto, M; Gotz, F. The D-alanine residues of *Staphylococcus aureus* teichoic acids alter the susceptibility to vancomycin and the activity of autolytic enzymes. *Antimicrob Agents Chemother*, 2000, 44, 2845-7.

[56] Plipat, N; Livni, G; Bertram, H; RB Thomson J. Unstable vancomycin heteroresistance is common among clinical isolates of methiciliin-resistant *Staphylococcus aureus*. *J Clin Microbiol*, 2005, 43, 2494–6.

[57] Plotkin, B; Bemis, D. Adherence of *Bordetella bronchiseptica* to hamster lung fibroblasts. *Infect Immun*, 1984, 46, 697-702.

[58] Plotkin, B; Erickson, Q; Roose, R; Viselli, S. Effect of Androgens and Glucocorticoids on Microbial Growth and Antimicrobial Susceptibility. *Current Microbiology*, 2003, 47, 514-20.

[59] Plotkin, B; Konaklieva, M. Possible role of *sarA* in dehydroepiandosterone (DHEA) - mediated increase in *Staphylococcus aureus* resistance to vancomycin. *Chemother*, 2007, 53, 181-4.

[60] Plotkin, B; Morejon, A; Laddaga, R; Viselli, S; Thijo, J; Schreckenberger, P. Induction of increased resistance to vancomycin in *Staphylococcus aureus* clinical isolates (MSSA, MRSA) by dehydroepiandosterone (DHEA). *Letters Applied Microbiol*, 2005, 40, 249-54.

[61] Retailleau, P; Ducruix, A; Ries-Kautt, M. Importance of the nature of anions in lysozyme crystallisation correlated with protein net charge variation. *Acta Crystallogr D Biol Crystallogr*, 2002, 58, 1576-81.

[62] Ruzin, A; Severin, A; Moghazeh, S; Etienne, J; Bradford, P; et al. Inactivation of *mprF* affects vancomycin susceptibility in *Staphylococcus aureus*. *Biochim Biophys Acta,* 2003, 1621, 117-21.

[63] Sergio, G. Exploring the complex relations between inflammation and aging (inflamm-aging): anti-inflamm-aging remodelling of inflamm- aging, from robustness to frailty. *Inflamm Res*., 2008, 57, 558-63.

[64] Spencer, N; Norton, S; Harrison, L; Li, G; Daynes, R. Dysregulation of IL-10 production with aging: possible linkage to the age-associated decline in DHEA and its sulfated derivative. *Exper Gerontol*, 1996, 31, 393-408.

[65] Stylos, W; Merryman, C; Maurer, P. Antigenicity of polypeptides (poly-alpha-amino acids). Distribution of sheep antibodies to polymers of alpha-L-amino acids of varying net electrical charge and lysozyme. *Int Arch Allergy Appl Immunol*, 1970, 39, 381-90.

[66] Tomasz, A; Waks, S. Mechanism of action of penicillin: triggering of the pneumococcal autolytic enzyme by inhibitors of cell wall synthesis. *Proc Natl Acad Sci U S A.* 1975, 72, 4162-6.

[67] Udo, E; Grubb, W. Molecular and phage typing of *Staphylococcus aureus* harbouring cryptic conjugative plasmids. *Eur J Epidemiol*, 1996, 12, 637-41.

[68] Umeno, D; Arnold, F. A C35 carotenoid biosynthetic pathway. *Applied and Environmental Microbiology*, 2003, 69, 3573-9.

[69] Umeno, D; Arnold, F. Evolution of a pathway to novel long-chain carotenoids. *Journal of Bacteriology*, 2004, 186, 1531-6.

[70] Urani, A; Privat, A; Maurice, T. The modulation by neurosteroids of the scopolamine-induced learning impairment in mice involves an interaction with sigma1 (sigma1) receptors. *Brain Res Brain Res Rev*., 1998, 799, 64-77.

[71] Walsh, T; Howe, R. The prevalence and mechanisms of vancomycin resistance in *Staphylococcus aureus*. *Annual Review of Microbiology*, 2002, 56, 657-75.

[72] Wong, S; Ho, P; Woo, P; KY, KY. Bacteremia caused by staphylococci with inducible vancomycin heteroresistance. *Clin Infect Dis*., 1999, 29, 760-7.

[73] Wu, J; Kusuma, C; Mond, J; Kokai-Kun, J. Lysostaphin disrupts *Staphylococcus aureus* and *Staphylococcus epidermidis* biofilms on artificial surfaces. *Antimicrob Agents Chemother*, 2003, 47, 3407-14.

[74] Xiong, Z; Kapral, F. Carotenoid pigment levels in *Staphylococcus aureus* and sensitivity to oleic acid. *J Med Microbiol*, 1992, 37, 192-4.

[75] Zimmermann, A; Katona, B; Plaisance, K. Association of vancomycin serum concentrations with outcomes in patients with gram-positive bacteremia. *Pharmacotherapy*, 1995, 15, 85-91.

[76] Zinder, O; Dar, D. Neuroactive steroids: their mechanism of action and their function in the stress response. *Acta Physiol Scand,* 1999, 167, 181-8.

In: Methicillin-Resistant Staphylococcus Aureus...
Editor: Charles L. Kolendi

ISBN: 978-1-60741-398-1

Chapter 12

RISK FACTORS FOR CARRYING METHICILLIN-RESISTANT *STAPHYLOCOCCUS AUREUS* AT HOSPITALIZATION IN A PEDIATRIC SURGICAL WARD

Koichi Ohno*

Department of Pediatric Surgery, Osaka City General Hospital, Osaka, Japan

ABSTRACT

Nosocomial infection caused by methicillin-resistant *Staphylococcus aureus* (MRSA) is one of the serious worldwide health problems. In order to investigate the risk factors for becoming an MRSA carrier and to prevent MRSA infections, bacteriological surveillance at the time of hospitalization is essential. Bacteriological examination of throat swabs and stool samples in the case of 508 males and 347 females, aged between 0 day and 24 years, who had spent more than 4 days in our pediatric surgical ward were performed within 48 h after hospitalization. In the case of patients with local infections and those with urological diseases, the discharge from the infectious regions and urine, respectively, were also examined. The patients were categorized on the basis of their history of previous hospitalization; neurological disorders, such as cerebral palsy; and age. The carrier rate of MRSA was compared among the groups in each category. We collected 854 throat swabs, 822 stool samples, and 127 other samples. At the time of hospitalization, 8.1% of 855 patients were identified as MRSA carriers. The carrier rate in the case of patients with a history of previous hospitalization within the past year (14.4%: 55/381) was higher than that in the case of patients without any history of previous hospitalization (2.6%: 8/309) and in those who were hospitalized more than a year ago (3.6%: 6/165) ($p < 0.01$). The carrier rate in the case of patients with a neurological disorder (20.9%: 23/110) was higher than that of patients without a neurological disorder (6.2%: 46/745) ($p < 0.01$). The carrier rates differed significantly with age ($p < 0.01$). The carrier rates in the case of patients aged less than 3 years (excluding neonates) (13.4%: 39/291) and patients aged more than 18 years (17.6%: 3/17) were higher than the average carrier rate (8.1%). In the former age group, 66.7%

* Corresponding author: Department of Pediatric Surgery, Osaka City General Hospital, 2-13-22, Miyakojima-hondori, Miyakojima-ku, Osaka 534-0021, Japan Tel.: +81-6-6929-1221, Fax: +81-6-6929-1091, E-mail address: k-ohno@r2.dion.ne.jp (K. Ohno).

(194/291) had a history of previous hospitalization within the past year. In the latter age group, 29.4% (5/17) had a neurological disorder. Bacteria are known to develop resistance in hospitals and cause serious nosocomial infections. However, MRSA carriers exist not only in hospitals but also in the general community. Therefore, it is very important to identify MRSA carriers at the time of hospitalization. Among our patients, 8.1% were MRSA carriers at the time of hospitalization. A history of previous hospitalization within the past year was identified as a risk factor for being an MRSA carrier. In contrast, the probability of being an MRSA carrier declines after a year of leaving the hospital. MRSA was, however, detected in 2.6% of patients without any history of previous hospitalization and in 3.6% of patients who were hospitalized more than a year ago. Community-acquired MRSA could also cause nosocomial infections. In the case of patients with neurological disorders, 1 out of 5 was an MRSA carrier. With respect to age, MRSA was frequently detected in patients aged less than 3 years (excluding neonates) since many of them had a history of previous hospitalization within the past year. In the case of patients aged more than 18 years, MRSA was frequently detected because many of them had neurological disorders. Bacteriological surveillance at the time of hospitalization is very effective in preventing MRSA infection in a ward. In the case of patients with a history of previous hospitalization within the past year, those with neurological disorders, those aged less than 3 years (excluding neonates), and those more than 18 years, it is particularly important to conduct bacteriological surveillance at the time of hospitalization.

Keywords: Nosocomial infection; Surveillance; Bacteriological examination; Methicillin-resistant *Staphylococcus aureus*.

1. Introduction

Nosocomial infection caused by drug-resistant pathogenic bacteria has been one of the serious clinical problems worldwide for a long time. Bacteria are known to develop resistance because of the overuse of antibiotics in medical facilities, and administration of antibiotics sensitive to the resistant bacteria has resulted in the emergence of many multidrug-resistant bacteria [1, 2, 3, 4, 5, 6, 7, 8]. The treatment of infectious disease caused by drug-resistant bacteria is therefore more difficult. Furthermore, it has also become clear that carriers of these bacteria exist not only in hospitals but also in the general community [9, 10]. Therefore, prevention of nosocomial transmission and nosocomial infections by taking the necessary precautions has been given more priority than the administration of sensitive antibiotics [11]. Moreover, the importance of the restricted use of sensitive antibiotics has been emphasized [12]. When an outbreak of nosocomial infection caused by drug-resistant bacteria occurs, it is usually believed that the bacteria develop resistance in a hospital, and medical workers in the hospital can usually be held responsible for this occurrence. However, it must be considered that admission of patients who are carriers of drug-resistant bacteria leads to the outbreak of nosocomial infections. Therefore, in order to prevent nosocomial infections, it is very important to identify patients who are carriers of drug-resistant bacteria at the initial stage of hospitalization [13]. In a pediatric surgical ward, especially, there are many weak, malnourished children as well as children with malignant disease. Furthermore, many patients are hospitalized repeatedly. Therefore, they are more prone to bacterial infections, and there is a high probability of them becoming carriers of drug-resistant bacteria. Bacteriological examinations at the time of hospitalization have been monitored in order to identify carriers and to prevent outbreaks of nosocomial infections and postoperative infections in our pediatric surgical ward. In this chapter, we have focused on methicillin-resistant *Staphylococcus aureus* (MRSA) and reported the carrier rate of MRSA, risk factors of becoming an MRSA carrier at the time of hospitalization, nosocomial and postoperative infections caused by MRSA during the study period, and antibiotics prophylactically administered to carriers who underwent surgery.

2. Patients and Samples

2.1. Patients

From June 2004 to November 2008, bacteriological examinations were conducted in the case of 855 patients at the time of hospitalization or transfer from a ward to the pediatric surgical ward. These patients had spent more than 4 days in the ward. They comprised 508 males and 347 females aged between 0 day and 24 years 6 months. They consisted of 103 patients who were less than 1 month old; 161, less than 1 year old; 130, less than 3 years old; 129, less than 6 years old; 122, less than 9 years old; 89, less than 12 years old; 76, less than 15 years old; 28, less than 18 years old; and 17, more than 18 years old. Of these patients, 12.9% (110/855) had neurological disorders such as cerebral palsy, Down syndrome, or West syndrome (Table 1).

2.2. Samples

Within 48 h after hospitalization, bacteriological examinations of throat swabs and stool samples were performed. In the case of patients with local infections, the discharge from the infectious regions was also examined. In the case of patients with urological diseases, a urine examination was also performed. A total of 1803 samples comprising 854 throat swabs, 822 stool samples, and 127 other samples, including 96 urine samples, 18 discharge samples from infectious regions, 11 discharge samples from a drain, and 2 skin swabs, were collected (Table 1).

3. Methods and Study Designs

3.1. Precaution for MRSA Carriers

All personnel were instructed to wash their hands before and after treating MRSA carriers. The surroundings of these patients were always kept clean with wipes containing 80% ethanol, and their excretions were segregated and disposed of separately. Sensitive antibiotics, such as mupirocin, were not administered for the purpose of MRSA elimination. When MRSA was detected in throat swabs of patients, they were instructed to wear surgical masks, and a screen was used to quarantine them from other patients. The carriers were confined to their beds and transfer was permitted only when necessary.

Table 1. Patients and Samples

Number of patients: 855 patients		
Gender: male: 508 Female: 347		
Age: from o day old to 24 years 6 months old		
0 – 30 days old	103	
– 1 year old	161	
– 3 years old	130	
– 6 years old	129	
– 9 years old	122	
– 12 years old	89	
– 15 years old	76	
– 18 years old	28	
18 years old –	17	
Neurological disorders (ND)		
Patients with ND	110	
Patients without ND	745	
Number of samples: 1803 samples		
Throat swabs	854	
Stool samples	822	
Other samples	127	
Urine:		96
Discharge from the infectious region		18
Drainage from a drain		11
Skin swabs		2

3.2. Relation between the Carrier Rates of MRSA and a History of Previous Hospitalization

The carrier rates of MRSA were determined by analyses of throat swabs, stool samples, and the patients. Further, the patients were categorized on the basis of their history of previous hospitalization, and the carrier rates in the case of the throat swabs, stool samples, and in the patients' groups were compared. In the case of patients with a history of previous hospitalization, they were divided into patients hospitalized within the past year and patients hospitalized more than a year ago, and the carrier rates in each sample and the patients' groups were also compared.

3.3. Relation between the Carrier Rates of MRSA and an Association with a Neurological Disorder

The patients were categorized on the basis of the presence or absence of a neurological disorder, and the carrier rates in the case of the throat swabs, stool samples, and in the patients' groups were compared. Further, the relationship between with the presence of a neurological disorder and a history of previous hospitalization was also investigated.

3.4. Relation between the Carrier Rates of MRSA and the Age

The patients were categorized on the basis of their age, and the relation between age and the carrier rates in the case of throat swabs, stool samples, and in the patients' groups were compared. In the age groups exhibiting a high carrier rate, the rates of patients with a history of previous hospitalization and rates of those with a neurological disorder were also investigated.

3.5. Community Acquired MRSA

MRSA that were identified in the case of patients without any history of previous hospitalization and in the case of those who were hospitalized more than a year ago were regarded as community-acquired MRSA (CA-MRSA). The clinical features of CA-MRSA were investigated.

3.6. Outbreaks of Nosocomial Infection and Postoperative Infection Caused by MRSA

The rate of infection caused by MRSA in the pediatric surgical ward was calculated in the case of all the patients during the study period. Prophylactically administered antibiotics to carriers who underwent surgery and postoperative infection caused by MRSA were retrospectively investigated.

3.7. Data Analysis

The data were analyzed using Fisher's exact probability test and the Chi-square test for independence; p values of less than 0.05 were considered significant.

4. Results

4.1. Relation between the Carrier Rates of MRSA and a History of Previous Hospitalization

MRSA was detected in 5.0% of the throat swabs (43/854), 4.5% of the stool samples (37/822), and 8.7% of the other samples (11/127); the carrier rate was 8.1% (69/855) (Table 2). Of 855 patients, 63.9% (546/855) had been hospitalized previously and 44.6% (381/855) had been hospitalized within the past year. MRSA was detected in 2.3% of the throat swabs (7/308) from patients without any history of previous hospitalization and in 6.6% of the throat swabs (36/546) from patients with a history of previous hospitalization ($p = 0.0052$). MRSA was detected in 0.7% of the stool samples (2/290) from the former patient group and in 6.6% of the stool samples (35/532) from the latter patient group ($p < 0.0001$). The carrier rate in the case of the former patient group was 2.6% (8/309) and that in the case of the latter group was 11.2% (61/546) ($p < 0.0001$) (Table 3).

Table 2. The carrier rates of MRSA in each sample groups

Throat swabs	5.0 %	(48 / 854 samples)
Stool samples	4.5 %	(37 / 822 samples)
Other samples	8.7 %	(11 / 127 samples)
Total carrier rate	8.1 %	(69 / 855 patients)

MRSA: methicillin-resistant *Staphylococcus aureus*

Table 3. Comparison of the carrier rates of MRSA between patients without any HPH and patients with a HPH

Throat swabs

	Negative	Positive	Carrier rate
HPH (-)	301	7	2.3 %
HPH (+)	510	36	6.6 %

($p = 0.0052$)

Stool samples

	Negative	Positive	Carrier rate
HPH (-)	288	2	0.7 %
HPH (+)	497	35	6.6 %

($p < 0.0001$)

All samples

	Negative	Positive	Carrier rate
HPH (-)	301	8	2.6 %
HPH (+)	485	61	11.2 %

($p < 0.0001$)

HPH: history of previous hospitalization

The 546 patients who had been hospitalized previously were divided into the 165 patients who were hospitalized more than a year ago and 381 patients who were hospitalized within the past year. MRSA was detected in 1.8% of the throat swabs (3/165) from patients who were hospitalized more than a year ago and in 8.7% of the throat swabs (33/381) from patients who were hospitalized within the past year ($p = 0.0022$). MRSA was detected in 1.9% of the stool samples (3/157) from the former patient group and in 8.6% of the stool samples (32/372) from the latter patient group ($p = 0.0035$). The carrier rate in the case of the former patient group was 3.6% (6/165) and that in the case of the latter patient group was 14.4% (55/381) ($p < 0.0001$) (Table 4).

The carrier rates were compared between patients who had never been hospitalized and patients who had been hospitalized more than a year ago. MRSA was detected in 2.3% of the throat swabs (7/308) from patients without any history of previous hospitalization and in 1.8% of the throat swabs (3/165) from patients who were hospitalized more than a year ago ($p > 0.9999$). MRSA was detected in 0.7% of the stool samples (2/290) from the former patient group and in 1.9% of the stool samples (3/160) from the latter patient group ($p = 0.3526$). The carrier rate in the case of the former patient group was 2.6% (8/309) and that in the case of the latter patient group was 3.6% (6/165) ($p = 0.5733$) (Table 5).

Table 4. Comparison of the carrier rates of MRSA between patients with a HPH more than a year ago and patients with a HPH within the past year

Throat swabs

	Negative	Positive	Carrier rate
HMTYA	162	3	1.8 %
HWPY	348	33	8.7 %

($p = 0.0022$)

Stool samples

	Negative	Positive	Carrier rate
HMTYA	154	3	1.9 %
HWPY	340	32	8.6 %

($p = 0.0035$)

All samples

	Negative	Positive	Carrier rate
HMTYA	159	6	3.6 %
HWPY	326	55	14.4 %

($p < 0.0001$)

HPH: history of previous hospitalization
HMTYA: history of previous hospitalization more than a year ago
HWPY: history of previous hospitalization within the past year

4.2. Relation between the Carrier Rates of MRSA and an Association with a Neurological Disorder

There were 745 patients without neurological disorders and 110 patients with neurological disorders. MRSA was detected in 3.6% of the throat swabs (27/744) from the former patient group and in 14.5% of the throat swabs (16/110) from the latter patient group ($p < 0.0001$). MRSA was detected in 3.6% of the stool samples (26/713) from the former patient group and in 10.1% of the stool samples (11/109) from the latter patient group (p =

0.0057). The carrier rate in the case of the former patient group was 6.2% (46/745) and that in the case of the latter patient group was 20.9% (23/110) ($p < 0.0001$) (Table 6).

Of the 745 patients without neurological disorders, 58.9% (439/745) had been hospitalized previously, and 97.3% (107/110) of the 110 patients with neurological disorders had been hospitalized previously ($p < 0.0001$). Of the patients in the former patient group, 39.7% (296/745) had been hospitalized within the past year, and 76.4% (84/110) of the patients in the latter patient group had been hospitalized within the past year ($p < 0.0001$) (Table 7).

Table 5. Comparison of the carrier rates of MRSA between patients without any HPH and patients with a HPH before more than a year ago

Throat swabs

	Negative	Positive	Carrier rate
WOH	301	7	2.3 %
HMTYA	162	3	1.8 %

($p > 0.9999$)

Stool samples

	Negative	Positive	Carrier rate
WOH	288	2	0.7 %
HMTYA	157	3	1.9 %

($p = 0.3526$)

All samples

	Negative	Positive	Carrier rate
WOH	301	8	2.6 %
HMTYA	159	6	3.6 %

($p = 0.5733$)

HPH: history of previous hospitalization
WOH: without any history of previous hospitalization
HMTYA: history of previous hospitalization more than a year ago

4.3. Relation between Carrier Rates of MRSA and the Age

The patients were categorized on the basis of age into groups of less than 1 month of age, less than 1 year of age, less than 3 years of age, less than 6 years of age, less than 9 years of age, less than 12 years of age, less than 15 years of age, less than 18 years of age, and more than 18 years of age.

MRSA was detected in 2.9% (3/103), 5.6% (9/160), 9.2% (12/130), 7.0% (9/129), 1.6% (2/122), 3.4% (3/89), 2.6% (2/76), 3.6% (1/28), 11.8% (2/17) of the throat swabs in each age group, respectively ($p = 0.1118$). MRSA was detected in 5.2% (5/97), 8.8% (14/159), 9.3%

(12/129), 0.8% (1/125), 0% (0/118), 3.5% (3/85), 1.5% (1/68), 0% (0/25), 6.3% (1/16) of the stool samples in each age group, respectively (p = 0.0009). The carrier rates were 6.8% (7/103), 11.2% (18/161), 16.2% (21/130), 7.8% (10/129), 1.6% (2/122), 3.4% (3/89), 5.3% (4/76), 3.6% (1/28), 17.6% (3/17) in each age group, respectively (p = 0.0007). In the case of the patients aged less than 1 year, less than 3 years, and more than 18 years, the carrier rates were higher than the average carrier rate (8.1%: 69/855) (Table 8).

Table 6. Comparison of the carrier rates of MRSA between patients without ND and patients with ND

Throat swabs

	Negative	Positive	Carrier rate
ND (-)	717	27	3.6 %
ND (+)	94	16	14.5 %

(p < 0.0001)

Stool samples

	Negative	Positive	Carrier rate
ND (-)	687	26	3.6 %
ND (+)	98	11	10.1 %

(p = 0.0057)

All samples

	Negative	Positive	Carrier rate
ND (-)	699	46	6.2 %
ND (+)	87	23	20.9 %

(p < 0.0001)

ND: neurological disorders

The rates of patients who had been hospitalized within the past year in each age group were 46.6% (48/103), 68.9% (111/161), 63.8% (83/130), 39.5% (51/129), 24.6% (30/122), 19.1% (17/89), 34.2% (26/76), 39.3% (11/28), 23.5% (4/17), respectively (p < 0.0001). In the case of patients aged less than 1 month, less than 1 year, and less than 3 years, the rates were higher than the average rates of patients who were hospitalized within the past year (44.6%: 381/855) (Table 9).

The rates of patients with neurological disorders in each age group were 1.9% (2/103), 8.1% (13/161), 19.2% (25/130), 21.7% (28/129), 7.4% (9/122), 14.6% (13/89), 10.5% (8/76), 25.0% (7/28), 29.4% (5/17), respectively (p < 0.0001). In the case of patients aged less than 3 years, less than 6 years, less than 12 years, less than 18 years, and more than 18 years, the rates were higher than the average rate of patients with neurological disorders (12.9%: 110/855) (Table 9).

4.4. Community Acquired MRSA

MRSA were detected in 8 of the 309 patients who had never been hospitalized and 6 of the 165 patients who had been hospitalized more than a year ago. MRSA was detected in 13 healthy patients; and these were considered to be CA-MRSA cases, except 1 patient who had stayed at an institution for mentally handicapped persons. Among the total patients, 1.5% of them (13/855) were CA-MRSA carriers, and 18.8%of the all MRSA carriers (13/69) were CA-MRSA carriers. They comprised 7 males and 6 females. There were 2 patients aged less than 1 month; 2, less than 1 year; 5, less than 3 years; 3, less than 6 years; and 1, more than 18 years. CA-MRSA was detected in 6 throat swabs, 8 stool samples, 2 umbilical discharge samples, and 1 discharge sample from a subcutaneous abscess (Table 10).

Table 7. Comparison of HPH between patients without ND and patients with ND

History of previous hospitalization

	HPH (-)	HPH (+)
ND (-)	306	439
ND (+)	3	107

($p < 0.0001$)

History of previous hospitalization within the past year

	HWPY (-)	HWPY (+)
ND (-)	449	296
ND (+)	26	84

($p < 0.0001$)

ND: neurological disorders
HPH: history of previous hospitalization
HWPY: history of previous hospitalization within the past year

4.5. Outbreaks of Nosocomial Infection and Postoperative Infection Caused by MRSA

During the study period, since an MRSA wound infection occurred after removal of the subcutaneous benign tumor in the case of a patient who was not an MRSA carrier, the rate of MRSA infection was 0.1% (1/855).

Of the 69 carriers, 34 (49.3%) underwent surgical treatment. Contaminated operations in 13 patients and noncontaminated operations in 21 patients were performed. Antibiotics sensitive to MRSA were prophylactically administered to 9 of the former patient group and 6 of the latter patient group (p = 0.0337). Thus, the all 69 carriers who underwent surgery were not infected with MRSA postoperatively (Table 11).

Table 8. The carrier rates of MRSA in each age group

Age group	Throat swabs (%)	Stool samples (%)	All samples (%)
0 – 30 days old	2.9	5.2	6.8
– 1 year old	5.6	8.8	11.2
– 3 years old	9.2	9.3	16.2
– 6 years old	7.0	0.8	7.8
– 9 years old	1.6	0.0	1.6
– 12 years old	3.4	3.5	3.4
– 15 years old	2.6	1.5	5.3
– 18 years old	3.6	0.0	3.6
18 years old –	11.8	6.3	17.6
p-value	0.1118	0.0009	0.0007

Table 9. The rates of patients with a HPH within the past year and the rates of patients with ND in each age group

Age group	HWPY (%)	Neurological disorders (%)
0 – 30 days old	46.6	1.9
– 1 year old	68.9	8.1
– 3 years old	63.8	19.2
– 6 years old	39.5	21.7
– 9 years old	24.6	7.4
– 12 years old	19.1	14.6
– 15 years old	34.2	10.5
– 18 years old	39.3	25.0
18 years old –	23.5	29.4
p-value	< 0.0001	< 0.0001

HPH: history of previous hospitalization
ND: neurological disorders
HWPY: history of previous hospitalization within the past year

Table 10. Community acquired MRSA in patients without any HPH and patients with a HPH more than a year ago

Case	Age	M/F	Diagnosis	Sample	HPN
1	10d	F	vaginal atresia	T	Non
2	24d	M	umbilical granulation	T, S, U	Non
3	1m	M	umbilical granulation	S, U	Non
4	2m	F	bronchotracheal fistula	S	HMTYA
5	1y00m	F	lymphagioma	S	Non
6	1y04m	M	perineal lipoma	T	Non
7	1y11m	M	hypospatiase	T	Non
8	2y08m	F	acute abdomen	S, T	HMTYA
9	2y11m	M	enteritis after Sove operation	S	HMTYA
10	3y04m	M	displastic kidney	S	HMTYA
11	3y05m	F	pre-pubic sinous	S	Non
12	5y08m	F	rectal polyp	T	Non
13	18y01m	M	wound abscess	D	HMTYA

M: male, F: female, T: throat swab, S: stool sample, U: unbilica discharge
D: discharge from a drain
HPH: history of previous hospitalization
HMTYA: history of previous hospitalization more than a year ago

Table 11. Antibiotics administered to the 34 surgical carriers

	Carriers administered antibiotics sensitive to MRSA	Carriers administered antibiotics nonsensitive to MRSA
Contaminated operations	9	4
Noncontaminated operations	6	15

($p = 0.0337$)

Sensitive antibiotics
Vancomycin: 12, Arbekacin: 2,
Sulfamethoxazole/trimethoprim: 1

Contaminated operations
Pena operation, Anorectoplasty, Perforated ulcer, Colostomy, Colopexy , Soave operation, Polypectomy, Abscess drainage, Gant-Miwa operation

Noncontaminated operations
Ramstedt operation , Fundoplication, Repair of the diaphragm, Ventriculoperitoneal shunt, Splenectomy, Nephrectomy, Tumor biopsy, Tumor resection, Repair of inguinal hernia, Vaginoplasty, Urethroplasty, OK-432 injection, Resection of prepubic sinus

5. Countermeasure against MRSA

Many efforts have been made to develop broad-spectrum antibiotics for treatment of bacterial infections. Many people formerly hoped and asserted that all bacterial infections could be eradicated. However, new antibiotics have resulted in the emergence of new harmful drug-resistant bacteria, including multidrug-resistant *Pseudomonas aeruginosa* [7, 8], vancomycin-resistant *Enterococci* [4, 5], extended-spectrum beta-lactamase *Escherichia coli* [14], and extended-spectrum beta-lactamase *Klebsiella pneumoniae* [15], which have caused serious infectious diseases recently. In the same way, since MRSA was detected in the early 1960s [16, 17, 18], people have been exposed to the menace of MRSA infections worldwide. Although vancomycin was reinstated as an antibiotic sensitive to MRSA, and new sensitive antibiotics have been employed [19, 20, 21, 22], MRSA has not been exterminated. In some countries, including Japan and America, MRSA accounts for 40–60% of *Staphylococcus aureus* detected by culture [23,24]. Furthermore, the first report on vancomycin-resistant *Staphylococcus aureus* (VRSA) was published in America [25]. After it was widely advocated that the development of new antibiotics alone would not eradicate MRSA, the usefulness of active surveillance and the importance of preventing MRSA outbreaks were realized [10, 26]. In particular, in acute care hospitals, neonatal intensive care units, and intensive care units where there are many patients in fragile conditions, the importance of precaution has been recognized in order to prevent nosocomial transmission and infections caused by resistant bacteria. Although the Center for Disease Control and Prevention (CDC) in America published the necessity of the observance of standard precautions in the case of all inpatients, it was not effective in the reduction of MRSA. On the other hand, in some north European countries, less than 3% of the *Staphylococcus aureus* strains were found to be methicillin resistant. One of the guidelines of the Dutch Working party on Infection Prevention (WIP) for prevention of nosocomial infections caused by MRSA should be especially considered. The guideline was laid down on the basis of a "search and destroy" policy, and it resulted in strict observance of repeated active surveillance and quarantining of all people with MRSA [12]. According to the guideline, category 1 consists of patients with MRSA infections and all patients and medical staff who were MRSA carriers. All the patients in whom MRSA was detected were quarantined in a single room. All the medical staffs who were MRSA carriers had to leave their work to be treated with nasal mupirocin. A bacteriological examination was conducted every 5 days, and after negative results were confirmed more than 3 times, the staff could be reinstated. Furthermore, if more than 2 people with same strain of MRSA were identified in a ward, MRSA was regarded to be prevalent in that ward and complete countermeasures, including closing the ward, were undertaken. Thereafter, America followed in the footsteps of The Netherlands; the Society for Health-care Epidemiology of America (SHEA) introduced another guideline in 2003 that included active surveillance [11]. The guideline stated that not only patients with MRSA infection but all carriers of MRSA could cause nosocomial transmission of MRSA and outbreaks of MRSA infection; further, active bacteriological surveillance at the time of hospitalization was recommended in the case of high-risk patients.

6. Risk Factors for Carrying MRSA

Since patients in a pediatric surgical ward are weak and hospitalized repeatedly, they are very susceptible to becoming carriers of pathogenic bacteria. Jernigam et al.. [27] reported that a history of hospital stay of more than 5 days within a year was a risk factor for colonization of MRSA. Among our patients, 63.9% had a history of previous hospitalization, and 44.6% had a history of previous hospitalization within the past year. In both throat swabs and stool samples, a higher level of MRSA was detected in the case of patients with a history of previous hospitalization than in the case of patients without any history of previous hospitalization. The carrier rate in the case of the former patient group was also higher than that of the latter group. Furthermore, in the case of patients who were hospitalized within the past year, an even higher MRSA level was detected in each sample. Moreover, the carrier rate in the case of this patient group was still higher; 1 of the 7 patients with a history of previous hospitalization within the past year introduced MRSA into our pediatric surgical ward. However, the carrier rate in the case of patients who were hospitalized more than a 1 year ago did not significantly differ from that in the case of the patients without any history of previous hospitalization. When more than a year passed after discharge, MRSA levels in the throat and stool samples decreased spontaneously.

Some patients with neurological disorders, such as cerebral palsy, were kept in the pediatric surgical ward prior to an operation. This study showed that 1 of the 5 patients with neurological disorders was identified to be a carrier. In patients with neurological disorders, the rate of patients with a history of previous hospitalization within the past year was significantly higher than in patients without neurological disorders. Many patients with neurological disorders stay not only in hospitals but also in facilities for handicapped persons for long periods. These factors could be the cause of the high carrier rate in the case of patients with neurological disorders.

Many patients from newborns to those aged more than 15 years stay in the pediatric surgical ward. In this chapter, the patients were categorized on the basis of age, and a high carrier rate was detected in some age groups. Patients who were less than 1year of age (excluding neonates), less than 3 years of age, and more than 18 years of age exhibited a higher carrier rate than the average rate. Among the patients who were less than 3 years of age (excluding neonates), 66.7% of them (194/291) had a history of previous hospitalization within the past year. Further, among the patients who were less than 1 year of age (excluding neonates) 19.2% of them (25/130) had neurological disorders. On the other hand, in the case of patients aged more than 18 years, 29.4% of them (7/28) had neurological disorders. In the case of patients with neurological disorders in this age group, their symptoms were stable, and many of them stayed in the facilities over a long period of time. These factors contributed to an increase in the carrier rate in the case of patients aged less than 3 years (excluding neonates) and more than 18 years.

7. Community-Acquired MRSA

Although overuse of broad-spectrum antibiotics has resulted in the emergence of multidrug-resistant bacteria in hospitals, the Center for Disease Control and Prevention reported that 4 healthy people who had never been hospitalized died due to MRSA infection in 1999 [28]. After this report, CA-MRSA was recognized, and it made an impact in society. Recently, CA-MRSA has spread worldwide, and has become one of the serious medical problems [29, 30, 31, 32, 33, 34, 35]. Initially, it was thought that MRSA colonizing inpatients spread in the general community after they were discharged from the hospital [13]. However, CA-MRSA possesses Panton-Valentine leukocidin genes, and it is more harmful than hospital-acquired MRSA (HA-MRSA) [22]. This strain causes serious pneumonia in healthy people [36]. Furthermore, many CA-MRSA produce an exfoliative toxin (ET), and cause impetigo in infants and staphylococcal-scalded skin syndrome in neonates [37, 38]. Because of the characteristic features of CA-MRSA, some MRSA detected from healthy people got resistance in the general community. In this chapter, 13 carriers were detected among patients without any history of previous hospitalization and those who were hospitalized more than a year ago; they constituted 1.5% of all the patients (13/855) and 18.8% of all MRSA carriers (13/69). Of the 13 CA-MRSA carriers, 12 carriers were less than 6 years of age, and comprised 21.4% of the carriers aged less than 6 years (12/56). The 4 patients described in the first report on CA-MRSA were also children [28]; caution should be exerted while detecting MRSA in infants. Although MRSA was detected in only a few patients without any history of previous hospitalization and patients who were hospitalized more than a year ago, active surveillance at the time of hospitalization is recommended for all patients in order to identify harmful CA-MRSA.

8. Prevention of Nosocomial Transmission of MRSA

New antibiotics sensitive to MRSA have been developed [19, 20, 21, 22, 39, 40]. However, a history of the MRSA treatment proves that antibiotics alone can not exterminate MRSA. The Dutch WIP guideline was put forth on the basis of a conviction that anyone could be an MRSA carrier [12], and this is one of the ideal strategies to combat MRSA. However, regular repeated bacteriological surveillance for all inpatients and medical staffs requires a huge capital, and it is very difficult for all hospitals to provide single rooms in order to quarantine all the MRSA patients. In some countries where MRSA has already spread nationally, including Japan and America, the application of the Dutch WIP guideline is unrealistic. Since many carriers require surgical treatment in reality, we need to develop an effective strategy for the prevention of nosocomial MRSA transmission and MRSA outbreaks besides providing single rooms to quarantine patients. Organizing infection control teams (ICT), regular ICT rounds, surveillance of infections, investigation of antibiotics used in a hospital, and standard precautions for all inpatients are clearly very important [41]. Each ward should also make an active effort to combat MRSA. The usefulness and economical effects of active surveillance at the time of hospitalization have been reported [42]. Therefore, we initially monitored all the patients by conducting active bacteriological surveillance at the initial stage of hospitalization and identified MRSA carriers. Strict precautions were taken while handling these patients instead of the administration of sensitive antibiotics such as nasal mupirocin. MRSA is contagious and therefore spreads easily, and MRSA colonizing the throat is transmitted while coughing or sneezing. Therefore, the surroundings in which the carriers resided were always kept clean with wipes containing 80% ethanol, and their stool and urine were disposed of separately. The MRSA carriers detected by analysis of throat swabs were instructed to wear surgical masks, and a screen was used to quarantine them from other patients. They were treated as noncarriers when they tested negative for MRSA twice by bacteriological examination. Although the Dutch WIP recommends nasal mupirocin treatment for MRSA carriers [12], it was reported that the effects of this drug on pharynx-colonized MRSA are limited [43], and mupirocin-resistant MRSA has also been identified [44]. Administration of sensitive antibiotics against MRSA could alter the intestinal bacterial flora. Therefore, we did not administer any sensitive antibiotic to the carriers in order to eliminate MRSA. This resulted in the rate of nosocomial MRSA infections being only 0.1% during the study period. Although we could not eliminate MRSA infection completely, our strategy was found to be effective in controlling the infection. Therefore, we do not recommend administration of antibiotics sensitive to MRSA for the extermination of it.

9. Postoperative Infection Caused by MRSA and Prophylactically Administered Antibiotics to the Surgical Carriers

The prophylactic administration of antibiotics to MRSA carriers who underwent surgery was retrospectively investigated. Sensitive antibiotics were prophylactically administered to 44.1% (15/34) of the carriers who underwent surgery. Sensitive antibiotics were used to a significantly greater extent in the case of the carriers who underwent contaminated operations. However, the effectiveness of prophylactic administration of sensitive antibiotics for the treatment of carriers who underwent surgery could not be validated because all the carriers who underwent surgery, including the 19 carriers who were administered nonsensitive antibiotics, did not experience any postoperative MRSA infections. Some reports recommend the administration of prophylactic sensitive antibiotics to carriers who underwent surgery [45, 46, 47]. However, the use of sensitive antibiotics should be reduced as far as possible [12] because it leads to the emergence of new drug-resistant bacteria [1, 2, 3]. The carriers undergoing noncontaminated surgeries, including urological surgeries, did not require the administration of sensitive antibiotics. Even in the case of patients with contaminated surgeries, the prophylactic administration of sensitive antibiotics should be restricted to weak carriers who are prone to infection, which could progress and result in a serious situation, e.g., patients with gastrointestinal perforation, pan-peritonitis, bone marrow suppression, immunological disorders, septic conditions, or congenital heart anomalies. Randomized prospective studies are required to examine the effectiveness of the prophylactic administration of sensitive antibiotics for the MRSA carriers who underwent surgery.

10. Conclusion

In a pediatric surgical ward, the carrier rate of MRSA at the initial hospitalization stage was 8.1%. A history of previous hospitalization within the past year and the presence of a neurological disorder were considered to be the risk factors for becoming an MRSA carrier. In the case of patients aged less than 3 years (excluding neonates) and more than 18 years old, the carrier rates were higher than those in the case of the other age groups. CA-MRSA that was more harmful than HA-MRSA was detected in 18.8% of all the MRSA carriers and many CA-MRSA carriers were less than 6 years of age. Active bacteriological surveillance at the time of hospitalization and taking selective and strict precautions while handling MRSA carriers could satisfactorily prevent nosocomial transmission of MRSA as well as an outbreak of MRSA infection. We recommend neither the administration of sensitive antibiotics to MRSA carriers in order to eradicate MRSA or the quarantining of MRSA carriers in single rooms. All carriers who undergo surgery do not require the prophylactic administration of sensitive antibiotics. This should be limited to selective carriers who are weak and are prone to infection. Randomized prospective studies are required to validate the effectiveness of prophylactic administration of sensitive antibiotics to carriers who undergo surgery. Because the use of antibiotics sensitive to MRSA promotes the emergence of multidrug-resistant bacteria, the administration of sensitive antibiotics to carriers in order to eradicate MRSA as well as the prophylactic administration of sensitive antibiotics to surgical carriers should be limited as far as possible.

REFERENCES

[1] Aligholi, M; Emaneini, M; Jabalameli, E; et al. Emergence of high-level vancomycin-resistant Staphylococcus aureus in the Imam Khomeini Hospital in Tehran. *Med Princ Pract.*, 2008, 17, 432-434.

[2] Sievert, DM; Rudrik, JT; Patel, JB; et al. Vancomycin-resistant Staphylococcus aureus in the United States 2002-2006. *Clin Infect Dis.*, 2008, 46, 668-674.

[3] Saha, B; Singh, AK; Ghosh, A; et al. Identification and characterization of a vancomycin-resistant Staphylococcus aureus isolated from Kolkata (South Asia). *J Med Microbiol*, 2008, 57, 72-79.

[4] Hota, B; Blom, DW; Lyle, EA; et al. Interventional evaluation of environmental contamination by vancomycin-resistant enterococci: failure of personal, product, or procedure? *J Hosp Infect*, 2008, 71, 123-131.

[5] Sethi, AK; Al-Nassir, WN; Nerandzic, MM; et al. Skin and environmental contamination with vancomycin-resistant Enterococci in patients receiving oral metronidazole or oral vancomycin treatment for Clostridium difficile-associated disease. *Infect Control Hosp Epidemiol*, 2009, 30, 13-17.

[6] Huycke, MM; Sahm, DF; Gilmore, MS. Multiple-drug resistant enterococci: the nature of problem and an agenda for the future. *Emerg Infect Dis.*, 1998, 4, 239-249.

[7] Kawazoe, H; Takiguchi, Y; Inoue, T; et al. Infection treatment caused by multiple-drug resistant Pseudomonas aeruginosa in a patient undergoing allergenic hematopoietic stem cell transplantation. *Yakugaku Zasshi*, 2008. 128, 657-661.

[8] Wang, J; Bo, R; Xu, L; et al. A CARB-like beta-lactamase gene from a multiple-drug resistant Pseudomonas aeruginosa. *J Med Microbiol*, 2006, 55, 1609-1610.

[9] Bloomfield, SF. Community-acquired MRSA. *Lancet*, 2006, 368, 1328.

[10] Gould, IM. Community-acquired MRSA: can we control it? *Lancet*, 2006, 368, 824-826.

[11] Muto, CA; Jernigan, JA; Ostrowsky, BE; et al. SHEA guideline for prevention nosocomial transmission of multidrug-resistane strains of Staphylococcus aureus and Enterococcus. *Infect Control Hosp Epidemiol*, 2003, 24, 362-386.

[12] Wertheim, HFL; Vos, MC; Boelens, HAM; et al. Low prevalence of methicillin-resistant Staphylococcus aureus (MRSA) at hospital admission in the Netherland: the value of search and destroy and restrictive antibiotics use. *J Hosp Infect*, 2004, 56, 321-325.

[13] Salgado, CD; Farr, BM; Calfee, DP. Community-acquired methicillin-resistant Staphylococcus aureus: a meta-analysis of prevalence and risk factors. *Clin Infect Dis.*, 2003, 36, 131-139.

[14] Song, W; Lee, H; Lee, K; et al. CTX-M-14 and CTX-M-15 enzymes are the dominant type of extended-spectrum beta-lactamase in clinical isolates of Escherichia coli from Korea. *J Med Microbiol*, 2009, 58, 261-266.

[15] Jemima, SA; Verghese, S. SHV-28, an extended-spectrum beta-lactamase produced by a clinical isolate of Klebsiella pneumoniae in south India. *Indian J Med Microbiol*, 2009, 27, 51-54.

[16] Jevons, MP. "Celbenin"-resistant staphylococci. *British Med J*, 1961, 1, 124-125

[17] Jevons, MP; Coa, AW; Parker, MT. Methicillin resistance in staphylococci. *Lancet,* 1963, *1*, 904-907.

[18] Eriksen, KR; Erichsen, I. Clinical occurrence methicillin-resistant strains of staphylococcus aureus. *Ugeskr Laeger*, 1963, 125, 1234-1240.

[19] Petersen, PJ; Bradford, PA; Weiss, WJ; et al. Projan SJ *In vitro* and *in vivo* activities of tigecycline (GAR-936), daptomycin, and comparative antimicrobial agents against glycopeptide-intermediate *Staphyrococcus aureus* and other resistant gram-positive pathogens. *Antimicrob Agents Chemother*, 2002, 46, 2595-2601.

[20] Rybak, MJ; Hershberger, E; Moldovan, et al. In vitro activities of daptomycin, Vancomycin, linezolid, and quinuspristin-dafopristin against staphylococci and enterococci, including vancomycin-intermediate and resistant strains. *Antimicrob Agents Chemother*, 2000, 44, 1062-1066.

[21] Lamb, HL; Figgitt, DP; Faulds, D. Quinuspristin/Dalfopristin: a review of its use in the management of serious gram-positive infections. *Drug.*, 1999, 58, 1061-1097.

[22] Yanagihara, K; Okada, M; Fukuda, Y; et al. Efficacy of quinuspristin-dalfopristin against methicillin-resistant Staphylococcus aureus and vancomycin-insensitive S. aureus in a model of hematogenous pulmonary infection. *Chemotherapy*, 2004, 50, 260-264.

[23] Weber, JT. Community-acquired methicillin-resistant Staphylococcus aureus. *Clin Infect Dis.*, 2005, 41, 269-272.

[24] Niki, Y; Hanaki, H; Yagisawa, M; et al. The first national surveillance of bacterial respiratory pathogens conducted by the Japanese Society of Chemotherapy. Part 1: a general view of antibacterial susceptibility. *J Infect Chemother*, 2008, 14, 279-90.

[25] Center for Disease Control and Prevention. Staphylococcus aureus resistant to vancomycin- United States. *MMWR*, 2002, 51. 565-567.

[26] Spicer, WJ. Three strategies in the control of stapylococcus including methicillin-resistant Staphylococcus aureus. *J Host Infect*, 1984, 5 Suppl A, 45-49.

[27] Jernigan, JA; Pullen, AL; Flowers, L; et al. Prevalence of and risk factors for colonization with methicillin-resistant Staphylococcus aureus at the time of hospital admission. *Infect Control Hosp Epidemiol*, 2003, 24, 409-414.
[28] Center for Disease Control and Prevention. Four pediatric deaths from community-acquired methicillin-resistant Staphylococcus aureus- Minnesota and North Dakota, 1997-1999. MMWR Morb. *Mortal Wkly Rep.*, 1999, 48, 707-710.
[29] Kale-Pradhan, P; Johnson, LB. Treatment and recurrence management of staphylococcus infections: community-acquired MRSA. *Expert Rev Anti Infect Ther.*, 2008, 6, 909-915.
[30] Bogut, A; Koziol-Montewka, M; Baranowicz, I; et al. Community-acquired methicillin-resistant Staphylococcus aureus (CA-MRSA) in Poland: further evidence for the changing epidemiology of MRSA. *New Microbiol*, 2008, 31, 229-234.
[31] Takano, T; Saito, K; Teng, LJ; et al. Spread of community-acquired methicillin-resistant Staphylococcus aureus (MRSA) in hospitals in Taipei, Taiwan in 2005, and comparison its drug resistance with previous hospital-acquired MRSA. *Microbiol Immunol*, 2007, 51, 627-632.
[32] Baranovich, T; Potapov, V; Yamamoto, T. The first isolation of Panton-Valentine leukocidin (PVL) positive community-acquired methicillin-resistant Staphylococcus aureus (CA-MRSA) in Russia. *Euro Surveill*, 2007, 12, E070315. 4.
[33] Krzyszton-Russjan, J; Tambic-Andrasevic, A; Bukovski, S; et al. First community-acquired methicillin-resistant Staphylococcus aureus (MRSA) strains in Croatia. *Clin Microbiol Infect*, 2006, 12, 697-698.
[34] Vourli, S; Perimeni, D; Makri, A; et al. Community acquired MRSA infections in a paediatric population in Greece. *Euro Surveill*, 2005, 10, 78-79.
[35] Takazawa, Y; Taneike, I; Nakagawa, S; et al. A Panton-Valentine leucocidin (PVL)-positive community-acquired methicillin-resistant Staphylococcus aureus (MRSA) strain another such strain carrying a multi-drug resistance plasmid, and other more-typical PVL-negative MRSA strains found in Japan. *J Clin Microbiol*, 2005, 43, 3356-3363.
[36] Francis, JS; Doherty, MC; Lopatin, U; et al. Severe community-onset pneumonia in healthy adults caused by methicillin-resistant Staphylococcus aureus carrying the Panton-Valentine leukocidin genes. *Clin Infect Dis.*, 2005, 40, 100-107.
[37] Yamaguchi, T; Yokota, Y; Terajima, J; et al. (Cloanl association of Staphylococcus aureus causing bullous impetigo and emergence of new methicillin-resistant clonal groups in Kansai district in Japan. *J Infect Dis.*, 2002, 185, 1511-1516.
[38] Magilner, D; Byerly, MM; Cline, DM. The prevalence of community-acquired methicillin-resistant Staphylococcus aureus (CA-MRSA) in skin abscesses presenting to the pediatric emergency department. *N C Med J.*, 2008, 69, 351-354.
[39] Korazono, M; Ida, T; Yamada, K; et al. In vitro activities of ME1036 (CP5609), anovel parenteral carbapenem, against methicillin-resistant Staphylococci. *Antimicrob Agents Chemother*, 2005, 48, 2831-2837.
[40] Ueda, Y; Kanazawa, K; Eguchi, K; et al. In vitro and in vivo antibacterial activities of SM-216601, a new broad-spectrun parenteral carbapenem. *Antimicrob Agents Chemother*, 2005, 49, 4185-4196.
[41] Kobayashi, H. National hospital infection surveillance on methicillin-resistant Staphylococcus aureus. *Dermatology*, 2006, 212 Suppl, 1-3.

[42] West, TE; Guerry, C; Hiott, M; et al. Effect of targeted surveillance for control of methicillin-resistant Staphylococcus aureus in a community hospital system. *Infect Control Hosp Epidemiol*, 2006, 27, 233-238.

[43] Hayakawa, T; Hayashidera, T; Katsura, S; et al. Nasal mupirocin treatment of pharynx-colonized methicillin-resistant Staphylococcus aureus: preliminary study with 10 carrier infants. *Pediatr Int.*, 2000, 42, 67-70.

[44] Hurdle, JG; O'Neill, AJ; Mody, L; et al. In vivo transfer high-level mupirocin resistance from Staphylo coccus epidermidis to methicillin-resistant Staphylococcus aureus associated with failure of mupirocin prophylaxis.*J Antimicrob Chemother*, 2005, 11, 1166-1168.

[45] Gemmell, CG; Edwards, DI; Fraise, AP; et al. Guideline for prophylaxis and treatment of methicillin-resistant Staphylococcus aureus (MRSA) infections in the UK. *J Antimicrob Chemother*, 2006, 57, 489-608.

[46] Thomas, S; Cantrill, S; Waghorn, DJ; et al. The role screening and antibiotic prophylaxis in the prevention of percutaneous gastrostomy site infection caused by methicillin-resistant Staphylococcus aureus. *Aliment Pharmacol Ther.*, 2007, 25, 593-597.

[47] Calleja Kempin, JI; Banares, R; Polo, JR; et al. Effect of antibiotic prophylaxis with vancomycin on methicillin-resistant Staphylococcus aureus infection following liver transplantation. *Rev Esp Enferm.*, 1993, 84, 22-25.

In: Methicillin-Resistant Staphylococcus Aureus...
Editor: Charles L. Kolendi
ISBN: 978-1-60741-398-1

Chapter 13

POPULATIONS AT-RISK FOR METHICILLIN-RESISTANT *STAPHYLOCOCCUS AUREUS* INFECTION

***Tze Shien Lo*[1]*, Augusto M. Alonto*[2] *and Stephanie M. Borchardt*[1*]**

[1]Fargo Veterans Administration Medical Center, Fargo, North Dakota
[2]MeritCare Medical Center, Fargo, North Dakota

Methicillin-resistant *Staphylococcus aureus* (MRSA) was identified as a nosocomial pathogen in the 1960s and later emerged as a cause of community-associated infections in the 1990s. Established risk factors for hospital-associated MRSA infection include older age, recent hospitalization or surgery, past antimicrobial therapy, residence in a long term care facility, dialysis, and indwelling percutaneous devices and catheters. Whereas, risk factors for infection with community-associated MRSA include age less than 2 years and greater than 65 years, participation in contact sports (e.g. rugby, football, and wrestling), injection drug use, residence in a correctional facility, and concurrent influenza-like illness. Identified at-risk populations for community-associated MRSA infection also include racial minorities (i.e., Pacific Islanders, Native Americans/Alaska Natives, and Pacific and Canadian aboriginal), homeless persons, and men who have sex with men. The dichotomy between these two risk factor profiles warrants further study in order to increase our understanding of populations at risk for MRSA infection.

Methicillin-resistant *Staphylococcus aureus* (MRSA) was identified as a nosocomial pathogen in the 1960s. [1] MRSA acquisition has traditionally been linked to health care settings and specific patient populations. Established risk factors for MRSA infection include older age; recent hospitalization or surgery; past antimicrobial therapy; residence in a long term care facility; dialysis; and indwelling percutaneous devices and catheters. [2, 3]

In the 1980s, however, MRSA infections were reported among persons in the community without traditional MRSA risk factors. [4] Because these infections appeared to have been acquired in the community, they have been termed community-associated (CA) MRSA infections. More widespread identification of CA-MRSA in the United States occurred during

* Corresponding Author: Stephanie Borchardt, Fargo VA Medical Center, 2101 Elm Street N, Fargo ND 58102, Phone: 701-239-3700 Ext. 9-3348, FAX: 701-237-2652, E-mail: Stephanie.Borchardt@va.gov

the 1990s following the deaths of four children due to severe MRSA infections in MN and ND. [5] Risk factors for infection with CA-MRSA include age less than 2 years and greater than 65 years; [6, 7] participation in contact sports (e.g. rugby, football, and wrestling); [8, 9] injection drug use [9]; residence in a correctional facility [8, 9]; and concurrent influenza-like illness. [10] Identified at-risk populations for community-associated MRSA infection also include racial minorities such as Pacific Islanders, Native Americans/Alaska Natives, and Pacific and Canadian aboriginals, [11] homeless persons [12], and men who have sex with men. [8, 9]

The changing epidemiology of MRSA infections from hospital-associated (HA) to those among persons with no healthcare contact has expanded the at-risk population and has; therefore, become increasingly important for primary care providers to be aware of the risk factors associated with each type of infection. Failure to recognize risk factors for MRSA and to use them as a diagnostic tool may lead to a delay in appropriate treatment and increased morbidity and mortality. [13]

Hospital-Associated MRSA (HA-MRSA)

In hospitals in the United States, MRSA is the most common nosocomial pathogen causing serious infections such as bacteremia, surgical site infection, prosthetic device-related infection and pneumonia. [14, 15] Kuehnert et al. reported the overall rate of MRSA among all *S. aureus* infections in U.S. hospitals to be 43.2%. [16]

HA-MRSA cases are defined as patients with MRSA infection identified after 48 hours of hospital admission; have a history of hospitalization, surgery, dialysis, or residence in a long-term care facility within the previous year; or a permanent indwelling catheter or percutaneous medical device present at the time of diagnosis. Those patients without any of these risk factors are defined as CA-MRSA. [17] The following are risk factors for acquiring or progressing to HA-MRSA infection.

Older Age

An elderly person is defined as a person $\geq$ 65 years of age. Elderly persons are more susceptible to infection because of decreased immune function, poor nutrition, underlying chronic conditions, such as diabetes mellitus, chronic kidney disease, or chronic obstructive pulmonary disease. [18, 19] They also accounted for 47% of inpatient days-of-care in nonfederal acute hospitals. [20] Therefore, elderly patients, because of their vulnerability and longer hospital stays, have a higher risk of acquiring various nosocomial infections including infections caused by HA-MRSA. [7, 21] A recent study conducted by Kuehnert et al. found that the rates of MRSA infection, including septicemia and pneumonia, increased with age ($p<0.05$). [16] The German Nosocomial Infection Surveillance System also demonstrated that older age contributed to mortality in patients with MRSA pneumonia. [22]

Colonization with MRSA

The anterior nares is the major reservoir for MRSA, however, these bacteria can also be found colonizing the axillae, perineum and open wounds, but to a lesser extent. [2, 23] Numerous studies have demonstrated that colonization with MRSA is an important risk factor for infection; such as skin and soft tissue infection, surgical site infection, device-related infection, bacteremia, and endocarditis. Patients on hemodialysis or peritoneal dialysis were found to have a higher rate of MRSA infection if they were MRSA carriers prior to their infection when compared with non-carriers. [24-30]

Hospitalization

Present or recent hospitalization is also a risk factor for acquiring MRSA infection. [31] However, patients who are admitted to the intensive care unit or burn unit are at an even higher risk of MRSA colonization and infection compared with patients admitted to other wards. [32-34]

Noskin et al. found that >55% of nosocomial infections identified among patients in the ICU were due to MRSA. [35] In United States ICUs, Klevens et al. found that MRSA increased by approximately 3% each year from 1992 to 2003. He also demonstrated that, in 2003, MRSA accounted for >60% of all *S. aureus* infections in most ICUs. [7, 21]

Long Term Care Facilities

Numerous studies have demonstrated that persons who reside in a long term care facility or nursing home are at an increased risk of being colonized or infected with MRSA. [36-39]

Antimicrobial Exposure

Prior antimicrobial exposure has been implicated as a risk factor for acquisition of MRSA infection. Several studies have demonstrated that use of fluoroquinolones promoted MRSA infection. [40-42] Muller et al. discovered that use of both a fluoroquinolone and a penicillin increased the risk of MRSA isolation from a clinical specimen among hospitalized patients. [43]

Post-Surgical Procedure

Surgical site infections (SSIs) are serious and common complications among hospitalized patients. Patients with SSIs are at an increased risk of developing MRSA infection. [44] Data collected from the National Nosocomial Infection Surveillance system demonstrated that *S. aureus* accounted for almost 20% of surgical site infections. [15] Studies on cardiothoracic surgery and gastroduodenal procedures revealed that MRSA accounted for a high percentage of postsurgical infections. [45, 46]

Prosthetic Devices

Chu et al. found that patients with prosthetic devices, including orthopedic, cardiovascular and intravascular devices, were at an increased risk of acquiring *S. aureus* bacteremia as a complication of device-related infection. MRSA accounted for 35% of staphylococcal bacteremia. [47]

Diabetes Mellitus

Because of abnormalities in cell-mediated immunity, phagocyte dysfunction secondary to hyperglycemia, and diminished vascularization; diabetic patients are at an increased risk for bacterial and fungal infections. [48] Although, diabetic patients are at risk of developing MRSA bacteremia, endocarditis and pneumonia; [49-51] the most fearful complications of diabetes mellitus are wound infections, especially diabetic foot ulcers, because of their high incidence and poor outcomes. Diabetic foot ulcers or infections are the most common reason for hospitalization among diabetic patients in the United States. [52] Tentolouris et al. demonstrated that *S. aureus* was the most common pathogen found in both infected and non-infected diabetic foot ulcers and 50% of the pathogens were MRSA. [53] Another study conducted by Roghmann et al. showed that clinical outcomes are worse and hospital length of stay is longer among diabetic patients with chronic ulcers caused by MRSA. [54]

Community-Associated MRSA (CA-MRSA)

Skin and soft tissue infections such as cellulitis, abscesses, folliculitis, and impetigo are the most commonly reported infections caused by CA-MRSA. However, other types of infections have been reported, including bacteremia, septic arthritis, osteomyelitis, pneumonia, urinary tract infections, sinusitis, and meningitis. [6]

Outbreaks

One of the first CA-MRSA outbreaks was reported in the early 1980s among intravenous drug users with skin and soft tissue infections. [4] Since then, outbreaks have been reported in other populations, including military personnel, prison inmates, men who have sex with men, child care centers, tattoo recipients, the homeless, and athletes. Through these observations, it is believed that MRSA can be transmitted through a variety of mechanisms.

Age

CA-MRSA infections have disproportionately affected children <2 years of age and adults >65 years of age. [6, 7] Older age is likely a risk factor for CA-MRSA for similar reasons it is a risk factor for HA-MRSA.

Contact Sports

Skin trauma from contact sports, intravenous drug use, or other injuries may transmit MRSA from one person to another. Sharing sports equipment, towels, and other personal items, such as razors for shaving, are other possible mechanisms of transmission. Interestingly, among athletes participating in contact sports, such as football, playing certain positions present more of a risk than others. An MRSA outbreak was reported among the St. Louis Rams professional football team in 2003. Only linemen and linebackers were affected, suggesting that players with frequent contact with other players were at a greater risk of infection. In addition, the MRSA skin infections occurred at sites that were not covered by uniforms, suggesting that skin-to-skin contact was the predominant form of transmission. [55, 56]

Household

Physical contact between family members or in child care facilities is another way that MRSA can potentially be spread. Family pets, such as cats and dogs, have also been reported to be carriers of MRSA and may be responsible for spreading infections to family members. [55-64]

In a large population-based surveillance study of MRSA infections in three United States cities between 2001 and 2002, characteristics that were potentially related to MRSA infection included any visit to a physician's office in the past year prior to infection, receipt of any antimicrobial agent in the year prior to infection, chronic non-infectious skin disease, and a crowded household; defined as a home with greater than 1 person per bedroom. [6]

Racial Minorities

In addition, there have been several reports of increased incidence of CA-MRSA infection among specific racial populations such as Pacific Islanders, Native Americans, and Canadian aboriginals. [11, 65] Recent reports have also shown that CA-MRSA infection is more prevalent among blacks compared to whites. [6, 7] MRSA prevalence has been associated with socioeconomic status, which might confound the association between race and incidence of CA-MRSA. [7]

Concurrent Influenza-Like Illness

Hageman et al. revealed, through analysis of data from influenza outbreaks which occurred during 2003-2004, that 17 probable influenza cases (12 cases had laboratory evidence of influenza) had concurrent *S. aureus*-associated community-acquired pneumonia (CAP). Eighty-eight percent of these CAP infections were due to MRSA. The authors concluded that, in communities with a high prevalence of MRSA, coverage for MRSA should be included in empiric therapy for severe CAP during periods of high influenza activity. [10]

It is concerning that there are patients with reported CA-MRSA infections without isk factors. This is demonstrated in a series from Los Angeles in which four of 14 patients who were found to have CA-MRSA necrotizing fasciitis did not have a serious coexisting condition or risk factor. [66]

Between 1996 and 2002, the Community Health Network of San Francisco reported a relatively stable number of nosocomial onset MRSA infection. In contrast, there was a significant increase in the number of community onset MRSA infections, such as in the pediatric clinics, primary care clinics, HIV ward, and urgent care clinics. These findings suggest that community based strains of MRSA are migrating into the hospital setting. [67] Indeed, there have been many reports of outbreaks involving CA-MRSA infections in the hospital, including skin and soft tissue infections and sepsis in otherwise healthy newborns, mastitis and breast abscesses in post-partum women, health-care associated blood stream infections, and nosocomial infections in orthopedic patients. [68-72] The proportion of CA-MRSA acquired in the hospital setting has increased from 17% to 56% between 1999 and 2004, in one recent review of 352 patients. [73] Another report has shown a significant increase (24% to 49%) in CA-MRSA strains responsible for MRSA bloodstream infections reported in hospitals between 2000 and 2006. [74]

At the same time, HA-MRSA is being seen more commonly in the community. For example, in one study of patients with MRSA infections acquired in the community, 29% were found to have HA-MRSA strains, as found by molecular typing. [75] A recent report by the United States Active Bacterial Core Surveillance Network has extrapolated that 58% of the 94,360 MRSA infections in the United States in 2005 are HA-MRSA infections acquired in the community. [7]

These observations imply that the distinction between hospital and community infections is blurring in regard to MRSA. We are seeing HA-MRSA infections in the community and CA-MRSA infections in the hospital. Because HA-MRSA and CA-MRSA infections are now difficult to distinguish from each other, it is important to obtain susceptibility testing in order to guide antibiotic therapy.

References

[1] Barrett, FF; McGehee, RF; Jr., Finland M. Methicillin-resistant Staphylococcus aureus at Boston City Hospital. Bacteriologic and epidemiologic observations. *N Engl J Med.,* 1968, 279(9), 441-8.

[2] Lowy, FD. Staphylococcus aureus infections. *N Engl J Med.,* 1998, 339(8), 520-32.

[3] Brumfitt, W; Hamilton-Miller, J. Methicillin-resistant Staphylococcus aureus. *N Engl J Med.*, 1989, 320(18), 1188-96.

[4] Saravolatz, LD; Markowitz, N; Arking, L; Pohlod, D; Fisher, E. Methicillin-resistant Staphylococcus aureus. Epidemiologic observations during a community-acquired outbreak. *Ann Intern Med.*, 1982, 96(1), 11-6.

[5] From the Centers for Disease Control and Prevention. Four pediatric deaths from community-acquired methicillin-resistant Staphylococcus aureus—Minnesota and North Dakota, 1997-1999. *JAMA*, 1999, 282(12), 1123-5.

[6] Fridkin, SK; Hageman, JC; Morrison, M; et al. Methicillin-resistant Staphylococcus aureus disease in three communities. *N Engl J Med.*, 2005, 352(14), 1436-44.

[7] Klevens, RM; Morrison, MA; Nadle, J; et al. Invasive methicillin-resistant Staphylococcus aureus infections in the United States. *JAMA*, 2007, 298(15), 1763-71.

[8] King, MD; Humphrey, BJ; Wang, YF; et al. Emergence of community-acquired methicillin-resistant Staphylococcus aureus USA 300 clone as the predominant cause of skin and soft-tissue infections. *Ann Intern Med.*, 2006, 144(5), 309-17.

[9] Moran, GJ; Krishnadasan, A; Gorwitz, RJ; et al. Methicillin-resistant S. aureus infections among patients in the emergency department. *N Engl J Med.*, 2006, 355(7), 666-74.

[10] Hageman, JC; Uyeki, TM; Francis, JS; et al. Severe community-acquired pneumonia due to Staphylococcus aureus, 2003-04 influenza season. *Emerg Infect Dis.*, 2006, 12(6), 894-9.

[11] Centers for Disease Control and Prevention. Community-associated methicillin-resistant Staphylococcus aureus infections in Pacific Islanders—Hawaii, 2001-2003. *MMWR Morb Mortal Wkly Rep.*, 2004, 53(33), 767-70.

[12] Young, DM; Harris, HW; Charlebois, ED; et al. An epidemic of methicillin-resistant Staphylococcus aureus soft tissue infections among medically underserved patients. *Arch Surg.*, 2004, 139(9), 947-51; discussion 951-3.

[13] Corriere, MD; Decker, CF. MRSA: an evolving pathogen. *Dis Mon.*, 2008, 54(12), 751-5.

[14] Diekema, DJ; BootsMiller, BJ; Vaughn, TE; et al. Antimicrobial resistance trends and outbreak frequency in United States hospitals. *Clin Infect Dis.*, 2004, 38(1), 78-85.

[15] Emori, TG; Gaynes, RP. An overview of nosocomial infections, including the role of the microbiology laboratory. *Clin Microbiol Rev.*, 1993, 6(4), 428-42.

[16] Kuehnert, MJ; Hill, HA; Kupronis, BA; et al. Methicillin-resistant-Staphylococcus aureus hospitalizations, United States. *Emerg Infect Dis.*, 2005, 11(6), 868-72.

[17] Garner, JS; Jarvis, WR; Emori, TG; Horan, TC; Hughes JM. CDC definitions for nosocomial infections, 1988. *Am J Infect Control*, 1988, 16(3), 128-40.

[18] Saltzman, RL; Peterson, PK. Immunodeficiency of the elderly. *Rev Infect Dis.*, 1987, 9(6), 1127-39.

[19] Norman, DC; Toledo, SD. Infections in elderly persons. An altered clinical presentation. *Clin Geriatr Med.*, 1992, 8(4), 713-9.

[20] Graves, EJ; Gillum, BS. National hospital discharge survey: annual summary, 1994. *Vital Health Stat 13.*, 1997(128), i-v; 128.

[21] Klevens, RM; Edwards, JR; Tenover, FC; et al. Changes in the epidemiology of methicillin-resistant Staphylococcus aureus in intensive care units in US hospitals, 1992-2003. *Clin Infect Dis.*, 2006, 42(3), 389-91.

[22] Gastmeier, P; Sohr, D; Geffers, C; et al. Mortality risk factors with nosocomial Staphylococcus aureus infections in intensive care units: results from the German Nosocomial Infection Surveillance System (KISS). *Infection*, 2005, 33(2), 50-5.

[23] John, JF; Berg, NL. *Staphylococcus aureus* In: Hospital Epidemiology and Infection Control. Mayhall CG, editor. 2nd ed. Philadelphia, PA: Lipincott, Williams & Wilkins; 1999 p. 325-345.

[24] Kune, GA; Moritz, V; Carson, P; et al. Postoperative wound infections: a study of bacteriology and pathogenesis. *Aust N Z J Surg.*, 1983, 53(3), 245-8.

[25] Ruef, C; Fanconi, S; Nadal, D. Sternal wound infection after heart operations in pediatric patients associated with nasal carriage of Staphylococcus aureus. *J Thorac Cardiovasc Surg.*, 1996, 112(3), 681-6.

[26] Sheagren, JN. Staphylococcal infections of the skin and skin structures. *CUTIS*, 1985, 36(5A), 2-6.

[27] Yu, VL; Goetz, A; Wagener, M; et al. Staphylococcus aureus nasal carriage and infection in patients on hemodialysis. Efficacy of antibiotic prophylaxis. *N Engl J Med.,* 1986, 315(2), 91-6.

[28] Luzar, MA; Coles, GA; Faller, B; et al. Staphylococcus aureus nasal carriage and infection in patients on continuous ambulatory peritoneal dialysis. *N Engl J Med.*, 1990, 322(8), 505-9.

[29] Lye, WC; Leong, SO; Lee, EJ. Methicillin-resistant Staphylococcus aureus nasal carriage and infections in CAPD. *Kidney Int.*, 1993, 43(6), 1357-62.

[30] Davis, KA; Stewart, JJ; Crouch, HK; Florez, CE; Hospenthal, DR. Methicillin-resistant Staphylococcus aureus (MRSA) nares colonization at hospital admission and its effect on subsequent MRSA infection. *Clin Infect Dis.,* 2004, 39(6), 776-82.

[31] Simor, A; Ofner-Agostini, M; Paton, S. The Canadian Nosocomial Infection Surveillance Program: results of the first 18 months of surveillance for methicillin-resistant Staphylococcus aureus in Canadian hospitals. *Can Commun Dis Rep.*, 1997, 23(6), 41-5.

[32] Rutala, WA; Katz, EB; Sherertz, RJ; Sarubbi, FA; Jr. Environmental study of a methicillin-resistant Staphylococcus aureus epidemic in a burn unit. *J Clin Microbiol,* 1983, 18(3), 683-8.

[33] Cook, N. Methicillin-resistant Staphylococcus aureus versus the burn patient. *Burns*, 1998, 24(2), 91-8.

[34] Farrington, M; Ling, J; Ling, T; French, GL. Outbreaks of infection with methicillin-resistant Staphylococcus aureus on neonatal and burns units of a new hospital. *Epidemiol Infect.*, 1990, 105(2), 215-28.

[35] Noskin, GA; Rubin RJ; Schentag JJ; et al. National trends in Staphylococcus aureus infection rates: impact on economic burden and mortality over a 6-year period (1998-2003). *Clin Infect Dis.*, 2007, 45(9), 1132-40.

[36] Lee, YL; Cesario, T; Gupta, G; et al. Surveillance of colonization and infection with Staphylococcus aureus susceptible or resistant to methicillin in a community skilled-nursing facility. *Am J Infect Control.*, 1997, 25(4), 312-21.

[37] Muder, RR; Brennen, C; Wagener, MM; et al. Methicillin-resistant staphylococcal colonization and infection in a long-term care facility. *Ann Intern Med.*, 1991, 114(2), 107-12.

[38] Hsu, CC. Serial survey of methicillin-resistant Staphylococcus aureus nasal carriage among residents in a nursing home. *Infect Control Hosp Epidemiol*, 1991, 12(7), 416-21.
[39] Strausbaugh, LJ; Jacobson, C; Sewell, DL; Potter, S; Ward, TT. Methicillin-resistant Staphylococcus aureus in extended-care facilities: experiences in a Veterans' Affairs nursing home and a review of the literature. *Infect Control Hosp Epidemiol*, 1991, 12(1), 36-45.
[40] Weber, SG; Gold, HS; Hooper, DC; Karchmer, AW; Carmeli Y. Fluoroquinolones and the risk for methicillin-resistant Staphylococcus aureus in hospitalized patients. *Emerg Infect Dis.*, 2003, 9(11), 1415-22.
[41] Bosso, JA; Mauldin, PD. Using interrupted time series analysis to assess associations of fluoroquinolone formulary changes with susceptibility of gram-negative pathogens and isolation rates of methicillin-resistant Staphylococcus aureus. *Antimicrob Agents Chemother*, 2006, 50(6), 2106-12.
[42] Graffunder, EM; Venezia, RA. Risk factors associated with nosocomial methicillin-resistant Staphylococcus aureus (MRSA) infection including previous use of antimicrobials. *J Antimicrob Chemother*, 2002, 49(6), 999-1005.
[43] Muller, A; Mauny, F; Talon, D; et al. Effect of individual- and group-level antibiotic exposure on MRSA isolation: a multilevel analysis. *J Antimicrob Chemother*, 2006, 58(4), 878-81.
[44] Anderson, DJ; Kaye, KS. Staphylococcal surgical site infections. *Infect Dis Clin North Am.*, 2009, 23(1), 53-72.
[45] Wells, FC; Newsom, SW; Rowlands, C. Wound infection in cardiothoracic surgery. *Lancet*, 1983, 1(8335), 1209-10.
[46] LoCicero, J; 3rd, Nichols, RL. Sepsis after gastroduodenal operations: relationship to gastric acid, motility, and endogenous microflora. *South Med J.*, 1980, 73(7), 878-80.
[47] Chu, VH; Crosslin, DR; Friedman, JY; et al. Staphylococcus aureus bacteremia in patients with prosthetic devices: costs and outcomes. *Am J Med.*, 2005, 118(12), 1416.
[48] Powers, AC. Diabetes mellitus. In: Harrison's Principles of Internal Medicine. Fauci AS; Braunwald E; Kasper DL; editors. 17th ed. New York, NY: McGraw-Hill; 2008 p. 2275-2304.
[49] Khatib, R; Johnson, LB; Sharma, M; et al. Persistent Staphylococcus aureus bacteremia: incidence and outcome trends over time. *Scand J Infect Dis.*, 2009, 41(1), 4-9.
[50] Rogers, BA; Drake, AK; Spelman, D. Methicillin Resistant Staphylococcus aureus Endocarditis in an Australian Tertiary Hospital: 1991-2006. *Heart Lung Circ*, 2008 [E-pub ahead of print].
[51] Rello, J; Torres, A; Ricart, M; et al. Ventilator-associated pneumonia by Staphylococcus aureus. Comparison of methicillin-resistant and methicillin-sensitive episodes. *Am J Respir Crit Care Med.*, 1994, 150(6 Pt 1), 1545-9.
[52] Singh, N; Armstrong, DG; Lipsky, BA. Preventing foot ulcers in patients with diabetes. *JAMA*, 2005, 293(2), 217-28.
[53] Tentolouris, N; Petrikkos, G; Vallianou, N; et al. Prevalence of methicillin-resistant Staphylococcus aureus in infected and uninfected diabetic foot ulcers. *Clin Microbiol Infect.*, 2006, 12(2), 186-9.

[54] Roghmann, MC; Siddiqui, A; Plaisance, K; Standiford, H. MRSA colonization and the risk of MRSA bacteraemia in hospitalized patients with chronic ulcers. *J Hosp Infect.,* 2001, 47(2), 98-103.

[55] Kazakova, SV; Hageman, JC; Matava, M; et al. A clone of methicillin-resistant Staphylococcus aureus among professional football players. *N Engl J Med.,* 2005, 352(5), 468-75.

[56] Nguyen, DM; Mascola, L; Brancoft, E. Recurring methicillin-resistant Staphylococcus aureus infections in a football team. *Emerg Infect Dis.*, 2005, 11(4), 526-32.

[57] Diep, BA; Chambers, HF; Graber, CJ; et al. Emergence of multidrug-resistant, community-associated, methicillin-resistant Staphylococcus aureus clone USA300 in men who have sex with men. *Ann Intern Med.*, 2008, 148(4), 249-57.

[58] Adcock, PM; Pastor, P; Medley, F; Patterson JE; Murphy TV. Methicillin-resistant Staphylococcus aureus in two child care centers. *J Infect Dis.*, 1998, 178(2), 577-80.

[59] Campbell, KM; Vaughn, AF; Russell, KL; et al. Risk factors for community-associated methicillin-resistant Staphylococcus aureus infections in an outbreak of disease among military trainees in San Diego, California, in 2002. *J Clin Microbiol,* 2004, 42(9), 4050-3.

[60] Centers for Disease Control and Prevention. Methicillin-resistant Staphylococcus aureus skin or soft tissue infections in a state prison—Mississippi, 2000. *MMWR Morb Mortal Wkly Rep.*, 2001, 50(42), 919-22.

[61] Gilbert, M; MacDonald, J; Gregson, D; et al. Outbreak in Alberta of community-acquired (USA300) methicillin-resistant Staphylococcus aureus in people with a history of drug use, homelessness or incarceration. *CMAJ*, 2006, 175(2), 149-54.

[62] Centers for Disease Control and Prevention. Methicillin-resistant Staphylococcus aureus skin infections among tattoo recipients—Ohio, Kentucky, and Vermont, 2004-2005. *MMWR Morb Mortal Wkly Rep.*, 2006, 55(24), 677-9.

[63] Baptiste, KE; Williams, K; Willams, NJ; et al. Methicillin-resistant staphylococci in companion animals. *Emerg Infect Dis.*, 2005, 11(12), 1942-4.

[64] Sing, A; Tuschak, C; Hormansdorfer S. Methicillin-resistant Staphylococcus aureus in a family and its pet cat. *N Engl J Med.*, 2008, 358(11), 1200-1.

[65] Groom, AV; Wolsey, DH; Naimi, TS; et al. Community-acquired methicillin-resistant Staphylococcus aureus in a rural American Indian community. *JAMA*, 2001, 286(10), 1201-5.

[66] Miller, LG; Perdreau-Remington, F; Rieg, G; et al. Necrotizing fasciitis caused by community-associated methicillin-resistant Staphylococcus aureus in Los Angeles. *N Engl J Med.*, 2005, 352(14), 1445-53.

[67] Carleton, HA; Diep, BA; Charlebois, ED; Sensabaugh, GF; Perdreau-Remington, F. Community-adapted methicillin-resistant Staphylococcus aureus (MRSA): population dynamics of an expanding community reservoir of MRSA. *J Infect Dis.*, 2004, 190(10), 1730-8.

[68] Kourbatova, EV; Halvosa, JS; King, MD; et al. Emergence of community-associated methicillin-resistant Staphylococcus aureus USA 300 clone as a cause of health care-associated infections among patients with prosthetic joint infections. *Am J Infect Control*, 2005, 33(7), 385-91.

[69] Seybold, U; Kourbatova, EV; Johnson, JG; et al. Emergence of community-associated methicillin-resistant Staphylococcus aureus USA300 genotype as a major cause of health care-associated blood stream infections. *Clin Infect Dis.*, 2006, 42(5), 647-56.

[70] Saiman, L; O'Keefe, M; Graham, PL; 3rd; et al. Hospital transmission of community-acquired methicillin-resistant Staphylococcus aureus among postpartum women. *Clin Infect Dis.*, 2003, 37(10), 1313-9.

[71] Reddy, P; Qi, C; Zembower, T; Noskin, GA; Bolon, M. Postpartum mastitis and community-acquired methicillin-resistant Staphylococcus aureus. *Emerg Infect Dis.*, 2007, 13(2), 298-301.

[72] Healy, CM; Hulten, KG; Palazzi, DL; Campbell, JR; Baker, CJ. Emergence of new strains of methicillin-resistant Staphylococcus aureus in a neonatal intensive care unit. *Clin Infect Dis.*, 2004, 39(10), 1460-6.

[73] Maree, CL; Daum, RS; Boyle-Vavra, S; Matayoshi, K; Miller, LG. Community-associated methicillin-resistant Staphylococcus aureus isolates causing healthcare-associated infections. *Emerg Infect Dis.*, 2007, 13(2), 236-42.

[74] Popovich, KJ; Weinstein, RA; Hota, B. Are community-associated methicillin-resistant Staphylococcus aureus (MRSA) strains replacing traditional nosocomial MRSA strains? *Clin Infect Dis.*, 2008, 46(6), 787-94.

[75] Davis, SL; Perri, MB; Donabedian, SM; et al. Epidemiology and outcomes of community-associated methicillin-resistant Staphylococcus aureus infection. *J Clin Microbiol*, 2007, 45(6), 1705-11.

In: Methicillin-Resistant Staphylococcus Aureus...
Editor: Charles L. Kolendi ISBN: 978-1-60741-398-1

Chapter 14

Staphylococcus Aureus: Risk Factors for Oronasal Transmission and Dental Aspect

E. Bahar Tuna, Nursen Topcuoglu, Koray Gencay and Guven Kulekci
Istanbul University Faculty of Dentistry, Istanbul, Turkey

Abstract

Staphylococcus aureus is a major cause of surgical wounds and infections associated with indwelling medical devices; it is found particularly on the nasal membranes and skin. Naso-oral transmission of *S. aureus* is an important risk factor for development of oral and dental infections. Although prophylactic and postoperative antibiotics are given to patients after oral surgery operations, most strains of *S. aureus* are reported to be resistant to penicillin. After the 1960s, methicillin was replaced by newer penicillin-type antibiotic that were not affected by ß-lactamase, and the resistant *S. aureus* strains became known as *methicillin resistant S. aureus* (MRSA).

There were insufficient data available to determine whether the *S. aureus* isolates were colonizing or infecting the oral cavity and regarding *S. aureus* colonization of the nasal floor in patients with oronasal fistula. Recently our research results showed that bacterial transmission was proven for large oronasal fistulas and a correlation was found with *S. aureus* counts in the children with cleft lip and palate. There were no MRSA strains among our patient population. It appears that as fistula size increases, significantly higher colony numbers of *S. aureus* are found in saliva samples. In addition, *S. aureus* tends to survive in the oral cavity as a result of transmission through the nasal passages as long as an unrepaired cleft exists. Colonization of this site by nasal bacteria seems to occur when the fistulas are sufficiently large to maintain a similar bacterial environment. Therefore, we reported that the high numbers of salivary *S. aureus* count are attributable to poor oral hygiene and dietary factors thereby favoring the outgrowth of *S. aureus,* which can cause wound breakdown. We also recommended that coordination of care and an understanding of special care issues in oral health and maintaining infection control are essential for the health care team for a patient after orodental operations.

Introduction

Staphylococcus aureus is frequently isolated from healthy individuals and causes a variety of diseases in humans. The external nares are almost certainly the main reservoir of *S. aureus* [Kluytmans *et al.*, 1997; Graham *et al.*, 2006]. Although the oral cavity harbors a complex microflora consisting of mostly non-pathogenic microorganisms, it is not usually isolated from the oral cavity. It is considered as a part of the transient microbiota when it is isolated from the oral cavity. However, a change in the oral microbiota may develop for several reasons. In immuncompramised individuals and patients with periodontitis, it may occur in high numbers and can be an infection source to other body sites and individuals [Dahlén and Wikström, 1995; Smith *et al.*, 2001].

Methicillin-resistant *S.aureus* (MRSA) which are generally resistant to multiple antibiotics have emerged among persons in hospitals, nursing homes and other healthcare institutions since the 1960s and are called hospital- or health care- associated MRSA (HA-MRSA) [Barrett *et al.*, 1968]. They account for a large part of nosocomial infections worldwide and are associated with longer hospitalization and higher lethality. Community-associated MRSA (CA-MRSA), which was first reported in the 1980s [Weber, 2005], differs from HA-MRSA.

The anterior nares are known to be the primary colonization site of *S. aureus*. Colonization sites other than the anterior nares, e.g oral colonization, may play an important role in the development of CA- MRSA infection. MRSA are most frequently spread via transiently contaminated hands of health care professions [Kurita *et al.*, 2006]. In dentistry, standard precautions are considered effective to prevent MRSA transmission to the dental team and from patient to patient. In particular, strict adherence to hand disinfection before and after every patient contact and wearing personal protection equipment are recommended [Siegel *et al.*, 2006; Zimmerli *et al.*,2008].

S. aureus is known as one of the major serious hospital-acquired pathogen in many countries and the prevalence of MRSA has increased dramatically worldwide. The prevalence of MRSA among healthy children varies very widely; 0.61% [Hussain *et al.*, 2001] or 11% [Adcock *et al.*, 1998] in the United States, 8.7% in Nigeria [Ako-Nai *et al.*, 1991] and 0.1% in Turkey [Soysal *et al.*, 2006]. The incidence of MRSA among healthy children peaked at 12.6% in 1992-93 and then decreased 4.0% in 2002 throughout a long term periodical examination in Japan [Suzuki *et al.*, 2007].

S. aureus has been well-documented to inhabit the nares, skin wounds, and respiratory tract, but little is known about its presence in the oral cavity. Köndell *et al.* [1984] examined 110 patients who visited a dental hospital, and the incidence of *S. aureus* was 38%, 21% and 11% in the nose, saliva and gingiva, respectively. Ohara-Nemoto *et al.* [2008] reported the occurrence of *S. aureus* was 46.4% in saliva and 44.4% in nasal swab in healthy adults. Smith *et al.* [2003] isolated *S. aureus* from 1,017 specimens among 5,005 specimens submitted to a regional diagnostic oral laboratory, which the MRSA rate was 5% in a three- year retrospective analysis. They reported that, *S. aureus incidence* was increasing with age and the infection was commonly associated with angular cheilitis.

There occurrence of *S. aureus* in subgingival samples also reported in few studies. Rams *et al.* [1990] observed in 22.3 % of the patients with chronic periodontitis while Dahlén and Winkström [1995] reported the rate as 8.2%. Loberto *et al.* [2004], isolated *S. aureus* in 4.5%

of the patients with chronic periodontitis, while the rate was 22.3% in the oral cavity. Zirnsli Fritschi *et al.* [2008] found *S. aureus* at a high prevalence in non-smoking individuals with aggressive periodontitis (60.5%). This was an agreement with results from studies of nosocomial infections showing higher levels of *S. aureus* in nasal colonization of non-smokers [Fang *et al.*, 2004].

The incidence of this organism in the mouths of children is of interest. Millar *et al.* [2001] isolated *S. aureus* from 37% (MRSA rate was 2%) of the mouthwashes of 539 healthy 7-8 year-old children. Miyake *et al.* [1991] reported the *S. aureus* incidence on the tongues was 32.6% (MRSA rate was 2%) in 307 healthy children with 0-5 years of age, where Suzuki *et al.* [2007] maintained the incidence of *S. aureus* between 33 and 44% throughout an examination period between 1987 and 2003. Mc Carthy *et al.* [1965] reported that the incidence of *S. aureus* in the mouth region was 45% in newborn babies and become 100% by the age of eight months.

Clinical experience has shown that oropharyngeal carriage of MRSA can be difficult to eradicate [Working Party Report, 1998]. Therefore, it is important that consideration to be given to the oral cavity if eradication of colonization by MRSA in children is clinically appropriate.

Staphylococci are common residents of the nasal flora [Wenzel and Perl,1995; Kluytmans *et al.*, 1997; Güçlü *et al.*, 2007], and the main reservoir for *S. aureus* is the anterior nares [Edmond *et al.*, 1996] thus they may consistently pass into the mouth. Although these micro-organisms are considered to be transiently resident in the oral cavity [Dahlén and Wikström, 1995; Percival et al., 1991; Smith et al.,2001], these may be continuously provided from the nasal cavity. Transmission of *S. aureus* usually occurs through close contact with a person who has either a draining lesion or asymptomatic carriage of *S. aureus*. Kawada *et al.* [2003] suggest that methicillin-resistant *S. aureus* or methicillin-sensitive *S. aureus* may be transmitted between healthy, lactating mothers without mastitis and their infants by breastfeeding. Contamination from skin flora is also possible; however, the PFGE genotyping provided evidence of nasal-oral trafficking of the *Staphylococcus* species [Ohara-Nemoto *et al.*, 2008].

The most probable transmission route seems to be direct contact with hands contaminated with nasal microflora. Apart from that route, nasal bacteria may be transmitted through an oronasal fistula to the oral cavity, and it may be able to survive in the oral environment in children with cleft lip and palate (CLP) [Tuna *et al.*, 2008].

Cleft lip and palate (CLP) affecting both the nasopharyngeal space and the oral cavity could predispose an alteration of normal flora at both sites [Pearce *et al.*, 1995; Arief *et al.*, 2005]. Primary closure of a CLP should result in an intact palate with separation of the oral and nasal cavities [Cohen *et al.*, 1991; Karling *et al.*, 1993; Murphy and Willmot, 2005]. As a result of failure in surgical repair and disturbed anatomy, an oronasal fistula, an epithelial-lined communication between the oral and nasal cavities, might occur in patients with CLP [Cohen *et al.*, 1991; Brennan *et al.*, 2001]. Oronasal fistulas are mostly seen after the surgical treatment of CLP, and as the child grows, it could become a larger defect.

In the presence of oronasal fistulae, the nasal passages are open to extensive contamination by aerobic and anaerobic bacteria in the saliva. Many authors described the problems of oronasal transmission related to fistulas. Crawford and Fischer [1971] reported that bacteriologic examination of patients with unrepaired clefts revealed bacteria in the exposed nasal passages that closely resembled the aerobic flora of the oral cavity. Bacterial infections after primary CLP surgery increase the risk of wound breakdown, palatal fistulas, poor growth and speech [Chuo and Timmons, 2005].

Little microbiologic data exist regarding *S. aureus* colonization of the cleft nasal floor in patients with oronasal fistula. Brennan *et al.* [2001] concluded that a lack of colonization of cleft nasal floor might reflect poor transmission of bacteria through the oronasal fistula. Recently, our research results showed that bacterial transmission was proven for large oronasal fistulas and a correlation was found with *S. aureus* counts in the children with cleft lip and palate [Tuna *et al.,* 2008]. It appears that as fistula size increases, significantly higher colony numbers of *S. aureus* are found in saliva samples. In addition, *S. aureus* tends to survive in the oral cavity as a result of transmission through the nasal passages as long as an unrepaired cleft exists.

We found *S. aureus* colonization in 53.1% and 40.6% of the patients with oronasal fistula in the saliva and nasal samples, respectively. Although there were no MRSA strains among this CLP patient population, we conclude that nasal carriage of *S. aureus* is an important risk factor during oral and dental treatments. The children in our study were not using any acrylic oronasal obturator prosthesis, while the prosthetic devices in the oral cavity may encourage colonization of MRSA (Tawara *et al.,*1996; Smith *at al.*, 2001). It is difficult to eradicate the mature MRSA denture biofilms than their planktonic counterparts [Lee *et al.*, 2009; Rossi *et al.*, 1996]. Therefore, a meticulous oral hygiene for patients with CLP and oronasal fistula is essential especially patients with obturator prosthesis. On the other hand, the dentist should evaluate carefully the growth, dental arch development, and overall dental management of patients with CLP.

Conclusion

S. aureus transmission from oronasal fistula in children with CLP is possible and increases with fistula size. Oral carriage of MRSA may serve as a reservoir for re-colonization of other body sites. Therefore, the bacterial transmission through oronasal fistula has to be intercepted by clinical foresight.

References

[1] Adcock, PM; Pastor, P; Medley, F; Patterson, JE; Murphy, TV. *Methicillin-resistant Staphylococcus J Infect Dis*., 1998 Aug, 178(2), 577-80.

[2] Ako-Nai, AK; Torimiro, SE; Lamikanra, A; Ogunniyi, AD. A survey of nasal carriage of *Staphylococcus Ann Trop Paediatr*, 1991, 11(1), 41-5.

[3] Arief, EM; Mohamed, Z; Idris, FM. Study of viridans streptococci and *Staphylococcus* species in cleft lip and palate patients before and after surgery. *Cleft Palate Craniofac J.* 2005 May, 42(3), 277-9.

[4] Barrett, FF, McGehee, RF, Finland. Methicillin-resistant *Staphylococcus aureus* at Boston City Hospital. Bacteriologic and epidemiologic observations. *N Engl J Med.*, 1968, 279(9), 441-8.

[5] Brennan, PA; Markus, AF; Flood, TR; Downie, IP; Uppal, R. Do oral *Cleft Palate Craniofac J*, 2001 Jul, 38(4), 399-400.

[6] Chuo, CB; Timmons, MJ. The bacteriology of children before primary cleft lip and palate surgery. *Cleft Palate Craniofac J.*, 2005 May, 42(3), 272-6.

[7] Cohen, SR; Kalinowski, J; LaRossa, D; Randall, P. Cleft palate *Plast Reconstr Surg.*, 1991 Jun, 87(6), 1041-7.

[8] Crawford, JJ; Fischer, ND. Oral and respiratory flora of individuals with normal and repaired palatal clefts. *Cleft Palate J*, 1971 Apr, 8, 166-76.

[9] Dahlén, G; Wikström, M. Occurence of enteric rods, staphylococci and Candida in subgingival samples. *Oral Microbiol Immunol*, 1995, 10, 42-46.

[10] Edmond, MB; Wenzel, RP; Pasculle, AW. Vancomycin-Resistant *Staphylococcus aureus*: Perspectives on Measures Needed for Control. *Ann Intern Med.*, 1996; 124, 329-334.

[11] Fang, S; Skeete, D; Cullun, JJ. Preoperative risk factors for postoperative *Staphylococcus aureus* nosocomial infections. *Surg Technol Int.*, 2004, 13, 35-8

[12] Güçlü, E; Yavuz, T; Tokmak, A; Behçet, M; Karali, E; Oztürk, O; Egeli, E. Nasal carriage of pathogenic *Eur Arch Otorhinolaryngol*, 2007, 264(1), 85-8.

[13] Graham, PL; Lin, SX; Larson, EL. A U.S. population-based survey of *Staphylococcus aureus* colonization. *Ann Intern Med.*, 2006, 144, 318-25.

[14] Hussain, FM; Boyle-Vavra, S; Daum, RS. Community-acquired methicillin-resistant *Pediatr Infect Dis J.*, 2001, 20(8), 763-7.

[15] Karling, J; Larson, O; Henningsson, G. Oronasal fistulas *Scand J Plast Reconstr Surg Hand Surg.,* 1993 Sep, 27(3), 193-201.

[16] Kawada, M; Okuzumi, K; Hitomi, S; Sugishita, C. Transmission of *Staphylococcus aureus* between healthy, lactating mothers and their ınfants by breastfeeding. *J Hum Lact,* 2003, 19, 411-17.

[17] Kluytmans, J; Van Belkum, A; Verbrugh, H. Nasal carriage of *Staphylococcus Clin Microbiol Rev.*, 1997, 10(3), 505-20. Review.

[18] Köndell, PA; Nord, CE; Nordenram, G. Characterization of *Staphylococcus Int J Oral Surg.*, 1984, 13(5), 416-22.

[19] Kurita, H; Kurashina, K; Honda, T. Nosocomial transmission of methicillin-resistant *Staphylococcus aureus* via the surfaces of the dental operatory. Br Dent J 2006, 201, 297-300.

[20] Lee, D; Howlett, J; Pratten, J; Mordan, N; McDonald, A; Wilson, M; Ready, D. Susceptibility of MRSA *FEMS Microbiol Lett.*, 2009, 291, 241-6.

[21] Loberto, JCS; De Paiva Martins, CA; Dos Santos, SSF; Cortelli, JR; Jorge, AOC. *Staphylococcus* spp. in the oral cavity and periodontal pockets of chronic periodontitis patients. *Brazilian Journal of Microbiology*, 2004, 35, 64-68.

[22] McCarthy, C; Snyder, ML; Parker, RB. The indigenous *Arch Oral Biol.*, 1965, 10, 61-70.

[23] Millar, MR; Walsh, TR; Linton, CJ; Zhang, S; Leeming, JP; Bennett, PM; ALSPAC Study Team. Carriage of antibiotic-resistant bacteria by healthy children. *J Antimicrobial Chemotherapy*, 2001, 47, 605-10.

[24] Miyake, Y; Iwai, T; Sugai, M; Miura, K; Suginaka, H; Nagasaka, N. Incidence and characterization of *Staphylococcus aureus* from the tongues of children. *J Dent Res.*, 1991, 70, 1045-7.

[25] Murphy, TC; Willmot, DR. Image analysis of oronasal fistulas *Plast Reconstr Surg.*, 2005 Jan, 115(1), 31-7.

[26] Ohara-Nemoto, Y; Haraga, H; Kimura, S; Nemoto, TK. Occurrence of staphylococci in the oral cavities of healthy adults and nasal-oral trafficking of the bacteria. *J Med Microbiol*, 2008, 57, 95-99.

[27] Pearce, C; Bowden, GH; Evans, M; Fitzsimmons, SP; Johnson, J; Sheridan, MJ; Wientzen, R; Cole, MF. Identification of pioneer viridans streptococci in the oral cavity of human neonates. *J Med Microbiol*, 1995 Jan, 42(1), 67-72.

[28] Percival, RS; Challacombe, SJ; Marsh, PD. Age-related microbiological changes in the salivary and plaque *J Med Microbiol*, 1991, 35(1), 5-11.

[29] Rams, TE; Feik, D; Slots, J. Staphylococci in human periodontal diseases. *Oral Microbiol Immunol*, 1990, 5, 29-32.

[30] Rossi, T; Peltonen, R; Laine, J; Eerola, E; Vuopio-Varkila, J; Kotilainen, P. Eradication of the long-term carriage of methicillin-resistant *J Hosp Infect*, 1996, 34, 311-20.

[31] Siegel, JD; Rhinehart, E; Jackson, M; Chiarello, L. The Healthcare Infection Control Practices Advisory Committee. Management of multidrug-resistant organisms in healthcare settings, 2006. www.cdc.gov/ncidod/dhqp/pdf/ar/mdroGuideline2006.pdf

[32] Smith, AJ; Jackson, MS; Bagg, J. The ecology of *Staphylococcus* species in the oral cavity. *J Med Microbiol*, 2001, 50, 940-6.

[33] Smith, AJ; Robertson, D; Tang, MK; Jackson, MS; MacKenzie, D; Bagg, J. *Staphylococcus aureus* in the oral cavity: a three- year retrospective analysis of clinical laboratory data. *Br Dent J*, 2003, 195, 701- 3.

[34] Soysal, A; Sahin, H; Yagci, A; Barlan, I; Bakir, M. The low rate of mehicillin resistant *Staphylococcus aureus* in Turkish children. *Jpn J Infect Dis.*, 2006, 59(3), 195-6.

[35] Suzuki, J; Yoshimura, G; Kodomoto, N; Kuramoto, M; Kozai, K. Long-term periodical isolation of *Staphylococcus aureus* and methicillin- resistant *Staphyloccoccus aureus* (MRSA) from Japanese children's oral cavities. *Pediatric Dental Journal*, 2007, 17, 127-130.

[36] Tawara, Y; Honma, K; Naito, Y. Methicillin resistant *Staphylococcus aureus* and *Candida albicans* on denture surfaces. *Bull Tokyo Dent Coll*, 1996, 37, 119-28.

[37] Tuna, BE; Topçuoglu, N; Ilhan, B; Gençay, K; Kulekçi, G. *Staphylococcus Cleft Palate Craniofac J.*, 2008 Sep, 45(5), 477-80.

[38] Weber, JT. Community-associated methicillin-resistant *Clin Infect Dis.*, 2005, 41 Suppl 4, S269-72. Review.

[39] Wenzel, RP; Perl, TM. The significance of nasal carriage of *Staphylococcus J Hosp Infect.*, 1995, 31(1), 13-24. Review.

[40] Working Party Report. Revised guidelines for the control of methicillin-resistant *Staphylococcus aureus* infection in hospital. *J Hosp Infect,* 1998, 39, 253-90.

[41] Zimmerli, M; Widmer, AF; Dangel, M. Methicillin-resistant *Staphylococcus aureus* (MRSA) among dental patients: a problem for infection control in dentistry? *Clin Oral Invest*, 2008 Dec 23 [Epub ahead of print].

[42] Zirnsli Fritschi, B; Albert Kiszely, A; Persson GR. *Staphylococcus aureus* and other bacteria in untreated periodontitis. *J Dent Res.*, 2008, 87, 589-93.

In: Methicillin-Resistant Staphylococcus Aureus...
Editor: Charles L. Kolendi
ISBN: 978-1-60741-398-1

Chapter 15

COMMUNITY-ASSOCIATED METHICILLIN-RESISTANT *STAPHYLOCOCCUS AUREUS*: ETIOLOGY, DIAGNOSIS, PREVENTION AND TREATMENT

H. Richard Weiner
University of Wisconsin-Milwaukee

ABSTRACT

Staphylococcus aureus has developed resistance to many antibiotic agents. Once a source of morbidity and mortality limited to hospitalized patients, drug-resistant *S. aureus* now appears among other populations – often in young people without significant medical conditions. Appropriate control measures, early diagnosis, and proper antibiotic choice can limit its spread and severity.

RESISTANT *STAPHYLOCOCCUS AUREUS*

When benzylpenicillin (Pen G) was introduced in 1940, *S. aureus* was uniformly susceptible; but within a decade, the majority of *S. aureus* isolates demonstrated resistance to penicillin.[1] Almost as soon as semisynthetic agents such as methicillin were introduced in the 1960s, resistant strains of *S. aureus* (methicillin-resistant *S. aureus,* MRSA) were identified. Multiply-resistant forms of the organism appeared in the 1970s, and spread globally in the 1980s. MRSA was found in patients with exposure to the health care system, such as hospitalization or nursing home care, chronic dialysis, previous antibiotic treatment, or exposure to invasive procedures or devices; however, these infections remained confined to hospitals or health-care related settings, and they were always susceptible to vancomycin.

The past decade has brought two disturbing developments. First, vancomycin can no longer be depended on to cure all *Staphylococcus* infections. *S. aureus* with reduced susceptibility to vancomycin (VISA) was first reported in 1997, and resistant forms (VRSA) were identified in 2002. Second, MRSA – previously a problem only for the seriously ill patient – has appeared among athletes and people in large, communal dwellings, such as college students in dormitories. [2, 3, 4, 5, 6, 7]

Definition and Laboratory Characteristics of Community-Associated Methicillin-Resistant *Staphylococcus Aureus*

The Centers for Disease Control and Prevention criteria for CA-MRSA state that it must be diagnosed in the outpatient setting, or by culture within 48 hours of hospital admission in a patient who has not been hospitalized, placed in a nursing home, skilled nursing facility or hospice, undergone surgery in the year prior to infection, or has a permanent indwelling catheter or transcutaneous medical device.

Community-associated *S. aureus* (CA-MRSA) differs genetically from healthcare-acquired *S. aureus* (HA-MRSA). The dominant American clones of CA-MRSA, US 300 and US 400, contain staphylococcal chromosomal cassette (SCC) *mec* IV and V, while HA-MRSA clones US 100 and 200 contain the larger SCC *mec* I, II, and III. Since SCC *mec* IV and V are smaller, they confers less resistance – thus the differences in antibiotic susceptibility. Healthcare-acquired MRSA has resistance to all but a few agents, while CA-MRSA is always resistant to beta-lactams and frequently to erythromycin, but remains susceptible to numerous other antibiotics, including ciprofloxacin, clindamycin, gentamicin, linezolid, mupirocin, rifampin, tetracycline, trimethoprim-sulfamethoxasole, and vancomycin. [8, 9]

CA-MRSA can carry genes for Panton-Valentine Leucocidin (PVL), which produces a necrotizing toxin. Panton-Valentine Leucocidin-producing MRSA can cause necrotizing pneumonia and necrotic skin lesions. [10, 11, 12] The necrotic lesions appear as darkened skin with little inflammation and may be mistaken for spider bites. Few strains of methicillin-susceptible or HA-MRSA carry the PVL gene.

Increased virulence leads to striking clinical presentations, such as larger boils which spread more quickly, skin infections which extend to fascia instead of coalescing into abscesses, bacteremia and pneumonia.

Another worrisome feature of CA-MRSA is inducible resistance. Clindamycin, long a viable treatment option, may show *in vitro* activity against MRSA, but fail clinically. The disk-diffusion test (D-test) detects this phenomenon. Erythromycin and clindamycin discs are placed on an agar disc streaked with the organism. If the zone around the clindamycin disc is blunted (forms a D shape), then the organism has a heightened mutation rate, and could develop resistance during treatment. [17]

Populations at Risk for CA-MRSA

In addition to history of MRSA infection or colonization, high community prevalence, recurrent skin disease, recent or frequent antibiotic use, risk factors for infection with community-acquired methicillin-resistant *S. aureus* are the "Four C's": contact, crowding, contaminated surfaces, and lack of cleanliness. The typical patient might be an athlete, especially a football player, wrestler and fencer. Groups of patients have been found in prisons and daycare centers, settings that bring socially active groups into small living areas.

Men who have sex with men have been noted to be affected by CA-MRSA, although there do not appear any factors which confer extra risk over heterosexual activity. Infection is spread predominantly by person to person contact, although it may also be transmitted by contaminated surfaces and objects.

Clinical Presentation: Signs and Symptoms

Healthcare-associated MRSA (HCA-MRSA) occurs in patients who have recent or chronic exposure to the health care system, and presents as intravascular catheter or hematogenous infections, in surgical sites. Dialysis access is a common setting.

By contrast, the common presentation of CA-MRSA is skin and soft tissue infection. Of those, about half present as abscesses and a third as cellulitis. Other, less frequent dermatologic manifestations include folliculitis, wound infection, impetigo, furuncles and carbuncles. Far less common is invasive disease, which can appear as bloodstream infection, most often without focus. Cases have been reported of postinfluenza necrotizing pneumonia [14], and infections of peritoneal fluid, bone and joints.

Prevention of CA-MRSA

Transmission of CA-MRSA results from contact with contaminated skin or contaminated surfaces, so infection can be controlled by appropriate hygiene. Patients and caregivers should wash their hands with soap and warm water after touching infected areas, soiled bandages and dressings, or infected body fluids. Soap dispensers should be used instead of soap bars, and paper towels instead of cloth towels. Alcohol hand sanitizers are a convenient alternative to soap and water when hands are not visibly soiled. Hand washing before meals and after using bathrooms should be encouraged. Common areas, such as cafeterias, rest rooms, locker rooms and shared work stations, should be cleaned at regular intervals.

Skin lesions should be covered completely, and showers with disinfectant soaps such as hexachlorophene or chlorhexidine should be taken before close physical contact. Activities which may involve chafing, abrasions and lacerations can increase risk of transmission.

Towels and personal items, including razors, towels, clothing and athletic equipment should not be shared. When equipment must be used by several athletes, it should be cleaned thoroughly with hot water and detergent, and dried at 180° F. Health care workers and athletic trainers should consider disposable gloves when performing wound care. Contaminated surfaces and equipment should be cleaned and disinfected with a household disinfectant or bleach solution (1:9 dilution), then allowed to air dry.

Colonization with *S. aureus* usually occurs in the nares. It creates a reservoir and is a risk factor for subsequent infection, both of the patient and contacts. Decolonization of the nares can be achieved with five to ten days of mupirocin applied to the anterior nares two or three times a day. In health care workers, this is effective for periods up to three months; in the community, decolonization may last longer, because of decreased exposure. More aggressive approaches include oral doxycycline with rifampin, trimpethoprim-sulfamethoxazole, ciprofloxacin, clindamycin or linezolid. Daily use of hexachlorophene or chlorhexidine showers assist in decolonization. In the absence of clinical findings, cultures after these measures is not warranted.

Community-associated MRSA is easily transmitted in sports settings, but routine precautions and safe practices can prevent it. Football players and wrestlers, two groups involved in reported outbreaks of CA-MRSA, should be taught to recognize the danger posed by cuts and abrasions, and to report them to coaches and trainers, who should enforce adequate coverage, use of topical antibiotics including mupirocin as indicated, and referral to medical providers when necessary. Physicians should obtain wound cultures more frequently, and prescribe therapy directed against the resistant skin organisms.

If it is not possible to cover skin lesions completely, coaches should consider removing players from practice and competition until the infections can be covered appropriately or have healed. Although this may compromise an individual's performance in a game, it protects the team and competitors from infections that may require intravenous antibiotics and surgical drainage – therapy that may end a season, or a career.

Treatment [15, 16]

Incision and drainage is an essential part of curing CA-MRSA infection. In small lesions, adequate debridement may remove the need for antibiotic therapy.

The emergence of vancomycin-resistant *S. aureus* has prompted the development of new agents for severe inpatient infections. Recognition of mild or moderate MRSA affecting skin and soft tissue in the community has changed prescribing practices. The available antibiotics, their advantages, limitations, potential for complications, and costs, are summarized below in Table 1. [17, 18]

CONCLUSION

In summary, community-associated methicillin-resistant *Staphylococcus aureus* has emerged as a nationwide problem over the past decade. [19, 20] Commonsense approaches to prevention, along with intelligent use of the laboratory and available antimicrobials, can protect patients from this new threat.

Table 1. Agents active against Methicillin-resistant *S. aureus*

Drug	Comments	Cost per day
Vancomycin	Drug of choice for severe MRSA infections. May cause neutropenia, fever, rash and ototoxicity	$18.28
Daptomycin (Cubicin)	Intravenous only. Cannot be used in pneumonia. Associated with myopathy and increased creatinine kinase.	$166.08
Linezolid (Zyvox)	Available in intravenous and oral preparations.	$156.00 (intravenous) $128.68 (oral)

Table 1. (Continued)

Drug	Comments	Cost per day
Quinupristin/dalfopristin (Synercid)	Intravenous only. Severe myalgias, thrombophlebitis, hyperbilirubinemia may occur.	$330.00
Tigecycline (Tygacil)	Intravenous only Minocycline derivative. Nausea and vomiting frequent.	$136.20
Minocycline and doxycycline	Well absorbed, excellent penetration. Nausea, vomiting and vertigo may occur.	$5.08 (minocycline) $2.20 (doxycycline)
Trimethoprim/sulfameth ox-azole (Bactrim, Septra)	Resistance is rare. Should be dosed as two DS tablets twice a day. *Streptococcus* often resistant.	$0.94
Clindamycin	Excellent penetration and bioavailability. Resistance may occur. D-zone test for resistance. *C. difficile* colitis may occur.	$9.51
Mupirocin (Bactroban)	Topical. Bacitracin is almost as effective, and much cheaper.	$30/30 grams

ACKNOWLEDGMENTS

The author thanks Brian P. Buggy, M.D., for his thoughtful insights and suggestions in the preparation of this paper.

This material appeared in somewhat different fashion in *Journal of American College Health.*

BIBLIOGRAPHY

Chambers, HF. The changing epidemiology of *Staphylococcus aureus. Emerg Infect Dis.*, 2001,7, 178-182.

Fridkin, S; Hageman, J; Morrison, M; Sanza, LT; Como-Sabetti, Jernigan, JA; et al. Methicillin-resistant *Staphylococcus aureus* in three communities. *N Eng J Med.*, 2005, 352, 1485-1487.

Herold, BC; Immergluck, LC; Maranan, MC; Lauderdale, DS; Gaskin, RE; Boyle-Vavra, S; et al. Community-associated methicillin-resistant *Staphylococcus aureus* in children with no identified predisposing risk. *JAMA,* 1998, 279, 593-598.

Ellis, MW; Hospenthal, DR; Dooley, DP; Gray, PJ; Murray, CK. Natural history of community-associated methicillin-resistant *Staphylococcus aureus* colonization and infection in soldiers. *Clin Infect Dis.*, 2004, 39, 971-979.

Centers for Disease Control and Prevention. Methicillin-resistant *Staphylococcus aureus* infections in correctional facilities – Georgia, California, and Texas, 2001-2003. *MMWR Morb Mortal Wkly Rep.*, 2003, 52, 992-996.

Centers for Disease Control and Prevention, Public health dispatch: outbreaks of community-associated methicillin-resistant *Staphylococcus aureus* skin infections. Los Angeles County, California, 2002-2003. *MMWR Morb Mortal Wkly Rep.*, 2003, 52, 88.

Lindenmayer, JM; Schoenfeld, S; O'Grady, R; Carney, JK. Methicillin-resistant *Staphylococcus qureus* infection in a rugby football team. *Br J Sports Med.*, 1998, 153-154.

Salgado, CD; Farr, BM; Calfee, DP. Community-associated methicillin-resistant *Staphylococcus aureus:* a meta-analysis of prevalence and risk factors. *Clin Infect Dis.*, 2003, 36, 131-139.

Naimi, TS; LeDell, KH; Como-Sabetti, K; et al. Comparison of community- and health care-associated methicillin-resistant *Staphylococcus aureus* infection. *JAMA*, 2003, 290, 2976-2984.

Gillet, Y; Issartel, B; Vanhems, P; et al. Association between *Staphylococcus aureus* strains carrying gene for Panton-Valentine leukocidin and highly lethal necrotising pneumonia in young immunocompetent patients. *Lancet*, 2002, 359, 753-759.

Baggett, HC; Hennessy, TW; Rudolph, K; et al. Community-onset methicillin-resistant *Staphylococcus aureus* associated with antibiotic use and the cytotoxin Panton-Valentine leukocidin during a furunculosis outbreak in rural Alaska. *J Infect Dis.*, 2004, 189, 1565-1573.

Francis, J; Doherty, M; Lopatin, U; Johnston, CP; Sinha, G; Ross, T; at al. Severe Community-onset pneumonia in healthy adults caused by methicillin-resistant *Staphylococcus aureus* carrying the Panton-Valentine leukocidin genes. *Clin Infect Dis.*, 2005, 40, 100-107.

Siberry, GK; Tekle, T; Carroll, K; Dick, J. Failure of clindamycin treatment of methicillin-resistant *Staphylococcus aureus* expressing inducible clindamycin resistance in vitro. *Cin Infect Dis.*, 2003, 37, 1257-1260.

Hageman, JC; Francis, J; Uyeki, TM; et al. Staphylococcus Pneumonia Working Group. Emergence of methicillin-resistant *Staphylococcus aureus* as a cause of community-associated pneumonia during the influenza season, 2003-2004 [abstract]. In: Program and abstracts of the *Infectious Diseases Society of America* 42nd Annual Meeting: Boston, Mass; September 30-October 3, 2004, 60-61. Abstract LB-8.

Dellit, T; Duchin, J; Hofmann, J; Gurnai Olson, E. Interim guidelines for evaluation and management of community-associated methicillin-resistant *Staphylococcus aureus* skin and soft tissue infections in outpatient settings. September 2, 2004. www.metoke.gov/health/prevcont/mrsa.htm.

Hepburn, MY; Dooley, DP; Skidmore, PJ; Ellis, MW; Starnes, WF; Hasewinkle, WC. Comparison of short (5 days) and standard (10 days) treatment for uncomplicated cellulitis. *Arch Intern Med.*, 2004, 164, 1669-1674.

Editors of The Medical Letter. Treatment of Community-Associated MRSA Infections. *The Medical Letter*, 2006, 48, 13-14.

Kowalski, TJ; Barbari, EF; Osmon, DR. Epidemiology, Treatment and Prevention of Community-Associated Methicillin-resistant *Staphylococcus aureus* Infections. *Mayo Clin Proc.*, 2005, 80(9), 1201-1208.

Jones, ME; Mayfield, DC; Thornsberry, C; Karlowsky, JA; Sahm, DF; Peterson, D. Prevalence of oxacillin resistance in *Staphylococcus aureus* among inpatients and outpatients in the United States during 2000. *Antimicrob Agents Chemother*, 2002, 46, 3104-3105.

Moran, GJ; et al. Methicillin-Resistant *S. aureus* Infections among Patients in the Emergency Department. *NEJM*, 2006, 355, 666-674.

In: Methicillin-Resistant Staphylococcus Aureus... ISBN: 978-1-60741-398-1
Editor: Charles L. Kolendi

Chapter 16

COMMUNITY-ACQUIRED METHICILLIN-RESISTANT *STAPHYLOCOCCUS AUREUS*: CHARACTERIZATION, RISK FACTORS AND TREATMENT

Hossam M. Ashour
Department of Microbiology and Immunology, Faculty of Pharmacy, Cairo University, Egypt.

COMMENTARY

Methicillin-resistant *Staphylococcus aureus* (MRSA) is commonly known for its resistance to methicillin and other antibiotics such as oxacillin, penicillin, and amoxicillin. Hospital-acquired MRSA (HA-MRSA) evolved due to the widespread use of antibiotics in hospitals during the twentieth century. HA-MRSA infections occur most frequently among persons with compromised immune systems in hospitals and other healthcare facilities. Traditionally, most of the attention of the scientific community has been focused on HA-MRSA. The last decade has witnessed the emergence of a genetically different form of MRSA, which has been called Community-acquired MRSA (CA-MRSA) due to the fact that it is acquired from the community and not from hospitals or other healthcare facilities. CA-MRSA infections started to receive attention because they can be highly virulent and can occur in otherwise healthy people, who have no apparent risk factors for MRSA acquisition. The emergence of the first community outbreaks was reported in the 90's [1, 2]. Since then, the prevalence of CA-MRSA infections has been rapidly increasing.

CA-MRSA is defined as an infection in which bacteria is isolated in an outpatient setting or from a patient within 48 hours of hospital admission. This patient must have a medical history free of MRSA infection or colonization, hospitalization, surgery, or dialysis in the previous year. Moreover, the patient must not have had permanent indwelling catheters or per-cutaneous medical devices. Contrary to HA-MRSA infections, CA-MRSA infections are acquired by persons who have not been recently hospitalized nor had a medical procedure.

There are genetic, clinical, and epidemiological characteristics that distinguish CA-MRSA from HA-MRSA. The emergence of methicillin resistance in *Staphylococcus aureus* is caused by a shift in cell wall constituents resulting from mutation of Penicillin-Binding Protein 2A (PBP2A) [3]. The mutated PBP2A is encoded by *mecA* gene and resides in a staphylococcal cassette chromosome *mec* (SCC *mec*) that causes site-specific excision and reintegration of resistance genes [4]. Although *mecA* is the mobile genetic element responsible for methicillin resistance in *Staphylococcus aureus*, CA-MRSA is genotypically distinct from HA-MRSA. Clones of CA-MRSA include USA 300 and USA 400, whereas clones of HA-MRSA include USA 100, USA 500, and USA 800 [5]. The USA 300 clone is the most predominant CA-MRSA clone in USA, and has recently emerged in European countries and Japan [6]. Its main molecular characteristics are the presence of Panton-Valentine leukocidin (PVL), the arginine catabolic mobile element (ACME), and SCC *mec* [5]. So far, there are five identified SCC *mec* types, which vary in size, genetic composition, and antimicrobial resistance patterns. CA-MRSA harbors the distinct SCC *mec* type IV and less commonly SCC *mec* type V. On the other hand, HA-MRSA carries SCC *mec* types I, II, or III [5]. The genetic distinction suggests that CA-MRSA infections were not simply the result of HA-MRSA moving into the community.

From the clinical standpoint, HA-MRSA is typically associated with pneumonia, urinary tract infections, blood stream infections, and surgical wound infections. On the other hand, CA-MRSA usually presents with skin and soft tissue infections. Increased prevalence of CA-MRSA within the community has posed a concern to healthcare providers. Although CA-MRSA infections usually present as skin and soft tissue infections that are irresponsive to typical therapy, they can also lead to serious systemic infections that can be fatal. An example is the CA-MRSA pneumonia, which is an acute necrotizing process that is typically associated with high rates of bacteremia. As a result of the widespread reservoir of CA-MRSA in the community, there has been an increase in the prevalence of CA-MRSA in the hospitals [7]. This will ultimately render the epidemiological characteristics of MRSA less valuable in identifying CA-MRSA from HA-MRSA. The isolation of CA-MRSA from hospitals emphasizes the importance of correct genetic and clinical characterization of MRSA.

Contrary to HA-MRSA which has well-defined risk factors, there are no clearly defined risk factors for CA-MRSA. Studies suggest that risk factors for CA-MRSA include overcrowding [8], intravenous and inhalational drug abuse [9], antibiotic use during the past twelve months [10], obesity [11], African-American race and female sex [12]. Several reservoirs for CA-MRSA have been described, among them are zoonotic reservoirs. These include opiate abusers [9], skin of children with atopic dermatitis [13], the district nurse patient (DNP) population [14], healthy dogs [15], horses [16], and pigs [17].

Clinicians need to be able to correctly characterize CA-MRSA, and identify the risk factors that contribute to its spread. This can help design strategies for control and treatment of CA-MRSA. Whereas HA-MRSA is frequently known to have multidrug resistance, CA-MRSA infections are not multidrug resistant. This is a key piece of information for a practicing clinician who might not have proper research settings that can help him characterize CA-MRSA molecularly by rapid PCR-based assays. Most of the characterized CA-MRSA strains have increased susceptibility to non-beta-lactam antibiotics, such as vancomycin, trimethoprim-sulfamethoxazole, clindamycin, tetracyclines, linezolid, and levofloxacin [18]. In contrast to HA-MRSA, different options of efficient therapies can be implemented once a CA-MRSA infection has been suspected. However, relapses occur frequently and must be dealt with promptly using thorough strategies such as total-body decolonization [19, 20]. For example, one safe and effective decolonization strategy employed a combination of intranasal mupirocin, body washes with chlorhexidine, and oral doxycycline [20]. Inspite of all that, it is probable that CA-MRSA antimicrobial resistance will develop over time. Thus, optimal treatments of MRSA infection and colonization have to be determined to avoid the emergence of this resistance. Attempts to prevent relapses by designing vaccines that offer long-lasting protection against *Staphylococcus aureus* have not been successful so far [21]. However, researchers should pursue research in this direction with the aim of providing long-lasting vaccine protection against MRSA, which will not lead to the emergence of antibiotic-resistant strains, as is the case with antimicrobial therapy. Based on the increased prevalence of CA-MRSA in many countries, there should be active routine screening tests for CA-MRSA carriage in all patients admitted with purulent skin and soft-tissue in countries which face high rates of CA-MRSA infection.

REFERENCES

[1] Herold, BC; et al., Community-acquired methicillin-resistant Staphylococcus aureus in children with no identified predisposing risk. *JAMA,* 1998, 279(8), p. 593-8.

[2] Collignon, P; et al., Community-acquired meticillin-resistant Staphylococcus aureus in Australia. Australian Group on Antimicrobial Resistance. *Lancet*, 1998, 352(9122), p. 145-6.

[3] De Jonge, BL; Tomasz, A. Abnormal peptidoglycan produced in a methicillin-resistant strain of Staphylococcus aureus grown in the presence of methicillin: functional role for penicillin-binding protein 2A in cell wall synthesis. *Antimicrob Agents Chemother,* 1993, 37(2), p. 342-6.

[4] Katayama, Y; Ito, T; Hiramatsu, K. A new class of genetic element, staphylococcus cassette chromosome mec, encodes methicillin resistance in Staphylococcus aureus. *Antimicrob Agents Chemother,* 2000, 44(6), p. 1549-55.

[5] Herman, RA; et al., Etiology and treatment of community-associated methicillin-resistant Staphylococcus aureus. *Am J Health Syst Pharm*, 2008, 65(3), p. 219-25.

[6] Shibuya, Y; et al., Emergence of the community-acquired methicillin-resistant Staphylococcus aureus USA300 clone in Japan. *J Infect Chemother*, 2008, 14(6), p. 439-41.

[7] Nathwani, D; et al., Guidelines for UK practice for the diagnosis and management of methicillin-resistant Staphylococcus aureus (MRSA) infections presenting in the community. *J Antimicrob Chemother*, 2008, 61(5), p. 976-94.

[8] Methicillin-resistant Staphylococcus aureus skin or soft tissue infections in a state prison—Mississippi, 2000. *MMWR Morb Mortal Wkly Rep*., 2001, 50(42), p. 919-22.

[9] El-Sharif, A; Ashour, HM. Community-acquired methicillin-resistant Staphylococcus aureus (CA-MRSA) colonization and infection in intravenous and inhalational opiate drug abusers. *Exp Biol Med (Maywood),* 2008, 233(7), p. 874-80.

[10] Lo, WT; et al., Risk factors and molecular analysis of panton-valentine leukocidin-positive methicillin-resistant Staphylococcus aureus colonization in healthy children. *Pediatr Infect Dis J.,* 2008, 27(8), p. 713-8.

[11] Anaya, DA; Dellinger, EP. The obese surgical patient: a susceptible host for infection. *Surg Infect (Larchmt)*, 2006, 7(5), p. 473-80.

[12] King, MD; et al., Emergence of community-acquired methicillin-resistant Staphylococcus aureus USA 300 clone as the predominant cause of skin and soft-tissue infections. *Ann Intern Med.*, 2006, 144(5), p. 309-17.

[13] Chung, HJ; et al., Epidemiological characteristics of methicillin-resistant Staphylococcus aureus isolates from children with eczematous atopic dermatitis lesions. *J Clin Microbiol*, 2008, 46(3), p. 991-5.

[14] Thomas, S; et al., Meticillin-resistant Staphylococcus aureus carriage among district nurse patients and medical admissions in a UK district. *J Hosp Infect*, 2007, 66(4), p. 369-73.

[15] Epstein, CR; et al., Methicillin-resistant commensal staphylococci in healthy dogs as a potential zoonotic reservoir for community-acquired antibiotic resistance. *Infect Genet Evol*, 2009, 9(2), p. 283-5.

[16] Weese, JS; et al., Community-associated methicillin-resistant Staphylococcus aureus in horses and humans who work with horses. *J Am Vet Med Assoc.*, 2005, 226(4), p. 580-3.

[17] Wulf, MW; et al., Prevalence of methicillin-resistant Staphylococcus aureus among veterinarians: an international study. *Clin Microbiol Infect*, 2008, 14(1), p. 29-34.

[18] Stefani, S; et al., Pathotype and susceptibility profile of a community-acquired methicillin-resistant Staphylococcus aureus strain responsible for a case of severe pneumonia. *Diagn Microbiol Infect Dis.*, 2009, 63(1), p. 100-4.

[19] Miller, LG; et al., A prospective investigation of outcomes after hospital discharge for endemic, community-acquired methicillin-resistant and -susceptible Staphylococcus aureus skin infection. *Clin Infect Dis.*, 2007, 44(4), p. 483-92.

[20] Simor, AE; et al., Randomized controlled trial of chlorhexidine gluconate for washing, intranasal mupirocin, and rifampin and doxycycline versus no treatment for the eradication of methicillin-resistant Staphylococcus aureus colonization. *Clin Infect Dis.*, 2007, 44(2), p. 178-85.

[21] Shinefield, H; et al., Use of a Staphylococcus aureus conjugate vaccine in patients receiving hemodialysis. *N Engl J Med.*, 2002, 346(7), p. 491-6.

In: Methicillin-Resistant Staphylococcus Aureus... ISBN: 978-1-60741-398-1
Editor: Charles L. Kolendi

Chapter 17

The Changing Pattern of Infections with MRSA and other Organisms after Major Reconstructive Surgery of the Maxillofacial Region

CME Avery
University Hospitals of Leicester
United Kingdom

Abstract

There have been few publications on the incidence of infection with methicillin resistant *Staphylococcus aureus (*MRSA) in the head and neck region. In an initial study, over two years, the pattern of infections and the acquisition of MRSA in two groups of patients who had undergone free flap reconstructions of maxillofacial defects were studied. The first group (n=31) were given a 5-day perioperative course of antibiotic prophylaxis (long-term) and the second group (n=33) a 24-hour course (short-term). MRSA was the main infecting organism detected. Seventeen of the 64 patients (27%) developed an infection with MRSA and all infections were acquired after operation. Serious infections delayed healing or discharge in 8% (5) as a result of MRSA and 6% (4) with infections other than MRSA. The median length of hospital admission was significantly greater if MRSA was acquired (p=0.005). There were significantly fewer patients with wounds infected by MRSA in the short-term antibiotic group (4/33 compared with 13/31, p=0.01).

The incidence of acquisition of MRSA and other organisms was then studied over a 6 year period up to 2008. One hundred and fifty-nine patients were reconstructed with a total of 187 free (155) or pedicled (32) flaps. The incidence of MRSA per flap procedure was 19% (36/187). The MRSA acquisition rate decreased from 31% (14/45) with a long-term antibiotic regimen to a rate of 15% (22/142) with the shortened regimen. An infective organism was detected after 85% (159/187) of procedures. The differing groups of organisms were divided in to MRSA alone (2%), "other organisms" (60%), MRSA with "other organisms" (23%), and no growth (15%). The range of organisms increased and these were mainly faecal flora. There was a significant decrease in the incidence of MRSA (p=0.02) and an increase in the detection of Pseudomonas (p=0.02). Age of 70 years or greater (p=0.01) and a pedicled flap (p=0.03) were significant predictors for acquisition of other infectious organisms. The median duration of hospital admission has remained similar throughout the period of the study.

The incidence of MRSA and other infections remains relatively high after major reconstructive maxillofacial surgery with free or pedicled flaps, although the incidence of MRSA is decreasing. A sustained reduction in infections should be achievable through the use of appropriate antibiotic prescribing policies, improved design of specialist wards including the provision of single bedded rooms and isolation facilities together with strict adherence to standard infection control measures. It is not yet known whether the earlier detection of MRSA with rapid screening and preventative isolation will further reduce the incidence of MRSA.

Introduction

The infection of surgical wounds with methicillin-resistant *Staphylococcus aureus* (MRSA) has become increasingly common across the world and often results in impaired wound healing (1-4). Hospital acquired infections lead to a prolonged stay in hospital and result in additional expense (5, 6). Infection with MRSA is especially serious and causes morbidity and mortality, particularly after vascular operations (7) and burns (8). In recent years infection with MRSA has not been uncommon at the Leicester Royal Infirmary, United Kingdom, particularly with major reconstructive vascular procedures, and has led to significant morbidity and attempts to minimise the risk of infection (9).

Perhaps surprisingly there have been relatively few reports of infection with MRSA after major operations in the head and neck region. Infection with MRSA and other organisms causes morbidity and occasionally mortality, reduces the quality of life and prolongs the stay in hospital, which is a major component of the resulting increase in additional costs to the health care service (10-12). Infection with MRSA has complicated wound healing after reconstructive head and neck surgery at the donor site of the radial (13) and other free flaps. Until recently there have been few substantive studies of the incidence of MRSA after major head and neck reconstructive surgery. In the first prospective study of infections with MRSA after major maxillofacial surgery Avery (14) reported on the effect of moving from a standard 5-day perioperative course of antibiotic prophylaxis to a shorter 24-hour perioperative course for patients undergoing reconstruction with a free flap. There were significantly fewer patients with wounds infected by MRSA in the short-term antibiotic group than the longer term group.

Following this report the routine use of a short-term antibiotic surgical prophylaxis regimen, whenever possible, was continued and later generation cephalosporin drugs were avoided. The use of cephalosporin antimicrobials and prolonged treatment, together with many other factors, increase the risk of acquiring infection with MRSA (15-17). During this time there has been an increased awareness of the need to improve cross-infection control measures on the head and neck ward and within the Leicester Royal Infirmary. Across the National Health Service (NHS) in the United Kingdom there has been a politically driven agenda to introduce multiple interventions, including improved cleaning practices and universal screening, to reduce the incidence of MRSA (18). This chapter reports on the initial analysis of a larger cohort of consecutive oncology patients managed with either a free or pedicled flap since the beginning of the original study. The current issues in the literature are also reviewed

PART A

The initial study: a comparison of short and long-term antibiotic prophylaxis regimens with free flap reconstruction.

i.) Patients and Data

The following data were recorded on consecutive maxillofacial patients undergoing reconstruction with a free flap; demographic details, medical history, and risk factors for infection, including the presence of MRSA, anaemia, diabetes, misuse of drugs, renal failure, steroids, immunosuppression, previous operations, radiotherapy, chemotherapy, or admission to hospital within the past year. The stage of the tumour, details of the operation, the antibiotics that were given, and infection of the wound were documented. Free flaps were used for reconstruction after surgical resections primarily for oral carcinoma.

The initial period of study was from June 2002 to May 2003 (n=31). In September 2003 there was a switch to shorter perioperative course of antibiotics and a further cohort of patients were studied until September 2004 (n=33). This shortened regimen was maintained as long as there was no clinical and microbiological evidence of infection necessitating additional antibiotic therapy. The incidence of MRSA and other infections before and after changing from the 5-day (long-term) to a 24-hour (short-term) perioperative prophylactic course was audited.

ii.) Isolation Measures

The maxillofacial patients shared a 14-bed high dependency type ward with patients who were attended by the departments of otolaryngology and plastic surgery, including those with minor burns. All of the patients were managed in single rooms with contact isolation measures regardless of whether MRSA was cultured or not (14). These measures included wearing gloves and gowns but not masks. The hands were washed and an alcohol gel advised after patient contact.

iii.) Classification of Wound Infection

Infection was defined as breakdown of the wound or purulent discharge with substantial microbial growth. Infections were graded using the classification of Szilagyi (19) that was developed for vascular surgery and is simple to apply. A type 1 infection was defined as involving the skin, type 2 the skin and subcutaneous tissues, and type 3 exposed the underlying graft, or in this particular study the deep major vascular structures, bone plate or other implanted material. Delayed healing was defined as a wound that required dressing after the thirtieth day. An infection that delayed healing or discharge beyond 30 days was defined as a serious complication.

Infection or colonisation with MRSA was treated using an MRSA eradication programme. This included aseptic nursing, and for patients, mupirocin nasal cream, daily body washes with triclosan soap (Aquasept, SSL, Cheshire, UK), and an appropriate antibiotic. Potentially infected catheters and intravenous lines were removed as soon as possible after surgery or if suspected as a focus of infection.

iv.) Detection of MRSA

Screening swabs were taken from all patients from the anterior nares, axillae, and perineum on admission, on the first and tenth day after operation and pooled. The donor site of the flap was swabbed during the operation and after removal of the dressing on the tenth day. In the absence of clinical infection no further swabs were obtained after the tenth day. If infection was suspected all of the wounds were swabbed and the screen was repeated. Swabs were inoculated into 7% salt nutrient broth and incubated for 16-18 hours at 30°C. The broth was then subcultured on to mannitol salt agar plates containing 2 mg/Lf Oxacillin and incubated for a minimum of 24 hours at 37°C. Yellow colonies on thee plates were identified using a combination of Gram stain, DNAse reaction, slide coagulase test, and oxacillin sensitivity.

v.) Regimens of Antibiotic Prophylaxis

Both groups were given 1.5 g cefuroxime and 500 mgs metronidazole at induction of anaesthesia with additional peroperative doses of 750mgs of cefuroxime and 500mgs of metronidazole after 3 hours. In the first group (long-term) the course was continued after operation at 8 hourly intervals for up to 5 days or longer if clinically indicated. In the second group (short-term) the final doses were given within 8 hours of completing the operation. Additional doses were given only if clinically indicated.

vi.) Statistical Analysis

The groups of patients who did and did not contract MRSA colonisation or infection were compared by the Mann-Whitney U test for age, length of operation, duration of stay in hospital,and treatment with antibiotics. Fisher's exact test was used to compare age groups (< 70 and 70 years or more), sex, and risk factors for infection, advanced tumour stage (T3 and T4), type of flap, and admission to the intensive care unit (ICU). The chi-squared goodness of fit test was used to look for evidence of a temporal relationship or "clustering" between individual patients that acquired MRSA.

PART A: RESULTS OF THE INITIAL STUDY

Patient Groups

The patient details of both the long and short-term antibiotic groups are listed in **Table 1**.

Table 1. Details of patients and admissions. Data are mean (SD) except where stated otherwise.

	Type of antibiotic prophylaxis			
	Long-term (n=31)		**Short-term (n=33)**	
	Acquired MRSA		**Acquired MRSA**	
	Yes (n=14)	No (n=17)	Yes (n=4)	No (n=29)
Age (years)	69 (13)	60 (21)	63 (19)	58 (14)
Duration of operation (hours)	10 (2)	11 (2)	12 (2)	10 (2)
Median (range) duration of antibiotics (hours)	144 (48-192)	144 (72-360)	30 (24-48)	24 (24-120)
Median (range) duration of hospital stay (days)	18 (9-81)	12 (8-112)	50 (9-71)	14 (6-60)

i.) Long-Term Antibiotic Group (n=31)

Long-Term Group: Colonisation and Infection with MRSA

Fourteen of the 31 patients (45%) were either colonised or infected with MRSA (**Table 2**). There was no significant difference in age, sex or stage of tumour between those who did and did not acquire MRSA.

MRSA was not isolated preoperatively, during the operation, or on the first postoperative day. It was first detected on the tenth postoperative day in 9/31 patients (29%), in 13 (42%) at the time of their discharge from hospital, and in 14 (45%) after discharge (**Table 3).** Thirteen patients (42%) developed an infection by MRSA, which was isolated from 43 sites. (**Table 2**). 4 patients had delayed healing.

There was no significant difference in duration of operation, antibiotics given and incidence of admission to ICU between the patients who acquired MRSA and those who did not. However, the median (18, range 9-81 days) duration of stay in hospital in those who acquired MRSA was significantly longer (p=0.01) (**Table** 1).

Long-Term Group: Potential Risk Factors for Infection

Ten of the 31 patients (32%) had potential risk factors for infection (excluding anaemia) (**Table 4**).

Long-Term Group: Infections at the Flap Donor Site

Seven donor sites became infected (23%). There was no significant difference in the incidence of infections between osseocutaneous (3/11) and soft-tissue donor sites (4/20).

Long-Term Group: Wounds Infected by MRSA

Wound infections were classified as type 1 in 19% (4), type 2 in 67% (14), and type 3 in 14% (3).

Long-Term Group: Infections other than MRSA

Swabs from 16 sites in 10 patients (32%) grew one or more organisms that were not MRSA, 8 (26%) developed one or more infections and one patient had delayed healing (**Table 5**). Cultures from two patients who developed fatal chest infections grew no pathogens.

Long-Term Group: Wound Infections not Caused by MRSA (Table 6).

iv.) Short-Term Antibiotic Group (n=33)

Short-Term Group: Colonisation and Infection with MRSA

Four patients (12%) acquired and became infected by MRSA and there were no significant differences between these four and the other 29 in age, sex, stage of tumour, duration of operation, or use of antibiotics. These four patients stayed in hospital longer than the other 29 but the difference was not significant because of small numbers (p=0.07). One patient had their discharge delayed because of chest and wound infections.

MRSA were first cultured on the tenth postoperative day in one patient, and in the other 3 at the time of discharge (**Table 3**).

Short-Term Group: Risk Potential Factors for Infection

Fourteen patients (42%) had potential risk factors for infection (excluding anaemia) (**Table** 4).

Short-Term Group: Infections at the Flap Donor Site

Two donor sites (6%) developed infections.

Short-Term Group: Wounds Infected by MRSA

Infections were classified as type 1 in 44% (4), type 2 in 33% (3), and type 3 in 22% (2).

Short-Term Group: Infections other than MRSA

Twelve patients (36%) had an appreciable growth of one or more other organisms at 26 sites, 10 (30%) developed one or more infections and 4 had delayed healing (**Table 5**).

Table 2. MRSA acquisition and infection. Data are number (%).

	Long-term group (n=31)	Short-term group (n=33)
Patients who acquired MRSA	14 (45)	4 (12)
Patients with MRSA infections	13 (42)	4 (12)

Table 3. Patients in either group that acquired MRSA before the 10th day. Figures are number of patients (%).

Day	Long-term group (n=31)	Short-term group (n=33)
Before operation	0	0
During operation	0	0
Postoperative day 1	0	0
Postoperative day 10	9 (27)	1 (3)
At discharge	13 (39)	4 (12)
After discharge	14 (42)	4 (12)

Table 4. Potential risk factors and infection by MRSA.

Infection risk	Long-term group (n = 31)	Short-term group (n = 33)
MRSA if risk factors*	5/10	4/14
MRSA if no risk factors*	9/21	0/19
P-value**	1.0	0.02
MRSA if diabetic	2/4	3/5
MRSA if non-diabetic	12/27	1/28
p–value**	1.0	0.01
MRSA if anaemic	5/14	1/8
MRSA if not anaemic	9/17	3/25
p–value**	0.47	1.0

*Excluding anaemia

**Fisher's test

Table 5. Acquisition of and infection by organisms other than MRSA. Figures are numbers (%).

	Long-term group (n=31)	Short-term group (n=33)
Patients with non-MRSA growth	10 (32)	12 (36)
Patients with non-MRSA infections	8 (26)	10 (30)

iv.) A Comparison of the Short and Longer Term Antibiotic Groups

The mean duration of antibiotic treatment was significantly shorter in the short-term group (p=0.001) than the long term group regardless of whether patients acquired MRSA or not. There was no significant difference in the length of operation or hospital stay (**Table 1**).

In the long-term group the median age was higher (69 compared with 56 years, p=0.08) and there were twice as many 70 years or older (48% compared with 21%, p=0.04). The incidence of acquisition (p=0.005) and infection (p=0.01) by MRSA decreased significantly between the long- term and short-term groups, and there were fewer infections at donor sites (**Table 6**). There was no evidence of temporal clustering of infections with MRSA but the numbers were small. The pattern of acquisition of and infection by organisms other than MRSA remained broadly similar (**Table 6**).

Table 6. Comparison of the two groups as regards infections. Data are number (%).

	Long-term group (n=31)	Short-term group (n)=33	p value
MRSA			
Acquisition	14 (45)	4 (12)	0.005
Infection	13 (42)	4 (12)	0.01
Infection of donor sites	7 (23)	2 (6)	0.08
Other infections			
Acquisition	10 (32)	12 (36)	0.80
Infection	8 (26)	10 (30)	0.78
Donor site infection	3 (10)	3 (9)	1.0

v.) Combination of the Two Cohorts

Overall, MRSA was acquired by 28% of patients and 27% developed an infection. Acquisition of MRSA was associated with a significantly longer median length of hospital admission (P=0.005). There was no evidence of clustering of patients that acquired MRSA. Infection developed at most sites that acquired MRSA (57%) and MRSA usually affected more than one site in an individual patient (72%). The majority of wound infections were type 1 (27%) or type 2 (57%). Substantial growth of non-MRSA organisms was detected in 34% of patients and 28% developed an infection. Infection developed at most sites (60%) with growth. The wound infections were mainly type 2 (44%) or type 3 (44%).

The incidence of patients with serious complications caused by wound or chest infection with MRSA was 8% (5/64) and by non-MRSA organisms was 6% (4/64). The incidence of donor site infections with MRSA (14%, 9/64) and non-MRSA (9%, 6/64) organisms was also similar. The numbers were too small to compare individual donor sites but no one site was particularly prone to infection. There was no significant difference in the number of soft-tissue only (4/46) and osteocutaneous donor sites (5/18) that developed infection with MRSA.

Several trends were identified. Type 2 and 3 wound infections were associated with a longer median hospital admission than type 1 infections (21 compared with 12 days, P<0.001). Diabetics had a higher incidence of 3 or more infections (3/9 compared with 5/55, P=0.08) and serious infections (3/9 compared with 9/55, P=0.35). Patients with risk factors for infection (excluding anaemia) had more multiple infections (6/24 compared with 2/40, P=0.04) and patients with anaemia had more serious infections (7/22 compared with 5/42, P=0.09).

PART B: THE CONTINUING STUDY: FREE AND PEDICLED FLAP RECONSTRUCTIONS USING A PREDOMINANTLY SHORT ANTIBIOTIC REGIMEN

i.) Patients and Data

The study was continued from September 2003 using a predominantly shortened antibiotic antibiotic regimen. Data was again prospectively collected on consecutive patients but in this analysis pedicled flaps used for head and neck reconstruction were also included up to March 2008. The detailed analysis of wound infection rates and the distinction between infection and colonisation will be reviewed in a later analysis when greater numbers are available. The current article assesses the overall changing pattern of infectious organisms detected between 2002 and 2008.

ii.) Isolation Measures

The isolation measures have remained the same during this period. During this time there has been an increased awareness of the need to improve cross-infection control measures on the head and neck ward and within the entire hospital. Therefore multiple interventions, including improved cleaning and "deep" cleaning practices, have been introduced to reduce the risk of infection.

iii.) Wound Infection

The same wound classification has been used and the detailed results will be reported in a later article. Infection or colonisation with MRSA was treated using the same MRSA eradication programme.

iv.) Detection Of MRSA

The same MRSA detection system has been used. It has become routine practice to screen all patients undergoing major or minor surgery for MRSA that are admitted or transferred to the head and neck ward.

v.) Regimens of Antibiotic Prophylaxis

The use of cefuroxime has largely been discontinued and replaced with augmentin in combination with metronidazole. The frequency of drug administration has remained the same. All patients, unless allergic, received 1 gram of augmentin and 500 mgs of metronidazole at induction of anaesthesia with additional peroperative doses of 500 mgs of augmentin and 500mgs of metronidazole after 3 hours. The final doses were given within 8 hours of completing the operation. Additional doses were given only if clinically indicated.

vi.) Statistical Analysis

The analysis was completed using appropriate tests as previously listed.

PART B: RESULTS OF THE CONTINUING STUDY

One hundred and fifty-nine patients underwent reconstruction for head and neck malignancy with a total of 187 free (155) or pedicled (32) flaps. The median operation time was 10 (IQR 6-12 hours). There were 90 males and 69 females with a median age of 62 years (range 19-91).

Continuing Study: Acquisition of MRSA

The overall incidence of MRSA per flap procedure was 19% (36/187). There was a substantial decrease of 16% from the initial rate of MRSA acquisition of 31% (14/45) with the long term antibiotic regimen to 15% (22/142) with the shortened regimen. This data is not exactly comparable with the initial study as these results now include pedicled flaps over a slightly longer period of time but confirm a similar pattern.

Continuing Study: Acquisition of MRSA and other Organisms

An infective organism was detected after 85% (159/187) of procedures. At least 27 differing species were identified. The differing groups of species were broadly divided in to MRSA alone (2%), "other organisms" (60%), MRSA together with "other organisms" (23%), and no growth (15%) (**Figure 1**).

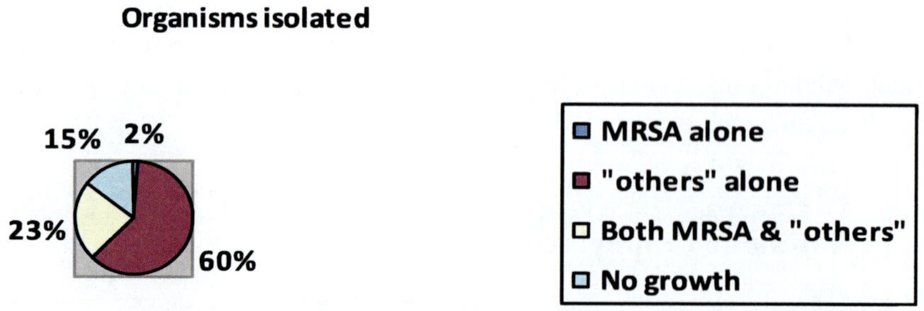

Figure 1. Main subgroups of organisms isolated.

The range of other organisms increased and these were mainly faecal flora. Prior to September 2003 there were 50 organisms affecting 45 patients but after the change of antibiotic regimens there were 249 differing organisms affecting 142 patients. The changes in the incidence of the 3 main organisms are shown in **Figure 2.** There is a significant decrease in the incidence of acquisition of MRSA (p=0.02) and increase in the incidence of Pseudomonas (0.02), with a trend towards an increase in *Candidiasis* as well (p=0.08). There was a 3 times greater odds ratio difference of acquiring MRSA before September 2003 than in the period afterwards up to 2008.

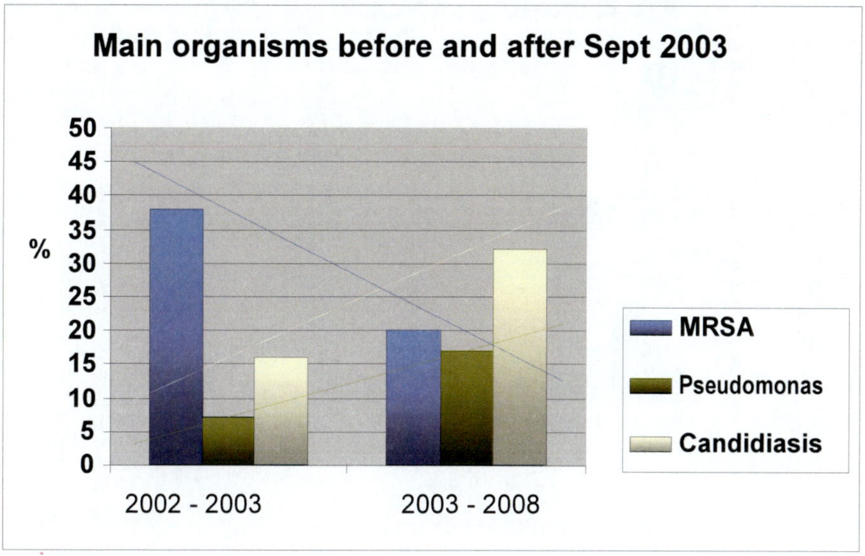

Figure 2. Main changes in organisms isolated.

Continuing Study: Potential Risk Factors for MRSA and "Other Organisms"

The incidence of MRSA acquisition was not influenced by the length of operation, age, or potential risk factors such as diabetes and anaemia but the incidence of wound infections has not yet been analysed. However, age of 70 years or greater (p=0.01) and a pedicled flap (p=0.03) were significant predictors for acquisition of other infectious organisms.

Continuing Study: Length of Hospital Admission

During the period of this study the median duration of hospital admission has remained similar. This may be due to the acquisition of "other organisms" delaying discharge rather than MRSA. The median hospital stay with MRSA alone was 29 days and with "other organisms" was 25 days **(Figures 3 & 4)**. The distribution of length of stay was skewed for both of these groups by several patients with a prolonged admission. The durations of hospital stay for both of these groups was significantly longer than the group without growth, which had a median stay of 11 days ($p<0.0001$) **(Figure 3)**. There was no statistically significant difference between the group with MRSA and the "other organism" group (p=0.92). There was no overall difference in the median length of stay between the long (23 days) and short term (21 days) antibiotic prophylaxis groups.

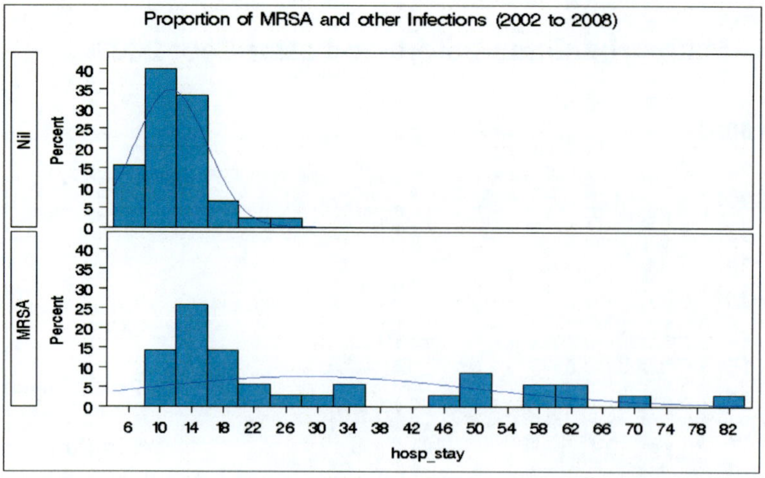

Figure 3. The duration of hospital stay for MRSA and "no growth organisms".

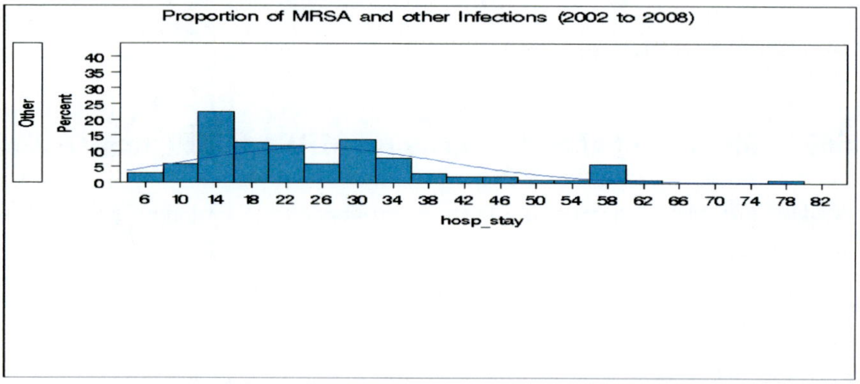

Figure 4. The duration of hospital stay for "other organisms".

Discussion

The principle mechanism of transmission of MRSA is direct interpersonal contact, but indirect transfer from contaminated environmental surfaces may occur (20, 21) as well as transmission through the air (22, 23). Shiomori (22) isolated MRSA from the air and inanimate surfaces in the rooms of 3 patients with head and neck carcinoma who had acquired MRSA. An increasing number of seemingly innocuous potential routes of transmission have been described including bed control handsets (24), mobile communication devices (25) and surgical marker pens (26). Multiple factors within the host, the quality of wound care and the environment also affect the rates of surgical wound infection (16). Intensive care units, high dependency units, and surgical units are high-risk environments for the acquisition of MRSA (5, 27). Scriven (9) reported that 17% of patients, also at the Leicester Royal Infirmary, United Kingdom, who had vascular operations acquired MRSA, mostly in the ICU or on the ward after the operation. The longer patients stayed in hospital the more likely they were to acquire MRSA.

There have been few substantive publications that have dealt with infections by MRSA after major head and neck surgery. Small case series and reports have described mortality as a result of serious systemic complications (28) and wound healing complications at the donor site of a free flap (13, 29). In the initial part of the study reported in this chapter, and which is based on the prospective study by Avery (14) in 2006, the overall incidence of MRSA acquisition (28%) and infection (27%) associated with the reconstruction of the maxillofacial region using free tissue transfer techniques was relatively high. However, the incidence decreased significantly from 45% to 12% after the introduction of a short-term antibiotic regimen. There was a trend to increasing frequency and severity of infections in patients with risk factors for infection and Type 2 and 3 wound infections were associated with a longer median hospital admission than type 1 infections (21 compared with 12 days, $P<0.001$). Diabetics had a higher incidence of 3 or more infections (3/9 compared with 5/55, $P=0.08$) and serious infections (3/9 compared with 9/55, $P=0.35$). Patients with risk factors for infection (excluding anaemia) had more multiple infections (6/24 compared with 2/40, $P=0.04$) and patients with anaemia had more serious infections (7/22 compared with 5/42, $P=0.09$). However, most infections caused relatively minor wound complications. Serious complications occurred with infections by both MRSA (8%) and other organisms (6%). MRSA was the predominant infective organism but Pseudomonas, coliforms, and methicillin-sensitive Staphylococcal aureus infections became more common as the incidence of MRSA declined. In the same year Morimoto (30) described a reduction in the incidence of MRSA following a similar change in antibiotic strategy. Within a total of 103 patients, treated over a 14 year period, there was a significantly reduced probability of acquiring MRSA with shorter surgical procedures and once the use of third generation cephalosporins and imipenem had been discontinued.

In the follow-on study, reported in this chapter, which included both free and pedicled flaps the incidence of MRSA acquisition has remained relatively low (15%), although this is still a significant rate. There was no significant difference in the incidence of MRSA acquisition at the donor sites of free or pedicled flaps. The incidence of unsuspected colonisation may have been higher as wounds were not swabbed without evidence of infection. Acquisition of MRSA alone was uncommon (2%) and MRSA was more usually detected together with "other organisms" (60%). There was a significant decrease in the incidence of acquisition of MRSA ($p=0.02$) once a shortened antibiotic regimen had been introduced. Only 15% of patients had no detectable growth. The range of other organisms detected broadened and these were mainly faecal flora. There was a significant increase in the incidence of Pseudomonas ($p=0.02$), with a trend towards an increase in Candidiasis as well ($p=0.08$). However, infection with Clostridum difficile has not been a problem.

In this latest study the incidence of MRSA acquisition was not influenced by the length of operation, age, or potential risk factors such as diabetes and anaemia but a more detailed analysis of wound infection rates will follow when greater numbers are available. However, elderly patients of 70 years or greater ($p=0.01$) and those with a pedicled flap ($p=0.03$) were at a significantly higher risk of acquiring other organisms. During the period of this study the overall median duration of hospital admission has remained stable. This seems to be due to the acquisition of other organisms rather than MRSA that have delayed discharge. The median hospital stay with MRSA alone was 29 days and with "other organisms" was 25 days. These durations were both significantly longer than the group that did not acquire an infectious organism, median 11 days ($p<0.0001$). The duration of admission is not an entirely independent outcome factor as the discharge of a few patients was further delayed because of the time taken to find a nursing home once an infection with MRSA had resolved. There have been several retrospective studies of infection after major head and neck surgery. In 2004 Watters (10) reported detection of MRSA in 45% (25/55) of patients and emphasised the additional morbidity incurred as a result of wound infections that required further surgery. The duration of hospital admission of patients with MRSA was almost three times longer than that of those without MRSA and the economic costs were similarly increased. More recently Shiomori (31) detected MRSA in 75% (46/61) of patients with head and neck cancers and the acquisition of other pathogens such as methicillin sensitive Staphylococcus aureus (MSSA) was also common (25%, 15/61). Patients with MRSA had a significantly longer duration of hospitalisation. It should be appreciated that the scope of head and neck surgery is wide. The range of surgical resections and reconstructions varies considerably and it should be ascertained whether a paper is reporting principally on oropharyngeal and laryngeal tumours, or oral and mandibular or maxillary tumours. In 2008 Rogers (32) reported on a cohort of patients which is comparable to the current study. Tumours of the maxillofacial region were reconstructed with primarily free and pedicled flap procedures. The incidence of MRSA during a five year-period was similar, 14% (39/287), but this only included new cancer patients. The incidence of infection was probably an underestimate as this was a retrospective study. The two main risk factors for acquisition of MRSA were advanced stage of cancer and reconstruction with a free flap. Patients with advanced neck disease (N2+) had an MRSA rate of approximately 40%. Within the "oncological" patient group, presumably those managed with medical treatment, the incidence of infection was extremely high 82% (60/73). A high number, just over 25%, of patients were MRSA positive at the time of admission. As a result of that study all patients are now screened on admission and managed in separate side rooms to reduce the risk of cross-contamination. In the current continuing study no potential risk factors for acquisition, as opposed to infection, with MRSA were identified but a more detailed analysis of wound infection rates will follow when greater numbers are available. It is likely that infection rates will vary between individual units at differing times as local factors such as the effectiveness of infection control measures alter.

In the initial part of this study patients were screened for MRSA if they had previously had an MRSA infection, had wounds or were transferred from high-risk environments. At that time no patients were identified as carriers of MRSA and there was no justification for routine screening before admission. All but one of the MRSA infections was acquired in the ward and inadequate compliance with measures to control cross-infection was at least partly responsible for the spread of infection. However, screening has been advocated for rates of carriage as low as 1.3% (33). More stringent methods of infection control have continued to be promoted, including a "search and destroy" strategy (4, 34, 35). According to Cooper (36) (37) isolation may be effective even when MRSA is endemic. These studies have advocated multiple interventions to control outbreaks of MRSA within the hospital environment but lacked control groups. The effectiveness of any screening programme will depend upon many factors including the sensitivity of the technique, the availability and effectiveness of treatment decolonisation protocols and compliance with standard cross-infection control measures (38). Any of these elements may be deficient and recent guidelines in the United Kingdom have supported targeted screening for patients in higher risk groups such as those being transferred or re-admitted and previous carriers or those thought to be infected. This applies to admissions to intensive care, high dependency wards, and the surgical specialities including trauma, orthopaedic and vascular surgery (38). It seems clear that surgical head and neck wards should be also be included as high risk wards. At the Leicester Royal Infirmary a standard chromogenic direct plating detection system is used that will give a negative result within 24 hours and a positive result within 48 to 56 hours. This means that ideally patients should be screened about two weeks prior to surgery to allow time for decolonisation if necessary. Most recently there have been increasing political and medical pressures to introduce screening for most surgical procedures, including low risk surgical interventions, although the value of universal screening is unproven (18). In the United States of America success with a universal screening policy in conjunction with strict contact precautions has been reported (39). The introduction of universal screening within the United Kingdom, although controversial, has now been introduced in 2009 for all elective admissions and will be mandatory for emergency admissions as soon as is practical (18).

The introduction of "rapid" nucleic acid amplification methods has been advocated to shorten the time to a laboratory result and hence improve the efficiency of screening programmes. Unfortunately logistical delays often negate this advantage and the techniques are significantly more expensive. A recent large study of universal screening with a rapid detection technique on a medical ward did not demonstrate a practical advantage over conventional techniques and it was thought unlikely that the additional expense could be justified when compared to alternative conventional control measures (40). In a large controlled study within a surgical environment with endemic MRSA, but relatively low rates of infection, the incidence of MRSA surgical site infection and nosocomial MRSA acquisition did not change significantly with universal screening (41). The latter study supported the hypothesis that transmission after surgery was a major cause of MRSA infection. At present there seems to be no substantial additional benefit from a rapid screening technique applied universally when adequate conventional cross-infection control measures are already in place. These results do not preclude the use of the technique in specialised units such as an intensive care department but an effective infection control system would limit any additional potential advantages (18). Some uncontrolled studies have shown a reduction in MRSA rates (42) (43), particularly when pre-emptive isolation is possible (41).

In our experience the incidence of MRSA infections is declining and other infective organisms are becoming increasingly important causes of serious wound infections. The current study is not controlled and other confounding factors such as the fluctuation in the rate of infection within the Leicester Royal Infirmary will have had an impact on rates of MRSA acquisition. Although there was a relatively small decline in the incidence of MRSA-bacteraemia within the University Hospitals of Leicester (UHL) at the beginning of this study period there has now been a more sustained and substantial reduction (**Figure 5**). This probably reflects a relatively high initial incidence of endemic MRSA within the institute when compared to the reduction in national rates of bacteraemia as reported by the Health Protection Agency (44) (**Figure 6**).

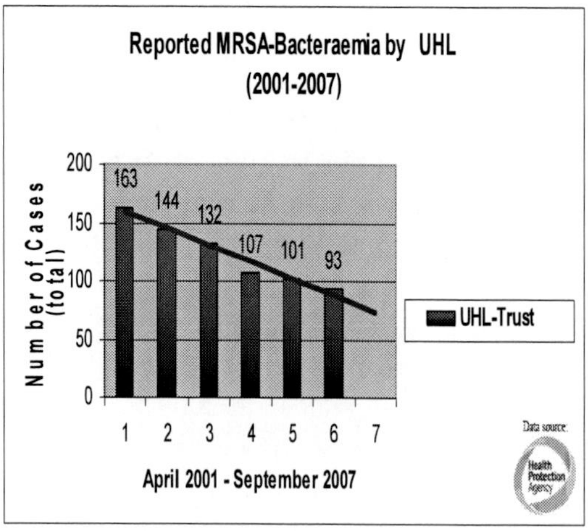

Figure 5.The decreasing incidence of MRSA-bacteraemia within the University Hospitals of Leicester (UHL).

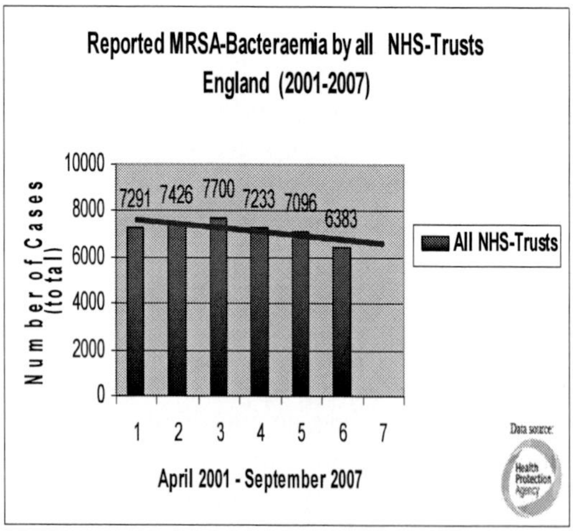

Figure 6. The decreasing incidence of MRSA-bacteraemia within the National Health Service (NHS) in England.

However, the monitoring of only MRSA-bacteraemia rates provides only an overview of the problem. The vast majority of patients in this study did not have a detected bacteraemia despite developing serious MRSA related wound complications and occasionally pneumonia. At present we continue to closely monitor for all infections.

Conclusion

There has been a dramatic reduction in the incidence of acquisition of MRSA over the last 6 years but a significant increase in other infective organisms. The incidence of MRSA after major maxillofacial surgery and other head and neck surgery, particularly when flap reconstruction is required remains appreciable. The majority of infections are acquired after surgery and the head and neck ward is a high risk area for acquiring MRSA. The current focus should be on achieving a sustained reduction in contamination with MRSA and other infective organisms together with a reduction in associated morbidity. This should be pursued through the use of appropriate antibiotic policies, the improved design of specialist wards including the provision of single bedded rooms and isolation facilities together with strict adherence to standard infection control measures. All patients should be screened on admission but the value of rapid MRSA screening techniques is not yet proven.

REFERENCES

[1] Mulligan ME, Murray-Leisure KA, Ribner BS, Standiford HC, John JF, Korvick JA, et al. Methicillin-resistant Staphylococcus aureus: a consensus review of the microbiology, pathogenesis, and epidemiology with implications for prevention and management. *Am J Med* 1993 Mar;94(3):313-28.

[2] Ayliffe GA. The progressive intercontinental spread of methicillin-resistant Staphylococcus aureus. *Clin Infect Dis* 1997 Jan;24 Suppl 1:S74-9.

[3] Reacher MH, Shah A, Livermore DM, Wale MC, Graham C, Johnson AP, et al. Bacteraemia and antibiotic resistance of its pathogens reported in England and Wales between 1990 and 1998: trend analysis. *Bmj* 2000 Jan 22;320(7229):213-6.

[4] Duckworth G. Controlling methicillin resistant Staphylococcus aureus. *Bmj* 2003 Nov 22;327(7425):1177-8.

[5] Plowman RG, N. Griffin, M. Roberts, JA. Swan, AV. Cookson, B. Taylor, L. *The socio-economic burden of hospital acquired infection.* London: Public Health Laboratory Service 2000.

[6] Plowman R, Graves N, Griffin MA, Roberts JA, Swan AV, Cookson B, et al. The rate and cost of hospital-acquired infections occurring in patients admitted to selected specialties of a district general hospital in England and the national burden imposed. *J Hosp Infect* 2001 Mar;47(3):198-209.

[7] Naylor AR, Hayes PD, Darke S. A prospective audit of complex wound and graft infections in Great Britain and Ireland: the emergence of MRSA. *Eur J Vasc Endovasc Surg* 2001 Apr;21(4):289-94.

[8] Gang RK, Sanyal SC, Bang RL, Mokaddas E, Lari AR. Staphylococcal septicaemia in burns. *Burns* 2000 Jun;26(4):359-66.

[9] Scriven JM, Silva P, Swann RA, Thompson MM, Naylor AR, Bell PR, et al. The acquisition of methicillin-resistant Staphylococcus aureus (MRSA) in vascular patients. *Eur J Vasc Endovasc Surg* 2003 Feb;25(2):147-51.

[10] Watters K, O'Dwyer T P, Rowley H. Cost and morbidity of MRSA in head and neck cancer patients: what are the consequences? *J Laryngol Otol* 2004 Sep;118(9):694-9.

[11] 1Penel N, Lefebvre JL, Cazin JL, Clisant S, Neu JC, Dervaux B, et al. Additional direct medical costs associated with nosocomial infections after head and neck cancer surgery: a hospital-perspective analysis. *Int J Oral Maxillofac Surg* 2008 Feb;37(2):135-9.

[12] Penel N, Mallet Y, Roussel-Delvallez M, Lefebvre JL, Yazdanpanah Y. Factors determining length of the postoperative hospital stay after major head and neck cancer surgery. *Oral Oncol* 2008 Jun;44(6):555-62.

[13] Avery CM, Harrop C. Rapid healing of MRSA infection at the suprafascial radial donor site. *Int J Oral Maxillofac Surg* 2002 Jun;31(3):318-21.

[14] Avery CM, Ameerally P, Castling B, Swann RA. Infection of surgical wounds in the maxillofacial region and free flap donor sites with methicillin-resistant Staphylococcus aureus. *Br J Oral Maxillofac Surg* 2006 Jun;44(3):217-21.

[15] Anonymous. Tackling antimicrobial resistance. *Drug and Therapeutics Bulletin* 1999 Feb;37(2):9-16.

[16] Bumpous JM, Johnson JT. The infected wound and its management. *Otolaryngol Clin North Am* 1995 Oct;28(5):987-1001.

[17] Schentag JJ, Hyatt JM, Carr JR, Paladino JA, Birmingham MC, Zimmer GS, et al. Genesis of methicillin-resistant Staphylococcus aureus (MRSA), how treatment of MRSA infections has selected for vancomycin-resistant Enterococcus faecium, and the importance of antibiotic management and infection control. *Clin Infect Dis* 1998 May;26(5):1204-14.

[18] 18Wilcox MH. Screening for MRSA. *BMJ* 2008 Apr 26;336(7650):899-900.

[19] 1Szilagyi DE, Smith RF, Elliott JP, Vrandecic MP. Infection in arterial reconstruction with synthetic grafts. *Ann Surg* 1972 Sep;176(3):321-33.

[20] Boyce JM, Potter-Bynoe G, Chenevert C, King T. Environmental contamination due to methicillin-resistant Staphylococcus aureus: possible infection control implications. *Infect Control Hosp Epidemiol* 1997 Sep;18(9):622-7.

[21] Boyce JM. Understanding and controlling methicillin-resistant Staphylococcus aureus infections. *Infect Control Hosp Epidemiol* 2002 Sep;23(9):485-7.

[22] Shiomori T, Miyamoto H, Makishima K. Significance of airborne transmission of methicillin-resistant Staphylococcus aureus in an otolaryngology-head and neck surgery unit. *Arch Otolaryngol Head Neck Surg* 2001 Jun;127(6):644-8.

[23] Shiomori T, Miyamoto H, Makishima K, Yoshida M, Fujiyoshi T, Udaka T, et al. Evaluation of bedmaking-related airborne and surface methicillin-resistant Staphylococcus aureus contamination. *J Hosp Infect* 2002 Jan;50(1):30-5.

[24] Brady RR, Kalima P, Damani NN, Wilson RG, Dunlop MG. Bacterial contamination of hospital bed-control handsets in a surgical setting: a potential marker of contamination of the healthcare environment. *Ann R Coll Surg Engl* 2007 Oct;89(7):656-60.

[25] Brady RR, Fraser SF, Dunlop MG, Paterson-Brown S, Gibb AP. Bacterial contamination of mobile communication devices in the operative environment. *J Hosp Infect* 2007 Aug;66(4):397-8.

[26] Tadiparthi S, Shokrollahi K, Juma A, Croall J. Using marker pens on patients: a potential source of cross infection with MRSA. *Ann R Coll Surg Engl* 2007 Oct;89(7):661-4.

[27] Duckworth GJ. Diagnosis and management of methicillin resistant Staphylococcus aureus infection. *Bmj* 1993 Oct 23;307(6911):1049-52.

[28] Parton M, Beasley NJ, Harvey G, Houghton D, Jones AS. Four cases of aggressive MRSA wound infection following head and neck surgery. *J Laryngol Otol* 1997 Sep;111(9):874-6.

[29] Lo RC, Burd A. MRSA - again. *Br J Plast Surg* 2001 Apr;54(3):279-80.

[30] Morimoto Y, Sugiura T, Tatebayashi S, Kirita T. Reduction in incidence of methicillin-resistant Staphylococcus aureus (MRSA) after radical surgery for head and neck cancer. *Spec Care Dentist* 2006 Sep-Oct;26(5):209-13.

[31] Shiomori T, Miyamoto H, Udaka T, Okochi J, Hiraki N, Hohchi N, et al. Clinical features of head and neck cancer patients with methicillin-resistant Staphylococcus aureus. *Acta Otolaryngol* 2007 Feb;127(2):180-5.

[32] Rogers SN, Proczek K, Sen RA, Hughes J, Banks P, Lowe D. Which patients are most at risk of methicillin resistant Staphylococcus aureus: a review of admissions to a regional maxillofacial ward between 2001 and 2005. *Br J Oral Maxillofac Surg* 2008 Sep;46(6):439-44.

[33] Papia G, Louie M, Tralla A, Johnson C, Collins V, Simor AE. Screening high-risk patients for methicillin-resistant Staphylococcus aureus on admission to the hospital: is it cost effective? *Infect Control Hosp Epidemiol* 1999 Jul;20(7):473-7.

[34] Verhoef J, Beaujean D, Blok H, Baars A, Meyler A, van der Werken C, et al. A Dutch approach to methicillin-resistant Staphylococcus aureus. *Eur J Clin Microbiol Infect Dis* 1999 Jul;18(7):461-6.

[35] Humphreys H, Duckworth G. Methicillin-resistant Staphylococcus aureus (MRSA)--a re-appraisal of control measures in the light of changing circumstances. *J Hosp Infect* 1997 Jul;36(3):167-70.

[36] Cooper BS, Stone SP, Kibbler CC, Cookson BD, Roberts JA, Medley GF, et al. Systematic review of isolation policies in the hospital management of methicillin-resistant Staphylococcus aureus: a review of the literature with epidemiological and economic modelling. *Health Technol Assess* 2003;7(39):1-194.

[37] Cooper BS, Stone SP, Kibbler CC, Cookson BD, Roberts JA, Medley GF, et al. Isolation measures in the hospital management of methicillin resistant Staphylococcus aureus (MRSA): systematic review of the literature. *Bmj* 2004 Sep 4;329(7465):533.

[38] Coia JE, Duckworth GJ, Edwards DI, Farrington M, Fry C, Humphreys H, et al. Guidelines for the control and prevention of meticillin-resistant Staphylococcus aureus (MRSA) in healthcare facilities. *J Hosp Infect* 2006 May;63 Suppl 1:S1-44.

[39] Robicsek A, Beaumont JL, Paule SM, Hacek DM, Thomson RB, Jr., Kaul KL, et al. Universal surveillance for methicillin-resistant Staphylococcus aureus in 3 affiliated hospitals. *Ann Intern Med* 2008 Mar 18;148(6):409-18.

[40] Jeyaratnam D, Whitty CJ, Phillips K, Liu D, Orezzi C, Ajoku U, et al. Impact of rapid screening tests on acquisition of meticillin resistant Staphylococcus aureus: cluster randomised crossover trial. *BMJ* 2008 Apr 26;336(7650):927-30.

[41] Harbarth S, Fankhauser C, Schrenzel J, Christenson J, Gervaz P, Bandiera-Clerc C, et al. Universal screening for methicillin-resistant Staphylococcus aureus at hospital admission and nosocomial infection in surgical patients. *JAMA* 2008 Mar 12;299(10):1149-57.

[42] Cunningham R, Jenks P, Northwood J, Wallis M, Ferguson S, Hunt S. Effect on MRSA transmission of rapid PCR testing of patients admitted to critical care. *J Hosp Infect* 2007 Jan;65(1):24-8.

[43] Keshtgar MR, Khalili A, Coen PG, Carder C, Macrae B, Jeanes A, et al. Impact of rapid molecular screening for meticillin-resistant Staphylococcus aureus in surgical wards. *Br J Surg* 2008 Mar;95(3):381-6.

[44] Health Protection Agency. Results of the quarterly reporting of mandatory surveillance data In: Health Protection Agency. MRSA Bacteraemia tables: *www.hpa.org.uk;* 2007.

In: Methicillin-Resistant Staphylococcus Aureus... ISBN: 978-1-60741-398-1
Editor: Charles L. Kolendi

Chapter 18

The Phenomenon of Antibiotic Resistance in an Evolutionary Perspective

Roberto Manfredi*

Department of Internal Medicine, Aging, and Nephrologic Diseases, Division of Infectious Diseases, "Alma Mater Studiorum" University of Bologna, S. Orsola-Malpighi Hospital, Bologna, Italy

Keywords: Antimicrobial resistance, spread, evolutionary perspective.

The extraordinary capacity of a number of bacterial organisms to develop resistance to antibiotic compounds, remains a hot topic on the minds of doctors, hospital staff, reporters, and the community since several years. It is also heralded as a textbook example of "Darwinian evolution" continuously in action [1, 2]. As a consequence, these multiresistant bacteria are being studied also by "evolutionary" scientists, with the hope that they will reveal secrets or trace a pathway as to how molecules-to-man evolution could have happened [2, 3].

Basically, antibiotics are natural substances secreted by bacteria and fungi in order to kill other bacteria that are competing for limited nutrients; in fact, the majority of present antibiotics are synthetic derivatives of these natural products. Scientists are dismayed to discover that some bacteria have become resistant to antibiotics, mostly through various genomic (chromosomal-, plasmid-, or transposon-carried) mutations.

Changes in natural echosystems, including the release of large amounts of antimicrobial agents in the environment as happened in the last decades, might act on the population dynanmics of microorganisms too, including the selection of resistance determinants, with the visible consequences for human health, in terms of spread of multiresistant organisms in the commumity and especially in hospital facilities [3].

Hospitals have become a breeding ground for antibiotic resistant bacteria, and all microorganisms as whole. In these settings, bacteria proliferate in an "artificial" environment filled with sick people who have some underlying cause of immunodeficiency and concurrent

* Corresponding auther: Associate Professor of Infectious Diseases, University of Bologna c/o Infectious Diseases, S. Orsola Hospital Via Massarenti 11 I-40138 Bologna, Italy Telephone: +39-051-6363355, Telefax: +39-051-343500, E-mail: Roberto.manfredi@unibo.it

diseases, and where antibiotics have eliminated competing bacteria that are not resistant to used compounds [3].

However, the history learns that bacteria which became resistant to modern antibiotics have even been found in the frozen bodies of people who died long before those antibiotics were discovered or synthesized [4].

As known, antibiotics were first discovered through a providential experiment conducted by Dr. Fleming in late 1920s. His work eventually led to the large-scale production of penicillin from the mold *Penicillium notatum* in the 1940s. Almost concurrently, as early as the late 1940s, resistant strains of penicillin-resistant bacteria began to appear [5]. Currently, it is estimated that over 70% of the different species of bacteria that cause hospital-acquired infections prove resistant to at least one of the antibiotics currently used to treat them [6].

Antibiotic resistance continues to expand for a very broad spectrum of reasons, including over-prescription of antibiotics by physicians, non-completion of prescribed antibiotic treatments by patients, self-prescription of antibiotics by patients themselves, use of antibiotics in animals as growth enhancers, increased international travel, and poor hospital hygiene [5].

In extreme synthesis, bacteria are known to develop antimicrobial resistance through two primary ways:.

- By gene mutation, and
- By using a built-in design feature to swap DNA (called horizontal gene transfer); by this last way, bacteria have the possibility to transfer resistance genes among different species which share the same environment.

An antibiotic kills a bacterial cell by simply disrupting an essential structure, or by blocking a critical metabolic function [2]. Antibiotic resistance of bacteria also encompasses a loss of structural and/or functional systems. Anyway, on a general basis the evolutionary mechanisms require a gain of functional systems for bacteria, to evolve into more adaptive organism.. For instance, an antibiotic may bind to a specific bacterial protein, so that this last protein cannot function properly. The normal protein is usually involved in copying the DNA, making proteins, or building the bacterial cell wall, all necessary mechanisms for bacteria, to grow and reproduce themselves. If the bacteria have a mutation in the DNA which encodes for one of those proteins, the compound cannot bind to the altered protein; as a direct consequence, the mutant bacteria survive and multiplicate. In the presence of antibiotics, the process of natural selection will occur, favoring the survival and reproduction of the mutant bacteria [2, 3]. Obviously, the mutant bacteria are better able to survive in the presence of the stated antibiotic(s), and will continue to cause illness in the patient, and to increase morbidity and mortality rates.

Although the mutant bacteria can survive well in the hospital environment, from an evolutionary point of view every change occurs at some cost. Actually, i.e. the altered protein is less efficient in performing its normal function, making the bacteria less fit in an specific environment, without the inducing antibiotics [2, 3]. Typically, the non-mutant bacteria are better able to compete for resources and reproduce faster than the mutant form. However, also adaptive mechanisms are activated by resistant microorganisms: the inhibition of some penicillin-binding proteins has been shown to elicit compensative responses in surviving colonies of *Pseudomonas aeruginosa* [7].

Bacteria can also become antibiotic-resistant by gaining mutated DNA from other bacteria. But this still is not an example of evolution in action. No new DNA is generated (a requirement for molecules-to-organism evolution), it is just DNA sequences moving around. However, this mechanism of exchanging DNA is necessary for bacteria to survive in extreme or rapidly changing environments like hospitals or nursing home facilities, or in experimental conditions [2, 8].

The mechanisms of bacterial mutations and natural selection aid microorganism populations in becoming resistant to antibiotics. However, mutation and natural selection also result in bacteria with defective proteins that have lost their normal functions. In general, the evolution process requires a gain of functional systems for bacteria to evolve [2, 3]. Mutation and natural selection, thought to be the driving forces of evolution, only lead to a loss of functional systems. Therefore, antibiotic resistance of bacteria is not an example of "evolution in action", but rather variation within a bacterial kind. It is also a testimony to the wonderful design "the destiny" gave to bacteria, master adapters and survivors in a sin-cursed world [2, 3].

Actually, evolutionary medicine is relevant also to family practice and family physician training, in a wide spectrum of perspectives [9]. Its incorporation into medical training and practice is underway, since its potential scientific burden may be supported by the generation of testable hypotheses and practical applications in relation to both disease pathogenesis, prevention, and treatment [1, 9].

REFERENCES

[1] Alpert, JS. In Medicine, signs of evolution are ever-present. *Am J Med.*, 2006, 119, 291

[2] Gilmore, MS. The molecular basis of antibiotic resistance: where Newton meets Darwin. *Int J Med Microbiol*, 2002, 292, 65

[3] Martínez, JL. Antibiotics and antibiotic resistance genes in natural environments. *Science*, 2008, 321, 365-7

[4] Bacterial antibiotic resistance: proof of evolution? www.apologeticspress.org/articles/439

[5] Antibiotic resistance: How did we get to this? flemingforum.org.uk/slides/antibiotic_resistance.pdf

[6] The problem of antimicrobial resistance, National Institute of Allergy and Infectious Diseases. www.niaid.nih.gov/factsheets/antimicro.htm/

[7] Blázquez, J; Gómez- Gómez, JM; Oliver, A; Juan, C; Kapur, V; Martín, S. PBP3 inhibition elicits adaptive responses in *Pseudomonas aeruginosa*. *Mol Microbiol*, 2006, 62,84-9

[8] Is natural selection the same thing as evolution? In "The New Answers Book", Ken Ham Ed., Master Books, *Green Forest, Arkansas*, 2006

[9] Naugler, CT. Evolutionary medicine: update on the relevance to family practice. *Can Fam Physician*, 2008, 54, 1265-9

INDEX

A

B

C

D

E

F

G

H

I

J

K

L

M

N

O

P

Q

R

S

T

Y

Z